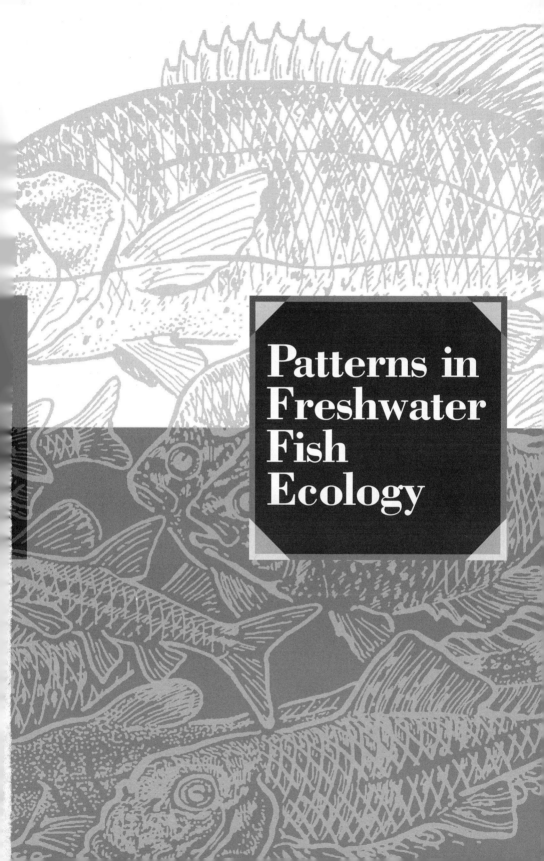

Patterns in Freshwater Fish Ecology

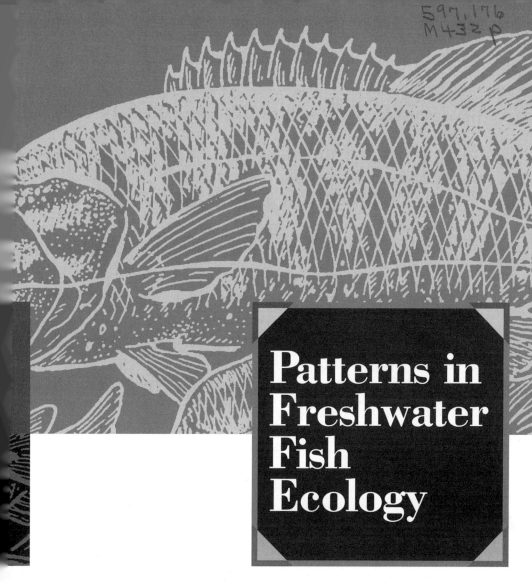

Patterns in Freshwater Fish Ecology

William J. Matthews

Professor, Department of Zoology
Curator of Fishes, Oklahoma Museum of Natural History
The University of Oklahoma
Norman, OK

CHAPMAN & HALL

INTERNATIONAL THOMSON PUBLISHING
Thomson Science

New York • Albany • Bonn • Boston • Cincinnati • Detroit
London • Madrid • Melbourne • Mexico City • Pacific Grove
Paris • San Francisco • Singapore • Tokyo • Toronto • Washington

Distributors for North, Central and South America:
Kluwer Academic Publishers
101 Philip Drive
Assinippi Park
Norwell, Massachusetts 02061 USA

Distributors for all other countries:
Kluwer Academic Publishers
Distribution Centre
Post Office Box 322
3300 AH Dordrecht, THE NETHERLANDS

Library of Congress Cataloging-in-Publication Data

Matthews, William J. (William John)
   Patterns in freshwater fish ecology / William J. Matthews.
     p. cm.
   Includes bibliographical references and index.
   ISBN 0-412-02831-X (alk.paper)
   1. Freshwater fishes--Ecology. I. Title.
QL624.M38   1998
597.176--dc21   97-21339
                                                CIP

Cover Drawing: Coral McCallister, Norman, OK
Cover design: Curtis Tow Graphics

Copyright © 1998 by Chapman & Hall
Copyright © Second Printing 1998 by Kluwer Academic Publishers

All rights reserved. No part of this publication may be reproduced, stored in a retrieval system or transmitted in any form or by any means, mechanical, photocopying, recording, or otherwise, without the prior written permission of the publisher, Kluwer Academic Publishers, 101 Philip Drive, Assinippi Park, Norwell, Massachusetts 02061

Printed on acid-free paper.

Printed in the United States of America

For my family

Contents

Introduction ... xvii

1. Overview of Fishes and Fish Assemblages ... 1
    1.1 Problems and Approaches in Fish Ecology ... 1
        "Reighard's Lament" ... 1
    1.2 Diversity of Fishes and Their Habitats ... 2
        Fish Versus Tetrapod Diversity ... 3
        Examples of Diverse Fish Faunas ... 5
        Diversity of Fishes by Geographic Region ... 5
        Diversity Within Taxonomic Groups ... 6
        Morphological Diversification ... 7
        Habitats of Fishes ... 10
    1.3 Questions about Fish Assemblages ... 11
        Focus of This Book ... 11
        Factors Potentially Influencing Assemblage Structure ... 12
        Macroecology ... 15
        Classes of Explanations for Fish Assemblage Structure ... 15
        Effects of Fish Species and Assemblages in Ecosystems ... 19
    1.4 Explanations? ... 22
        Correlative Versus Mechanistic Explanations for Observed Patterns ... 22
        Time Scales for Evolutionary and Ecological Thought ... 24
        Influence of Phylogeny on Ecology ... 25
        Classification and Evolutionary Systematics ... 26
    1.5 Guideposts ... 28

2. Structure of Fish Assemblages ... 30
    2.1 Introduction ... 30
        What Is a Fish "Assemblage" in the Context of a "Locality"? ... 30
        What Is Fish Assemblage "Structure"? ... 32

viii / Patterns in Freshwater Fish Ecology

2.2 Number of Species, Families, and Species per Family	33
Numbers of Fish Species in an Assemblage	33
The Empirical Evidence: Numbers of Species per Locality	38
Local Assemblages in Individual Stream Pools?	44
Number of Families in Local Assemblages, and Species per Family	45
2.3 Trophic and Functional Groups, and Predator/Prey Species	54
Composition of Assemblages by Trophic Groups	54
Relative Abundance of Piscivore to Prey Species in Assemblages	60
Functional Groups of Fish in Assemblages	66
2.4 Abundance, Body Size, and Mouth Size	73
Distribution of Abundance of Species Within Assemblages	73
Body and Mouth Size Structure Within Whole Assemblages	77
2.5 Summary	84
3. Discrete Versus Overlapping Assemblages and Assemblage Stability	86
3.1 Discrete Versus Overlapping Assemblages	86
Background	86
Fish Assemblages—Discrete Versus Overlapping?	88
Distinct Fish Assemblages in Streams?	90
Distinct Fish Assemblages Among Lakes?	91
Random Versus Discrete Groupings of Fish Species—	
A Multivariate Model	93
Detecting Discrete Fish Assemblages in Streams and Lakes—A	
Statistical Test of the Evidence	99
3.2 Stability of Fish Assemblages	104
Background	107
Classical and Descriptive Studies	108
Qualitative Changes in Fish Assemblages	109
Quantitative Studies—Temperate Streams	110
Assessment of Long-Term Samples by Similarity Indices and	
Coefficients of Variation	120
Multivariate Assessment of Temporal Variation in Fish Assemblages	122
Stability Across Combined Spatial and Temporal Scales	126
3.3 Summary	127
4. Stream Ecology and Limnology as Related to Fish Assemblages	130
4.1 Introduction	130
4.2 Physical and Chemical Limnology and its Effect on Fish and	
Fish Assemblages	131
Physical Properties of Water	133
Light in the Aquatic Environment	135
Heat Budgets and Temperature in the Aquatic Environment	136
Effects of Light and Heat on Fish Assemblages in Lakes	136
Thermal Effects on Fish Distribution in Lakes	137
Water Chemistry: Oxygen	138
Effects of Oxygen on Fish in Lakes	138
Water Chemistry: Alkalinity and pH, Carbonate Systems	140

	Effects of Environmental pH on Fish	141
	Physical and Chemical Factors in Streams and Effects on Fish	142
	General Physicochemical Effects on Habitat Selection by Fish	146
4.3	Lentic Versus Lotic Environments as Fish Habitats	149
	Characteristics of Lentic Versus Lotic Environments	149
4.4	Characteristics of Lakes that Influence Fish Assemblages	151
	Lakes	151
	Lake Formation, Life Span, and Demise	152
	Lake Formation and Sources of Fishes	155
	Eutrophication: Trophic Status of Lakes	157
	Stratification Patterns and Effects on Fishes	159
	Wind and Water Movements in Lakes	163
	Morphometry of Lakes	166
	Summary: Limnology of Lakes	171
4.5	Characteristics of Streams that Influence Fish Assemblages	171
	Watersheds and Ecosystem Concepts in Streams	171
	Formation of River Basins and Drainage Patterns	172
	Terminology for Streams of Various Sizes	174
	Drainage Patterns	175
	Headward Cutting of Channels—New Habitat for Fishes?	180
	Sources of Water for Streams	181
	Location of Flow Within Stream Channels and Configuration of Channels	182
	Stage rises, Floods, and the Hydrograph	185
	Woody Debris in Streams	187
	Channel Formation and Maintenance	188
	Anthropogenic Effects on Hydrographs and Flood Frequency	188
4.6	Summary	189

5. Influence of Global to Regional Zoogeography on Local Fish Assemblages 191
   5.1 Introduction 191
   5.2 Global Zoogeography of Freshwater Fishes 192
      Zoogeographic Realms 192
      Testing Patterns of Fish Families on a Worldwide Zoogeographic Template: Was Wallace Right? 194
      Family Richness Among Basins, and "Basin Richness" Within Families 196
      Primary, Secondary, and Tertiary Freshwater Fishes (Myers) 197
      Continental Movements and Freshwater Connections 202
         Northern Landmass 205
         Southern Landmass 206
         Attaining Modern Configuration 207
      Occurrence of Freshwater Fish Families Relative to Continental Movements 207
      Zoogeography of Ostariophysians—Alternative Hypotheses 213
      Application of Global Biogeography to Local Fish Assemblages? North America as an Example 219

5.3 Subcontinental Zoogeography of Freshwater Fishes	227
Glacial Periods of the Pleistocene—Effects Worldwide, "Land Bridges," and Beringia	227
North American Fish Regions as an Example of Subcontinental Zoogeographic Effects	235
Eurasia	238
Southeast Asia	239
Africa	240
South America	240
Australasia and New Zealand	240
5.4 Regional Biogeography of Fishes	241
Importance of River Basins in Fish Biogeography	241
Local Endemics	243
Vicariance Biogeography and "Historical Ecology"	245
Directional Affinities of Species in a Fauna	249
"Ecoregions"	254
Regional Versus Local Species Richness	255
Zoogeography in Individual Species: Range Sizes and Body Sizes	260
5.5 Merging Concepts of Geological, Zoogeographic, and Basin-Level Effects in Ecological Time	261
Interactive Ecology	262
6. Physical Factors Within Drainages as Related to Fish Assemblages	264
6.1 Introduction	264
6.2 Area Effects at "Drainage" Level	264
Drainage Area and "Insular" Effects	264
6.3 Local Habitat Size	268
Stream Width	268
Stream Depth and "Pool Development"	271
Depth of Lakes	273
Volume	274
Discharge	276
6.4 Habitat Structure, Cover, Complexity, and Productivity	279
Structure or Cover	279
Woody Debris	279
Rock Structure in Streams	282
"Soft" Structure: Macrophytes, Algae, Leaf Packs	282
Cover	283
Habitat "Heterogeneity"	283
Heterogeneity at Increasing Scales: Connectivity of Habitat Units	287
Primary Productivity in Local Habitats	289
6.5 Zonation of Fishes in Lakes and Streams	290
Habitat Zonation in Lakes	290
Coves as Special Habitats?	293
Longitudinal "Zonation" or "Continuum" of Stream Fishes?	296
A Test of the "Continuum" Versus "Zonation" Hypothesis for Stream Fishes	305
Stream Order and Longitudinal Distribution of Fishes	307

Potential Problems with the Horton–Strahler System of Stream Order for Fish Ecologists	310
Alternatives to Stream Order?	312
6.6 Landscape Ecology: Looking Laterally, Instead of Just Up and Down the Stream?	312
6.7 Multivariate Analyses of Fish Distributions, and Influence of Environmental Variables	315
7. Disturbance, Harsh Environments, and Physicochemical Tolerance	318
7.1 Introduction	318
Abiotic Versus Biotic Regulation of Fish Assemblages	319
7.2 Definition and Time Scales of Disturbance	320
What Is "Disturbance"?	320
Theoretical Implications of Disturbance	322
Time Scales of Disturbance	323
7.3 Floods	326
Effects of Floods on Streams	326
Potential Long-Term Effects of Floods on Stream Ecosystems and Fish Assemblages	331
Field Studies of Flood Effects on Fishes	333
Habitat Use by Stream Fish During and After Floods	334
A Major Flood Event in a Typically "Benign" Warm-Water Stream: Effects on the Fish Fauna	336
7.4 Drought	341
Known Changes in Fish Assemblages During and After Drought	344
Phases of Effects of a Drought	349
7.5 Physicochemical Stress	353
Tolerance of Individual Species for Physicochemical Stress in the Environment	353
Effects of High Temperature	354
Effects of Cold Temperature	357
Oxygen	357
pH and Acidification, Natural "Stained" Water	360
Intraspecific Variation	361
Sublethal or Indirect Effects of Stressors	363
"Harsh" Conditions—Detriment or Benefit to Individual Species?	365
An Example of Salinity Structuring a "Freshwater" River Assemblage	365
Physicochemical Tolerances and Selectivity of Stream Fishes as Related to Ranges and Local Distributions	366
Hypothetical Tolerance and Selectivity Strategies	368
Influence of Phylogeny on Physicochemical Tolerance?	370
Fish in Microrefugia from Physicochemical Stress: When to Stay and When to Go?	374
7.6 Disturbance and Physicochemical Effects: Combined	377
Integrating Disturbance and Physicochemical Stress into Distribution of Fishes in a River/Lake System	377

8. Morphology, Habitat Use, and Life History	380
8.1 Introduction	380
8.2 Water as a Medium: Fluid Drag	381
8.3 Morphology	384
Body Regions and Relation of Their Morphology to Ecology	384
Approaches to Morphological Studies of Fish Ecology	385
Generalizations from Descriptive Morphology	386
Swimming or Holding Position Against Fluid Drag	388
Function Morphology in Swimming	389
Body and Caudal Fin Shape Relative to Hydrodynamics	390
Axial Fins: Role in Propulsion or Controlling Movements	391
Paired Fins: Control of Motion in Swimming or Their Use to Hold Position in Flowing Water	393
Additional Morphological Features of Benthic Fishes	394
Shapes of Fishes in Nonflowing Habitats	395
Trophic Morphology of Fishes	396
Descriptive Studies of Morphology Relative to Food Use	399
Functional Analyses	400
"Other" Morphological Features Relative to Ecology	403
Ecomorphology	403
Overview	410
Intraspecific Variation in Morphology	410
Summary: Morphological Approaches and Fish Ecology	411
8.4 Hydraulics, Morphology, Microhabitat, and Food Use	412
Larval Fish Success Relative to Hydraulics of Habitat	416
Habitat Selection, Patch Choice, Effect on Feeding or Reproduction	417
8.5 Reproduction and Life History	419
Life History of Fishes	420
Overview of Life-History Patterns	421
Age at First Reproduction and Reproductive Life Span	425
Factors Regulating Length or Timing of the Reproductive Season	426
Single or Repeated Spawning?	427
Prespawning Phase of Reproduction	429
Mating Systems or Mate Selection	429
Mating patterns	429
Sexual selection	429
Alternative life histories	431
Spawning and Clutch Parameters	432
Egg Size Versus Egg Number	433
Intraspecific Variation in Clutch Parameters	434
Individual Variation	435
Effect of Reproductive Status on Morphology and Hydrodynamics of Individuals	435
Parental Care and Guarding	435
Early Life History: Survival of Eggs and Growth of Larvae	437
Finding Food as Free-Living Larvae	438
Growth, Recruitment, and Mortality	439

Evolutionary and Environmental Constraints in Reproductive Traits	440
Intraspecific Differences in Reproductive Traits	441
Interspecific Differences in Life-History Traits	442
Life History Versus Habitat Features	442
Effects of Predators on Life History	443
Genetics or Heritability of Life-History Traits	444
Different Life-History Strategies May Work Within a Single Place	445
8.6 Movement and Migration	448
Home Pools or Limited Home Ranges	448
Short-Term Movements	450
Rheotaxis	451
Spawning Movements	451
Nonspawning Seasonal Movements and Migrations	452
Diadromy	453
9. Interactive Factors: Competition, Mixed Species Benefits, and Coevolution	455
9.1 Interspecific Competition and Resource Partitioning	455
Background	455
Niche Segregation or "Resource Partitioning"	458
Competition Among Fishes: Historical Perspective	460
Is Competition Important in Fish Communities?	464
Empirical Evidence for Competition in Fish Assemblages	466
Experimental Evidence of Competition Between Fish Species	467
Resource Partitioning in Fish Assemblages: Throug 1970	470
Empirical Evidence of Resource Partitioning: Early 1970s	474
Experimental Work—Werner, Hall, and Colleagues	476
Other Studies of Resource Partitioning: Late 1970s	478
Resource Partitioning: 1980 to the Present	480
Conclusions: Competition and Resource Partitioning in Fish Assemblages	491
Remaining Questions About Competition and Resource Partitioning in Fish Assemblages	496
9.2 Intraspecific Competition, Density Effects, and Resource Partitioning	497
Density Dependent Effects	499
Competition Between Size Classes of a Species	502
Intraspecific Competition: Cold-Water Fishes	503
Intraspecific Competition in Warm-Water Stream Fishes	506
Intraspecific Resource Partitioning?	507
Summary: Intraspecific Competition or Resource Partitioning	512
9.3 Mixed-Species Effects in Assemblages	512
Background	512
Potential Benefits	513
Evidence of Mixed-Species Phenomena	515
Case Histories: Mixed Species Groups	515
Examples from Streams and Reservoirs	521
More Complex Mutualisms	522

xiv / Patterns in Freshwater Fish Ecology

Conclusions: Mixed-Species Groups in Assemblages	523
Optimal Numbers of Species in Mixed Groups?	524
9.4 Coevolution in Fish Assemblages?	525
10. Interactive Factors: Predation Effects in Fish Assemblages	532
10.1 Predation and a Hypothetical Model	532
Hypothetical Model of Interactions of Numbers of Prey and Piscivore Species	533
10.2 Empirical Studies of Piscivory	536
Background on Piscivory in Fish Communities	536
Case Histories: Introduced Piscivores in Streams	538
Case Histories: Piscivory in Lakes	539
Localized Effects of Piscivores in Streams	542
Piscivore Effects Throughout Longer Stream Reaches	544
Piscivore Influence on Distribution, Habitat, or Behavior of Prey Species	546
10.3 Experiments with Piscivore–Prey Systems	547
Experimental Evidence: Effects of Predators in Streams and Lakes	547
Effects of the Environment on Outcome of Predator–Prey Contests	549
Effects of Other Animals on Piscivory	550
10.4 Theoretical Aspects of Predation	552
Balancing the Risk: Eat Well and Die Young(er)?	552
Theoretical Prey Behavior Under Predator Threat	554
Nonlethal Effects of Predators on Growth, Reproduction, or Interactions of Prey Species	557
Fragmentation of Populations by Predators?	559
Quantifying Effects of Piscivory in Streams?	560
Growth as a Survival Strategy?	563
10.5 Conclusions: Effects of Piscivory in Fish Assemblages	563
11. Effects of Fish in Ecosystems	565
11.1 Introduction	565
11.2 Direct Effects of Fish in Ecosystems	567
Effects of Fish on Abundance of Macroinvertebrates	567
Effects of Fish on Life History or Behavior of Macroinvertebrates	573
Effects of Fish on Large Macroinvertebrates: Crayfish	574
Effects of Fish on Vertebrates: Frogs and Salamanders	574
11.3 Planktivory	575
Effects of Fish on Zooplankton	575
Effects of Fish on Phytoplankton	581
11.4 Herbivory	583
Algivory	583
Effects of Fish on Vascular Plants	589
11.5 Nutrient Effects and Ecosystem Engineering	590
Nutrient Changes by Fish in Ecosystems	590
Ecosystem "Engineering"	593

11.6	Indirect Effects of Fish in Ecosystems	594
	Second-Order Effects	595
	Trophic-Level Cascade Effects	597
	Fish Effects in Food Webs	601
	Biomanipulation	604
11.7	Keystone Species and Strong Interactors	610
	Strong Interactors	611
	An Example of a Strongly Interacting Fish Species	611

Literature Cited	617
Subject Index	733
Taxonomic Index	744
Locality Index	752

Introduction

Nearly a decade ago I began planning this book with the goal of summarizing the existing body of knowledge on ecology of freshwater fishes in a way similar to that of H. B. N. Hynes' comprehensive treatise *Ecology of Running Waters* for streams. The time seemed appropriate, as there had been several recent volumes that synthesized much information on a range of topics important in fish ecology, from biogeographic to local scales. For example, the "Fish Atlas" (Lee et al., 1980) had provided range maps and basic entry to the original literature for all freshwater fishes in North America, and in 1986 Hocutt and Wiley's *Zoogeography of North American Fishes* provided a detailed synthesis of virtually everything known about distributional ecology of fishes on that continent. Tim Berra (1981) had summarized in convenient map form the worldwide distribution of all freshwater fish families, and Joe Nelson's 1976 and 1984 editions of *Fishes of the World* had appeared. To complement these "big picture" views of fish distributions, the volume on *Community and Evolutionary Ecology of North American Freshwater Fishes,* edited by David Heins and myself (Matthews and Heins, 1987), had provided an opportunity for more than 30 individuals or groups to summarize their work on stream fishes (albeit mostly for warmwater systems).

For tropical waters, *The Fishes and the Forest* (Goulding, 1980) had intrigued ecologists with the intricate ecological relationships of South American freshwater fishes to seasonally flooded forests, and Lowe-McConnell's (1987) *Ecological Studies in Tropical Fish Communities* expanded an earlier volume (Lowe-McConnell, 1975) to provided a valuable synthesis both of her own work and a diverse, multilingual primary literature on the ecology of fishes in the Neotropics and Africa. Many other books or works in the primary literature set the stage for an attempt to write a comprehensive book on freshwater fish ecology, but those volumes were particularly influential for me.

In addition to the recent availability of these syntheses, there was also a maturing of fish ecology into a discipline linked to aquatic ecology and to ecology in general, as reviewed in Heins and Matthews (1987). Much of the history of "fish ecology" is found in the all-important works on exploration and discovery of species, alpha-level taxonomy and subsequent revisions of groups, distributional museum catalogs, and "state" or regional fish books produced by many ichthyologists from the 1800s to mid 1900s. By late in this century, fish distributions, evolution, and zoogeography worldwide seemed sufficiently well-known to support a synthesis at that level. The other extreme in work fundamental to present-day fish ecology is the voluminous primary literature on biology, natural history, autecology or management of individual species, dating to the late 1800s when individuals like Stephen Forbes began basic and management-oriented studies of fishes. By the 1970s there was a huge amount of information on individual species, summarized in Kenneth Carlander's two-volume *Handbook of Freshwater Fishery Biology*.

Against this huge background of information on freshwater fish distribution, taxonomy, and biology, modern "fish ecology" began to emerge in the 1960s and 1970s. During this era two very important changes in studies of freshwater fishes began. First, there was a renewed interest in testing general ecological theory with freshwater fishes. For example, individuals like Alan Keast, Earl Werner, Thomas Zaret and others began quantitatively testing general precepts of population ecology or MacArthurian community ecology with freshwater fishes. Secondly, there was a renewal of interest (which had largely lapsed in the mid-1900s) in the effects of fish on the ecosystems in which they occurred. With the 1965 paper by J. L. Brooks and S. I. Dodson many limnologists became interested for the first time in the potential effects of fish on lake plankton and productivity, and by the early 1980s several groups began to show that fish also affected structure or function of stream ecosystems. Finally, by late in the 1980s it was apparent that biodiversity issues were not limited to tropics or rain forests, and that a "biodiversity crisis" included freshwater aquatic systems in much of the world. As a result, there was a great sharpening of interest in freshwater fish during the 1980s as barometers of the condition of ecosystems.

Thus, when I began outlining this book about 1988, there was not only a very substantial background against which to cast such a volume, but a sharp interest in ecology of fishes by individuals who would not typically consider themselves "ichthyologists". Ichthyology, natural history, general ecological thought, and management all melded during the 1970s and 1980s into something that can properly now be called freshwater "fish ecology", and many extant scientists now label themselves as fish ecologists.

Soon after undertaking the writing of a book on "freshwater fish ecology", it became apparent that a single book summarizing everything known in the discipline was impossible, but, thankfully, not needed. Since I began writing this book, several others have fortuitously appeared that summarize large blocks of information on freshwater fishes. Wootton's (1990) *Ecology of Teleost Fishes* summarized fish ecology in general, but with emphasis on ecology of individual fishes, their energetics, and their reproduction. Gerking's (1994) *Feeding Ecology of Fish* and the volume on fish feeding ecology edited by Stouder et al. (1994) have thoroughly summarized that topic, also freeing the present book from a need to repeat much of the known details on trophic ecology of fishes. Finally, the 1992 volume *Systematics, Historical Ecology and North American Freshwater Fishes* edited by R. L. Mayden, has, in concert with many other publications in the primary literature, begun to make it possible for ecologists interested in freshwater fishes to place their work in an evolutionary, phylogenetic perspective, so that a mature view of "adaptation" as a paradigm can be tested. Finally, Nelson's (1994) third edition of *Fishes of the World* appeared, summarizing latest thinking on fish systematic relationships.

Freed from the requirement to summarize "everything" in freshwater fish ecology, I have chosen to address "patterns in freshwater fish ecology", with focus on (1) the structure of local fish assemblages, i.e., those seen in a single location in freshwater lakes or streams, and (2) on the effects of those fish in the systems where they live. This book is an attempt to assess, for graduate students, upper-division undergraduates, or professionals wanting a review, the major patterns in ecology of freshwater fishes, from the perspective of factors ranging from "deep-evolutionary/global" to "recent/local" that influence the composition of local fish assemblages in lakes and streams. In this, I borrow freely from the paradigm of Smith and Powell (1971) of a series of "screens" from very large to very small scale that influence composition of local assemblages.

This book has three parts. Chapters 1–4 assess general patterns in diversity of fishes worldwide; diversity, structure and stability of local fish assemblages; and the limnological and physical features that influence local fish assemblages. Chapters 5–10 focus at increasingly small spatial scales on factors that influence the structure of local fish assemblages. These range from a review of global and regional zoogeography and distributional history of freshwater fishes through evolutionary time (Chapter 5), to local physical structure or environmental stress factors (Chapters 6 and 7), to morphology and life history (Chapter 8), to interactive factors like competition (Chapter 9) and predation (Chapter 10). The third part of the book is a single chapter (Chapter 11) that summarizes known influences of fish species or local fish assemblages on ecosystem structure or processes where they do occur.

Numerous caveats about this book are in order. First, I have not attempted the primary literature in languages other than English, which is particularly important to some areas of Africa or the Neotropics. Similarly, most of this book draws on reprints and journals most available to me, or systems with which I am most familiar, thus North American streams are most used as examples (although I incorporate work from all areas of the world, when possible). Additionally, there will be no "great synthesis" at the end of the book proclaiming primacy of one scale of effects over another (e.g., zoogeographic vs. local interactions) in determining structure of local fish assemblages. Instead, I hope that this book may help a fish ecologist standing in any lake or stream to contemplate the *range* of factors, from those in the deep evolutionary past to those that happened only moments ago, that influence the fishes they see or collect on a given day. Parts of this book should be used with caution, because I have in places indulged in speculation or potential explanations that exceed the limits of the peer-reviewed literature, or I have included syntheses that have *not* been previously published in any peer-reviewed journal. Such new materials, whether presented as specific examples or as proposed syntheses (e.g., fish functional groups, Chapter 2), have *not* had anonymous peer-review, although they have been scrutinized by volunteer reviews of chapters that I thank below. The reader should simply be aware that in many sections I present opinions or ideas supported by some field data or observations, but that may be rejected if they are subjected to more detailed testing in the future.

Finally, some readers will be disappointed that I have not devoted more of this book to anthropogenic damage to fish assemblages or aquatic ecosystems. There is indeed great need for a synthesis of the history of the terrible effects of thoughtless actions of humankind on freshwater fishes and on aquatic systems, and for recommendations for saving these systems for the future. However, this aspect of fish ecology has such a voluminous literature that to give it adequate treatment would have swamped other aspects of this volume. Suffice to say that probably none of the aquatic systems described in this book really contain assemblages identical to those that existed before encroachment of "civilization" into a region. In most cases the pre-settlement fish assemblages were not documented or were poorly studied at best, so that few living fish ecologists have ever seen a completely pristine fish assemblage with all its original components. All we can do at present is hope to retain, through careful basic study and understanding, fish assemblages and aquatic systems in as nearly intact a form as possible. A great service of and challenge to fish ecology has been and will be to identify, within the framework of the last two centuries, the degree to which variation in assemblages is natural, and thus identify magnitudes of change that signal or reflect an exceeding of natural change (typically as a result of acts of humankind).

I owe a huge measure of thanks to many individuals who have discussed ideas, helped with field work that has influenced my thinking, shared unpublished data, or served as reviewers of chapters or sections of chapters. Many of the ideas presented in this book relate to my studies in the central United States in Piney Creek, Izard County, Arkansas; Brier Creek, Marshall County, Oklahoma; Lake Texoma, Oklahoma-Texas; or at long-term sampling sites throughout the Midwest and Southwest. My brother Bob, who was instrumental in starting the Piney Creek project, urged completion of this book for years. Bob Cashner, Fran Gelwick, and Henry Robison also have been long-term collaborators and supporters of this writing. Others who have helped in the field, many of whom are co-authors or co-holders of the data on those systems, include: Ken Asbury, Irene

Camargo, Betty Cochran, Tony Echelle, David Edds, George Harp, Bret Harvey, Thomas Heger, Loren Hill, Jan Hoover, Steve Kashuba, Mike Lodes, Susan Matthews Jones, Andrew Marsh, Rebecca Marsh, Edie Marsh-Matthews, Doug Martin, Amy Matthews, Scott Matthews, Roland McDaniel, Mary Power, Steve Ross, Scott Schellhaass, Bill Shepard, Art Stewart, Jeff Stewart, Barbara Taylor, Chris Taylor, Bruce Wagner, and many students or technicians at the University of Oklahoma Biological Station. For help in sampling the Roanoke River during flood conditions, I thank Eric Surat, Jeff Bek, Bridgett Lambert, Jill Stockett, John Styron, Don Cherry, Tim Jesse, David High and Jody Hershey.

It is difficult to know where many of the thoughts presented in this book originated, but I suspect that I have either stolen or sharpened many ideas as a result of discussions with the persons named above, or with Ned Andrews, Cynthia Annett, Art Brown, Ken Beadles, Alan Covich, Frank Cross, Mike Douglas, Ray Drenner, Alex Flecker, Carter Gilbert, Jim Gilliam, Moshe Gophen, Owen Gorman, Nick Gottelli, Nancy Grimm, Gary Grossman, David Heins, Clark Hubbs, Vic Hutchison, Robert E. Jenkins, Mike Kaspari, Hiram Li, Mark Lomolino, Mike Meador, W. L. Minckley, Gary Meffe, Peter Moyle, Jimmie Pigg, Mark Pyron, Gary Schnell, Bill Shelton, Steve Threlkeld, Caryn Vaughn, Matt Winston, Kirsten Work, Earl Zimmerman, present or past members of the graduate student fish ecology group at the University of Oklahoma, and many visitors to the OU Biological Station or Department of Zoology.

I cannot thank enough the colleagues who reviewed entire chapters or sections of chapters. Individuals who critically reviewed one or more whole chapters were: Bob Cashner, Ray Drenner, Fran Gelwick, Moshe Gophen, Bret Harvey, Roger Lemmons, Edie Marsh-Matthews, Mary Power, Chris Taylor and several anonymous reviewers engaged by Chapman & Hall. Others who critically reviewed, offered technical comments or shared additional information on particular sections of chapters included Bill Dietrich, John Lundberg, Richard Mayden, Peter Moyle, Steve Ross, Royal Suttkus, and Gary Wellborn. C. R. Scotese provided technical interpretation of maps of continental movements. Numerous graduate students in my fish ecology classes have provided excellent input on various chapters, particularly Keith Gido, Jacob Shaefer, Karl Polivka, Bruce Stewart, Phil Gaines, and Roger Lemmons. I particularly thank Edie Marsh-Matthews for detailed involvement in all aspects of the completion of this book. Edie suggested rearrangements of several chapters, read the final draft of every chapter for editorial detail, checked all literature cited, did much of the indexing, criticized all syntheses and ideas, suggested several new analyses, pointed out references, provided expert input on life-history concepts, and in some cases "saved me from myself" by convincing me to delete speculations that exceeded the bounds of credibility. I thank all of the reviewers above, and note to the reader that any remaining errors of logic are my own.

Numerous individuals shared unpublished data, collecting records, and similar information. Chapter 2 draws heavily on collection records and/or personal communications provided by Bob Cashner, Alex Flecker, Rosemary Lowe-McConnell, Chris Peterson, Don Stewart, Donald Taphorn, Kirk Winemiller, and Robin Welcome for tropical or Australian systems. Others sharing important collection records, unpublished data, or impressions from their field work were Janalee Caldwell, Keith Gido, Mark Lomolino, Peter Moyle, Darrell Pogue, Henry Robison, Steve Ross, and Laurie Vitt. Finally, I thank the many individuals whose published data I have drawn on, reanalyzed, or used in some similar way in compiling this book. Without the field sampling and/or experimental work of literally hundreds of ichthyologists and fish ecologists during the last century there would be no book to write in "fish ecology".

The University of Oklahoma Biological Station (Loren Hill, Director) and Department of Zoology (James Thompson, Chair) provided intellectual and logistical support during writing of this book. I owe special thanks to Von Pevehouse and Donna Cobb of the Biological Station for typing some tables, and for a huge effort at compiling the literature cited from my piles of reprints and scratch notes. Coral McCallister, Illustrator in the Department of Zoology, made heroic efforts to convert my hand-drawn figures to ones suitable for publication, produced original art for the book cover, and gave permission to reprint two of her original drawings, and Caroline Tawes, Gloria Stephens, and Shalia Newby were helpful in many ways. Dan Hough, Oklahoma Biological Survey, helped in many ways with computer work during the last 15 years. Phil Lienesch helped check literature, Kirsten Work helped index and proof pages, and Becky Ziebro drew one of the original graphs.

Many individuals at Chapman & Hall were important in production of this book. Greg Payne was instrumental in initiation of the project. Henry Flesh, Lisa LaMagna, and Kendall Harris played important editorial roles.

Finally, I thank my entire extended family for their patience while this book or the field work underlying it was in progress, and for forgiving many missed holidays, family gatherings, trips and similar activities. Without their tolerance and encouragement this book would not have been possible. The book is thus dedicated "to my family", particularly to Edie, Andy and Becky, to Scott and Amy, to Bob, and to Louise P. Matthews on the occasion of her 90th birthday.

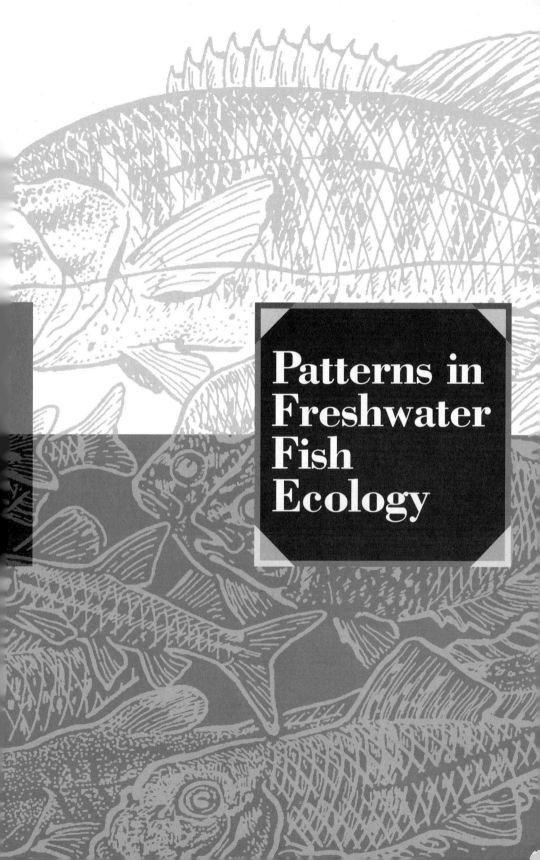

Patterns in Freshwater Fish Ecology

1

Overview of Fishes and Fish Assemblages

The mass of men lead lives of quiet desperation.

—Thoreau, in *Walden*, 1854

If there were no bad weather, no university duties, and no human interference with breeding environment and breeding fish ... work might be carried on with as little interruption as that of the laboratory ...

—Reighard, *Biological Bulletin*, 1920

1.1 Problems and Approaches in Fish Ecology

"Reighard's Lament"

The mass of women and men who study ecology of fishes do indeed "lead lives of quiet desperation"—but not the world-weary despair of Thoreau. Most, like Reighard, exist in a state of chronic tension between deep satisfaction with their work and frustration from the fact that fishes are so complex that no individual can understand all that could be known about their ecology, even if there were no "university (or agency or museum) duties." There are so many kinds of fishes, occupying such diverse habitats, evolving in such complex ways, and with such complicated ecological traits, that a total synthesis of "understanding fishes" will always elude ichthyologists and ecologists. This book is an attempt to synthesize what is known about ecology of freshwater fishes, to make sense of their distribution and abundance across the Earth, their membership in local assemblages, and the ways they, in turn, influence the ecosystems where they live. These are some of the "patterns in freshwater fish ecology" addressed in this book.

With more than 55,000 total names (including those no longer considered valid) that have been used for fish species or subspecies, worldwide (Eschmeyer, 1990), and more than 25,000 extant species (Nelson, 1994), it is obvious that there are more kinds of fishes than anyone can learn in a lifetime. I first realized the hopelessness of "knowing" all fishes well at the Shedd Aquarium in Chicago in 1982. At the annual meeting of the American Society of Ichthyologists and Herpetologists, I was standing at a tropical freshwater display next to one of the grand old men (GOM) of ichthyology. Being fascinated by some tropical species whose identity I did not know, I asked the GOM what kind of fish it was. The GOM looked at me, shrugged without hesitation and harumphed, "I'll be damned if I know!" It was then that I first realized the truth: Nobody, no matter how long they study, can "know" intimately all the fish in the world. The best that most of us can hope is to know well the fishes in the geographic regions we study and to gain broader insight into larger patterns and generalities about groups of fishes and their ecology on a worldwide basis without intimate knowledge of individual species.

Potentially more tractable than "knowing all fishes" are questions about what or how many kinds of fish occur together in assemblages in a small lake, a specific part of a larger lake, or a particular stream reach, and whey they live there. These questions can be explored with ecological tools that are available. However, some lakes and streams have such a mind-boggling diversity of fishes that understanding all of their ecological relationships is probably impossible. The champions of freshwater fish diversity are the Great Lakes of Africa, where generations of ichthyologists have attempted to clarify taxonomy, speciation processes, and comparative ecology (Fryer and Iles, 1972; Greenwood, 1984; Witte, 1984) of the hundreds of endemic cichlids.

This book will focus on local fish assemblages. My operational definition of a fish assemblage is "fish that occur together in a single place, such that they have at least a reasonable opportunity for daily contact with each other." This working or "field" definition lacks some scientific rigor but is used for two reasons. First, unlike the word "community," which often implies predictable interactions among species, the term "assemblage" means only that the fish were, in fact, found together by observers or collectors using standard techniques such as snorkeling or seining, in an area that would typically be considered a "sample site." Second, in a more formal ecological sense, the term community often has been used more broadly, including more than one taxon (i.e., not restricted to "fish communities," "bird communities," etc.), to encompass all of the organisms that occur together in a particular place (e.g., the community comprises the living component of the ecosystem).

Problems that confound any studies of structure and function of freshwater fish assemblages include the following: the high diversity of species and habitats; the lack of adequate taxonomic work in some regions; the complexity of or variability in interactions among fishes; time scales for explanations; gaps in fossil or geological records; correlative versus mechanistic explanations for patterns; and understanding present-day ecological traits of species in the context of the evolutionary history of a clade. In this book I will seek explanations for the assemblage structure by factors that range from zoogeographic to local, which act to determine the presence or absence of a given species in an assemblage. I will also describe ways that fishes as members of local assemblages feed back on ecological processes to affect or modify their ecosystems.

1.2 Diversity of Fishes and Their Habitats

Fish are probably more diverse at all taxonomic levels and have more species than all other vertebrate groups combined. Nelson (1994) recognized 57 orders of living fishes, with 482 families. These numbers are in sharp contrast to numbers of orders and families of amphibians (8, 27), reptiles (4, 49), birds (29, 165), and mammals (23, 122) worldwide (Orr, 1971). Nelson (1994) suggested that the number of described living fish species may eventually reach about 28,500. In North America alone, the number of described species has increased from about 770 in the *Atlas of North American Freshwater Fishes* (Lee et al., 1980) to 971 (Mayden, 1992b). Mayden suggested that more than 1000 species of fish are actually known in North America, because some forms known to experts to be distinct species are awaiting formal description. Page (1993) suggested that the number of described species in two major North American groups (the minnows, Cyprinidae, and the darters and their allies, Percidae) could increase from the approximate 400 species today to as many as 800 in a few decades. In Table 1.1, from Nelson (1994), are 23 orders of fishes with a total of more than 10,000 species that occur, at least in part, in freshwater.

Fish Versus Tetrapod Diversity

In contrast to the estimates for fishes, the numbers of species of all other vertebrates combined on Earth are estimated at no more than 25,000. Nelson (1994) totaled 24,618 species of fishes, worldwide, compared to 23,550 tetrapods. May (1988), reviewing biodiversity worldwide, estimated the numbers of described species of vertebrates as follows:

Fishes—19,000

Amphibians—2800

Reptiles—6000

Birds—4500

Mammals—4500

Even if May's estimates are conservative (see fishes, for example), this estimate of all nonfish vertebrate groups would total 22,300 species. It becomes obvious, therefore, that a student of fishes has the opportunity to "know" approximately as many individual species as students of all other vertebrate groups combined.

In spite of fishes being more speciose overall than any other tetrapod group, with very high species richness in some tropical assemblages, the numbers of fish species in many local assemblage may not differ greatly from those of other vertebrate groups in areas comprising local assemblages. Whereas a 300-meter reach of a typical North American stream might include 20–30+ fish species, a 300-meter line transect bird survey might reveal 10–15 species, possibly with a maximum of 30 species in a very diverse habitat. A 300-meter transect of small mammal traps in grassland or woodland in the north temperate zone might produce as few as 5, but more typically 15–20 mammal species; a survey of

Table 1.1. Teleost Orders with Freshwater Species

Number and Order[a]	Typical Freshwater Examples	Number Freshwater Species
20 Osteoglossiformes	*Arapaima, Arawana,* mooneyes, knifefishes, mormyrids	217
21 Elopiformes	Ladyfishes, tarpons	7
22 Anguilliformes	Freshwater eels (in part)	26
25 Clupeiformes	Shads, herrings	80
26 Gonorhynchiformes	Milkfish, Kneriidae	29
27 Cypriniformes	Minnows, suckers, loaches	2,662
28 Characiformes	Characins, pike-characids, hatchet-fishes	1,343
29 Siluriformes	Catfishes (of 34 families)	2,287
30 Gymnotiformes	Knifefishes, weakly electric fishes	62
31 Esociformes	Pikes, pickerels, mudminnows	10
32 Osmeriformes	Freshwater smelts, galaxiids, lepidogalaxiids	71
33 Salmoniformes	Trouts, salmons, whitefish, chars, graylings	66
40 Percopsiformes	Troutperch, pirateperch, cavefish	9
42 Gadiformes	Burbot	2
45 Mugiliformes	Mullets	7
47 Beloniformes	Medakas, ricefishes, needlefishes, halfbeaks	56
48 Cyprinodontiformes	Rivulines, killifish, topminnows, poeciliids, goodeids, pupfishes	805
52 Gasterosteiformes	Stickelbacks, pipefishes	41
53 Synbranchiformes	Swamp eels, spiny eels	87
54 Scorpaeniformes	Sculpins, Baikal oilfishes	62
55 Perciformes	Temperate bass, sunfish, black bass, darters, luciopercinids, perch, drums, cichlids, gobies, anabantoids, many others	2,185
56 Pleuronectiformes	Soles, some flounders	20
57 Tetradontiformes	Some puffers	20
Total freshwater fish species		10,000+

Note: Several orders with only a few species that wander into freshwater have been omitted.

[a]Order number follows the assignment by Nelson (1994).

Source: Modified from Nelson (1994).

amphibians and reptiles in a typical midwestern North American habitat might produce 15–20 species. However, L. Vitt (personal communication) pointed out that in the southeastern United States, the diversity of amphibians and reptiles is much greater and that a thorough survey of several hundred square meters might yield as many as 100 species. Caldwell (1996) and Duellman (1978) reported a total diversity of tropical frogs ranging from 33 to 81 species in local assemblages. Vitt (1996) reported, from a large number of studies, a range of 5–41 lizard species in local tropical sites in South and Central America, with an average across all studies of 21 species per site. Thus, although local fish assemblages can be quite diverse, the number of fish species in many local assemblages may not be more than the number of bird, mammal, or reptile species.

Examples of Diverse Fish Faunas

The extremes of local fish diversity are found in the famous endemic cichlids of the East African rift or Great Lakes (Fryer and Iles, 1972). Estimates vary, but the deep and large lakes Malawi, Tanganyika, and Victoria have been reported to have as many as 242, 247, and 238 species, respectively (Lowe-McConnell, 1987, p. 29). Hori et al. (1983) found 38 species in a 20 × 20 m quadrat in Lake Tanganyika. Most of the species in the African Great Lakes are endemic to a single lake, and most belong to the family Cichlidae (Meyer, 1993), comprising a total of more than 700 endemic species. Other highly speciose lakes or lakes with high numbers of endemic species include Lake Chad (Africa, 93 species), Lake Baikal (Russia, 50 species, with a "species flock" of sculpins), the Black Sea (156 species), Caspian Sea (74 species), the North American laurentian Great Lakes, with more than 110 species each in lakes Erie, Michigan, and Ontario (Barbour and Brown, 1974).

For stream fishes, maximum assemblage complexity is in tropical streams, where 100+ can occur within a short reach of river (K. Winemiller, pers. comm). The Amazon River basin has the world's richest freshwater fish fauna, with more than 1,300 species (Lowe-McConnell, 1987, p. 123). In North America, greatest stream fish species diversity is centered in the Appalachian and Ozark uplands, with some rivers systems having as many as 100 to 200 species (Robison and Beadles, 1974; Hocutt and Wiley, 1986; Robison and Buchanan, 1988; Etnier and Starnes, 1993). By comparison, European streams are depauperate, as are streams west of the Continental Divide in North America (Moyle and Herbold, 1987). The entire Nida River (Poland) has about 25 known species, and the Colorado River basin of western North America has about 32 native species, in contrast to about 375 species in the Mississippi River basin (Burr and Mayden, 1992).

In contrast to whole-system biodiversity, a typical reach of 200–300 m in medium or large streams in eastern North American streams contain 20 or more species of fishes, with substantially fewer in the semiarid to arid lands from midwestern to western North America [e.g., 8–12 species at stream sites in central Oklahoma (Matthews and Gelwick, 1990)]. Moyle and Herbold (1987) suggested that in the Mississippi River basin of North America, a typical sample of fishes might include 10–30 species, whereas a similar sample of fishes in rivers of western Europe or western North America would contain fewer than 10 species.

Diversity of Fishes by Geographic Region

Moyle and Cech (1988), like Darlington (1957) and Nelson (1994), and Wallace a century ago (Berra, 1981), divided the world into six zoogeographic regions

6 / Patterns in Freshwater Fish Ecology

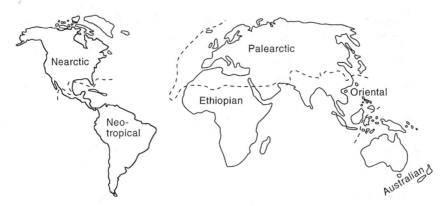

Figure 1.1. The faunal realms of the world. [Modified from *Fishes of the World, Third Edition* by J. S. Nelson, Copyright © 1994 by John Wiley and Sons, Inc. Redrawn with permission of John Wiley and Sons, Inc.]

or realms: the African, the Neotropical (= Central and South America), the Oriental, the Palearctic (= Europe and Asia north of the Himalaya Mountains), the Nearctic (= North America), and the Australian (including Australia, New Zealand, and oceanic islands of the region) (Fig. 1.1). The total numbers of freshwater fish species per region (Moyle and Cech, 1988) are at least:

African	2000
Neotropical	2600
Oriental	700
Palearctic	420
Nearctic	950
Australian	~225 (including marine fishes that enter fresh water)

Even if those numbers are slightly low, due to recent species descriptions, it is clear that ecologists in tropical freshwaters deal with a tremendous diversity of fishes relative to other regions of the world, that the Nearctic and Oriental fauna is also quite complex, and that only in northern Eurasia and in the Australian realm (including New Zealand and oceanic islands) are there "simple" fish assemblages in lakes and streams.

Diversity Within Taxonomic Groups

Taxonomically, the most species-rich groups are relatively few. Nelson (1994) tallied more than 10 families for only 11 teleost orders, including Anguilliformes, Characiformes (recently separated from Cypriniformes), Siluriformes, Osmeriformes (recently separated from Salmoniformes), Aulopiformes (all marine),

Gadiformes, Lophiiformes (all marine), Gasterosteiformes, Scorpaeniformes, Perciformes, and Pleuronectiformes. For orders with the greatest number of freshwater species (note that not all species in these orders are completely freshwater), Nelson (1994) reported the following numbers of species: Cypriniformes (2662), Siluriformes (2287), Perciformes (2185), Characiformes (1343), Cyprinodontiformes (805), Osteoglossiformes (217), Atheriniformes (171), Synbranchiformes (87), Clupeiformes (80), Osmeriformes (71), Salmoniformes (66), Gymnotiformes (62), and Scorpaeniformes (62).

It is, thus, obvious that freshwater fish ecologists will spend much of their time working with cypriniforms (minnows, suckers, loaches), siluriforms (catfishes), perciforms (perches, sunfishes, darters, and allies), characiforms (characins and allies, with much taxonomic uncertainty), cyprinodonts (topminnows, killifishes, live-bearers, pupfishes), and osteoglossids (hiodontids in North America and bonytongues, knifefishes, and elephantfishes in tropical regions). Students of high-altitude waters or cold regions will largely encounter salmoniforms.

Morphological Diversification

Nelson (1994) depicted body forms for most families of fishes. Lagler et al. (1962, p. 53) defined basic fish morphologies, standardizing terminology for cross-sectional shapes of the body. Summaries of diversity of fish body forms are in Wootten (1990, p. 4, general comparisons of diverse body forms), Lowe-McConnell (1987, pp. 30–31 and 120–121, for African and South American fishes, respectively), and Mahon [1984, and reproduced in Allan (1995), pp. 246–247, for stream fishes in North America and Europe. Aleev (1963), Gosline (1971), and Alexander (1967) provide overviews of morphological adaptations and diversity of body forms of fishes.

As an example of typical fish assemblages of temperate waters, Keast and Webb (1966) depict a wide array of basic body forms of fishes in a Canadian lake. Body shapes and sizes vary widely (Fig. 1.2), as do positions and structure of the mouth and fins (Fig. 1.3). Emphasizing the body (minus fins, or the "fuselage") suggests differences in streamlining and in hydrodynamics of all these fishes that occupy one northern lake (Fig. 1.2). Even greater differences can exist in body form for fishes of a temperate stream, where fishes are adapted for benthic as well as water-column microhabitats (Mahon, 1984). Finally, Keast and Webb (1966) detail the differences in mouth position, shape, and structure (Fig. 1.3) within this single assemblage of lake fishes. Note differences not only in position of the mouth (from dorsal to ventral) but also in angle of the mouth, size of the mouth relative to the head, protrusability of jaws (e.g., in *Pimephales*), and the snapping jaws of predators like *Esox* and *Labidesthes*.

Lowe-McConnell (1987, pp. 30–31 and 120–121) showed diversity of fish forms in freshwaters of Africa and in South America. There is strong similarity in the morphology of many of the fishes of the two continents: There are examples

8 / *Patterns in Freshwater Fish Ecology*

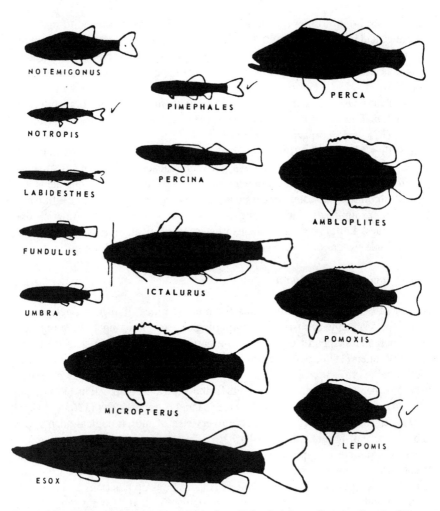

Figure 1.2. Typical body shapes of fishes from Lake Opinicon, Ontario, Canada. [From Keast and Webb (1966)].

in each group of a slim-bodied lungfish with tiny fins relative to body size; there are forms with flattened dorsal profile and upturned mouths (*Pantodon* and *Osteoglossum*) adapted for surface swimming and feeding; numerous species in each of Lowe-McConnell's figures exhibit downturned or fully ventral mouths, adapted for benthic feeding; barbeled forms appear prominently in both groups, and large predatory fishes with apparently strong jaws. The most noticeable difference in morphologies between the two continents is the tendency for fishes in South America to have heavy, crushing jaws (e.g., in *Collosoma*). These specialized structures equip numerous fishes of South American tropical streams for

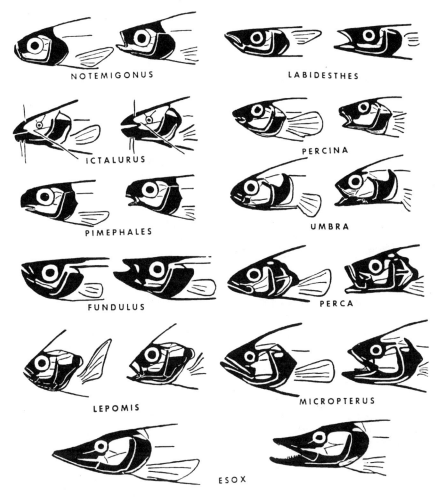

Figure 1.3. Typical mouth morphology of fishes from Lake Opinicon, Ontario, Canada. [From Keast and Webb (1966)].

crushing and consuming large tree-produced fruits, providing an essential link between "the fishes and the forest" (Goulding, 1980).

The morphology of a fish species is influenced by at least three factors: (1) phylogentic "baggage" that may constrain morphological diversification within the group; (2) adaptation of the body and fins for the hydrodynamic conditions where it lives; and (3) adaptation of head, jaw, and propulsive musculature for obtaining food. Additionally, for many species there is an adaptation of body size, shape, or flexibility (Schlosser and Toth, 1984) to permit use of specialized, space-limited microhabitats (e.g., interstitial spaces between stones of a stream bed), or to allow requisite life-history traits (e.g., larger body = larger clutch volume). The basic shape of the body is also modified in some species by acquisition of adornments of body or fins, often with value in reproductive activities or in sexual selection [e.g., knobs on dorsal spines of one group of North American darters (Page, 1983), or exaggerated tails of swordtails).

Within orders of fishes, there can be a tremendous diversification of body form; however, in many fish families, the majority of species have bodies of relatively similar shape. In most cases, families seem a readily observable and definable unit, and there has been greater stability in boundaries of fish families in the last century than in the taxonomic limits of orders (at the next higher level), and certainly than at the level of genera (at the next lower taxonomic level). In other words, most fish families consist of fishes that somewhat "look alike" in body form and in fundamental placement of fins. North American catfishes (Ictaluridae) vary widely in body size (e.g., the diminutive madtoms, *Noturus*, compared to large blue catfish, *Ictalurus furcatus*, which have exceeded 150 kg (Pflieger, 1971), but they all possess pectoral fins with hardened rays, massive heads with strong bone and jaws, and a lack of scales—traits that are common to the entire family (and for the most part, the order). As Archie Carr suggested (tongue in cheek) in the "Carr Key" issue of *Dopeia* (American Society of Ichthyologists and Herpetologists annual meeting, 1941), "any damn fool knows a catfish." Despite the jovial tone of this publication, it captured a fundamental truth about fish families—most observers can sort fishes to family groups, on the basis of fundamental morphology, even if they lack formal training in taxonomy.

Within a family, evolutionary radiation commonly resulted in modifications of fin shape or size, body depth or cross-sectional shape, longitudinal profile of body in position and structure of the mouth and alimentary canal, or in adult size. For example, within the family Cyprinidae (minnows), there is a wide array of mouth positions, varying from almost vertical to very downturned for feeding on benthic organisms, and some species have unique structures for scraping algae from surfaces of stones. Alimentary tracts vary from the very simple "S-shaped" guts of genera like *Notropis* to highly convoluted tracts like those of *Hybognathus*, or even winding around the swim bladder in many loops in *Campostoma anomalum* (Jordan, 1905, p. 33). However, virtually all cyprinids share the basic traits of relatively elongate, cylindrical bodies, with pelvic fins placed well forward on the body.

Habitats of Fishes

The range of habitats occupied by fish is also amazingly varied, perhaps best summarized by Carl Hubbs' famous statement that, "where there's water there's fish...." Almost any freshwater habitat will contain fishes, if it does not desiccate too often or exceed basic physiological limits for metazoans. Most lakes or ponds contain fish, although there are fishless lakes, particularly at high altitude, such as montane glacial lakes. Other lakes lacking fishes are "playas" and ephemeral ponds (Cole, 1966) of the North American southwest and lakes with unusually high salinity or pH. Mono Lake, California, lacks fishes because of high salinity (Moyle, 1976), and Edmondson (1966) describes the chemistry of numerous other lakes in the American West that have salinity or pH too high for fishes. Only the most resistant species of fish tolerate a pH below 4.0 or above 10.0 for long (Doudoroff, 1957). For example, Lahonton cutthroat trout of Pyramid Lake, Nevada (Wilkie et al., 1993) or *Tilapia grahami* of Lake Magidi, Kenya (Reite et al., 1974) survive chronic pH near or slightly above 10.0. At the other extreme, increases in anthropogenic acidity in the Northern Hemisphere continue to make more lakes inimical for fishes (Haines, 1981).

Few streams are naturally devoid of fish, although extreme headwaters may lack fishes, particularly if waterfalls prohibit recolonization if the upper stream occasionally goes dry. A full understanding of which lakes and streams contain fish and those relatively few that do not requires detailed knowledge of the zoogeography and behavior of fishes. To be in a given aquatic habitat, fishes (with a few rare exceptions) must have arrived by direct aquatic connections. In streams, the most likely route is direct migration (e.g., when fish rapidly repopulate a formerly dry reach of stream). Matthews (1987a) described the rapid return of fish into more than 2 km of dried creek bed after the onset of spring rains in Brier Creek, Oklahoma, and Cross (1967) noted the incredible abilities of some fishes of the American Great Plains to move into and occupy stream reaches after they are rewatered. Lacking anthropogenic assistance, movement of fishes among lakes is a much slower process (Barbour and Brown, 1974; Magnuson, 1976), and colonization of any closed body of water requires a longer time scale than that of fishes swimming back upstream after rains. However, there are natural processes by which fish are transported from one lake to another, including the likely accuracy of "folk tales" of raining of fishes in storms, and "transport on duck's feet" of eggs, and so forth, as well as the connectivities that may form and be lost between lakes over geological time or following immense floods. Kornfield and Carpenter (1984) documented transport of small fish in waterspouts in the Philippines. In that many fish species have existed more than a million years, they have had abundant time to move from one closed system to another, so no system should be considered completely "closed" to immigration.

1.3 Questions About Fish Assemblages

Focus of This Book

The fundamental question as to why there are so many kinds of fish is outside the scope of this book and is explored in detail in volumes such as Echelle and Kornfield (1984) for fish species flocks, Fryer and Iles (1972) for fishes of African

12 / Patterns in Freshwater Fish Ecology

Great Lakes, and Mayden (1992a) on evolution within major taxonomic lines. This book will accept that a diverse and complex array of fishes has evolved on Earth and will not attempt to focus on mechanisms of speciation. This book will focus on two large themes in freshwater fish ecology.

1. Why are fishes where they are, and what controls the structure of the fish assemblage in any particular locality?
2. What are the effects of fish species or assemblages on ecosystem structure and processes?

These two questions incorporate a great deal of what is important in understanding ecology of fishes, both about the fishes themselves and their interactions with the living and nonliving components of their ecosystems. These questions do not denigrate the importance of understanding phenomena that affect individual fish, e.g., individual behaviors, but places such individualistic questions in the context of understanding how individual traits influence the fish we find in a particular stream or lake.

Factors Potentially Influencing Assemblage Structure

Figure 1.4 is an attempt to summarize some of the major factors that potentially affect the presence or abundance of species in local assemblages. Arrows in the figure suggest unidirectional influence of one factor or phenomenon on another; that is, continental land mass formation and movements influence climate and the evolution of major taxa of fishes. At first glance, there seems to be an almost bewildering array of alternative hypotheses and interlinkages of phenomena that might regulate structure or processes in local fish assemblages (Fig. 1.4). However, the view of control of local fish assemblage structure and function becomes more tractable if (1) there is no assumption of any "one" primary controlling factor and (2) the contribution of various causal factors is viewed across spatial and temporal scales that range from the deep evolutionary past when continents were very different than they now are, and major lineages of fish evolved, to the daily behavioral interplay of individuals found within a local habitat.

For the deepest evolutionary past that directly concerns fishes, consider (at the top of Fig. 1.4) plate tectonics, including movements and collisions of continental land masses. As seen in Chapter 5, such global zoogeographic phenomena can explain why some major lineages of fish are present or absent on some continents. Major tectonic events have had a direct influence on which orders or families of fish are present on a land mass, hence as potential members of local assemblages. Moving to the upper left of Figure 1.4, continental movements have resulted in changes in climate for aquatic systems borne on the major land masses and, as plates have collided mountains, have been pushed upward. Such events, combined in some parts of the world with volcanic or local tectonic

Overview of Fishes and Fish Assemblages / 13

FRESHWATER FISH: LOCAL ASSEMBLAGES

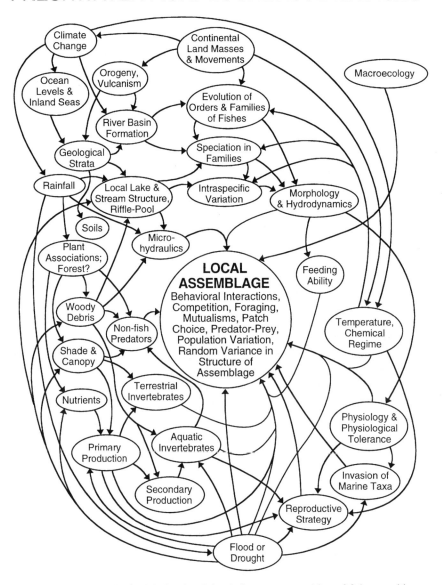

Figure 1.4. Factors from global to local that influence composition of fish assemblages.

activity (Minckley et al., 1986), have been major shapers of landscapes, establishing boundaries of large river basins. Obvious examples include the Andes and Appalachian mountains, where fishes on one side of a divide differ substantially from those on the opposite. Climate change worldwide also has resulted in periods of major inland seas and deposition of calcareous rock, sandstones, and other ocean-mediated formations in what is now the interior of continents. Fundamental differences of surface geologies in a region, combined with rainfall and continued orogeny versus downward weathering of mountains (hence, steepness of stream gradients), play a role not only in formation of drainage networks in river basins but also in the kinds of structure within a stream or lake: riffle-pool formation, size of substrate materials, upthrust shale beds, boulder fields, and the like—all of which contribute to the kinds of underwater structure (Chapter 6), hence "microhydraulics," within a stream (Fig. 1.4).

Isolation of ancestral fishes by continental movements had a strong role in the establishment or proliferation of orders and families, and speciation within families (Fig. 1.4). Evolution at this level established morphological and physiological diversification within major lineages, and local variation in stream or lake formations led to intraspecific variance within species. This intraspecific and interspecific variation in fishes resulted in differences in limits and opportunities for feeding, habitat use, and reproductive strategy. For example (upper center of Fig. 1.4), it is the details of the morphology of a fish or a life stage within a species (streamlining, fin structure, etc.) that combine with the microhydraulics of a given stream habitat to largely determine if a particular patch of microhabitat is advantageous for an individual—hence whether or not that individual will be a part of the local assemblage. Climatic, geologic, and soil factors (upper left, Fig. 1.4) that combine to produce certain plant associations, woody debris, and nutrient inputs to a particular stream or lake will, in extant time, influence the kinds of algal and invertebrate assemblages that are present. Algal and invertebrate production (lower left) will strongly influence extant food availability to fish, which, combined with their abilities to use habitat and their feeding morphology and feeding abilities, help to determine what fishes will be successful in a local assemblage.

To the right in Fig. 1.4, evolved physiological traits and tolerances (presumably influenced by continental movements, climate, etc.), combined with the modern physicochemical regime of a region (potentially overlain by flood or drought patterns), can influence the occurrence and reproduction of species in a given habitat—hence the persistence of species in a local assemblage. Also to the right in Figure 1.4, feeding ability and microhabitat use are strongly influenced by morphology, hydrodynamics (Chapter 8), and interspecific or intraspecific interactions (Chapter 9). In addition, dispersal ability and propensity for movement (not shown in Fig. 1.4) also influence local assemblage composition or variation in the assemblage over time.

Macroecology

At the interface between zoogeography and local phenomena is the concept of "macroecology" (Brown and Maurer, 1989; Brown, 1995), emphasizing the regulation of small-scale or local phenomena by large-scale regional patterns. Macroecology provides potential explanations for fish assemblage structure (not shown in Fig. 1.4); that is, shifts in geographic ranges of species, patterns of local invasion versus extinction, body size, and the related problems of energetic efficiency, all of which are the foci of macroecology (Brown, 1995). Much of the approach in macroecology (Brown, 1995) is extrinsic to the local community and is based on phenomena similar to my zoogeographic explanations. The reason for considering macroecology as a potentially separate explanatory approach is that macroecology focuses less on mechanisms (e.g., why a given phenomenon influences fish assemblages) than on seeking patterns indicating that a given large-scale phenomenon is indeed linked (even though causality may be unknown) to some local phenomenon.

Classes of Explanations for Fish Assemblage Structure

As an overview of Figure 1.4, four broad classes of explanations relate to the question of why particular fishes are found in a local assemblage: (1) zoogeography (both global and regional) and "deep past" evolution of fishes, (2) local abiotic phenomena, (3) autecology of individual species, and (4) biotic interactions among fishes (perhaps as mediated by ecosystem phenomena). The general domain of each of the four basic classes of explanations (with some overlap at interfaces, like "morphology") is depicted in bold outline in the modified Figure 1.4 that is presented in Figure 1.5.

I am not the first to suggest a hierarchy from large-scale to local influences on the composition of fish assemblages. Simpson (1953) argued that land bridges acted like filters to influence the composition of local terrestrial faunas. For aquatic organisms in general, Macan (1963, p. 7) pointed out that a species may be absent from a place simply because, in the course of zoogeographic history, it never reached a given body of water. As stated by Macan, "This simple possibility must obviously be examined before more involved ones are investigated." Macan (1963) then argued that, assuming an organism could have reached a particular place, there were more localized biological, chemical, and physical factors that influenced its success.

Smith and Powell (1971) were the first to suggest a series of successively more localized "screens" to explain the composition of local fish assemblages (Fig. 1.6). Tonn (1990) suggested a similar scheme to account for fish species present in individual northern lakes. Moyle (1994) depicted the screening of the "potential fish fauna" by a range of factors from Pleistocene events to local community interactions and anthropogenic impacts.

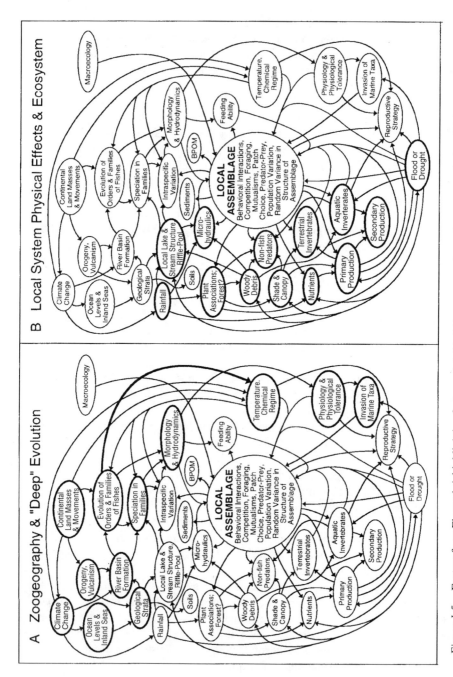

Figure 1.5. Factors from Figure 1.4, highlighting factors related to (A) zoogeography or deep-past evolution, (B) local phenomena, (C) ecology of individual species, and (D) local biotic interactions between fish species.

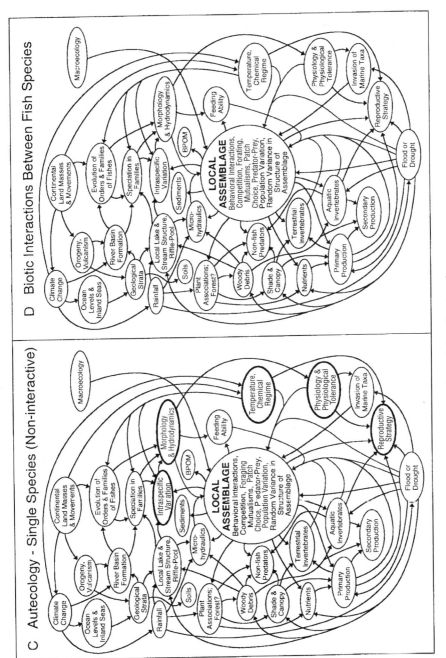

Figure 1.5. Continued.

18 / *Patterns in Freshwater Fish Ecology*

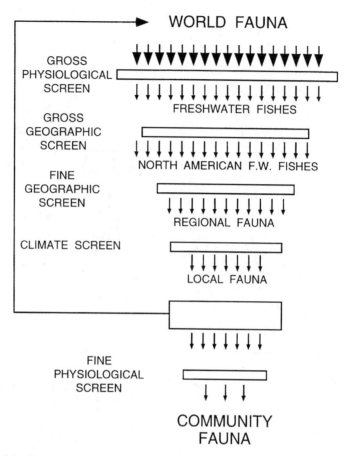

Figure 1.6. Progressively more localized "screens" to occurrence of fishes in local assemblages. [From Smith and Powell (1971)]. (Courtesy of the American Museum of Natural History.)

Under zoogeography and "deep-past" evolution, I consider questions about geological phenomena, river basin boundaries, historical biogeography, glaciation, and the history of the evolution of major groups of fishes. "Local" physical phenomena that can directly impact assemblage composition include size of habitat (e.g., water depth, stream width), pool development (Schlosser, 1987b), permanency of water (Wellborn et al., 1996), longitudinal zonation in streams, habitat zonation within lakes, habitat complexity or heterogeneity (Gorman and Karr, 1978), and habitat structure, as well as microhabitat phenomena such as flow patterns, oxygen concentrations, temperature gradients, and the like. Autecology of individual fish species offers distributional explanations based on phenomena like body structure, hydrodynamics, innate habitat preference, physicochemical tolerances, resistance to disturbance or drought and flood, bioenergetics, behavior, food acquisition, reproduction, larval success, productivity of populations, and shapes of mortality curves. Interactive, or "community," biological phenomena that may explain details about fish distribution and coexistence include competition, predation, mutualism, or resource partitioning. At the interface among "community," "physical," and "ecosystem" phenomena, "bottom-up" factors (in contrast to "top-down" factors like predation) like primary productivity, nutrient availability, and food availability exert control on a fish assemblage. Finally, although a detailed description of all anthropogenic factors is beyond the scope of this book, humans have affected (mostly negatively) a huge number of the factors that influence fish assemblages, in addition to directly manipulating the fishes themselves. In Figure 1.7, approximately half of the causal elements that I have identified to have control over fish assemblages are indicated to have some known impact of man, at scales from global "climate change" to local changes in "intraspecific variation." It seems unlikely that such pervasive anthropogenic changes in factors that influence fish assemblages could fail to have serious consequences for native fishes of a region.

Effects of Fish Species and Assemblages in Ecosystems

Where fish occur and why is only one side of the ecological coin. The other is the question of how, in direct and indirect ways, those fishes influence ecosystem structure or function. Some influences on ecosystems can be attributed to individual "keystone" species or "strong interactors" (Paine, 1980). I would also like to know how whole assemblages of fishes influence the ecosystems where they occur.

The relationships suggested in Figure 1.8, depicting herbivorous *Campostoma* attempting to be "cryptic" while eating benthic algae and maintaining vigilance with regard to a predatory bass, are indeed real, with strong direct and indirect effects throughout local stream systems (Gelwick and Matthews, 1992). Relationships become amazingly complex if all of the direct and indirect connectivities among various fishes and other levels of the ecosystem are considered with respect to both structural (components) and functional ecosystem phenomena (Winemiller, 1995). However, even looking for such fish effects in ecosystems is relatively new in aquatic ecology.

Through much of the middle of this century, limnologists appeared to consider fish relatively unimportant, and the impression is that they thought that if all fishes disappeared from a lake, it would have little effect on structural or functional properties of the system. Brooks and Dodson (1965) changed all that, and there are now a huge number of studies showing that fish can alter structure of lower trophic levels in lentic environments (e.g., zooplankton size structure) or life-history patterns of prey, with potential cascading effects to lower levels of ecosystems (Threlkeld, 1987; Kitchell, 1992; Wellborn, 1994; Wellborn et al., 1996).

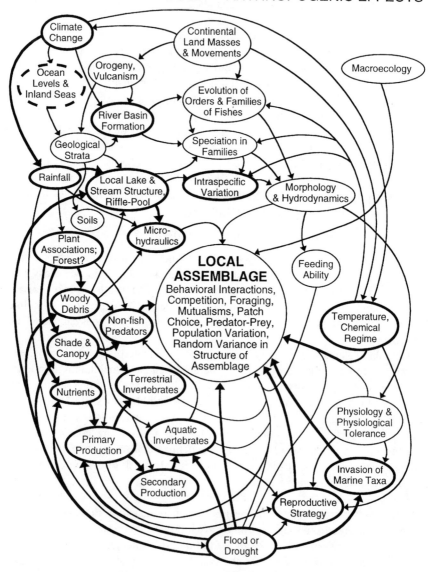

Figure 1.7. Factors from Figure 1.4 that influence composition of local fish assemblages, with factors having a demonstrated or highly likely effect of human activities outlined in bold. (Dashed line around one factor suggests that human effects are speculative.)

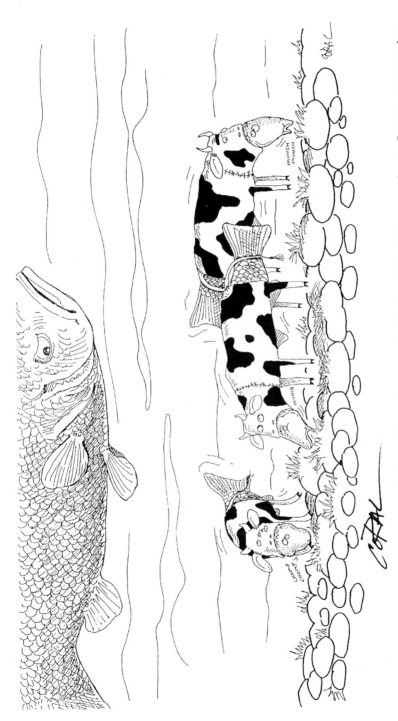

Figure 1.8. Depiction of the potential for piscivorous bass to influence the behavior of the herbivorous minnow *Campostoma anomalum*, producing the cascading trophic effect on attached algae described in Power and Matthews (1983). (Artwork courtesy of Coral McCallister.)

In spite of a few articles (e.g., Hall, 1972; transport of nutrients by fish in streams in the eastern United States), stream ecologists were slower to consider the many roles of fish in running-water habitats. For example, in his classic book on stream ecology, Hynes (1970) described adaptations of fish in detail, but included little about fish influence in streams. The River Continuum (e.g., Vannote et al., 1980) gave relatively little attention to the effects of fishes. It is now well known that many of the phenomena reported for fish in lake ecosystems also apply in streams, with important direct or indirect effects of fish in stream food webs, cascading trophic effects, death and decay, insectivory, herbivory, detritivory, nutrient transport, nutrient cycling or spiraling, breakdown of particulate organic matter, uptake of dissolved organic carbon, downstream transport of materials like coarse particulates, or changes in bacterial, algal, or invertebrate communities (e.g., Grimm, 1988; Gilliam et al., 1989; Power, 1990a, 1996b; Harvey, 1991a,b; Gelwick and Matthews, 1992; Flecker, 1996a, 1997).

1.4 Explanations?

Correlative Versus Mechanistic Explanations for Observed Patterns

Explanations for structure of fish assemblages, effects of fish in ecosystems, or similar questions, are of two general types:

1. "Correlative" explanations, for which an actual mechanism is not known or tested, or in which hypotheses are untestable
2. Explanations for which an actual mechanism has been tested and confirmed as likely

Neither type is inherently "good" or "bad." R. H. Peters (1991), in his *Critique for Ecology*, argued strongly that the first class of explanations was quite appropriate and useful for ecologists; that is, that finding a pervasive empirical pattern could be more applicable to real-world problems than academic testing of detailed mechanisms. An example of this kind of correlative explanation is the empirically known relationship between depth and total dissolved solids (combined as the "morphoedaphic" index) and production of fish in some lakes. Dunbar (1980) argued that even untestable hypotheses can be heuristic and lead to significant scientific advances. However, it seems intuitively more satisfying to be able to understand how a phenomenon is regulated or controlled—particularly if we ever wish to exert control over a system, help one recover from perturbation, or implement similar management actions.

"Macroecology" has been proposed (Brown and Maurer, 1989; Brown, 1995) as an overview of the ways in which larger-scale processes or patterns correlate with local phenomena. One of the best known macroecological patterns is that, all things being equal, more species exist in larger areas. This pattern has been observed worldwide for many taxa, and in both island and mainland situations. There has been also been an experimental demonstration of the species versus area effect (e.g., Simberloff, 1976), in which island size was manually reduced and the numbers of species subsequently shown to be smaller than for the original island. However, the cause of the pervasive species–area relationship remains controversial (Rosenzweig, 1995), and nobody has shown unequivocal evidence for any one out of all possible alternative hypotheses about mechanisms (more habitat "niches," higher rates of invasion, proportional sampling of the regional fauna, etc.) in larger areas that underlie the species–area phenomenon.

In contrast to explanations lacking confirmed mechanisms are the many examples in which (1) potential causes and (2) applicability to real-world conditions have been tested and causal factors identified. For example, causality is "known" or demonstrated statistically in many experiments showing that specific factors like habitat structure, physicochemical conditions, or biotic interactions influence assemblage. As another example, the algivorous North American minnow *Campostoma anomalum* can control standing crops of algae (see, e.g., the discussion in Flecker (1996). This began as a possible explanation for an observed phenomenon, which now has been tested and confirmed at various times and places and in various ways.

In 1981, M. E. Power and I first observed that pools with schools of *Campostoma* in Brier Creek, Oklahoma (USA) had scant standing crops of algae, in contrast to pools lacking *Campostoma*, which has profuse algal growth. When Power first pointed out this observation to me in the field and suggested that it was the grazing of the *Campostoma* that controlled the algae, I was skeptical. However, in a pilot trial, we introduced algae-covered substrates into one pool with a school of *Campostoma*, which swarmed to and ate most of the algae in a matter of minutes (Power and Matthews, 1983). Later, Power et al. (1985) tested the "*Campostoma* control algae" hypothesis in two ways: dividing a pool into grazed and nongrazed halves (with major differences in algal standing crop in a few days), and introduction of piscivorous largemouth bass (*Micropterus salmoides*) into a pool containing *Campostoma* but little algae. Emigration or refuging of *Campostoma* (because of bass threat) resulted in a few weeks in the establishment of dense algal growth in the pool. Subsequently, Power et al. (1988) showed that *Campostoma* could control algal density in a different stream system, and Gelwick and Matthews (1992) found that the effect of *Campostoma* controlling algal standing crop could be extrapolated among many stream pools in Brier Creek and differentially affect pools in a larger stream (Gelwick et al., 1997). In addition, Vaughn et al. (1993) and Gelwick (1995) showed that *Campostoma* could control algal standing crop in experimental streams where the potential confounding effects of other grazers (macroinvertebrates, crayfish) were excluded.

Thus, over a decade, a putative mechanism for regulation of algal standing crop in temperate stream pools ("algivorous fish control standing crop") was confirmed as a likely explanation by observation, field experiments, and controlled experiments in artificial streams. I suggest that in this case an observed phenomenon has an understandable cause (without, however, insisting that *Campostoma* always are "the" controlling factor for stream algae), which seems more satisfying than merely knowing that a pattern exists.

Time Scales for Evolutionary and Ecological Thought

Miller (1979) suggested separate time dimensions important in fish ecology to include "evolutionary," "short term," and "ontogenetic." Sugihara (1980) suggested the importance of evolutionary explanations for community structure, and King (1964) reinforced the need to consider "ultimate" or evolutionary versus "proximate" or recent ecological explanations for fish distributions or abundance. In considering the composition of a local fish assemblage, the interactions therein, or the effects of the assemblage on the ecosystem, I view potential explanations at three temporal scales: (1) "deep evolutionary past," (2) "recent evolutionary past," and (3) extant "ecological" or "proximal" explanations. The first two are evolutionary, but subdivided to distinguish between factors relating to the distant versus more recent past.

By "deep evolutionary past," I refer to geological/evolutionary events in relatively distant time [e.g., perhaps 50 millions of years ago (mya) or more] as Pangaea broke up, continents were moving toward their present configurations, northern continents were variously fragmented and rejoined (Chapter 5), and fishes were evolving at the family level. Because morphologies of fishes tend to be conservative within families, attaining the family level of evolution effectively locks some fundamental traits into the "phylogenetic" explanations for why some fishes do what they do (bulky, bony heads versus slim or lightweight heads, for example). As "recent evolutionary past," I consider geological/evolutionary events at the level of formation of major or minor drainage configurations within continents, continued dispersal, vicariance events, and evolution within families at the genus or species level and the acquisition of characteristics that may have been adaptive at the species level for the environmental milieu of that epoch. Thus, the recent evolutionary past would include events since about 50 mya and of the Pleistocene glaciation and postglacial events that have resulted in extant distributions of fishes. (Although there is evidence that in some areas, for example, southern Canada, fish faunas are still changing in response to postglacial recolonization from southern refugia). This plastic view of postglacial events also implies that some fishes now occur in environments quite different from those in which they evolved. For example, there is fossil evidence that pickerels and pikes (*Esox*) evolved in habitats quite different from those occupied at present by most members of the genus (Wilson and Williams, 1992).

This view of time also has a direct bearing on the question of coevolution within assemblages, considered in Chapter 9. Two kinds of coevolution often are considered—"stepwise" coevolution in which two species interact in a reciprocating evolutionary pattern, and "diffuse" evolution, implying only that a suite of species has existed together for a sufficiently long time that they have all more or less influenced the evolution of each other (Fox, 1988). That *Esox* evolved in habitats different from the present suggests that they may have not evolved with their extant prey and implies that coevolution of predator and prey in this case would be unlikely. Additional new evidence from fossil mammals (Graham et al., 1996) suggests that the present-day "communities" of these tetrapods may be very recent (only a few thousands of years) and that ranges of taxa responded individually during the Pleistocene—making the notion of coordinated, sustained (in geological time) communities unlikely. Will this prove to be the case for freshwater fishes if enough fossils are ever recovered to allow such an analysis?

Finally, there is a large class of extant or "ecological" explanations that focus on the ways in which existing fishes interact with each other or with the environment on a day-to-day basis, with factors that occur in the present regulating community structure (e.g., Wellborn et al., 1996). In extant ecological interactions, fish incorporate features acquired during both deep and recent evolutionary past, but in the context of their fitness in the present environment. No one scale of time-thought alone to provide an answer to the question of why any given suite or assemblage of fish species occurs together, at present, in any modern environment. All levels of thought must be integrated for understanding of many extant phenomena of fishes, as with all organisms.

Influence of Phylogeny on Ecology

In evaluating any of the above traits that influence local fish assemblages, it is desirable to consider phylogeny or environmentally induced plasticity as potential explanations. Many ecological traits of fish in an assemblage (behavior, food use, gamete production, etc.) could be due as much to the evolutionary history within a taxon as by ongoing "ecological" events. When possible, patterns in observed traits of species should be tested against a phylogenetic "null model," to ask if the observed pattern is congruent with accepted phylogeny of the clade (Mayden, 1992b). If the answer is "yes," then there may be nothing to explain in terms of adaptation to extant conditions, and the observed pattern among species may be a result of ancestral patterns within the clade. At the other extreme, observable traits, including morphology (e.g., feeding structure), propagule number or size, and others, are known to be highly variable or developmentally plastic, with phenotypic expression differing under different environmental conditions, and not due to any genetic differentiation. Between these two extremes, it is likely that much of the variation among species in ecological traits is attributable to a combination of phylogenetic history and recent fine-tuning of species or populations to a particular environment. Additionally, the variation in a trait within a species or population may be as or more ecologically important than the mean or average expression of the trait, and variability could itself be treated as a "trait" in populations.

Classification and Evolutionary Systematics

Classification and systematics of fishes have changed markedly in this century. For many years, the suggestion of a single lineage from "lower" to "higher" teleosts was accepted. Such a classification by Berg (1940), which he based on many earlier works, especially that of C. T. Regan (1909) had wide acceptance, and remained the basis for G. A. Moore's classification in 1968. Berg's classification included a huge order Clupeiformes that included not only clupeids but also many taxa no longer aligned with clupeids, like salmonids, esocids, smelts, and osteoglossids. Berg's Cypriniformes included characins and catfishes (now separate orders), and anguillid eels (now aligned with elopomorphs) were considered more advanced than cyprinoids. Gosline (1971) also retained the monophyly of teleosts, with advancement from lower to higher taxa in a single large lineage. However, in 1966, Greenwood et al. (1966) recognized that teleosts did not comprise a single lineage, and dramatically revised the assignment of teleost fishes into three "divisions" that did not all lead unidirectionally from "lower" to "higher" fishes. In Greenwood et al.'s classification, anguillid eels and tarpons (and possibly clupeids) were considered Division 1, and not in a direct lineage to "higher teleosts"; Division 2 included the osteoglossiforms, or "bony-tongue" fishes (including the genus *Hiodon* of North America, and mormyriforms, knifefishes, and others in the tropics). All other teleosts comprised Division 3, with evolutionary advancement from the most primitive groups, like Salmoniformes and the Ostariophysi (minnows, characins, catfishes), to the higher Acanthopterygii, with the perciform radiation having greatest diversity. Although the classification of Greenwood et al. (1966) has undergone considerable modification as knowledge accumulates on fish relationships, their classification still provides the basic framework for most recent summaries (e.g., Nelson, 1994; Moyle and Cech, 1988).

Transitions in taxonomy can influence interpretations of comparative ecologists. Consider the position of salmonids and related taxa as summarized in three editions of *Fishes of the World*, by J. S. Nelson (1976, 1984, 1994). In the first edition (Nelson, 1976), the order Salmoniformes was considered to be the "basal stock of euteleostean evolution" (and "lower" evolutionarily than the Ostariophysi, including minnows, suckers, catfishes, and related families. Salmoniformes in Nelson (1976) included Salmonidae (trouts and salmons), Osmeridae (smelts), Esocidae (pikes and pickerels), and numerous other families of related fishes. In the second edition (Nelson, 1984), the Ostariophysi were recognized at "the start of Euteleosti," followed by Salmoniformes, although Nelson (1984) noted that both groups likely had "equally primitive beginnings." By the second edition, some groups had been removed from the order Salmoniformes, and the close relationship of escocids to the rest of the order was in question. The third edition (Nelson, 1994) reflects the continued evolution of thought by experts on these groups, with the former Salmoniformes now separated into three orders (Salmoniformes, Osmeriformes, and Esociformes) to incorporate the most recent information on similarities and differences among these ecologically important groups of fishes. Nelson (1994) also cautioned that this assignment of groups is tenuous and that future research will no doubt result in continued change in comparative positions of these groups.

Another example of the ways changes in taxonomy can influence comparative ecological analyses [particularly for South America (e.g., Lowe-McConnell, 1987)] is the separation of the single family Characidae (Gosline, 1971; Weitzman, 1962) into as many as 16 families (Greenwood et al., 1966) or 10 families (Nelson, 1994). Regardless of the number of recognized families, relationships within and among those families for characins and characin-like fishes remain very uncertain. However, it seems clear that not all former "characins" are characins. Comparative ecologists must adjust their perspectives accordingly.

Ecologists working with fishes should remember that ongoing revisions of formal taxonomies and historical hypotheses among and within major fish groups reflect the efforts by systematists to improve knowledge of evolution within clades. Updating taxonomies to reflect the most recently available evolutionary information is sometimes viewed by nontaxonomists as an inconvenience (Cashner and Matthews, 1988), but, in reality, it is the ultimate contribution of systematics to the discipline of ecology, as those updated taxonomies provide ecologists with increasingly good information on which to base comparative and evolutionary hypotheses (Mayden, 1992b). In fact, the strong focus of systematists working with all vertebrate groups in recent decades on the formulation of evolutionary hypotheses, accepting as valid only monophyletic clades and rejecting or revising clades that are paraphyletic or polyphyletic (Etnier and Starnes, 1993), has provided comparative ecologists the much needed framework for designing studies or testing hypotheses.

The large number of phylogenies by systematic experts in many families compiled in Mayden (1992a) is a major resource for ecologists working with fishes in North America. At a much finer scale, there has been much recent revision within species complexes, with recognition of cryptic taxa as valid species in their own right. Such findings can make a huge difference in interpretations by ecologists. Within North American minnows, as examples, the former species *Notropis rubellus* recently yielded recognition of the new *Notropis suttkusi* from southern Oklahoma and western Arkansas (Humphries and Cashner, 1994).

Additionally, Mayden (1988a) recognized within the former single minnow species *Luxilus pilsbryi* two valid species: the nominal form from the White River drainage in Arkansas/Missouri, and the newly described cardinal shiner *Luxilus cardinalis*. In drainages of eastern Oklahoma and western Arkansas, at least two, and possibly more, as yet unnamed cryptic species exist within the *Luxilus chrysocephalus* complex (T. Dowling, Arizona State University, unpublished data and personal communication). Individuals carrying out comparative, interspecific ecological studies of minnow species throughout the region [e.g., Matthews (1987a), physiological tolerances of species] might find their original hypotheses or conclusions modified by knowing that two former "populations" deserved treatment as valid species. In summary, it would seem that fish ecologists have two major problems with regard to diversity and evolution of fishes: first, to find ways to deal with the incredible diversity of freshwater fishes, overall, and second, to remain aware of ongoing changes in evolutionary hypotheses for the groups with which they work. Only in this way—and not at arm's length from taxonomy—can ecologists work most successfully.

1.5 Guideposts

The goal of this book is, therefore, to sharpen awareness of why freshwater fish are where they are and what they do in those ecosystems. Some approaches to understanding fishes begin with the individual at a given time and place and ask how the external milieu of stimuli and resources influence that individual (e.g., Wootton, 1990). This approach begins with the fish and works outward. I prefer to begin at the other extreme, narrowing from a geological and evolutionary view to an increasingly localized view in time and space, ending with the presence of a given suite of fishes living together in daily contact. In viewing the composition of any particular fish assemblage or its effects in the ecosystem, explanatory power may be enhanced by thinking about that assemblage at all levels, from "ultimate" evolutionary and geological causes to "proximate" ongoing local mechanisms. Through such a multifaceted perspective, the local fishes that pose the day-to-day practical problems for conservation or management may best be understood.

This book is organized roughly as follows. The fundamentals of fish assemblage structure, function, and stability are described in Chapters 2 and 3. Basic physical and chemical properties of lentic (standing water) and lotic (running water) systems are described, as are some of the responses of fish to physical conditions in lakes or streams, in Chapter 4. The next chapters (Chapters 5–8) ask "why" freshwater fish live where they do from the perspective of largely abiotic factors, including zoogeographic history (Chapter 5), local physical patterns (Chapter 6), disturbance and physicochemical stress (Chapter 7), and single-species, "autecological" traits (like morphology and life history), that influence where a particular species can exist on Earth (Chapter 8). Chapters 9 and 10 address complex interactions of species or individuals within species, including competition, mutualisms, and predation. I also consider coevolution in fish assemblages. Thus, Chapters 9 and 10 describe how the fishes that could be present in "fundamental assemblages" (due to zoogeographic history or local physical conditions) are winnowed by biotic interactions to comprise the "realized assemblages" (with apologies to G. H. Hutchinson) of fishes that actually exist together in different places and times. Finally, there is a single chapter (Chapter 11), in which direct and indirect interactions of fishes with and their impacts on ecosystems are summarized. Where possible, I have tried to find generalizations in the large bodies of research that exist in each of these large subdisciplines of fish ecology. More often, I have found only that firm generalizations and absolute answers are elusive for a group so diverse as freshwater fishes. Clearly, a huge amount of research remains to be done before ecologists can "know" all about freshwater fishes and their roles in ecosystems. Enjoy.

2
Structure of Fish Assemblages

2.1 Introduction

What Is a Fish "Assemblage" in the Context of a "Locality"?

This chapter addresses empirical facts about structure of local freshwater fish assemblages. The evolutionary (ultimate) or mechanistic (proximate, ecological) explanations for that structure are the focus of later chapters. In Chapter 1, I defined a local "assemblage" as the fishes that would be found together in one particular place or "locality." What, then, is a "locality"? Although difficult to quantify rigorously, a locality, operationally, is a place in a stream or lake that would be included in a single typical collection or observational sample by an ichthyologist (e.g., one to a few hundred meters of a stream or a shoreline). Individual fishes in a "locality" have at least a reasonable chance of encountering each other during normal daily activities, although some may be nocturnal versus diurnal (Helfman, 1981a). Some fishes probably spend long periods of time in much less space than suggested above [e.g., individual darters living their lives in one riffle (Scalet, 1973), or some fishes moving less than 20 m in a stream between captures (Hill and Grossman, 1987)]. However, many stream or lakeshore fishes have the potential to and apparently often do move as much as 100 m or more in a given day (e.g., cruising lake shorelines to feed, or moving between pools in a stream). Therefore, an assemblage will include the individual fish that occur together in one locality at a given time or over a brief period of ecological time, having reasonable probability of encountering each other within the course of feeding, resting, movements, and so forth in a given day.

Spatial limits to a locality, hence to the fish assemblage, can be difficult to define. Physical limits to assemblages are rather easy to envision in many small streams but are more difficult to describe in large rivers or lakes. Many small streams, particularly in medium- or high-gradient regions, have riffle-pool differentiation that is reflected in different suites of fishes in the two microhabitats (Gelwick, 1990). Additionally, even within a relatively homogeneous region, rather different suites of species can occur in different local riffle assemblages (Blair, 1959). However, M. E. Power (personal communication) notes that in some regions [e.g., bedrock, bedded, canyon-bound western (USA) rivers], riffle–pool differentiation is less distinct. Other small streams (in low-gradient alluvial habitats) consist of series of pools connected by narrow points in the stream channel, although some prairie streams may lack such constrictions (R. Lemmons, personal communication). The ends of pools at riffles or where the channel is constricted often comprise a natural "leaky" barrier to upstream or downstream movement of water-column fishes. Pool-dwelling fish often turn back at a riffle or channel constriction/shallows, remaining within their resident pool (although this phenomenon is only now being critically examined, for example, in work by D. and M. Lonzarich in high-gradient streams in the Ouachita Mountains of Arkansas, and ongoing experimental work by J. Schaefer in artificial pool–riffle stream systems). Pools and riffles may comprise distinct physical limits to movement of many fishes, breaking up a small stream into understandable or biologically meaningful physical units (Matthews et al., 1994).

In large rivers and lakes, there are few observable (to the ichthyologist) physical boundaries to daily activities of most fishes. Nevertheless, the innate biology of many fish species of large rivers or lakes results in their remaining mostly within distinct compartments of the water body (e.g., Rutherford et al., 1987). Three subsets of fishes exist in many large freshwater systems. In both large rivers and lakes, there is a littoral fish assemblage (e.g., Werner et al., 1977; Keast et al., 1978; Gelwick and Matthews, 1990), consisting of taxa that spend most of their time in shallow water where they have some contact with or affinity for substrates or structures. Many are small-bodied taxa like the North American minnows (Cyprinidae) or darters (Percidae). Next, there is a true limnetic (or mainstream channel, in large rivers) fauna of deep, open waters, such as the ciscos of the North American Great Lakes, the white bass (*Morone chrysops*) in North American streams and reservoirs, or shads (Clupeidae). Finally, many species of fishes in lakes move onshore and offshore in a diel pattern (Hall et al., 1979; Hubbs, 1984; Matthews, 1986a) and thus are, at times, members of different spatially defined assemblages. The boundaries to these three compartments are fuzzy. The first compartment (littoral fishes) is probably tightly coupled with habitat, and in littoral assemblages, we might expect fish–fish interactions to be strong and frequent [although composition of the littoral-zone fish assemblage can change seasonally (e.g., Hall and Werner, 1977; Gelwick and Matthews, 1990)]. In more open waters, there is much variation in the parts of a large lake occupied by some species in response to seasonal or daily patterns or to water-quality variation (Matthews et al., 1989). In spite of the difficulty in defining boundaries to a fish assemblage in larger freshwater habitats, I still ask, "How many kinds of fishes occur together?"

For small lakes (e.g., kettle lakes) in which all fish potentially could encounter each other in a day, I considered the entire fauna part of one "assemblage" (for ennumeration of species). However, as lake size increases, species segregate more among habitats, forming inshore and deeper-water fauna (e.g., Keast and Harker, 1977; Werner et al., 1977). Therefore, in all but the smallest of lakes, I consider an "assemblage" to include only those fish encountered in a particular (usually homogeneous) part of the lake where ichthyologists take a sample (e.g., shoreline seining of a beach, snorkeling transect in weedbeds, gill nets set in a homogeneous patch of habitat, etc.). Again, the intent is not to rigorously define "microhabitats" in this chapter but to gain insight into the empirical question, "How many species live together in a place?", both in lotic (flowing water) and limnetic (standing water) habitats.

Like spatial limits, temporal boundaries to an assemblage also can be difficult to define: There are seasonal changes in the fish assemblage at a given site in many streams or lakes (Gelwick and Matthews, 1990). However, for practical purposes, consider an assemblage to consist of all of the fish collected or observed together in one sample or in samples repeated in a brief period of ecological time (up to a year) by typical ichthyological methods (e.g., seining, electrofishing, passive netting, snorkeling, etc.). A related issue is that of temporal limits to "a time." Is a "time" rigorously equal to one day or one fish collection, or should data on fish collected at a site be pooled when collections are repeated? Realizing that it is usually impossible in any one collecting or observing bout to capture or observe all of the fish species that are actually present, in some summaries that follow, I included the total number of fish species found at a single locality in repeated samples for up to a year, in order to incorporate the annual cycle of occurrence of fishes in that place.

What Is Fish Assemblage "Structure"?

The "structure" of an assemblage includes the following:

1. Number of species
2. Number of families
3. Species richness within families
4. Numbers of prey species versus piscivorous species
5. Proportional composition of the assemblage by trophic groups or by functional groups
6. Distribution of abundance of species at a locality
7. Body-size patterns for the whole assemblage
8. Distribution of "trophic potential" (mouth size) for the whole assemblage.

2.2 Number of Species, Families, and Species per Family

Numbers of Fish Species in an Assemblage

Humans are fascinated by numbers—particularly large ones. Whether home runs per season, total biomass in a fishing catch, or personal salary, "biggest" or "most" is often perceived as best. Ecologists are no exception to this rule, as we tend to celebrate or be impressed by high "diversity," "species richness," or (recently) "biodiversity" in systems—and with some cause, as there is a long-debated suggestion that ecosystems with more species are somehow more stable (Goodman, 1975; Auerbach, 1984), in better "health" (Karr et al., 1986), or otherwise "better." Consequently, ichthyologists and fish ecologists often give special attention to those places on Earth that have very high numbers of species in local assemblages, like the rift lakes of Africa (Fryer and Illes, 1972), other lakes with fish species flocks (Echelle and Kornfield, 1984), or certain tropical streams (Lowe-McConnell, 1975). This fascination is not without justification, because these very species-rich locations can be valuable laboratories for investigating processes that control speciation or shape the evolution of fishes. Therefore, great emphasis and interest has been placed on localities with, say, 200+ species in a given lake, 100 species in a single stream reach, or 50–60 species associated in a patch of underwater structure, such as a boulder in the African rift lakes.

Although such locations are fascinating to students of fishes, they do not appear to be the norm in the "big picture" of the world. Far more often, one or a few repeated samples in a given place yields a much lower number of taxa. Matthews and Robison (1997) summarized species richness in 404 seining collections in the Interior Highlands (Ozark and Ouachita mountains) in the state of Arkansas (Fig. 2.1). The greatest total number of species in local assemblages was in 2 Ozark upland collections with 38 and 33 species and in a Ouachita upland collection with 31 species. Mean species per local assemblage in the two uplands was much lower. For 222 Ozark collections (taken in small- to medium-sized streams in the uplifted northern portion of Arkansas, north of the Arkansas River), there was an average of 14.7 (standard deviation = 6.2) species per collection. The 182 collections in streams of the Ouachita Mountain uplift in southern Arkansas averaged 12.9 (standard deviation = 5.5) fish species per collection.

In a more limited portion of central United States (i.e., prairie or plains streams within the state of Oklahoma), Matthews and Gelwick (1990) found substantially fewer species per locality than in the upland streams to the east (in Arkansas, as above). For 53 collections in these typically low-structure, low-gradient systems (all outside the more species-rich Interior Highlands that are in eastern Oklahoma), Matthews and Gelwick reported a range of 3–15 species per collection and an average of fewer than 10 species per site. In those 53 collections plus 4 additional collections in a more recent study (Matthews and Gelwick, 1990), there was an average of 7.8 species per collection. For another 17 collections

34 / Patterns in Freshwater Fish Ecology

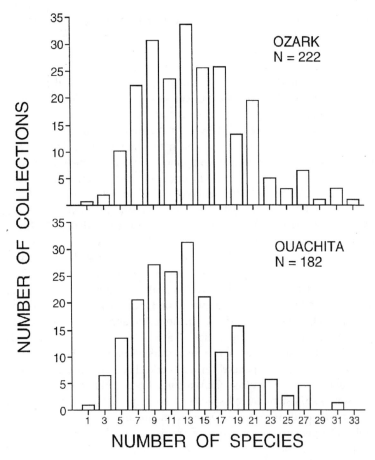

Figure 2.1. Frequency distribution of number of samples containing the indicated number of species (*x* axis) for 404 local collections in the Ozark and Ouachita mountains, Arkansas (USA). (unpublished data).

taken in urbanized streams in central Oklahoma, Matthews and Gelwick (1990) found an average of 7.6 species per collection (not significantly lower than for rural streams). Therefore, richness per local site typically exceeds 15 species in the mesic parts of the eastern United States, but in the more arid west-central region outside the uplifted regions (e.g., from about 100° west longitude to the Rocky Mountains), the number of species per assemblage may be substantially less.

A tally of information available in English language publications, or by personal communications from colleagues, suggested, worldwide, a general tendency for a range of about 5–18 species of fish in a locality (Fig. 2.3). This number seems curiously low, relative to the notions of the incredibly diverse fish assemblages that are often the focus of popular or scientific media. However, in considering these much lower "typical" numbers of species in fish assemblages, several facts about availability of habitats to freshwater fishes may be important.

First, even in tropical regions that are the undisputed champions of fish diversity (Fryer and Illes, 1972; Roberts, 1972; Lowe-McConnell, 1975; Winemiller and Leslie, 1992), many of the very-high-diversity fish collections are either from lakes with species flocks, large rivers, or brackish-water sites where freshwater and marine fishes mix. In the tropics, there are many small lakes with relatively few fish species, or kilometers of small streams where, just as in temperate streams, the number of species rapidly diminish toward the headwaters. For example, in a coastal stream in Costa Rica, Winemiller and Leslie (1992) found 80 species at a site near its confluence with the ocean. However, less than 13 km upstream, in foothill streams, they collected only 13 and 19 species—numbers comparable to those in headwaters streams in uplands of North America. In one isolated forest pool near this river, Winemiller and Leslie found only seven species.

Ibarra and Stewart (1989), sampling beach-zone faunas in the Napo River basin (Ecuador) from its headwaters at an elevation of 400 m to an elevation of 100 m at confluence with the Amazon, reported 208 species. The richest single collection included 84 species, but many were locally "rare," represented by only one individual. Thus, even at sites with great diversity, many species are represented by only a few individuals.

Despite the tremendous maximum diversity (1 site had 84 species) in the study by Ibarra and Stewart, which relates in part to the high total continental diversity in South America (Chapters 1 and 5), the average number of species collected per site was much less (mean = 39.2; their Table 1), and 8 sites had 30 or fewer species. Therefore, in spite of the tremendous continental diversity in South America, local diversity of species in Ibarra and Stewart's collections was, on average, only modestly greater than averages for speciose streams in eastern North America, which can often have 25 or more species per locality (Matthews, 1986b, 1986c).

In the Kretam Kechil River of Borneo, Inger (1955) collected only 1–19 species per site, even with the use of rotenone. In tropical streams of Panama, Angermeier and Karr (1983) reported from 6 to 25 species in assemblages within modest stream reaches. In streams of Sri Lanka, Wikramanayake and Moyle (1989) found 5–10 species per site. In the Luongo River, Africa, Balon and Stewart (1983) collected 5–27 species per site, with a strong influence of longitudinal position in the system on the numbers of species—much as in temperate parts of the world. Finally, on the basis of many samples throughout ten west African coastal rivers, Hugueny and Paugy (1995) reported local species richness averaging 5.0–27.2 species per site. Therefore, although tropical stream and lake systems clearly have the richest local fish assemblages on Earth, a great many localities within tropical systems appear to have, on average, no more, or not many more, species in local assemblages than might be found in temperate regions of the world.

Patterns of variation in richness of assemblages are influenced by longitudinal position in a watershed, by regional diversity, and by size of local habitat (see also Chapter 6). First, the numbers of fish species in local assemblages, worldwide, reflect the strong influence of position in the watershed. As an example, Edds (1989) depicted the numbers of species collected at 148 sites in the Kali Gandaki River, Nepal, at locations ranging from high altitude in the Himalayas to low-gradient conditions of the Gangetic Plain. The strong relationship between distance downstream from the headwaters and the numbers of species is apparent (Fig. 2.2), but the pattern is not strongly linear. Instead, it is clear that in headwaters [e.g., upper 100 km (Fig. 2.2)], there are only one or a very few species, whereas further downstream (e.g., 300 km or more below the headwaters), there is wide variance in the numbers of species, from fewer than 10 to more than 30 at any given site.

Table 2.1 summarizes distribution of species richness from headwaters to the lower mainstreams of 17 streams in eastern North America that were included in the analysis of longitudinal zonation by Matthews (1986b). One obvious pattern in Table 2.1 is the well-known increase in species richness from headwaters to downstream, producing apparent longitudinal zonation of species richness (as

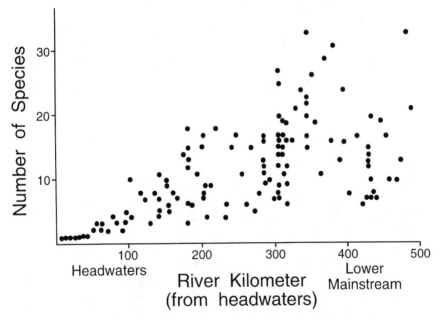

Figure 2.2. Number of species per collection versus distance from headwaters in the Kali Gandaki River, Nepal. [Modified from Edds (1989)].

Table 2.1. Number of Fish Species in Local Assemblages in Streams East of Rocky Mountains in North America

Stream	Species per location[a]										
Kiamichi River, OK	6	9	12	12	12	15	14	19			
Paddy's Run, WV	3	8	11	15	12	21	20	21	16	7	11
Otter Creek, OK	5	5	6	9	8	5	13	12	13	11	5	8
Tenmile Creek, OH	12	13	6	11	10	9	11	7	9	5	
Swan Creek, OH	13	9	11	10	11	13	13	5			
Big Sandy Creek, TX	22	22	28	30	31	31	33				
Janes Creek, AR	18	26	22	27	20	23	35				
Piney Creek, AR	19	17	15	23	18	17	22	12	18	27	
Big Creek, AR	14	16	15	14	16	16	16				
Dry Fork, OH	13	18	14	16	20	17	24				
Brier Creek, OK	8	12	11	16	16	18					
Mill Creek, OK	9	15	13	16	18	15	12	23	36	38	
Roanoke River, VA	24	26	25	20	22	22					
Mason Creek, VA	6	10	13	16	19	24	31				
Ouachita River, AR	17	10	18	30	32	28	18	32	24	16	
Caddo River, AR	12	10	22	26	31	27	30				
East River, WV	1	3	3	6	8	12	15	37			

$N = 149$ sites; mean = 16.6 species; standard deviation = 9.3

Note: Locations and references are in Matthews (1986b).

[a] Species per location are ordered from upstream to downstream sites.

discussed in more detail in Chapter 6). However, there is an asymptote or even a decrease in diversity in the lower mainstream of some of these streams. Possible causes include a decrease in habitat diversity in wide, relatively homogeneous lower mainstreams and cumulative anthropogenic effects such as buildup of silt, pollutants, and so forth. In other cases (e.g., Janes Creek, Arkansas, or Mill Creek, Oklahoma), the number of species is greatest at or near the lowest station. This can occur if the creek flows into a larger river, from which typical "big river" species (e.g., suckers, *Ictiobus* or *Carpiodes;* large catfishes, *Ictalurus;* gars, *Lepisosteus*) migrate into the lower parts of the creek proper.

The total number of species in local assemblages also is obviously influenced by large-scale patterns of diversity at the level of continents or large subcontinental regions. Europe and North America west of the Continental Divide have low numbers of native species (Moyle and Herbold, 1987), as do streams in isolated land masses like Australia (Pusey et al., 1993; Pusey et al., 1995; R. C. Cashner, personal communication), the British Isles (Penczak et al., 1991) or New Zealand (McDowall, 1968, 1990; Minns, 1990). Moyle and Herbold (1987) compared the fish fauna of eastern North America, western North America (west of the Continental Divide, where faunas generally are depauperate), and Europe. Moyle and Herbold suggested that the headwater fauna is similar taxonomically and in species richness in all three areas, consisting mostly of salmonids and other cold-water taxa. However, in eastern United States, the fauna becomes much more speciose downstream, whereas in western United States and in Europe, a cold-water fauna (with forms like trouts, sculpins, suckers—in North America—and dace) continues to dominate. Both in Europe and western North America, the numbers of species present in a local stream reach are relatively low, usually fewer than 10. For 183 localities in the cold deserts of the intermountain American West, Smith (1981) found from 1 to 11 species, with fewer than 7 species at most sites.

There is also, worldwide, a general relationship between habitat size and numbers of species (e.g., Smith, 1981). If collecting sites were spaced at regular intervals within any catchment, there would be substantially more sites in small tributaries or headwaters than in large, lower mainstreams. Worldwide, narrow or shallow stream headwaters contain far fewer fish species than exist further downstream (Schlosser, 1987). Furthermore, the source of water [spring fed or not (Hubbs, 1996)] can have a tremendous effect on the stability of a headwaters stream, hence its number of fish taxa. Therefore, it seems appropriate to focus less on the coexistence of high numbers of species at unique sites and to focus more on the many local fish assemblages that have substantially fewer coexisting species. The latter appear to represent the more usual situation for freshwater fish assemblages in lentic and lotic waters worldwide.

The Empirical Evidence: Numbers of Species per Locality

Probably the most asked question about structure of fish assemblages is, "How many species typically occur together at a locality?" Figure 2.3 summarizes numbers of fish species found in samples at 1657 locations in streams and bodies of standing water (lakes, reservoirs, ponds, lagoons, etc.), worldwide. Some of these articles were based on repeated sampling (e.g., seasonally) at a given site, whereas others were based on only a single sample, so there is some discrepancy among studies as to the likelihood of detecting rare species. Unpublished comparisons by myself, E. Marsh-Matthews, and H. W. Robison for several streams in Arkansas suggested that a second sample at a physically complex site could increase by about four the number of species detected in a single sample (from about 20 to 24 species, in the localities we analyzed). In a few articles (e.g., Whiteside and McNatt, 1972), only the average number of species detected for a group of sites was reported. Such numbers were tallied only once in Figure 2.3. In others (e.g., Meffe and Minckley, 1987), a large number of collections (10–15) from a single site was averaged and the mean value tallied only once. In some, species richness was determined from figures, which could introduce some error. However, given the large number of sites worldwide that are summarized in Figure 2.3, a modest number of errors in the counting of species is not likely to change overall results. Figure 2.2 includes sites on streams of all sizes from very small headwaters to portions of large rivers, which typically contain different numbers of species (Karr et al., 1986, p. 11; and as noted by numerous scientists, below).

For all 1657 site records from lentic and lotic freshwaters worldwide, there was a preponderance of cases with fewer than 20 species per local assemblage (Fig. 2.3). Across all kinds of freshwater habitats in this summary, including small or depauperate headwaters and species-poor continents as well as faunally rich tropics, the mode is at only seven to eight species per local assemblage (Fig. 2.3). Clearly, this mode is only a rough measure of central tendency, and many collection localities contained a total number of species ranging from about 5 to 17 or 18. Above these values, however, the numbers of samples with larger numbers of species decline markedly, with relatively few records in the literature I have reviewed of sites with more than 30 species present. It should be noted that Figure 2.2A, for all samples pooled worldwide, probably underrepresents the tropics and is strongly influenced by my including more records from temperate North America than elsewhere.

For North American streams, there are many studies documenting the numbers of species or composition of fish assemblages at a given locality (Fig. 2.3B). Fish faunas are much more diverse in the eastern part of North America, particularly in the vast Mississippi–Missouri–Ohio basin. Figure 2.3B shows numerous sites with more than 20 species in temperate North American streams, most of which are east of the Continental Divide. Sites for temperate North America with relatively few species (e.g., fewer than 10 species) are most common west of the Continental Divide, in cold-water montaine regions or in very small headwaters of streams. With all habitats combined, the mode for the numbers of species in North American streams is at about 11–12 species per local assemblage.

The compilation for temperate streams outside North America (Fig. 2.3C) includes sites in the British Isles and Europe, northern Asia, Nepal, temperate South America, and Australia–New Zealand. Figure 2.3C suggests that, worldwide, there are many sites ranging from 1 to about 16 species, and the proportion of local assemblages exceeding that range is relatively small.

With regard to tropical streams, I benefitted greatly from generous personal communications and/or sharing of unpublished data from colleagues. Robin Welcomme (personal communication) noted with respect to his own work in Africa that, "commercial catches (from a single site) in the tropics, even those from one canoe, can consist of tens or even over a hundred species." Welcomme also pointed out (personal communication) that "fish assemblage sizes in tropical systems are generally much larger than those of temperate systems, although these all have to be seen in the light of the position of the river reach in the system." (The subject of longitudinal zonation of stream fishes will be considered in detail in a later chapter, but it seems clear that the pattern holds as well in the tropics as in temperate systems.) From his data, Welcomme reported sites in African rivers having (in gill nets, cast nets, or traps) as few as 9 to as many as 58 species locally.

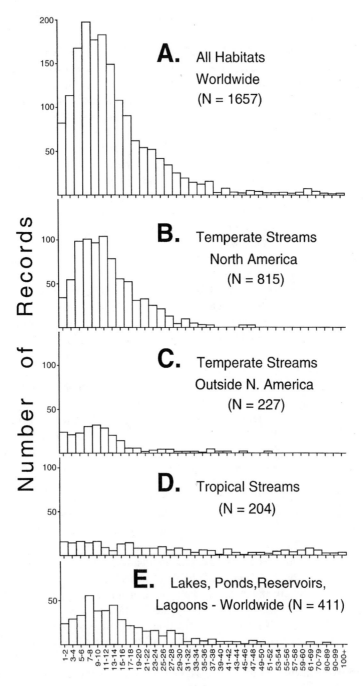

Figure 2.3.

◄ *Figure 2.3.* Frequency distribution of number of local samples containing the indicated number of species (x axis) for (A) worldwide, (B) temperate streams in North America, (C) temperate streams from elsewhere in the world, (D) tropical streams, and (E) lentic habitats. (See the Appendix for sources of information for this figure.)

Appendix

Sources of information tallied in Figure 2.3 were as follows:

Streams, eastern North America: Angermeier and Smogor, 1995; Barton, 1984; Bauer et al., 1978; Bell and Hoyt, 1980; Binderim, 1977; Bramblett and Fausch, 1991a; Cashner et al., 1994; Cowley and Sublette, 1987; Cross, et al., 1985; Dewey and Moen, 1978; Evans and Noble, 1979; Fausch and Bramblett, 1991; Fowler and Harp, 1974; Freeman et al., 1988; Garman et al., 1982; Hansen and Ramm, 1994; Harima and Mundy, 1974; Harrel et al., 1967; Harris and Douglas, 1978; Hastings and Good, 1977; Hocutt and Stauffer, 1975; Hoyt et al., 1979; Jackson and Harp, 1974; Jenkins and Freeman, 1972; Kuehne, 1962; Lyons, 1992; Mahon, 1984; Matthews, 1986c; Matthews and Harp, 1974; Meffe and Sheldon, 1990; Osborne and Wiley, 1992; Paller, 1994; Rahel and Hubert, 1991; Reash and Pigg, 1990; Richards, 1976; Rister, 1994; Schlosser, 1982; Simonson and Lyons, 1995; Smale and Rabeni, 1995a; Stauffer et al., 1975; Stewart et al., 1992; Swaidner and Berra, 1979; Taylor et al., 1991; Thompson and Hunt, 1930; Tramer and Rogers, 1973; Warren et al., 1994; Whiteside and McNatt, 1972.

Streams, western North America: Beecher and Fernau, 1982; Hughes and Gammon, 1987; Meffe and Minckley, 1987; Moyle and Vondracek, 1985; Saiki, 1984; Smith, 1982.

Streams, Eurasia: Hartley, 1948; Jones, 1975; Jungwirth and Schmutz, 1995; Kirchhofer, 1995; Lobon-Cervia et al., 1986; Mahon, 1984; Mann and Orr, 1969; Penczak et al., 1986; 1991; Sokolov et al., 1994.

Streams, Neotropics: Angermeier and Karr, 1983; Flecker, 1992; A. Flecker, personal communication of unpublished data; Galacatos et al., 1996; Goulding et al., 1988; Ibarra and Stewart, 1989; Lowe-McConnell, 1975, 1991, personal communication; Lyons and Navarro-Perez, 1990; C. Peterson, personal communication, unpublished data; Smith, 1981; K. Winemiller, personal communication, unpublished data; Winemiller and Leslie, 1992; Wootton and Oemke, 1992; Zaret and Rand, 1971.

Streams, tropical Africa: Balon and Stewart, 1983; Hugueny and Paugy, 1995; Lowe-McConnell, 1975; R. Welcomme, personal communication, unpublished data; K. Winemiller, personal communication, unpublished data.

Streams, Asia–Indonesia: Edds, 1989; Inger, 1955; Moyle and Senanayake, 1984; Wikramanayake and Moyle, 1989.

Streams, Australia–New Zealand: R. C. Cashner, personal communication, unpublished collections on Nymboida River, Australia; Hanchet, 1990; Pusey et al., 1993, 1995.

Lakes, reservoirs, ponds, lagoons—worldwide: Beecher and Ferneau 1982; Borges, 1950; Cady, 1945; Coke, 1968; Coulter, 1981; A. Flecker, personal communication, unpublished data; Galacatos et al., 1996; F. P. Gelwick, W. J. Matthews, and I. Camargo, unpublished data on Lake Texoma (USA); Giussani, 1989; Hall and Werner, 1977; Heard, 1962; Henderson and Fry, 1987; Hori et al., 1983; Keast, 1978a; Keast and Fox, 1990; Lowe-McConnell, 1964, 1975; Lyons, 1989a; Mahon and Balon, 1976; Pierce, et al., 1994; Reinthal, 1993; Sal'nikov and Reshetnikov, 1992; Spangler and Collins, 1992; Sylvester and Broughton, 1983; Vaux et al., 1988; Vigg and Hassler, 1982; Werner et al., 1977, 1978; Wiener et al., 1984.

Donald Taphorn (personal communication) likewise noted the importance of longitudinal position of sites in the tropics relative to numbers of species in local assemblages. Taphorn noted that in one creek in the state of Apure, Venezuela, there were 40–60 species present on any given date. However, Taphorn summarized (personal communication) for the Apure River, flowing from the Andean piedmont, "high mountain streams, with cold water, have very few species (3–8), and the diversity increases . . . downstream. Small streams around 200 m elevation have from 10–30 species. Just a bit further down into the llanos (plains) the number jumps considerably to between 30–80 species . . . between 30–50 on any given day (at a site)."

Eigenmann and Allen (1942) also noted the small number of species in some stream sites in western South America: "With the annual subsidence of the (annual flood) water there is . . . everywhere a local sorting of species according to preferred habitat. Thus in a given stream one may not obtain more than two, three or half a dozen species at the same time." Alex Flecker (personal communication) wrote that, "Most of my stream fish collections have been made . . . in the Andean piedmont (Venezuela). I generally collect 20–40 species (per site) on a sampling date)." Christopher Peterson (personal communication), sampling in "slightly acidic streams on the verge of the black water tributaries" (Venezuela) noted "a total of seventy species of fish for the year, and my average was 45 species (per collection)." Kirk Winemiller (personal communication) reported from his records a range from about 20–80 species locally in streams in Venezuela, and as few as 12 to as many as 41 species per locality in streams of Zambia.

Rosemary Lowe-McConnell (personal communication) cautioned that, with regard to considering "assemblages" of fish in the tropics, it is necessary to "distinguish between pool communities of fish, and concentrations of fish from a large drying floodplain which happened to be trapped together (but mostly not feeding)." This concern is certainly real, as it is well known that when the waters that have innundated the floodplain during the annual flood recede, fish can be trapped in very large numbers in shrinking pools (Lowe-McConnell, 1987). Likewise, Roberts (1972) noted that, "During any intensive collecting in the Amazon or Congo, a good proportion of species obtained in a given locale are likely to be either recently arrived or in the process of moving away from habitats that became unfavorable." Roberts (1972) also noted that although a typical rotenone collection in one place might "yield 70–80 species," only a third or less were likely to be permanent residents, with most others occurring in that particular place due to chance, and that the species would soon disappear from the site unless additional individuals migrated there. Lowe-McConnell (personal communication) also noted that in tropical areas, the problems are "twofold, taxonomic ones, and logistics of transporting . . . when you have to carry everything." This seems very different than typical field work in North America, where one can drive within easy hiking distance of most collecting sites and virtually every fish species is well known (even if not formally described). Thus, it might be that some reports of very large numbers of species in tropical systems could come from ephemeral congregations of fishes, but any overestimate of richness from such an error might, in the large picture of the tropics, be somewhat balanced by the potential underestimate of local richness where cryptic species have not yet been separated taxonomically and where collecting conditions are difficult at best.

Even with these caveats, Figure 2.3D shows many sites in the tropics with large numbers of species in the local assemblages. In a large proportion of stream sites in the tropics, the local assemblage is much richer than it would be for a comparably sized stream in temperate regions, with many sites exceeding 40 species locally and a substantial number with greater than 60 species locally. Two of the sites for which information was available actually had more than 100 species in the local assemblage. The mode that existed at low numbers of species in temperate streams was virtually lacking in tropical streams. However, as noted by several colleagues (above), there are many tropical sites (small streams, headwaters) that contain no more fish in a local assemblage than would be found in similar temperate streams.

For lentic habitats, including lakes, ponds, reservoirs, lagoons, or other standing bodies of water, worldwide, there was a mode at 7–8 species in local assemblages and a substantial decrease in sites with more than 14 species present. In spite of the very large numbers of species found in limited parts of some lakes [e.g., where rich species flocks exist (Echelle and Kornfield, 1984)], it seems much more typical, worldwide, to have relatively few species living together on a shoreline, in open waters of the limnetic zone, or in a small natural lake (e.g., in kettle lakes of glaciated northern areas). However, it is in the tropics where the greatest numbers of species occupy a similar habitat (i.e., occur as an assemblage). Alex Flecker (personal communication, unpublished data) found more than 40 species of fish in numerous artificial ponds in the Venezuelan floodplain, and Hori et al. (1993) consistently found 40 or more species of cichlids in 20 × 20-m quadrats in Lake Tanganyika. Figure 2.2E shows numerous lentic habitats with more than 45 species, and even some with more than 80 species. However, much as seen for temperate streams, the majority of lentic assemblages had fewer than 20 species (Fig. 2.3E).

To summarize the information in Figure 2.3, it appears that regardless of whether lentic or lotic habitats are considered, and across all landmasses, there is a preponderance of far fewer species in local assemblages than exist in the spectacular tropical sites that have been so widely publicized in popular and scientific writings. Although the importance of such species-rich sites is not in question, it would appear that fish ecologists working in many freshwater habitats face the problem not of explaining "why are there so many kinds of species" [to borrow from Hutchinson (1959)] living together but in understanding the evolutionary and extant processes that result in approximately 20 species (or fewer) occurring together in an average location.

Local Assemblages in Individual Stream Pools?

Probably the smallest spatial scale at which it is profitable to ask about the numbers of species in an assemblage is that of a single pool or riffle. Freeman et al. (1988) and Matthews et al. (1994) assessed the stability of fish assemblages in single pools. In Brier Creek, Oklahoma (USA), there is a known fish fauna of 31 species, dominated numerically by minnows (Cyprinidae) and sunfishes (Centrarchidae). For 14 pools surveyed in a 1-km reach of Brier Creek by snorkeling seven times in a year (Power et al., 1985; Matthews et al., 1994), there was a mode at 4–5 species per pool, but numerous pools had 6–8 species (Fig. 2.4). Thus, in any given pool at any time, the assemblage included about one-fourth to one-third of the fishes known from the creek. These pools ranged in length from 6 to 120 m [see detailed map in Power and Matthews (1983)], were about 2–10 m wide at base flow with depths to 1.5 m, and were in a permanently flowing, spring-fed section of this creek.

Although I have no quantitative data on the degree to which shallow riffles between pools (Power and Matthews, 1983) functioned to retain species within

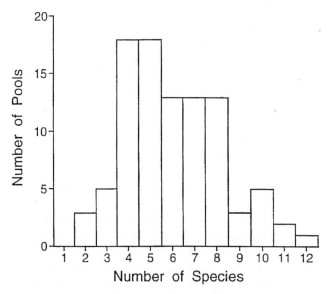

Figure 2.4. Frequency distribution of number of pool samples [14 pools across repeated snorkel surveys, summarized in Matthews et al. (1994)] versus number of species per pool, for Brier Creek, Oklahoma (USA).

pools, it was my impression from these surveys that riffles did comprise at least partial barriers to fish movement and that an assemblage within a given pool represented a biologically important and potentially interactive grouping (Matthews et al., 1994). Jacob Schaefer (unpublished data) has found in large artificial streams at the University of Oklahoma Biological Station that minnows typical of small streams (*Notropis boops, Cyprinella venusta, Campostoma anomalum*) can readily cross riffles less than 1 cm deep, but such shallows should largely restrict movement of larger species (e.g., predators) among pools. My observations suggest [as have many other studies, Baltz and Moyle (1984), for example] that different species use different microhabitats within pools, such as *Notropis boops* occurring mostly at heads of pools feeding on drift or *Fundulus notatus* at pool surfaces, but many of the species in these Brier Creek pools "cruise" the entire pool at times, and it is not possible to rule out their presence in any portion of the pool at any time.

Number of Families in Local Assemblages, and Species per Family

Other interesting metrics of local fish assemblage structure include the number of families and the numbers of species per family. A few families can incorporate most of the species-level diversity at a site. For example, in the study by Ibarra and Stewart (1989), characoids made up 51% of the species and 82% of the individuals, and catfishes made up 39% of the species and 15% of the individuals collected. Numerous authors (e.g. Roberts, 1972; Lowe-McConnell, 1975) have noted the high local diversity of fish species in tropical waters and posed mechanisms that allow or facilitate the coexistence of so many species. Explanatory mechanisms that typically have been offered relate to active or extant resource partitioning among species (e.g., Zaret and Rand, 1971), often ones that are closely related (e.g., within the same family). However, if families represent distinct, ancient lineages, each with a characteristic general morphology, an alternative view may be tenable. If streams have large numbers of species, but these are drawn from many families (i.e., if the species per family ratio is low; Fig. 2.5A), then the "resource partitioning" in diverse fish assemblages may be attributable as much to "primitive" family-level evolutionary differences that date to the "deep past" as it is to interspecific differences in recently evolved, "derived" ecological traits (as might be the case in Figure 2.5B, where the large number of species are all from two families).

To determine if high local species numbers are related more to enrichment of numbers of families or to enrichment in numbers of species per family, both relationships are depicted in Figure 2.6. In Figure 2.6, as the number of species increases, there is a gradual, distinct increase in the numbers of families, but not an overall pattern of increased species per family. This could mean, overall, that increased species richness in local stream fish assemblages results at least in part from a greater number of families per locality, as well as proliferation of species

46 / Patterns in Freshwater Fish Ecology

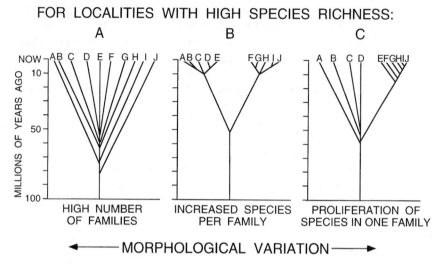

Figure 2.5. Three idealized patterns for regional diversification as related to relatedness of fish species in a single locality: (A) high diversification at the family level, with a single species representing each of 10 families in a local assemblage; (B) high diversification within two families, with five species from each family in a local assemblages, and (C) high diversification within one family (six species), with the remaining four species each from a different family.

within the families that are present. If this is the case, then high local species richness would relate both to evolutionary events in the "deep past" (i.e., when families were emerging) and to those in the more recent past (when extant species evolved). Note that an increased number of families should mean greater morphological diversity in an assemblage, relative to the morphological diversity that would exist if there were a large number of species within only one or a few families.

The pattern suggested above may not be true for streams in eastern North America. Figure 2.6A suggests that there are more species per family, on average, in stream systems in eastern North America than elsewhere. Furthermore, in eastern North American streams, there seems to be a longitudinal effect on the number of species per family. For example, in Byrd's Mill Creek (Stewart et al., 1992), the numbers of species per family increased markedly from headwaters (1.0) to downstream (3.7) stations. It is at the lowermost stations that the numbers of species proliferate in some families (e.g., particularly Cyprinidae and Centrarchidae) in Byrds Mill Creek. Kuehne (1962, Table I) also demonstrates this point: His headwater stations ranged on average from 1.0 to 4.0 species per family, whereas the most downstream stations ranged from 3.8 to 6.3 species per family.

Results for 20 streams in North America east of the Continental Divide and 14 streams elsewhere in the world (including Arivaipa Creek, Arizona) are compared in Figure 2.6. The question being asked is, "Where there is a high

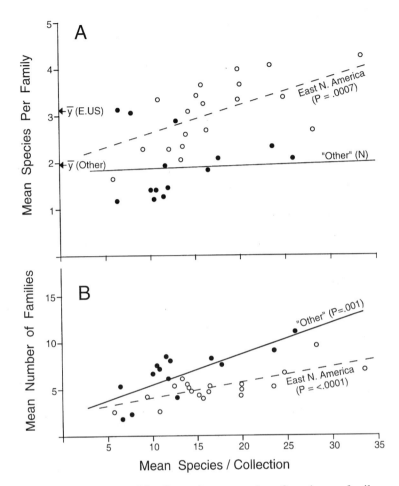

Figure 2.6. Mean number of families and mean number of species per family, versus mean number of species (averaged across collecting sites within each system) for 20 stream systems of eastern North America and for 14 stream systems elsewhere in the world. Delimitation of families follows Eschmeyer (1990). Sources of information tallied into the analysis were as follows: Eastern North America (Barton, 1984; Bauer et al., 1978; Bell and Hoyt, 1980; Evans and Noble, 1979; Fowler and Harp, 1974; Garman et al., 1982; Hansen and Ramm, 1994; Harima and Munday, 1974; Kuehne, 1962; Matthews, 1986b, Roanoke River, VA and Brier Creek, OK; Matthews and Gelwick, 1990; Matthews and Harp, 1974; Simonson and Lyons, 1995; Smale and Rabeni, 1995a, prairie sites only; Stewart et al., 1992; Swaidner and Berra, 1979; Taylor et al., 1991; Warren et al., 1994); western North America (Meffe and Minckley, 1987); Costa Rica (Winemiller and Leslie, 1992); Panama (Angermeier and Karr, 1983; Zaret and Rand, 1971); Guyana (Lowe-McConnell, 1964, for floodplain "ponds" after seasonal floods); Sri Lanka (Wikramanayake and Moyle, 1989); Australia (Pusey et al., 1993, 1995; R. C. Cashner, unpublished data on Nymboidea River); Borneo (Inger, 1955); Africa (Balon and Stewart, 1983); Austria (Jungwirth and Schmutz, 1995); England (Hartley, 1948).

mean richness of species, does this relate to (a) increased numbers of families, (b) increased numbers of species within families, or (c) both?" For streams outside eastern North America (including Africa, South America, Central America, Asia, Europe, and western United States), there was a very significant increase in the mean number of families with the mean numbers of species (Fig. 2.6B). However, in those streams, there was no significant increase in the mean numbers of species per family (grand mean = 1.95 species per family) as total species richness increased. In other words, it appears that in these 14 diverse streams worldwide, greater local species richness was related to the presence of large numbers of families, not to a proliferation of species within families.

Eastern North America appeared to contrast with this pattern. Figure 2.6B shows that for 20 stream systems in eastern North America, there is a larger mean number of families present where mean species richness is greater. However, there also was a significant increase ($p = 0.0007$, explaining 48.4% of the variance) in mean number of species within families (grand mean = 3.11 species per family) where total species richness was higher (Fig. 2.6A). If the analysis is reversed and the mean total number of species regressed onto mean families and mean species per family, the "number of families" enters the model first, accounting for 73.7% of the variance in total species. However, the mean number of species per family also was significant in the model, improving variance accounted for to 96.4%, or representing 22.7% of total variance in number of species accounted for by number of species per family, for streams of eastern North America.

To summarize, in streams of eastern North America, where a few families often dominate assemblages taxonomically (including Cyprinidae, Centrarchidae, and Percidae), systems with greater total species have higher species richness, not because more families are present but because of an increase in the numbers of species within families, on average. In streams elsewhere in the world, it appears that proliferation of species within families is less important to local species richness, and increased local richness is, on average, accompanied more by the increased number of families being present.

Another question that needs to be considered is whether, in more speciose local assemblages, there is a disproportionate number of species from a few families (e.g., Fig. 2.5C, for eastern North America). Thus, in contrast to the models above, a local site might have a large number of species because of the presence of numerous families and the proliferation of species in one or two families. This question suggests the examination of "balance" or "evenness" in the distribution of numbers of species among the families that are present. For nine stream systems (with multiple sites on each stream) in North America and the tropics, E. Marsh-Matthews (personal communication) calculated Hill's Modified Index of Evenness (Oakleaf Software) for the numbers of species per family. The following streams were included: eastern North America—Piney Creek, Arkansas (Matthews and Harp, 1974); Brier Creek, Oklahoma (Matthews, unpublished data); Janes Creek, Arkansas (Fowler and Harp, 1974); Panama—

R. Frijolito and R. Juan Grande (Angermeier and Karr, 1983); Costa Rica—Rio Tortuguero and its tributaries (Winemiller and Leslie, 1992); Africa—Luongo R. (Balon and Stewart, 1983); Australia—Mary River (Pusey et al., 1993). Hypothetically, if all families present contributed near equally to the total number of species in the assemblage, "evenness" should be high. Conversely, if one or two families, of the larger number typically present, contributed most of the species to the assemblage, "evenness" should be lower. For 50 local assemblages with a range of 6–55 species, Marsh-Matthews (personal communication) found a generally negative relationship (with a possible threshold effect; line fitted by eye) between total species richness and evenness in distribution of species among families (Fig. 2.7). These preliminary findings suggest that for an array of quite different stream sites, those with the largest numbers of species attained that higher number of species because of the proliferation of only one or a few families [i.e., species-rich sites have overrepresentation of a few families (e.g., Fig. 2.4C)].

Another way to examine species per family, locally, is to plot the numbers of species per family within a site, with the families arrayed on the x axis in order of numbers of species. [This is analogous to a "ranked-abundance" list (Pielou, 1975, p. 20), but ranked by species per family instead of individuals per species.] A site with a large number of species in one or two families would show a sharp peak at the extreme left and probably a "hollow curve" marking the rapid decrease in numbers of species per family beyond the first one to two families. Such a hollow-curve pattern was shown for species per family in plants and is common

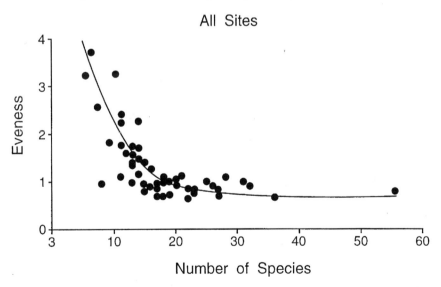

Figure 2.7. Evenness in distribution of species across families, versus number of species per local assemblage.

50 / *Patterns in Freshwater Fish Ecology*

at various taxonomic levels (Dial and Marzluff, 1989). At the other extreme, an assemblage with a linear decrease in species per family across families (e.g., an assemblage with 6, 5, 4, 3, 2, and 1 species per family) would exhibit a regular, linear downward slope.

To test this notion for some real assemblages, Figure 2.8 depicts the numbers of species per family for the two most species-rich local sites in each of seven stream systems included in the analysis of evenness versus total species numbers by Marsh-Matthews (Fig. 2.7). The results suggest (1) typical dominance of only one or two families, worldwide, and (2) a markedly greater tendency for one family to dominate assemblages in two of the three North American streams. (The dominant family was Cyprinidae in every case.) Comparison of the North

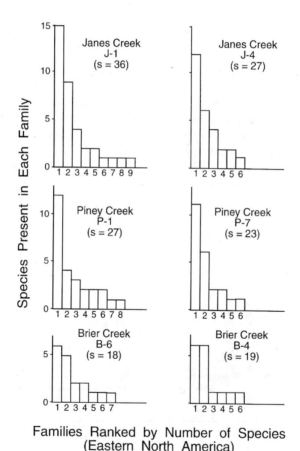

Figure 2.8. Number of species per family, with families ranked by number of species present, for local assemblages at (A) six locations in temperate North America and (B) eight locations elsewhere worldwide.

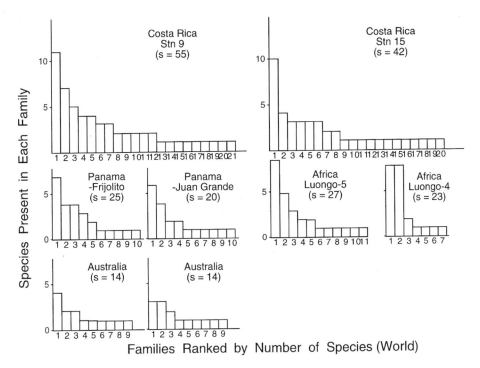

Figure 2.8. Continued.

American temperate streams (Fig. 2.8A) to tropical streams (Fig. 2.8B) also confirms the earlier suggestion that tropical streams tend toward more families, whereas North American streams may have more proliferation within one or two families, and relatively fewer families, overall, in local assemblages.

What causes this apparent pattern? One obvious hypothesis is that this is merely an artifact of there being more families present, overall, in tropical regions, with the local number of families comprising a low but proportional sampling of the "regional" families. I have not tested this hypothesis formally, but even in the regions (e.g., drainage basins) containing the North American streams, there are many more families available that could be present than actually occur in any one of the sites. For example, Janes Creek (with the greatest number of families in Fig. 2.8A) is in the White River basin of North Arkansas, where 21 native families occur (Robison and Buchanan, 1988) and hypothetically could be present in a given stream site.

Another possible hypothesis to account for differences between temperate and tropical streams in the distribution of species among families is better illustrated if Figure 2.8 is redrawn as a series of curves (all on the same scale) representing the tops of the histograms from Figure 2.8. The resulting Figure 2.9 shows the shape of the decrease in species per family. The number of species per family

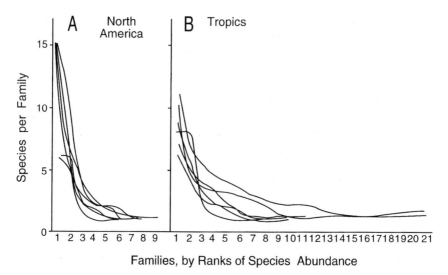

Figure 2.9. Tops of histograms from Figure 2.7(A) and Figure 2.8(B), redrawn to emphasize slopes of line for each locality.

is greater (for at least four sites) in North America than for any in the tropics, and the curves for those North American sites drop more steeply from left to right than do the curves for tropical streams in general. The steepness of the decline for North American streams suggests a possible explanation that is analogous to the family of resource-apportioning models (Pielou, 1975) often invoked to account for numbers of individuals per species in local assemblages. The niche-preemption model of May (1975), as detailed by Pielou (1975, pp. 21–22), accounts for the abundance or success of species in a resource-limited environment, by apportioning resources among species in a stepwise fashion, with each species starting with the dominant, obtaining a fixed percentage (denoted as k) of the available resources. Thus, the dominant species sequesters $k\%$ of 1.00 (total) resources, the next species sequesters $k\%$ of $1.00 - 1.00\ k$, and so on. If the abundance of a species is proportional to its sequestering of resources, the result will be a distribution of species among progressively smaller percentages of the total number of individuals in a community [as suggested by Fig. 2.1 of Pielou (1975)].

Now, if k is fixed arbitrarily as $= 0.5$, then the dominant species in an assemblage will acquire one-half of the resources, the next most successful species will acquire one-half of those remaining (i.e., one-fourth of the original resources), and so on, thus their abundances would equal one-half, one-fourth, one-eighth, and so on of the total number of individuals in the assemblage. If a gigantic leap of speculation is made that in some way families might, over time, divide the available resources (e.g., primary productivity), hence "success," in a local setting, and if "success" were measured in numbers of species that were present within a given family, a model could be contrived that is analogous to the niche-preemption model (Pielou, 1975) for numbers of species distributed among families (Figure 2.10). In this speculative figure, I have depicted the most "successful" family as having 16 species present locally (possibly in streams of eastern North America), the next most "successful" family having 8 species, the next 4, then 2, and, finally, 1. (To make the model "break" by one-half in each step, I have cheated and allowed the fifth and sixth families to each have one species—bringing the total to 32. Because no family can have only part of a species, I have also allowed any additional families that might be present to have one species each—noted by the dashed line in Figure 2.10.)

If this speculative model for family success at a local site is overlain on the curves for temperate streams (Fig. 2.9A) and, separately, on the curves for the tropical streams (Fig. 2.9B), the fit of the hypothetical curve is quite close for four of the six temperate stream sites, but it does not approximate the situation in tropical streams. The curves for the actual tropical streams begin too low, but even if we took the maximum value for tropical stream species per family, halved that, halved it again, and so on, it would be obvious that the curves for the tropics are not as steep as those for temperate streams and do not fit the $k = 0.5$ resource-preemption pattern well.

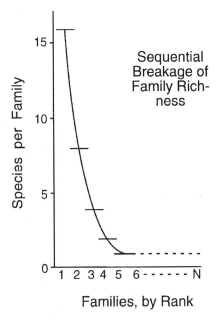

Figure 2.10. Speculative model of the pattern of species per family that would be produced by a "sequential breakage" model of family enrichment.

If the above speculation should prove to approximate patterns for temperate versus tropical stream sites in general (if many more examples were included), how might this be interpreted with respect to ecology of local fish assemblages? If, indeed, a niche- or resource-preemption pattern was evident [for species abundance (e.g., Pielou, 1975) or for family success (e.g., this analysis)], it would be consistent with (although obviously not proving) the notion that some families may contain species that, on average, are better competitors than others for a limited supply of some needed resources. If this is the case, we might speculate that fish species in temperate streams are resource limited, with one or two families having species most successful at obtaining critical portions of perhaps less diverse resources, whereas fish assemblages in tropical streams are characterized by more equitable success of more families, with a wider array of foods, habitats, and so forth, available, and a wider array of fundamental, family-level body plans being successful. Clearly, there is huge speculation here, but even with this limited suite of examples, it seems interesting that these stream fish assemblages in eastern North America are so different from those in the tropics with respect to species-per-family patterns.

2.3 Trophic and Functional Groups, and Predators/Prey Species

Composition of Assemblages by Trophic Groups

Hartley (1948), Nikolsky (1963, p. 263), Keast (1977), Gerking (1994, pp. 42–52), and virtually every other author who has studied fish diets note the influence of ontogeny and changing opportunities (Ross and Baker, 1983) or environmental conditions on feeding by fishes. Helfman (1994) provides an overview of the strong propensity of fishes to switch feeding "modes." For fishes in general, highly specialized feeding habits like those of some cichlids [eye-eaters, scale-eaters (Fryer and Iles, 1972)] or Amazonian herbivores [fruit-eaters (Goulding, 1980)] appear to be the exception rather than the rule, and some of species formerly considered trophic specialists [e.g., cichlids of African Great Lakes (Fryer and Iles, 1972)] are proving to be more omnivorous upon closer study. As a result, it is difficult to classify fishes in an assemblage into exclusive trophic groups or "guilds." Two factors need to be taken into account in any useful scheme distinguishing trophic groups in fish assemblages. First, almost all species change trophically during ontogeny (Miller, 1979), most consuming microinvertebrates or microalgae as postlarvae, increasing in invertebrate prey size as juveniles, and ultimately achieving their adult range of feeding repertoires. Second, adults of many species are opportunists, with stomach contents on a given day reflecting availability of foods or extant environmental conditions that influence feeding [Mendelson (1975), for northern minnows; Matthews et al. (1978), for *Luxilus pilsbryi;* Matthews et al. (1988), for *Morone saxatilis*]. Some species "switch" prey in a relatively predictable manner when levels of prey availability change (Murdoch et al., 1975).

Gerking (1994) reviewed several trophic or guild classifications for freshwater stream fish (his Table 3.1) and concluded that all lacked utility. Gerking objected to such classifications because of the versatility of feeding by most fish species ["trophic adaptability," which he attributed to Hartley (1948) but that could be traced at least to Forbes (1988a and b), who described a high degree of variability in feeding within species of fish]. Gerking (1994) considered guild classifications for freshwater fish to lack "maturity as a concept," experimental tests of assumptions, and criteria for delineating guilds. Gerking (1994) concluded that the guild concept has "not yet matured to the point where it offers fundamental feeding categories. . . ."

One tractable solution to at least part of the "trophic groups problem" (Gerking, 1994) is to recognize that different-sized classes within a species may predictably use different foods. In some species/habitats, the ontogeny of food use or place of feeding is highly predictable, such as the well-known pattern for bluegill (*Lepomis macrochirus*) in northern lakes to occur near shore as small juveniles, move offshore to live and feed in deeper water (avoiding shoreline piscivores) as larger juveniles, and return to shoreline weed beds or structures as adults (Keast, 1977). Thus, it is practical to classify some fish species into size-classed ecological (rather than taxonomic) species (Polis, 1984) as did Matthews et al. (1994) for centrarchids in Brier Creek, Oklahoma. The data do not yet exist to allow cataloging all age classes of all freshwater species to "ecospecies" or a trophic group, but an ecologist should be able to determine in a relatively straightforward manner the desirability of so treating common species (particularly larger-bodied taxa) in a given system.

A final caveat regarding trophic classification of fishes is that such groupings should not be confused with "functional groups" like those recognized for aquatic insects (Cummins and Klug, 1979). In a narrow sense, trophic classification should reflect *what* a species eats (e.g., insects), whereas functional groups should reflect *where* or *how* the species obtains food, or how it affects the ecosystem. Some recent "trophic" classifications do include aspects of both trophic and functional phenomena (e.g., Grossman et al., 1982; Schlosser, 1982), and the "trophic" categories of Keenleyside (1979) as summarized by Wootton (1990, p. 33) include where as well as what a species eats. However, a rigorous classification of fishes into functional groups has not yet been attempted, although I pose a preliminary classification later in this chapter.

By late in the last century, Forbes (1883) had divided the smaller fishes in Illinois (USA) into four "trophic" groups on the basis of diet analyses and the morphology of pharyngeal teeth and the alimentary canal. Without formally naming trophic groups, Forbes remarked on the high proportion of "mud eating" by the some groups and the contribution of insects to others. Forbes (1988) summarized his studies of food use by Illinois fishes by placing species approximately into the following named trophic categories based on foods of the adults: piscivorous fishes, mollusk (sic) eaters, insectivorous species, scavengers, vegetable feeders, and mud eaters. Forbes also noted marked differences between foods of juveniles and adults, and the substantial variability in food use among adults of a given species.

Hartley (1948) divided freshwater fishes in England into four groups: (1) specialist predators upon a single prey, (2) those taking "a wide variety of foods," (3) species with a diet in which "insects and plants predominate," and (4) those in which "insects and crustaceans predominate." Hartley's groups might correspond in part to more recent trophic classifications using categories like piscivores, omnivores, insectivore–omnivores, and insectivore–planktivores.

Since Hartley (1948), numerous authors have devised schemes for trophic classification of local or regional fishes. Grande (1984) has even classified the extinct fishes of the Green River formation (western United States) by trophic groups. Table 2.2 compares some that have been most followed. It is clear in Table 2.2 that individual authors have used schemes of trophic classification that fit the fishes in the system they studied, providing increased precision for the local system but making extrapolation to other systems problematic. However, in Table 2.2, several of the more general categories seem repeated often enough to suggest their broad applicability in new systems, including herbivore–detritivores, planktivores, omnivores, benthic invertivores, midwater–surface feeders (largely on insects), and piscivores. My own preference might be to separate the herbivore and detritivore categories, given the very different effects in the ecosystem of grazers of live attached algae versus consumers of detritus (e.g., Bowen, 1983; Flecker, 1996). Perhaps a good general approach would be to begin with the often-used categories above and add specialized categories (e.g., "snail-eaters," "aerial feeders") where such species are a component of the system.

Probably the most used system at present for North American streams is the general categorization by Karr et al. (1986) which was not included in Table 2.2. Karr et al. included, on the basis of defined proportions of diet, the relatively simple categories of invertivores, piscivores, herbivores, omnivores, and planktivores. Although this scheme ignores methods or microhabitats of feeding, it does match some of the most used categories in Table 2.2 and has broad applicability. Karr et al. (1986) provided a classification of 90 species of eastern North America by trophic groups, serving as a useful guide to the dominant fishes in many systems east of the Continental Divide.

Brown and Matthews (1995) classified the fishes of four streams in midwestern United States by the trophic categories of Karr et al. (1986) and found relative abundance of individuals by trophic groups in these streams to reflect a difference between upland and low-gradient systems. Two clear, permanently flowing upland streams had many more insectivores, and two low-gradient, slow-flowing or intermittent streams had fewer insectivores and more omnivores and piscivores (Fig. 2.11).

Table 2.2. Trophic Categories Recognized by Various Authors

Trophic Group	N	LM	H	ML	S	G	W	ML	WL
Herbivorous–detritiphagic	X		X	X	X				X
Detritivores							X		
Scavengers							X		
Mud- or detritus-feeders		X							
Ooze feeders						X			
Herbivores		X					X		
[with 3 subcategories: grazer, browser, planktivores, in Wootton (1990)]									
Algae feeders						X			
Plankton-feeders			X	X	X		X	X	
Zooplanktivores								X	
Neustonivore								X	
Omnivores		X	X	X	X	X			X
Carnivores	X	X					X		
(invertivores)									
General invertivore			X		X				X
Benthic invertivore			X	X	X	X		X	
[separated re. soft versus hard substrates by Grossman et al. (1982)]									
Zoobenthivore								X	
Midwater–surface feeders			X	X	X	X			
[separated by Grossman et al. (1982)]									
Predators (piscivores)	X	X	X			X	X	X	X
Aerial feeders							X		
Snail-eaters			X						
Lampreys (?)			X						
Predators (on fish and invertebrates)				X	X				

The term first used is retained, hence "plankton-feeders" (Lowe-McConnell, 1975) rather than "planktivores." N = Nikolsky (1963); LM = Lowe-McConnell (1975); H = Horwitz (1978); ML = Moyle and Li (1979); S = Schlosser (1982); G = Grossman et al. (1982); W = Wootton [1990, following Keenleyside (1979)]; ML = Magnuson and Lathrop (1992); WL = Winemiller and Leslie (1992).

That longitudinal zonation of fish species from headwaters to mainstreams occurs in streams worldwide is well known (Matthews, 1986b; Balon and Stewart, 1983). However, less has been shown about longitudinal or local distributions of trophic groups (although Morin and Naimen (1990) predicted downstream increases in detritivore-herbivores). Additionally, Chipps et al. (1994) demonstrated differences in fish assemblages by trophic group in upland streams of eastern North America, with more insectivors and insectivore-piscivores in pools, and more benthic insectivores and herbivore-detritivores in riffle habitats. For both an upland stream (Piney Creek) and a low-gradient stream (Brier Creek), there was a distinct trend for an increase in the proportion of insectivores from headwaters to downstream (Fig. 2.12). Herbivores and omnivores were scarce at the most downstream site, but increased moderately in proportion at several

58 / Patterns in Freshwater Fish Ecology

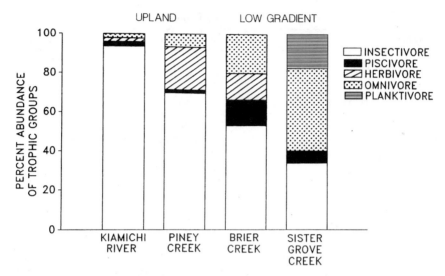

Figure 2.11. Percent abundance of five defined trophic groups of fishes, for four streams ranging from high-gradient, upland streams (Kiamichi River, Oklahoma; Piney Creek, Arkansas) to low-gradient prairie-margin streams (Brier Creek, Oklahoma; Sister Grove Creek, Texas). [Reprinted from Brown and Matthews (1995) with kind permission from Elsevier Science-NL, Sara Burgerhartstraat 25, 1055 KV Amsterdam, The Netherlands].

upstream sites in collections in Piney Creek in August 1982 (Fig. 2.12A). (Piscivores were not included for Piney Creek, as too few were captured to allow discussion of patterns.)

Most of the upstream sites in Piney Creek have large rock substrates, largely exposed to sunlight. Even the smallest sites in Piney Creek are 4–6 m wide with at least a partially open canopy. Under those conditions, growth of periphyton is substantial, providing foraging opportunity for herbivores like *Campostoma anomalum, Notropis nubilis,* and *Phoxinus erythrogaster.* Further downstream (e.g., P-5 to P-1 in Figure 2.12A), pools are larger and deeper, with more sand or silt in the bottoms, and comprise more of the area of the streambed than do rock riffles. Although large cobble–boulder riffles exist in downstream reaches, there is proportionately much less surface area available for periphyton attachment than at upstream sites where a preponderance of rock substrates and clear water promote periphyton growth. The large downstream riffles are suitable for the abundant darters (*Etheostoma* species), which feed mostly on insect larvae, comprising a substantial proportion of the "insectivore" category. The larger and deeper downstream pools with substantial currents are ideal for drift-feeding and water-column minnows (Cyprinidae), topminnows (Fundulidae), and other insectivorous taxa; thus, both pools and riffles in lower Piney Creek are numerically dominated by insectivorous individuals, and herbivores and omnivores are in relatively low numbers.

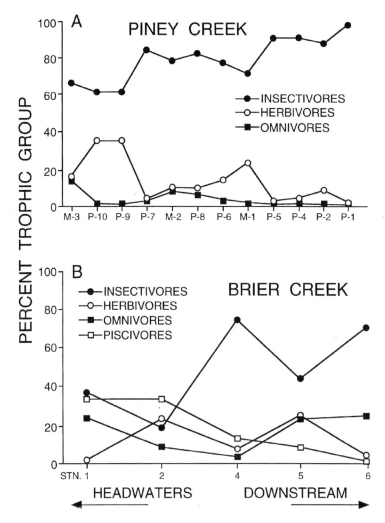

Figure 2.12. Percent abundance of fishes in defined trophic groups, by locality, in Piney Creek, Arkansas and Brier Creek, Oklahoma. [From Matthews (unpublished data)].

Longitudinal patterns in trophic groups are less obvious in Brier Creek (Fig. 2.12B), but they can be detected. As in Piney Creek, insectivores are most abundant downstream. Lower Brier Creek, from Station 4 downstream, flows year-round (Matthews, 1987a), but headwaters sites typically cease flow in later summer, creating pond-like conditions that favor sunfish species (which as a family are typically fishes of slow-moving or lentic waters). Presumably, permanent flow with drifting food items and substantially flowing riffles at downstream sites favor insectivorous minnows, topminnows, and juvenile sunfishes in pools, and insectivorous darters (Percidae) in riffles. Herbivores (*Campostoma anomalum*) are of relatively low abundance throughout most of Brier Creek (Ross et al., 1985; Matthews et al., 1988), occurring mostly where solid substrates permit growth of periphyton, and relatively scarce in pools with silty or soft bottoms. Piscivores, mostly green sunfish (*Lepomis cyanellus*) and black basses (*Micropterus salmoides* and *M. punctulatus*) increase in relative abundance at upstream sites, which are dominated by deep, slow, or nonflowing pool environments much of the year. Here, physical conditions are harsh and extreme [low oxygen conditions, dramatic temperature changes (Matthews, 1987a)].

The extreme headwater (Station 1) consists of one deep and several smaller pools dominated by piscivorous–insectivorous large sunfishes (*Lepomis*) and two species of insectivorous minnows. In these headwaters, species like green sunfish, *Lepomis cyanellus,* which is very tolerant of physicochemical stress (Matthews, 1987a), is the dominant species. Thus, longitudinal patterns in trophic groups do exist for fish assemblages in Brier Creek, but they are likely controlled mostly by local variation in flow, pool development (Schlosser, 1987), and physicochemical conditions (Matthews, 1987a). For both Piney and Brier creeks, evaluation of local assemblage structure from the perspective of trophic groups can provide more information about ecological processes than mere enumeration of species, families, or other taxonomic categories.

In summary, several of the trophic classifications systems that have been proposed can be used (cautiously) for many freshwater fishes—particularly for species that mature rapidly to use of adult foods or if size-specific "ecological species" are recognized (e.g., Polis, 1984; Matthews et al., 1994). These schemes may not yet be perfected, but even with their flaws (Gerking, 1994), they seem to be useful steps in the direction of examining general patterns in the ecological as well as the taxonomic structures of fish assemblages.

Relative Abundance of Piscivore to Prey Species in Assemblages

The numbers of piscivorous species relative to numbers of potential prey species in local assemblages can have important implications for system dynamics. It is known that in localized habitat patches in streams (e.g., within a single pool), some piscivores can control fish assemblages (Power and Matthews, 1983; Matthews et al., 1994). There are many examples of changes in fish communities after the addition of exotic piscivores to lakes or artificial impoundments [e.g., peacock bass in Lake Gatun, Panama (Zaret, 1989), Nile perch in Lake Victoria (Witte et al., 1992; Stiassny, 1996), and brown trout in New Zealand streams, (McDowall, 1984; Crowl et al., 1992)]. Numbers of piscivore versus prey species are of considerable theoretical interest, with implications for pervasiveness of "top-down" versus "bottom-up" controlling factors in ecosystems (e.g., Power, 1992) or changes in dynamics of predation when more than one kind of predator attacks a prey base.

Structure of Fish Assemblages / 61

Interactions of piscivores and their prey are addressed in Chapter 10. Here, I provide an empirical view of the numbers of prey species typically found in local assemblages with given numbers of piscivorous species, drawing examples from the literature and my samples from the midwestern United States. I focus on *numbers of species* of piscivores versus prey in local assemblages, rather than abundance of those species. The question to be asked from field studies of assemblages is whether there is a positive, negative, or no relationship between numbers of prey and piscivorous species. Causality will be considered in Chapter 10. Here, I only establish the empirical information on structure of fish assemblages from the perspective of numbers of predator versus prey species.

In Figure 2.13, the number of prey species in local assemblages is plotted against the number of piscivore species for 188 locations sites in 21 streams of eastern North America. I include only samples that were apparently thorough and comprehensive (data from: Matthews and Harp, 1974; Jackson and Harp, 1973; Bauer et al., 1978, two streams; Binderim, 1977; Fowler and Harp, 1974; Matthews, 1986b, three streams; Hocutt and Stauffer, 1975; Harrel et al., 1967; Tramer and Rogers, 1973, two streams; Evans and Noble, 1979; Jenkins and Freeman, 1972; Harris and Douglas, 1978; Dewey and Moen, 1978; Gelwick and Gore, 1990; Stewart et al., 1992; Stauffer et al., 1975; Kuehne, 1962). This analysis is focused on individual sampling sites rather than averages for entire

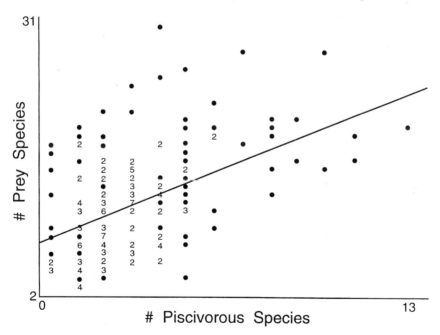

Figure 2.13. Number of prey fish species versus number of piscivorous fish species (as defined in text) in 178 local fish assemblages in eastern North America.

streams. This results in pseudoreplication in a strict statistical sense (sites within a stream may not be independent), but to average prey–piscivores across sites would mask the local variation within stream systems that is important. Analysis was limited to North American streams, from which I have most sources of data, but similar questions could be asked of diverse systems worldwide.

"Piscivores" included all species that as adults include fish as a substantial component of their diet. Thus, for eastern North America, some of the taxa scored as piscivores included the following: creek chub, *Semotilus atromaculatus;* catfishes of genus *Ictalurus* and *Pylodictis;* all black bass (*Micropterus*); *Morone,* except *M. mississippinesis;* most sunfish, *Lepomis,* but excluding small-bodied taxa like *L. humilis* and *L. megalotis;* all gars (Lepisosteidae); all esocids; and large percids (*Stizostedion*). I excluded taxa like *Cottus,* which has been reported to eat small fish (Phillips and Kilambi, 1996), but not as a major part of the diet. I scored as prey all other species having adult body size sufficiently small to be eaten by at least some of the piscivores above. Hence, all topminnows, darters, artherinids, poeciliids, clupeids, and minnows [except *Semotilus atromaculatus* and large-bodied taxa like carp (*Cyprinus carpio*)] were included as prey. Large bodied, nonpredatory taxa like catostomids, *Polyodon,* and sturgeons were considered neither piscivore nor prey.

Across the 188 sites, there were, as would be expected, many more prey than predator species per site on average (Figs. 2.14 and 2.15). For all sites, there was a mean of 11.4 prey species per site, with wide variance (Fig. 2.14) and as many as 31 prey species in one locality. No localities lacked prey species. The mean number of piscivore species detected per site was 3.1, with much less variance than for prey species, and only a small proportion of the sites with more than 5 piscivore species present (Fig. 2.15). Sites with high numbers of piscivores (e.g., 10–13) generally had a substantial number of large sunfish species (*Lepomis*), scored as piscivores. It is generally recognized that the kind of sampling (usually seining) on which these studies were based can underestimate the numbers or biomass of large adult piscivores, as they may be difficult to collect, but thorough sampling by seining at a given site will usually detect juveniles of those species (i.e., allowing their inclusion in the data set as "present"). Thus, I suspect that the number of piscivore species indicated for these streams in Figure 2.15 is representative; that is, that most stream locations in eastern North America have one to five species of piscivores.

Overall, there was a positive relationship between number of piscivorous species and prey species for all 188 sites on 21 streams in eastern North America (Fig. 2.13). The numbers of prey and piscivorous species were correlated significantly ($r = 0.49$, accounting for 24% of the variance). There was evidence for greater numbers of both prey and piscivore species in some complex habitats. For example, in Mill Creek, Oklahoma, Binderim (1977) found the greatest numbers of prey and piscivore species at the two most downstream, complex sites (19 prey: 11 piscivores, and 20 prey: 13 piscivores). However, for the Mill

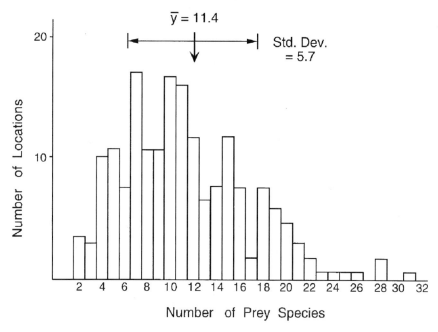

Figure 2.14. Frequency distribution of locations with specified number of prey species per local assemblage for 178 local fish assemblages in eastern North America.

Creek sites, collections were pooled across time. (Thus, it is possible that not all piscivore and prey species were present simultaneously—as some could have migrated in and out of the sites at various times.)

Instead of absolute numbers of prey or piscivore species per site, it may be desirable to focus on the ratio of prey to piscivores to ask if there are any regularities of pattern having biological meaning. The numbers of prey and piscivore species were converted to a prey/piscivore ratio for the 21 streams in eastern North America (above). (Ten sites lacking piscivores were omitted, because a ratio with denominator zero is not useful.) For the remaining 178 sites, there was an average of 4.8 prey species for each piscivore species present in local assemblages (Fig. 2.16), with many locations having a prey/piscivore ratio from slightly more than 1:1 to about 5:1. The mode at about 3:1 and steepness of the histogram (Fig. 2.16) suggests that for many local stream sites, there are about three to four times as many prey species as piscivore species. This value, of about three prey species for every piscivore species in local fish assemblages, is very different from the ratio of prey to predator species reported for 14 "community food webs" by Cohen (1977). Cohen found *more* predator than prey species on average in communities, with only about three prey species for every four predator species. However, my 3:1 ratio (above) is virtually identical to the ratio of prey to predators (average predators/prey = 0.36) reported by Jeffries

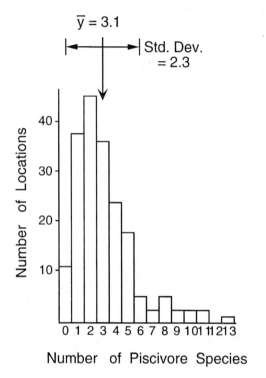

Figure 2.15. Frequency distribution of locations with a specified number of piscivorous fish species in 178 local fish assemblages in eastern North America.

and Lawton (1985) for stream invertebrate assemblages at 96 sites, worldwide. The implication seems to be that predator–prey dynamics could be very different in aquatic versus "other" systems.

Another question about the relationship between prey and piscivore species is if sites with a greater overall number of species exhibit either larger or smaller ratios of prey to piscivore species than do sites with few species. In other words, as the total number of species in a local assemblage increases, is the increase related to the disproportional addition of either prey or piscivore species, or could some approximate ratio of prey to piscivore species prevail across sites with widely differing total species richness?

Although there is a weak correlation ($p = 0.03$; but only 2.6% of the variance accounted for) between the prey/piscivore ratio and the total species richness in Figure 2.17, the relationship appears distinctly nonlinear. Although a wide range of prey/piscivore values was possible for many levels of local species richness, there were no high prey/piscivore ratios in assemblages with either very low or very high numbers of species. Part of this is a statistical artifact: if the total number of species in an assemblage is very low (four, for example), the maximum

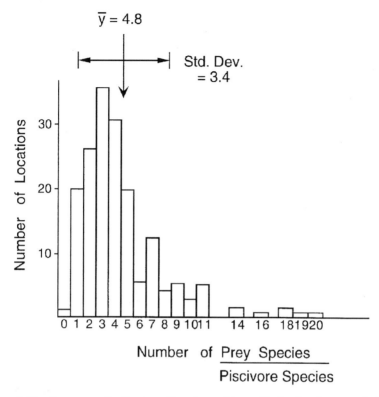

Figure 2.16. Frequency distribution of locations with specified ratio of prey to piscivore species, for 178 local fish assemblages in eastern North America.

value that could result for prey/piscivore ratio (assuming at least one piscivore) would be 3/1 = 3.0. Thus, on the left side of Figure 2.17, it is logical that values would be small with the potential in increase toward the middle of the figure, which they do. However, it is less intuitive why ratios decrease toward the right. Beyond approximately 25 total species, none of the ratios exceed about 6:1. This suggests that in all of the more speciose local assemblages, there are substantial numbers of piscivores, which (being in the denominator of the ratio) helps to keep the ratio values low (e.g., 30 prey/5 piscivorous species = ratio of 6.0). Thus, there is in none of the 21 streams a situation with a very large number of prey species without some concomitant increase in the numbers of piscivore species. To answer the original question, there is no regular pattern of increased or decreased ratio of prey to piscivore species at any given total species richness. However, it is obvious from the envelope of space enclosed by all of the points in Figure 2.17 that for much of the midrange of values for species richness (e.g., 10–25 total species), there is a very wide range of possibilities of relationships between number of prey and piscivore species. Some sites in this range have

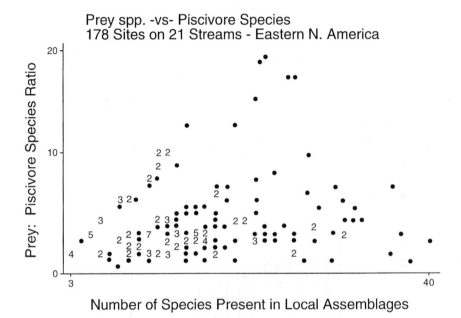

Figure 2.17. Prey to piscivore fish species ratio versus total number of fish species present in 178 local assemblages in eastern North America.

only one successful piscivore, others have many; hence, the wide range in prey/piscivore ratios. There seems to be no single "solution" to the number of prey species available to or needed to support a single piscivorous species across the range of environments in our sample of streams.

Functional Groups of Fish in Assemblages

Tropic groups describe "what" a fish eats, with some schemes (e.g., Keenleyside, 1979) including, in general, "where" (e.g., water column, benthic) a fish eats. Functional groups should reflect how a fish obtains food. There are two important reasons for considering the "how" of food-getting (i.e., "functional groups" of fishes).

1. Knowing "how" fishes feed is linked closely to differences in abilities of species to use similar food items (e.g., if a common, abundant or widely distributed prey were available, such as zooplankton) and may help us understand the outcome of competition between species (e.g., McComas and Drenner, 1982).
2. "How" fish feed may have strong impact on an ecosystem (e.g., burrowing feeders resuspending benthic organic matter, detritus, nutrients, and so on, or exposing benthic prey to other fishes).

I will compare three "functional" (or partly so) classifications: the functional group classification of aquatic insects by Cummins (1978, p. 31), the fish trophic classification of Keenleyside [1979, his chapter 2, summarized by Wootton (1990, p. 33)], and a proposed scheme—speculative and untested—for fish functional groups.

In Table 2.3, the invertebrate functional groups of Cummins (1978) are compared to the trophic groups of Keenleyside (1979), as summarized by Wootton (1990). The invertebrate functional groups of Cummins (1978) are used widely in aquatic ecology and are intimately tied to the River Continuum concept (Vannote et al., 1980), which forms the basis for a prevailing school of thought about functional relationships in streams. More than any other fish "trophic group" system, the Keenleyside–Wootton scheme incorporates function (e.g., how feeding is accomplished) with categories of foods that are used. Some of the "trophic" categories of the Keenleyside–Wootton scheme for fishes match Cummins' functional groups for aquatic insects rather well (Table 2.3).

Grazing fishes (e.g., *Campostoma* or juveniles of some suckers, Catostomidae) remove periphyton by a scraping lower jaw function, much like the invertebrate "scrapers." Fishes that eat macrophytes by chewing and ingesting are analogous to invertebrate "shredders" that feed by fragmenting large pieces of plant material [living plant material or coarse particulate organic matter (e.g., leaf litter)]. In both cases, the effect of fish and of insect is similar—altering the ecosystem in like manner—even though a single fish may remove or process much larger quantities of material in a single "bite" than any invertebrate can. The "engulfer"

Table 2.3. *Comparison of Aquatic Invertebrate Trophic (Functional) Groups (Cummins, 1978) with Fish Trophic Groups of Keenleyside (1979) as Summarized by Wootton (1990).*

Invertebrate Functional Groups (Cummins)	Fish Trophic Groups (Keenleyside–Wootton)
Shredders	Browsers
Living plants and CPOM	
Collectors	Detritivores
FPOM[1]	Scavengers
	Filter-feeders
	Benthivorous pickers
	Disburbance pickers
Scrapers	Grazers
Attached algae	
Piercers	None
Living animal tissue, pierce, and suck	
Engulfers	Particulate feeders
Predators on whole animals or parts	Piscivores (ambush, lure, stalk, chase)
	Grasping benthivores
Parasites	Ectoparasites (incl. scale and fin eating)

[1]Fine particulate organic matter

insects of Cummins, which prey on whole animals or animal parts, seem much like piscine particulate feeders (e.g., on zooplankton, other small invertebrates), grasping benthivores, or piscivores, all of which are carnivorous and "engulf" all or parts of living animal prey.

The "piercers" among the aquatic insects do not seem to be represented by freshwater fishes. More invertebrates probably live as "parasites" than do fishes, but Keenleyside and Wootton note that some fishes feed as ectoparasites, by eating scales, fins, and so forth of other fishes. Several important categories of feeding by fishes do not seem to be represented in the insect functional groups of Cummins (1978), including phytoplanktivores and species that pick or sort benthic animal prey from sediments or gravel substrates. However, there remains a relatively good match of fishes and aquatic invertebrates in what are actually functional groups—describing the ways aquatic insects or freshwater fishes obtain food and carrying the implications that they might fit similar categories in their effects on the ecosystem (Chapter 11).

In spite of the reasonably good match of Cummins' and Keenleyside–Wootton's schemes for functional or trophic classification, I would like to classify the feeding of fishes more from the perspective of their potential effect on the ecosystem. Freshwater fishes affect their ecosystems in three primary ways: (1) killing or removing smaller animals or plants from the population, (2) actively producing fragments or feces containing the bodies of their prey, and (3) mechanically disturbing substrates, increasing the resuspension and/or transport of materials on water currents. Although these effects are explored fully in Chapter 11, it is important to consider them here to complete a picture of the functional structure of fish assemblages.

In Table 2.4, freshwater fishes (with examples) are divided into two major categories: those species that mechanically disturb the substrates from which they feed, and those that feed without disturbing substrates. This distinction is important, because of potential effects of a fish species on the ecosystem if, by virtue of their feeding activities, materials associated with the substrates (biofilms, attached algae, detritus, nutrients, particulate organic matter) are returned from a "closed" compartment (e.g., locked in place in sediments or some other "inactive" form) to an "active" compartment (e.g., exposed for uptake or processing by other components of the stream or lake ecosystem). Some of the distinctions within the two major categories may seem trivial with respect to ecosystem effects, but they are not. For example, piscivores feeding by suction/engulfing smaller fishes versus biting to kill and then returning to swallow all or parts of the prey would seem initially to have the same effect: removing a live prey individual from its population. However, the piscivore that engulfs and swallows prey whole and intact may remove the prey individual from the water column with little "leftovers," whereas a piscivore killing by biting or slashing, then returning to consume pieces of the prey would release prey body fluids, mucus, and perhaps bits of flesh into the water column. These materials, in turn, could

Table 2.4. *Proposed Fish Functional Groups, with Emphasis on Interactions of Fish Feeding with Ecosystem Processes; Examples from North American Fauna*

I. PHYSICALLY DISTURB SUBSTRATES:
  1. Grazers (scrape, pick, or "shovel" materials from hard substrates—stony or woody debris)—*Campostoma*
  2. Benthic detritivores—*Prochilodus*
  3. Deep burrowers in soft substrates—*Cyprinus*
  4. Gravel-disturbers—*Hypentelium*, some *Moxostoma*
  5. Mud- or sand-eaters (consume soft sediments, biofilms, sand)—*Hybognathus*, adult *Dorosoma cepedianum*, *Pimephales*, *Hybognathus*, *Luxilus pilsbryi*
  6. Stone-turners—some *Percina* (log perch)
  7. Scavengers of animal material—*Amieurus*
  8. Egg-eaters—many small cyprinids or small *Lepomis*

II. DO NOT DISTURB SUBSTRATES
  On invertebrate prey
  9. Surface-feeders (*Lepomis*, many minnows)
  10. Drift-feeders (*Notropis boops*)
  11. Benthic-pickers (*Cyprinella venusta*), many riffle-dwelling darters (*Percina, Etheostoma*)
  12. Water-column particulate feeders (suck—*Menidia*; bite or snap jaws—*Labidesthes*)
  13. Snail crushers (*Lepomis microlophus*)
  14. Filter feeders (*Dorosoma*)
  On fish as prey
  15. Suction piscivores (*Micropterus*)
  16. Overrun or biting piscivores (*Morone saxatilis*)
  On vascular materials
  17. Seed crushers (none in North America; see *Colosoma* and others in South America, from Goulding (1980)

attract smaller fishes, invertebrates, and so forth, and release nutrient-bearing fluids, with an overall different effect in the system than that of a "clean" engulfing of prey. The "microbial loop" (i.e., direct uptake of nutrients by bacteria in the water column) could be enhanced if piscivores directly release nutrients from prey by virtue of mechanical actions of feeding.

The proposed functional groups for fishes (Table 2.4) have not been tested for general usefulness. Like all "trophic" classifications, this scheme suffers from the fact that many species cross boundaries at times. However, this variation in feeding–functioning in the ecosystem, rather than being "noise," is an important component of any study of the relationship between fishes and their ecosystem. There are obvious effects of ontogeny in this scheme. For example, our observations in streams of the midwestern United States show that although adult *Erimyzon oblongus* (creek chubsucker) feeds by removing insects from gravel as an adult, small juvenile *Erimyzon* feed largely by grazing attached algae. Thus, this species, like many others, moves among functional groups as it grows.

The scheme in Table 2.4 needs much refinement to categorize the structure of fish assemblages by the functional roles of their members in ecosystems, and it focuses only on feeding activities. Other activities (e.g., nest building) obviously have the potential to change ecosystems. This scheme of fish "functional groups" is presented as a start toward quantifying effects of fishes in ecosystems by "how" as well as "what" or "where" they feed. At this point, I only wish to establish that the structure of a fish assemblage should include a view of how fish in the assemblage feed, as well as what items they use.

As an example of applying this scheme of functional groups to a moderately complex fish fauna, I have classified, by functional group, the common species that I have collected in Piney Creek, Izard County, Arkansas (USA). The classification is not perfect because exhaustive underwater observations combined with information on stomach contents have not been made for all of these species. Accordingly, I classified each species on the basis of my limited snorkeling observations and extensive above-water observations of their feeding activities, with reference (when in doubt) to information in Robison and Buchanan (1988), and my 20+ years of noting the behaviors and microhabitats of species in Piney Creek (Matthews, 1986c; Matthews et al., 1988a).

Of 44 species from Piney Creek, 54.5% are benthic pickers of invertebrates (including all darters that feed on aquatic insects in riffles), 36.4% are surface feeders on invertebrates, 29.5% are grazers of attached algae, 29.5% feed on drifting invertebrates in the water column, and 25% are suction piscivores, with water-column particulate feeders (e.g., on zooplankton), mud/sand/biofilm users, benthic detritivores, and several other less common functional groups also represented (Fig. 2.18). In all, 12 of the 17 postulated functional groups of fishes occur in Piney Creek. Lacking are burrowers of soft substrates, scavengers of dead animal material, filter feeders, biting piscivores, and seed crushers (which are generally lacking from North America). Clearly, more information on the way that the structure of the fish fauna can affect the ecosystem might be gained by quantifying on the basis of abundance of each species. However, the overview of numbers of species that are known to play the functional roles shown in Figure 2.18 suggests diverse roles of fishes in Piney Creek, and the expectation of an ecologist for fish–ecosystem interactions would be different in Piney Creek from that in a simple fauna dominated by fewer functional groups. Of the 44 Piney Creek species, exactly 50% (Table 2.5) are included, at least in part, in functional groups 1–8 (Table 2.4); that is, they acquire food in some way that has the potential to disturb the substrates where they feed, thus affecting materials *other than* the food items (e.g., by returning them to the water column, making them available to water-column predators, etc.).

Brier Creek, in southern Oklahoma (USA), has an environment very different from that of Piney Creek (Ross et al., 1985), with the former physicochemically very harsh relative to the latter. As noted above and in Brown and Matthews (1995), the fish fauna of the two creeks differ in proportion of trophic groups [based on the classification of Karr et al. (1986)]. However, the overall distribution of species across functional groups appears rather similar for the two systems. For Brier Creek (Fig. 2.18; raw data not shown) as well as Piney Creek (Fig.

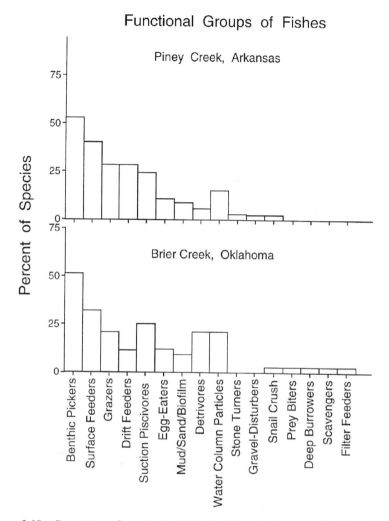

Figure 2.18. Percentage of species in 16 defined functional groups for Piney Creek, Arkansas and Brier Creek, Oklahoma (USA).

2.18), the two most common groups are benthic pickers and surface feeders. Brier Creek fauna includes 14 of the functional categories in Table 2.4 relative to 12 in Piney Creek. Brier Creek, with less consistent flow, has fewer drift feeders but more detritivores and water-column particulate feeders. Brier Creek also contained a filter feeder (*Dorosoma cepedianum*), a scavenger (*Amieurus melas*), and a deep burrower (*Cyprinus carpio*), all of which Piney Creek lacked. However, Brier Creek lacked the species that feed by stone-turning or gravel-disturbing, which were present in Piney Creek. In spite of differences in less

Table 2.5. Classification of 44 Fish Species in Piney Creek, Arkansas, by Functional Group as Adults, Following Categories in Table 2.4.

Cyprinidae			Salmonidae	
Campostoma spp.	1		Oncoryhnchus mykiss	9, 10, 15
Cyprinella galactura	10, 9		Fundulidae	
Hybopsis amblops	11		Fundulus catenatus	11, 9
Luxilus chrysocephalus	11		Fundulus olivaceus	9
Luxilus pilsbryi	10, 9, 11, 5		Poeciliidae	
Nocomis biguttatus	9, 10, 11, 1		Gambusia affinis	9, 1
Notemigonus crysoleucas	9, 12, 10		Atherinidae	
Notropis boops	10		Labidesthes sicculus	12
Notropis greenei	10		Centrarchidae	
Notropis nubilus	1		Ambloplites constellatus	15
Notropis rubellus	10, 9, 1		Lepomis cyanellus	9, 15, 10, 8
Notropis sabinae	11, 10		Lepomis macrochirus	9, 12, 15, 8
Notropis telescopus	10, 9		Lepomis megalotis	12, 9, 15, 8
Phoxinus erythrogaster	1		Lepomis microlophus	13, 11, 1, 2
Pimephales notatus	11, 12, 1		Micropterus dolomieu	15, 9, 10
Pimephales promelas	2, 12, 11		Micropterus punctulatus	15, 9
Semotilus atromaculatus	11, 10, 15, 1		Micropterus salmoides	15, 9
Catostomidae			Percidae	
Carpiodes carpio	1, 5, 11		Etheostoma blennioides	11
Erimyzon oblongus	5, 11, 1		Etheostoma caeruleum	11, 8
Hypentelium nigricans	4, 6, 11		Etheostoma spectabile	11, 8
Moxostoma duquesnei	12, 11, 4		Etheostoma zonale	11
Moxostoma erythrurum	5, 1, 11		Cottidae	
Ictaluridae			Cottus carolinae	11, 15
Ameiurus natalis	11, 2		Cottus hypselurus	11, 15
Noturus albater	11			
Noturus exilis	11			

Categories from Table 2.4, with number of Piney Creek species in parentheses: grazers (15), benthic detritivores (3), deep burrowers (none), Gravel-disturbers (2), Mud- or sand-eaters (4), stone-turners (1), scavengers of animal material (none), egg-eaters (5), surface-feeders (15), drift-feeders (13), benthic pickers (24), water-column particulate feeders (12), snail crushers (1), filter feeders (none), suction predators (11), overrun or biting piscivores (none), seed crushers (none)

common functional groups, both creeks present a rather similar distribution of species among the more common functional group categories (Fig. 2.18). Whether or not this pattern holds for more streams in North America remains to be seen, but these preliminary comparisons suggest that evaluating functional roles of fish species across a variety of streams could address generalities in structure of fish assemblages and their effects in ecosystems. Finally, remember that the emphasis from the perspective of functional groups is to provide an estimated overview of the ways that fish in an assemblage can alter the ecosystem (removing items, producing new items such as feces, changing nutrients from one compartment to another, moving materials downstream or back into the water column, etc.).

This is very different from merely tallying what kinds of food items a species uses or where (microhabitat) it takes food.

2.4 Abundance, Body Size, and Mouth Size

Distribution of Abundance of Species Within Assemblages

In typical fish assemblages, how many species are abundant? How many are rare (represented by only one or a few individuals)? Theoretical constructs (Pielou, 1975) and empirical studies suggest that in most animal assemblages, there will be a few very abundant species and many rare ones (Brown, 1995a). Sheldon (1987) analyzed rarity for fishes in Oswego Creek, New York and found that numerous species were represented by small numbers of individuals. Sheldon (1987) commented on the biological effects of being rare; for example, lowering the likelihood of interspecific or intraspecific encounters and even risking extinction if populations are too small and locally isolated. However, it is clear to anyone who has collected fish widely that some species usually occur in small numbers relative to other, taxonomically related species (such as *Phenacobius mirabilis* or *Notropis rubellus*).

Ibarra and Stewart (1989) found a substantial proportion of the sandbar fish species in the Napo River, Ecuador to be represented by only a single individual per site. If such samples are plotted with the number of species on the y axis and the number of individuals per species on an arithmetic x axis (a species-abundance distribution), the result often is a hollow curve, with a peak on the left indicating that many species are locally rare and a tail to the right representing the few species that are locally abundant (Fig. 2.19A). Preston's (1962) assessment of "commonness and rarity" across a variety of taxa suggested that if the x axis is converted to a log scale (he preferred base 2), there will be a "log-normal" bell-shaped curve with a peak somewhere on the x axis at relatively low numbers of individuals (Fig. 2.19B), but with the curve falling away from the peak in both directions (or often truncated at the left). This lognormal distribution of species abundances within assemblages results if there are a large number of relatively "rare" species and few "abundant" species in the assemblage. The lognormal distribution of abundance of species in natural communities has been considered a ubiquitous ecological property across many taxa (Sugihara, 1980), although the underlying biological meaning of the lognormal has alternately been suggested to be an artifact (May, 1975) or biologically meaningful (Sugihara, 1980), and numerous alternative mathematical series ("geometric," "log series," and "broken-stick") have been proposed as better-fitting alternatives under various assumptions [reviewed by Pielou (1975)]. Finally, for assemblages with relatively few species, it may be better to place the species into a "ranked-abundance" curve, depicting the abundance of each species on the y axis relative to its rank

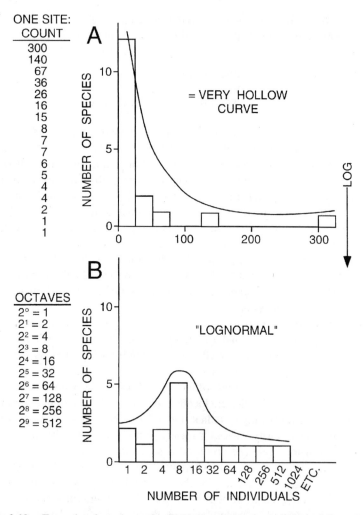

Figure 2.19. Example of numbers of individuals per species in a typical (hypothetical) local assemblage, plotted as number of species versus raw number of individuals per species on the x axis (A) and plotted versus an exponential (octaves) x axis (B).

of abundance on the x axis (Pielou, 1975). Unfortunately, as seen earlier in this chapter, local fish assemblages can range from only a few to as many as 100 species, so neither the species-abundance distribution nor the ranked-abundance approach may work across all sites that might be considered for freshwater fishes. King (1964) used a ranked-abundance approach to the abundance of fish within families (Percidae and Cyprindae), finding that for well-made collections there was a reasonably good fit to MacArthur's original "random-breakage, broken-stick" model [although Pielou (1975) shows that this model is largely untenable, as did Sheldon (1987) for his stream fish collections].

Using a ranked-abundance (log abundance versus ranked abundance) approach for creek segments with 6–23 species, Sheldon (1987) found curves for species-poor and species-rich sites that fit relatively well the patterns that had been shown for plants, birds, and other taxa (Whittaker, 1972). For Sheldon's Oswego Creek samples and for samples by Boschung and O'Neil (1981) that Sheldon (1987) used for comparison, the lognormal model best-fit distribution of abundance of stream fishes, followed by the log series [of Williams (1964)]. The random-breakage model of MacArthur (1957) did not fit the data from either sets of collections. Regardless of which [if any—see Pielou (1975, p. 61) on the difficulty of testing the fit of curves with empirical data] mathematical constructs best fit empirical data or theoretical implications of the numbers of rare versus abundant species, it is certainly of empirical interest to know what kind of rarity-commonness to expect in typical fish assemblages in order to predict future persistence of species or to pose hypotheses about interactions among fishes.

The most used assessment of the distribution of abundance of species plots the numbers of species (x axis) occurring in intervals or "octaves" of abundance (y axis) on a scale of log base 2 (as suggested by Preston), so that successive octaves represent doubling of abundance of the species (May, 1975). This scaling of the x axis (abundance) results in the abundance categories increasing as 2, 4, 8, 16, and so forth (Fig. 2.19B). A large body of evidence across diverse taxa has shown that there typically is a lognormal distribution of species across the octaves of abundance, forming a bell-shaped curve on either side of the octave with the maximum number of species (Preston, 1962; May 1975; Sugihara, 1980).

In Figure 2.20, species-abundance distributions are shown for six freshwater systems (four North American streams, a tropical stream, and a suite of northern lakes). For each system, 6–13 sampling sites were included, with the averages in numbers of species plotted for each octave in Figure 2.20. Figure 2.20 depicts the number of species per site for each system, averaged across sites (y axis), falling within successive $\log_2$ abundance octaves (x axis). In four of the systems (Piney Creek, Arkansas; Roanoke River, Virginia; Buffalo River, Mississippi; Rio Tortuguero system, Costa Rica), there is a distinct maximum in the mean number of species per site in the first octave (i.e., many species in those systems are "rare") represented by only one to two individuals in a site. In the remaining two systems (Brier Creek, Oklahoma, and six small lakes in Wisconsin), there also is a maximum in the number of species in the first octave although the "peak" is less distinct. For all six of the systems, there were more species occurring as only one to two individuals per site than at any other abundance.

Assuming that these systems are "typical" for moderately complex to complex fish assemblages, it appears that at a given site about three to nine species are "rare" (one to two individuals). In the next several octaves of abundance (3–4, 5–8, 9–16, 17–32, 32–64, 65–128), responses varied among systems, but in

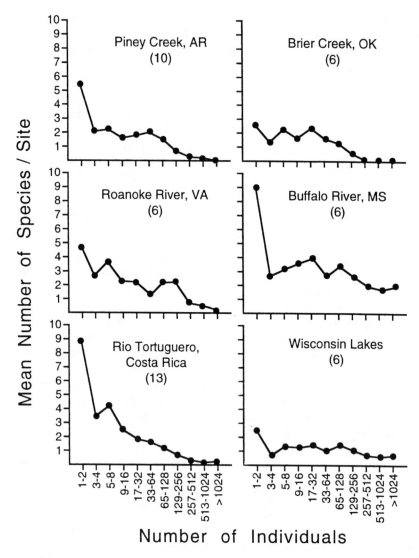

Figure 2.20. Number of fish species in each of 11 octaves of abundance (*x* axis) for local assemblages at 6–13 sites, for 4 streams in temperate North America, a Costa Rican river, and lakes in Wisconsin (USA).

most there were 1–4 species in each interval. In the next octave of abundance (129–256 individuals), Piney Creek, Brier Creek, Rio Tortuguero, and the Wisconsin lakes all had, on average, one or fewer species per site. At or above the octave with 257–512 individuals, only the Buffalo River system, in Mississippi, had, on average, more than one species per octave of abundance.

Overall, across six very dissimilar freshwater systems, the distribution of species per site across $\log_2$-based octaves of abundance (Fig. 2.20) show that fish assemblages have, on average, a large number of species that are detected as only 1–2 individuals per site, a modest number of species for which 3 to 100 or so individuals are present, and relatively few species that number in the hundreds of individuals per site. There are, even more than for many other taxa, many rare species and few highly abundant ones. The patterns in Figure 2.20 do not appear particularly lognormal, but I have made no attempt to formally fit the patterns to mathematical models like the lognormal, log series, or sequential breakage (Sheldon, 1987). Brown (1995a) points out that the most important message, regardless of the closest-fitting model, remains the "qualitative one"; that is, that in most ecological communities, there are a few relatively abundant species and many rare species. Fish assemblages seem to fit this pattern, overall.

Caveats are in order in evaluating results in Figure 2.20. First, the six systems selected for evaluation were based on three for which I have personal knowledge (i.e., made the collections) and others that seemed likely to capture a wide range of environmental conditions. Second, the numbers of species falling into the high-abundance octaves clearly will be influenced by the number of collections per site. Collections at the six sites in Figure 2.20 ranged from monthly for more than a year (e.g., Buffalo River), to a single sample per site (Rio Tortuguero); hence, they are not rigorously comparable. However, in each case, the sampling was by competent investigators using either a single "best" gear or a combination of collecting gears to attempt to collect all species at the site, and the similarities among patterns in Figure 2.20 seem greater than the differences, in spite of widely different efforts per site. I conclude that ecologists going to freshwater lakes or streams to sample fish should expect to find many rare and few abundant species in typical assemblages. To the extent that this is wrong, there may be fascinating local phenomena that regulate abundances of species. For now, I consider "many rare and few common" species as "expected" for the structure of fish assemblages, much as has been shown for a variety of taxonomic groups for decades (Fisher et al., 1943; Williams, 1964; Preston, 1948, 1962; May 1975; Sugihara, 1980).

Body and Mouth Size Structure Within Whole Assemblages

Many studies in ecology have addressed differences in body sizes among animal species. At a continental scale, taxa as diverse as mammals, birds, and fishes show more small than large species, but with fewer species either above or below a particular (typically sharp) mode in body size (Brown et al., 1993). Rosenzweig (1995) pointed out that Lindsey (1966) showed three exceptions to the "mode" rule, finding (as well as deep-sea fishes and frogs) a peak in number of species in the smallest size category for tropical freshwater fishes. Brown (1995a) and Brown and Nicoletto (1991) pointed out that samples of vertebrates (mammals) taken from smaller local patches of a relatively homogeneous habitat tended to have a more flattened curve, representing a more nearly equivalent distribution of body sizes among logarithmic size classes. Peters and Raelson (1984) and Peters (1991) reviewed global relationships between body mass and population density for mammals, finding a near-linear decrease in population density for mammals with body mass. Cotgreave (1993) also pointed out that, in a local area, the population density of small species is much higher than that of large-bodied species. Although this has been challenged by Blackburn et al. (1993), the literature overall suggests a general pattern of "many small and few large" animals in local assemblages.

There have been many studies of local animal assemblages, dating to Hutchinson (1959), empirically demonstrating rather regular spacing among species with respect to body size or marked differences in body size of coexisting, potentially competing taxa (Schoener, 1970; Bowers and Brown, 1982). Spacing of body size hypothetically could ameliorate competition within an assemblage, particularly if body size is related closely to food size. However, this idea has been challenged by Wiens and Rotenberry (1981), Wilson (1975), and others [reviewed in Case et al. (1983)] either because of a general lack of evidence for competition being important, because larger animals, with a potential for eating large-sized prey, were shown to eat a wide spectrum of prey sizes, or because of a lack of "universality" of the interspecific ratios earlier suggested. For pumpkinseed sunfish (*Lepomis gibbosus*), Osenberg and Mittlebach (1989) showed that prey-size selection related not only to prey (snail) size but also to the ways that relative body sizes of predators and prey influenced foraging behavior, such as attack probability. Many studies (e.g., Werner, 1977) have shown that optimal prey size may be different from maximum prey size predicted by gape limitation and related to capture probability, handling time, and so forth. However, there is at least an approximation for many fish with larger mouths eating larger prey. Hence, a consideration of body- or mouth-size distribution in a fish assemblage should reflect predation potential, if not actual consumption of a given size prey.

For animals with determinant growth, like birds and mammals, it is possible to focus on relative sizes of adults within local assemblages. For fish, as ectotherms with indeterminant growth and no "typical" adult body size within a species, it has been more common to estimate body-size relationships on the basis of maximum adult body size. For example, Schlosser (1990) showed, based on maximum adult body size, that the size of fish species generally increases from headwaters to downstream (Fig. 2.21), and Taylor and Gotelli (1994) compared range sizes to maximum body sizes for minnows of the genus *Cyprinella*. However, for fish having indeterminate growth, there is much less certainty within a species than for mammals or birds as to (1) "typical" adult body size or (2) time in months or years to reach a "maximum" (actually asymptotic) body size. Although much is known about some size-structured fish populations (Keast,

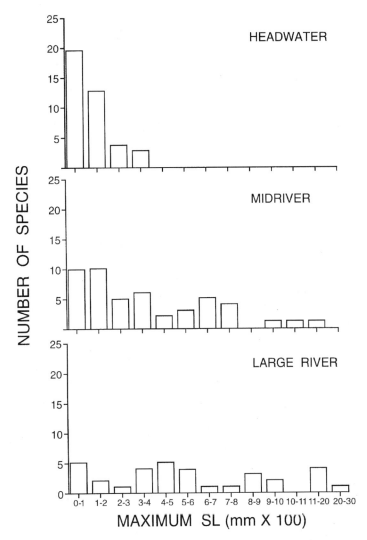

Figure 2.21. Number of fish versus maximum body size (standard length) in headwaters, midriver, and large river, for midwestern United States. [After Schlosser (1990)].

1977), less is known about size structure or relative size structure for whole assemblages of fishes considered across taxonomic lines.

I suggest that it might be desirable to focus not on size structure within fish populations, but to examine the spectrum of sizes represented across all species of a local assemblage. The only attempt to measure the size of all individuals in all species of a whole assemblage of which I am aware is that of Poff et al.

(1993), who included all fishes that were present in a documentation of the size distribution of all metazoans in an upland stream. Griffiths (1986) presented summaries of total numerical abundance (regardless of species) versus body size for a number of bird communities, finding strong suggestions of bimodality or polymodality of size distributions within communities. Thiebaux and Dickie (1993) provided an example of "body-size spectrum" of the biomass in Lake Michigan [from Sprules et al. (1991)] which included planktivorous and piscivorous fish. Thiebaux and Dickie (1993) also summarized total biomass per increment of body size from the data of Chadwick (1976), who measured the length of every fish recovered from total rotenone kill of the fishes of a small Canadian lake. Because (1) typical adult body size (Matthews, 1987b) and time to maturity (Keast, 1977; Matthews, 1986d) may vary widely across the geographic range of a species, (2) average or typical body sizes of individuals in a population is strongly seasonal (production of juveniles, which may dominate the population at certain times), and (3) so many (almost unlimited) combinations of species exist in local assemblages, the result for local assemblages is an endless array of possible body sizes of individuals present at a given time.

The ecological importance of body size for fish is well known. Different-sized individuals within fish species may function trophically or in other interactions with the environment as "ecological species" (Keast, 1977; Matthews et al., 1994), playing different roles as they mature. There are not only qualitative differences in diets as fishes mature (e.g., switch from invertebrate to fish prey for many piscivores), but also, because of gape limitation or size selectivity, larger fishes typically eat larger prey, or at least have a greater maximum prey size (Matthews and T. Crowl, unpublished data for striped bass in Lake Texoma). There is a long-standing argument about whether most fishes are gape-limited predators (taking the largest possible prey up to a maximum size, where intake drops to zero) or size-selective predators (taking the optimally sized prey, with prey election gradually declining at prey sizes larger or smaller than optimal) (Schmitt and Holbrook, 1984; Bence and Murdoch, 1986). However, for the assemblage-level summaries, it matters only minimally which model prevails across most fish species, so long as there is, indeed, a relationship among body size, mouth size, and prey size, on average. (However, note that in an assemblage of fishes if some species were gape limited and others size selective, the mere measurement of body or mouth size might overestimate or underestimate use of some size classes of prey.) Additionally, many aspects of fish life history are related to size—fecundity, reproductive behaviors, and so on—suggesting a whole different set of reasons that body size within an assemblage can be important.

Basic facts about body size for most fish assemblages are well known to anyone who has sampled fish by any technique: far more small than large individuals are present in a typical fish assemblage. However, documentation of this phenomenon is largely lacking and, for any one assemblage, would clearly be "gear dependent" (e.g., one has less chance of catching a large than a smaller individual with a small seine, because of differences in swimming speed or evasive abilities of the two). Nevertheless, "structure" should include a determination of how the sizes of individuals are distributed within any assemblage.

To evaluate the body size of individuals for a whole assemblage, a class under my direction measured the standard length (tip of snout to terminus of hypural plate) of all individual fish (mostly insectivore–omnivores, with a few piscivores; the algivorous *Campostoma anomalum* omitted) collected by seining in the pools and riffles at one location [Station 6 of Ross et al. (1985); Matthews et al., 1988a] in the lower mainstream of Brier Creek, Oklahoma in June 1985 (not including obvious young-of-year) (Fig. 2.22). This site was relatively easy to seine, and even if we did miss some of the larger individuals (e.g., *Lepomis, Micropterus*) that might have been present, there is still, the suggestion of a very wide array of mouth sizes. If, in fact, size of prey is linked to size of mouth for small stream fishes, it is apparent that this fish assemblage was capable of eating invertebrates or small fishes across a wide array of sizes. Most of the assemblage ranged in body length from about 40 to 60 mm Standard Length (SL). Peaks in the size distribution existed (Fig. 2.22), but they were not sharply defined, and a wide range of prey sizes appears likely to present optimal prey size (Pianka, 1978) for some of the individual fishes.

Closely related to body size (length or mass) is mouth size (e.g., gape width). Most articles on gape size and prey size are limited to one species. I know of no published attempts to quantify the array of mouth sizes in a whole fish assemblage. To illustrate one example of the spectrum of mouth sizes that existed in one stream fish assemblage, I measured to the nearest 0.1 mm the width of the lower jaw at its widest point for all individual fish ($N = 558$ specimens)

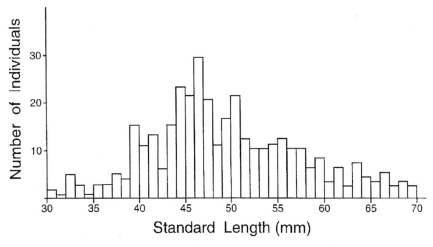

Figure 2.22. Number of individuals per increment of standard length, for one local assemblage sampled in midsummer in Brier Creek, Oklahoma.

collected by F. P. Gelwick and myself by seining pools at Station P-2 (see Matthews, 1986c) of Piney Creek, Arkansas in August 1986. At this site, there were swift main-channel pools with many drift-feeding minnows, and several backwater pools with minimal flow. This sample was dominated numerically by the minnows *Luxilus pilsbryi, Cyprinella galactura, Notropis nubilus, Notropis telescopus,* and the topminnow *Fundulus catenatus.* Other taxa included in the sample were as follows: *Nocomis biguttatus, Notropis rubellus,* and *Pimephales notatus* (Cyprinidae); *Lepomis* species and *Micropterus* species (Centrarcidae); *Etheostoma* species (Percidae); *Fundulus olivaceus* (Fundulidae); *Moxostoma* species and *Hypentelium nigricans* (Catostomidae). These species are all insectivorous, omnivorous, or piscivorous as adults; I omitted the algivorous *Campostoma anomalum* from measurement.

Figure 2.23 shows the distribution of individuals in this assemblage across 0.3-mm categories of mouth size. With all of the above species included, there are at least two obvious peaks (at about 2.5 mm and 4.7 mm mouth width, and a minor peak at about 7 mm). Because of the 5-mm mesh size used in sampling, the assemblage collected may underrepresent small young-of-year fishes (i.e., the truncation of the curve at a mouth size of 0.9 mm may not reflect some unknown number of very small individuals), but the declining limb of the histogram from about 2.5 mm mouth width is probably real—our seines perform very

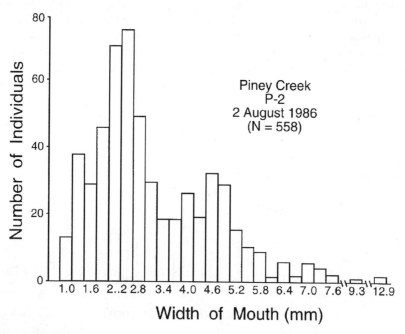

Figure 2.23. Number of individuals per increment of mouth width, for one local assemblage sampled in late summer in Piney Creek, Arkansas.

well in capturing individuals with a body size reflecting these mouth sizes. It appears that viewed without regard for taxonomic classification, this insectivore–omnivore–piscivore assemblage had many more individuals capable (assuming gape limitation) of engulfing small than large prey items, but the distribution of maximum prey sizes that could be consumed by this assemblage is at least biomodal, or trimodal. It would be of interest to know how this pattern changes with time of year, or from place to place within or among watersheds, or how the gape size of fishes relates to the size of potential prey (mostly aquatic invertebrates) found in the stream. This latter question is related to the concept of "species stacking" posed by Griffiths (1986); that is, that the existence of an abundance peak in size distribution of food organisms could result in accumulations of more than one species of predator (fish, in our case) at a size corresponding to efficient feeding on the peaks in size of food items. Thiebaux and Dickie (1993) present a model including the feature that predatory fish of a given size range feed on prey of a given size range; thus, an increase in numbers of prey will result in a corresponding number of predator individuals. "Species stacking" of stream fish species at sizes corresponding to peaks in available benthic or drifting invertebrate prey remains untested with field data.

Finally, although it is easy to note peaks in maximum potential feeding sizes in Fig. 2.23, it may be profitable to view these same data from the perspective of the proportion of the fish assemblage that can eat invertebrates of a given size. Figure 2.24 shows, for the 558 individual fish, the proportion in the assemblage that could potentially engulf prey across the range of prey sizes, assuming that for each size class of mouth width, all individuals in the class could eat prey items at least equal to the smallest mouth width in the class (i.e., for mouth width class from 1.5 to 1.7 mm, I assumed all members could engulf a prey item as wide as 1.5 mm).

Therefore, in Figure 2.24, 100% of the fish in our sample can eat a prey item 0.9 mm wide, and roughly half of the individuals could eat a prey item 2.7 mm wide. The curve falls off rather sharply, because there are relatively few large individuals in our sample, so that at a prey width of 4.1 mm, only about one-fourth of the fish could be potential predators. A prey item as wide as 5 mm would escape gape-limited predation from all but about 10% of the fish in these pools. This analysis is preliminary and is based on only one sample of fishes from one site, but this approach might be useful for asking what potential effect the entire suite of fishes in an assemblage might have on invertebrate or other prey—instead of asking (experimentally or empirically) what effect a single species might have on the potential prey in a system. It might also be possible to speculate on levels of potential competition among fishes in this assemblage with varying mouth widths. It must be remembered that for these species it is likely [at least occurs for one species, *Luxilus pilsbryi* (Matthews et al., 1978)] that a given individual will eat prey from a wide array of sizes up to the size that begins to present difficulty with handling of prey items. However, Matthews

84 / Patterns in Freshwater Fish Ecology

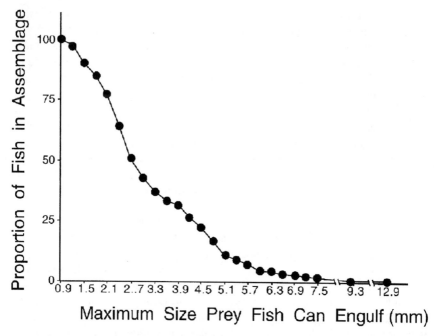

Figure 2.24. Data from Figure 2.23, rearranged to depict the proportion of fish in the Piney Creek assemblage capable of engulfing prey of a given size.

et al. (1982) showed for three Roanoke River, Virginia, darters (Percidae) that there was a substantial relationship between mouth width and the size prey actually eaten; thus, estimating the prey-impact potential based on an array of multispecies mouth sizes may be profitable, suggesting a direct way that the structure of the fish community (distribution of mouth sizes) can impact another ecosystem component (invertebrates).

2.5 Summary

A local assemblage of fishes includes those individuals that occur together in a relatively short period of time and in a "place" with dimensions such that individual fish have a reasonable probability of encountering each other in the course of daily activities. Although it is difficult to rigorously define, with absolute boundaries in time or space, such a locality, a huge literature in fish ecology is based on the presumed operational definition that an "assemblage" comprises the fishes that would logically be included in a single sample using conventional ichthyological methods. Within a single carefully taken field sample, using any of a variety of methods, it should be possible to characterize fish assemblage "structure," although it is well known that seasonal patterns may make it necessary to sample repeatedly to actually understand all of the fishes that occur in a single place during a whole year.

Structure in fish assemblages has a variety of meanings. The number of species, which is arguably the most used metric, varies from only a few species in small headwater streams worldwide to very high diversity (as many as 100 or more species) in some tropical settings. However, on average, the typical or modal number of species in an assemblage is probably more on the order of 15 or fewer at a given time. The number of families represented in an assemblage is usually greater at sites with greater overall richness, but the numbers of species per family vary geographically. In eastern North America, there tends to be a marked increase in species per family at species-rich sites; in much of the rest of the world, the proliferation of total species is by addition of substantially more families.

Trophic and functional group schemes are still being developed, and no single approach works best worldwide. By combining several of the most used trophic designations, I have proposed a trophic classification that may be useful and I offer a largely untested view of classifying fish by functional group (which includes physical effects on the environment, as well as feeding habits). Whereas taxonomic classification is critical for understanding details of interactions among species in assemblages, functional approaches may help to provide broad views of the ways that groups of fish taxa affect their habitats in common. A large number of assemblages in eastern North America averaged about 11 prey species and 3 predator species. Predator/prey ratios varied widely among assemblages, with low-richness assemblages constrained in predator/prey ratios, but with intermediately diverse assemblages (e.g., 15–30 species) showing wide variation in predator/prey species ratios, from about 20:1 to 2:1. With respect to the abundance of individual species per site, many localities showed a preponderance of "rare" species (occurring as only 1–2 individuals) and a small number of species that were "abundant" (e.g., > 500 individuals in a sample). Body- or mouth-size distributions for all fishes in samples from streams in eastern North America suggested a preponderance of small individual fish in assemblages, and that invertebrates or juvenile fishes could outgrow the gape limitations of many of their potential predators.

Overall, a review of patterns in structure of fish assemblages worldwide suggested that some of the general patterns that have been well established (e.g., greater richness in the tropics) held reasonably well at the level of local assemblages, but that such broad regional patterns did not necessarily predilect local assemblage phenomena. Many of the patterns in community structure from North American examples described in this chapter have not been tested elsewhere.

3

Discrete Versus Overlapping Assemblages, and Assemblage Stability

3.1 Discrete Versus Overlapping Assemblages

> *An animal society is composed of animals habitually occurring together in the same locality and the same class of situations. Such an association is, of course, composed of many species, variously related to their special environment, some attracted to it by one set of conditions and some by another.*
>
> —S. A. Forbes, 1907, "On the Local Distribution of certain Illinois Fishes: an Essay in Statistical Ecology"

Background

A fundamental question about any taxon of plants or animals is whether species occur in discrete, repeatable (presumably interactive) groupings or, alternatively, if local assemblages are merely aggregations of individual species, each occurring where the environment favors its growth, reproduction, and fitness (McIntosh, 1995; Scott, 1995). Observation of discrete and repeatable assemblages across the landscape has been considered evidence that groups consist of interacting species that "work well" together, or at least do not violate interactive (usually competition based) "assembly rules" (e.g., Diamond, 1975). Lack of discrete or repeatable assemblages would argue for less regulation of co-occurrence by interspecific interactions and for the formation of site-specific assemblages on the basis of optima for each individual species. There is a rich history of debate in community ecology about "individualistic" versus "interactive" views of assemblages, as summarized by Brown (1995a) and McIntosh (1995). Fishes have had little attention in this regard.

Early in this century, Frederick Clements dominated thinking in plant community ecology with the "community unit" view of species distributions across landscapes. Clements (1916) viewed plants as distributed in discrete assemblages or communities occurring in a repeatable and predictable fashion. Although his more extreme concepts of plant communities as "superorganisms" with their own life span, emergent properties, and so forth, are no longer accepted, the core of his argument—namely that organisms form discrete communities—remains, with modification, an important question. [However, Kindscher and Wells (1995) note that, "Traditional Clementsian plant community associations have not been very useful to plant ecologists who study prairies because of the complexity of the mix of species in these communities".]

H. A. Gleason (1917, 1926) countered with the idea that observed plant assemblages represented the accumulation of individual species at places in the landscape where conditions favored each, with little influence from other co-occurring species. In this view, there are no emergent, interactive properties of communities to be found—each species merely responds to the suite of environmental conditions that causes it to do more or less well at a site. Brown (1995a) argued that the views of Gleason and Clements are not necessarily in conflict, but that they were viewing plant associations at two different scales, with repeatable, emergent properties evident at a larger scale.

Finally, the plant ecologists Curtis and McIntosh (1951), Whittaker (1956, 1967, 1972), and McIntosh (1967) focused on plant assemblages on a continuum or along environmental gradients, with interest in positions on environmental gradients (e.g., altitude, temperature, moisture) where individual species appeared and then disappeared. If the "starting" and "stopping" points for individual species on environmental gradients were nonrandom and clumped, this was considered evidence that discrete communities prevailed. Alternatively, random spacing of distribution limits of species would suggest individualistic responses of each species, forming a "passive," nonstructured (by interactions) aggregation of species at a given site. In a variation on this idea, Whittaker (1970) suggested that common responses of individual plant species to environmental gradients might, at first, be expected to result in groups that consistently coexist. However, he concluded that competition resulted in displacement of species' centers of abundance from each other—producing a continuum of species peaks on a gradient. To quote Whittaker (1970): "A plant community is a system of interacting, niche differentiated, partially competitive species. These species have evolved toward scattering of their population centers along environmental gradients ... undisturbed communities ... in most areas, intergrade continuously." These ideas are summarized by Hoagland (1995) and McIntosh (1995).

In general, the view that plant associations typically form a continuum along environmental gradients, with individualistic occurrences of plants along gradients, has prevailed, with little evidence of distinctly bounded "community units" (Auerbach and Shmida, 1993). Patten and Ellis (1995) showed that dominant plant species in Kenya varied individually in distribution and abundance along gradients, resulting in a continuum of intergrading patterns. However, Patten and Ellis (1995) also noted that along this continuum, identifiable association "types" existed, which were repeated where environmental conditions were similar. This underscores a problem with the "discrete" versus "individualistic" view of assemblages: Discrete assemblage patterns could result either from integrated linkages among species or similar responses of many individual species to abiotic conditions. Either would result in the observer finding repeated patterns in coexistence of species in field studies.

Brown (1995a) emphasized that interactive, competitive "MacArthurian" views of assemblage composition are not inconsistent with Gleason's views of individualism or the views of more recent plant ecologists (Whittaker, MacIntosh, Auerbach) of species occurring on a continuum. As Brown noted, "the interactions of a species with other organisms are just as individualistic as its relations with the physical environment." Therefore, it would seem that individualism can apply not only to dispersal ability or tolerances/optima of abiotic conditions, but also to the individual response of species to other members of the assemblage—all combining to produce the assemblages seen in snapshot samples at a site at one time.

Animal ecologists have given less attention than plant ecologists to questions about discrete versus continuous or overlapping assemblages, although there are many examples of work with animal taxa (McIntosh, 1995). Diamond's (1975) assessment of forbidden combinations of species and assembly rules for birds on islands argued against passive, individualistic accumulation of species and supported the occurrence of bird species in structured (competition-limited) assemblages. In contrast, Brown and Kurzius (1987) suggested that composition of local assemblages of desert rodents "reflected primarily the ability of local patches to meet the *individualistic* requirements of different species rather than highly structured interactions between particular combinations of species." One additional proposal is that many taxa occur in a nonrandom fashion, with "breaks" in distribution according to hierarchically nested patterns in habitats (Kolasa, 1989), but Matthews (1986b) found little evidence of predictable locations of breaks in fish species assemblages in streams of the eastern United States.

Fish Assemblages—Discrete Versus Overlapping?

"Fish people" do not talk much to "plant people," or so it would seem, based on their different views of "communities" or questions about structure of assemblages. Fish ecologists typically pose questions having a different dimensionality than do ecologists working with terrestrial plants or animals, both of which have more ability than fish (which are limited to stream courses or lakes) to move across the landscape. Terrestrial plant or animal ecologists search for distribution patterns in multidimensional space; fish distributions in streams are strongly influenced by and restricted to (at one scale) longitudinal variation. Similarly, ecologists working with fish in small lakes may consider that the boundary for the assemblage is obvious (i.e., the lakeshore). However, most of the questions asked by plant or terrestrial ecologists can apply and are important to understanding fish assemblages. Despite the physical boundaries to the aquatic medium, we need to know if freshwater fish assemblages (in streams or lakes) exist as repeatable, discrete kinds (Tonn and Magnuson, 1982) or, alternatively, if local fish assemblages exist on a continuum of types if observations are made across substantially large regions of the landscape (e.g., within a lake district, throughout a river basin, etc.). Bear in mind that the question being asked in this section is not whether broad fish faunal boundaries exist. A huge number of surveys in North America at the scale of river basins or whole watersheds show that fish faunal provinces exist at a continental scale (Hocutt and Wiley, 1986; Burr and Mayden, 1992) and fish faunal regions (Pflieger, 1971; Matthews and Robison, 1988) exist within smaller zoogeographic areas. The question here is whether *local* fish assemblages exist as distinctive and repeatable types or, alternatively, if local assemblages form a continuum over a wide range of overlapping kinds of grouping of species.

It is well known that fish species do not occupy habitats at random (e.g., Keast and Fox, 1990). Our problem is, accepting the fact that distributions of individual species are typically nonrandom, how does the nonrandom occurrence of all species in a region aggregate to the resultant community or assemblage structure— organized into finite assemblage types, or comprising local aggregations of spatially overlapping species?

Forbes (1907) apparently accepted that individual species formed "animal societies" on the basis of their individualistic needs or responses. On the same page of his 1907 essay, he noted that, "A pike and a minnow may be members of the same associate group, to whose habitat, however, the pike is especially attracted by the minnow, and the minnow by the facilities which are offered there for concealment or escape from the pike." Thus, even before phytosociologists began to argue plant associations in North America, Forbes was considering assemblages of fish from the perspective of individualism of species, yet the predator–prey implications suggest that interactions might also characterize "animal societies." Forbes (1907) explicitly asked, "What Illinois fishes are habitually found in each others' society, and what is the relative frequency of their associations?" In this work, limited to darters (Percidae) as his example, Forbes used "coefficients of association" of species pairs, forming "associative tables." Forbes (1907, p. 288) further suggested his belief in individualistic accumulation of species in local associations, stating that "species which are equally attracted to some local situation and unequally attracted to others, will be less frequently associated than those whose local preferences (for habitat) are altogether similar." However, Forbes (1907) also noted that local coexistence of darters should be affected by "many things," including "variations in the mere instinct of segregation," the kind of food preferred, physicochemical preferences, and seasonal migration. The remark on variations in the "instinct of segregation" is particularly interesting, in that Grinnell (1917), Gause (1934), Hutchinson (1957a), MacArthur (1972), and Hardin (1960) would not formulate concepts on niche, competitive exclusion, or resource partitioning until years later!

Other early fish ecologists believed in the individuality of the occurrence of species. Thompson and Hunt (1930, p. 66), acknowledging the work of Forbes (1907), indicated that "most instances of the association of different species of fishes (in Illinois) are explained satisfactorily by similar environmental preferences." However, on the next page, Thompson and Hunt (1930, p. 67) described the strong association of certain minnow species by mutualism, which we now know is likely mediated by spawning of one species in the pebble nests of the other. The concept of individualism of fishes in Illinois was amplified by Larimore and Smith (1963) in discussing miscellaneous associations of species:

> This lack of similarity or consistency in associations (of species pairs in different streams) seems to suggest little interdependence between species but rather dependence of certain species on certain ecological factors. These factors may occur together in one stream system, thus bringing two species together, or they may be separated in another stream system, thus separating species.

Distinct Fish Assemblages in Streams?

This chapter is not specifically concerned with longitudinal patterns within a single stream (see Chapter 6), but it should be pointed out that Smith and Powell (1971) were among the first to use multivariate analyses of fish assemblages and showed in Brier Creek, Oklahoma, the distinct separation of upstream and downstream fish assemblages in multivariate space. An important contribution to the question of distinctiveness of fish species associations (if not whole assemblages) was that of Echelle et al. (1972) in which repeated associations of species pairs or triads were detected in samples broadly spaced in the upper Red River basin in Oklahoma. They also noted "three major complexes of positively associated fishes," including two, four, and six species, respectively, that related to gradients of physical conditions (salinity). However, showing the repeated occurrence of a particular species in associations is not the same as showing the overall separation of local Red River fish assemblages into distinctive types. For example, a local assemblage that contained their "*lutrensis–affinis*" association (including *Cyprinella lutrensis, Gambusia affinis, Lepomis cyanellus, Lepomis megalotis, Pimephales promelas,* and *Pimephales vigilax*) would, in most cases, also include other species. Perhaps, as suggested later by the "core–satellite" hypothesis of Hanski (1982) or core fish species of Matthews (1982), the predictable associations detected by Echelle et al. are the "core" of the upper Red River assemblages, but with the assemblages "filled in" by less predictable additions of other species from the pool of regionally available taxa.

Echelle and Schnell (1976) published a factor analysis of species associations on the basis of 131 fish collections (Pigg and Hill, 1974) in the Kiamichi River basin in Oklahoma. Echelle and Schnell (1976) detected discrete associations of species, not all of which were longitudinally controlled. They noted that for two of the defined groups, "certain species were rather widely distributed over the drainage," and they emphasized ecological flexibility of some species. For example, their "steelcolor shiner group" (named for most common species in group) was prominent in main channels and large tributaries throughout the drainage. Echelle and Schnell (1976) considered that both habitat preference and competitive interaction accounted for distribution patterns in this drainage, but that it would be "difficult, if not impossible, to assess their relative importance."

Grady et al. (1983) used discriminant function analysis to test the structure of fish assemblages throughout the Bayou Sara drainage of the lower Mississippi River. Their distribution of fish species along a discriminant function based on physical and chemical measurements associated with capture locations sorted fish species into discrete assemblages (their Fig. 3) that also related to the position of stations in the watershed. Pyron and Taylor (1993) collected dominant fish species (15-min samples) at 45 lowland sites, ranging from flowing streams to oxbow lakes, sloughs, and so forth, and found "breaks" in distribution and abundance of species, suggesting the existence of distinct assemblage types. A multivariate hierarchical procedure (TWINSPAN) placed the fish species into five primary groups.

The model of Schlosser (1987) predicts changes in fish assemblages in streams along gradients of flow stability or of pool development. In a test of Schlosser's model (Capone and Kushlan, 1991) in dry-season pools of the flow-variable Sulphur River, northeast Texas (USA), three distinct fish assemblages were detected by cluster analysis of species' abundances and were predictable on the basis of the physical gradients of depth and persistence of water. Schlosser's (1987) model could be interpreted to predict gradual change in the structure of fish assemblages (e.g., overlapping or continuum assemblages) downstream as pools become better developed and flow continuous. However, Capone and Kushlan (1991) found three distinct assemblages. Both Schlosser (1987) and Capone and Kushlan (1991) are probably correct, but it would be interesting to determine if sampling in the Sulphur River system when pools are not isolated (e.g., during flowing conditions) would yield the impression of a more gradual differentiation among local assemblages. Finally, as in all of the cited studies, it would be of interest to know more about seasonal or flow-related changes in assemblages of species. Matthews and Hill (1980) showed that fish species in the South Canadian River, Oklahoma, formed only transient microhabitat assemblages that changed with season and with flow (i.e., a lack of consistently discrete assemblage types). Starrett (1950a) noted long ago the propensity of minnows to form assemblages that changed with season and with the hour of the day, arguing against rigid boundaries to discrete species assemblages in the Des Moines River (Iowa) drainage.

Distinct Fish Assemblages Among Lakes?

Among lakes of a particular district, distinct and repeatable fish assemblages have been reported (e.g., Tonn and Magnuson, 1982). For 18 small (2–90 hectares)

Wisconsin lakes, Tonn and Magnuson found sharply distinct assemblages in summer, but not in winter. Overall, however, these small lakes were separable into "mudminnow–cyprinid" assemblages and "centrarchid–*Esox*" assemblages. Tonn and Magnuson (1982) indicated that differences between the two assemblage types related primarily to winterkill (lower oxygen levels in some lakes did not allow persistence of large predators) and to connectedness of individual lakes to nearby streams (which could serve as refugia in winter or allow recolonization of predators in warm weather). Tonn and Magnuson summarized that these "ecologically striking" differences provided a good example of assemblage consistency due to mechanistic (extant) processes.

Rahel (1984) identified three distinct fish assemblages (centrarchid assemblage, cyprinid assemblage, mudminnow–yellow perch assemblage) in 43 northern Wisconsin lakes (38 had fish) that ranged from bog ponds to small oligotrophic lakes, with pH and winter oxygen levels critical to determining and maintaining differences among assemblages. Direct gradient analysis of winter oxygen concentrations completely separated the three types of assemblages. Where piscivores were present, small-bodied cyprinids were excluded, with the exception of one species (*Notemigonus crysoleucas*).

For another 100 northern Wisconsin lakes ranging in size from about 5 to 1500 hectares, Rahel (1986) reported a core assemblage present in all lakes, including *Catostomus commersoni, Micropterus salmoides,* sunfish (*Lepomis* spp.), bullheads (*Amieurus* spp.), and *Perca flavescens*. In some lakes (with higher alkalinity), a sculpin plus several species of minnows and darters were added to this core group (Rahel, 1986). Although Rahel (1986) showed six of these "other" species more common in small lakes and nine species less common in small lakes, the existence of the core species throughout the entire lake district suggests that, in this region, fishes of lakes did not, overall, form distinct, nonoverlapping assemblage types. Rahel (1986) did not emphasize winterkill of piscivores in his suite of lakes, as did Tonn and Magnuson (1982) and Rahel (1986), but the lakes included by Rahel were, on average, much larger than the small lakes studied by Tonn and Magnuson—hence, less likely to exhibit winterkill.

A more speculative proposition by Ryder and Kerr (1978) for fish communities in Canadian lakes was discussed by Carline (1986). Ryder and Kerr (1978) suggested that there were three distinctive and stable community types described as "salmonid," "percid," and "centrarchid" on a gradient from oligotrophic to euthophic lakes. According to this model, lakes with mixed assemblages would likely be less stable than those having one of the distinctive types, although, as Carline (1986) pointed out, data do not exist to rigorously test this postulate.

Bendell and McNicol (1987) found the cyprinid assemblages in 58 lakes in Ontario divisible into two types, based on their requiring or not requiring hard substrates for spawning. (However, they did not ask whether or not the lakes could be divided into two or more groups based on their entire fish assemblages.)

Random Versus Discrete Groupings of Fish Species—A Multivariate Model

It would be desirable to test the results of various studies to determine if fish assemblages show, overall, evidence of consisting of accumulations of individualistic species versus discrete associations of species, as has been asked historically for plant associations. Before attempting to test actual data, I pose a simple model (Fig. 3.1) of placement of species (of fish for our purposes) along an environmental gradient. This gradient could be any condition or suite of conditions that vary from one extreme to another. It should also be emphasized that in contrast to often-studied plant situations, this "gradient" for fish need not be (and should not be) in any one stream, that is, this is not intended to be a longitudinal gradient (which is a special case considered in Chapter 6). Instead, assume that this "gradient" reflects the range of conditions that would exist if a large number of sites were sampled within a region with a common fish faunal pool (e.g., a river basin or a lake district).

Along this gradient of conditions, I placed 15 species with the peak of their abundance determined by randomization (upper part of Fig. 3.1). Each species has a maximum abundance of 100 individuals per sample and an identical, symmetrical curve of abundance about the peak (ignoring "tails" that would probably occur in nature). This model mimics the extreme "Gleasonian" condition, with each species occurring independent of any other along an environmental gradient. Although each species was placed on the gradient randomly, there are regions of the gradient that appear to have more or fewer species.

In the lower part of Figure 3.1, 15 species were placed nonrandomly to represent very strong associations in three groups of five species each. This is intended to mimic the Clementsian perspective of "community units," although this distribution of species could occur either from positive interactions among species in a group or from a common response of all five species in a group to the range of environmental conditions.

The objective of this simplistic model is to ask if a commonly used multivariate technique can indeed be used to detect differences in patterns if "samples" are taken from the two artificial distributions of species. Assume that in both the upper and lower (random versus grouped placement of species) parts of Figure 3.1, fish are "sampled" by an adequate collecting technique at 37 locations equally spaced along the environmental gradient (numbered locations on the horizontal axis). Keep in mind that this does not mean that the samples are actually sequentially adjacent to each other in space—the locations are scattered throughout a river basin or some other reasonable area for sampling but lie in a particular position on the hypothetical environmental gradient.

For each of the 37 hypothetical sampling locations (in the upper and lower parts of Fig. 3.1), I estimated, from the vertical axis, the abundance of each species, entering the abundance data in a typical 15 species × 37 location matrix

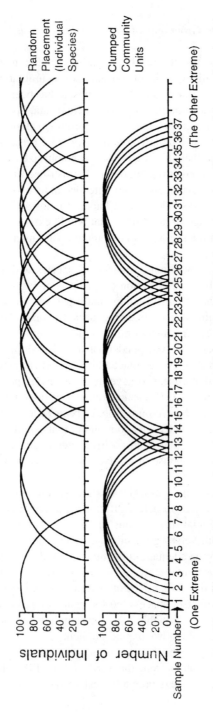

Figure 3.1. Fifteen fish species with peaks of abundance curves placed on a hypothetical environmental gradient (x axis) at random (upper) or placed in three nonrandom groups of five species each (lower).

(Tables 3.1 and 3.2). Detrended correspondence analysis (DCA, by the DECOR-ANA module of the Cornell Ecology Programs), which has had frequent use in fish ecology (Matthews and Robison, 1988; Edds, 1993), was used to carry out an ordination of both matrices. If DCA (or related multivariate analyses) is effective at detecting structure among the 37 assemblages, we should find a random placement of samples in the ordination for the randomly assigned species, and a nonrandom distribution of samples into discrete community "types" for the samples from the grouped distribution of species on the environmental axis. For example, note that in the lower part Figure 3.1 and in Table 3.2 that samples 5–11 contain identical species in relatively similar abundances; samples 16–22 and samples 28–33 do likewise. We should expect to find these groups of samples clumped in an ordination, with others (e.g., samples 12–14 or samples 24–26) forming transitions between the three dominant assemblage types.

Both DCAs accounted for a very high percentage of the total variance on the first axis. For the model with randomly placed species, DCA axes 1 and 2 had eigenvalues of 0.948 and 0.085, respectively. For the model with a grouped placement of species, DCA axes 1 and 2 had eigenvalues of 0.987 and 0.044, respectively. With such a large amount of the variance accounted for on the first DCA axis, there is, for these models, no need to examine the second axis, and for both models the rest of analysis can be limited to axis 1. [Note that although this was carried out as a DCA, there is no detrending of the first axis; thus, the first axes are identical to a reciprocal averaging (RA) analysis (Gauch, 1982)]. For the random model, DCA axis 1 was slightly less than 10 standard deviations long; for the grouped species model, the first axis was slightly greater than 10 standard deviations. Figure 3.2 summarizes the placement of numbers of samples along increments of one standard deviation for DCA 1 for both models. For the random model, there is a maximum in numbers of samples in the eight to nine standard deviation increment (corresponding to samples 8–13 which contained identical species; Fig. 3.1 and Table 3.1), but there is no increment on DCA 1 lacking samples. In contrast, the placement of samples on DCA 1 is highly grouped in Fig. 3.2 for the grouped species model: There are three sharp peaks, as would be expected from inspection of species placement in the original distribution (Fig. 3.1) and there are three standard deviation increments with no samples at all.

By visual inspection, therefore, it appears that the DCA of samples from the two artificial (random versus grouped) distributions of species detected patterns that would be expected in assemblages sampled from such distributions of species. However, instead of evaluating Figure 3.2 by eye, it is desirable to test statistically whether the distributions of samples on DCA axis 1 are nonrandom. The Poisson distribution (Zar, 1984) of "rare and random" events (i.e., the occurrence of samples in an increment) provides an appropriate test. The numbers of samples per standard deviation increment (10 increments for randomized species model; 11 increments for grouped species model) were tallied [with tails of the distributions

Table 3.1. Abundance of Species 1–15 in 37 "Samples" Taken from Artificial Distributions of Species by Random Placement on the Environmental Gradient in Figure 3.1 (Upper)

SPECIES	RANDOM SAMPLE #																																				
	1	2	3	4	5	6	7	8	9	10	11	12	13	14	15	16	17	18	19	20	21	22	23	24	25	26	27	28	29	30	31	32	33	34	35	36	37
1	95	100	95	92	82	70	42																														
2				5	50	72	84	92	99	100	98	91	80	65	35																						
3						38	65	81	91	96	100	99	92	85	71	50																					
4														42	68	82	91	96	100	99	91	81	70	40													
5															1	72	84	92	99	100	98	91	80	62	38												
6																48	68	81	92	99	100	99	92	81	74	41											
7																			50	70	86	94	99	100	99	92	84	82	40								
8																			35	64	80	91	96	100	91	92	85	82	45								
9																						45	70	81	75	97	100	99	91	80	65	40					
10																							1	49	42	82	91	99	100	96	90	80	61	38			
11																										68	80	92	98	100	99	91	81	67	44		
12																											35	82	80	90	99	100	99	95	92	72	50
13																														35	62	80	89	96	100	97	92
14																																	30	61	88	88	96
15																																		36	60	78	86

Table 3.2. Abundance of Species 1–15 in 37 "Samples" Taken from Artificial Distributions of Species by Nonrandom "Clumped" Placement on the Environmental Gradient in Figure 3.1 (Lower)

CLUMPED

SPECIES	1	2	3	4	5	6	7	8	9	10	11	12	13	14	15	16	17	18	19	20	21	22	23	24	25	26	27	28	29	30	31	32	33	34	35	36	37
1	25	60	80	90	95	98	100	96	90	78	58	20																									
2	1	50	72	84	90	95	99	97	92	81	64	40																									
3		25	60	75	85	93	96	98	94	82	72	46	10																								
4			45	68	80	90	95	99	96	90	80	65	38																								
5			20	48	75	88	92	100	98	94	88	74	58	20																							
6													50	70	85	95	99	100	98	92	83	70	42														
7													35	60	80	92	98	99	99	94	90	76	47	20													
8													5	50	72	88	96	98	99	96	95	82	68	40													
9														30	62	80	92	96	100	99	98	86	72	55	10												
10															50	75	88	95	99	100	100	95	88	70	42												
11																								18	58	75	88	95	99	99	96	89	76	58	25		
12																									42	65	81	92	97	100	97	92	80	65	40		
13																									22	58	75	88	93	98	98	95	86	74	52	20	
14																									1	40	62	80	87	96	100	96	88	80	64	38	
15																										20	52	76	85	95	99	100	94	88	76	59	20

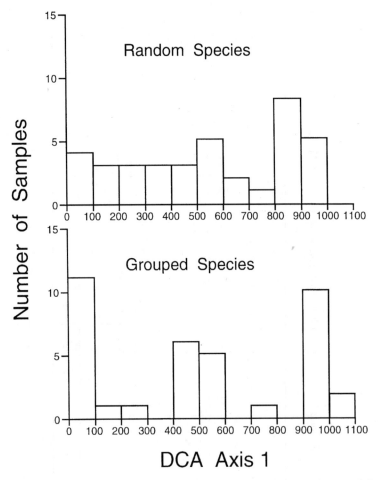

Figure 3.2. Frequency distribution of samples for random (upper) and grouped (lower) species on first axis of a detrended correspondence analysis of the data from Figure 3.1, as described in text.

pooled so that no increment contained less than 1 occurrence (Zar, 1984)] and actual frequencies were compared to expected frequencies by a chi-square test (Sokal and Rohlf, 1969). For randomly placed species, $\psi^2 = 7.8$, df = 4, and $p > 0.05$ (i.e., not significant); thus, the frequency of occurrence of samples per increment of the DCA axis 1 did not differ from random. The original random placement of individual species on the environmental gradient resulted in a DCA axis lacking any statistically significant grouping of samples (i.e., there were no discrete "community units").

For species artificially placed into three well-defined groups (Fig. 3.1) the distribution of samples on DCA axis 1 was distinctly nonrandom ($\psi^2 = 22.6$, df = 3, $p < 0.005$), indicating that the groupings of samples on the axis (Fig. 3.2) are significant [i.e., discrete "community units" (cf. Clements) were detectable by the detrended correspondence analysis].

Obviously, none of the above is intended to reflect anything about realistic distribution of fish assemblages along environmental gradients or their grouping into discrete assemblage types or "community units." The exercise was merely to ask, given an initial condition of known random or highly nonrandom placement of species, if we could use a multivariate analysis (DCA) followed by statistical test (Poisson distribution) of placement of samples on one or more axes to detect random versus nonrandom patterns in assemblages. The results suggest that it would be appropriate to test results of multivariate analyses of actual fish assemblages with a Poisson (or similar) approach to ask if scatter of the assemblages along axes is nonrandom.

Detecting Discrete Fish Assemblages in Streams and Lakes—A Statistical Test of the Evidence

Following the model above, I tested the results from several multivariate analyses to determine if fish assemblages occur in distinct groups (i.e., clumped or clustered in multivariate space) or if they occur randomly in multivariate space. I analyzed some of my own unpublished data on midwestern (USA) fishes and results from seven published studies. All of the stream studies included multivariate analyses of local fish assemblages at individual sampling stations and were displayed on a scatterplot of the first two axes [usually DCA or principal components analysis (PCA)]. Three studies of small lakes also were included, in which the entire known lake fauna was included as the "local assemblage." Streams included were (1) my samples at 96 stream sites in midwestern United States in June 1978 [unpublished data, but see map in Matthews (1985), for localities]; (2) 123 sites throughout the Kali Gandaki River drainage, Nepal (Edds, 1989); (3) 47 sites in South Carolina (USA) coastal streams (Paller, 1994); (4) 45 of Paller's sites, excluding two outliers, (5) 48 assemblages in streams of the upper Red River basin in Oklahoma (Taylor et al., 1993); (6) 49 sites in streams across Oregon (USA) (Whittier et al., 1988). Lake studies were (1) 39 lakes in southcentral Ontario (Canada) (Jackson and Harvey, 1993) and (2) 40 lakes in the Bruce Lake District, Ontario (Jackson and Harvey, 1989).

Unlike the model data in the earlier example, real data on fish assemblages rarely result in a single axis accounting for so much variance. Therefore, most ordinations of actual data use at least two axes to display differences among fish assemblages. For each real data set, the original scatterplot of samples on the first two multivariate axes was redrawn as necessary from the published figure (or generated originally for my unpublished data) to omit all author-imposed indications of structure (e.g., ellipses, polygons, codes reflecting drainages, or similar indicators of "groups"). On the "bare" scatterplots (Fig. 3.3), a rectangular grid was superimposed (adjusted in size to an appropriate number of grid quadrants to permit Poisson analysis—typically averaging from one to two samples per quadrant, Zar, 1984). Furthermore, any quadrants in the rectangular grid lying completely outside a polygon enclosing *all* samples were omitted (i.e., to eliminate regions of the scatterplot where there was no evidence that any assemblage would be found). With the number of quadrants thus reduced, the number of quadrants containing 0, 1, 2, 3, ... samples were tallied and the frequency distribution of quadrants containing a given number of samples was compared to a Poisson distribution of random placement of samples by a chi-square test to determine if the samples differed significantly from random placement on the scatterplot. If the frequency distribution of actual samples per quadrants is not significantly different from that predicted by Poisson distribution, the distribution of samples on the scatterplot is random, overall; and, there would be no need to consider the fish assemblages to exist in discrete groups (i.e., there is no contagious or clumped distribution of samples in multivariate space). However, if the Poisson results are significant, the distribution of samples across the scatterplot is different than random [i.e., with contagious or clumped distribution (none were "regular" or "hyperdispersed")], suggesting that the samples of fish assemblages may exist in discrete groups. Note that this procedure does not tell us how many such groups exist on the scatterplot—merely that statistically significant clumping or contagion of samples exists. It is still necessary to identify the actual groups by eye—as no satisfactory procedure seems to exist as yet for setting boundaries to the groups when the number of samples is relatively small. For very large numbers of samples, spatial geostatistics (Issacs and Srivastava, 1989; Rossi et al., 1992) procedures might be useful, but none of the studies available at present had enough data points for geostatistical approaches. [See Gelwick (1995) for an example of application of spatial geostatistics to aquatic systems.]

By eye, it appeared that my samples in 1978 from streams throughout midwestern United States formed at least two distinct clusters, with a high density of samples in the lower left part of the scatterplot (Fig. 3.3A) and a disjunct cluster of about 22 samples in the lower right corner [these are samples from the Texas hill country—a unique region of uplifted limestone and clear streams known to have a distinctive fish fauna including several local endemics (Hubbs et al., 1991)]. With 85 samples that can be counted (a few are hidden in the display) and 52 quadrants, mean samples per quadrant is 1.635. More quadrants (26) than expected at random contained no samples, and 3 quadrants contained 8, 9, and 10 samples, respectively—far more than expected at random. Overall, the ψ^2 comparing the actual distribution of samples among quadrants to the expected from a Poisson distribution was 52.3, which, with four degrees of freedom (with tails pooled; $6 - 2 = 4$) is very highly significant ($p < 0.005$). Therefore, the distribution of Midwest samples among quadrants in the DCA scatterplot is nonrandom, with strongly (statistically) clumped or grouped samples. Such a grouped pattern suggests the existence of discrete fish assemblages in streams

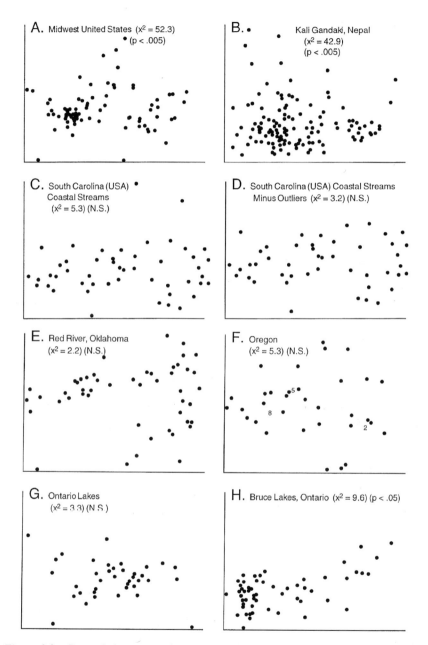

Figure 3.3. Detrended correspondence analysis (DCA) of eight sets of samples from streams or lakes in North America (A, C, D, E, F, G, H) and Nepal (B), with ψ^2 values from a Poisson analysis of placement of samples in two-dimensional cells, as described in text. [Modified from Edds (1993), Paller (1994), Taylor et al. (1991), Hughes and Gammon (1987), Whittier et al. (1988), Jackson and Harvey (1989, 1993).]

across the midwestern United States (although there clearly are several outliers and at least two samples that are not assignable easily to either of the two main groups).

Figure 3.3 also depicts seven published studies similarly analyzed by overlaying a grid on scatterplots of multivariate axes, followed by a chi-square comparison of the actual frequency distributions of samples per quadrant to those expected from a Poisson distribution. Edds (1989) study of the Kali Gandaki River, Nepal, included samples from the extreme upstream, mountainous region, which were markedly different from all other samples in the river systems (Edds, 1993). The upstream sites are omitted from the present analysis as are two other outliers in Figure 5 of his dissertation (Edds, 1989). For the remaining 123 samples, ψ^2 = 42.9, df = 4, and $p < 0.005$ (i.e., indicating a very nonrandom and clumped distribution of samples on the scatterplot). Edds' (1989) scatterplot appears to have numerous scattered outliers at the top of the figure (Fig. 3.3B), but within the remaining suite of samples, there are regions of greater and lesser density that the Poisson analysis suggests to represent distinct fish assemblages in the lowlands, "low hills," and "high hills" in agreement with Edds (1989, 1993).

The PCA scatterplot of South Carolina streams by Paller (1994) allowed testing sensitivity of this Poisson approach to inclusion or exclusion of outliers. Including outliers can introduce substantial "empty space" to the analysis, making the distribution of sample frequency per quadrant more likely to be nonrandom. For 47 stream sites (Fig. 3.3C), including two apparent outliers gives $\psi^2 = 5.3$, df = 2, $p > 0.05$, suggesting random distribution of samples among quadrants, that is, no discrete clumping. Elimination of two outliers (and the previously enclosed empty region of the scatterplot; Fig. 3.3D) leaves a region of samples lacking in any strong apparent clusters. However, for this reduced suite of 45 samples, $\psi^2 = 3.2$, df = 3, and $p > 0.05$. Therefore, including or excluding outliers had no substantial effect on the outcome. Paller's analysis addressed distributions of fish assemblages among stream order for a number of coastal drainages. His Figure 1 with sites coded by stream order (Paller, 1994) suggests a general blending and continuity of assemblages from headwaters to downstream (left to right in the figure) in the systems, with distinct differences at the ends of the gradient but considerable overlap of assemblages among adjacent stream orders. Thus, both Paller's analysis and the Poisson analyses of his scatterplot supported the view of these assemblages as lying along a continuum.

A canonical correspondence scatterplot of 48 samples from the upper Red River basin, Oklahoma, USA (Taylor et al., 1993) showed no difference in distribution of samples across quadrants from that predicted by Poisson analysis ($\psi^2 = 2.2$, df = 3, $p > 0.05$) (Fig. 3.3E). Taylor et al. (1993) noted that although species and sites could be classified into discrete groups by a hierarchical clustering procedure, site groups showed "considerable overlap" on a habitat template, and there was "considerable continuity, blurring the discreteness of groups." Taylor et al. (1993) noted that "groups of sites showed much overlap on the environmental template, suggesting that assemblages do not form highly integrated units." It appeared both from my Poisson analysis and the interpretations of Taylor et al. that fish assemblages in the upper Red River basin did not exist in distinct, discrete units, but tended to form a continuum of changing assemblages across environmental gradients (driven largely by salinity in this system).

Whittier et al. (1988) provided a DCA ordination of 49 sites throughout all parts of Oregon (USA), with emphasis on distinctiveness of faunas among river basins. Although they had eight locations with only one species (rainbow trout) present, the overall pattern of distribution of samples among quadrants was random ($\psi^2 = 5.32$, df = 3, $p > 0.05$) (Fig. 3.3F). Although their DCA axis 1 provided separation of fish samples from montane and nonmontane regions (Whittier et al., 1988), the present Poisson analysis suggested that, overall, there was a continuum of stream fish assemblage types if all of Oregon is considered.

Finally, for two lake districts in Ontario (Figs. 3.3G and 3.3H), Jackson and Harvey (1989, 1993) used correspondence analysis of fish presence–absence to provide an ordination of 39 and 40 lakes, respectively (Figs. 3.3G and 3.3H). For one set of lakes, the distribution of samples among grid quadrants was nonrandom, with clumped dispersion of samples on the multivariate axes (Bruce Lakes; Fig. 3.3H). This appears to result from a region of relatively high sample density to the left side of Figure 3.3H.

In summary, use of a Poisson distribution to estimate expected distributions of samples among superimposed grids in published multivariate studies of fish assemblages produced mixed results. For only three of the eight studies (my Midwest samples included), was there a nonrandom placement of samples on two-dimensional axes from multivariate analyses of assemblage. For the other five, the dispersion of samples on the grid did not differ from a random (Poisson) pattern. There is as yet no best way to examine a scatterplot to determine if there actually are well-defined and rigorously bounded clumps or clusters of species. It is clear that a Poisson analysis could produce a finding of "nonrandom" for a variety of reasons, including a strong gradient from dense to nondense parts of a figure. However, if a Poisson analysis indicates that dispersion of samples is nonrandom, this at least gives the investigator a reason to look further for patterns in the dispersion—including the segregation of samples into discrete community types. In other words, a nonrandom dispersion (by Poisson analysis) seems to be a necessary but not sufficient condition to "prove" that there are discrete fish assemblage types across space. Until a better test is devised, it may be the best we can do to (1) establish that a perceived pattern in random or nonrandom and (2) if the pattern is nonrandom, use inspection by eye to estimate the number of and limits to clusters of assemblage groups.

One obvious conclusion is that there is no simple answer for fishes to the Clementsian–Gleasonian question, "Do fish assemblages consist of discrete types, or do fish assemblages grade one into another along a continuum?" Additionally, it seems likely that samples taken at different spatial scales have very different probabilities of being found "random" or "nonrandom" in multivariate space. For example, samples taken from the entire midwestern United States, which obviously include quite different sites (e.g., clear upland streams of the Texas Hill Country versus low-gradient prairie streams in Kansas) seem intuitively more likely to include samples on either side of environmental boundaries, and thus exist in two or more clusters in multivariate space. In contrast, a large number of samples, all from the Kansas prairie region, would seem less likely to fall out in discrete groupings. As long as investigators are sensitive to such confounding effects, this line of inquiry seems potentially fruitful for work defining the basic nature of freshwater fish assemblages—which may become possible as more and more good examples of fish assemblage similarities and differences appear in the literature.

3.2 Stability of Fish Assemblages

In all types of waters, fish populations appear to fluctuate in abundance and in species composition from year to year.
—Starrett, 1951, re. the Des Moines River, Iowa

... one can return to a locality at the same time of year and collect essentially the same species ... in the same relative numbers.
—Smith and Powell, 1971, re. Brier Creek, Oklahoma

Both are correct. Individual fish species do fluctuate in abundance from year to year, but, at most locations, it is possible to find essentially the same assemblage year after year with respect to the presence of species. This means simply that although abundances of individual species may rise and fall in response to both abiotic and biotic (density dependent) factors, most species are able to persist, barring unusual (often anthropogenic) changes in the environment. Clearly, this is not universal for all species—some are more sensitive to environmental changes than others. Further, as detailed below, some common species are highly persistent over time, being taken in virtually every sample made at a location. Others we have referred to as "come and go" species (Marsh-Matthews and Matthews, unpublished data) in that they frequently are absent in a given sample, yet present in a subsequent one. It is worth noting also that although the "come and go" species in streams are not usually the most abundant ones, they often are not the locally very rare species either. In this section, I explore patterns of qualitative persistence of species and quantitative stability of assemblages in a wide variety of habitats and determine if a consensus about constancy of fish assemblages can be synthesized from the literature.

Like many phenomena in fish ecology, stability of assemblages across time is both spatially and temporally scale dependent. At one extreme, the fish species or numbers observed over a given meter-square quadrant of substrate or hovering near a particular boulder in a pool might change minute to minute, or at least day to day. At the other extreme, for an entire catchment, if undisturbed and with stable habitat (like spring-fed headwater runs) the fish assemblage found a century ago might remain relatively unchanged today. The length of time considered is obviously important. No stream fish assemblage at any one place can remain stable "forever" because of the dynamic nature of river basin erosion and change. Although changes in river basins are typically thought of as taking place over evolutionary time, the change in the characteristics of a given stream segment can be rapid. Patrick et al. (1991) document changes in the location of the downcutting "nickpoint" in a Mississippi (USA) stream within this century, with resulting changes in distribution of riffle-dwelling benthic fishes in the system. Similar views were espoused long ago by Shelford (1911), who drew parallels between succession of stream morphology and fish assemblages as rivers cut headward. From another "long" perspective of change in fish assemblages with evolutionary time, Roberts (1975, p. 264) suggested a "taxon cycle" for African fishes, such that

> By taxon cycle is meant the succession of taxa that inhabit a place as it becomes available for colonization and gradually acquires a richer fauna, and then loses taxa as it becomes unfavorable. From the long view of geological and evolutionary time, it is trivial to ask at a given place on the face of the earth "do fish assemblages change?"

Of course they do. However, our concern in this section is with changes in fish assemblages that we, as ecologists, see at a given place in a stream or lake within our lifetime, or at least within the span of a few generations of our species. We have little impact or control beyond that time frame.

At the other extreme in span of time, it is well known that fish assemblages at a given site often change seasonally or monthly, or over short spans of time associated with floods or drought. For example, Gido et al. (1997) showed the magnitude of baseline variation in detrended correspondence analysis scores for fish assemblages at stream sites in San Juan River, New Mexico (USA) from month to month over a span of 1.5 years. The magnitude of change increased in late summer as juveniles joined the assemblage (Gido et al., 1997). Harrell (1978) and Matthews (1987a) quantified rapid changes in local fish assemblages or their habitat associations in harshly fluctuating southwestern (USA) streams in response to flood and drought, respectively. The effects of such events on fish assemblages are detailed in Chapter 7. The dramatic seasonal migrations of fishes in tropical rivers (Roberts, 1972; Welcomme, 1985; Lowe-McConnell, 1987) are well known. Within North American streams seasonal local movements or migrations of individual species are well understood (Hall, 1972), and the fish assemblage at individual sites in a stream may change from month to month (Gelwick, 1990) or by season (Mendelson, 1975; Matthews and Hill, 1980; Orth and Maughan, 1984). Gelwick (1990) also made clear that changes can be at differing tempos in contrasting habitats (e.g., more temporal change in riffles than in pools). Taylor et al. (1996) showed for 10 sites in the upper Red River (Oklahoma) basin that the magnitude of monthly or seasonal change could depend on stream size (mainstem versus tributaries). Matthews et al. (1994) also showed moderate to high levels of change in the composition of fish assemblages in individual stream pools from month to month in Brier Creek, Oklahoma. However, the well-known seasonal changes in fish assemblages in many parts of the world are not the primary focus of this section.

Primary focus in this section is on fishes in a local assemblage, as previously defined. However, I also consider the evidence for temporal change at several spatial scales, from individual pools or riffles to the fauna of whole catchments. The focus is only on the empirical findings of studies (i.e., if a fish assemblage changed, or how much). Mechanisms influencing change in fish assemblages will be the subject of much of the second part of this book. There are many well-known examples of pervasive changes in fish assemblages because of human influence (introduction of sea lampreys to the North American Great Lakes; introduction of exotic piscivores to Lake Gatun, Panama, and Lake Victoria, Africa; construction of dams, artificial waterways, and the like), but the question in this chapter is limited to assessing levels of apparently natural change in fish assemblages where there is no single, major anthropogenic event known in the system.

Why is stability of fish assemblages important? Two categories of answers are possible. First, there is the urgent need in environmental management to know how much fish assemblages in streams or lakes change naturally over time, as a baseline for detecting any anthropogenic effects in the system. Second, from a more theoretical view, many of the great concepts in ecology depend on the assumption that populations achieve stability or that communities operate in a regular and predictable fashion. For example, in population ecology, stable age distributions, density-dependent population regulation, and all optimality predictions relate to stability of the population. In community ecology, classical concepts about competition, resource partitioning, or coevolution assume that systems will eventually reach either a steady state or a predictable cycle in their components. Change, whether caused by stochastic, abiotic factors (floods, droughts, apocalyptic events), or by unexplained vagaries in reproduction, survival, and so forth of the biota, upsets the predictions from such orderly models and makes them less than fully applicable to a system. Our task is to determine how well fish assemblages in the real world fit assumptions of stability or predictability over time. There are, as suggested above, two major components of interest. The population biologist and the fish manager often want to know if the numbers of size structure of individual species are constant or predictable. The community ecologist or biogeographer wants to know if the total assemblage exhibits properties of stable structure. It is with this latter question that we will be most concerned, although also examining views on changes in fish populations in this section.

Background

Long-term records of populations or communities are common for some groups, e.g., plankton (Glover, 1967; Gray, 1977; Colebrook, 1979; Makarewicz and Baybut, 1981; Dugan and Livingston, 1982). Some of the longest records of population fluctuations are for commercially important fishes, mostly marine (Bannister, 1977; Skud, 1982). In contrast, detailed records of changes in assemblages of stream or lake fishes over periods of a decade or more are scarce. Even scarcer are records in which a single individual has taken part in or directed all of the sampling (e.g., Meffe and Minckley, 1987). When a single individual has physically participated in making every sample, as I did (Fig. 3.4) for 12 fixed stations on Piney Creek, Izard County, Arkansas, from 1972 through 1996 (Matthews and Harp, 1974; Matthews 1986c; Matthews et al., 1988; Matthews and E. Marsh-Matthews, unpublished data), or Clark Hubbs did for 129 sites in Texas sampled in 1953 and 1986 (Anderson et al., 1995; Hubbs et al., in press, it is possible to really know that the stream reach being sampled is precisely the same segment as in previous years (even decades earlier) and that the individual can subjectively and realistically know if sampling efforts are comparable.

Figure 3.4. Depiction of a given investigator participating in sampling sites repeatedly across several decades. (Original artwork by Coral McCallister.)

Stability or persistence of natural fish assemblages has received much empirical and theoretical attention in recent years (Grossman et al., 1990; McIntosh, 1995). Assemblages with stable and persistent composition have been considered regulated by "deterministic," equilibrium, or interactive biotic processes (Moyle and Vondracek 1985). Assemblages lacking significant stability or persistence are purported to be influenced more by "stochastic," nonequilibrium, random, or abiotic processes (Grossman et al., 1982; Schlosser, 1982). Such fundamental differences between assemblages at opposing ends of a stability–instability continuum suggest the importance of carefully evaluating a variety of fish assemblages at different spatial or temporal scales before generalizations are suggested (Ross et al., 1985). Characterizing fish assemblages as typically stable or unstable carries important ramifications for the appropriate application of much equilibrium-based ecological theory, not only to fishes but to aquatic biota, in general, and all animals in dynamic environments (Grossman et al., 1982; 1985).

Classical and Descriptive Studies

The debate about stability of fish assemblages in temperate streams is not new. Starrett (1951), Hubbs and Hettler (1958), Smith and Powell (1971), and others presented contrasting views, but most early fish ecologists considered fish assemblages in natural streams relatively stable, barring gross cultural disturbance. Some of the now-classic studies were stimulated by changes in conditions within watersheds, including flood and drought, general environmental degradation, or pollutants. Studies in response to cultural impact (e.g., pollution, dredging, channelization, etc.) have been conducted, such as those by Thompson and Hunt (1930), Larimore and Smith (1963), and Mills et al. (1966).

An additional kind of study that sheds light on persistence of fish assemblages is that of natural, experimental, or accidental defaunation and recolonization (Larimore et al., 1959; Berra and Gunning, 1970; Olmsted and Cloutman, 1974). Whereas extrapolation of defaunation studies to changes in nonimpacted conditions is problematic, they nevertheless suggest some stability in fish assemblages. However, such studies may merely show that fishes in areas adjacent to defaunated reaches are similar to those in the defaunated area, and that because of a source of nearby colonists, they give an unrealistic impression of structured organization of the fish assemblage.

Hubbs and Hettler (1958) considered most fish populations (in Texas, USA) to "maintain approximately the same relative abundance," but, on the basis of many repeated collections, they reported several "notable exceptions" related to flooding or drought during the 1950s. Species showing marked interyear changes in abundance at particular localities included the following: *Percina sciera* (reduction in stream volume from drought); *Etheostoma lepidum, Campostoma anomalum, Gambusia geiseri, Cyprinella venusta,* and *Notropis amabilis* (floods, modified spring flow, introduced species); *Phenacobius mirabilis* (factor unknown); and *Etheostoma spectabile* (flooding of riffle habitat).

Cross and Brasch (1968) chronicled changes in the fish fauna in the upper Neosho River (Kansas, USA) system from 1952 to 1967, where pollutants from cattle feedlots had caused known fish-kills and severe drought of the mid-1950s also had impact on fishes. Six species common in 1952 were not found in 1967, and 15 other species were reported as "depleted" at mainstream locations. In general, Cross and Brasch (1968) found that bottom-dwelling fishes "declined strikingly" and that larger-bodied species (from reservoirs) or small surface-dwelling species (typical of more sluggish waters) increased in abundance.

King (1973) compared minnows and darters at more than 100 locations in Boone County, Iowa (USA) between 1947 and 1972. The same species dominated in abundance in the two different periods. However, *Etheostoma flabellare* disappeared from one creek in the 25 years between collections. Several minnow species (*Pimephales notatus, Notropis stramineus, Campostoma anomalum,* and *Cyprinella spiloptera*) increased in streams where bottom types changed from rubble-gravel to more silty-sand (apparently due to altered agricultural practices and siltation).

Trautman and Gartman (1974) summarized samples in one Ohio (USA) creek in 1887, 1929–1938, 1954–1973, and 1973, where channelization had been carried out for a century. Seven species were reported depleted or extirpated due to habitat loss or general environmental changes. (This article is also valuable in that it provides semiquantitative summaries of the abundance of species throughout the watershed for a span of almost 90 years.)

Recently, Fausch and Bestgen (1996) have documented apparent postsettlement changes in rivers of the Great Plains in Colorado (USA), with noteworthy losses of numerous species. However, Fausch and Bestgen (1996) also point out the critical need for fish surveys in little-known areas, in that, for numerous species, it is now impossible to know (due to the lack of early surveys) exactly what the presettlement conditions or fauna were like.

Qualitative Changes in Fish Assemblages

If we compare qualitative composition of fish assemblages from the earliest to the most recent records at a given location, how similar are they? For example, in 1977–1979, I made fish collections at a location in the Roanoke River, Virginia, where David Starr Jordan (Jordan, 1889) had sampled almost a century earlier. I found the fish assemblage little changed (at least qualitatively) from the species reported by Jordan. Of the 25 species collected by Jordan, 22 were present in my samples (= 88% persistent).

Numerous recent articles have addressed the question of persistence of fish species at particular localities. From this perspective, the interest is in the continued existence of the species regardless of fluctuations in its abundance. There is ecological justification for this view, in that a species that is present, even in low numbers, is at least available to increase in numbers if environmental conditions change.

Over a span of 11 years in Brier Creek, Oklahoma, 16 of 17 species from collections in 1969 remained present in 1981, and in Piney Creek, Arkansas, all species taken in 1972 were again taken in 1981 (Ross et al., 1985) if collections within each system are pooled. Matthews et al. (1988a) showed from an expanded database on these two streams and in the Kiamichi River, Oklahoma, that common or abundant species were typically persistent and that only some rarer species were nonpersistent. Many sites in temperate streams appear to have a core of abundant species that are almost always found in samples from a locality, with additional species present or not from one sample to the next. In Piney Creek, Arkansas, for 12 samples at the most downstream site (P-1, Matthews et al., 1988a), of 41 total species, 8 have occurred in all or all but 1 of the samples, 20 species have appeared and disappeared in samples at various schedules across time ("come and go"), and 13 have been "rare" (occurring only 1–2 times in samples spanning 24 years (E. Marsh-Matthews and Matthews, unpublished data).

Gunning and Suttkus (1991) presented raw data on the abundance of 28 species (each comprising > 1% of the total for a given family) taken in collections over 16 years at 1 site on the Pearl River at Monticello, Mississippi. Of those 28 species, 19 were taken in every year, and 5 others were taken in all but 1 year. This local assemblage therefore consisted largely of very persistent species. Similarly, for 22 samples in the Pearl River at Bogalusa, Louisiana by Gunning and Suttkus (1991), 23 of 37 species were present in every year.

In long-term samples where species disappear, there are often one or more anthropogenic effects that can be linked to the declines (Burr, 1991). Cross and Moss (1987) chronicled historical loss of species from sites on river mainstreams in Kansas, with declines worsening as water withdrawal for agriculture increased and stream flows seriously decreased. For two sites on the Smokey Hill or Solomon rivers, Kansas, Cross and Moss (1987) document 23 species that have become locally extirpated in this century. However, there is evidence from historical exploration records compared to recent conditions that it is actually the small streams of the central United States that have been altered even more than the river mainstreams (Matthews, 1988).

Quantitative Studies—Temperate Streams

One of the first modern long-term studies of changes in local stream fish assemblages was Gard and Flittner's (1974) analysis of annual samples in Sagehen Creek, California from 1952 to 1961. Sagehen Creek is a trout stream with nine known species, several of which have distinct upstream boundaries linked to elevation. For samples pooled across 10 stations, Gard and Flittner (1974) reported overall for the 10 years a significant decline in abundance of two species, an increase in two species, and no trend in abundance for five species. However, individual species varied as much as an order of magnitude in abundance among years, apparently in response to floods. Erman (1986) continued the analysis of structure of fish population in Sagehen Creek with data through 1983 and concluded that, overall, most fish populations had been stable over the long term, in spite of years with widely varying flow regimes. However, construction of an impoundment low in the main channel of the creek appeared to result in altered fish assemblages in the lower mainstream [e.g., an increase in suckers (Catostomidae) from the reservoir].

Moyle and Vondracek (1985) suggested that relatively simple fish assemblages in another cold-water stream (Martis Creek, California) were stable over a 5-year period ending in 1983, despite vagaries of stream flow, and concluded that segregation of species by habitat, microhabitat, or diet provided ecological structure to the assemblages. However, for at least one site on Martis Creek, subsequent studies for an additional 5 years (ending in 1989) showed that major flooding could result in an alteration in the relative density of fish species (Strange et al., 1992). On the basis of the 10-year span of data, Strange et al. (1992) concluded that stream fish communities might show more than one relatively stable state, with the community shifting between alternate states in response to severe environmental events. However, they emphasized that "community assembly between states will not be random" and that "the community that emerges at any given time will be determined by how particular high discharge events influence the existing biotic context."

Grossman et al. (1982) analyzed 12 years of samples at one location on Otter Creek, Indiana (USA) (in Whitaker, 1976), using a nonparametric multisample analysis (Kendall's W) to test for concordance of rank abundance of the 10 most abundant species (seasonally) across all years. They concluded that assemblage structure was nonpersistent, as there were no significant multiyear correlations for any season. From these findings and a review of the literature, Grossman et al. (1982) argued that in many moderate-sized streams of the North American midwest, local fish assemblages are unstable and likely dominated by stochastic processes such as floods and droughts. Rebuttals that followed (Herbold, 1984; Rahel et al., 1984; Yant et al., 1984; Grossman et al., 1985) provided a variety of views about interpretation of the data set analyzed by Grossman et al. (1982), but no single answer to the question, "Are fish assemblages of streams stable?"

Schlosser (1985) showed large differences between 2 years in the abundance of juveniles of minnow and sunfish species relative to discharge in a small Illinois (USA) stream, expanding the view of stochastic regulation of fish assemblages (e.g., Grossman et al., 1982, 1985) and showed ontogeny to be a critical factor. Schlosser (1982) also considered local fish assemblages to vary in stability, depending on longitudinal position in a watershed, and generalized these findings to a model of fish assemblage structure and stability as a function of pool size or depth (Schlosser, 1987).

Ross et al. (1985) concluded that the fish faunas of two midwestern streams (Piney Creek, Arkansas, and Brier Creek, Oklahoma) were persistent (qualitatively) over a decade, but the fauna was quantitatively stable (based on abundance data) only in the stream with more stable environmental conditions (Piney Creek). Matthews et al. (1988a) concluded from rank concordance tests and quantitative similarity indices for data sets spanning 14, 17, and 5 years at multiple sites on Piney and Brier creeks and on the Kiamichi River (Oklahoma) that the fish faunas (of the whole watersheds) were stable and persistent [*sensu* Connell and Sousa (1983)]. However, stability was greater at individual sites in the more environmentally stable stream.

Additionally, the fish assemblages and the total fauna of Piney Creek recovered rapidly (within 8 months) after a catastrophic flood (Matthews, 1986c). Individual sites varied in their response to a "hundred-year" flood, with rock-bound reaches returning rather rapidly to former physical conditions and sand-gravel reaches reforming riffle and pool configurations more slowly. Consequently, the return of fish assemblage structure to preflood conditions was more rapid at some sites than others, but by 8 months after the flood (August 1993), all but 2 of 12 sites exhibited substantial resemblance to that of the previous summer (August 1992). Peter Moyle (personal communication) has suggested that the timing of the flood in December (outside the reproductive season for these species) may have contributed to the minimal long-term effect on the fish assemblages.

Several other noteworthy studies of stream fish assemblage stability appeared in the late 1980s. Meffe and Minckley (1987) showed that the fish assemblages of a desert–canyon stream (Aravaipa Creek, Arizona, USA) were both persistent qualitatively and stable quantitatively (based on concordance or ranks) over decades, in spite of massive flash floods. [W. L. Minckley (personal communication) and his graduate students at Arizona State University are now analyzing more than 30 years of annual data from this system.] Ross et al. (1987) concluded from rank abundance across multiple sites and a multivariate principal components analysis of local assemblages in Black Creek, Mississippi that the fauna was highly concordant across a span of 9 years and that assemblages at individual sites were relatively distinctive. For Cedar Fork Creek, Ohio (USA), Meffe and Berra (1988) found for 38 samples over 9 years at 1 location that the assemblage was persistent and stable (ranks concordant both for the entire suite of 30 species and for the 12 most abundant species), but that there were species that existed infrequently or as transients. Meffe and Berra (1988) noted that absolute population sizes fluctuated markedly but that relative abundances of species were consistent.

Freeman et al. (1988) examined the stability of fish assemblages at three sites on small streams in the Appalachian Mountains, with seasonal samples over a 40-month period. A total of 13 species was collected; thus, this was a relatively simple fauna more like that in California streams (Erman, 1986) than the more diverse streams studied by Ross et al. (1987), Matthews et al. (1988a), or Meffe and Berra (1988). Freeman et al. found three groups of species they considered "resident," "seasonal," and "occasional" in short reaches of stream (30-m each). They found that relative abundances of species were constant at two of their three sites, based on coefficients of variation of individual species. However, they found sharp differences in year–class strength among years. They concluded that density-dependent mechanisms regulated populations of the abundant species, whereas environmental variation had a substantial role in abundance of young of year or survival of some species (density independent).

Grossman et al. (1990) reviewed all recent multiyear studies of stability of stream fish assemblages in North America and reanalyzed some of the original data (provided by authors) from those studies. Grossman et al. (1990) assessed the stability in nine assemblages for which sufficient quantitative data existed to permit calculation of coefficients of variation (CV) in the abundance of individual species. They and Freeman et al. (1988) used the following criteria to describe variation in populations, with an overall view of the assemblage stability based on "examining CV values for all assemblage members":

Coefficient of Variation	Interpretation
< 25%	Highly stable
25–50%	Moderately stable
50–75%	Moderately fluctuating
> 75%	Highly fluctuating

Grossman et al. (1990) concluded that CVs were a "better estimator of population/assemblage stability" than was Kendall's W (nonparametric rank correlation) as used by numerous previous authors. Ebeling et al. (1990) also criticized the use of rank correlation as a measure of temporal concordance of (marine) fish communities as "overly simplistic and misleading" for any conclusions about regulation of assemblage structure. Grossman et al. (1990) found that across all reanalyzed species and data sets, the most typical CV was about 0.85–1.05 (i.e., "highly fluctuating"). On the basis of reanalyses of numerous authors' data by CV of individual species, Grossman et al. (1990) concluded that many stream fish assemblages are "composed of species that vary substantially in population size." They interpreted this as evidence that there is high natural variability in lotic fish assemblages.

Coefficients of variation appear to be a useful tool for detecting differences among species in their variation in samples over time. Grossman et al. (1990) discussed their limitations, the most severe of which I would suggest to be "decomposing an assemblage into its component populations", thus "no longer examining assemblage level behavior." Additionally, another problem may exist with the interpretation of the CV of individual species as an index of variability of whole *assemblages*. What most ecologists or managers working with stream fish need seems to be a quantitative tool that allows the use of existing data (25–30 years for some systems) for assessment of changes in assemblages as a potential gauge of future changes that could occur from anthropogenic effects.

As pointed out clearly in a symposium at the annual meeting of the American Fisheries Society in 1995, many of these original data were not gathered with the intent of detecting fine details in the change in abundance of individual species among samples. In most cases, data from these studies were collected as typical "ichthyological" samples, in which competent investigators used seines, electroshockers, or other devices to attempt to capture (1) all of the species present in a stream reach in (2) proportions reflecting their relative abundance. In almost no case has there been an attempt to rigorously quantify density per unit area in such samples (and, indeed, such attempts may be in conflict with the first goal—to capture all of the fish species present at a site).

The reality is that most ichthyologists in long-term sampling have applied an approximate criterion of "equal effort," timewise, among collections at a given site, usually working about 45 min to 1.5 h and until all possible habitats have been sampled approximately in proportion to their representation at a site (Gunning and Suttkus, 1991). Sampling generally ceases when all habitats have been sampled exhaustively and additional sampling seems not (after numerous seine hauls or kick sets) to be producing any additional species or marked changes in apparent abundances of species. With such sampling (or any other technique of which I am aware), the absolute numbers of a given minnow species could easily depend on encountering one school, more or less, containing the species. The bottom line is that taking of, say, 200 versus 300 individuals of a common species in a 1.5 h seining collection in a large creek site is likely trivial. In other words, it is asking too much of real-world sampling of long stream reaches (e.g., 100–300 m) that any team visiting a site on two occasions (even if numbers of actual fish were *identical*) could capture a highly similar number of individuals. In the scenario above, it is the sampling "error" (from a statistical sense) that at least in part makes the CV appear large, even though the real abundance of this species is unchanged. Finally, I agree with Grossman et al. (1990) that similarity indices (particularly percent similarity, PSI) are potentially useful for assessing levels of change in fish assemblages, and E. Marsh-Matthews and I are now assessing the performance of this measure in more detail. From a large number of samples in four stream sites, we find a mean of approximately 0.60 similarity among all possible pairs of samples.

As a test of using coefficients of variation (Grossman et al., 1990) to detect the magnitude of change in a fish assemblage, I calculated CVs for all 28 species across 16 years of collections in Table 1 of Gunning and Suttkus (1991). These are probably some of the best and most reliable data in existence on the abundance of fishes at a single site over a long period of very regular sampling. However, Suttkus (personal communication) informs me that sampling was, at some times, during high water and that physical conditions at the site were changed from some years to another. For example, Suttkus (personal communication) wrote that, "during a particular spring survey a particular bar may have a firm gravel substrate that is available for sampling, while the next spring it may be inundated by four to eight feet of water or it may have been eroded away to winter floods."

Inspection of their original data makes me believe, subjectively, that this assemblage has properties that I consider to reflect "stability" in the long term. (The fact that physical conditions in this river differed from year to year, as described above, would tend to promote variation in composition of the local assemblage and, thus, should make any detection of "stability" more conservative. I consider this site to have had relatively stable properties in that most species were present in all years. Although raw abundances obviously vary a great deal within some species (e.g., *Cyprinella venusta*—their *Notropis venustus*—from 2246 to 22,578 individuals collected in a year) the pattern of abundant species largely remaining abundant in this diverse assemblage suggests strong stability.

The coefficients of variation ranged for the 28 species from 0.40 to 3.87. Suttkus (personal communication) indicated that an unusually high value for *Elassoma zonatum* in one year (191 individuals) was due to sampling a spawning aggregate on a newly seeded slope during high water and that in other years the species was only present in low numbers. This single large sample of *E. zonatum* resulted in a very large CV for this species. If this large value is omitted as an outlier, the range is from 0.40 to 1.45, with a mean of 0.92 in CV values. However, the frequency distribution of CV is distinctly bimodal (Fig. 3.5), with a peak at about 0.4–0.6 and another at about 1.2. Clearly, in this data set, some species have a high CV; others, distinctly low CV. Correlation of CV with mean abundance for species (with two outlier species omitted) was significant ($r = -0.395$) and negative, indicating a tendency for the most abundant species, overall, to have a low CV's (i.e., appear more stable) and the scarce species to have a high CV. This is an undesirable property of CV for use as a measure of overall stability of an assemblage (i.e., if "stability" of individual species is linked to its overall abundance). Grossman et al. (1990) did not encounter this difficulty in using CV for smaller data sets, but in this particular case [Pearl River (Gunning and Suttkus, 1991)] with massive and meticulously taken collections, it would appear that the CV is linked to abundance (Fig. 3.6). It also seems linked to families of fishes; that is, the very abundant minnows have substantially lower CV than species in most other groups (Table 3.3).

The studies cited above suggest that there is likely no single, simple answer to the question of stability of stream fish assemblages, and no one simple quantitative tool for the detection of change. The outcome of studies has varied depending on the kind of streams or environments under consideration, the numbers and kinds of fish species included in analysis, and the spatial or temporal scale of the work. Most fish ecologists now view fish assemblages as occurring on a gradient from stable to unstable (G. Grossman, personal communication), with parallels to Barbara Peckarsky's (1983) "harsh-to-benign" hypothesis for aquatic organisms in general. Rather than prolonging the debate in the literature on

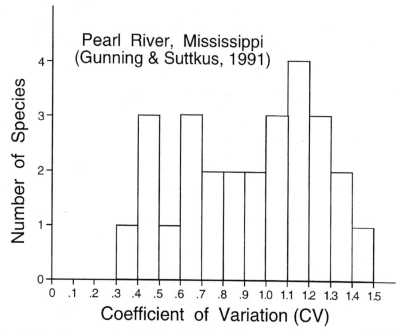

Figure 3.5. Frequency distribution of species versus the coefficient of variation in their abundance at one locality in the Pearl River, Mississippi. [Calculated from data in Gunning and Suttkus (1991)].

"whether fish assemblages are stable," I suggest "whole assemblage" analyses using "community index" measures (Jaccard's, for qualitative similarity; percent similarity, PSI, for quantitative similarity) and multivariate analyses (principal components analysis, PCA; detrended correspondence analysis, DCA) to attempt to place some of the available long-term studies for stream fishes into a context of magnitude of change in the assemblage from small to large. For comparative purposes of change in actual abundance, I also computed a CV for each common species, as suggested by Grossman et al. (1990).

First, a few comments about basic properties of each of the proposed measures, which have been used commonly in the literature to analyze something about changes in fish (or other) assemblages and whose properties are rather well known (e.g., Wolda, 1981). In choosing a measure to assess long-term changes in fish assemblages, the investigator implies the parts of the assemblage in which greatest interest lies. Use of a Jaccard's index is only for changes in presence–absence of species. With Jaccard's index, all taxa that are included are equal in influence on outcome, but the investigator can choose to include virtually every species that has ever been detected or to include only species for which there is reason to believe that they are real "members" of the assemblage—and

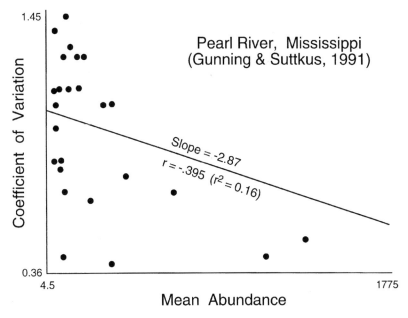

Figure 3.6. Coefficient of variation versus mean abundance for species from Gunning and Suttkus (1991), with two outlier species omitted (see text for details).

then ask how persistent they are. Alternatively, it is possible to exclude a priori very rare species (e.g., taxa not occurring at least twice over a long series of collections, or some other rule to reduce emphasis on rare species).

If the investigator selects the PSI, emphasis is not only quantitative (abundance) but also is focused on the most abundant species. Because PSI works on proportional abundance of each species within the sample, summing the minimum value for each species in the pair of samples, it results in a value of zero for each species not present in *both* samples. Hence, rare species contribute nothing to a PSI, and species present but in low abundance can contribute only a small increment to the total. Nevertheless, it is intuitively understandable to most ecologists to have a measure reflecting the similarity of two assemblages based largely on relative abundance of dominant taxa. It is also important at this point to note that I have abandoned use of an index that I and others used much in the past, the index of Morisita (e.g., Matthews et al., 1988). Linton et al. (1981) compared the properties of Morisita's and PSI indices and showed that for most of the range of values, Morisita's index produced a score modestly higher than the PSI. In that I have often considered stream fish assemblages "stable," it should strengthen the test of this hypothesis to use a measure (PSI) that lacks this inflation of low values, even if minor. Actual values calculated for fish assemblages by both indices are in Matthews et al. (1988a). Overall, the PSI

Table 3.3. Coefficient of Variation of Fish Species Sampled at Pearl River at Monticello, MS, by Gunning and Suttkus (1991) in 16 Yearly Samples (Calculated from Raw Data in their Table 1)

Species	Mean Abundance	Coefficient of Variation
Dorosoma cepedianum	60.4	1.45
Dorosoma petenense	149.4	1.29
Cyprinella venusta	10,628.7	0.55
Hybognathus nuchalis	1,775.1	0.92
Notropis longirostris	1,264.4	0.47
Pimephales vigilax	1,087.8	0.40
Notropis volucellus	601.2	0.66
Notropis texanus	352.9	0.76
Notropis atherinoides	302.0	1.05
Macrhybopsis aestivalis	237.3	1.08
Carpiodes velifer	133.1	1.26
Hypentelium nigricans	10.1	0.95
Ictalurus punctatus	126.9	1.15
Fundulus notatus	80.6	1.14
Fundulus olivaceus	20.9	1.13
Fundulus catenatus	10.9	0.81
Lepomis megalotis	286.6	0.36
Pomoxis annularis	78.4	1.31
Micropterus punctulatus	58.1	0.41
Lepomis macrochirus	55.1	0.67
Micropterus salmoides	17.9	0.77
Elassoma zonatum	12.3	3.87
Lepomis humilis	6.1	1.39
Ammocrypta beani	175.2	0.65
Percina vigil	53.8	1.26
Percina sciera	27.0	0.81
Percina caprodes	6.9	1.12
Ammocrypta vivax	4.5	1.07

seems more readily understood and to better reflect values in the range typical of many comparisons of fish assemblages.

An important choice is whether to use measures that respond to absolute numbers or to the relative abundance of species in an assemblage. As indicated above with respect to sampling, even the best of field techniques commonly used probably lack precision in the determination of true abundance or density of species. However, most of the techniques commonly used (seining, electrofishing, snorkeling) are quite good at estimating relative abundances of species when carried out by skilled individuals. However, even for experienced individuals, sites and times vary substantially in "catchability" of fishes. For example, the same site on two dates, one with rock substrates bare from recent flooding and another with rocks slippery from algal growth, will result in differences in ability of investigators to move about to sample fishes. Moderate differences in water levels, current speeds, or structure in the water all influence the ability of investigators to capture fishes by seine, shocking, or any other active devices. E. Marsh-Matthews and I scored (independently, after making the sample) a large number of field sites in the midwestern United States on a scale from 1 (poor) to 10 (ideal) for seine sampling conditions on a sampling trip across six states. We rarely scored sites as either extreme, and many were in the range of about 5–8. With these differences in "seinability" of sites, even careful documenting of start–stop times, use of block nets, and so forth, will not assure identical effort. As a result, the same fish assemblage at a site, being sampled on two different dates, would not produce absolute numbers of individuals that are equal.

Even with all of these caveats, long-term data sets are nevertheless invaluable in fish ecology and ichthyology, providing an irreplaceable record of historical patterns in structure of fish assemblages over years. With the known vagaries of sampling, I consider that sites with similar relative abundances of species across time are stable. For example, examine the structure in species abundances in the following artificial matrix, assuming for one field site a total of five samples, each a year apart:

SAMPLE NUMBER:	1	2	3	4	5
Species A:	15	20	30	60	25
Species B:	30	40	60	120	50
Species C:	45	60	90	180	75
Species D:	60	80	120	240	100
Species E:	150	200	300	600	250

I would consider the assemblage as a whole at this hypothetical site to be very stable across time, because the relative proportion of each species was identical among samples and all species were present in all samples. Differences in absolute numbers per species per sample, or in total numbers, could realistically occur due to sampling differences at this site on various occasions. Comparison of consecutive pairs of samples from this set would result in a Jaccard's index of 1.00 and a PSI of 1.00. In contrast, comparison of these samples by a coefficient of variation for each species results in an estimate of considerable change across time by the criteria of Grossman et al. (1990). In this case, each species (and, therefore, the mean for the assemblage) exhibits a CV of 0.59, which would be "moderately fluctuating" by the criteria of Grossman et al. Use of the CV, therefore, seems not to reflect the components of similarity that I consider important in characterizing "changes" in this assemblage over time. However, it should be kept in mind that the PSI focuses on relative abundance, whereas the CV focuses on absolute abundance (which is more sensitive to sampling error) (S. T. Ross, personal communication).

Assessment of Long-Term Samples by Similarity Indices and Coefficients of Variation

To assess the magnitude of assemblage changes at several sites in North American streams for which long-term collection data exist, I calculated two similarity indices comparing samples and CVs comparing abundances of individual species across time. Several of the sites were included in Grossman et al. (1990), and I have added my own more recent samples to sites on Piney and Brier creeks. Table 3.4 summarizes the sites; each is a single reach on a stream. Data for Otter Creek, Indiana are from Whitaker's (1976) original paper [i.e., the data on which Grossman et al. (1982) was based]. I retained all of Whitaker's 18 "major species" in these analyses. The Pearl River data were published by Gunning and Suttkus (1991) and represent their intensive sampling as described earlier. Raw data for Black Creek (Ross et al., 1987) were provided by S. T. Ross. Otter Creek, Pearl River, and Black Creek data consisted of raw abundances of species collected by seining, with the Pearl River data within each year representing pooled seasonal samples. All three streams have a relatively rich and diverse fish fauna, with 16 or more species retained in analyses (Table 3.4). Two western streams with relatively simple faunas (Aravaipa Creek, Arizona, and Martis Creek, California) were included on the basis of tables or figures published in Meffe and Minckley (1987) and Strange et al. (1992). The last four sites are headwaters (Brier S2

Table 3.4. Comparisons of Studies: Variation of Assemblages Among Years, Using Jaccard's Index, Percent Similarity, and Coefficient of Variation

Site	No. of Samples	No. of Species Included	Span of Years	Mean Jaccard	Mean PSI	Mean CV
Otter Ck., IN (Whitaker, 1975)	12	18	13	0.795	0.469	1.37
Pearl River, MS (Gunning & Suttkus, 1991)	16	28	16	0.925	0.803	1.03
Black Ck 1, MS (Ross data)	7	16	9	0.787	0.558	0.98
Aravaipa Ck, AZ (Meffe & Minckley, 1987)	14	7	30	0.866	0.796	N/A
Martis Ck, CA (Strange et al., 1992)	10	5	11	0.911	0.628	N/A
Piney P-1, AR (Matthews et al. data)	7	24	24	0.740	0.793	1.07
Piney P-9, AR (Matthews et al. data)	7	21	24	0.727	0.619	1.29
Brier S-5, OK (Matthews et al. data)	9	18	26	0.666	0.558	1.56
Brier S-2, OK (Matthews et al. data)	9	15	26	0.628	0.523	1.43

and Piney P9) and downstream (Brier S5 and Piney P1) sites in two midwestern creeks that differ in environmental stability and have been chronicled in detail in Ross et al. (1985), Matthews (1986), and Matthews et al. (1988a). Samples in Piney and Brier creeks span 24 and 26 years, respectively.

For each of the sites for which abundance data were available, I calculated a coefficient of variation (CV), following Grossman et al. (1990), across all samples. Also, to compare each site to itself in consecutive collections, Jaccard's index of qualitative (presence–absence) similarity and a percent similarity index (PSI) based on relative abundance of species were calculated. Very rare species, either excluded by original authors from calculations of stability or those that occurred only once in all samples at the Piney or Brier creek sites, were omitted. (If these rare species had been included, the Jaccard's index values would have been slightly lower than those presented in Table 3.4.)

By the criteria of Grossman et al. (1990), every site included in the present calculations would be considered "highly unstable" for means of CV scores across all species. These calculated values are similar to those obtained and discussed by Grossman et al. (1990) (i.e., with means of CV at or above 1.00). Similarity indices for these streams presented a rather different picture. For Jaccard's index, based on presence–absence of species in consecutive samples, the means for sites ranged from 0.628 (Brier Creek S2) to 0.925 (Pearl River). Values tended to be lower in Brier Creek, which is overall a rather harsh environment (Ross et al., 1985), particularly in the headwaters where average daily temperature fluctuation can be as great as 10°C and headwaters are dewatered in dry years (Matthews, 1987a; Matthews et al., 1988a). Jaccard's index values (Table 3.4) were very high in the Pearl River (a large, permanent river with a complex fauna) and in Martis and Aravaipa creeks (which are much smaller streams with very-low-diversity faunas, but in which most species are persistent).

Values for percent similarity (PSI), based on abundances of species, were lower (Table 3.4), ranging from 0.469 at the Otter Creek site to 0.803 at the Pearl River site. Across all nine sites included in this analysis, the grand mean (mean of means) for PSI values representing consecutive collections was 0.639, a value similar to that reported by Marsh-Matthews and Matthews (MS) for all possible interyear comparisons for four stream sites in Arkansas. In these and similar streams, it seems likely that from one time to another, the proportional similarity of fish assemblages at a given site will be about 60–70% between samples. Although not suggesting that such sites are "stable" across time, these values do suggest that fish assemblages remain rather similar, barring disturbances. The use of index values, particularly those that equate to a minimum "percent" of similarity from one time to another, seem intuitively reasonable as approximations of the magnitude change from one time to another. There is, however, one distinct weakness in calculating intersample similarity indices unless the data are inspected carefully. It is possible to determine how much an assemblage differs from Time 1 to Time 2 by inspecting a PSI; however, from

Time 2 to Time 3, a new and similar PSI could be obtained *either* if the site diverged *more* from the original (Time 1) condition *or* if it changed completely *back* to the original condition. In other words, inspection of PSI or other index values can tell us how much a given site is expected to change in any one time interval, on average. A series of PSI values cannot, as noted by Grossman et al. (1990), explain how much in a particular direction the whole assemblage is changing over a series of samples in time. For that much more difficult question, multivariate ordination approaches have much promise. They have been used commonly in recent years to assess differences among fish assemblages at different sites (Edds, 1993) or to assess changes in fish communities over short periods of time [e.g., monthly (Gelwick, 1990); before and after manipulation (Meffe and Sheldon, 1990)], but they have had relatively little use in assessments of long-term variation in stream fish assemblages [but see Ross et al. (1987) and Hansen and Ramm (1994)].

Multivariate Assessment of Temporal Variation in Fish Assemblages

For all nine sites in Table 3.4, I carried out detrended correspondence analysis (DCA) or (if DCA appeared inadequate due to low eigenvalues) principal components analysis (PCA) based on the covariance matrix. PCA based on a correlation matrix may overemphasize rare species or those of only modest abundance, and recent analyses by E. Marsh-Matthews and myself showed that scores on the first PCA axis sometimes are highly correlated with total abundance of fish in individual samples.

If a DCA produced a reasonably high eigenvalue on the first axis [e.g., 0.20 or higher is acceptable (M. Palmer, Oklahoma State University, personal communication) and with a value of only 0.15 retained for the large Pearl River data set], it was used as the descriptor of changes within a site over time. For sites with low eigenvalues in a DCA, a PCA based on a covariance matrix (or correlation matrix, in one case) accounted for > 30% of sample variance on the first PC axis. For all cases, locations of samples in two-dimensional ordination space are shown on scatterplots (Figs. 3.7 and 3.8).

The most important information in the scatterplots is an overall impression of displacement or movement of sample points across time. At two possible extremes (neither met by any of the real data), a sample could remain in or cycle precisely about a point in multivariate space, or it could move progressively further away from the original condition in a virtually straight vector. The sites from five published studies (Fig. 3.7) exhibited a range of movements in multivariate space: Otter Creek, Indiana, and Black Creek, Mississippi, showed moderate change over time [DCA axes exceeding 1.5 standard deviations in length, suggesting at least partial turnover of assemblages with time (Gauch, 1982)] but appeared to return twice or more to near their original condition. The Pearl River site showed complex movement in multivariate space, with samples from the early 1970s

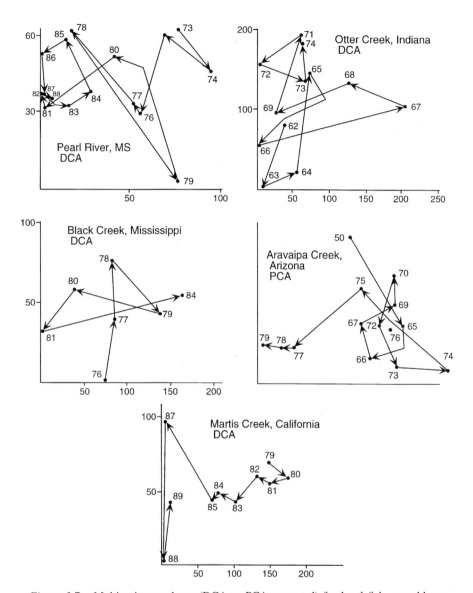

Figure 3.7. Multivariate analyses (DCA or PCA, as noted) for local fish assemblages sampled across time from nine sites in temperate North America. For each site, the x axis is the first multivariate axis and the y axis is the second multivariate axis; numerical values on the x and y axes is the length of the gradient (for DCA); arrows connect samples from first to last, with the years adjacent to the dots. [DCA or PCA calculated by Matthews from data in Gunning and Suttkus (1991); Pearl River, Grossman et al. (1982), Otter Ck., IN; Ross et al. (1987) and original data provided by S. T. Ross, Black Ck., MS; Meffe and Minckley (1987), Aravaipa Ck., AZ; Strange et al. (1992), Martis Ck., CA.]

moving little on the first DCA axis, intermediate samples in time showing marked displacement to the right and left in the figure, and ultimately (1980 on) cycling in the left portion of the scatterplot. Comparing the earliest to latest samples at the Pearl River site, there appears, overall, to have been displacement of the assemblage, but with great variation in samples between some years.

The Aravaipa Creek, Arizona canyon reach showed cyclic movement in multivariate space from 1950 to about 1975, then a sharp and persistent displacement away from the early years on PCA axis 1, with 1977–1979 distinctively changed from the original assemblage. It will be very interesting to learn if the samples taken since 1979 by W. L. Minckley and his students have cycled again to near the position of earlier samples or remained altered.

The most striking study with respect to displacement of an assemblage in multivariate space was that of Martis Creek, California (Fig. 3.7). An earlier analysis of 5 years of data in Martis Creek (Moyle and Vondracek, 1985) suggested that assemblages at several sites remained relatively stable. Based on a total of 10 years of sampling at one site, Strange et al. (1992) found that after 1983 the assemblage changed markedly in response to severe flooding that eliminated recruitment of some species. A DCA (based on percentages taken from a figure of Strange et al.) reflected and agreed with their conclusion: After 1983, the assemblage was very different in multivariate space than before 1983, and there was evidence (1987–1989 samples) that this assemblage had settled into another (alternative) stable point as suggested by Strange et al. (1992). However, continued sampling [through 1996 (Strange, 1995; P. B. Moyle, personal communication)] now shows that the assemblage in Martis Creek is shifting back to a structure similar to that described in Moyle and Vondracek (1985). Moyle pointed out (personal communication) that in 1994–1996, trout comprised < 15% of total fish (i.e., similar to earliest samples) but that Lahontan redside (*Richardsonius egregius*) appear to be absent, hence not recovered to earlier conditions. This entire case underscores the need for very-long-term samples of benchmark fish assemblages, so that divergences from original condition can be determined to be permanent or transient, and the fact that different components of fish assemblages can change in abundance at differing tempos.

My samples on Piney and Brier creeks (Fig. 3.8) showed examples both of cycling near original conditions and of displacement in multivariate space. For Brier Creek Station 2 (a harsh headwaters site), the assemblage varied substantially over 26 years, but if the first sample (1969, by C. L. Smith) is omitted, most of the remaining samples have cycled in the left portion of the multivariate space. In fact, the greatest displacement in space was from the 1969 sample to the next, taken in 1976. We lack knowledge of events on the creek during that time period, so an explanation remains wanting. The more downstream Brier Creek site (Station 5) is a midreach site characterized by perennial flow and relatively benign (for Brier Creek) conditions (Matthews, 1987a; Power et al., 1985; Gelwick and Matthews, 1992). At this site, which has distinct riffle–pool

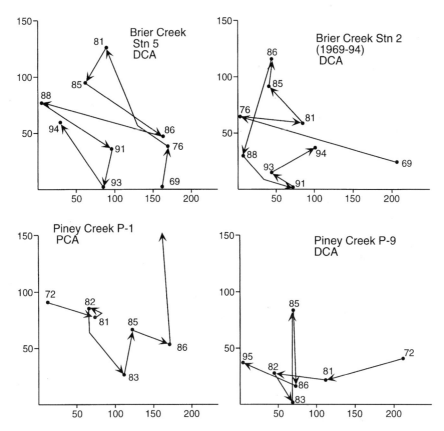

Figure 3.8. Same as for Figure 3.7, but with unpublished data of Matthews and collaborators, Brier Ck., OK, and Piney CK., AR.

zonation (Matthews et al., 1994) and is intermediate in overall conditions between the intermittent headwaters and the deeply incised downstream sites with mud substrates, the fish assemblage has varied substantially over the years but has moved back and forth along DCA axis 1 several times. The most recent collection (1994) is substantially displaced from the first ones (1969 or 1976), but the marked fluctuation along DCA axis 1 in the past suggests that this site may well "recycle" in the future.

Both Piney Creek sites showed evidence of some directed changes in time. At Piney Creek P-1, the DCA provided little discrimination among sites, with a low eigenvalue (0.09) and a short DCA axis 1 (less than one standard deviation). In part, this suggests that the variation, overall, has been relatively small (and note in Table 3.4 that both Jaccard's and PSI indices were rather high for this station). Within the multivariate space occupied by this site, movement of sequential samples has been, generally, from negative to positive on PCA axis-

1, with little evidence of cyclicity. In fact, in this PCA, there is some evidence of a relationship between total numbers of fishes collected and scores on the first axis. For Piney Creek P-9, a small headwaters site, there was evidence of marked change overall from the samples in summer 1972 to summer 1995 but with considerable cycling of assemblage structure on PCA axis-1 during the 1980s. Interestingly, the greatest known disturbance in the watershed (at both sites) was in December 1992, when a massive "hundred-year" flood had devastating effects on stream channels and riparian zones (Matthews, 1986c). However, in agreement with Matthews (1986c), it appeared that this major flood had only modest impact on the composition of fish assemblages at these stations (see displacement of points from 1982 to 1983, which is not remarkable relative to other intersample intervals).

Stability Across Combined Spatial and Temporal Scales

The fundamental question of how stable or unstable fish assemblages are at the local scale also needs to be viewed in the context of various levels of spatial or temporal consideration (Table 3.5). Although most analyses in this chapter have focused on stability of fish assemblages in a given stream reach, questions about relative stability of fish groupings are of interest at scales ranging from whole river basins down to the assemblage of fish in an individual stream pool or riffle. I suggest in Table 3.5 that not only the degree of stability but also the mechanisms influence that stability differ at the various spatial scales. At one extreme, the fish fauna of zoogeographic regions, or whole river basins, tends to be very stable across years (barring massive anthropogenic changes of the systems or large-scale species introductions), by virtue of evolutionary and geologic processes that have resulted in particular fishes being present in particular abundances in those regions (see Chapter 5). However, it is clear (e.g., in the Colorado River basin of the North American west) that the fish fauna of an entire basin can be massively altered through incautious acts of man (Minckley, 1991).

At the other extreme of space and time, the fish fauna in a single stream riffle or pool can change rapidly (i.e., with schools of fish immigrating or emigrating in a matter of seconds, or a mixing of fishes among pools in response to a small spate). Perhaps the surprising thing is that fish assemblages within small pools do tend to be at least relatively stable (e.g., PSI 50–60%), in many cases at a scale of months or across seasons. The stability of fishes within a single small land-scale unit like a pool or a riffle may depend largely on environmental stability of that habitat (e.g., Chapter 6), but at this scale, fish are also vulnerable to displacement by predators (Power et al., 1985) or competitors (Taylor, 1996). Overall, most local fish assemblages, which lie midway on the spatial scale of Table 3.5, tend to be stable across time, but it is clear that investigators can influence the outcome of measurements of stability by choice of size of habitat units under study. It is equally important that fish managers, seeking information

Table 3.5. *Spatial Scale and Stability of Stream Fish Faunal Patterns in Eastern North America*

Spatial Scale Considered	Degree of Stability	Proposed Mechanisms
Zoogeographic (regional or basin fauna)	Stable	Long-term evolution of faunas, some barriers to migration, physicochemical tolerances of members of faunas for harsh environments.
Whole-stream fauna	Typically stable	Migration within stream, movement to refugia during environmental extremes, recolonization of dewatered areas; availability of stable environments or stable structural microhabitats. Low rates of whole-stream extinction, low immigration rates from other systems. Elasticity following perturbation.
Longitudinal patterns	Mostly stable, but may shift if conditions are harsh	General headwaters to mainstream patterns in fish distribution exist, but environmentally variable (harsh) years may result in displacement, (e.g., of headwaters fauna downstream). Patterns not strongly linked to stream order; more influence by local availability of microhabitats for particular fish.
Single stream reach (200–400 m)	Very stable to unstable	Often stable across time if no environmental perturbation or microhabitat modification, but readily unstable. Changes in availability of microhabitats (e.g., riffles or pools) at local scale may alter fish assemblage. Predators and competition can also influence assemblage structure in a given small reach of stream. Stream features constrained by bedrock may have more permanent microhabitats, and thus fish assemblages, than in geomorphically unstable streams.
Single pool	Stability typically (PSI=0.5–0.6)	Fish assemblages in individual pools often stable across time intervals of weeks to months, but floods, movement, or predators and prey can rapidly alter assemblage of a single pool. Some individual species or life-history stages more prone to flood displacement than others. Physical integrity or stability of pool is important, as is depth or heterogeneity.

on baseline variation in fish assemblages, need to carefully define the scales that are of interest for the stocks they manage.

3.3 Summary

There is a long history in plant ecology of asking whether or not individual species occur in discrete community units or in local assemblages determined by the individualistic responses of each species to gradients of environmental conditions. In general, the individualistic view had prevailed. Animal ecologists historically focused more on competition or on limiting similarities among species to limit their occurrence in communities or assemblages. However, many of the questions are actually the same and can be summarized as, "Can we expect to find discrete, repeated groupings of animals (fish) occurring broadly across the landscape?" To answer this question, one approach is to sample broadly across kinds of habitats, and then estimate similarities and differences among samples based on the ordination of the species that are present in each. Such an analysis typically produces a two- (or three-) dimensional display of samples, where investigators can look for clumping of samples into discrete groups.

It should be possible to use objective statistical tools to ask if there is clumping of samples into discrete groups in a multivariate display. One potential approach is a Poisson analysis to ask if there is significant clumping of samples. In this chapter, I first used a simple model of random versus clumped placement of individual species on an environmental gradient to test the ability of a frequently used multivariate analysis (detrended correspondence analysis, DCA) to detect differences between clumped and randomized placement of species. The DCA, followed by Poisson analysis, properly discriminated random from clumped data, suggesting that this approach may be useful for detection of patterns in real data. For five of eight real multivariate analyses of fish samples over large areas, there was nonrandom clustering of samples, suggesting repeated patterns in distributions of species in assemblages. However, this approach cannot determine the mechanism for these nonrandom patterns—mechanisms need to be sought on the basis of life-history or natural-history traits of individual species, combined with assessments of their interactions.

Stability or persistence of fish assemblages over both short and long periods of time has a substantial history of debate. Most fish assemblages, when viewed from the magnitude of variation observed in the abundance of individual species in repeated samples, appear quite variable, with some unknown degree of effect of variation in the sampling effort contributing to the effect. However, when viewed from the perspective of the assemblage as a whole, most stream fish assemblages that have been sampled over multidecade scales show persistence in presence–absence of most species (barring human disturbance) and most also show relative continuity in abundance of at least the more common species in the assemblage. Many rare or modestly abundant species may not be detected in a single sample, but such species often appear to "come and go" in samples over time at a given place, suggesting, for these rare taxa, the difficulty of adequate detection in a single sample. Thus, if a species has been shown to occur earlier at the site, it should not be considered "gone" unless it fails to appear in several consecutive samples. Finally, there is growing recognition that all fish assemblages do change, with the interest focusing more on "how much" or "under what conditions" is the change, and if there is a directional trajectory of change in a given assemblage. A review of numerous published and unpublished multivariate analyses for stream fish assemblages in North America showed that although there was substantial change in some assemblages from time to time, there was little evidence of long-term, persistent change from one state of the assemblage to another. In contrast, the dominant pattern seemed to be a relatively random movement of local fish assemblages in multivariate space over periods of years, often with a given assemblage changing in some particular direction, yet returning to near its original conditions. On balance, and at the time scale of many years, most stream fish assemblages seem relatively resistant to change or oscillate about some relatively stable condition, barring human intervention or introductions of exotic species.

4

Stream Ecology and Limnology as Related to Freshwater Fishes

4.1 Introduction

Due to the heavy rainfall 29–30 August 1996, Brier Creek (Marshall County, Oklahoma, USA) rapidly rose about 1.3 m above base flow. Before this event, a prolonged summer drought had lowered the water level so that riffles were less than 1 m wide, with little or no surface flow, and pools were connected mostly by water flowing through gravel bars. The creek was very clear during the summer low-flow period, and dense masses of attached and floating algae had become the dominant feature of many pools in the stream. Small- and medium-sized fishes were intimately associated with the algal masses, apparently finding both invertebrate food items and shelter from their own predators in and near the floating columns or mats of *Spirogyra, Rhizoclonium,* and *Chara*. Sediments and detritus from senesced algae and leaf litter had accumulated on the bottoms of deeper pools, creating a "loose" and richly organic benthic layer as much as several centimeters deep and harboring many varieties of aquatic invertebrates. A week after the stage rise, water levels in the creek were again at base flow. The clarity of water permitted light penetration (and human sight) to the bottom of shallower pools, although a clay suspension remained in deeper pools, blocking vision. Minnows, sunfish, and small bass could be observed swimming actively in many of the shallow pools. Pools and riffles were markedly different than before the flood. The detritus and live mats of algae were largely gone, scoured and transported to some unknown location downstream. Deeply carved channels in the bedrock of the streambed were now apparent, as the pool-bottom debris and detritus had been removed.

After the stage rise, the course of one riffle had been completely changed within the streambed so that the flow of water now connecting the adjacent pools was over previously dry gravel. An entirely new pool had been created adjacent to one of the original pools, leaving a gravel bar in midstream. Fish in this reach of Brier Creek now occupied microhabitats that either (1) did not exist before the flood, such as the new pool and riffle, or (2) were markedly altered from pre-flood conditions with respect to underwater shelter (algal mats) or substrates (exposed bedrock). In addition, fishes were, at least temporarily, foraging in an environment with fewer visual cues because of turbidity, in contrast to the very clear conditions that had existed a week earlier. This event no doubt affected invertebrates: Although they were not sampled in association with this spate, sampling before and after other stage rises in this creek showed that the quantity and quality of invertebrates available as foods for fish (both in benthos and the drift) were quite different after a spate (M. R. Meador, W. J. Matthews, F. P. Gelwick, and T. Gardner, unpublished data).

This stage rise, which was relatively small in comparison to others I have seen in Brier Creek, may have had only minimal long-term impact on the distribution of fish assemblages among pools (Matthews et al., 1994). However, it probably had strong immediate effects on activities, distribution, or movements of fish in the stream during the event, and the spate, in general, illustrated the dependence of day-to-day ecology and behavior of fish on events and processes that collectively are referred to as limnology and/or stream ecology. To understand the ecology of freshwater fishes, the fundamental structure and physical or biotic processes of their aquatic environments must be understood.

4.2 Physical and Chemical Limnology and its Effect on Fish and Fish Assemblages

Limnology and physical stream ecology affect local fish assemblages by (1) formation of major habitat features over geologic time (e.g., lake formation and drying, drainage network modifications or flow reversals due to tectonic uplifting), (2) establishment or maintenance of physical or chemical habitat structure or zonation on a day-to-day basis during "nonevent" periods, and (3) abrupt modification of habitat structure or chemistry during periods of active change or "events." The two latter effects can occur rapidly, whereas the first typically requires geological time (although there can be "instantaneous" formation of lakes by the isolation of oxbows or landslides during a single flood event). The intent of this chapter is not to review all of limnology or stream ecology, as numerous texts summarize this body of knowledge (e.g., Hutchinson, 1957b, 1967, 1975; Wetzel, 1983; Cole, 1984; Hynes, 1970; Whitton, 1975; Allan, 1995), but to summarize important principles of limnology/stream ecology as they relate to fishes and fish assemblages. Information in this chapter is a synthesis from the volumes cited above if no specific reference is given.

The phenomena considered in limnology are strongly interrelated both in extant and geological time. Figure 4.1 suggests the more obvious of those linkages.

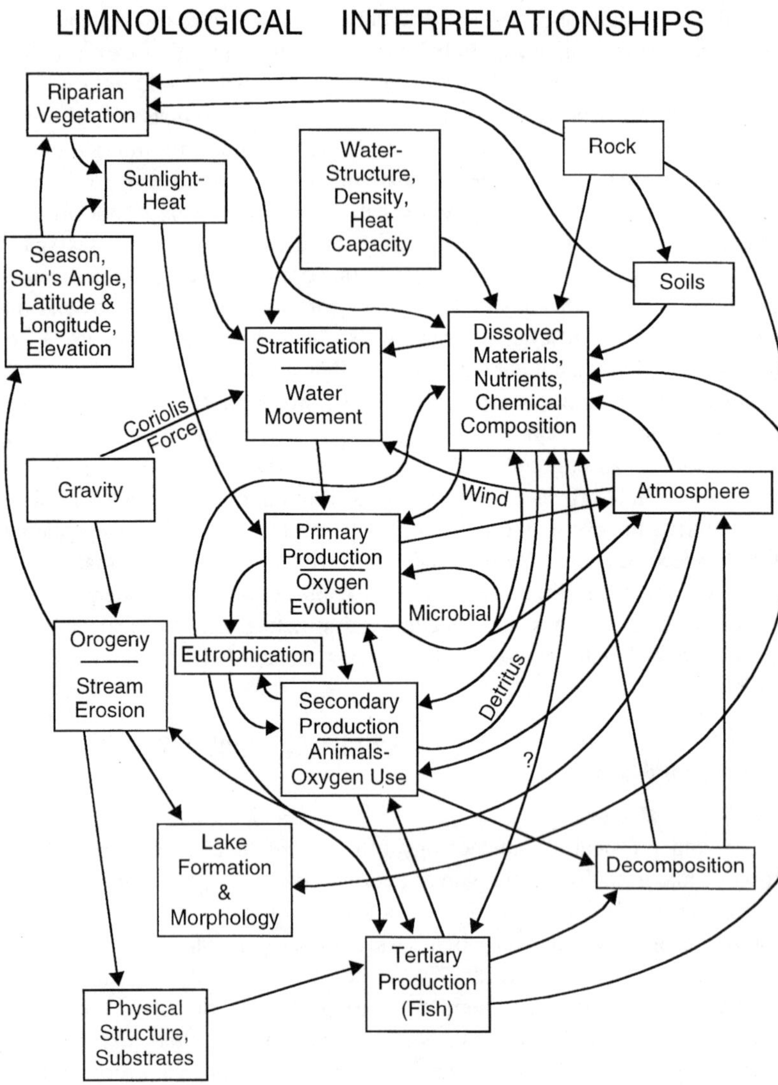

LIMNOLOGICAL INTERRELATIONSHIPS

Figure 4.1. Interrelationships of factors that influence ecology of individual lakes or streams.

Factors of most importance in extant, ecological time include the following: heating or cooling of water, light penetration, and the movement of water, which collectively establish the gradients of physical and chemical conditions that fish encounter and that influence fish movements or biology; physical extremes that actually kill fish or limit their habitat use; and the primary production (aquatic *or* terrestrial) that ultimately supports the fishes in an assemblage. Some of the variables with most pervasive influence on the limnological parameters of aquatic systems are very large-scale factors (Fig. 4.1) like (1) geology (= "rock" in Fig. 4.1), which determines water chemistry, streambed substrates, rates of downcutting, formation of individual lakes or drainage basins, as well as soils and nutrient inputs, (2) orogeny or uplifting of a region, which influences stream gradients and helps establish drainage patterns, and (3) climate. Before considering more complex or indirect linkages that can influence fishes, fundamental properties of water must be understood, such as composition and structure of water, density of water, light penetration in water, and heat relationships within water bodies, as well as differences between lentic and lotic habitats. In general, classical limnology has assumed "bottom-up" control of ecosystems, with productivity at higher trophic levels regulated by nutrients and primary productivity. Conversely, there is a growing body of knowledge that suggests that organisms in secondary or tertiary trophic levels, like fish, can influence ecosystems through "top-down" trophic effects cascading from predators to producers (i.e., algae or macrophytes) or by influencing nutrient availability to algae/macrophytes (Chapter 12). All of the biological phenomena that occur in a body of water take place in the context of the physicochemical milieu that is collectively studied in physical limnology. In the sections that follow, I first outline general principles about limnological factors like specific gravity of water, light penetration, thermal effects, or chemistry of freshwaters, using (in keeping with most limnology texts) lakes as the focus. Then, in a separate section, I relate physical/chemical limnology to streams.

Physical Properties of Water

A water molecule consists of one oxygen and two hydrogen atoms, with a 104.5° bond angle between the two hydrogen atoms and a strong dipole moment that is negative at the oxygen end of the molecule and positive at the hydrogen end (Wetzel, 1983). Water molecules in ice exist in a tetrahedron configuration, with each individual molecule both comprising the center and of one tetrahedron, and helping to form the "points" of adjacent tetrahedra. Each water molecule is hydrogen-bonded to its four nearest neighbors (i.e., those forming the "points" of the tetrahedron at which it is at the "center"). The tetrahedral configuration has a rigid lattice with a considerable amount of open space; thus, ice is about 8% less dense than, and floats on, liquid water. This accounts for the freezing of lakes from the surface downward [i.e., leaving liquid water (where fish live) under the surface ice]. Pools in some shallow streams freeze solid, with the joining of anchor ice on the streambed and surface ice and the resulting depletion of liquid water (and obvious death of all fish). Even at relatively southern latitudes (e.g., Oklahoma, 35°N), small headwaters or ephemeral pools may freeze completely, killing all fish trapped therein.

As ice warms, there is increased movement of molecules within the open lattice, which becomes semistable then collapses (Wetzel, 1983). At melting, the collapse of the rigid lattice allows denser packing of water molecules; thus, liquid water has a higher specific gravity than ice. Liquid water exists as clusters of individual water molecules (monomers) weakly combined by cooperative hydrogen bonding into polymers of various sizes (i.e., different numbers of water monomers forming the polymer) (Liu et al., 1996a). All of the polymers, at least up to the size of pentamers, appear to have a quasiplanar "puckered-ring" configuration that is "somewhat flexible and distortable" (Liu et al., 1996a). Larger water clusters (hexamers and larger) appear to lose the planar cyclic structure in favor of three-dimensional structures that have a lower energy state (Liu et al., 1996a). Liquid water is dominated by pentamers (Liu et al., 1996b), but at increasingly high temperatures, some hydrogen bonding is disrupted, resulting in more of the smaller polymers.

After melting of ice, continued warming from 0°C to 3.98°C results in increasingly close packing of water molecules in the liquid state (i.e., an increase in specific gravity of water). Maximum packing of molecules, hence the greatest specific gravity of water, is at 3.98°C; thus, water at this temperature is denser than that at any other, and it sinks downward toward the lake bottom. As water warms above 3.98°C, there is increased molecular vibration and interatomic spacing, causing water molecules to be spaced further apart and specific gravity to decrease. As a result, "warm" water has a lower specific gravity and floats on top of "cold" water in lakes at any temperature above 4°C. These basic physical properties of water set the stage for the well-known vertical stratification patterns (Section 4.4) that occur in many freshwater lakes (Wetzel, 1983; Cole, 1994). During summer, many temperate-zone lakes are directly stratified, with a nearly complete separation of water masses into an upper epilimnion and a lower hypolimnion, often separated by a well-defined metalimnion or thermocline. The epilimnion is characterized by warm water, circulation and mixing of the water column within the epilimnion (but *not* extending downward into the hypolimnion), predominance of photosynthesis and oxygen production, and the presence of most of the live algal biomass. The hypolimnion in directly stratified lakes is cooler, lacks or has much less photosynthesis than the epilimnion, and is characterized by oxidation of materials "raining" down from the epilimnion, which can deplete oxygen in the summer. In winter, as described in Section 4.4, "reverse stratification" can occur; that is, if water directly under an ice cover is actually colder, then water of greater density (e.g., at about 4°C) lies deeper in the water. Winter stratification generally occurs only where there is an ice cover; in warm monomictic lakes (lacking ice cover), winds cause circulation of the entire water column all winter, and stratification occurs only in the summer.

Another critically important effect of the physics of water is that change in specific gravity per degree centigrade is nonlinear and is much greater at higher temperatures within the environmental range for fishes (Wetzel, 1983). The difference in specific gravity between two parcels of water at 8°C and 9°C, respectively, would only be about 0.00007 g/cm, whereas at 28°C versus 29°C, the difference would be 0.00028; that is, there is almost four times as much difference in density per one degree difference in temperature at the hotter range. Wetzel (1983, p. 11) notes that for a 1°C difference at 29–30°C versus at 4–5°C, the difference in specific gravity is 40-fold. As a result, stratification in summer, due to thermal profiles, is much stronger in lakes than is the winter stratification, even though the absolute number of degrees differing from surface to bottom of a lake might be the same.

Light in the Aquatic Environment

Sunlight provides most of the energy that moves through the food web to fishes in lakes or streams. Lakes and streams receive various portions of their energy by input of allochthonous materials from the terrestrial environment, typically more in streams (Vannote et al., 1980). However, in both lentic and lotic environments, direct influx of solar radiation allows the fixing of energy in carbon compounds. The vertical distribution of different wavelengths of light plays a major role in determining the exact location of autochthonous primary production in a lake or a stream pool. Vertical profiles of light quantify and quality in a lake strongly influence not only the site of primary production but also the location of organisms (zooplankton, some fishes) that feed on that primary production.

Light entering the upper atmosphere is delivered at the rate of the solar constant, about 1.94 cal/m^2 per minute, but it arrives at very different schedules at the poles versus the equator, making light availability more seasonal at higher latitudes (Wetzel, 1983). The tropics receive more total solar energy per year than do the polar regions, its delivery is spaced more equitably across time, and (because of greater reflectance at low sun angles) more of the light entering the upper atmosphere in the tropics actually penetrates to reach the water's surface.

After entering the top of the atmosphere, light is lost in the atmosphere by scatter, reflection, and absorption by atmospheric carbon dioxide, ozone, and water vapor. At the surface of the water light is lost to reflection (but with ripples actually increasing the penetration of light). Within the water column, light is absorbed and converted to heat (more than half of the light entering the water), lost to scattering, and used in photosynthesis.

The attenuation of light in water is closely linked to dissolved and suspended materials. Clear water absorbs red and infrared wavelengths (about 700 nm and greater) rapidly and converts the energy to heat; thus, most solar heating is in the upper meter of the water column. In clear water, blue light penetrates most deeply (Wetzel, 1983). For example, in clear water at a depth of about 40 m, almost 80% of the blue light penetrating below the surface of the water would still remain, but the remaining percentages of green, violet, yellow, orange, and red, would be about 60, 50, 25, 0, and 0, respectively (Wetzel, 1983). However, dissolved organics like humic acid can markedly increase the absorption of violet and blue wavelengths (e.g., 400–480 nm) (Wetzel and Likens, 1991).

Photosynthetically active radiation (PAR) is at wavelengths from about 390 to 710 nm, with peaks in the blue (~ 460 nm) and orange (~ 620 nm) ranges. Hence, the color and organic matter content of water will have strong influence on (1) the depth of penetration of various wavelengths of light (Wetzel, 1983) and (2) the contribution of that light to primary production at a given depth.

Heat Budgets and Temperature in the Aquatic Environment

Although there are minor ancillary sources of heat in lakes, and perhaps in streams (e.g., reradiation from warmed sediments), it is the conversion of solar radiant energy to heat in the upper few meters of water that is the source of the most significant heating in freshwater. Heat input and its dissipation and distribution determines the annual "heat budget" [defined as the heat storage capacity of a lake (Wetzel, 1983, p. 89)] of a system and the local temperature at any given time and place within the system. The temperature at a given stratum in a lake changes rather slowly unless wind forces water downward that has temperature different from that at a lower level (e.g., during a wind-induced seiche or during spring or autumn turnover).

The heat budget of a body of water is, very simply, the amount of heat that must be input to that body to raise it from its coldest average (for the entire body of water) winter temperature to its hottest average summer temperature. For lakes that lack ice cover in winter, the heat budget can be calculated directly from knowledge of the maximum and minimum annual temperatures and the volume of the water body. For lakes having ice cover in winter, an additional factor must be included: the heat of fusion required to actually melt the ice (i.e., the input of kilocalories that, at a temperature of 0°C, must be added to water to melt ice, although not raising the temperature).

Effects of Light and Heat on Fish Assemblages in Lakes

As a result of the greater strength of stratification in warm lakes, fish at latitudes further from the equator or at high elevations may experience shorter periods of summer stratification than do fish in warmer climates, where lakes can be strongly stratified for months (Matthews et al., 1985; Matthews and Hill, 1988). Therefore, summer anoxia in the hypolimnion may present more difficulty for fish assemblages at low latitudes, whereas winter anoxia during reverse stratification (discussed in a later section) may be a more substantial problem for northern lake fish assemblages.

Light penetration is one of the most important parameters directly and indirectly influencing the kinds of fishes which will occur in a local assemblage in lakes (Rodriguez and Lewis, 1997) and streams. The penetration of light into water is critically important to fishes because of its role in (1) producing heat, (2) driving photosynthesis, which directly or indirectly provides feeding opportunities, (3) providing behavioral or reproductive cues, and (4) allowing sufficient illumination for permit visual activities like foraging or finding mates.

Thermal Effects on Fish Distribution in Lakes

Thermal properties of water impact fish distribution. Because they are ectotherms, the internal temperature of freshwater fishes generally follows that of the external medium, and most fish thermoregulate behaviorally. However, the core temperature of a fish can lag considerably behind that of the external medium (Stevens and Sutterlin, 1976), with important effects in thermal energetic models for fishes (Erskine and Spotila, 1977; Kubb et al., 1980).

In that temperature influences virtually all physiological and behavioral processes in fish (Hutchison, 1975), it also is critical to the understanding of a fish assemblage to know how heat passes through the system they occupy and how members of the assemblage respond to various thermal regimes. In lakes, vertical distribution of heat, reflected in temperature profiles, is a key regulator of fish habitat selection. Magnuson et al. (1979) showed that fish in lakes were importantly influenced by spatial distribution of temperature gradients, with different temperatures predilecting habitat selection of different species. Magnuson et al. (1979) considered fish in lakes to have "thermal niches" and that availability of advantageous temperatures represented a measurable "ecological resource," and Brandt et al. (1980) showed thermal habitat partitioning in Lake Michigan. Elrod and Schneider (1987) showed that depth distribution of lake trout in Lake Ontario corresponded to temperature, with interannual differences in fish distribution relating to annual differences in weather.

In addition to avoiding harmful temperatures, some fish use the best available temperatures for a variety of physiological functions. Brett (1971) found that sockeye salmon move to warmer temperatures to feed (where they were benefitted by the ability to be more active to capture prey), then to cooler temperatures to metabolize the meals and shunt energy into growth (with less metabolic demand for maintenance physiological activities, there remains more of a given meal to be devoted to growth of somatic tissues or reproductive products).

The heat budget of a lake also is important from the perspective of fishes that ripen ova or spawn after appropriate temperatures (or degree-days). Because of the requisite input of heat to melt ice (without raising the temperature until the ice is melted), two identical lakes—one with and one lacking ice cover—given equal input of heat from the atmosphere and solar radiation, would warm at different schedules, with the ice-covered lake warming less rapidly, delaying fish reproduction in the ice-covered system. In lakes, fishes often follow closely the temperature profile [although not always seeking the "optimal" temperature if it is coupled with harmful conditions (e.g., ideally cool temperatures being so low in a lake that they are within the anoxic hypolimnion) (Matthews et al., 1985; Coutant, 1985)].

Water Chemistry: Oxygen

Over the range of environmental temperatures occupied by fishes (0°C to or slightly in excess of 40°C), there is, at saturation, sufficient oxygen for fish to live without stress. As for any gas dissolved in liquid, cold water at saturation holds more oxygen by weight than does warm water. Air-saturated freshwater at 4°C has about 13.1 parts per million (ppm) of oxygen, whereas at 38°C, it contains about 6.6 ppm (Wetzel, 1983), which is sufficient for respiration for virtually any fish. It is, therefore, the limnological variables resulting in departures from saturation that are of the most concern relative to distributions of freshwater fishes in local assemblages. Departures from saturation result primarily from chemical and biological processes (respiration) that decrease oxygen concentrations, and photosynthesis, which increases oxygen concentrations. Phenomena that exacerbate the effects of either include the isolation of parcels of water from exposure to air (e.g., the hypolimnion of a stratified lake) and increased oxygen evolution or demand due to the greater availability of nutrients, increased productivity, and the subsequent increased decomposition or respiration in the water column. Barica and Mathias (1979) found an approximately 50% increase in the rate of oxygen depletion under ice for stratified compared to nonstratified lakes.

In lakes or deep, slow-moving streams, the direct input of atmospheric oxygen is slow (unless there is wind-mixing), relative to the diel changes in oxygen concentrations that result from algal photosynthesis and total respiration of the biota. In such bodies of water, the oxygen availability is driven in part by diurnal photosynthesis of phytoplankton (or attached or floating, filamentous algae in stream pools), countered by respiration of animals throughout the day and both animals and algae at night, combined with oxidation of dead animal and plant materials. It is common to find the oxygen concentration in well-lighted, near-surface waters of a lake in midsummer to increase from about 7 or 8 ppm in the early morning to 12 ppm or more (supersaturation) by mid-afternoon. Whitney (1942) reviewed previous work on diel fluctuations of oxygen concentration in standing waters, with a typical fluctuation of about 3 ppm, and oxygen maxima in mid to late afternoon. Littoral zones, with aquatic macrophytes, show the greatest oxygen fluctuation. Wiebe (1931) showed changes as great as 6 ppm in a day in a pond dominated by an algal bloom. Diurnal changes in dissolved oxygen of as much as 4–6 ppm are common, and oxygen fluctuations can be much greater in macrophyte-dominated littoral zones than in pelagic surface waters (Wetzel, 1975, pp. 134–135). Diurnal oxygen changes can be summarized in oxygen curves that allow the calculation of net production and respiration in lentic waters (Wetzel and Likens, 1991).

Effects of Oxygen on Fish in Lakes

No limnological factor is more important to fishes than oxygen, with anoxia causing outright death of individuals and low oxygen concentrations causing a wide array of stress-related responses. Distribution of oxygen in freshwaters and deviations of freshwaters from oxygen saturation (in space or time) have serious consequences for fish assemblages. Outright lethal limits will be addressed in more detail in Chapter 7 as an influence structuring fish assemblages. However, winter oxygen concentrations strongly influence the composition of northern lake fish assemblages (Rahel, 1984; Tonn and Magnuson, 1982; see Chapter 7). These effects are not limited to northern lakes: In Lake Texoma (Oklahoma–Texas), Matthews et al. (1985) and Matthews and Hill (1988) found strong stratification in July and August, with anoxia below 13 m during much of that time. Under those circumstances, most fish avoided the anoxic hypolimnion (Matthews et al., 1985; Matthews and Hill, 1988). Matthews (1987a) showed a correspondence between oxygen tolerances of various fish species and their upstream limits in Brier Creek, Oklahoma, which has a progressively more stressful oxygen regimen in headwater pools.

Low oxygen does limit distributions or activities of fishes. Davis (1975) reviewed the oxygen requirements of fishes and the effects of hypoxia on fish physiology, energetics, behavior, and growth, as well as the ability of fish to acclimate to low oxygen and the interaction of oxygen with toxic materials. Moore's (1942) review of the earlier literature and his tests with numerous species showed that, in general, oxygen concentrations less than 3.5 ppm are lethal in 24 h, but that above 5.0 ppm, fish were not killed. Spoor (1977) showed that larval largemouth base (*Micropterus salmoides*) go through a window at about day 6, when they are most vulnerable to low oxygen, but that their tolerance subsequently increases. Carlson and Siefert (1974) noted a delay in first feeding at low oxygen concentrations for largemouth bass and for lake trout (*Salvelinus namaycush*).

Some fishes have remarkable tolerance for low dissolved oxygen, such as fishes in low-oxygen spring heads (Hubbs and Hettler, 1964). Lowe et al. (1967) showed that differential resistance of desert fish species related to their behaviors and to their morphology (as this affected the ability to use surface oxygen). Campagna and Cech (1981) showed that Sacramento blackfish (*Orthodon microlepidotus*) of western United States have a high capacity to survive in low-oxygen environments because of their blood oxygen saturation characteristics and compensatory ventilation rates that allowed respiration even at very low oxygen concentrations. Ultsch et al. (1978) found that some darter (*Etheostoma*) species from slow-water habitats with summer hypoxia decreased their oxygen uptake below a critical minimum value. Klinger et al. (1982) showed that central mudminnows (*Umbra lima*) could use oxygen from gas bubbles, thus surviving periods of anoxia under ice cover. This species is the core of the assemblage of fishes "winterkill" lakes in the northern United States (Rahel, 1984). Gee et al. (1978) showed that 24 of 26 species in the northern United States extended habitat use ability by respiration at the oxygenated surface film.

Suthers and Gee (1986) found that juvenile yellow perch (*Perca flavescens*) balanced diel microhabitat choice between avoiding hypoxia overnight in a cattail bed (where oxygen demand was high) and using the cattail bed during the day (when oxygen was not limiting) as a foraging site. Saint-Paul and Soares (1987) showed that many fish species abandoned macrophyte cover in favor of aquatic surface respiration (ASR) during periods of anoxia in an Amazon floodplain lake, but that two species remained within weedbeds and reduced locomotor activity instead of migrating to open water for ASR. In that the macrophyte zone is important as shelter, source food, and avoidance of predators, the ability of the two non-ASR species appeared advantageous. The authors had no direct information on mechanisms for survival of species in the reduced-oxygen environment of macrophyte beds but suggested the possibility of anaerobic metabolism. Burton and Heath (1980) showed a capacity for anaerobic metabolism at low-oxygen tensions for three species of fish.

Kramer (1987) discussed oxygen as a limiting resource for fishes and reviewed the optimality of the response of fish to low oxygen, indicating that they should use strategies that included changes in activity, aquatic surface respiration, air breathing, and spatial changes in habitat. Kramer showed that some species combined two or more of the strategies to survive low-oxygen conditions. His review included effects of lowered oxygen on activity levels, feeding, reproduction, and predator avoidance.

Water Chemistry: Alkalinity and pH, Carbonate Systems

One of the most important distinctions with respect to chemistry of various bodies of water relates to their geology, [i.e., whether or not they occur in calcareous strata (possessing carbonates), producing alkaline, well-buffered water, or in noncalcareous strata (granite, shales, quartzites, etc., lacking carbonates)]. Streams or lakes in calcareous strata, dominated by limestone (calcium carbonate) or dolomite (calcium–magnesium carbonate) are typically well buffered, maintain a pH at or above 8.0, and resist a change in pH. Waters lacking calcareous buffering typically exhibit lower pH, more extremes in pH fluctuation, and have less predictable chemical features overall.

The details of the carbon dioxide–bicarbonate–carbonate buffer system are beyond the scope of this book and are found in any standard limnological text. In brief, pure water exposed to the atmosphere (with no background buffering of carbonate rock) will equilibrate at about pH 4.3, as a balance of carbon dioxide dissolving from the atmosphere, producing carbonic acid, which undergoes a first dissociation to produce bicarbonate ions (and release one hydrogen ion) and a second dissociation to produce carbonate ions and free another hydrogen ion. Limnologists estimate the buffering capacity of water by its "alkalinity" (i.e., the combination of bicarbonate and carbonate ions), measured by the ability of the water to resist titration by weak acids. Until the "alkalinity" is exhausted by the addition of acid, it will continue to buffer the water against a decrease in pH. Where streams flow over or through limestone or dolomite, they receive additional carbonate and bicarbonate ions and, hence, are "alkaline" and well buffered. A typical calcareous stream will have an alkalinity as high as 180–200 "units" [expressed as milliequivalents per liter of calcium carbonate (Wetzel, 1983)].

By the well-known principle of LeChatlier, any removal of a chemical species from this reversible equilibrium system will result in a reaction of the system to ameliorate the change. Thus, in a poorly buffered stream or lake, removal of carbon dioxide from the water by photosynthesis of aquatic plants or algae is followed by a shift of the chemical equilibrium to the left, compensating for the loss of a carbon dioxide molecule by using up free hydrogen ions to shift carbonate to biocarbonate, or bicarbonate to carbonic acid (which is unstable and may not really exist in water). Because the system "uses up" hydrogen ions in response to photosynthesis, the process of photosynthesis is, in effect, removing hydrogen ions from the water, hence raising the pH. This is, in fact, the phenomenon commonly observed in weedbeds in lakes or slow-moving streams, where the pH may increase from slightly alkaline (e.g., pH = 7.5) to as much as more than 8.5 or even more than 9.0 in the afternoon of sunny days. Conversely, respiration, producing free carbon dioxide in the water, will shift the equilibrium bicarbonate buffer system to the right, with the end result being the release of hydrogen ions into the water and a lower pH (as is commonly observed at night in the water column or near the bottom of lakes in vertical profile).

In a sun-lit lake with high levels of photosynthesis in the summer, it is common to find the pH varying by as much as an entire unit from the surface to near the bottom (e.g., Matthews et al., 1985). For example, in the downlake portion of Lake Texoma in summer 1982, we often found the pH to vary from about 8.5 at the surface to 7.3 or lower at the bottom, with an average difference from surface to bottom of about 1.2 pH units in 10 weekly samples at one station [data from Matthews et al. (1985)]. Thus, both spatially and temporally, fish can be exposed to a substantial range of pH conditions in natural habitats, particularly if buffering capacity is limited. In softwater streams, with poor buffering and often with high concentrations of humic acids, or in lakes in granitic formations, the pH is often as low as 4.5 or 5.0, which can cause stress for many fishes (Haines, 1981).

Effects of Environmental pH on Fish

One important role of pH with respect to fish physiology is (as in all vertebrates) that a relatively homeostatic internal pH is requisite for many physiological systems (Schmidt-Nielson, 1975). Across a wide range of environmentally encountered pH values, fish can maintain a relatively stable internal pH, even compensating over time for a lowered environmental pH (Eddy, 1976). However, below an external pH of about 5, fish lose their ability to adjust internal pH

(Packer and Dunson, 1970), which decreases. A loss of body sodium follows (Packer and Dunson, 1970), but the more serious consequence for lowered internal pH may be in the exacerbated Bohr shift (decreased loading of oxygen to the blood) that is known in fish as the "Root effect" (Root, 1931; Börjeson and Höglund, 1976), which can result at a low pH in an inability of fish blood to be saturated with oxygen [regardless of the level of oxygen available in the external water (Schmidt-Nielsen, 1975)].

A low pH can have a variety of subtle effects other than directly killing individuals. Values below about a pH of 6 can result in a marked decrease in some fishes in oogenesis (Ruby et al., 1977), egg fertility or growth of fry (Craig and Baksi, 1977), or egg hatchability and growth (Menendez, 1976). Fromm (1980) reviewed a wide array of effects of acid stress on freshwater fishes, including the following as factors that eliminate fish from acidic waters: (1) reproductive failure from failed calcium metabolism and lack of protein deposition in oocytes, (2) harm to gill mucus and gill membranes, (3) loss of salts, and (4) lowered capacity of hemoglobin to transport oxygen. Fromm (1980) noted that in spite of the potential stress of water with low pH, most fish failed to discriminate the pH within a range of about 5.5–10.5 in laboratory tests. In addition to physiological responses (Fromm, 1980), fish may exhibit altered behaviors at a low pH, including decreased activity and decreased feeding (Jones et al., 1985). Finally, even among closely related members of a genus or within different strains of a single species, fish may differ markedly in their ability to tolerate a low pH (Robinson et al., 1976; Gonzalez and Dunson, 1989), and general debilitation from other sources (e.g., starvation) may make fish much more vulnerable to episodic lowered pH (Kwain et al., 1984).

Physical and Chemical Factors in Streams and Effects on Fish

In contrast to lakes, whose physical and chemical properties have been studied in detail for more than a century, streams received fewer comprehensive studies until later in this century. Some of the landmark studies of physical and chemical factors in streams include Stehr and Branson (1938) and Neel (1951). Probably the most comprehensive single study of physics, chemistry, and biota of a stream in the early literature is the classic work by Minckley (1963) on Doe Run, Kentucky. In his monograph, Minckley documented in detail the variation in chemical values, turbidity, light, and other physical variables, as well as longitudinal distribution and variation in algae, aquatic invertebrates, and fishes. Essentially all work in stream ecology, worldwide, was reviewed in Hynes' (1970) *Ecology of Running Waters,* which remains the benchmark in descriptive studies for lotic systems. Recently, Allan (1995) has synthesized stream research since Hynes, with emphasis on functional phenomena. Minckley (1963), Hynes (1970), and Allan (1995) all provide integration of fishes into stream ecosystems and/or provide details on ways fish function in streams. In this section, I relate physical and chemical properties of water (above) as they uniquely relate to streams.

A century of studies have documented vertical stratification in lakes, and many patterns are well known (Wetzel, 1983). It is less well known, or less commonly reported, that pools in small streams also stratify and that, particularly in ephemeral or isolated pools or pools with modest flows, there can be marked vertical profiles of temperature. Neel (1951) found differences more than 3°C between surface and bottom in pools less than 1 m deep in a small upland stream. In pools as shallow as 1.2–1.5 m in Brier Creek, Oklahoma, I have measured as much as a 5°C difference in water temperature from surface to bottom. Particularly during hot weather, these differences can be very significant to survival of fish, as surface temperatures exposed to sunlight can exceed 35°C.

Although deep streams or stream pools can exhibit vertical thermal stratification, but thermal mixing in most streams is rapid, and because of exposure to air in riffles and turbulent areas (below waterfalls, etc.), the temperature of a small stream can rapidly mimic that of air, even when the change in air temperature is rapid. On one January afternoon, I measured an increase from 10°C to 20°C in Brier. The day was unseasonably warm, with full sun, after air temperatures had been cold. Through one summer, continual temperature measurements in upper Brier Creek reflected an average daily thermal change of about 10°C, with very little variation in the magnitude of daily warming from day to day.

The rapid temperature changes that can occur in small streams must be important in the ecology of fishes, although I know of no study that has evaluated the changes in behaviors or interspecific interactions that would occur during such a drastic alteration of temperature. However, in numerous experiments in outdoor artificial streams at the University of Oklahoma Biological Station, we have observed that some species of minnows only become active as temperatures warm during the day, and Jacob Schaefer (unpublished data) has found that in these streams, fish migrate between pools much more if temperatures are warmed artificially within a given day.

Given the rapid temperature changes that can occur in streams like Brier Creek, the influence of thermal changes on internal physiology of fishes could strongly affect their well-being. The physiological Q_{10}, reflecting the change in reaction rates with a change of 10°C, is approximately 2.5–3.0 for many physiological processes (Schmidt-Nielson, 1975), much as it is for chemical reactions in nonliving systems. Hence, a fish experiencing an increase of 10°C in Brier Creek or elsewhere should experience at least a doubling of its overall metabolic rate and that of many internal enzyme reactions. To be able to accommodate such rapid changes in internal physiology suggests that some fishes are plastic in their abilities and tolerances of rapid changes in environmental conditions [e.g., Red River pupfish, *Cyprinodon rubrofluviatilis* (Renfro and Hill, 1971)]. Recent awareness of "heat-shock" proteins as continuously present in fish (see Chapter 7) may relate to the propensity for rapid temperature change in their environment.

In lakes, the unequal penetration of various wavelengths of light causes vertical zonation of phytosynthesis, which results in vertical zonation in the distribution of phytoplankton and zooplankton, which, in turn, influences the depths of feeding zones occupied by planktivorous fishes and their predators. In contrast, sunlight penetrates to the bottom of clear, shallow pools, and virtually all underwater substrates in small, clear streams are potential sites for feeding by algivorous or omnivorous fishes, or for invertivorous fishes (that may feed on macroinvertebrates that use the attached algae as shelter, feeding sites, etc.). However, in deeper rivers, light can be attenuated within the water column, reducing or eliminating phytosynthesis at the bottom. For example, in the great depths (80+ m) of the main-channel Orinoco or Amazon rivers, which must lack light, the most common fishes caught in deep-water trawling are the weakly electric gymnotiform fishes (E. Marsh-Matthews and J. Lundberg, personal communication) that find food and mates by olfactory or electric organ sensing (Kramer, 1990).

Oxygen variations, both in a diel cycle and spatially at a given time, are also common in streams. Whitney (1942) cited earlier authors as finding diel fluctuations of 3 ppm in rivers. Thompson (1925) provided a valuable summary of oxygen concentrations under ice in the Illinois River in winter, which were sufficiently low to stress fish. Minckley (1963) showed the patterns in annual variation in dissolved oxygen in Doe Run, Kentucky, and demonstrated diel variation of more than 2 ppm, as well as changes in oxygen in the stream on cloudy versus clear days. Davis (1975) documents the variations in dissolved oxygen concentration that can be encountered in streams (as well as lakes). Hynes (1970) reviewed previous studies of oxygen in streams and noted that (1) typically, fluctuations may not be as great as in lakes, (2) that the fluctuations are caused by photosynthesis and respiration of biota—as well as by changes in temperature and gas saturation—and (3) that in some situations as much as 10 ppm fluctuation in dissolved oxygen concentrations have been noted in stream waters in a day.

Simonsen and Harremoës (1978) showed diel fluctuations as great as 20 ppm in the River Havelse, Denmark, a reach impacted by human and farming wastewater. However, both Hynes (1970) and Allan (1995) indicated that in "small, turbulent" streams (where reaeration is greatly enhanced by rapids, small waterfalls, etc.) oxygen is usually near saturation. Simonsen and Harremoës (1978) and Allan (1995) noted that in large rivers, there is much less reaeration of water than in small streams. Simonsen and Harremoës (1978) concluded that plant production and total respiration are major factors driving diurnal oxygen values in some streams. Simonsen and Harremoës (1978) derived a mathematical model for oxygen fluctuations in rivers, driven by photosynthesis, total community respiration, and reaeration. Their model assumes that respiration is a constant across a 24-h period, but that photosynthesis is symmetrical about and peaks at noon. Their model predicted that oxygen maxima in rivers will vary from slightly after noon if reaeration is rapid (e.g., rapids, entrainment of air), to late in the afternoon or near sunset (depending on length of day) for rivers with minimal reaeration. Rivers with high reaeration coefficients also were predicted to have much less diel fluctuation in oxygen concentration (Simonsen and Harremoës, 1978).

Measurements in the wide, shallow South Canadian River near Norman, Oklahoma (Matthews, 1977) provide an example of variations in oxygen concentration that could be encountered by fish in one reach of stream within a day and at different seasons of the year. Dissolved oxygen was measured near the surface (by Winkler method, electronic meter, or both) at fixed sites in the river or in the lower part of a small tributary in the morning from about 8 to 10 A.M., and in the afternoon from about 1 to 3 P.M. The habitat was diverse, with a mainstem shallow channel, deeper channel-edge pools, and sluggish or nonflowing habitat in the creek (Matthews and Hill, 1979a). On 14 February 1976, for 13 sites the mean increase from morning to afternoon was only 1.4 ppm, and within morning and afternoon samples, oxygen concentrations ranged from 9.2 to 11.4 and 10.7 to 13.6, respectively (i.e., very modest temporal or spatial variation in oxygen available to fishes). On 10 August 1976, the spatial and temporal variation in oxygen concentration at the same 13 sites was much greater, with a mean morning–afternoon increase of 3.6 ppm, and a range in oxygen values in the morning hours from 3.3 to 10.8 ppm and from 8.6 to 12.9 ppm. Some individual sites (e.g., in backwaters with algae) increased more than 6 ppm from morning to afternoon. On 17 October 1976, there was substantial encrusting blue-green algae growing on the shallow river bottom. The mean increase in oxygen at that time, for the same 13 sites, was only 1.8, but there was dramatic difference between individual points in the stream, with some increasing as much as 6 ppm to a supersaturation of 19.0 ppm and others decreasing as much as 3.3 ppm from morning to afternoon, in backwaters where respiration appeared substantial.

Figure 4.2 shows, for 2 days in October 1976, the oxygen measurements made in a complete array of 24 sampling points per day, each sampled in the morning and in the afternoon. Figure 4.2 indicates that (1) a general increase in oxygen concentrations in the water column of South Canadian River and Pond Creek (in response to photosynthesis) but that (2) there is a very marked local variation among points, all of which were within a 1-km reach of this shallow river–creek complex and (3) that dissolved oxygen increased as much as 6 ppm from morning to afternoon at some points. Thus, fish in this habitat might hypothetically be exposed to a wide range of oxygen conditions both temporally and spatially within a single day.

Long ago, Coker (1925) showed that streams with a low pH had a very low number of species in assemblages that included mostly trout and a few minnows and sculpins. In more than 20 years of sampling fish in the Ozark and Ouachita mountains of Arkansas, I typically find greater apparent densities of fish in the Ozark streams, which are calcareous with a stable, high pH (usually 8.3–8.5), than in the Ouachita streams, which are in noncalcareous shales, sandstones, and other noncarbonate rock and are less alkaline than streams in the Ozarks. Still, these streams are all well above the truly low pH conditions that are encountered in "stained water" or as a consequence of acidification in many poorly buffered water systems of the world.

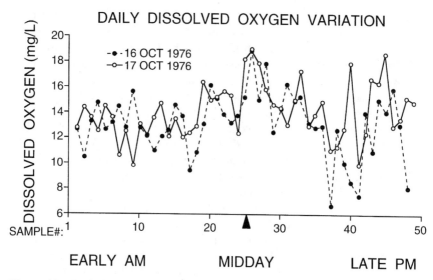

Figure 4.2. Variation in dissolved oxygen (at various points within 1 km of stream during 2 consecutive days in the South Canadian River and a tributary creek, in central Oklahoma (USA). [Graphed from data in Matthews (1977).]

General Physicochemical Effects on Habitat Selection by Fish

There is a huge literature based on field and laboratory studies to suggest that gradients of temperature, oxygen, or pH can be strong cues to fish in habitat selection or avoidance. Two examples of combined studies of field and laboratory habitat selection by fishes in response to physicochemical variables are Neill and Magnuson (1974), for a northern lake fish assemblage, and Matthews and Hill (1979a) for habitat selection by one species (*Cyprinella lutrensis*) across gradients of physicochemical variables in a southwestern United States river.

One issue that is not clearly resolved about habitat selection by fishes in physicochemical gradients is whether they have the capacity to detect and follow gradients of temperature, oxygen, or pH to some ideal or "optimal," or if they fit a kinetic physiological model consisting of increased nondirected movements if conditions are "uncomfortable," with decreased movement in microhabitats where they become "comfortable."

Höglund and Harding (1969) suggested differences between "avoidance homeostasis" and "orientational homeostasis" in fishes in gradients of oxygen, pH, or other physical factors. Jones et al. (1985b) showed that fish can differentially respond to gradients of hydrogen ion and carbon dioxide (although the two are usually confounded in natural waters), primarily relying on carbon dioxide for discrimination in hydrogen ion–carbon dioxide gradients but able to detect and use hydrogen-ion gradients alone to avoid lethal pH conditions. Whereas many tests have focused on the avoidance of low pH by fish, Serafy and Harrell (1993) showed that fish avoided pH levels above 9.5. However, when such high environmental pH values were associated with very high oxygen concentrations (as can occur in dense weedbeds with high rates of photosynthesis, fish in their experiments did not avoid the combined high pH (9.5–10.0) and supersaturated oxygen conditions.

Davis (1975, p. 2307) noted that, "it is not clear whether avoidance behavior in response to low oxygen constitutes a highly directed form of behavior. It may result simply from increased locomotor activity with more random movement, which is satisfied by discovery of improved oxygen conditions." Davis (1975) indicated that it was not known if fish have oxygen receptor systems (e.g., associated with the gills), but Daxboeck and Holeton (1978) suggested the location of oxygen-sensitive chemoreceptors on the first gill arch of rainbow trout (*Oncorhynchus mykiss*). Jones (1952) showed that at 13°C or lower, sticklebacks (*Gasterosteus aculeatus*) did not sense entry to low-oxygen environments, but after a few minutes in hypoxic conditions, they increased random movement until locating improved conditions. However, Jones found that at 25°C, sticklebacks rapidly responded to and avoided water with less than 2 ppm of oxygen. Whitmore et al. (1960) suggested that fish sensed and actively avoided oxygen levels that were low but well above concentrations that were directly harmful. In contrast, Hill (1968) found that the spring cavefish (*Chologaster agassizi*) was relatively nonselective in any gradient of oxygen conditions above about 6 ppm, suggesting that they only avoided detrimental conditions. Logperch (*Percina caprodes*) actively avoided dissolved oxygen less than 2 ppm, whereas a bimodal breather (African reedfish, *Erpetoichthys clalbaricus*) did not avoid dissolved oxygen concentrations even as low as 0.5 ppm (Beitinger and Pettit, 1984).

Temperature selection has probably been studied more than any other stimulus to fish movement or habitat selection [reviewed in Hutchison (1976) and Hutchison and Maness (1979)]. A large literature shows that fish have both diel and seasonal cycles of thermal "preference" and that acclimation or recent prior exposure to high temperatures influence temperatures occupied by fish in thermal gradients. Young fish tend to select higher temperatures than older individuals of the same species (McCauley and Huggins, 1979). A variety of approaches to the study of temperature selection in fishes includes (1) acute selection during periods of time too brief for fish to begin acclimating to new temperatures, in an acute gradient (Matthews and Hill, 1979a; Matthews, 1987a), (2) "behavioral" thermoregulation, in which movements of the fish control heating or cooling in a "shuttlebox" (Neill et al., 1972; Beitinger et al., 1975), and (3) long-term selection of temperatures by fish in a thermal gradient, in which the temperature selected and the acclimation temperature of the fish merge into one "final preferendum" (Fry, 1947; Crawshaw, 1975; Beitinger and Fitzpatrick, 1979; Reynolds and Casterlin, 1979a). McCauley and Casselman (1981) found that the final thermal preferrendum corresponded to the temperature at which growth was optimum for numerous species of fish. Ingersoll and Claussen (1984) found differences in thermal preferences of two darter species that matched their habitat (riffle versus pool) temperatures in the wild.

Most thermal selection studies have been in laboratory gradients, but wild fish in field environments have shown thermoregulatory behaviors (Brett, 1971; Spigarelli et al., 1982; Zimmerman et al., 1989). Ferguson (1958) showed a relationship between preferred temperature and the distribution of fish in streams and lakes. Spigarelli et al. (1983) found that brown trout (*Salmo trutta*) tracked in the field selected a modal annual temperature corresponding to optimal metabolic temperature for the species. Brown (1971) showed that the desert pupfish (*Cyprinodon*) very precisely thermoregulates in very hot microhabitats, remaining very near their critical upper limits in order to feed most effectively on algae, but virtually never entering hotter, potentially lethal, parts of the habitat. Bulkley and Pimentel (1983) used thermal preference and avoidance tests for razorback suckers (*Xyrauchen texanus*) to make recommendations for thermal controls related to dam operations in the upper Colorado River basin.

As with oxygen, pH, or other stimuli, there remains uncertainty about how fish detect and respond to thermal gradients [i.e., if they are actually selecting optimal temperatures or merely avoiding harmful conditions (Otto and Rice, 1977; Cherry et al., 1977; Barila et al., 1982; Matthews, 1987a)]. Neill (1979) contrasted predictive (directed) movements of fish versus reactive (undirected) movements and concluded that increased movement with increased turning behavior could accomplish thermoregulation in fish. Reynolds and Casterlin (1979b) found a decrease in activity at temperatures approximating a final preferendum; that is, with decreased activity, fish would tend to remain at the appropriate environmental suite of conditions. Crawshaw and Hammel (1974) showed the brainstem sensitive to temperature and potentially playing a role in behavioral regulation of internal temperature by habitat selection in fish. Nelson and Prosser (1981) located thermally sensitive preoptic neurons in green sunfish (*Lepomis cyanellus*). Fish also have the capability to physiologically influence their internal heating and cooling rates by shunting blood to or away from points of heat exchange (Reynolds, 1977; Hutchison and Maness, 1979).

It is clear from a large number of studies that fish are responsive, often very precisely, to differences in temperature in laboratory gradients and that that thermoregulatory or thermoselection behavior is important in their ecology in the wild. Whether fish are, in the wild, actively avoiding harmful temperatures or selecting optimal or "best" temperatures for various physiological or behavioral functions cannot be categorically determined. However, although most studies have been of one or a few related species, thermal selection in natural streams and lakes could play a substantial role, at least where temperatures differ sharply within the available habitat, in determining which species actually coexist in local assemblages (Stauffer et al., 1976). In the context of natural environments, it remains to be seen how much of the coexistence of fish species in local assemblages is influenced by individualistic or species-specific thermal responses, compared to biotic interactions or other abiotic cues. Hlohowskyj and Wissing (1987) shed some light on this question, as they found differences among three coexisting darter species (*Etheostoma*) in the degree their habitat selection were potentially influenced by temperature gradients.

4.3 Lentic Versus Lotic Environments as Fish Habitats

Characteristics of Lentic Versus Lotic Environments

Lakes and ponds ("standing" waters; lentic habitats) and a variety of other named "nonflowing" aquatic habitats (bogs, playas, sinkholes, etc.) typically have both vertical and horizontal habitat patterns that are important to the ecology of fishes. Streams ("running" waters; lotic habitats), broadly defined as any watercourse that contains water moving downhill at any time of year, are of many shapes and sizes, and have in common that all expose fish to flow at least during parts of the year and have a preponderance of horizontal (including longitudinal) habitat patterning (in spite of known vertical stratification in some pool environments). The horizontal nature of streams is at several scales. At the continental scale, many stream systems have montaine, piedmont, and coastal plain regions, as is typical of much of the southeastern United States (Fig. 4.3A). At an intermediate scale (Fig. 4.3B), stream ecologists, particularly in Europe, refer to the upper portion of a basin as the rhithron, and the lower portion as the potamon, with the annual 20°C mean isocline in water temperature separating the two zones (roughly into "trout" water and "warm" water). At the local scale, many streams have well-defined microhabitats within a reach, resulting in turbulent run or riffle areas with coarse substrates, and pools or glides with nonturbulent flow over a variety of substrates from very fine silt to large boulders (Fig. 4.3B). At the microhabitat scale, flowing water is characterized by interspersion of erosional zones (where materials are entrained into the water column and removed) and depositional zones (where transported materials fall out of the water column) (Fig. 4.3B). In streams, these zones often correspond to riffle and pool microhabitats, respectively. There often is distinct zonation of physicochemical variables between riffle and pool environments. Classic articles quantifying riffle–pool differences or physicochemical differences over short distances, horizontally or vertically, in a small stream are Stehr and Branson (1938) and Neel (1951). However, the distinction between riffles and pools is not complete: Within "riffles," there are microdepositional zones (behind rocks), and within "pools," there can be a complex array of erosional and depositional areas that depend on the vagaries of flow through the pool, formation of eddies, and so forth. All of these horizontal zonations within streams at both scales may, but does not always, result in differing fish assemblages, as described in detail in Chapter 6.

In most streams (except in springs or spring runs), fishes are exposed to a

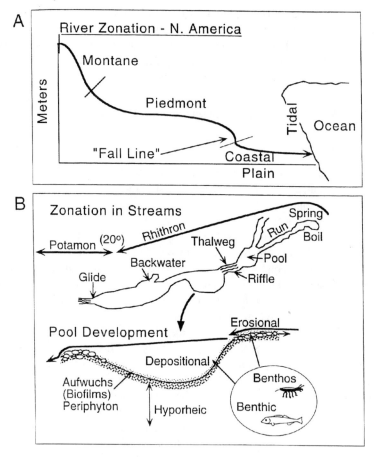

Figure 4.3. Stream zonation in major rivers of eastern North America (A) and local reaches within streams in a typical small stream (B).

relatively broad range of physicochemical conditions that result from the interaction of flow, terrestrial inputs, and exports. Streams typically are more well mixed than lakes, and as a result, there often are fewer local gradients of chemical (e.g., oxygen, conductivity) or physical variables (e.g., turbidity, temperature) than exist in lentic waters of lakes or reservoirs. In spite of the fact that the boundary can be fuzzy between lentic and lotic habitats for fishes, some generalizations are possible. Table 4.1 suggests some physical and biotic phenomena, all of which influence fishes present in the system, that can differ between "typical" lentic and lotic habitats.

However, the distinction between lentic and lotic habitats is not always as clear-cut as is sometimes suggested (Table 4.1) because large deep rivers may have some biotic components that are like those in shallow or turbid lakes (i.e., standing crops of plankton, vertical stratification of physicochemical conditions,

Table 4.1. Characteristics of Typical Lentic and Lotic Habitats

Lentic	Lotic
Water relatively stationary, or flows variable	Unidirectional flows
Closed system—materials recycle	"Open" system—materials pass through one way
Energy mostly autochthonous	Allochthonous or autochthonous
Much phytoplankton and zooplankton	Little plankton, much benthic invertebrates and attached algae
Little "drift," except at surface due to wind	"Drift" an important component
Vertical zonation dominates (littoral, limnetic, profundal benthic)	Longitudinal zonation (erosional–depositional; riffle–pool)
Vertical stratification of temperature and oxygen	Usually well mixed
Less water-level fluctuation	More water-level fluctuation
Deep	Shallow
Oxygen depletion often a problem	Low oxygen less common
Smaller temperature range	Larger temperature range
Organisms hold place vertically	Organisms hold place horizontally
Less long-distance fish migration	Long-distance migration of many fishes
Fish nesting sites limited to littoral zone	Fish nesting sites limited more by velocity and substrates

slow water movements). Thus, the physical environment encountered by a fish in a large, slow-moving river may be more like the environment it would encounter in a lowland lake (or a reservoir) than in a small, rapidly changing headwater stream. Additionally, in harsh environments (e.g., wide, shallow, minimally shaded rivers of the American Midwest (Matthews, 1988; Brown and Matthews, 1995)], solar and chemical inputs may result in sharp local gradients (even within a few meters, horizontally) in temperature, dissolved oxygen, conductivity, pH, or other physicochemical variables (Matthews and Hill, 1979a; 1979b).

Just as stream environments may have some lentic characteristics, lake environments may have properties similar to those typical of streams. Many lakes (particularly artificial reservoirs) have inflows and outflows, or consistent gyres or internal currents, resulting in relatively predictable currents at velocities that could transport or give behavioral cues to fish, zooplankton, or other prey for fishes. For example, larval fishes may be particularly vulnerable to transport on currents or washout of particular regions within lakes (Matthews, 1984). Finally, artificial impoundments, particularly on reservoir mainstems, have many properties that place them as a hybrid of river and "lake" conditions (Thornton et al., 1990).

4.4 Characteristics of Lakes that Influence Fish Assemblages

Lakes

Initially, the term *limnology* was restricted to studies of standing bodies of water and was often referred to as "the study of physics and biology of lakes," or "inland oceanography." Only more recently has "limnology" included at least the physical and chemical aspects of streams. Much of the original knowledge of structural and functional properties of lakes came from Europe or from small northern lakes in North America, but there is now a huge body of empirical information on lake systems worldwide, including the tropics, and the subset of limnology that focuses on artificial reservoirs has achieved a rather separate status as a discipline or at least a subdiscipline of limnology (e.g., Thornton et al., 1990) in recent years. I include reservoirs under general consideration of lakes, unless differences are noted. Finally, there has been a major effort in North America in this century to manage ponds and small lakes for fish (Swingle, 1950; Bennett, 1970). Many ponds are now artificial entities, but small lakes or ponds are common features of post-Pleistocene environments in formerly glaciated northern regions. There is no rigorous dividing line between a "lake" and "pond," but a working delineation of the two may be to consider a lake as a lentic water body sufficiently large that wind can have substantial influence on its ecology, whereas a pond is sufficiently small that fetch (the distance wind can "push" water, creating wave action) does not allow major surface waves, and Langmuir circulation (spiraling, horizontal motion of water across a lake) may not be sufficiently well developed to result in any significant lateral transfer of materials (including water) downwind. As a result, ponds would lack seiches (rocking motion of water in the basin), except in unusual conditions. From the perspective of importance to fish, I first review formation of lakes, then describe segmentation of fish habitat because of limnological features in extant time, and finally note some of the kinds of disturbance events that affect lakes or reservoirs.

Lake Formation, Life Span, and Demise

Lakes form naturally in various ways, and their life span can strongly influence development or composition of local fish assemblages. In fact, permanency of lakes, once formed, and their sources of water (inflowing rivers or not) are at the heart of some of the most important concepts about the evolution of fishes. Four examples of lacustrine faunas are illustrative: the pluvial lakes of the North American Great Basin, the former lakes of the northeastern United States, Lake Chichancanab, Mexico, and Lake Victoria, east Africa.

During the Pleistocene, the Great Basin of the interior of the American west contained at least 21 pluvial lakes ranging in size from a few to more than a thousand square kilometers (Hubbs et al., 1974). At present, all that remains of these vast lakes is a number of small, relict aquatic habitats, mostly as springs or marshes. The detailed account of the scanty present-day waters in these basins by Hubbs et al. (1974) underscores the extreme reduction of available water, leaving fishes only in widely isolated habitats. To quote Hubbs et al. (1974), "Here, bare remnants of fish fauna have survived in the extreme isolation that has resulted from the almost complete desiccation of an area that not more than a few thousand years ago was one-fifth covered by lakes fed by streams of ample flow." These fishes typify the range of variability in morphology and ecology of fishes that can occur when vicariance events separate formerly continguous populations. Vicariance is often followed by local differentiation in morphology, physiology, or life-history traits. Allopatric speciation can result, producing sister species.

In the northeastern United States, the Newark Supergroup of lakes existed during discrete windows in time (perhaps 21,000 years or so each), after which the individual lakes dried completely (McCune et al., 1984). The typical limnological/ geological cycle for these lakes appears to have involved lake histories characterized by a period of expansion of the lake over former swamps and dry land, a period of maximum lake size, depth, and stratification, and a period of contraction and complete evaporation (McCune et al., 1984). A well-preserved fossil record indicates that species "flocks" (*sensu* Echelle and Kornfield, 1984) of semionotid fishes developed repeatedly in individual lakes in this region (McCune et al., 1984). The repeated appearance of different semionotid species flocks, with a very wide diversity of body form, in these lakes (Fig. 4.4) was probably due to the ephemeral (in geological time) nature of the lakes.

Lake Chichancanab, Mexico, has consisted of variously filled, desiccated, and interconnected smaller lake basins. Humphries (1984) described the extreme modifications of head structure among five very closely related (as determined genetically) species of pupfishes (*Cyprinodon*) that comprise a species flock in Lake Chichancanab. Although Humphries argued against a model of intralacustrine "microallopatric" speciation based on water-level fluctuations and ephemeral isolations of water bodies within the Chichancanab basin (in favor of a model of sympatric speciation resulting from competition), it is true that this lake has undergone periods of fragmentation due to water-level fluctuation. Whether or not this fragmentation led to speciation in a direct (vicariance) manner, it seems plausible that if competition did cause the speciation events, it would have been exacerbated among populations isolated or crowded (over brief geological time) into shrinking, fragmented remnant basins within the present-day lake.

Finally, one of the most baffling of all lakes for evolutionary biologists and ichthyologists is Lake Victoria, in east Africa—with its tremendous "flock" of more than 200 endemic cichlid species (before their decimation by the introduced piscivorous Nile perch, *Lates nolitica*), interpreted as "explosive speciation" by P. H. Greenwood (1984). Many of the historical accounts and arguments about the fishes of this lake and speculation about their speciation were based on interpretations that the lake was "young," having been formed about 750,000 millions of years ago (mya) (Greenwood, 1984). As of 1984, Greenwood was uncertain whether or not the Victoria cichlids were monophyletic but was clear that regardless of number of lineages, there was extensive speciation either from one or more ancestral types. Stager et al. (1986) reported maximal drying at 15,000 to 13,000 years ago, but "no evidence that the lake level fell low enough

154 / *Patterns in Freshwater Fish Ecology*

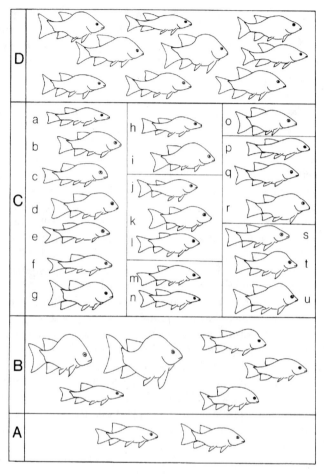

Figure 4.4. Semionotid fishes from lakes in the Newark Basin, New Jersey. [From McCune et al. (1984).]

to confine fish to refugia in small isolated ponds or around river mouths." However, all arguments about gradual versus rapid or "explosive" speciation in these fishes now must undergo close scrutiny in the light of the recent finding that Lake Victoria apparently was completely dry during a period from about 17,000 to 15,000 years ago (Johnson et al., 1996b). Evidence from paleolimnological and palynological coring evidence shows a horizon of terrestrial vegetation that is 15,000 years old, in what is presumed to have been the deepest part of the lake basin. This argues against isolated peripheral basins or persistent small rivers that could have harbored fishes. Johnson et al. (1996b) concluded that after this period of complete desiccation, one or more species reinvaded Lake Victoria and underwent speciation to literally produce hundreds of species in 15,000 years.

However, this argument depends on the assumption that the core by Johnson et al. (1996b) was actually taken in what would have been the deepest part of Lake Victoria 15,000 years ago (M. Gophen, personal communication).

Therefore, from four widely separated examples of lakes with complex assemblages of fishes now or in the past, it is possible to conclude that the formation, life span, and demise of lake basins can be critical in the evolution of fishes. Often, the core of the argument in favor of sympatric versus microallopatric species is an understanding of the geological or paleolimnological evidence for complete desiccation of a given basin at a particular time in the past. Clearly, lakes that undergo periods of fragmentation into isolated peripheral waters (like the lakes of the Great Basin of North America) offer ample opportunities for allopatric speciation in isolation by relict populations of a formerly widespread fauna. In this light, how do lakes form?

Lake Formation and Sources of Fishes

Given that lake formation and persistence is critical to fish assemblage structure, how do lakes form? Although Hutchinson (1957b) and Wetzel (1983) review a wide range of possibilities, most can be categorized as (1) tectonic, (2) volcanic, (3) glacial, (4) landslide, or (5) fluviatile lakes. Other minor processes not assignable to the above include coastal lagoons cut off by oceanic sand deposit, solution lakes, and even speculation about formation of lakes (e.g., Carolina Bays) by impact cratering of extraterrestrial objects. How does each category of lake formation relate to the likelihood of its subsequent invasion or evolution of species assemblages?

The key to answering this question is whether or not the lake was formed in a place or manner that allowed rapid colonization of an existing fish fauna or, alternatively, if it had to await slower colonization from elsewhere (e.g., Magnuson, 1976; Browne, 1981; Mandrak, 1995). The other obvious questions are whether or not, after formation, the lake had connections to other bodies of water or to streams, and whether or not paleontological evidence indicates strong anoxia of a lake [with likely winterkill (e.g., Rahel, 1984)]. Some lakes seem likely to have been formed in a way that traps or allows rapid colonization by existing fishes (Table 4.2). Examples include oxbow lakes, whose instant of formation (often one major flood) would readily trap a preexisting fauna, possibly quite rich, from a main river, and landslide lakes, in which the collapse of a major landform (side of a mountain, etc.) might dam an existing stream with its resident fishes. Another mode of lake formation is the tectonic uplifting of large portions of the seafloor, elevating what formerly was a depression in the ocean floor to above sea level and trapping water as a lake. Lake Okeechobee, Florida, appears to be an example. Colonization of freshwater fish would be slow, with the original marine fauna largely being eliminated over geological time while rainfall gradually results in "freshening" of the lake water.

156 / Patterns in Freshwater Fish Ecology

Table 4.2. General Processes Forming Lakes [Compiled from Wetzel (1983)], and Potential Sources of Fishes for Colonization

Process	Sources of Fish	Time Required
I. Tectonic Processes		
Fault block (grabens) uplifting or depression	Existing streams or small lakes	Rapid
Uplifting of seafloor depressions (epeirogenesis)	Marine fauna	Slow
II. Volcanic Lakes		
Crater lakes	None	Slow
Caldera collapse within volcanic cone	None	Slow
Lava flows—dam valleys	Dammed stream	Immediate
III. Glacial Lakes		
Terminal moraine—dam valley	River below glacier	Moderate (initially too cold?)
Lateral moraine—dam inflowing streams	Inflowing streams	Moderate (too cold?)
Ice scour (e.g., cirque lakes)	Isolated, high elevation	Slow
Ice block melting—kettle lakes	Isolated to semi-isolated?	Slow
IV. Landslide Lakes		
Landslides block canyon	Existing stream	Immediate
V. Oxbow Cutoff (flood)	Existing stream	Immediate
VI. Solution Lakes		
Gradual solution of material	None	Slow
Sinkhole collapse (doline)	None	Slow
VII. Shoreline Lakes		
Sand bar dynamics—trap water		
a. Along large inland lakes	Existing lake	Immediate
b. Alone marine coasts	Marine/estuary	Slow (for freshwater taxa)
VIII. Manmade Lakes		
Reservoirs	River and tributaries	Immediate
Ponds	Tributary streams, or none	Immediate to slow

Other forms of lake formation would seem to result in an initially fishless lake, or at least one with a depauperate fish fauna. Glacial processes, such as glacial scour lakes, take place in very cold climates, where apparently few species could or would exist in close proximity to a glacier. Some glacial lake formation [e.g., kettle lakes formed from slow (centuries) melting of large fragmented ice blocks, buried in till as glaciers retreated] obviously would have no preexisting supply of fishes and would have to await colonization. For such lakes, the pattern of connectivity by small streams would be critical. Volcanic lakes often form by filling of extinct or inactive cones by rainfall—with no apparent opportunity for natural colonization of fishes from outside—due to the cylindrical downslope of the normal crater of a volcano. Each lake is unique, and sometimes extant lakes are the product of more than one tectonic or geologic event; thus, complex patterns of initial colonization by fishes should be expected. Nevertheless, it would seem that the first questions to ask of a lake fish assemblage would be "How was the lake formed?" and, secondarily, "How long ago?" This is not to imply that the fish species initially trapped during lake formation or those that first colonize will remain the dominant taxa in the assemblage. After lake formation (e.g., reservoir construction), some preexisting species will benefit and many others will be lost, as spawning areas or other natural habitats are flooded (Hubbs and Pigg, 1976). However, knowing the potential for the presence of fishes over the geological time of formation of natural lakes may make it easier to predict the richness of the natural fauna. For example, lakes formed in association with Pliestocene ice sheets (Chapter 5) may have had little potential for colonization by fishes at the time of their formation, and richness of their faunas now depends strongly on postglacial colonization events. There is evidence that lakes in southern Canada have been in the process of recolonization since the last glacial retreat and that the location of a lake with respect to glacial refugia has an important influence on their richness (Mandrak, 1995).

Eutrophication: Trophic Status of Lakes

Many lakes change gradually from oligotrophic to eutrophic conditions. However, there is a huge variance in the time scale of this process, with some lakes having remained oligotrophic for millions of years; others exhibiting gradual, natural increases in eutrophication as nutrients accumulate from the basin; yet others suffering cultural eutrophication from anthropogenic activities in a few decades. Additionally, humans have great potential to influence or reverse the process of cultural eutrophication, such as by phosphorus control (Stoermer et al., 1996). Cole (1994) reviewed the interplay of original conditions at time of lake formation with the morphometry of the lake and the cultural activities in the basin as revealed by the techniques of paleolimnology. By coring of sediments, careful inspection of layers (varves), and an array of dating techniques (Cole, 1994), it is possible to reconstruct much of the history of a lake. (Paleontological techniques that focus on organisms take advantage of the relatively good preservation of pollen and of diatoms; unfortunately, fish scales or other parts seem less appropriate for their work.)

The overriding effects of trophic status of a lake on its fishes are on (1) the availability to fishes of primary and secondary production as food and (2) oxygen availability in summer and winter (for those lakes with ice cover). Although there are many intermediate stages on the trophic scale for lakes, the extremes of typical oligotrophy and eutrophy offer very different environments for fish assemblages, as summarized in Table 4.3.

Oligotrophic lakes typically have low concentrations of nutrients that control primary production (phosphorus, nitrogen), are relatively clear and well oxygenated, and often occur in high-altitude or cold conditions. In cold high-elevation

Table 4.3. Features of Oligotrophic Versus Eutrophic Lakes

Oligotrophic	Eutrophic
Water poor in nutrients	Water rich in nutrients
Inorganic N < 200 µg/L	Inorganic N-700 µg/L or more
Total phosphorus < 10 µg/L	Total phosphorus = 30 + µg/L
Deep, steep sided	Shallow, broad littoral zone
Small epilimnion, large hypolimnion	Small hypolimnion
Marked transparency	Limited transparency
Blue to greenish water	Green to yellow or yellow-brown
Sediments low in organics	Sediments rich in organic matter
Oxygen abundant at all levels	Oxygen depleted in summer hypolimnion, or throughout water column in winter (for lakes with ice cover)
Few littoral plants	Littoral plants abundant
Limited phytoplankton	Abundant phytoplankton, often with large blooms of bluegreens
Diverse benthic fauna, forms intolerant of low oxygen	Depauperate benthic fauna, mostly forms tolerant of low oxygen
Low biomass of benthos	High biomass of benthos
Cold-water fish—salmonids and whitefish	Few fish in summer hypolimnion; many "tolerant" fish species
Often in mountains, with granite or rock substrates, and thin soils	More common in lowlands or on alluvial strata

Source: Modified from Cole (1994) and Wetzel (1983).

or high-altitude oligotrophic lakes, salmonids are typically the dominant taxa, with sculpins present in some regions; "forage" fish species may be scarce or lacking, and relatively few large piscivores are present. The most important feature from the perspective of fishes is that production is sufficiently low that the demand for oxygen from decomposition processes (mostly hypolimnetic) does not deplete oxygen in any portion of the water column; thus, essentially all depths of the lake are available as habitat. Because oxygen remains available throughout the water column, organic materials that rain from the epilimnion into the hypolimnion are well broken down, and there is no net accumulation of organic matter in the sediments. There can be, for a given lake, a relatively unchanging balance of production and decomposition, without substantial increases in organic matter content in the layers revealed by sediment cores. Hutchison (1957b) referred to this as "trophic balance" that could remain stable for long periods of time. Cole (1994) summarized differences in the oligotrophic–eutrophic status of a lake related not to the lake origin but to the ratio of volume of epilimnion to hypolimnion (E/H ratio). Where E/H is small, there is a relatively large hypolimnion, which may entrap sufficient oxygen during spring circulation to allow oxygenated breakdown of materials all summer, ameliorating any tendency toward anoxia. Hence, deeper lakes with a large hypolimnion also tend to be secondarily oligotrophic, whereas shallow lakes with a small hypolimnion begin the summer stratification period with relatively little total oxygen in the hypolimnion to facilitate the breakdown of organic materials.

Changes in a drainage basin (climate, plant associations, anthropogenic activities, etc.) can rapidly alter the trophic status of a lake. A diagnostic marker in sediment cores signaling the onset of eutrophication is any sharp increase in percent organic matter, suggesting that anoxic conditions have occurred in the water column. Once a lake reaches a point at which oxygen is depleted near the bottom, there can be a feedback whereby (1) decomposition of organic materials raining from the epilimnion is incomplete in the hypolimnion and (2) retained material results in greater demand for oxygen in the next year's stratified period, followed by (3) even less effective decomposition of the materials coming from the epilimnion in the next year. Additionally, oxidation-reduction processes become very different once oxygen is depleted, and nutrients such as phosphorus that may have been effectively locked in sediments can be released back into the water column, to be potentially carried upward into the euphotic zone, exacerbating production in the epilimnion.

Thus, a lake moving from oligotrophy to eutrophy would have a greater primary production and, therefore, a potentially greater production of planktivorous fish species (e.g., shads, *Dorosoma*) that comprise prey for piscivores, but would offer much more limited habitat space for fishes during summer stratification, as the hypolimnion can become mostly or entirely anoxic. There may be a substantial replacement of sensitive species in the fish assemblage with new species [e.g., green sunfish, largemouth bass, shads (*Dorosoma*), bullhead catfish] that are tolerant of less oxygenated waters. Within historic time, there also has been an obvious tendency to stock such fishes into waters wherever productivity is sufficiently high to support forage and sport fishes. Furthermore, the stocking of planktivorous fishes, particularly in small lakes, may have important direct and indirect effects at lower trophic levels (Chapter 11), and nutrients from the fishes (e.g., Drenner et al., 1986) could promote increased algae blooms, hence eutrophication.

There is one trophic condition that can be important to fish assemblages that does not fit conveniently into an oligotrophic-to-eutrophic scale. This is the "dystrophic" condition found in waters of particularly low calcium content. Dystrophic waters have few bacteria; thus, plant material dies, fragments, but does not decompose, and gradually fills a lake to create a bog. Such habitats may have sufficient humic materials to make them relatively acid, and fishes of such bog lakes or their outflowing springs and streams may be particularly resistant to low pH conditions (Matthews, unpublished data for *Phoxinus* species).

Stratification Patterns and Effects on Fishes

There are many patterns of stratification in lakes, with literally dozens of named fundamental or combined patterns. Most often, stratification is the result of thermal differences between layers of water, but salinity differences can also cause or exacerbate stratification patterns. The classic patterns typical of many of the lakes of the world that contain fish are (1) dimictic and (2) warm monomictic. Dimictic lakes stratify twice in a year—in summer and in winter—with periods of complete circulation in autumn and spring. Warm monomictic lakes stratify only once a year—in summer—with wind-driven circulation of the water column throughout autumn, winter, and spring. The fundamental difference between the two types is in that lakes at southern latitudes, lacking ice cover, cannot stratify strongly enough at relatively low temperatures to resist wind-driven movements of water, hence almost continual mixing until summer stratification occurs. A third major category is that of meromictic lakes—which do not mix completely from top to bottom. Such lakes often are very deep or may fail to mix because of deep salt layers that confer permanent density differences between the epilimnion and hypolimnion. As a result, meromictic lakes offer a particularly limiting environment for fishes—if their deeper waters are anoxic, high in salinity, or both.

Typical, thermally mediated stratification has a strong seasonal cycle. Beginning in spring, assume the water column is well mixed, homothermal, and with abundant oxygen from surface to bottom. Wind, sufficient to overturn the lake against any incipient stratification, is the driving force in circulation. For a typical temperate-zone lake, spring circulation might be in progress from a mean water temperature of 6°C or 8°C up to 15–18°C. At these temperatures, density differences per degree difference are less than at warmer temperatures. If, in late spring, there is a period of calm with little wind and substantial heating of the surface, there can be many incipient periods of stratification; but by early summer, the thermal differences can become sufficient that two "lakes" form (the epilimnion and the hypolimnion) with little mixing between the two water masses. As continued heating of surface waters takes place, the strength of stratification increases, until a point is reached at which no windstorm can readily overturn the lake. During this classic period of summer stratification, water can circulate within the epilimnion and there can be some water movement within the hypolimnion, but the two do not mix (and often are separated very abruptly by a narrow thermocline or (in the case of salt-mediated stratification) halocline.

Furthermore, the vertical stratification or zonation of lakes may differ markedly from one part of the lake to another (e.g., uplake versus deeper areas downlake), providing a cooler available habitat regugia in deeper stratified waters and strongly influencing seasonal uplake or downlake migration of some species (Matthews et al., 1989). For reservoirs on rivers, the location of the "plunge point" (where inflowing river water "plunges" under the mass of lake water in a density-dependent fashion) can make a substantial difference in vertical physical or chemical profiles in the water column at two points even a short distance apart (upstream versus downstream of the plunge point) (Ford, 1990).

In the warm surface waters [as high as 30°C in Lake Texoma, Oklahoma–Texas, on a regular basis (Matthews et al., 1985)], primary production can be intense, producing blooms of phytoplankton, peaks in zooplankton densities, and peaks in larval fish production and growth. All of this biological activity produces a steady rain of sinking phytoplankton, castes of zooplankton and aquatic insects, fecal pellets, and all manner of dead or dying material into the hypolimnion. This organic rain is oxidized or broken down by aerobic bacteria until the oxygen demand exhausts the supply. The hypolimnion then becomes anoxic and inhospitable to fishes not only because of the lack of oxygen but because of the buildup of toxic by-products of anaerobic activity. Proteins, carbohydrates, and fats in or near the sediments are acted on by anaerobes to produce organic acids, which may, in turn, be acted upon by methane-producing bacteria to produce methane. This methane can then begin to rise in the water column, where it is oxidized by methane-oxidizing bacteria, thus further depleting any oxygen in the hypolimnion. The overall results are a hypolimnion lacking oxygen and having high concentrations of toxic products of fermentation such as hydrogen sulfide, ammonia, methane, and acetic and other organic acids. This combination can be outright toxic to fishes and, in the case of hypolimnetic-release reservoirs, can cause substantial fish kills in the receiving river below the dam.

With the onset of autumn, surface waters eventually cool until the thermal difference between epilimnion and hypolimnion is minimal, and some wind event will result in mixing of the lake throughout the water column. Autumnal circulation allows reaeration of the whole lake, redistribution of materials, and oxidation of the toxic anaerobic products in the hypolimnion. This event can ephemerally lower the oxygen concentration of the whole lake, result in noticeable odor, and facilitate an autumnal plankton bloom. Circulation continues until winter cooling of surface waters increases their density so that they move downward and tend to remain deep in the water column. In warm monomictic lakes, the water temperature rarely reaches that of the maximum density of water (4°C), there is no ice formation, and water circulates all winter, with adequate oxygen for fishes throughout the water column. In more northern climates, cooling of surface water to 4°C confers on that water the maximum possible density; hence, it may move to and remain at the bottom of the lake. Wind may, however, keep the entire lake circulating at 4°C, until some rapid drop in temperature (usually on a calm night) freezes surface water into what will become a permanent, thickening layer of ice that remains all winter. If this ice cover is formed at about 4°C, warmest water will be deepest in the lake, and near-surface waters can continue to be cooled from 4°C to near zero, resulting in a "reverse" thermal stratification with the coldest liquid water above and the warmer 4°C water below. This is the classic dimictic pattern, with winter stratification now in place. See Wetzel (1975, Fig. 6-4, and pp. 72–76) for details on the autumnal and winter patterns in stratification. There are many variations on the stratification patterns described above, with additional periods of turnover, incipient stratifications that are broken up, unusual years without stratification, and so forth, all described in detail in Wetzel (1975, 1983), but the basic patterns are sufficient to explain most effects on fish.

Some aspects of summer stratification impose stress on fish, restricting their use of habitat and, for species needing cooler water, confining them to too-warm upper layers in some situations or crowding them into small cool-source refugia (Coutant, 1985). However, lake fishes generally have evolved in such environments and, thus, should be adapted to deal with stratification at a natural level. However, due to the actions of man, many lake fishes no longer live in an environment matching that in which they evolved. The greatest potential for stress on fish might hypothetically be in situations where the lake was oligotropic over evolutionary time but has been anthropogenically converted to mesotrophy or euthophy in historical times—resulting in more prolonged stratification or more rapid oxygen depletion than in the lake environment in which the fish were evolved. Various phenomena during summer stratification may influence feeding or habitat use by fish. At typical summer temperatures, there is not only a marked difference per degree in specific gravity of water but also in viscosity (warmer water being less viscous). These differences in viscosity can have a critical influence on distribution of phytoplankton in the water column. Phytoplankton generally sink and remain in the upper water column only by floatation adaptations and by reproducing sufficiently fast that the population is maintained despite sinking of individuals. Phytoplankton cells sinking through the water column of a lake will move downward more rapidly in warm (e.g., epilimnetic) water than in cooler (more viscous) metalimnetic (transition zone) or hypolimnetic layers. Hence, as phytoplankters sink from warmer to cooler layers, there is a decrease in downward velocity. This phenomenon can result in the accumulation of cells at the warm–cool boundary or in any layer with differential density or viscosity of water, which, if in an oxygenated zone, can become a feeding site for planktivorous fishes.

Winter stratification is a period of lowered metabolic activity for many fishes, with decreased feeding or opportunities for feeding, some taxa entering almost a torpor-like state at very cold temperatures, and others forming pods of relatively inactive individuals that feed little or are quiescent (Johnsen and Hasler, 1977; Kolok, 1991). Photosynthesis can continue substantially under clear ice in winter, but a snow cover can eliminate as much as 99% of light transmission (Wetzel, 1983), resulting in little photosynthesis and the potential for depletion of oxygen by decomposition processes. The resulting winterkill from oxygen depletion has a strong effect on the composition of the local fish assemblages in many northern regions (Rahel, 1984). Survival of fish in many winter lake systems may depend substantially on the length of ice cover, the trophic status of the lake (hence, the level of biological and chemical oxygen demand in winter), and the local drainage networks that allow fish to move or not into stream refugia to escape winter anoxia.

Wind and Water Movements in Lakes

Water moves in lakes as a result of wind driven surface waves and the horizontal motion of Langmuir circulation [spiraling cells of water, moving across a lake (Wetzel, 1983)], currents, and periodic movements such as seiches (rocking of water within the basin, because of accumulation of water masses downwind). Most movement of water is in the epilimnion, but water can move in the hypolimnion because of density currents or heating from sediments, which can redistribute nutrients or move organisms toward the light. Here, I address movements of water in the epilimnion as they influence fishes or their prey.

Beginning with a calm water surface, the onset of a breeze creates surface ripples, pushing downward on the facing surface of a ripple, and pulling upward on the lee surface by virtue of the shear stress between the air and water. This eddying effect of air moving over water trends to enhance the amplitude of ripples, such that as wind increases, they grow into waves. Although an observer above the surface sees a succession of waves (e.g., approaching a beach), individual water molecules make no net horizontal movement in small waves, merely oscillating in a spiral fashion in one place at the surface of the lake. There also is rapid diminishment of size of oscillations with depth; hence, a 1-m-tall surface wave can result in oscillation with a diameter of about 25 cm at a depth of 4 m, and as little as 2 mm at a depth of 18 m. Thus, even rather large wave action that might have substantial impact on surface fishes or those in littoral zones would have essentially negligible hydraulic effect on fish deeper in the water column. Thus, fish can escape wave-induced turbulence by merely moving offshore and downward. In daily samples for three summers in the littoral zone of Lake Texoma, we have found the most abundant small fish species (*Menidia beryllina*) to essentially abandon the littoral zone when wave heights exceed about 25 cm (W. J. Matthews, D. R. Edds, and P. Lienesch, unpublished data).

However, consider a fish suspended or swimming in the water column at a depth of 4 m below a wave passing overhead and setting up an orbit of water as described above. Assuming a surface wave height of 1 m and movement of water at 4 m depth in an orbit with a 25-cm diameter, what is the effect on the fish? Although I know of no direct observations, the answer would seem to depend on the size of the fish and its activity. When I have snorkeled in the ocean, with swells of 1 m or more at the surface, I have noted that if I was not actively swimming, the elliptical motion was essentially undetectable by me (except by noting my proximity to the coral heads a meter or so below—which is attention-capturing, indeed) because I rose and fell in the water with the water mass. I suggest that a small fish motionless (i.e., without active forward swimming) in the water column would be little affected by the orbiting of the water mass in which it was entrained. This would also include larval fishes. Furthermore, this "motionless" fish would seem motionless with respect to other objects floating in its near proximity in the water column (e.g., food items).

In contrast to a small nonswimming fish, a large-bodied fish would seem more likely, depending on its location, to be impinged upon by more than one cell of water in motion in 25-cm-diameter orbits beneath successive wave crests and troughs at the surface. Intuitively, this would seem to subject the fish to considerable turbulence, hydraulic "confusion," or similar notable effects. Finally, for a fish swimming actively and horizontally through the water at the hypothetical depth of 4 m, there would seem to be a greater possibility of crossing through regions of the water with upward, downward, or sidewise motion, all potentially displacing the fish from a straight-line forward path, much like an aircraft buffeted by variable winds at angles to its flight path. Additionally, the forward-swimming fish subject to erratic motion of the water would also perhaps be pursuing prey items that were also being moved in variable dimensions due to the motion of the water. Testing the effect of such motion (e.g., in wave tanks?) on predation efficiency or other behaviors of water-column fishes would seem desirable.

The maximum wave height that can be sustained at the surface of a lake depends on the distance that the wind works on the water, or "fetch." Hypothetical maximum wave height (in centimeters) will be 0.105 times the square root of the fetch (in centimeters) (Wetzel, 1983). For the beach at the OU Biological Station, on the downwind (north) side of Lake Texoma, the fetch is about 3 km, which would result in a calculation of about 99.6 cm (or 1 m)—which approximates my observations of maximum wave height at this location during windstorms in the last 20 years.

From ripples to formation of small waves, there is no net horizontal movement of water across the surface of the lake. As wind increases, however, there is the establishment of Langmuir circulation cells about 5–20 m wide, indicated visually by Langmuir "strips" at the surface. At this point, water moves downward in convergence zones and upward at divergent zones, and there is net movement of water downwind across the surface of the lake. At the zones of convergence, there is an accumulation of floating materials (e.g., external tests of dead zooplankton, exudia of aquatic insects, foam, etc.) forming the visible Langmuir strips and potentially offering surface fishes microlocations of concentrated prey.

The movement of water across the surface can alter distribution of surface or near-surface organisms, cause water to pile up at the downwind side of the lake, and initiate a seiche (i.e., a periodic whole-lake water movement). In the simplest form in a stratified lake, a seiche is initiated when wind-driven water piles up at the downwind end of the lake, resulting in a downward displacement of the epilimnion on the hypolimnion (without much mixing between the two layers). As epilimnetic water is forced deeper on the downwind side of the lake, the epilimnion is pushed upward on the upwind side. After reaching maximum displacement of the thermocline from horizontal, there is a gravity-caused rebound and reverse rocking of the epilimnion on the hypolimnion, with many oscillations (up to several hours long, each) possible and only damping out over a period of days. Seiches are explained more fully by Wetzel (1983), with details of their dependence on whether or not a lake is stratified, the width of the lake, and the schedule of wind moving over the lake.

Matthews et al. (1985) described such an internal seiche in Lake Texoma on 23 August 1982 and its effect on fishes (Fig. 4.5). In summer 1982, we made weekly vertical profiles of physicochemical conditions (Hydrolab) and fish distribution (echolocation) at six sites throughout the main, deep downlake basin of Lake Texoma. From early July until mid-August, the lake had been strongly stratified, with an anoxic hypolimnion where very few fish were detected. Furthermore, during the July–August period, vertical profiles in any given week were almost identical throughout the downlake basin (i.e., the thermocline—actually a "chemocline" in Lake Texoma—was essentially level). Strong southwest winds prevailed on 22–23 August. On 23 August, we found that on the downwind (north) side of the basin, the anoxic hypolimnion had been driven downward to a depth of 18 m (Fig. 4.5), whereas on the upwind (south) side anoxic water was at a depth of only 10 m. As we noted (Matthews et al., 1985), it was only on this date that we found substantial numbers of fish in anoxic water below the discontinuity layer—possibly because the vertical "rocking" of the anoxic hypolimnion resulted in a loss of cues for habitat selection by the fish. This seiche illustrates that fish can be at least temporarily entrapped in, or fail to move from, anoxic water if a formerly calm stratified lake [with fish concentrated just above the hypolimnion (Matthews et al., 1985)] is suddenly subjected to strong vertical displacements of epilimnetic and hypolimnetic water masses. Moshe Gophen (L. Kinnerett Laboratory, personal communication) suggests that it is possible for a rapid vertical elevation of the anoxic water mass (e.g., on the

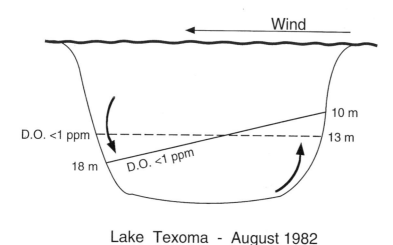

Figure 4.5. Schematic of a seiche detected by measurement of vertical oxygen profiles in Lake Texoma, Oklahoma (USA).

upwind side of a lake, in response to downward movement of the epilimnion on the downwind side) to actually result in massive kills of oxygen-sensitive fishes.

Winds and water movement appear capable not only of transporting poorly swimming biota (zooplankton, larval fishes) but also of changing their concentrations or patchiness at the lake surface. In summer 1983 and 1984, along with S. Kashuba and M. Lodes, I made a series of small (~ 15 m long) tows on north–south transects across Lake Texoma, using a 1-m^2 Tucker trawl. On 9 June 1983, after a period of sustained south winds that had prevailed for at least a week, we found that larval shad (*Dorosoma* spp) were very patchy, with a coefficient of variation [CV = standard deviation/mean; values substantially greater than unity indicate clumped or patchy distributions (Sokal and Rohlf, 1969)] equal to 59.5 and also concentrated more toward the north (downwind) side of the lake (Fig. 4.6). On 14 June 1983, after a strong windstorm with prevailing winds from the north, the clumping of larval shad in patches was substantially less distinct (CV = 16.3) and, although fewer individuals were collected overall, the previous pattern of concentrations toward the north had been broken up (Fig. 4.6). Similarly, for larval *Menidia beryllina* (not shown), the degree of patchiness was less after (CV = 7.2) than before (CV = 20.2) the change in wind direction and magnitude.

Morphometry of Lakes

Much can be inferred about potential fish assemblage structure in a lake from its morphometry. Many morphometric indices have been devised to facilitate limnological comparisons among lakes (Cole, 1994), but their relatedness to fish assemblage structure should not be assumed a priori. For example, a very large or a very small lake might have a similar index value comparing, say, depth to volume but have very different fish faunas. Nevertheless, among lakes of generally similar size, latitude, or geologic setting, morphometric indices may help fish ecologists to select suites of lakes for comparative studies or design of multilake experiments.

One of the simplest measures of the morphometry of a lake is "fetch," that is, the distance that wind can do work in pushing water across a lake. The maximum possible fetch in a lake is obviously identical to the longest open-water, straight-line distance across the lake, but from a practical basis, the fetch of a lake should, on average, be estimated as the distance from shore to shore on an axis approximating that of prevailing winds. In that fetch determines the maximum wave height that can be attained on a lake and in that a wave shock is one of the critical features affecting the quality of a littoral zone habitat for many fish species, it becomes important to know what the fetch, hence likely wave conditions, is in order to predict the composition of a littoral zone assemblage.

As an example, along the north (downwind) shore of Lake Texoma, we made weekly beach seining collections at four fixed sites during much of 1981–1986 (Matthews et al., 1992). These four sites were, from east to west (Fig. 4.7) (1)

Stream Ecology and Limnology as Related to Freshwater Fishes / 167

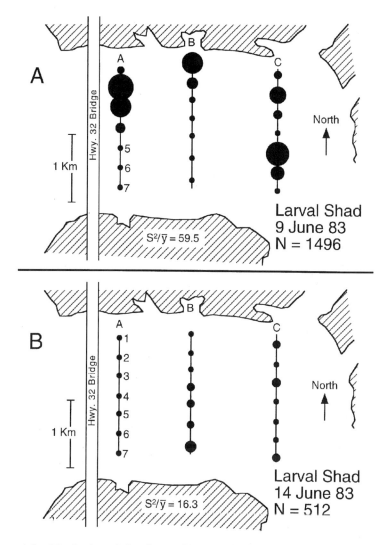

Figure 4.6. Distribution of abundance of larval shad (*Dorosoma* spp.) in six 15-s tows of a 1-m² Tucker trawl on each of three transects across Lake Texoma, Oklahoma–Texas, on 9 June (A) and 14 June (B) 1983. Size of darkened circle at each tow on transect is proportional to number of larval shad captured.

"rocky beach" (fully exposed to ~ 4 km or more of fetch from prevailing winds), (2) "sandy beach" (partially exposed as above, but protected by a point of land that effectively reduced the fetch and wave height on days with southwest wind, but allowed full fetch and wave shock on days with southeast winds, (3) "mouth of Mayfield Flat" (mostly protected from south winds by a point of land that

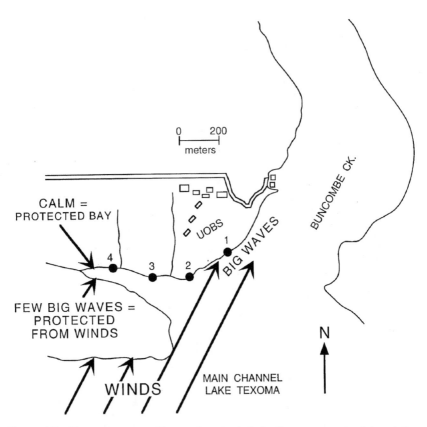

Figure 4.7. Exposure to prevailing southwest winds for four permanent seining stations (1–4) on the north shore of Lake Texoma (Oklahoma). UOBS = University of Oklahoma Biological Station.

reduced fetch to about (½) km at typical water levels, and (4) "upper Mayfield Flat" (in the upper part of a cove, fully protected from south winds by the point of land and having only a few hundred meters of fetch, at most, across which wave height could increase. As a result, the average wind exposure and wave shock was far greater at Stations 1 and 2 than at Station 3 or Station 4 (Fig. 4.7). Station 1, where waves on windy days often were 50 cm or higher, was dominated by large rock riprap (part of which was artificial); Station 2, where waves driven by southeast winds also reach 50 cm or more, consisted mostly of clean, wave-swept sand over hard clay; Station 3, where wave heights rarely reach 30 cm, had more clay and less sand, with some soft sediments; Station 4 had a bottom consisting of soft sediments, organic debris, and depositional materials rarely disturbed by wave action.

Local fish assemblages were very different across the gradient from full wave shock to minimal wave action (Matthews et al., 1992; Matthews, unpublished data): In the exposed Stations 1 and 2, some black bass (*Micropterus* spp.) and sunfish (*Lepomis* spp.) were present around the largest boulders, but the primary species were highly mobile atherinids (*Menidia beryllina*) and shads (*Dorosoma* spp.), as well as juvenile striped bass (*Morone saxatilis*) that could readily enter and leave the area as wave levels changed daily. Station 3 was also dominated by *Menidia* and shad but had, on average, far more of the littoral zone minnows such as *Cyprinella lutrensis*, *Cyprinella venusta*, and *Pimephales vigilax*. Only at the protected upper end of the cove were "quiet water" taxa, such as *Gambusia affinis, Pimephales vigilax,* and *Fundulus notatus* present in substantial numbers, along with small individual *Lepomis* of various species and some *Micropterus*. Chapter 8 will address the relationship between body form and hydrodynamics of fishes, but it seems obvious that the differences in local fish assemblages were marked in this part of Lake Texoma in response to local wave shock and resultant differences in habitat quality. However, some species, [e.g., white bass (*Morone chrysops*)] used the four microhabitats in approximately equal numbers (Matthews et al., 1992).

One of the most used metrics in limnology is "shoreline development," which compares the actual length of shoreline of a lake to the shoreline length that would exist if a lake of the same surface area were a perfect circle. Lakes with high values for shoreline development typically have many coves, convoluted shorelines, or other irregular features, and a correspondingly less open-water or deep limnetic area. Accordingly, natural lakes with large shoreline development values often are shallow, with a high percentage of the surface area existing over water not sufficiently deep to escape surface heating, wave action, or high primary productivity, as well as the opportunity for extensive development of beds of rooted aquatic macrophytes. Such lakes often have a high proportion of "littoral zone" fishes and/or large populations of planktivorous fishes that feed by grazing zooplankton or phytoplankton from near-surface waters. Because such convoluted, shallow lakes often have a high exposure of water to sediments, they may be rich in nutrients, not stratify deeply, and have a small hypolimnion that becomes anoxic early in summer. Hypothetically, there could be correspondence between shoreline development and the tendency for a lake to be or become eutrophic, to support more benthic or water-column omnivores, or to have fewer large, deep-water piscivores. However, such lakes do favor "inshore" piscivores, like largemouth bass (*Micropterus salmoides*), that are particularly efficient predators near or within structure. Finally, there are some "lakes" that have very high shoreline development values but little shallow littoral zone. This condition may be typical of large man-made reservoirs, such as Lake Norfolk or Bull Shoals Lake in north Arkansas or Broken Bow Lake in Oklahoma (USA). These lakes were all formed by the construction of large dams across a deep river valley, resulting in flooding of rugged terrain with steep hillslopes and many tributary streams. These artificial lakes would all have high values for shoreline development, but because their normal water level now lies on what was a terrestrial hillslope or mountain side, they typically have deep water near shore.

A better predictor of size of hypolimnion to epilimnion and the amount of deep to shallow water in a lake may be "relative depth," which is a measure of maximum depth divided by the average diameter of a lake. The maximum depth is obtained by direct measurement. The average diameter could hypothetically be determined by taking an infinite number of measurements from shore to shore across the center of the lake, but this is impractical. However, given that the surface area is known from mapping, it is possible to calculate relative depth as

$$Z_r = \frac{88.6 \ Z_{max}}{SQRT(A)}$$

(Cole, 1994) or the identical

$$Z_r = \frac{50 \ Z_{max} \ SQRT(\pi)}{SQRT(A)}$$

(Wetzel, 1983). Where Z_r is the relative depth, Z_{max} is the maximum depth, and SQRT(A) is the square root of the lake area. Both formulas for Z_r give the same answer. A large value indicates a relatively deep lake. Such values may be most appropriate for comparing natural lakes and are of perhaps little utility in comparing large, shallow reservoirs. For example, the dimensions of Crater Lake result in a Z_r of about 7.5%, implying that it is a relatively deep lake, that littoral zones would be limited, that oligotrophy might be expected more than eutrophy, and so forth. In contrast, Lake Texoma, assuming a maximum depth of 25 m and an area of 377 km^2, would have Z_r = 0.0114% (a value so small that its use seems questionable for assuming much about the fish fauna). However, this very small Z_r value does suggest that much of Lake Texoma is very shallow, and this is, indeed, reflected in its extensive littoral zone development. However, I doubt that comparisons among shallow reservoirs based on these very small values for Z_r would allow much prediction about the fish faunas.

Bathymetric mapping of lakes can be very useful as a predictor of their potential trophic status (Wetzel, 1983) and, hence, the composition of the fish fauna. Bathymetric data are typically presented as hypsographic curves depicting either the area or the volume of a lake that exists within a particular range of depths [see Wetzel (1983, Fig. 3.11)]. Comparing hypsographic curves can have predictive value about the proportion of a lake that is shallow versus deep or the shape of the slope of the lake bottom (e.g., evenly sloped, concave, convex) which will relate to how much of the lake can support rooted macrophytes, serve as a shallow habitat for fishes, its trophic status, or its potential productivity.

One of the curiosities of lake morphometry is the very predictive "morphoedaphic index" (MEI) that has been developed empirically by fisheries biologists and limnologists. This value, equal to total dissolved solids (TDS) in milligrams per liter, divided by mean depth in meters,

$$\text{MEI} = \frac{\text{TDS (mg/L)}}{Z_{mean}\ (m)}$$

has a very strong potential to predict the production of fish biomass in the lake. Although predictive basis for this measure has been much debated and modifications in its form have been proposed (Jackson et al., 1990; Downing and Plante, 1993), this value is one that appears empirically useful for predicting actual values for the standing crop of fish.

It is important to remember that for all the indices of lake morphometry, local factors not accounted for in the index can result in limnological or ichthyofaunal conditions being very different between lakes that have very similar morphometric scores. However, morphometric indices can help fish ecologists to generally categorize lakes that they are studying, relative to others in the region. Finally, great caution should be used in applying standard limnological morphometry measures (developed in natural lakes, with geological factors responsible for their shape) to artificial impoundments (in which the deepest point is typically at or near the dam, and the configuration reflects the terrestrial landscape that has been flooded more than anything about natural features of the body of water).

Summary: Limnology and Lake Fishes

Investigators interested in the composition of lacustrine fish assemblages need to (1) very carefully define the physical dimensions and location of the "assemblage" they are studying and (2) have fundamental information on the general limnology and "behavior" of the lake in which they are working. Obviously, the best detail is possible where a large, multifaceted research team (including physical limnologists) has been working in a given system for a long period of time [e.g., Lake Kinneret, Israel or Lake Mendota, Wisconsin (Kitchell, 1992)]. However, in many lakes throughout the world, fish ecologists lack such a massive assimilation of background information and must make do as best they can to understand ecology of local fish assemblages. Lacking such information, it is necessary, at a minimum, to acquire good information on extant chemical and physical conditions in the lake if there is to be any progress at all in understanding fish associations, assemblage structure, or interactions.

4.5 Characteristics of Streams that Influence Fish Assemblages

Watersheds and Ecosystem Concepts in Streams

One of the most fundamental units for understanding stream ecosystems and the ecology of fish in streams is the total watershed (i.e., the stream proper and all of the terrestrial area that drains into it). The importance of land–water interactions (Bisson and Sedell, 1984) and whole watersheds as conceptual units has repeatedly been demonstrated in stream ecology (e.g., Fisher and Likens, 1973; Minshall, 1988). Recently, there also has been increased emphasis on the role of the hyporheic zone in streams—that zone in which water percolates below the streambed and where the exchange of materials, nutrients, and biofilms have an important influence on the environment of a fish in the water column or on the streambed (Bretschko and Klemens, 1986).

Vannote et al. (1980) drew together in the River Continuum Concept (RCC) the trend for at least the previous decade toward focusing on the continuity of biota and stream ecosystem phenomena from headwaters to lower mainstems. Although Vannote et al. (1980) included little about functional roles of fish in streams, stream fish have since that time become increasingly integrated in paradigms for stream ecology (Power et al., 1985; Minshall, 1988; Grimm, 1988; Gelwick and Matthews, 1992; Allan, 1995). Although the RCC emphasized continuity, complementary views of streams have emerged in which patch boundaries, creating a mosaic of habitats within stream reaches, have been emphasized (Naiman et al., 1988; Pringle et al., 1988; Townsend, 1989). From the perspective of fishes, both the views of streams as large continua or as patchy mosaics seem important. There is no question that individual fish respond strongly to details of habitat or microhabitat within a stream reach (i.e., patches). Alternatively, many fish species are highly mobile, and a given species in a watershed typically occurs in or moves through a wide array of patches—possibly playing major roles in integrating stream ecosystems in the large longitudinal view of the system as a continuum (Flecker, 1996). Finally, human impacts play huge roles in river systems (Welcomme, 1995). The challenge for fish ecologists seems to be to keep both perspectives in mind in formulating questions or solutions to questions about stream fish or their role in the ecosystem. To understand the effects of stream systems on fish and the role of fish in stream systems, it is necessary to understand the basic structure or stream ecosystems.

Formation of River Basins and Drainage Patterns

Allan (1995) emphasizes the difficulty of knowing when an extant river system formed or began, because existing rivers have replaced earlier drainages over time, often with rapid or gradual changes in major flow patterns. As examples, until tectonic elevation of the central shield and orogeny of the Andes, the Amazon River basin appears to have drained westward to the Pacific Ocean (Sioli, 1975), and virtually all of the now eastward-flowing streams of the American Great Plains appear to have flowed through a north-to-south Ancestral Plains Stream in the Pliocene (Metcalf, 1966), until the growing Arkansas River cut through the Interior Highland during the Sangamonian interglaciation of the late Pleistocene, bisecting the Ozark and Ouachita regions and capturing much of

Stream Ecology and Limnology as Related to Freshwater Fishes / 173

the flow of the ancestral river (Mayden, 1985). Jenkins et al. (1972) and Hocutt et al. (1986) describe apparent prehistoric drainage connections and stream captures during the Pleistocene to Recent in the Appalachian Mountains in the eastern United States and their effect on composition of specific local faunas. Burr and Page (1986) summarized much of the history of the evolution of drainage basins in the central United States, and their map (reproduced as Fig. 4.8) outlines hypothetical drainage patterns of the Pliocene that have had a major effect on distributions of fishes in local assemblages.

Fundamental changes in drainage history of a region are critical to zoogeographic interpretation of fish faunas and, hence, to understanding the origins of species in existing fish assemblages. In fact, much of the activity of ichthyologists

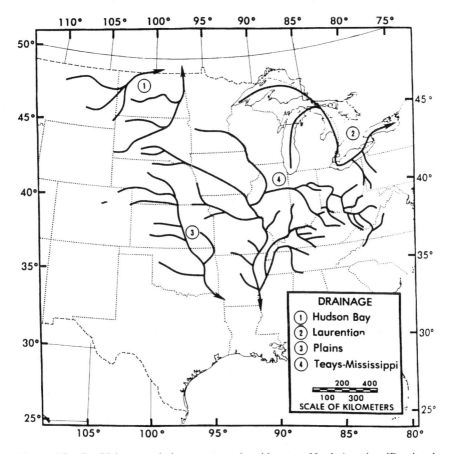

Figure 4.8. Pre-Pleistocene drainage patterns in midwestern North America. [Reprinted from Burr and Page (1986) in the Zoogeography of North American Freshwater Fishes, CH Hocutt and EO Wiley (editors). Copyright © 1986 by John Wiley and Sons, Inc. Reprinted with permission of John Wiley and Sons, Inc.]

during the last two centuries has been tied to understanding major distributions of fish relative to basin history. Virtually all articles in Hocutt and Wiley (1986) emphasize the history of river basin relationships as key to understanding extant fish faunas or species distributions. Mayden's (1987, 1988b) vicariance or historical biogeography of fishes in North America is based on detailed knowledge of drainage events and cladogenesis.

Terminology for Streams of Various Size

Streams are many different sizes, with the longest, the Mississippi–Missouri River with mainstream 7000 km long, and the largest on an areal basis, the Amazon Basin, with a watershed of 7,000,000 km^2 (Hynes, 1970). Drainage density, defined as the total length of streams per area drained, is greater in wet climates. Overall, total stream length and drainage density in a catchment are related to interactions of stream water with the terrestrial and geological components in the streambed: The long travel time allows more opportunity for physical and chemical interactions within the stream or between both surface and hyporheic water of the channel and the substrates. Thus, high drainage density may result in a more intimate local interplay of a stream with its landscape.

There is no single standard terminology in use for drainages of various sizes, but many ichthyologists use that recommended by Jenkins et al. (1972), which I will attempt to follow in this book. Jenkins et al. (1972) definitions, from smallest to largest patterns, are as follows:

1. System—a group of interconnected streams within a drainage
2. Drainage—an interconnected major group of streams, or systems entering the marine habitat (such as the Roanoke or Cumberland drainages)
3. Basin—a group of interconnected drainages (such as the Ohio Basin or Mississippi Basin)
4. Slope—refers collectively to all drainages on their surfaces, such as "Atlantic slope" or "Gulf (of Mexico) slope."

The terminology of Jenkins et al. (1972) clearly is at the zoogeographic scale and does not help with the definition of terms for local stream patterns. For standardization at a smaller scale, we could use the terminology (in part) of Frissel (1986) with progression in size from "microhabitat" to "pool–riffle sequence" to stream "reach." The worst overlap in terminology for streams of various sizes seems to be in the use of the terms "watershed" or "catchment," which have largely overlapped in usage. In general, "catchment" seems to have been used most often for a small stream (e.g., a brook or creek and its surrounding terrestrial drained area), whereas "watershed" has been used to refer to all of the area drained by a stream, ranging in size from the smallest creeks to large rivers. In this book, I use both "catchment" and "watershed" to refer to the area drained by a specific small stream up to the size of a large creek. Hence, I refer to the Piney Creek or Brier Creek watershed, but I would not use "watershed" for a stream as large, say, as the Buffalo River of North Arkansas (~ 6000 km^2). Following the terminology of Jenkins et al. (1972), it would seem most appropriate to refer to the Buffalo River "system," in that it includes numerous small to large named creeks, each with its own "watershed."

Drainage Patterns

Understanding basic forms of drainage patterns is critical to understanding fish distribution and potential dispersal in a region. In most regions, small- to medium-sized watersheds are of two basic forms: dendritic and trellised. Dendritic drainage patterns would arise in random fashion if a relatively level region (e.g., a plateau) with uniformly hard rock erodes downward. The existence of rocks of various hardness influence erosion patterns to depart from fully random, but the Ozark Highlands (central United States) province offers excellent examples of dendritic watersheds (Fig. 4.9), as upland plateaus have been dissected into steeply sloping hills and valleys, which exhibit a dramatically branched pattern (e.g., Piney Creek, Izard County, Arkansas). Typically, such patterns form in regions without severe macrolevel constraints of rock-bound phenomena. Because of the joining of many stream branches, dendritic streams achieve relatively high values for "stream order," per the Horton–Strahler (Horton, 1945; Strahler, 1957) system (Fig. 4.10). In this much used classification of streams, smallest headwater tributaries are first order; where two first-order streams join, the system becomes second order; where two second-order streams join, they become third order, and so on. Note that joining of two streams of *unequal* order does not result in an increase in ordinal status of the stream. Hypothetically, the branching of dendritic drainages into many tributaries of varying sizes offers much heterogeneity of habitat and could increase the number of species occurring in the watershed [although Matthews (1986b) cautioned against uncritical acceptance that stream faunas increase significantly in richness with increase in order].

Alternatively, where geology of a region is primarily "ridge and valley," there can be series of long, parallel mountain ridges (in some places 100 km long or greater) with valleys in between. In these regions, such as the Ridge and Valley Province north of Roanoke, Virginia, and in the northern Ouachita Mountains of southwest Arkansas, creek or river mainstreams may run for many kilometers without inflow of any other major tributary, and the drainage exhibits a "trellised" pattern (Fig. 4.11A). In a trellised pattern, there are typically many small tributary streams, often flowing down sides of mountains parallel to each other but not joining to form large tributaries. Instead, each is a small tributary "run" that increases the total flow of the mainstream but does not increase its stream order (Fig. 4.11A). Finally (Fig. 4.11B), in arid lands (e.g., the American West), many kilometers of stream channel that would appear on maps to increase in size and

176 / *Patterns in Freshwater Fish Ecology*

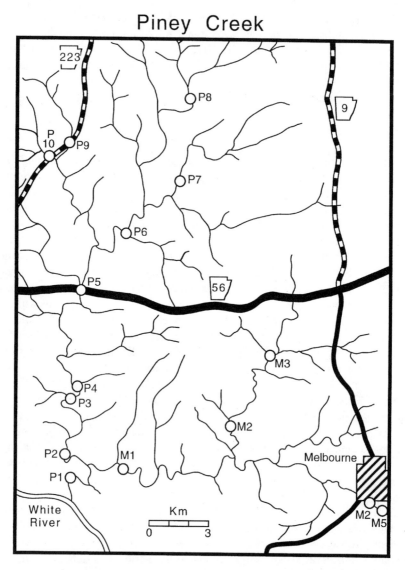

Figure 4.9. Twelve permanent stations on Piney Creek, Izard County, Arkansas, sampled from 1972 to 1995 by W. J. Matthews and colleagues.

in stream order may, in reality, consist mostly of isolated pools (e.g., Fausch and Bramblett, 1991). Another way of considering the distribution of flowing streams within a region is to calculate its "drainage density," which is merely the total length of all streams (in kilometers) divided by the total area drained (km^2). In general, streams in wetter regions have a greater drainage density,

Stream Ecology and Limnology as Related to Freshwater Fishes / 177

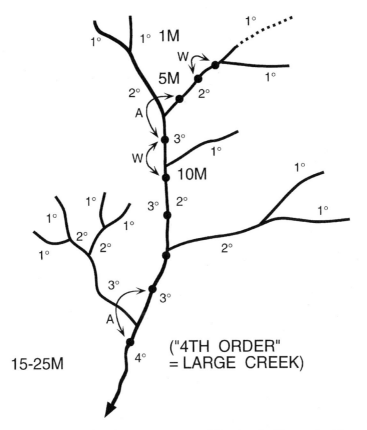

Figure 4.10. Stream order of stream segments, following the Horton–Strahler system, depicting reaches with typical width of 1, 5, and 10 m and larger. Two-headed arrows with W = two sites within a given stream order; two-headed arrows with A = two sites differing across stream orders.

which hypothetically results in a greater number and kinds of microhabitat for fishes that exist in regions with low drainage density and rather straight stream networks with minimal convolution or branching across the face of the landscape.

Highly complex drainage patterns can evolve from "stream capture," merging or dividing original river basins in an intricate fashion. Stream capture in its simplest form occurs if, for two streams in close proximity, the headward erosion of one breaks through to reroute the headwaters of the other so that its outflow now follows the channel of the "capturing" system. In terrain with high relief, major river basin formation or reshaping can occur as a result of stream capture, with subsequent entry of one species into a system that previously was out of its range. In this way, "new" taxa can be introduced into an "old fauna" of the capturing river. As an example, consider the hypothetical multiple captures across

178 / *Patterns in Freshwater Fish Ecology*

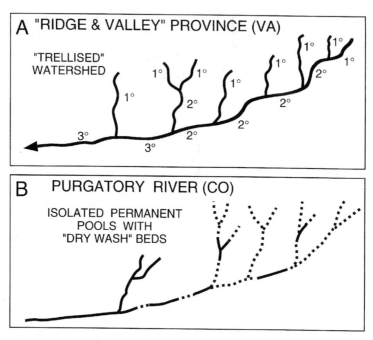

Figure 4.11. Typical Horton–Strahler stream order for a ridge and valley, trellised watershed (A), and for a largely intermittent watershed (B).

the Atlantic Divide of the eastern United States, as postulated by Ross (1969; cited in Jenkins et al., 1972) to account for the present-day configuration of the New, Roanoke, and James drainages in Virginia (Fig. 4.12).

According to Ross (1969), the original Teays River flowed westward and the prehistoric James and Roanoke systems flowed eastward, from the Blue Ridge divide. First, in geologic history, the Roanoke breached the Blue Ridge to capture and divert eastward much of the upper Teays, giving the Roanoke a large, new headwater west of the Blue Ridge. Subsequently, however, the James also breached the Blue Ridge and beheaded what was for a time the uppermost northern Roanoke River drainage. This final event resulted in an approximation of modern configurations of the three systems, with the former Teays (now New River) flowing westward and north to the Ohio River, and the Roanoke and James rivers with their separate systems, each including channels west of the original Blue Ridge divide. This set of events offers a plausible explanation for why a substantial number of fish species occur at all in the Roanoke system (e.g., *Luxilus albeolus*, closely related to and apparently derived from ancestral *Luxilus cornutus/chrysocephalus* stock that entered the Roanoke system from the west via this stream capture). However, Jenkins et al. (1972) also noted substantial differences between the fish fauna of the James and Roanoke drainages (e.g.,

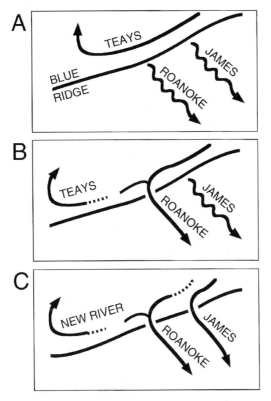

Figure 4.12. Hypothetical sequence of stream capture between James, Roanoke, and Teays (New) rivers, Virginia.

many more catostomids in the Roanoke) that Ross' hypothetical stream capture cannot account for. Understanding extant distributions of fish faunas is complex (Mayden, 1987), and stream capture is only one of many explanations that must be considered in the zoogeographic history of fishes in an assemblage.

Such stream captures can also lead to the reestablishment of contact between evolutionarily separate lineages within a single species (or species complex). The blacknose dace, *Rhinichthys atratulus,* widespread in eastern North America, has at least two named subspecies. The nominal *R. a. atratulus* has a native range almost entirely east of the Atlantic Divide, whereas *R. a. obtusus* (distinctive from the nominal subspecies in nuptial coloration and in several morphometric or meristic characteristics) occurs west of the Atlantic Divide (New River tributaries) (Matthews et al., 1982b). However, in Upper Meadow Creek, a tributary of the James River in Craig County, Virginia, was "probably gained by the James drainage by a capture of the upper section of Sinking Ck., a major tributary of the New River drainage" (Matthews et al., 1982a). By this stream capture, the putative subspecies *obtusus* gained entry to the James drainage, where, in a stream reach below a cataract, it comes into contact with the nominal subspecies *atratulus*. [In this particular case, this situation permitted a test of whether or not the two "subspecies" retained integrity in this contact zone, finding that 58–72% of individuals exhibited coloration, morphology, or both, suggesting interbreeding of the two stocks in the contact zone (Matthews et al., 1982a)].

This example of a local event illustrates the fact that a stream capture could place two highly similar forms in intimate contact where not only reproduction could be affected but many ecological phenomena like competition, mutualism, or complex indirect interactions in an assemblage could be altered by such a natural introduction of a new taxon. To the extent that some species may be "keystone" interactors in a system (Power et al., 1996), such a natural introduction could be very important to an assemblage. Furthermore, such an introduction by stream capture could be very rapid; that is, once a capture occurred during some event like a final major flood to break down geological barriers between two streams, the fish in a local assemblage might readily be swamped with individuals of the invading taxon in brief, ecological time. Stream fish ecologists, particularly those working in mountainous regions, must be fully aware of local geology and sensitive to potential stream captures as an explanation of local assemblage structure.

Headward Cutting of Channels—New Habitat for Fishes?

Drainage patterns ultimately depend on headward cutting of new channels. Moving water plays the primary role in extension of drainage networks into headwaters. Dietrich and Dunne (1993) review channel head formation. The following summary of the process was suggested by W. E. Dietrich (personal communication):

> The headward extension of lowland rivers is typically driven by overland flow where it concentrates and cuts through the ground surface. This overland flow may result from ground saturation in the valley bottom upslope of the channel head, or from precipitation intensities that exceed the infiltration capacity of the ground surface. The flowing water exerts a substrate shear stress on the ground surface, the magnitude of which depends on the slope of the ground surface and the depth of overland flow, which in turn varies with the amount of excess precipitation as well as the resistance to flow offered by grass, reeds or other. This vegetation and the cohesive properties of the soil surface also offer resistance to shear erosion by the overland flow. Channel extension typically occurs where the shear stress exceeds the shear resistance of the ground surface.

Hypothetically, if the substrate were a completely homogeneous material, without strain points or cracks, it should always fracture or slump at even intervals, related to shear stress of flowing water and a coefficient for composition of the material. However, during repeated cycles of wetting, infiltration, and drying, the substrate becomes cracked, so that slumping to form headcutting of channels comes at the points weakened by previous events. Dietrich emphasized that the critical shear stress necessary for channel formation is quite different for a watershed covered with natural, undisturbed vegetation, and one with vegetation altered or removed by overgrazing or similar deleterious land-use practices. This may explain the observation that headward cutting of small streams seems now to be proceeding at a rather rapid rate (i.e., too fast to be sustained over geologic time). This may indeed be the case, with rates of small stream expansion (e.g., pasture creeks) having increased markedly over presettlement rates. This would seem to have the potential to expand defined watercourses with pools and shallows upstream perhaps beyond the capacity of natural rainfall to maintain flowing channel, which could be important to long-term success of fish "lured" upstream during spates.

Sources of Water for Streams

A second major feature of streams, in addition to size, is the source of water. At one scale, all streams can be placed into two general categories: those originating from perennial spring sources and those that originate mostly from runoff [and typically have ephemeral headwaters (Matthews and Styron, 1981; Hubbs, 1995)]. With respect to actual movement of rainfall into streams, there also are two primary categories: runoff and groundwater (Beaumont, 1975). Runoff transports allochthonous material into the stream, which can influence the chemistry and nutrients of the water and also provide (especially in forested, canopied streams) much of the raw material (coarse particulate organic matter, CPOM) on which aquatic invertebrates (and some fishes) feed. As the CPOM is further broken down by shredders and other invertebrates, there is a releasing of nutrients, detritus (which some fish eat) is produced, and the invertebrates are food for many omnivorous/insectivorous fishes.

In contrast to runoff water, groundwater inputs to the stream have passed through any of a variety of underground reservoirs and typically carry less particular matter, and fewer living organisms are transported to a stream by surface runoff. Groundwater inputs, however, typically comprise water in chemical equilibrium with soil or rock and have a strong influence on chemistry of the stream. "Groundwater" also may consist of two basic types: that which is trickling through soil or rock layers in small interstitial spaces (and thus may be in most intimate contact with the substrates) and that which is actually flowing underground in a larger stream [e.g., the underground rivers of karst regions (Hobbs, 1992)]. This latter water may be the source of large springs, which have their own unique water chemistry, often release large amounts of carbon dioxide, and lead to deposition of marl.

Three major factors control water movement in soil: size and distribution of soil pores, attraction of soil particles for water molecules (adhesion), and cohesion between water molecules (which "pull" molecules along, following one another). Dry soils hold available water more tightly, as adhesion is much greater at the soil–water interface. The infiltration rate of water into soil is a function of number and size of pores in the soil and the amount of water already in them. (In arid soils, surface particles may form fine particles when hit by raindrops, actually packing the soil and inhibiting infiltration). Water movement is of two kinds in soil: saturated (percolation) and unsaturated flow from wetter to drier parts of the soil (even upward). The dry soil strongly pulls initial layers of water molecules, then other water molecules follow due to cohesion.

Where does water go when it rains? If the precipitation rate exceeds the attainable rate of infiltration, water flows overland, even if the soil is unsaturated, directly into stream channels. At lower rainfall intensities, water may also flow overland if soil layers are saturated. Water that does move into the soil will move downward until low-permeability rock layers are encountered, then it will move laterally ("throughflow") to eventually enter a stream as "groundwater." After a rain, throughflow down slopes may take days, such that the upper slopes are not saturated but lower slopes are. Another rain at that time will rapidly cause saturation of soil and initiate resumption of overland flow as runoff. This is often observed in periods of the year with repeated rains and may result in relatively sustained flow of creeks in areas that at other times of year may be dry or intermittent (e.g., in central Oklahoma).

Location of Flow Within Stream Channels and Configuration of Channels

Hynes (1970, p. 7) depicts the expected location of various current speeds in a cross section of an ideal stream channel. Current is fastest at midstream and usually slightly below the surface (Fig. 4.13). In such an idealized stream channel, one-half of maximum current speed would be at about two-thirds the depth of the stream (Hynes, 1970), leading to the commonly used convention of measuring "average current speed" at two-thirds the depth of the stream in field studies. However, this relationship does not apply in rapids or other shallow, turbulent parts of streams. Matthews et al. (1982b) found current velocities did not differ significantly from surface to bottom at 50 points in the riffle habitat.

Where a stream bends, higher current speeds are on the outside, with slower speeds inside (Fig. 4.13). This results in the alternation of erosional and depositional zones upstream and downstream, as each meander has a slow "inside" of the curve and a faster "outside" (Fig. 4.14). In addition, water flowing on the outside of a curve tends to also move outward in a helical spiral (Allan, 1995). This pattern of current speeds results in the continued lateral cutting of the stream into parent material on the outside of curves, resulting in undercutting the vertical bluffs of rock or earth, often accompanied (for rock substrates) by large boulders falling into the stream on the outside of the curve (Fig. 4.13). There, the swift water on the outer side of the curve can further deepen the pool by hydraulic

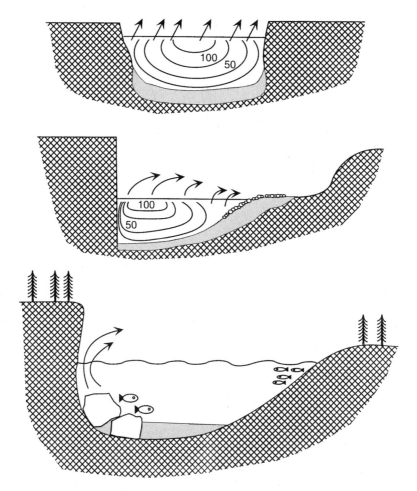

Figure 4.13. Isopleths of current speed (in cm/s) within a cross section of a straight stream channel (upper) and a curved section of stream channel (middle). Lower panel depicts results of increased speed and undercutting of stream banks on the outside of the curved channel.

scour around the boulders. The result, in meandering streams in rocky terrain, is often the presence of vertical or undercut walls of rock or earth on the outside of curves and the deposition of sand or gravel bars on the inside of the curves, where current speeds lessen and drop their load of materials (Pflieger, 1971, p. 249).

Fish use these habitats very differently, even within a given species. For example, at Station P-1 in Piney Creek, Arkansas (Matthews and Harp, 1974), one of the most abundant species is the minnow *Luxilus pilsbryi*. The largest

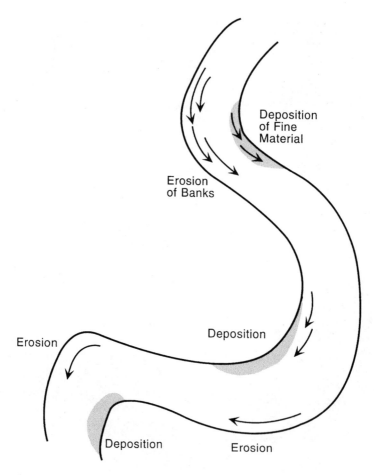

Figure 4.14. Spatial distribution of erosion of stream banks and of deposition of fine materials in a sequence of meanders in a hypothetical stream.

individuals are often in water 1.5 m or deeper in pools near large boulders on the outside of the channel at the base of a vertical dolomite bluff (Fig. 4.13). On the inside of this curve, there is a large, shallow bar of sand and gravel, over which we often capture large numbers of small- to medium-sized *Luxilus pilsbryi* in seine sampling. Other species are also differentially distributed across this cross section of the stream as a result of the outside-to-inside current gradient, with large-bodied smallmouth bass (*Micropterus dolomieu*) and suckers of the genus *Moxostoma* in deeper waters on the outside of the channel, and small minnows or topminnows (*Fundulus*) of various species in far greater numbers in slow-flowing water over gravel and sand on the inside of the channel. Thus, on a single cross-stream transect at this location (or in many stream sites), one would find a very differing set of physical structures, current speeds, bottom material size, and fish species (Fig. 4.13).

Such a hypothetical or idealized location of various current speeds in a channel cross section (e.g., Hynes, 1970, p. 7) would occur in a hydraulically perfect stream channel with regular meanders of flow in a sinusoidal pattern (Leopold and Langbein, 1966). This ideal pattern allows the stream to do the least total work in turning and is probabilistically and theoretically the form that streams should all take (Leopold and Langbein, 1966), resulting in a very predictable alternation of straight and curved segments and erosional and depositional zones. See Ritter et al. (1995, pp. 215–218) for a review of the physics of stream meandering. However, in real streams, there is much variability in location of flow, with locations of microhydraulic zones depending locally on positions of riffles, chutes, inflow to pools, and so forth, which, in turn, depend on the position of perched gravel pools, bedrock structures or constrictions of channel, and boulders. Finally, there is evidence that fish assemblages are strongly interdependent on the hydraulics of a river and its floodplain (Power et al., 1995a) and that the geomorphology of the floodplain and the interaction of the stream proper with the floodplain has major consequences for the life histories of fishes or the stream system food web (Power et al., 1995b).

There is also alternation of deep and shallow locations within any reach of stream, and alternation of flow from side to side (Leopold and Langbein, 1966). In gravel-bottomed streams, this alternation of erosional and depositional current speeds and change of greatest flow from side to side of the stream channel results in the deposition of gravel with a steep downstream slope (i.e., with a slope just less than that which will cause gravel-slides at that point in the stream). The downstream slope of such gravel bars is an important microhabitat for small fishes. At the upstream end of one pool in Brier Creek, Oklahoma, the maximum slope in a gravel bar deposited by a modest flood was nearly 45°. In sand-bottomed streams, similar microhabitats are produced, with "sand ridge" habitats [often used by *Notropis girardi* and other minnows (K. Polivka, unpublished data)] forming where shallow sand bars "break" downstream to depths of 0.5 m or more.

Stage Rises, Floods, and the Hydrograph

The time course of effects by a rainfall event on stream fishes can be described by a hydrograph (Fig. 4.15). Flooding intensity (as fluctuation of water levels in the hydrograph) is inverse to the size of the basin (e.g., local storms can affect small watersheds more than large basins). Thus, upland creeks can flood more rapidly and dramatically than large rivers. Typically, a small stream may rise rapidly from base flow to flood peak, then decline to base flow over several days. In undisturbed watersheds, where natural vegetation and leafy debris retain water, slow runoff, increase infiltration, and delay release into the stream, the hydrograph

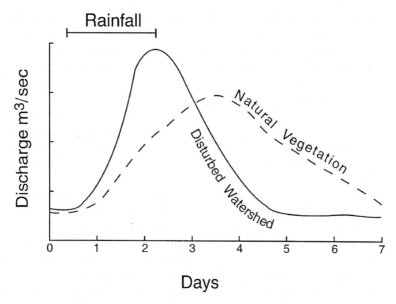

Figure 4.15. Example of a hydrographs naturally vegetated versus a disturbed watershed (= "flashy" hydrograph).

rises and falls more slowly and is of lower amplitude than a "flashy" hydrograph in watersheds where natural vegetation has been removed (Fig. 4.15).

During one scouring flood in Brier Creek, Oklahoma (Power and Stewart, 1987), I recorded stage rise on a bridge abutment. Near peak flow, the water was rising so rapidly that I could see it moving up on the elevation markers, and the peak was a single, dramatic cresting of water, followed by a visibly obvious onset of falling water on the stage gauge. In another case that I studied a posteriori (as soon as travel into the area was possible), Piney Creek (Izard County, Arkansas) exhibited (debris lines) a stage rise of 12 m vertically, in a single day, in response to a 15-in. rainfall (Matthews, 1986c).

Very rapid rises of water occur in arid lands. Flash floods are common in canyon and arroyo country of the American Southwest, where sudden rainstorms can create essentially a wall of water moving within confined stream reaches. However, in noncanyon topography, rather "ordinary" streams can, at times, exhibit essentially flash flooding like that in the West. Such floods create very high current speeds (e.g., in excess of 200 cm/s; Matthews, Brier Creek, unpublished data) and carry large amounts of woody debris and sediment down the channel.

A flashy hydrograph, resulting in more episodes of flood and drought than in the past, has direct consequences for the physical structure of a stream. In Brier Creek, the bottom of many pools actually consists of a sculptured bed of sandstone, with many small channels 10–15 cm deep carved by flow of water. During low-flow periods, these bedrock channels fill with silt and sediments, so that the pool bottoms appear like flat, unstructured habitats. However, a scouring flood sufficient to remove these sediments occurs every year or so, revealing the true nature of the channel bedrock. The sculpted rock exposed after a flood offers a very different microhabitat for fishes, with respect to hydraulic environment, shelter, hiding places from predators, and the like. In spite of the substantial erosive nature of floods in Brier Creek, adult fishes seem only minimally relocated after such an event (Matthews et al., 1994).

In Piney Creek, Arkansas, major floods such as that in December 1982 both remove from and input to the stream channel massive amounts of sand. Where the creek overflowed its banks to cut across the inside of a bend in the watercourse, decreased current speeds caused deposition of sand in layers as much as 1 m thick. However, in parts of the Piney Creek watershed, the 1982 flood also caused severe erosion of sandy banks, contributing then, and later, to input of sand into the stream channel. The total balance to the stream (i.e., negative or positive mass balance of sand into or out of the channel) remains unknown. At the microhabitat level, many cobble riffles in Piney Creek become partly filled with sand (increasing embeddedness) during nonflooding years. Immediately after the 1982 flood, it was obvious that many gravel/cobble bars had been churned and reworked and that the new clean gravel/cobble had far more interstitial spaces (e.g., for aquatic macroinvertebrates or small fishes) than before the flood.

Woody Debris in Streams

Wood comprising usable sized structure for fishes is classified as "woody debris" (Wallace and Benke, 1984; Andrus et al., 1988; Bilby and Ward, 1989). Floods input massive amounts of large woody debris, as riparian trees are either broken and uprooted by the force of the water (Piney Creek, 1982) or are undercut and fall into the stream (Brier Creek, frequent observations, WJM). The result is the input of many trees into the creek channels, many of which lodge to later provide structure in the stream. These microhabitats are among the richest for numbers and diversity of fishes in the streams. In addition, woody debris has the effect of deepening stream pools because it increases hydraulic scour (Andrus et al., 1988).

Flooding also can also remove large woody debris, as it can refloat logs or other large material that is within the stream channel and deposit it farther from the creek or above the normal water level, effectively isolating that woody debris from the stream, at least until the next major flood. Brown and Matthews (1995) provide a more detailed discussion of the dynamics of woody debris input and removal by floods.

Because large woody debris in many streams comprises important sources of primary and secondary productivity (Benke et al., 1984), as well as physical habitat for fish, floods, as related to the flashiness of the hydrograph, are important to stream fish assemblages. Golladay and Hax (1995) found greater reduction in meiofauna (very small invertebrates) on wood than on sediments in a Texas stream during an artificially engineered flow disturbance.

Finally, transport and deposition of large woody debris during floods can have very-long-term effects in streams and habitat for fish. An October 1981 flood in Brier Creek, Oklahoma, deposited many large trees with limbs intact into a pool at Station 6 [most downstream site (Ross et al., 1985; Matthews et al., 1988)] in Brier Creek, and deepening of pools by scour near the woody debris followed, providing both physical structure (wood) and deeper pools for fish. Those tree trunks (with many limbs now gone) remained in Station 6, embedded in sand/gravel of the stream bed, as recently as August 1996 (Matthews, personal observation). However, smaller woody debris can be rapidly moved within even a small stream. Golladay and Hax (1995) reported the transport of most marked pieces of woody debris (wooden dowels, 3 cm in diameter and 0.5 m long, water saturated before being placed in the stream) more than 500 m downstream in Sister Grove Creek, Texas, during an artificially engineered flood.

Channel Formation and Maintenance

Finally, at a larger scale, the flow of water may be critical to channel formation and maintenance, with direct impact on fish assemblages. For example, in the upper Colorado River basin, regulation of the river by dams has resulted in less frequent channel-forming/maintaining floods. As a result, midchannel bars are being stabilized with vegetation, and fringing backwaters are being cut off from the mainstream by sediment deposits that form sills across the backwaters. Ultimately, these sills can deny fish access to backwaters, which function as important nursery sites for young and/or or resting and feeding places for adults (including the endangered Colorado squawfish, *Ptychochelius lucius,* and the razorback sucker, *Xyrauchen texanus* (Minckley, 1991). Wuerthrich (1996) summarized a large experimental flood in the Colorado River, designed to restore natural channel features. (Results remain unclear in 1997). Unnatural flow regimes [e.g., abrupt daily release of water from dams (e.g., Glen Canyon)] also flush backwater habitats abruptly, killing or removing fish larvae or fry. As a result, numerous species like the humpback chub, *Gila cypha,* and razorback sucker are in danger of extinction in the American west (Minckley 1991; Minckley and Deacon, 1991). Andrews (1994) clearly showed, for Sagehen Creek, California, that bedload transport was dependent on periods of elevated flow, at or near bankfull.

Anthropogenic Effects on Hydrographs and Flood Frequency

John D. Black (1954), in his classic *Biological Conservation,* noted the effects of the stripping of natural vegetation and anthropogenic disturbance on the flow of small streams in his native Ozark Mountains. A stream that in his youth had flowed perennially, was altered thusly:

> The hillside changed owners, and was clear-cut to provide a good pasture for goats. In three short years the slope was rent by ugly gashes into the bright red clay and completely through the thin, brown, forest topsoil. The brook runs now only six or seven months of the year; sometimes only four or five. A bone-dry creek bed is now a matter of accepted fact. Roaring floods are expected. (Black, 1954)

A flashy hydrograph often results from anthropogenic activities such as stripping of vegetation (e.g., timber cutting), overgrazing, or urbanization (with many streets and parking lots that make runoff rapid). As an example, Big Bend National Park (Texas, USA) now has creeks (Terlingua and Tornillo creeks) with little flow during much of the year. In presettlement conditions, the watersheds were much different than today. Late in the 1800s, cattle operations moved in, and severe overgrazing occurred. The land now appears arid and flows are very irregular, whereas the greater vegetation in the presettlement condition apparently resulted in more stable flows in those creeks.

In Oklahoma, much land has been converted to grazing for cattle. Although the pastures often appear to be densely vegetated to a passing motorist, they are only a weak remainder of the dense and tall vegetation that naturally existed in the native tall-grass prairie. This comparison becomes dramatic when one finally sees a true prairie, with big bluestem growing to head high or taller, or attempts to walk through this dense prairie grass. It then becomes easy to envision the differences in infiltration and water release time to streams in the tall-grass prairie versus those in most rural grazing lands of the lower Great Plains. Accordingly, there have apparently been major changes in the water quality and physicochemical conditions, particularly in smaller tributary streams (Matthews, 1988), which likely have strong consequences for the fish assemblages.

4.6 Summary

Fundamental physical and chemical aspects of lake and stream ecology all relate to local assemblages of fishes. Water provides habitat for all fishes; thus, the physical properties of that water (temperature, oxygen, pH, etc.) have strong effects on individual fishes, summing to a strong influence on which species or life stages will occur in a particular habitat. Long-term and large-scale phenomena (e.g., geological processes forming lakes or river basins) set the stage for the kinds of aquatic environments that will be found in a region of the Earth. Within ecological time in those aquatic environments, it is solar input, water clarity, and wind-driven changes that most influence daily conditions in lakes, and rainfall and flow probably have most influence on daily conditions in streams. These changes in conditions, on daily, seasonal, or interannual scales, have great influence on the physical conditions encountered by a fish or an assemblage of fishes, hence those fish that can or will occur locally. Variation in physical and chemical conditions is normal in most lakes or streams, and fish are generally well adapted to meet natural challenges. Anthropogenic changes that influence rates or schedules of processes in these systems or that change their chemical or physical structure are most likely to have serious long-term consequences for fish assemblages.

5

Influence of Global to Regional Zoogeography on Local Fish Assemblages

5.1 Introduction

To understand the occurrence of fish in local assemblages, it is necessary to consider the overriding, global influence of zoogeographic history and drainage basin relationships on present-day assemblages. G. R. Smith (1981) concluded that barriers to dispersal (basin boundaries) were of paramount importance in controlling species density and patterns of evolution of fishes. Without zoogeographic information, experimental assessments of present-day ecological phenomena run the risk of attributing present-day functioning or composition of fish assemblages to extant, adaptive processes, even though composition of those assemblages may relate as much or more strongly to historical events (e.g., Smith, 1981; Mayden, 1987) in either the "deep" or "recent" past that influence which species can occur in a stream or lake. Understanding the evolutionary origins [as much as is known for some groups, cf. Fink and Fink (1981)] or distributional affinities of the individual species in a local assemblage or regional fauna is of importance, as is an understanding of geological events affecting the location, in making any predictions about comparative ecology of species or assemblages.

Information on continental movements (Briggs, 1986; Golonka et al., 1994), age of taxonomic lineages (Cavender, 1986), and local geology or geomorphic history (Gilbert, 1980; Minckley et al., 1986) may provide more insight into the coexistence of certain species combinations than do hypotheses based on extant species interactions, present-day flow patterns or discharge schedules, and the like. An ecologist working on a local fish community problem might view rather differently the predicted interactions between two species in a habitat to which they have dispersed from two different directions in recent geological history as compared to two species whose history suggests the likelihood of millions of years of coexistence. Whereas some species interactions (like predator–prey or competition) could develop rapidly in ecological time, other interactions between species, such as mutualisms or interactive "resource partitioning," might require longer periods of contact. It may be helpful, for example, for a fish ecologist in the Mesa Central of Mexico to realize that some of the species in an assemblage may be of older, more northern origin and some of recent Central American origin.

This chapter thus describes some of the zoogeographic factors that influence coexistence of fishes at the local level, particularly continental associations, origins of faunas, intracontinental faunal regions, drainage basin-boundary effects, and similar large-scale phenomena. Of necessity, I draw largely on the North American fauna for examples and amplification, because much of the history of the common taxa is well known. However, ecologists working with freshwater fish elsewhere in the world should, in most cases, be able to locate detailed ichthyological accounts that help to provide at least an overview of the zoogeographic factors that influence the regional species pool (fauna), hence the local assemblages. As a descriptive model, the "faunal screens" of Smith and Powell (1971) (see Fig. 1.6) as further modified by Tonn (1990) or Moyle (1994) are useful (Fig. 5.1). The concept of zoogeographic filters dates at least to Simpson (1953), but Simpson's emphasis was on the filtering role of barriers or corridors like land bridges and island archipelagos, mostly on movement of mammal faunas. The concepts of Smith and Powell, Tonn, and Moyle incorporate many biogeographic through local biotic factors to "filter" the available fishes of the world down to those found in one locality (much as is the intent of this entire book!).

This chapter emphasizes the influence of zoogeography and evolution at three different scales: (a) "deep" global-evolutionary events, (b) subcontinental, geologic, and related phenomena, and (c) more recent regional or basin-level effects on local assemblages. This chapter builds on the knowledge of continental movements and the history of evolution of modern, major groups of freshwater fishes to help explain the appearance of those groups in the fossil or recent records in different land masses. To this end, extant distributional similarities of freshwater fish faunas, worldwide, are considered at the family level. Finally, the sections on subcontinental and regional zoogeography examine effects of glacial and geological events, underscore the importance of river basin boundaries, review the "historical ecology" of Mayden and other workers, test some ideas about the influence of regional zoogeographic patterns on local assemblages, and review use of "ecoregions" in fish ecology.

5.2 Global Zoogeography of Freshwater Fishes

Zoogeographic Realms

Berra (1981) describes the establishment of the six zoogeographic realms by Alfred Russell Wallace. These regions, perceived by Wallace as major delimiters

Influence of Global to Regional Zoogeography on Local Fish Assemblages / 193

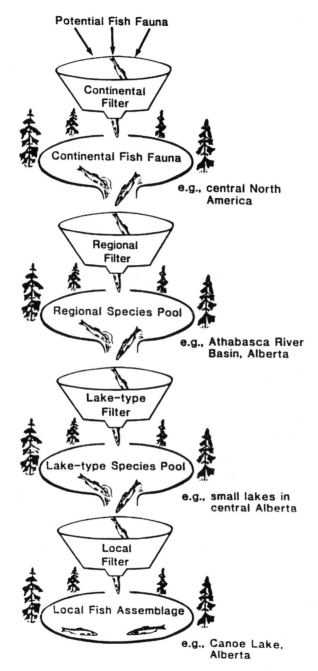

Figure 5.1. Series of "filters" influencing composition of local fish assemblages in small north temperate lakes (Canada). [From Tonn (1990).]

of the Earth's biota, remain today, with some minor modifications, important as defined regions in zoogeography of fishes (Darlington, 1957; Berra, 1981; Moyle and Cech, 1988; Nelson, 1994) and other vertebrates. Moyle and Cech (1988, pp. 320–332) provide details on the distributions of major fish groups and on freshwater fish diversity within the six regions. Darlington (1957, Chap. 2), although now outdated with respect to taxonomy of some groups, also is valuable as a catalog of many fish families among the zoogeographic realms.

From the perspective of fish ecology, the established zoogeographic realms or regions continue to be used as a first-order predictor of the numbers or kinds of fishes to expect in any local assemblage. I test the veracity of those realms, below. However, as shown in Figure 2.2 of this book, location in a faunally rich realm does not by itself dictate that all local assemblages will be species rich: Many small or simple sites, worldwide, have only a few fish species. Nevertheless, a fundamental awareness of similarities and differences among the faunas of these major regions would clearly be of paramount importance in, say, an attempt to conduct a comparative ecological study on a global scale. Something as conceptually simple (albeit logistically challenging) as evaluating niche overlap in the field for two to three closely related species in various tropical and temperate sites, worldwide, might be predicted to have very different outcomes depending on the composition, richness, or species "packing" of a region. With a range of approximately an order of magnitude in species richness (from about 225 freshwater fish in the Australian to about 2600 in the Neotropical), it is clear that experimental assumptions or protocols applied in one region might be less than fully appropriate if applied in an identical manner elsewhere in the world.

These differences in species richness at the level of entire realms is clearly related to the degree of connectedness or isolation of each region from others during the last 200 million years or so, when many of the present-day teleosts were evolving (Cavender, 1986; Moyle and Cech, 1988). Darlington (1957) was the last major biogeographer to tacitly reject continental drift and seek explanations in prehistoric land bridges [including some incredibly long ones (e.g., from extant Africa to South America)]. Although now-drowned land bridges (e.g., Beringia) clearly were important to the distribution of freshwater fishes, it is now universally accepted that continental movements (Golonka et al., 1994) provide the explanations for much of the dispersal of primary and secondary freshwater fishes over landmasses of the Earth (Moyle and Cech, 1988).

Testing Patterns of Fish Families on a Worldwide Zoogeographic Template: Was Wallace Right?

Students of fish biogeography have for a very long time basically followed Wallace's designations of faunal realms or regions, and many experts on regional faunas have attested to boundaries generally matching those of Wallace (e.g., Myers, 1966). Details (e.g., position of "Wallace's LIne") may be argued (Berra,

1981), but, in general, the regions have withstood the test of time as defining large fish faunal regions for fish as well as terrestrial vertebrates. However, so far as I am aware, nobody has, until now, tested the boundaries of freshwater fish faunal regions in an "uninformed" or "naive," objective quantitative analysis of similarities and differences in fishes among smaller units of geographic space— to ask if those smaller units would indeed aggregate into the faunal realms of Wallace.

To test the hypothesis that Wallace's realms faithfully reflect known modern concepts about distributions of major fish taxa (families), I used the maps of 97 primary and secondary freshwater fish families in Berra (1981) as follows. First, I arbitrarily divided the terrestrial regions of Earth into 52 regions (mostly large river drainage basins, hereafter called "basins") (Fig. 5.2), *without* consulting any references on details of range boundaries for any fish groups (and attempting to ignore those that I already knew) and roughly following the outlines of the major river basins of the world as provided by Figure 1 of Berra (1981).

I then used Berra's range maps for families to score as present (1) or absent (0) each primary or secondary fish family in each of the 52 basins, resulting in a rectangular (1–0) matrix with the 97 families across the regions. [Although aware that some of the families in Berra's compilation have not been split into more than one family (e.g., Fundulidae separated from Cyprinodontidae), I followed families as mapped, assuming that this analysis, if useful, should be sufficiently robust to tolerate a few such discrepencies.] Similarities among all possible pairs of basins [as "OTU"s (Sneath and Sokal, 1973)] with respect to the presence and absence of families ["characters" (Sneath and Sokal, 1973)] were estimated by Jaccard's index and a simple matching coefficient (Sneath and Sokal, 1973) to produce two triangular matrices of basin similarity values. Each of the matrices was clustered by both UPGMA and single-linkage clustering (Sneath and

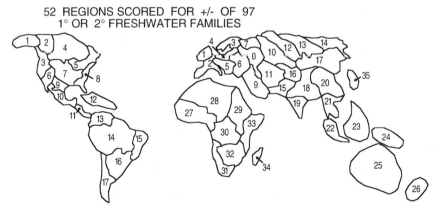

Figure 5.2. Map of the world indicating location of 52 regions scored for presence or absence of primary and secondary freshwater fish families.

Sokal, 1973) by the NTSYS computer programs (Rohlf, 1990) to produce a total of four dendrograms depicting clusters of basins at various levels of similarity. Evaluation of all four phenograms suggested that the most interpretable structure was produced by the Jaccard–UPGMA procedure; hence, it is used below.

The Jaccard–UPGMA dendrogram resulted in (with all cutoff levels uniform) nine identifiable clusters of drainage basins, worldwide. Five clusters were relatively large, with 5–17 drainage basins included, and the remaining 4 clusters included only 1–2 basins each. The results of this analysis are depicted in Figure 5.3, indicating the positions of the nine regions having similar fish faunas, worldwide.

The results were striking. Several of the major clusters, derived objectively and "naively" from presence–absence of families in basins, matched almost perfectly the subjectively or intuitively defined zoogeographic realms of Wallace. Cluster 1 of Figure 5.3A is equivalent to the Nearctic realm. Cluster 2 (Fig. 5.3B) coincides almost perfectly with the Palearctic of Wallace, with the exception that far-north Africa was not scored as a separate area in my analysis, hence could not have been included in the Palearctic. Cluster 3 (Fig. 5.3C) is tantamount to the Oriental region. Cluster 4 (Fig. 5.3D) is Africa, or Etheopia, minus Madagascar. Cluster 5 (Fig. 5.3E) of my analysis is interesting in that it matches the Neotropical, minus South America south and west of the Paraguay–Parana basin (which comprises independent Cluster 6, Fig. 5.3F), which, as a region, has long been known for its depauperate fish fauna.

Interestingly, my Cluster 5 (Fig. 5.E) does include the Magdalena River basin of northwest South America and much of Central America. Lundberg et al. (1986) noted that the Magdelana basin apparently had a long history (15 million years) of the presence of fish species identical to some of those present there today, but that the Magdelana had apparently suffered noteworthy local extinctions. My Cluster 7 (Fig. 5.3G) links Madagascar with the West Indies, on the basis (for these depauperate freshwater faunas) of sharing Cichlidae and Cyprinodontidae [per Berra (1981)]. Parenti (1981, p. 462) removed Aplocheilidae from the family Cyprinodontidae which is recognized by Nelson (1994). However, because aplocheiloids (albeit two subfamilies) occur in both Madagascar and the West Indies, alteration of the taxonomy by Parenti (1981) should not alter the relatedness of those two regions, as shown in my analysis (Fig. 5.3G). Cluster 8 (Fig. 5.3H) matches Australia and island regions southeast of Wallace's Line, and Cluster 9 (not shown) consisted of New Zealand, with no primary or secondary freshwater families. In summary, the "naive," numerical analysis of similarities and differences among 52 major river basins or regions of the world resulted in a view of world fish faunal regions that strongly supports most of the century-old suggestions of faunal regions by Wallace.

Family Richness Among Basins, and "Basin Richness" Within Families

From the summary of Berra's (1981) maps of family distribution divided into 52 drainage basins or areas, worldwide, it was also possible to examine the numbers of basins having few versus many primary and secondary freshwater families (Fig. 5.4). There is an obvious strong bimodality in Figure 5.4, with one peak at low family richness (21 of the 52 regions containing only 4–6 primary or secondary freshwater families) and another peak at relatively high family-level richness (18 of the 52 regions having 20–35 such families). Relatively few regions were intermediate. The general pattern, reflecting larger patterns well known to biogeographers, appears to be that river basins in some regions of the world have a very high diversification in the numbers of freshwater families, including basins largely in the Neotropical, Ethiopian, or Oriental realms, and that others (Nearctic, Palearctic, Australian) do not (Berra, 1981). The high numbers of families per region also coincide with those realms (Neotropical, Etheopean, and Oriental) with a high number of endemic primary freshwater fish families (Berra, 1981).

It is also possible from my summary of Berra's maps to ask how many families occupy only one or a few basins and how many occur widely among many basins over the face of the Earth. Figure 5.5 (plotting number of families versus increasing numbers of regions occupied, on a $\log_2$ scale) indicates that of the 97 primary or secondary families included by Berra, only 9 are limited to one of the drainage areas I defined. Clearly, such families are "endemic" to the extreme! There was a sharp peak (Fig. 5.5) at 5–8 regions, with 43 (44%) of the families occurring across that moderately wide distributional pattern. Far fewer families occurred across a large number of drainages (e.g., 20 or more), and only one family occurred in the highest octet (including 33–52 regions). This family, Cyprinidae (noted earlier to be most nearly cosmopolitan of all freshwater families), was present in 41 of the 52 drainages, absent only in South America, east of Wallace's Line in the Australian, and in extreme north Asia. Therefore, it would seem that in evolution of fishes worldwide, there have been two relatively different patterns among the higher taxa that are broadly important in fresh water. Some, like Siluriformes (catfishes) and Characiformes (characins and allies) have experienced a proliferation of families within an order, with many endemic families restricted to one continent. At the other extreme, the order Cypriniformes (as presently known) is characterized by fewer families, but by families like Cyprinidae and Cobitidae (loaches) that are widely distributed among drainage basins over large parts of the world.

Primary, Secondary, and Tertiary Freshwater Fishes (Myers)

Fish species differ markedly with respect to their ability to cross salt water (Moyle and Cech, 1988), and it is possible to describe families on the preponderance of salt tolerance of their species. Smith and Powell's (1971) "freshwater filter" and "continental filter" purported, for any given stream site to screen out, from the worldwide fish fauna at large, marine fishes that do not tolerate fresh water and freshwater fish on continents other than the one under consideration. Naturally,

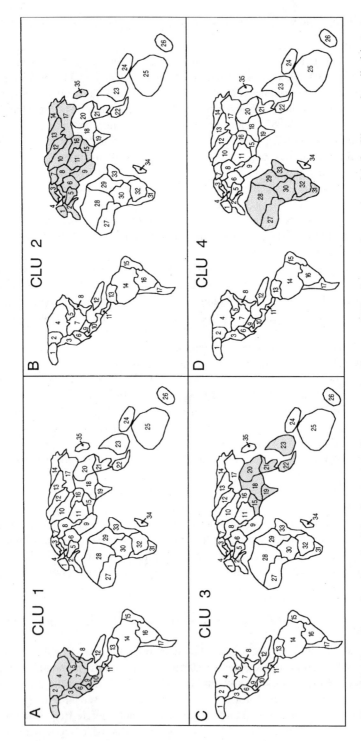

Figure 5.3. Locations of regions in eight clusters, based on presence—absence of 97 primary and secondary fish families, as described in text.

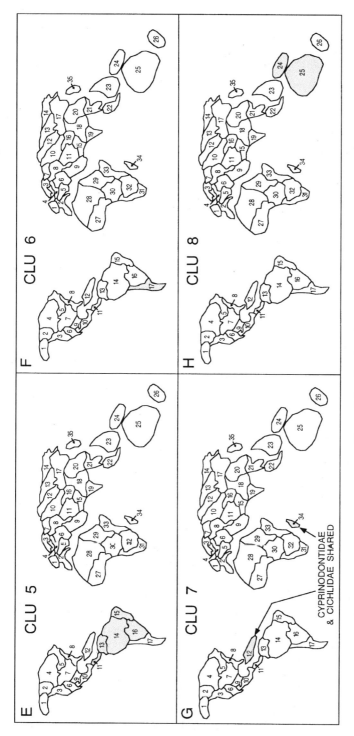

Figure 5.3. Continued.

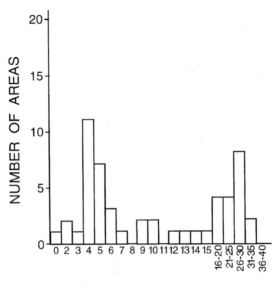

Figure 5.4. Frequency distribution of number of areas with specified number (*x* axis) of primary and secondary freshwater fish families, worldwide.

neither of these criteria are absolute in geological or even ecological time. Some marine taxa do make excursions into fresh water, and there are numerous brackish-water estuarine forms that move considerable distances into water with salinities as low as 3–4 ppt [e.g., fishes of marine/estuarine origin found a considerable distance upstream into freshwater bayous at the western end of Lake Pontchartrain, Louisiana (Cashner et al., 1994), or even less than 1 ppt (R. C. Cashner, pers. comm.)].

In three influential papers, G. S. Myers (1938, 1949, 1951) developed a classification of fishes that occur in fresh water on the basis of their tolerance for salt water (summarized by Berra, 1981 and Moyle and Cech, 1988). In Myers' classification, amplified by Darlington (1957), primary freshwater fish families are those that generally cannot tolerate any elevated salinity above that of "fresh water," thus ocean is a major barrier to their dispersal. Secondary freshwater families can tolerate limited excursions into saline waters (hence might migrate short distances through marine or brackish habitats). Peripheral freshwater families are of marine origin, having invaded freshwater recently, or moving back and forth from ocean to fresh water. Within families, some species vary in their salt tolerance. Gunter (1942) compiled a list of all known fish species in North and Middle America that were known to occur both in fresh water and sea water, which makes clear that within many "freshwater" families, a few of the species

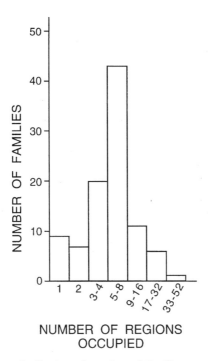

Figure 5.5. Frequency distribution of number of families versus number of regions occupied (from Fig. 5.2).

may enter brackish or salt water. Gunter noted that, "Several catostomids, cyprinids, centrarchids and other strictly freshwater fishes enter slightly salt or even strongly salt water in North America, some of them going into water that is over 12.00 parts per thousand saline . . . ," and offered the opinion that, "Probably most freshwater fishes can be accustomed to low concentrations of salt water." As an example, although a "freshwater" species, some populations of largemouth bass (*Micropterus salmoides*) live in brackish coastal waters (Meador and Kelso, 1990).

In general, however, a family can be assigned to "primary," "secondary," or "peripheral" on the basis of its distribution and history (Berra, 1981). Moyle and Cech (1988) generally follow Myers' classification, but lump together Myers' primary and secondary families as "freshwater dispersants" (with members "by and large not capable of traveling for long distances through salt water") and "saltwater dispersants" ("whose distribution patterns can be explained in a large part by movements through salt water"). Note that in considering dispersal of freshwater fishes across "salt water," not all oceanic waters can be assumed to be at a salinity of "seawater" at all times. Crossman and McAllister (1986, p. 91) note that during periods of active melting of glaciers, there may have been a freshwater surface layer accumulated on northern oceans, which could have provided corridors of dispersal for freshwater fishes. Crossman and McAllister also cite Barber and Murty (1977) as suggesting that freshwater bridges could have existed *under* perennial sea ice in the postglacial period, offering dispersal of fish populations. Lindsey and McPhail (1986, p. 642) also noted that temporary freshwater or brackish-water bridges could have existed in the sea between adjacent river confluences, during periods of high runoff. Conversely, Novacek and Marshall (1976) argued that shallow intrusions of ocean water into continental regions (during rising sea levels) could result in salinities exceeding those of normal ocean water, because of evaporation in the shallow-water intrusions. In considering all potential crossing of apparent oceanic salt barriers, it is necessary to keep in mind that even primary and secondary families may have had some opportunities to traverse what is normally considered "ocean," and that narrow inlets of seawater only a few meters wide and deep could be formidable barriers to primary freshwater fishes (Myers, 1966). Therefore, a consideration of the history of connectivity of landmasses and their freshwater drainages is requisite to any understanding of the biogeographic history of primary or secondary families. For example, Howes (1991) pointed out that the distributions of subfamilies of Cyprinids in Eurasia matched the lithospheric plates of those continents and subcontinents very closely.

Continental Movements and Freshwater Connections

The "theory" that landmasses change position and connectivity ("continental drift") has a history somewhat analogous to that of the "theory" of organic evolution, in that no serious scientist has doubted the fundamental facts of either in recent decades. The debate is over details, although highly important ones (Dickins, 1994). At present, geologists rely most on magnetic anomaly data, fracture zone locations, evidence of locations of seafloor subduction and spreading, to place boundaries of prehistoric landmasses; paleomagnetic data and "hot spot tracks" to determine orientation of continents relative to the Earth's spin axis; and paleomagnetic data, along with biogeographic and paleoclimate indicators to determine past latitudes of continents (Golonka et al., 1994). McFarland et al. (1979, pp. 794–795) provide a readable review of continental drift, emphasizing the movement of continents with plates, in a very long-term pattern of "fragmentation–coalescence–fragmentation" and give a succinct account of changes in vertebrate life on Earth in the context of geological changes. Nance and Murphy (1994) suggest a semiregular, half-billion-year cycle of continental coalescence and fragmentation and provide scenarios for existence of supercontinents up to 750 million years ago (mya), long before the existence of Pangaea as the most recent supercontinent.

Cracraft (1974) provides a detailed analysis of the evolution and distribution of all vertebrate groups in the context of continental drift as it was known at the time of his writing, including the hypothetical origins in either Laurasia or Gondwanaland of many freshwater fish families. Darlington's (1957) detailed summary of fish distribution suffered from his necessity to hypothesize complicated scenarios of movements of fish along land routes or land bridges that have never existed. Darlington's work remains a useful summary of the distributions of freshwater fish families, but many of his hypotheses about long movements of fishes are unnecessary when continental movement is accepted. Myers (1966) clearly rejected any hypothesis other than continental drift to explain relationships between the rich African and South American fish faunas, whereas Darlington (1957) required a land bridge spanning the present Atlantic Ocean.

Although a given species may now occur on a particular existing continent, it is quite possible (particularly in the case of the Holarctic faunas) that the species or its immediate ancestor evolved (and acquired fundamental ecological traits) on a different continent, with dispersal as an extant species (or a recent shared ancestor) to the continent of interest. Because of repeated mergers and separations of western North America, eastern North America, Europe, and northern Asia (to use the extant names) during the last 150 million years, many fishes that now occur on two continents formerly may have been widespread on one. Freshwater fish ecologists studying in the Northern Hemisphere need to understand the evolutionary history and derivation of the Laurasian fauna. However, in placing fish zoogeography in the context of continental movements, it is important to remember that, much like cladograms provide only hypotheses of the true evolution of fish groups, estimates of dates (millions of years ago) and positions of continental connections provided by geologists are also hypotheses and subject to change (e.g., Dickins, 1994). Below, I review distributions of some major fish groups in light of previous zoogeographic work, taking much information from Briggs (1986), but with updated details on continental positions from Golonka et al. (1994).

Briggs (1986) and Moyle and Cech (1988) summarized continental affinities and movements since the existence and subsequent breakup of the worldwide "supercontinent," Pangaea, some 200 mya. Although geological and geographic experts may disagree on details, the basic patterns described by Briggs (1986) and Moyle and Cech (1988) provide an overview of potential pathways of evolution and movements of freshwater fish lineages. A much more detailed and updated presentation of continental landmasses, locations of mountain ranges, and paleoclimates during the last 550 million years is in maps of Golonka et al., (1994), and there are some differences (below) between their maps and those of Briggs (1986) that could influence the interpretation of zoogeography of some fish groups.

To facilitate the understanding of the ebb and flow of continental connections since the breakup of Pangaea, it is convenient to reduce continents to an abstract series of blocks (Fig. 5.6), as Cracraft (1974) did to review distributional histories of vertebrates in general. In the Triassic, ~ 230–200 mya, all of the large land-

204 / *Patterns in Freshwater Fish Ecology*

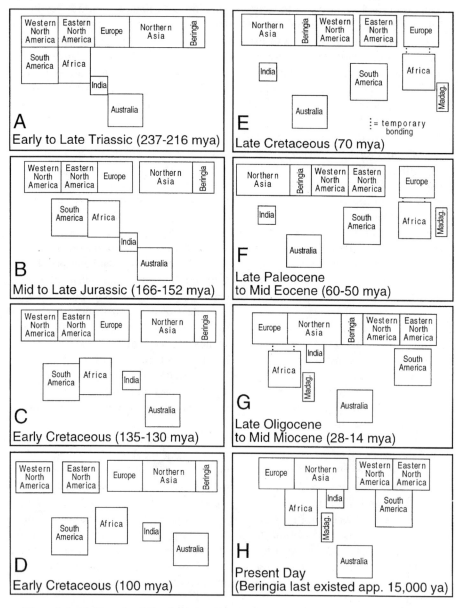

Figure 5.6. Schematic of distributions, movements, and connections of continental landmasses from Triassic to the present.

masses of the Earth were more or less connected into one large continent, Pangaea (Fig. 5.6A). By about 160–150 mya, Pangaea had separated into a northern group of landmasses (Laurasia) that included present-day Asia, Europe, and North America [with North America and Europe (Euramerica) separated from the Asian continent by the Turgai Sea] and a southern continent (Gondwanaland) comprising South America, Africa, Antarctica, Australia, and India (Fig. 5.6B). This basic north–south megacontinent separation would persist until almost modern times, although intermittent connections of Africa with southern Europe (Modern Spain) may have taken place occasionally from about 100 mya to Recent (C. Scotese, Dept. of Geology, U. Texas at Arlington). In Figure 5.6, Antarctica is omitted, although it remained as a land connection between South America and Australia as recently as 100 mya (Cracraft, 1974), because paleoclimate maps (Golonka et al., 1994) depict the temperatures of Antarctica as too cold to generally support freshwater fishes.

One of the first rifts in the landmass of Pangaea separated present-day North America from Africa, opening the germinal North Atlantic. Separation of the northern and southern megacontinents has been considered complete by the Early Cretaceous, approximately 135 mya (Fig 5.6C), and North America–Europe (Euramerica) and Asia remained as two separate landmasses. However, recent evidence from dinosaur material in Africa suggests that a land bridge between Africa and Europe may have remained into the Early Cretaceous (e.g., 150–130 mya), keeping open a route of dispersal between the northern and southern supercontinents (Morell, 1994). The events described to this point were largely before evolution of modern fish genera, but some modern families, such as sturgeons, paddlefish, gars, and bowfins were in existence (see below). Below, I consider the northern and southern landmasses separately.

Northern Landmass

By the Middle Cretaceous (100 mya), the North American midcontinent seaway (Fig. 5.6D) split western North American from eastern North America. Here, the maps of Golonka et al. (1994) and Briggs (1986, Fig. 1.4) differ. Briggs' Figure 1.4 depicts for the Mid-Cretaceous, a continued connection of eastern North America with the European continent, whereas Golonka et al. (their Fig. 61) show the separation of Europe from eastern North America (followed in Fig. 5.6D) and a connection of Europe with Asia which continued to at least 94 mya (Golonka et al., 1994, Fig. 67). However, the maps by Golonka et al. are global in scope; they were not intended to capture all fine details of connections of continents (C. Scotese, personal communication). Therefore, it is possible that intermittent connectivities between closely positioned continents occurred during periods of sea-level lowering (cf. Novacek and Marshall, 1976) that do not appear on maps of Golonka et al. (1994).

By Late Cretaceous, about 90–70 mya, a land connection (Beringia) existed between eastern Asia and western North America across the present-day Bering Strait (Fig 5.6E). McPhail and Lindsay (1986) detail several appearances and disappearances of this land connection (Beringia) and its potential role in dispersal of freshwater fishes between Asia and North America (and redispersal of fishes into Alaska–Canada after glacial retreats (below). By about 65–60 mya (Paleocene) the North American mid-continental sea was gone, and Briggs (his Fig. 1.6) suggested that the Beringia connection now provided a continuous connection of the entire northern landmass of Europe–to–North America–to–Asia (with Europe still connected to North America, instead of to Asia). However, Golonka et al. depict Europe from about 70 to 50 mya to be separated by seawater from the rest of the northern landmass, as shown schematically in Figure 5.6F. By about 65 mya (below), fossil Hiodontidae, Clupeidae, and Esocidae were present in North America (Cavender, 1986), and salmonids and ictalurid catfishes appeared in North America or shortly thereafter.

Southern Landmass

At about 135 mya, South America and Africa remained one landmass, but India had separated (Fig. 5.2C) and was moving northward toward its collision with south Asia. A rift appeared between southern South America and south Africa about 130 mya (Golonka et al., 1994, Fig. 55), with the separation of the two continents by a narrow sea by 119 mya depicted by Golonka et al. (their Fig. 58). Thus, by about 100 mya (Early Cretaceous) the southern continents were separated into the basic elements of which they consist today, except that India remained isolated. However, and this is important in selecting among some of the competing hypotheses of Ostariophysian zoogeography, it can be argued that some degree of land bridge remained in place between South America and Africa as the continents moved apart. Novacek and Marshall (1976) cite various geological sources as indicating that the break between Africa and South America was as recent as 90 mya and also cite estimates from seafloor spreading to argue that final separation of the two continents was not complete until early Santonian, about 80 mya. This postulated that a late separation of Africa and South America could have facilitated transfer of characoid fishes (below).

By 70 mya, the African continent was divided into a western and eastern element (Fig. 5.6E) by the incursion of a large inland tongue of the sea (Golonka et al., 1994, Fig. 70), or may have been even more fragmented (Briggs, 1986). In the southern hemisphere at 65 to 50 mya, India was traveling north toward its collision with Asia, which would cause the uptrusting of the Himalayas. Metcalfe (1994) depicts the collision of India with the Asian continent in Late Eocene (i.e., perhaps 35–40 mya). At least by the Late Oligocene (28 mya), India had joined the Asian landmass, and Africa neared its juncture with southwestern Asia (Golonka et al., 1994, Fig. 79).

Attaining Modern Configuration

By the Miocene, Briggs' Figure 1.8 shows all of the continents essentially in their modern configuration, with the Beringia connection of Asia and western North America persisting. Although Golonka et al. (their Fig. 82) show the separation of Africa from the Laurasian landmass as late as 14 mya, which could have implications (below) for evolutionary history of important groups like cyprinids, C. Scotese (personal communication) indicated that intermittent land connections of Africa to the Laurasian landmasses could have occurred during the last 100 mya in the vicinity of present-day Spain, and that an intermittent connectivity of Arabia and Iran has been possible since about 30 mya. Thus, there have been opportunities for dispersal of freshwater fishes between Africa and the northern landmasses during that time. North and South America remained separated as recently as 20 mya in the Late Tertiary (Briggs, 1986) but were joined by 14 mya (Golonka et al., 1994, Fig. 82) as Central America was uplifted (Fig. 5.6G). By this time, virtually all of the modern North American fish families were present on the continent, with the exception of Atherinidae, Poeciliidae, and possibly others of Central or South American origin. By 15–20 mya, many modern genera existed in North America, such as *Ictiobus, Ictalurus, Amierus, Fundulus,* and *Morone* (Smith, 1981; Cavender 1986).

Occurrence of Freshwater Fish Families Relative to Continental Movements

Berra (1981) lists a total of 97 primary or secondary freshwater fish families, worldwide, all of which have distribution patterns affected in some way by the events described above for the last 200 million years. [The number of families have been increased by recent systematic revisions; e.g., by division of family Cyprinodontidae by Parenti (1981).] It is beyond the scope of this book to attempt a review of the zoogeographic history of each of these families. However, it seems important to review general patterns in freshwater fish distributions that coincide with the worldwide continental events that have been considered, and to evaluate families that are most widely distributed or play roles in many local fish assemblages.

The global patterns in distribution of primary and secondary freshwater fish that can be placed into simplified categories follows [with approximate number of families, taken from maps of Berra (1981) and modified per advice of Nelson (1994)—in parentheses].

1. Endemics to a single continent or region (75 families): Many families [e.g., the electric eels (Electrophoridae) in South America, the sunfishes (Centrarchidae) of North America, the gymnarchids (Gymnarchidae) of Africa, or the hillstream loaches (Homalopteridae) of Asia] occur on a single present-day continent, with no evidence of any wider former distribution. For most such families, there is no need to invoke

continental "drift" to explain their present distribution within a continent, but in some areas (e.g., North America), the former schism of the continent into two parts (e.g., in the Early to Late Cretaceous) may influence the distribution of a family within the continent (e.g., Lepisosteidae limited to east of the Continental Divide in North America).

The numbers of endemic families vary widely among the present-day faunal regions or continents. South America, which was isolated for 100 million years (since its separation from Africa about 120 mya, until its connection to North America less than 14 mya) with respect to potential exchange of freshwater fishes, has by far the greatest number of endemic families, about 30 [maps in Berra (1981)]. Many of these are families of the order Siluriformes (catfishes). The Oriental region and Africa have 19 and 15 endemic families, respectively. For Africa, this level of endemism apparently reflects its own long period of isolation from other continents (similar to that of South America, or not, depending on interpretations of a possible land connection of Africa to south Europe and Asia about 60 mya). For the Orient, there is geological evidence of a long and convoluted period of rifting of relatively small blocks of land [described as continental "slivers" and "fragments" by Metcalfe (1994)], each of which could have carried its own freshwater fishes, with the isolation and repeated vicariance events apparently facilitating evolution of family-level endemics. For India, included in the Orient in my count above, there was the obvious long period (~ 100 million years) of oceanic isolation from any other continent, and some families apparently arose during that period.

Fewer endemic families (8) are in North America, and only one primary or secondary freshwater fish family is endemic to Eurasia [Valenciidae, separated from Cyprinodontidae by Parenti (1981); distributed only in a small part of Europe just north of the Mediterranean Sea]. In addition, the "peripheral" family Cottocomephoridae, endemic to Lake Baikal and area rivers in Russia, provides an example of a family of marine origin undergoing extensive radiation in isolation in fresh water (Berra, 1981). Australia has only three endemic primary/secondary freshwater families, in spite of its very long isolation from the rest of the continents.

2. Holarctic (2 families): Two truly Holarctic families are widely distributed throughout the Laurasian landmasses, showing influence of the former links between North America, Asia, and Europe. The Esocidae (pikes and pickerels) occur throughout temperate to cold fresh waters of the Nearctic and Palearctic (Berra, 1981, p. 38). Umbridae (mudminnows) also show a broadly disjunct Holarctic distribution, with members in eastern Asia, Europe, Eastern North America, and Western

North America. This pattern suggests a former, much wider distribution for mudminnows throughout the Laurasian landmasses. However, their absence from much of the northern Asian continent suggests that dispersal of this family could have been, during various connections of landmasses, by transfers involving eastern Asia–Beringia–western and eastern North America–Europe.

3. Transpacific (3 families): This distribution pattern reflects the connection of Asia and North America via Beringia, but without extension of the family into Europe. The ancient family Polyodontidae (paddlefishes), with only two extant species (*Polyodon spathula* in eastern North America and *Psephurus gladius* in the Yangtze River drainage of China) depicts a trans-Pacific biogeographic distribution, with fossil representatives in North America west of the Continental Divide validating the North American–Asian link for this family in the Early Tertiary (Grande and Bemis, 1991). The sucker family (Catostomidae) follows a similar transpacific patterns, with modern representatives of several genera throughout North America, one species (*Catostomus catostomus*) in eastern Siberia, and one species (*Myxocyprinus asiaticus*) in China. In addition to these living Recent fishes, Grande and Bemis (1991) considered that five separate Eocene teleost fish groups of the Green River Formation, Wyoming, have trans-Pacific distributions, and none have trans-Atlantic distributions. Grande and Bemis (1991) use all of this evidence to argue for a much closer affinity of North American fishes to Asia, than to Europe.

4. Transatlantic (1 family): The only primary or secondary freshwater family to show an extant trans-Atlantic biogeographic pattern is Percidae (perches and darters). The lack of this family in eastern Asia and in western North America argues against its connection through Beringia; thus, the most likely avenues of transfer would have been during periods of connectivity of eastern North America and Europe. However, this presents a biogeographic conundrum, because recent geological evidence suggests that Europe and eastern North America have not been connected since 100 mya (Golonka et al., 1994), whereas Berra (1981) and Gilbert (1976) presumed that the family did not reach North America until probably the Pliocene (beginning 12 mya). Well-defined subgroups of this family occur on one or the other continent but not both (e.g., darters of North America). This all suggests old divisions between subgroups of the Percidae and begs the question of how a relatively recent transfer of this primary freshwater family took place since 12 mya, if not via Beringia.

5. Eurasian–Oriental (2 families): Cobitidae (loaches) occur throughout present-day Europe, Asia, and the Oriental region, but nowhere else,

with the exception of two species in far-north Africa—one at the very northern tip of Africa (in Morocco) and one in northern Ethiopia (Berra, 1981). These cypriniform fishes, although very common in the warm waters of Asia, apparently never moved sufficiently far northward to cross Beringia into North America and must have entered Europe after the last connection of that continent to eastern North America (more than 100 mya). The catfish family Siluridae also has a Eurasian–Oriental distribution, although it is not as wide ranging as the loaches (Berra, 1981).

6. Neotropical–African (1 family): The diverse and ecologically important family (Characidae) is limited in basic distribution to South America and Africa (with extension into southern North America after the relatively recent connection of North and South America via Central America. This pattern apparently results from the distribution of the family across the South American–African landmass before its separation about 120 mya, as noted above, but without further transport of species with movement of the Indian subcontinent. It should be noted that the characins of South America and Africa are now divergent, with the long period of separation leading to differences within the family. Nelson (1994) suggested that only one other completely freshwater family (Osteoglossidae, below) is shared by Africa and South America, although the osteoglossids are not limited to those two continents. Nelson (1994) also notes the continuing need for taxonomic revisions in this family—hence its position as the only truly South American–African "family" could change with systematic decisions in the future.

7. African–Oriental (7 families): Numerous ecologically important tropical fish families with large numbers of species share occurrence in central to southern Africa, southern Asia, and/or the Orient. These include the Notopteridae (featherbacks, Osteoglossiformes); the catfish (Siluriformes) families Bagridae, Schilbeidae, and Clariidae; the Channidae (snakeheads; Channiformes); and the perciform families Anabantidae (climbing perches) and Masticembelidae (spiny eels). In each of these cases, the family is distributed disjunctly in Africa and in southern Asia–Orient, with a hiatus between their eastern and western occurrences. One explanation for this pattern is the movement of ancestral taxa within each group from south and west to the north on the Indian subcontinent, but this landmass broke away from the African landmass by at least 130 mya; thus, ancestral taxa "riding" the Indian subcontinent northward would have had to be present in the relatively deep past. The earliest fossils reported for snakeheads (Berra, 1981) are from the Pliocene (within 12 mya) in India and Java, suggesting

that this taxon might have been represented on the earlier Indian subcontinent at the time of its collision with Asia at least 28 mya.

8. Broadly Tropical (5 families): Aplocheilidae (rivulines), formerly included in Cyprinodontidae (Parenti, 1981; Nelson, 1994) fit the "broadly tropical" biogeographic pattern, with distribution in Central and South America, Africa, Madagascar, the southern Oriental region, and the Indo-Malaysian archipelago (Nelson, 1994). Cyprinodontidae [*sensu stricto*, per Parenti (1981)] also fits the broadly tropical distribution represented in North and South America, central Africa, and the Mediterranean. The cichlids (Cichlidae), broadly tropical, comprise one of the most evolutionarily diverse and successful freshwater fish families on Earth, with large numbers of species throughout South and Central America (with extensions into southern parts of North America), Africa, Madagascar, and parts of India. Leaffishes (Nandidae) occur in South America, west Africa, and India–Orient. For this group, it could be hypothesized that ancestral forms were distributed across South America–western Africa before their schism, with transport via India to the Orient, if ancestral forms were available at a sufficiently early time. Berra (1981) suggested that leaffishes could reflect a formerly wide distribution among the connected southern Gondwanian continents, with subsequent extinctions over wide areas giving the family its present distribution. Berra (1981) also suggested a former Gondwanian distribution for the family Osteoglossidae, which now has a wide dispersal across northern South America, Central Africa, and the southeastern Orient, and into northern Australia (pointed out by Berra as the only family to cross Wallace's Line from the Orient into the Australian to any great extent).

9. Holarctic, Africa, and Orient (1 family): Only the very widely distributed minnow family Cyprinidae (Fig. 5.7) fits this pattern, occurring as prominent members of fish assemblages in North America, Europe, Asia, the Orient, and Africa (but lacking in South America). The origins and distributional history of the Cyprinidae have been the subject of much debate and can only be understood in the context of the history of the superorder to which cyprinids belong, the Ostariophysi. This complicated scenario, contrasting a cyprinid origin in the northern (Laurasian) landmasses versus a Gondwanic (southern landmass) origin, is outlined in a subsequent section.

10. Nearctic–Neotropical (1 family): Only one freshwater fish family shares extensive distribution in both North America and South America (excluding those of recent transfer across the Central American land bridge). The Poeciliidae (live-bearers) are widely distributed in fresh and brackish waters of eastern North and South America, Central

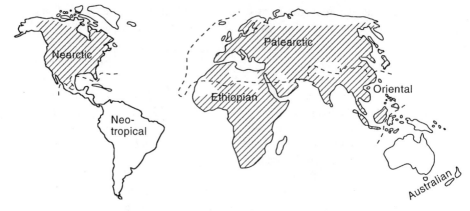

Figure 5.7. Distribution of the family Cyprinidae.

America, and islands of the Carribbean. Many members of the family are quite salt tolerant—hence dispersal of the family has been facilitated. They are characterized by internal fertilization and high tolerance for harsh environments, both traits that have likely facilitated their dispersal in the Western Hemisphere. Myers (1966) considered dispersal of Poeciliidae across the Panamanian land bridge to be an important zoogeographic event connecting the two continents.

11. South Circumpolar–Gondwandian (1 family): Only the Galaxiidae exhibit this very southern pattern of distribution. The Galaxiidae occur only in the southern parts of South America, Africa, Australia, and New Zealand (Berra, 1981). Berra notes that this distribution pattern likely reflects a wider distribution of this family during prehistoric Gondwanaland, but that McDowall (1978, original not seen) suggested dispersal of this peripheral family across seawater. Berra suggested that Osteoglossidae might also be considered a Gondwandland-dispersalist, but because of its affinity with tropical areas, I considered it under "broadly tropical," above.

12. Ancient Families of Reduced Distribution (2 families): Although there is only one extant species of the family Amiidae (*Amia calva*, in Eastern North America), the family is well known in Europe from Eocene to Miocene fossils (Berra, 1981), and fossils exist from Brazil. Gars (Lepisosteidae) now occur in fresh and brackish waters of eastern North America, Cuba, and Central America. Fossils (earliest in South American Early Cretaceous) have been found in North America, South America, Europe, the Eocene of Africa (Patterson and Longbottom, 1989), and India (Nelson, 1994). The family Hiodontidae now also appears to be reduced from a formerly more widespread distribution.

Zoogeography of Ostariophysians—Alternative Hypotheses

The highly successful and largely freshwater fishes of the superorder Ostariophysi (Nelson, 1994) are an excellent example of the difficulties that can be encountered in trying to find a single best explanation for the extant distributions of old lineages. [See also Lundberg (1993) for a summary of the difficulties in trying to explain extant distributions of fishes among two regions by any single, even very large, vicariance event.] Ostariophysi, as now defined, includes the order Gonorhynchiformes [with families Channidae, Gonorhynchidae, Kneridae, and Phractolamidae (Nelson, 1994)] as "Anotophysi," sister to the other major clade, Otophysi (Fink and Fink, 1996). Otophysi includes all of the groups previously considered "ostariophysans" (i.e., fishes with a Weberian apparatus, including minnows and their allies, characoids, catfishes, and the South American weakly electric fishes). The origin and dispersal of the minnows (Cyprinidae) has been particularly debated, underscoring the difficulty of knowing with confidence the biogeographic history for some old and complicated groups (Fink and Fink, 1981; Howes, 1991). However, this difficulty does not negate the need for best estimates of zoogeographic and phylogenetic affinities of species for use in framing or testing ecological hypotheses (Mayden, 1992b).

There are two conflicting basic arguments about cyprinid origin and geographic history contrasting (1) a northern origin in Asia or Eurasian landmasses with subsequent dispersal through Laurasia and (later) southward into Africa [generally supported by Briggs (1979), Cavender (1991), and Banarescu and Coad (1991)] and (2) a southern or Gondwanic origin in South America–Africa, with subsequent dispersal into Eurasia, then to North America [supported, in general, by Novacek and Marshall (1976), Moyle and Cech (1988), and Howes (1991) in part)]. Lundberg (1993) is relevant here, as he reviewed in detail the putative effects of a South American–African vicariance on numerous families and higher taxa of fishes. Although no totally acceptable resolution to the problem of cyprinid origins appears available, the basic arguments can best be considered in light of the evolutionary history of the entire group Ostariophysi (*sensu lato*) as follows.

The most recent phylogeny for the ostariophysans is that of Fink and Fink (1996), only slightly modified from Fink and Fink (1981). Figure 5.8 is a simplified version of their hypothesis for ostariophysan relationships. Within the Otophysi (Fig. 5.8) from oldest to most highly derived in this phylogeny are an extinct group represented by the fossil genus *Chanoides*, then Cypriniformes (minnows and carps), Characiformes (characins and characin-like fishes), Siluriformes (catfishes), and Gymnotiformes (weakly electric fishes). The latter two groups are sister taxa (Nelson, 1994), with no inference that one is "ancestral" to the other— only that there was a common ancestor. Fink and Fink (1996) note very close phylogenetic relationship between Siluriformes and Gymnotiformes, and consider them to form one monophyletic group, the Siluriphysi. See also Lundberg's (1993) review of zoogeography of the ostariophysans.

214 / Patterns in Freshwater Fish Ecology

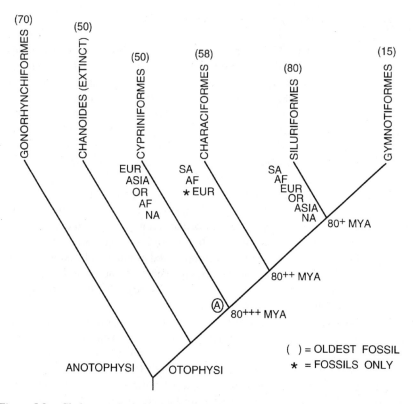

Figure 5.8. Cladogram depicting evolution of the Ostariophysi, modified from Fink and Fink (1996), with date of oldest known fossils in parentheses, and approximate minimum number of millions of years ago for cladogenesis giving rise to each other.

If the phylogeny of ostariophysans of Fink and Fink (1996) is accepted, then answers about age and distributional history of the cyprinids must not be inconsistent with the overall evolutionary and distributional history of any of the groups in Otophysi. One of the most fundamental evolutionary principals must be kept in mind: sister taxa diverge only once. If the phylogeny of Fig. 5.8 is accepted, any group diverging lower in the clade than any more derived group must, by definition, be older than the latter (e.g., Lundberg, 1993). For example, assuming all of the clades in Fig. 5.8 monophyletic, the order including the minnows (Cypriniformes) *must* be older (as a group, not necessarily the family Cyprinidae) than the Characiformes, which, in turn, have an older origin than the Siluriformes or Siluriphysi, and so on. The phylogeny of Fink and Fink (1996), therefore, has an influence on the acceptable dating of minimum ages of the groups as follows.

Fossils can fix the *minimum* age of a group or its least age in a particular region—although they never provide an absolute *first* or oldest member. Lundberg

(1992) noted the "unique contribution of paleontology" being the direct estimation of minimum ages of taxa. As stated by Lundberg (1992), " . . . the oldest fossil taxon of a monophyletic group provides the best evidence for the age of its own group, the stem of its sister group, and so forth." Lundberg (1993) outlines, in more detail, the use of paleontological data to date a clade, indicating that,

> It is perhaps tediously obvious that the oldest fossil of a group provides a direct estimate of its own minimum ages of origin and differentiation. Further, because sister groups originate simultaneously by cladogenesis of their common ancestor (a fossil) provides an estimate of the minimum age of origin of its sister groups. . . . It also follows that the common ancestor . . . of this sister group pair . . . existed and had differentiated prior to the minimum age set by (the fossil).

Parenti (1981, p. 534) pointed out that "a fossil can only give the minimum estimate of the age of a group, and that *most groups must be older than their oldest fossil representatives*" (italics mine). Thus, fossils in an advanced otophysian group can provide a starting point for dating otophysian history. Oldest fossils of the gymnotiforms are known only from the Upper Miocene (Fink and Fink, 1996, p. 220) (perhaps 10–5 mya), too recent to help date Otophysi in general. Catfishes (siluriforms) are much more informative; see Lundberg's (1992) dating of ictalurid origins. The earliest known catfish fossils date to very late Cretaceous, about 80 ± 5 mya (John Lundberg, personal communication) from marine strata in the midwestern United States, and material of approximately a similar age is known from South America. Lundberg indicated (personal communication) that the otoliths of the midwestern U.S. material appear much like those of ariid catfishes, whose otoliths are very distinctive.

If 80 mya is used as the conservatively earliest confirmed date for a catfish, all groups sister to the Siluriphysi (Fig. 5.8) must have origins more than 80 million years old. Although fossils of such antiquity are lacking for the Characiformes [oldest is probably Paleocene in North Africa, at least 58 mya, Patterson (original not seen, cited by Cavender (1991, p. 40)] and Cypriniformes [oldest is about 50 mya, in Asia (Cavender, 1991)], we *must* revise the minimum dates for those groups to 66+ mya and older, as indicated in Figure 5.8.

Thus, at a minimum, the stem of characoids must be 80++ mya, and the stem of cypriniforms at 80+++ mya or older (Fig. 5.8). It would seem reasonable, assuming that some geologic time passed between divergences of these different lineages, to estimate a minimum date for cypriniform divergence as at least 85 mya or so. Novacek and Marshall (1976) argued for a possible origin of the ostariophysans [the Otophysi of Fink and Fink (1996) and in Fig. 5.8] as long ago as Early Cretaceous (e.g., possibly as old as 130 mya). This estimate seems somewhat older than other reports in the literature, but even if all of my estimates above are in error, it must be in favor of older divergence of the major groups within Otophysi. Is it clear that the otophysans are old—sufficiently old to have experienced substantial dispersal by the time of separation of Africa and South

America [final separation 106–84 mya (Lundberg, 1993)] and also old enough to have dispersed (even as freshwater or brackish-water forms) across available post-Pangaean land connections that likely existed sporadically since 100 mya (C. Scotese, personal communication) between Africa and Eurasia.

The "southern" hypothesis for cyprinid origins was detailed by Novacek and Marshall (1976) as follows. [Note that the taxonomic relationships shown in Novacek and Marshall (1976) do not coincide with the phylogeny of Fink and Fink (1981, 1996), as the former have characoids, gymnotids, and cyprinoids grouped as Cypriniformes, and less derived than siluriforms.] Ancestral ostariophysans were presumed to be present in South America in the Cretaceous (Novacek and Marshall, 1976), with primitive cypriniforms and primitive siluriforms on that landmass, and moving into western Africa by Middle Cretaceous before separation of South America and Africa. From these primitive forms, modern South American catfishes, gymnoitids, and characiforms were derived in the Neotropics, and the primitive cypriniforms and siluriforms in Africa provided the stock for more advanced cypriniforms, siluriforms, and characiforms on that continent.

Novacek and Marshall (1976) postulate subsequent movement of cyprinoids into Europe via a tertiary connection (perhaps 60 mya) with southwest Europe, then to radiate throughout Laurasia. Moyle and Cech (1988, p. 333) seemed to concur with Novacek and Marshall (1976) by indicating that Africa came into contact with Laurasia near the end of the Cretaceous (~ 65 mya?), introducing ostariophysans to Laurasia. However, Moyle and Cech suggested that cyprinids evolved either in northern Africa or Europe in the early Paleocene (~ 60 mya?), then spread through Laurasia (including North America). Note that a "southern" hypothesis depends on the extinction of the primitive cypriniforms from South America subsequent to movement of part of that group to Africa—no cypriniforms live today in South America. The "southern" hypothesis for ostariophysan evolution and dispersal would fit available evidence on continental movements, but it is difficult to envision extinction of cyprinoids from all of South America, in light of their incredible success on most other landmasses (Cavender, 1991).

One "northern" hypothesis of cyprinid origins by Briggs (1979) depended on the principle that "centers of origin" coincide with the region of greatest species richness for a group—placing the origin of cyprinids in southeast Asia. However, the notion of "most speciose region = evolutionary origin" has been challenged by many recent zoogeographers. However, there is better or additional evidence for possible origins of cypriniforms in Asia, and this hypothesis can be congruent with problems in modern distribution of more derived groups—the characoids and the catfishes.

Returning to the point that a group can originate only once, placing cypriniform origins in east Asia also must place the stem ancestral group for the characiform–siluriform–gymnotiform lineage at that same location. How reasonable is such a postulate, in light of modern distributions and fossil records for those groups?

Extant characiforms occur only in South America and Africa, which superficially would argue for their Gondwanian origin within those continents. However, characiform fossils from about 50 mya are known from *brackish* water formations in Europe (Cavender, 1991, pp. 39–40), and Cavender cites other authorities relative to saltwater tolerance of early characiform fishes. If early characiforms had substantial tolerance for brackish or salt water, their dispersal from eastern Asia (location in common with primitive cyprinids) via marine or estuarine waters to Europe, then to northern Africa (site of the oldest known characoid fossil, ~58 mya) during intermittent connectivity of Europe to Africa (C. Scotese, personal communication) should have been possible. The westward dispersal of early somewhat salt-tolerant characoids could then have placed this lineage into northeastern South America [which has earlier been argued to have a potential connection to western Africa until 80 mya (Novacek and Marshall, 1976)]. To summarize, the placement of the earliest stem characoids in Asia does not preclude the dispersal of this group along brackish waters of southern Eurasia, leading to its present-day distribution in South America and Africa. It is also possible that the divergence of cypriniforms from the stem of characoids and higher group (Node A, Fig. 5.8) took place further west in Eurasia, with the high diversity of cypriniforms in eastern Asia attributable to eastward dispersal to that region. This would still allow the origins of Otophysi in the northern, not southern, landmasses in the late Cretaceous.

What about siluriforms and gymnotiforms relative to the "northern" origin of the otophysian lineage? Higher phylogeny of catfishes is uncertain (Lundberg, 1992; Nelson, 1994) and under continuing revision. However, ariid or ariid-like catfishes and some other early forms (Lundberg, 1996; ASIH abstract) were marine and/or estuarine, hence not constrained by fresh waters and/or continental movements. Although the distribution of completely freshwater catfish lineages may be explained in terms of continental movements (Lundberg, 1996), an early location of proto-siluriforms in the Eurasian landmass does not preclude the appearance of catfishes in Africa–South America–North America by marine lineages. Obviously, for the South American–African separation to have placed freshwater catfish families into isolation on the two continents by vicariance (Lundberg, 1996) requires that the ancestors to the groups were indeed on the South American–African landmass by (at latest) about 80 mya (if Novacek and Marshall's postulate of continued connection of the continents is accepted).

Lundberg (personal communication) points out that an additional argument for a southern location of ostariphysan radiation is the existence of the highly pleisiomorphic catfish family Diplomystidae (probably sister to all other catfishes) in southern South America. However, Lundberg also notes that an extinct catfish of the family Hysidoridae, known only from North America (Wyoming, Early Middle Eocene), was also very pleisiomorphic among catfishes [probably sister to all catfishes other than Diplomystidae, and sharing with the Diplomystidae the trait of teeth of the maxilla, unlike any other catfishes (Nelson, 1994)]. To summarize, catfishes and the gymnotiforms could have attained their present-day distribution from a stem Siluriphysi ancestor in the Eurasian landmasses, dispersing via saltwater, estuarine, or freshwater connections into the African, then South American continents. To accept that gymnotiforms *are* catfishes that diverged from the siluriform lineage after stem taxa reached South America, which would facilitate acceptance of a "northern" origin for the Otophysi (including cyprinoids), requires either "considerable parallelism" or reversals of "solid morphological synapomorphies for catfishes" (J. Lundberg, personal communication). Otherwise, ancestral, "stem" Siluriphysi (i.e., silurid–gymnotid line) must have reached South America, and then diverged.

Finally, what are the arguments that the Otophysi line could have diverged from Antophysi (the gonorhynchiform lineage) somewhere in Asia? The most primitive extant family in Anatophysi (Channidae) is marine–brackish–occasional fresh water (Nelson, 1994), and known from the lower Cretaceous (possibly 100–135 mya), from Africa, Brazil, and southern Europe (Berra, 1981). The very wide distribution of this family, which is sister to all other living gonorhynchiforms, suggests that primitive Anatophysi–Otophysi divergence could easily have been on the northern continental mass (while obviously not ruling out a southern location for this divergence).

If nothing with respect to related groups rules out the possibility of a cyprinid ancestor in the northern rather than the southern landmasses of the late Cretaceous, what does the evidence from cyprinid fossils and modern species distributions and richness suggest? At present, Eurasia has 155 cyprinid genera, with 70 in China and 42 in Southeast Asia (31 of which are endemic to the latter region) (Howes, 1991). Africa has 23 cyprinid genera, with 13 endemics, with only 3 genera shared with Eurasia (Howes, 1991). North American cyprinids are phoxinins (Howes, 1981; Cavender and Coburn, 1992) probably derived from Europe by a trans-Atlantic connection (Howes, 1981). The real question with regard to cyprinids is "Did they move from Africa to Eurasia, or from Eurasia to Africa?"

Cavender (1981) emphasized the lack of early fossil cyprinids from Africa, in spite of the tremendous effort that has gone into searches for fossil organisms on that continent in this century. The earliest cyprinid fossils in Africa date only to late early Miocene (~ 20 mya at most) in Kenya (Cavender, 1991). VanCouvering (1977) summarized her fossil finds at eight sites throughout north and east Africa to include Bagridae by about 45 mya, other catfishes by 23 mya, and characids by 23–36 mya, but no cyprinids until 17 and 12–13 mya. Cavender indicated that in central Africa, the earliest cyprinid fossils are Pleistocene (in Zaire) and that large areas of the western and southern parts of Africa apparently lack any cyprinid fossils. Additionally, African cyprinids consist of only three clades, all of which *share* a distributional pattern between Africa and southeast Asia. None of this evidence suggests Africa (or the South American–African, or Gondwanian landmass) as the origin for cyprinids. Cavender summarizes suggestions of previous authors that the fossil record of Africa is best explained by assuming that cyprinids migrated into Africa from Southeast Asia, through the Near East, about 18 mya in the late early Miocene, which coincided with a land mammal exchange of that same period (Cavender, 1991). VanCouvering (1977) suggested that, "Cyprinids probably entered Africa after the collision of the African and Eurasian continents (~ 18–16 mya) at the same time that Eurasian–African mammal migrations took place. . . ."

To summarize all of the above, Cyprinidae is the most widespread modern family of primary or secondary freshwater fishes. Its origins and dispersals over evolutionary time cannot be understood except in the context of a scenario that is acceptable within limits set by a phylogenetic hypothesis for the Otophysi. If the phylogeny of Fink and Fink (1996) and all available evidence of continental movements and connections are combined, it is possible to reconstruct a scenario for movement of all the otophysan groups, worldwide, from a common ancestor in the northern landmasses (i.e., probably Asia). It is not, however, my intent here to try to strongly support this "northern" hypothesis, but to

1. Point out the evidence by previous authors for a northern origin of the cypriniforms and their sister taxa, and suggest that within the framework of a recent phylogeny a northern origin is possible

2. Suggest that a southern origin also seems possible, but requires extinction of proto-cyprinids from South America, or their restriction to Africa as the point of origin, and an explanation for the absence of early cyprinid fossils from the African continent

3. Underscore the great difficulty in ascertaining a best hypothesis for origins for some groups (cf. Fink and Fink, 1981).

Application of Global Biogeography to Local Fish Assemblages? North America as an Example

What does all the above have to do the ecology of fishes in local assemblages? Whether the focus of a local ecological study is distributional, "macroecological," responses of species to abiotic conditions, or interactions among species, there should be predictions about outcomes of the study that relate to the very broad biogeographic affinities of the fishes in the local assemblage. For example, a comparative study of oxygen, temperature, or salt tolerances of local species might have a predictably different outcome if it is known that most of the families that are represented are (1) recently derived from marine taxa, (2) Holarctic, northern-affinity families, (3) families invading from the tropics (in geological time), or similar facts. The needs will vary with the investigation, but I illustrate the kind of information that can come from zoogeography by examining the common families represented in the central United States, where I have made a variety of comparative studies.

The information that follows for each family is largely from Moyle and Cech (1988), Briggs (1986), and Cavender (1986), all of whom present overviews of ages, origins, or derivations of major North American freshwater families. Cavender (1986) reviewed all known information on fossils to the Triassic for North American families. Other important reviews of North American fish origins or fossils include those of Gilbert (1976), or Myers (1966, for fish related to Central American connection) and G. R. Smith (1981, for the Late Cenozoic = Miocene through Holocene, with complete compilation of fossils and references for those epochs; many extant species are represented). The freshwater fish fauna of North America was depauperate through the Mesozoic (ending about 70 mya) (Cavender, 1986, p. 702), but some of the Cenozoic fauna of North America was derived from resident Cretaceous fishes (Cavender, 1986, p. 707). However, lampreys or their close relatives were present in North American as long as 300 mya. Other ancient families of extant fishes (gars, bowfins, sturgeons) were present in North America by the Cretaceous (perhaps 135 mya). The following summary, by family, suggests the history and possible origins of fish families in North America as related to common fishes in the central Great Plains. However, Cavender (1986) cautions that the knowledge of fish fossils relating to North American taxa are, on a worldwide scale, too poorly known for confident use of fossils to place origins of families. For a succinct time line of the presence of fish families in freshwaters of North American, see Wilson and Williams (1992, p. 232).

> Petromyzontidae (lampreys) is an ancient group found in temperate waters of both the Eastern and Western Hemisphere. Cavender reports freshwater fossils of lampreys very like (and perhaps included in) Petromyzontidae by the Pennsylvanian (~ 310–280 mya) in Illinois or the Mississippian (~350–310 mya) in Montana.
>
> Acipenseridae (sturgeons) is an archaic, widely distributed family in the Northern Hemisphere. A definite *Acipenser* species is known from Upper Cretaceous (perhaps 80 mya) in Canada (Cavender, 1986). *Acipenser fulvescens* is known as a modern species from the Late Pleistocene (perhaps 1 mya) in Pennsylvania (Cavender, 1986). Grande and Bemis (1991) diagnosed Acipenseridae as part of a phylogeny containing all Polyodontidae (below), and considered sturgeons sister to all paddlefish. Wilson and Williams (1992) show a hiatus in sturgeon fossils in North America from Eocene through Miocene (~ 50–30 mya).
>
> Polyodontidae (paddlefishes) has a fossil record from the Upper Cretaceous (~ 80–60 mya), but paddlefishes now exist only in Asia and North America, with extant genera in the Mississippi River (*Polodon spathula*) and Yangtse River basins (*Psephurus gladius*), China; fossils are known from deposits in Wyoming and Montana. Grande and Bemis (1991) reviewed all known

paddlefish on the basis of newly acquired materials and established that *Polyodon spathula* is a highly derived, not a primitive or "living fossil," member of the family. The genus *Polyodon* has been in Montana for at least 60 million years (Grande and Bemis, 1991), with the fossil paddlefish genus *Paleopsephurus* in Montana since the Late Cretaceous (at least some 70 mya or more?) (Grande and Bemis, 1991). The phylogeny of Grande and Bemis (1991, p. 100) indicates that sturgeons are a sister group to all paddlefish and that the sturgeon–paddlefish lineage was widespread in the Northern Hemisphere by Late Cretaceous, including both China and North America. The biogeographic distribution of paddlefish is trans-Pacific, not trans-Atlantic, and all known fossil paddlefish are North American (Grande and Bemis, 1991, p. 115), implying an origin for the group in the Western Hemisphere (my own conclusion, not so stated by Grande and Bemis). Wilson and Williams (1992) show gaps in the fossil record in Paleocene and Oligocene–Miocene but assume they were present on the continent.

Earliest fossils of gars (Lepisosteidae) are from west Africa [assignable to the genus *Atractosteus* (Wiley, 1976)] and there is an extensive fossil record of this group from the Early Cretaceous (135 mya). They were widespread in North America, Europe, Africa and India, mostly in freshwater deposits (Cavender 1986). Wiley (1976) provided a detailed account of the biogeographic tracks of the two genera (*Atractosteus* and *Lepisosteus*), demonstrating the presence of each genus on both the Laurasian and Gondwanian landmasses of the Mesozoic. From a presumption that the vicariance event resulting in speciations may have been the breakup of Pangaea, Wiley (1976) hypothised a minimum age for the genera of gars to be 180 mya. Wiley hypothesized also that the *Atractosteus* vicariance producing an African–North American track dated to the Jurassic, which would coincide with the fissure of African continent from eastern North America before 166 mya (Golonka et al., 1994).

Amiidae (bowfins) were common in North America and Europe in the Early Tertiary, and fossils are known from Greenland. Bowfins have an "extensive record" (Cavender, 1986) in North America since the Cretaceous. Fossil amiids are known since the Cretaceous in China and Brazil, and from Jurassic to Miocene in Europe. Amiids apparently had a worldwide distribution in South America, Asia, and the African continent dating to the Jurassic (Patterson and Longbottom, 1989; Nelson, 1994).

Hiodontidae (goldeyes) were an important member of North American Paleocene fauna, with the earlies fossils of the family from Alberta, Canada about 65 mya (Cavender, 1986). They now occur only in eastern North America but clearly were in western North America, and the closet related family, known only from fossils, is in China and Siberia. Other recently discovered fossils from the Cretaceous in Asia (Wilson, 1992) are very closely related to Hiodontidae (Wilson and Williams, 1992), suggesting that hiodontids in North America could be remnants of a former widespread northern hemisphere group, although with the genus *Hiodon* restricted to North America (Wilson and Williams, 1992). Guo-Qing and Wilson (1994) have described, from recently discovered material in Montana, a third species of *Hiodon* for North America, which bridges the evolutionary gap between *Hiodon* and the extinct *Eohiodon*, and suggests that "the Hiodon lineage had evolved as early as the late Eocene (e.g., 40 mya)." All of this evidence argues for the presence of hiodontids in eastern North America via a trans-Pacific connection–western North American connection. During the period 70–50 mya, a broad terrestrial connection from North America to Asia via Beringia was in existence (Briggs, 1986; Golonka et al., 1994).

Esocidae (pikes) are of Laurasian origin, with a very old separation of species groups (Cavender, 1986). The family probably originated in North America, given that the oldest known esocid fossils are Cretaceous (Wilson, 1992), more than 65 mya, from Alberta and Saskatchewan (Wilson and Williams, 1992). Wilson and Williams (1992) noted the continuous fossil record of Esocidae in North America from Paleocene to present argues for a North American origin of the family.

Clupeidae (shads): Cavender (1986) indicated that the oldest freshwater clupeid is *Knightia vetusta* from the Middle Paleocene (e.g., about 60 mya) in Montana (per Grande, 1982). *Knightia eocaena* from Early Eocene deposits 55–50 mya in Wyoming is the "Green River fossil fish" that is widely sold (Grande, 1984). It is clear that freshwater clupeids were well established in western North America from 50 to 40 mya (Grande, 1984). However, the record of clupeids in North American fresh water is in the Middle and Late Tertiary (from about 30 to 15 mya), but they reappear in fresh water at the Plio-Pleistocene horizon (Miller, 1982a; Wilson and Williams, 1992). Perhaps, as a family of marine origin, they persisted in the ocean from which they reinvaded fresh waters.

Salmonidae (trouts and salmons) has long been thought to be a group of apparently marine origin (Briggs, 1986), but it is now known that the oldest salmonid fossils are Eocene (~ 50 mya) from freshwater localities in British Columbia and northern Washington (including series of all size classes), supporting the hypothesis that the earliest salmonines were primarily fresh water (Wilson and Williams, 1992). Cavender (1986) indicated that since the Eocene (53–38 mya), salmonids have been known from the Pacific slope of North America.

The superorder Otophysi includes fishes with weberian apparatus, including minnows, suckers, catfishes, and characins. See the phylogeny in Fink and Fink (1996). The oldest fossil ostariophysans are catfishes in North America and South America, but no minnows or catostomids are in South America at present. Fink and Fink (1981, p. 347) noted evidence that ostariophysans comprise a very old group, dating to *before* the upper Cretaceous (origin possibly 100 mya?).

Cyprindae (minnows): North American cyprinids, with the exception of *Notemigonus*, are all phoxinins, comprising a very derived and highly evolved group (Cavender and Coburn, 1992, p. 295). The oldest known cyprinids in North America and Oligocene (31 mya); by Miocene (24 mya) and Pliocene (15 mya), some modern genera of minnows appear in the North American fossil record (Cavender 1986), and feeding morphology was highly differentiated. Lacking other hypotheses, on the basis of the extreme diversity of cyprinids in Europe, the older fossils in Europe than North America (Novacek and Marshall, 1976), and the clearly derived phylogenetic relationships of North American to Laurasian taxa (Cavender and Coburn, 1992), it would appear likely that the family entered North America by way of a trans-Atlantic connection (Howe, 1991). Further, the monophyly of North American cyprinids has yet to be established, but would require two, instead of one, invasions from Europe or Eurasia (Cavender and Coburn, 1992).

Catostomidae (suckers) fossils are known in North America from ~ 50 mya (Paleocene) in Alberta. Their origin (supported by Briggs) was Asia, athough Moyle and Cech (1988, p. 335) argued for a European–North American connection. At present, catostomids occur mostly in North America with one species in China, with essentially the same trans-Pacific biogeographic pattern as paddlefish. The genus *Chasmistes* dates to the Miocene (30 mya); *Ictiobus* is known from both sides of the Continental Divide by about 20 mya.

Ictaluridae (catfishes) may have been early invaders of North America from Asia (based on the possible close relationship of the Asian family Bagridae to North American Ictaluridae). Although fossil marine catfish were deposited in inland seas of North America about 80 mya (J. Lundberg, personal communication), ictalurid fossils are known in North America from about the late Paleocene (~ 60 mya) (Lundberg, 1992), including occurrences in the Pacific Northwest until near the Pleistocene. Cavender (1986) indicated that the oldest extant genera of the family date to Oligocene (about 40 mya), including *Ictalurus* and *Amieurus*. The oldest flathead catfish is Middle Miocene (20–25 mya). John Lundberg (1996, ASIH abstracts) recently reported the finding of a gigantic nonictalurid catfish species in present-day Arkansas that dates to about Middle Eocene.

Cyprinodontidae (pupfishes) has a poor fossil record in North America, but one pupfish from Death Valley may be as old as Miocene (~ 20 mya) (Cavender, 1986).

Fundulidae (killifishes) are widespread in the Old World tropics and in North and South America. The oldest North American fossils are Middle Miocene (~ 20 mya), with Miocene sites in Nevada, California, and Montana. By Pliocene (15 mya), the genus *Fundulus* was common in fossil sites of the Great Plains but disappeared in the Rocky Mountain area (Cavender, 1986). Although not old in North America, it is obvious that this family had in the past a very wide dispersal, perhaps aided by a warm climate maximum during the Miocene (Cavender, 1986).

Poeciliidae (gambusias) occur naturally only in the Western Hemisphere. They apparently originated in Central America and dispersed to North America (Myers, 1966). Wilson and Williams (1992) list no fossil poeciliids in North America.

Atherinidae (silversides) is a family of warm marine origin, with no fossil record in North American fresh water until Late Tertiary (Cavender, 1986). Cavender noted a Pliocene atherinid from Oklahoma assigned to *Menidia*, and the existence of atherinids from the Pliocene in Arizona and central Mexico. Fossil atherinids also are known from the Pleistocene in Central Mexico (Cavender, 1986).

The families Amblyopsidae (cavefishes), Percopsidae (trout perches), and Aphredoderidae (pirate perches) all evolved in and are endemic to North America (Briggs, 1986). All probably evolved from marine ancestors, although pirate perch (Aphredoderidae) were known from fossils by 25 mya in North America.

Centrarchidae (sunfishes) evolved in North America from marine ancestors and remain endemic to North America (with the exception of introductions of the largemouth bass as sport fish throughout world—often with undesirable effects). Earliest fossil centrarchids are Eocene (53–38 mya) from Montana (Cavender, 1986; additional fossils are from Oligocene (~ 30 mya) in South Dakota and from Lower Miocene (25 mya) in South Dakota. Centrarchid fossils west of the Continental Divide date to Middle Miocene (~ 20 mya?) (Cavender, 1986). Wilson and Williams (1992) show centrarchids as occurring continuously in North American since the Eocene.

Moronidae (or Percichthyidae) (temperate basses) are noted by Cavender (1986) to include, as a family, both marine and freshwater genera. Fossil percichthyids are known from North America during the period from about 50 to 40 mya (Cavender, 1986), but there is a gap in the fossil record for this family from the Miocene (~ 30 mya) until Recent (Wilson and Williams, 1992).

Percidae (perches and darters) are considered by Briggs (1986) to have originated in Europe and dispersed by land route to North America. I noted above the lack of percids in northeastern Asia, which argues for a trans-

Atlantic connection for this family between Europe and North America. Cavender (1986) notes that the earliest North American records for *Perca* are Pleistocene from Texas and Oklahoma. Fossil *Etheostoma* and *Percina* are known from Late Pleistocene (maybe 1 mya or less). Wilson and Williams (1992) also note Percidae fossils in North America only from the Pleistocene to Recent; hence, it seems likely that percids are a very recent arrival to North America through some unknown route. Such a recent arrival would obviously argue against the eastern North America–Europe transfer, and in favor of a trans-Pacific transfer, through Beringia. Gilbert (1976) argued for an apparent recent arrival of percids in North America "possibly not before the Pliocene" (12–2 mya). Oldest fossil darters are from the Late Pleistocene in South Dakota (Cavender, 1986), which argues that the incredible speciation that has occurred in this tribe has been in a very short period of time.

Cottidae (sculpins) are a peripheral freshwater family of marine origin. The oldest North American freshwater fossils are from the Late Miocene (12–15 mya?) from Oregon (Cavender, 1986). Wilson and Williams (1992) include cottids in North American fresh water from Pliocene (12 mya) to Recent.

What does such a cataloging of fossil fishes in North America offer to an ecologist wanting to study local assemblages of fishes? A time line (Fig. 5.9), adapted from Wilson and Williams (1992) with information from all the sources above, can help to put all of our modern assemblage members into historical and zoogeographic perspective. From Figure 5.9 [which has the times (in mya) expanded on the right side to emphasize detail], it is possible to obtain estimates of the *minimum* potential members of families present during given periods of evolutionary time for the most-encountered extant families in middle North America. Whereas we now have about 21 families present (excluding some mostly marine taxa, like drums) to form assemblages, at a time 10 mya only about 17 families with extant representatives would be confirmed. Moving back to 25 mya and to 50 mya, about 12 modern families were known as potentially present in central United States. Longer ago (~ 70 mya), about 11 of the now-present families are known to have been present (Fig. 5.9).

Although there are obvious gaps in the fossil record and scarcer members of other families will surely show up in the future fossil finds in North America, it may be reasonable to assume that at least the better represented families (in numbers of species, or numbers of individuals in prehistoric North America) would have been detected in the extensive searches for fish fossils that have been conducted. Knowing, for a region, the number of families that likely could have been in contact (Fig. 5.9), forming assemblages, and eating, competing, or reproducing as do modern fishes, gives the ecologist some ability to think in terms of the potential coevolution of fish species or fish assemblages (which we will address in a later chapter). At a minimum, identification of particular families

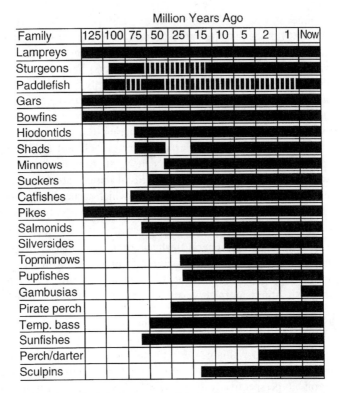

Figure 5.9. Time lines depicting presence of major fish families in freshwater of North American at a given number of mya. (Dashed lines indicate "likely" present but fossils lacking. Open space (shads) indicate apparent true lack of presence in North America, followed by reinvasion. [Modified from Wilson and Williams (1992).]

as "known present" or "not likely present" may be helpful in hindcasting the likelihood of predator–prey interactions, competitions, and the like in the past, when fishes that shape modern assemblages were evolving.

Additional help with this approach is in G. R. Smith's 1981 review, in which he provides, in tabular form, the known existence of particular fish species in the fossil record of North America since the Miocene (~ 25 mya). Also, Grande (1984) provides one of the few detailed comparisons of possible community-level interactions among extinct fossil fishes in the American West. McCune et al. (1984) similarly has drawn extensive conclusions about the ecology of now-extinct species flocks of semionotid fishes in prehistoric lakes of North America. A difficult thought process may be that of trying to work with the extant species and assemblages while placing assumptions in the context of realistic scenarios of past distributional relationships of the faunas. To this end, the awareness of the biogeographic history of the extant species may be immensely helpful in assessing the ecology of modern fish assemblages.

5.3 Subcontinental Zoogeography of Freshwater Fishes

Glacial Periods of the Pleistocene—Effects Worldwide, "Land Bridges," and Beringia

Four major named glacial periods during the Pleistocene had profound effects, that exist to the present, on the distributions of fishes in the Northern Hemisphere. The four periods can actually be divided into six or more, without sharp boundaries, (Conner and Suttkus, 1986). McFarland et al. (1979, p. 799) separate the most recent glacial period, the Wisconsinan, into early, middle, and later periods, with two warm "interstadial" periods in between. McFarland et al. (1979, p. 799) also provide a cross-reference of glacial stages and nomenclature that differ between Europe and North America. McPhee (1980, inside front cover) provides a different scheme for Pleistocent geological stages in Europe, and Colinvaux (1973, pp. 98–99), from the pollen record, summarizes two periods of late glacial advance in Europe, including the Older Dryas, an intervening warm period of glacial retreat (the Allerod), and the Younger Dryas, which ended with the "Pre-Boreal" forest period after the last glacial retreats.

Colinvaux (1973) surmises that vegetation of much of Europe was destroyed by ice sheets that advanced south from Scandinavia and north from the Alps, with few available refugia. Moyle and Herbold (1987) cites Wheeler (via personal communication) that freshwater fishes persisting in central Europe during glaciation could only have been cold-water forms, and that warm-water fishes in central Europe must have reinvaded from refuge in the south-flowing Danube system. Moyle and Herbold (1987) noted that warm-water fishes in Europe did survive glacial periods in southern areas like the Iberian Peninsula, the Balkans, and the Caspian and Black sea regions. In general, Moyle argues that the Pleistocene must have been a highly stressful period for European fishes, leading to its impoverishment in species to a level similar to that of North America west of the Continental Divide. In central Asia, glaciation, or the intervening First Interglacial period, may have promoted speciation and specialization of the Himalayan torrent fish fauna (Menon, 1954). Edds (1989) suggested that in southern Asia, the Ganges River could have provided a glacial refuge for fishes much as the Mississippi River did in North America, promoting a present-day fish fauna in the Gangetic region similar to that of eastern North America with respect to species richness, body size, reproductive traits, and trophic specialization (*sensu* Moyle and Herbold, 1987).

Most important of the Pleistocene ice sheet effects on fishes and freshwater systems were the following:

1. Extirpation of fishes in areas actually covered by the massive continental ice sheets or glaciers
2. Alteration of drainage patterns, by changing the flow of river systems to the south (e.g., in upper Great Plains and the northwest of North America), increasing discharge of rivers, and creating large glacial lakes at ice sheet periphery
3. Movement of fishes to nonglaciated refugia, from which they are to the present recolonizing formerly ice-covered regions
4. Alteration of landforms by glacial scour, widening or deepening river valleys, and leaving "hanging" valleys (forming waterfalls) adjacent to the larger river valleys
5. Creation of moraine lakes, and many thousands of smaller "kettle" lakes as ice blocks melted after glacial retreat

The most obvious effects of the glacial maxima were in close proximity to the continental ice sheets, but floras and faunas were affected worldwide. McFarland et al. (1979) noted the latitudinal shifting (farther south during glacial maxima) of entire continental biomes and the contraction of Amazon and Congo basin rain forests during the glacial periods (when much of Earth's water was ice bound, and even in the tropics the climate was more arid than at present). The glacial maxima also lowered sea levels, exposing more land and extending continental margins. Beringia was exposed as a terrestrial connection between Asia and North America during glacial maxima, most recently until about 15,000 years ago. The Pleistocene was also marked by many medium to very large inland lakes that played a major role in the distribution of fishes in western North America, in the "Pluvial" Great Basin of North America, far south of the glaciated areas (Hunt, 1974; Hubbs et al., 1974).

It should be noted that although humans may have first migrated to North America from Asia within the last 30,000 years—the exact date is disagreed on by authorities—the Pleistocene emergences of Berengia as a "land bridge" between the two continents did not likely have any major effect on the fish faunas of either continent at the family or genus level. All of the primary and secondary freshwater families in North America today were already on the continent before the Pleistocene. The importance of Beringia during the glacial maxima, from the perspective of freshwater fishes, was a major nonglaciated refuge from which they recolonized the northwestern parts of North America.

A second major potential effect of glacial sea-level lowering is to allow coastal river systems that at present are separated (e.g., along the Texas or Atlantic coasts of North America) to have had freshwater connections at substantial distances from the present shoreline of the ocean. Interconnecting of drainages due to sea-level lowering has been noted by Jenkins et al. (1972) and reviewed by Conner and Suttkus (1986) as avenues for prehistoric dispersal of fishes among pr̶ ently nonconnected river basins. (To envision this increased connectivity of coastal drainages during sea-level lowering, try the opposite "experiment": On any drainage map of a coastal region, draw a new coastline with the sea level *raised*. You will see many rivers that now connect near the ocean will lose their connectivity as their juncture is drowned by your new line. Reverse the process and those streams are rejoined by sea-level "lowering.")

An additional effect of glaciation in North America was substantial rerouting of many northern drainages. For example, the upper Missouri River flowed, preglacially, to the north, probably into Hudson Bay (Cross et al., 1986; Burr and Page, 1986). By the Nebraskan glaciation, the upper Missouri basin was diverted southward, introducing some northern fishes southward, and the Kansan glaciation further diverted to the south the streams of the Great Plains (Cross et al., 1986). Metcalf (1966) and Cross et al. (1986) document geological and biological evidence for a north–south transplains stream that many have crossed the present Red River (Oklahoma–Texas border) and flowed directly into the Texas Gulf Coast. Cross et al. (1986) document in detail the fish species whose distributions were affected by these events. I found high similarity in thermal tolerance (Matthews, 1986d) and in morphometrics and meristics (Matthews, 1987b) of populations of red shiner minnows (*Cyprinella lutrensis*) from drainages spanning the central and southern Great Plains. Such homogeneity of fundamental physiological or morphological traits across a 1200-km north–south gradient and among many extant river basins is much more understandable in light of the knowledge that all of those populations of *C. lutrensis* could have been united in one transplains river basin only half a million years ago.

Swiss naturalists appear to have been the first to notice, early in the 1800s, that some exposed landforms and deposits in ice-free valleys matched closely the arrangements of materials deposited by glaciers (Longwell et al., 1948). According to Longwell et al., it was Louis Agassiz who announced the opinion, in 1837, that glaciers in the past had covered a large part of Europe. Glacial periods on earth occurred in the late Precambrian, the Carboniferous, the Permian, and in the Pleistocene (Futuyama, 1986). McFarland et al. (1979) note three glacial periods in addition to that of the Pleistocene since the evolution of vertebrates at about 550, 415, and 300 mya. These three earlier glacial periods were mostly confined to the southern continents, where they would have directly affected only very ancient lineages of fish; extensive glaciation potentially affecting most modern fishes in the Northern Hemisphere has only been in the Pleistocene (McFarland et al., 1979). Broadly, from the perspective of life on Earth, Futuyama (1986) makes note that glacial periods (at least the most recent one) had, in spite of "violent fluctuations of the Pleistocene climate," little effect on extinction rates or rates of evolution. Futuyama notes that much of the Pleistocene fauna consisted of forms that still exist today, which certainly is true for fishes relative to the fossil record (Cavender, 1986). In fact, extant distribution patterns for many North American species can be attributed to either (1) movement to and remaining in refugia outside glaciated areas [e.g., the disjunct distribution of *Notropis nubilus* (Ozark minnow) in the Ozark uplands of Arkansas–Missouri at a substantial distance from its northern populations in the Driftless Area of Wisconsin (Lee et al., 1980)] or (2) reinvasion of glaciated areas by species "returning" from refugia (Crossman and McAllister, 1986; Mandrak and Crossman, 1992; Mandrak, 1995).

In summary, the Pleistocene glaciation in the Northern Hemisphere and the locations of refugia from the ice sheets had immense effects, many of which remain today on distributions of freshwater fishes throughout the Paleactic and Nearctic regions (Darlington, 1957; G. R. Smith, 1981; McAllister et al. 1986; Crossman and McAllister, 1986; Underhill, 1986; McPhail and Lindsey, 1986; Lindsey and McPhail, 1986; Moyle and Herbold, 1987; Burr and Mayden, 1992). The importance of refugia from glacial ice and the role of glaciation in facilitating dispersal in some regions is underscored by Lindsey and McPhail (1986, p. 642) who indicated that, "Fish distributions in northern North America make sense only in the context of Beringia" (as a refugium). Hocutt et al. (1986, pp. 172–182) document in detail the extensive effects of each period of Pleistocene glaciation on river drainages and fishes of eastern North America.

Since the beginning of the Pleistocene about 2 million years ago, there have been four major periods of glacial advance, interspersed by three warmer periods (Hunt, 1974). Those periods and some more recent events of interest, with approximate dates (which vary among locations) adapted from McFarland et al. (1979), Lindsey and McPhail (1986), and other sources, are listed below. However, it is important to keep in mind that there was much variance locally, and no single set of names for glacial periods can be used with firm dates in a cosmopolitan manner. Cross et al. (1986, pp. 372–373) show the considerable overlaps in the times included in the traditional nomenclature for North American glacial epochs (particularly the Nebraskan, Kansan, and Illinoian) as a result of local variation in glacial history.

Epoch		Years ago (ya) when stage began
Nebraskan	Cold	1,800,000 ya
Aftonian	Warm	
Kansan	Cold	1,300,000 ya
Yarmouth	Warm	
Illinoian	Cold	500,000–170,000 ya (varied with location)
Sangamon	Warm	
Wisconsinan	Cold	120,000–90,000 ya
Recent	Warm	About 12,000 ya
"Medieval Warm Period"		a.d. 800–1400
"Little Ice Age"		a.d. 1500–2850

The Wisconsinan consisted of three periods of glacial advance, beginning about 120,000, 60,000, and 30,000 years ago (McFarland et al., 1979) but varying in onset of glacial advance in various parts of the Northern Hemisphere. McPhail and Lindsey (1986) point out that in northwestern North America, the "classic" sequence of Pleistocene glacial periods was not as clear as elsewhere, because of local landform constraints on ice movements. McPhail and Lindsey note that, "it is clear that the last major glaciation in Cascadia was not exactly synchronous with the last major glaciation in the east." Conner and Suttkus (1986) note that early interpretations suggested that the four "classic" glacial periods were brief, separated by lengthy "warm" periods, but the opposite is now considered true. Conner and Suttkus (1986) also make clear that correlations of glacial periods with ecological and zoogeographic events elsewhere in the world are imprecise, and that pre-Wisconsinan geomorphic history must be viewed "broadly."

The two Recent intervals, the "Medieval Warm Period" (Hughes and Diaz, 1994) and the "Little Ice Age" (Fritz et al., 1994) are included to suggest (1) that climate variation is not a thing of the past and (2) that embedded within major Pleistocene climate periods, there could have been many lesser periods of global thermal variation and ice advance/recession [as indicated by Lindsay and McPhail (1986)] with potential effects on local fish assemblages, or on location (latitude/longitude) of particular kinds of fish assemblages. Many kinds of flora and fauna have been shown to respond by north–south migrations to changes in global temperatures (e.g., McFarland et al., 1979; Root, 1993). Hughes and Diaz (1994) offered evidence from tree rings and montaine glaciers that the Medieval Warm Period was regional, not synchronized globally, further suggesting the role that meterological/climatological variance can play at the scale of decades to centuries in localized episodes. However, Ritter et al. (1995, p. 471) show a minor but distinct "blip" of increased sea level, globally, that would correspond to the Medieval Warm Period.

It is unknown if the Little Ice Age was global, although there were glacial advances both in Europe and North America during that time (Fritz et al., 1994). However, there is well-supported evidence from paleolimnological reconstruction from Devils Lake, North Dakota (upper Great Plains of North America) for a more arid climate in the plains throughout the Little Ice Age, with very steep climatic gradients between the Great Plains and regions to the east and west. Fritz et al. (1994) summarized that during the Little Ice Age, glaciers expanded in western North America, floods were more frequent in the upper Mississippi Valley, and mesic trees expanded distributions in the east, whereas the northern Great Plains experienced more arid conditions. Such an increased gradient in climatic conditions could have exacerbated (or helped lead to) the variation that we still see between fishes of the Great Plains and the uplands to the east (Matthews, 1987a).

Hunt (1974, p. 91) shows the lower limits of the glacial advances in the eastern two-thirds of North America, and Burr and Page (1986) map the locations of glacial drift of the various ice ages. Glacial limits followed roughly the route of the present Missouri River, to which Longwell et al. (1948) attribute the position of the river—flowing at the foot of the ice sheets. Ice sheets extended across Illinois, Indiana, and Ohio almost to the present-day Ohio River, then across upper Pennsylvania, all of New York and New England (Hunt, 1974). The northern Appalachian Mountains were glaciated, but further south they were not.

The situation was more complicated in the Northwest and Alaska, as indicated by McPhail and Lindsey (1986, pp. 623–624) and Crossman and McAllister's map (1986, p. 57). The immense Laurentide Ice Sheet did not extend to the west coast of North America, and western icing may have had different times of maxima than did ice in the east. Instead, along the Pacific coast and into southern Alaska, there was a Cordilleran Ice Sheet, and the last western glacial period, the Fraser Maximum, did not reach peak glaciation until about 15,000 years ago (McPhail and Lindsey, 1986). Much of Alaska and the wide, exposed Beringia was free of ice during the glacial periods.

Lindsey and McPhail (1986, their Fig. 17.2) provide details on the coverage of ice sheets in Alaska, Siberia, and the broad, exposed terrestrial Beringia (1500 km north to south—not a narrow land bride) with now-drowned major river drainages that flowed into the Arctic Ocean or Bering Sea. Most of the information that follows is from their summary. Lindsey and McPhail (1986, p. 643) also show detailed patterns in the depth of the Bering Strait or elevation of Beringia above sea level, as ocean levels rose and fell during the last 250,000 years. At its most recent maximum about 18,000 years ago (Lindsey and McPhail, 1986) Beringia was a broad, exposed lowland approximately 1500 km wide from south to north, completely spanning the region between present-day Alaska and Siberia and extending north of Siberia. Numerous large rivers drained this lowland, and there appear to have been ample opportunities for temporary freshwater or brackish-water connections between the river mouths (Lindsey and McPhail, 1986). The climate of Beringia was sufficiently warm that cold-water fishes could survive, and ice-free routes generally existed, making Beringia along with western Alaska an important refugium for freshwater fishes when most of northern North America was glaciated. Lindsey and McPhail (1986, pp. 643–645) describe in detail the various potential connectivities of Beringian–Alaskan rivers and routes of access by freshwater fishes.

Crossman and McAllister (1986) show the major refugia of North American fishes during glaciation, including: (1) Beringian (= Beringia + central Alaska), (2) Columbian (Cascadia), (3) Missouri River basin, (4) Mississippi River basin, (5) southern Atlantic slope, and (6) possibly a refugium in the Banff–Jasper region of southwest Canada. Schmidt (1986) also notes the existence of Northeast Coastal and Atlantic Coastal refugia on the now-drowned but extended coastal plain due to sea-level lowering and in the Gulf of St. Lawrence region. Each of these refugia has important bearing on the reinvasion of fishes into the northern portions of the continent. Mandrak (1995) showed in a detailed analysis that fish species richness in lakes in Ontario was dominated by variables including (1) length of time since deglaciation, (2) length of time covered by glacial lakes, (3) distance from a dispersal corridor, and (4) mean annual air temperature. From this analysis and details on postglacial dispersal (Mandrak and Crossman, 1992), it seems clear that historical factors centered on Pleistocene glaciation and redispersal have been the overriding important variables relative to regional species pools in much of north–central North America. In addition, there is evidence that isolation of populations in different glacial refugia led to the evolution of local races (e.g., Foote et al., 1992).

Figure 5.10 is a composite of the information on postglacial dispersal routes taken from maps and text in Crossman and McAllister (1986), McPhail and Lindsey (1986), Lindsey and McPhail (1986), Schmidt (1986) and Mandrak and Crossman (1992). The map is a composite of major refugia and presumed redispersal corridors; details on the relative influence of particular refugia on a given region can be found in the original sources cited above. Bear in mind that the Great Lakes (and to their west) as they exist today were instead the approximate site of very large glacial lakes (e.g., glacial Lake Agassiz) that played a substantial role in altering drainages and producing connectivities for fishes from the Mississippi or Missouri refugia to the north. During maximum periods in

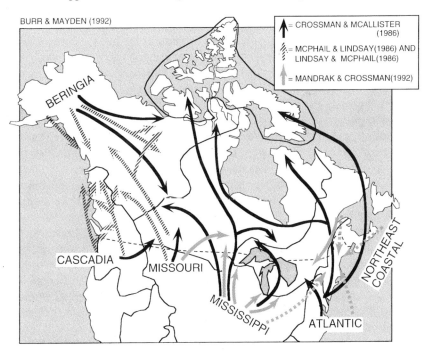

Figure 5.10. Major pathways of postglacial reinvasion from glacial refugia (synthesized from sources in figure legend).

size of the glacial lakes, some of their waters flowed directly southward into the Mississippi basin.

In general, the Mississippi Refugium (Fig. 5.10) accounted for an overall majority of reintroductions of fishes in northcentral Canada, the Hudson Bay region in part, and the Arctic Archipelago [terminology on regions follows Burr and Mayden (1992)]. For example, the Mississippi refugium has contributed as much as 85% of the species reinvading the Hudson Bay and Arctic Archipelago drainages (Crossman and McAllister, 1986). The Atlantic slope and perhaps coastal refugia have made major postglacial contributions to the Hudson Bay and Arctic Archipelago regions, but the process of reinvasion there continues, with the fauna remaining depauperate. The Mississippi refugium contributed a large percentage of the species recolonizing the northern Appalachians, which were entirely glaciated, and the now-drowned Atlantic and Northeast Coastal Plain regugia, along with the Gulf of St. Lawrence refugium, contributed the remaining species to recolonization of the northern Appalachians (Schmidt, 1986). To the west, the Missouri refugium was a major contributor to reestablishment of fishes in the interior of western Canada and into the Yukon–Mackenzie region. The Cascadian (or Columbian) refugium contributed species both by movement along the coast (through brackish connections) or by an inland corridor aided by the presence of glacial lakes that provided freshwater connections for species invading from the south (McPhail and Lindsey, 1986). Alaska was able to retain many of its fishes in situ during the glacial maxima, as ice-free regions remained there and in Beringia, and the Alaskan–Beringian refugium has contributed significantly to reintroduction of fishes to northern Canada throughout the Yukon–Mackenzie region and along the northwestern coast (Fig. 5.10). Even now, the massive Wrangell–St. Elias ice field that extends for hundreds of kilometers across coastal eastern Alaska probably blocks dispersal of freshwater fishes. This awesome ice field, viewed from the highway between Tok and Anchorage, Alaska, can give a "southern" fish ecologist a stunning glimpse of conditions that fish might have encountered in close proximity to continental ice sheets!

Moyle and Herbold (1987) considered western North America and Europe both reduced in richness by Pleistocene glaciation, when conditions favored large-bodied, long-lived species. Thus, western United States and Europe lost small-bodied fishes during Pleistocene glaciation, whereas the central United States and Canada retained fish of all sizes in the Mississippi–Missouri refugia. As a result, patterns in composition of fish assemblages, including body size, reproductive traits, and diet, are shared between western United States and Europe but not by eastern North America (Moyle and Herbold, 1987). Moyle and Herbold also cite Mahon (1984) to indicate fundamental differences in composition of fish assemblages in central Europe and eastern North America.

For ecologists interested in local fish assemblages, how did all of the subcontinental glacial and postglacial events influence fish species assemblages in North America (or elsewhere)? It is obvious from all the zoogeographic sources summarized above that there have been very major rearrangements of fish species distributions in at least the northern two-thirds of North America in the last 2 million years, with particularly important changes in the Wisconsinan with effects lasting to the present. Throughout the Pleistocene, some fish species probably did remain in contact with other particular species in glacial refugia, where some remain now as relict populations [e.g., in the Ozark uplands of Arkansas and Missouri (Black, 1940; Robison, 1986; Gorman, 1992)]. However, it seems very likely, based on resolved histories of postglacial dispersal (e.g., Cross et al., 1986; Mandrak and Crossman, 1992), that many fish species reinvaded different regions at varying schedules; hence, I speculate that the history of freshwater fishes in North America may be more a history of varying composition of local assemblages over geologic time than of very long periods of contact of particular species in basin-level or even local assemblages. Such an amorphous shaping of fish assemblages through Pleistocene–Recent (or longer) time would seem to have an important influence on the question of whether or not fish assemblages are coevolved. Freshwater fishes might have a pattern of coexistence during the environmental fluctuations of the Pleistocene analogous to that recently deduced for mammals (Graham et al., 1996), which showed that extant modern mammal communities may have existed only a few thousand years and that, in the past, the ranges of individual species changed individualistically, without integrated movements of whole faunas. Hinch and Collins (1991) showed that postglacial biogeographic processes had been an important factor in the present-day composition of lake assemblages in Ontario and suggested that differences in distributions of some fish species related to differences in their postglacial colonization rates (i.e., individualistic responses of species to the Pleistocene and Recent climatic events).

As an alternative hypothesis, we cannot disprove that fish faunas might shift geographic locations with the species intact as a group (i.e., having similar assemblage compositions) but at different locations. It would be fascinating to revisit some of the areas sampled near to or more than a century ago in North America (e.g., Jordan, 1889; Meek, 1894; Hubbs and Ortenburger, 1929) to determine if the same *species associations* exist that existed before damming and drainage alteration, although in *different* places.

North American Fish Regions, as an Example of Subcontinental Zoogeographic Effects

Burr and Mayden (1992) defined Nearctic fish faunal provinces, partly on the basis of very large and zoogeographically dominant river basins, but with coastal river drainages (e.g., Atlantic, eastern Gulf Coast, western Gulf Coast) combined on the basis of faunal similarities and geological affinities. The provinces of Burr and Mayden are summarized below, as an indication of the tremendous differences in the kinds and richness of fish faunas that can exist within one continent.

Similar contrasts in species diversity or patterns of occurrence exist among fish faunal regions in South America (Eigenmann, 1920; Eigenmann and Allen, 1942; Lundberg et al., 1986), Africa (Roberts, 1972; Hugueny, 1989, 1990; Hugueny and Lévêque, 1994), Asia (Smith, 1945; Jayaram, 1977; Chereshnev, 1990; Rainboth, 1991; Reshetnikov and Shakirova, 1993; Zakaria-Ismail, 1994), Europe (Maitland, 1969), or the Australian region (Coates, 1993).

The North American fish faunal provinces of Burr and Mayden (1992) include the following (with numbers of native species in parentheses, and a summary of their comments, with additional information from various references previously cited):

> Yukon–Mackenzie (65): Mostly salmonids, plus 13 cyprinids. Primary freshwater fishes invaders from the Mississippi basin, but there were unglaciated portions of Alaska that may have been refugia during glaciation.
>
> Arctic Archipelago (8): Most depauperate—includes 5 salmonids. No known primary freshwater fishes.
>
> Hudson Bay (101): Strong affinity to Mississippi and Great Lakes—but many species occur only at the southern edge of the Hudson Bay province, perhaps an area still being recolonized by southern survivors of Pleistocene glaciation. Fauna less than 14,000 years old because glaciers extirpated all fishes in area.
>
> Great Lakes/St. Lawrence (168): Connected to Mississippi basin during glacial times. Entire area ice covered until 15,000 years ago, so fauna consists of reentry of dispersalists from south. Mandrak and Crossman (1992) detail glacial refugia, routes, and approximate rates of redispersal into this region.
>
> Northern Appalachian (106): Cyprinidae = 29% of species. Entire region covered by glaciers, so freshwater fishes have recolonized by postglacial dispersal.
>
> Cascadia (60): Mostly salmonids, cyprinids, cottids, catostomids. Most glacier covered, but Columbia River was refuge for redispersal.
>
> Mississippi (375): Major center of fish evolution and served as major refuge for northern fishes during glacial advance. Also refuge for ancient, relict species (e.g., paddlefish, gars, bowfin). Mississippi is the "mother fauna" for North America and is a "cradle of temperate freshwater fish diversity" (Burr and Mayden, 1992, p. 28). Many endemic species: 16 in Ozarks, 30 in Tennessee River basin.
>
> Central Appalachian (177): East of Atlantic divide—mostly unglaciated = refugium for northern fishes during glaciation. Similar to Mississippi–Ohio fishes at genus level, but many species differ across the Atlantic divide. However, stream capture across the divide has secondarily mixed faunas.

Southeastern (Gulf slope) (268): Each river drainage in the province is distinctive. Important also is mixing of brackish-water species and differences in species above and below the "Fall Line." Forty endemic species in Mobile Bay drainage.

Colorado (32): West of Continental Divide, with 22 of the 32 species endemic. Now much invaded by introduced species from the east, major reductions in native fauna via dams, introduced species effects. Noteworthy are "big river" Colorado squawfish, humpback chub, and razorback sucker, all now endangered or in decline.

Great Basin–West Coast (98): Endemic, impoverished fauna, including relict minnows, catostomids, and pupfishes from large interior lakes during Pleistocene (~ 2 mya). Many isolated endangered species (e.g., pupfishes in desert springs).

Western Gulf Slope (132): Louisiana–Texas coastal rivers, to but not including the Rio Grande. Relatively depauperate subset of Mississippi fauna (to the north) and extensions from the south (like *Astyanax*). Blind catfishes of Edwards Aquifer deep underground highly unique. Faunal dispersal among drainages likely during low-water-level periods for ocean, providing freshwater routes via river connections (out into the present ocean).

Rio Grande (134): High percentage of endemic species, especially in closed basin in interior Mexico in isolated springs. Fauna includes 21 families, including five Nearctic or Holarctic families, and only two Neotropical families (Characidae and Cichlidae).

Sonoran–Sinaloan (45): Mexican Pacific slope and Baja California. Substantial desert areas; cyprinids, peociliids, and catostomids make up most species. Depauperate freshwater fauna, but in lower parts of rivers at least 125 marine species invade.

Central Mexican (205): Rich fauna with 25 families; cyprinids, catostomids, and ictalurids make up 20% of fauna, but most diversity is in Goodeidae (39 species), Poeciliidae (43 species), and Atherinidae [30 species, including a "species flock" endemic to the Mesa Central, diagnosed by (Echelle and Echelle, 1984)]. Eight species of the atherinid genus *Chrirostoma* in Lake Chapala (Echelle and Echelle, 1984). Overall = unusually high endemism, recognized long ago (Regan, 1906–1908) as a unique biogeographic subregion of North America.

In addition to the fish provinces of Burr and Mayden (1992) above, Miller (1966, 1982b) defined four fish provinces in Central America (Usumacinta, Chiapas–Nicaraguan, San Juan, and Isthmian) and suggested recognition of four fish provinces in Mexico (Rio Lerma, Rio Grande, Balsas, and Californian Sub-Region) (Miller, 1982), noting, however, that the faunal provinces in Mexico should not be considered firmly established until a more comprehensive analysis of Mexican fishes was completed.

Elsewhere in the world, subcontinental fish faunal provinces or regions have been diagnosed as follows.

Eurasia

Berg (1949) reviewed the freshwater fishes of the entire Soviet Union and adjacent countries, and described in detail (pp. 303–349) the distribution of more than 500 taxa among three major "sections" (Circumpolar, Meseurasian, and Sonoran), and a total of more than 40 "subregions," "provinces," and "districts" in a partially hierarchical scheme. He also documented in detail the "great similarity" between some of the river basins, noting the propensity for connections of the "upper reaches" of major basins in the Russian Plain. Seasonal fluctuations of waters of bogs in the headwaters of these major basins apparently permitted transfer of fishes among basins (more so than occurs in much of the rest of the world, where river basins tend to be have more distinct fish faunas).

Chereshnev (1990) detailed the distribution of freshwater fishes for 29 river basins in the northeastern Soviet Union from the Chukostsk Sea to the Sea of Okhotsk, based on extensive recent collecting. Five major ichthyofaunal regions were distinguished: Eastern Chukotka, the Anadyr River basin and adjacent areas, the Penzhina River basin and rivers flowing into Penzhina Bay, southern coastal rivers of the Sea of Okhotsk, and the Okhotsk group of rivers in the southwestern part of the study region.

Reshetnikov and Shakirova (1993) reviewed fishes from 25 water bodies of Central Asia and Afghanistan and proposed boundaries for fish regions and provinces. On the basis of a multivariate analysis of the presence–absence of species and a dendrogram of Reshetnikov and Shakirova, it appeared that about seven distinct regions could be identified, within which individual basins had moderately high overlap.

Exclusive of Southeast Asia, Banarescu and Coad (1991, map p. 128) described seven "major areas" of Eurasia based on geology, geography, and ichthyfaunal relationships, including the following (1) Europe and far-north Africa, (2) Southwest Asia (or West Asia), (3) Siberia, including Arctic and Pacific drainages, (4) Central or High Asia, (5) Western Mongolia, (6) South Asia including the Indian Subcontinent, and (7) East Asia, including Japan, northernmost Vietnam, and Taiwan. Banarescu and Coad (1991) specifically address cyprinid diversity and distribution among the regions, but their geographic information and bibliographic sources provide a detailed overview of this landmass.

In a review of the zoogeography of Indian freshwater fishes, Jayram (1977) did not define explicit boundaries that could be considered "fish provinces" but concluded that the fishes of India had three primary components: Indian (autochthonous, Gondwanian origin), Indo-Chinese, and Malayan. Jayram concluded that most of the present fauna originated in Indo-China, with 63% allied with the Oriental Realm. Jayram considered the Indian peninsula one of the oldest and most stable landmasses, comprising a "natural compact unit," now reduced to less than half its original size due to subduction under the Himalaya and Tibet following collision. Jayram (1977) indicated that little can be confirmed about the indigenous pre-Pleistocene fish fauna that existed on the Indian Subcontinent prior to its collision with Asia, because there have apparently been substantial extinctions.

Southeast Asia

Rainboth (1991) used phenetic clustering techniques to assess faunal regions in Southeast Asia, based on the presence–absence of cyprinid genera. Because Cyprinidae, with 1260+ species of 205 genera (Rainboth, 1991) is a dominant family in Southeast Asia, an analysis of their distribution can provide a relatively robust view of regional faunas overall. From this analysis, Rainboth provided a cluster with about seven major faunal elements, some hierarchical within larger elements. The strongest distinctions were between faunas of central Asia and southern Asia and, within central Asia, a dichotomy between "East Asia" and "high Asia." In addition to diagnosing cyprinid faunas, Rainboth (1991) provides an exhaustive review of and entry to the literature on geologic and drainage history, as well as fish explorations, for the Southeast Asian region.

Bornbusch and Lundberg (1989) noted from studies of catfish distributions that there is evidence of a post-Pleistocene vicariance between mainland Southeast Asia and the islands of Indonesia, as the Sunda Shelf (exposed by sea-level lowering during Wisconsinian glaciation) was reflooded to isolate the mainland from Recent insular elements. Yen (1985) divided the rivers in the North of Vietnam (Tonkin Gulf drainages) into five identifiable ichthyofaunal regions and found the greatest faunal affinities between these drainages and those of southern China and of Hainan Island, further suggesting the close ties of freshwater fishes across what are now oceanic expanses by virtue of prior movements of subcontinental plates of exposure of ocean shallows during glacial lowering of sea levels. Smith (1945, pp. 13–16) reviewed the zoogeographic regions of Thailand as applied to fishes, agreeing with Central, Peninsular, Eastern, and Southeastern divisions previously proposed for terrestrial animals and establishing newly defined fish regions (following river basins) in northern and western Thailand.

The most recent definition of zoogeographic regions for freshwater fishes of Southeast Asia is that of Zakaria-Ismail (1994), who included five regions: Indo-Chinese peninsula, Indo-Malayan archipelago, Mindanao (Philippines), Malay Peninsula, and Salween basin (the later being the westernmost, largely distinct ichthyofaunally from the rest of Southeast Asia, with a majority of species of Indian Subcontinent origin).

Africa

Lowe-McConnell (1987, pp. 35–39) summarized fish distributions in Africa and mapped approximately 10 fish faunal provinces form the combined work of Roberts (1975) and Greenwood (1983) for the continent, noting that these faunal provinces largely correspond to major coastal or interior lake drainage basins. On the basis of fish faunas in 52 river basins of West Africa, Hugueny and Lévêque (1994) recognized three fish faunal regions, including the Sudanian, the Upper Guinean, and the Lower Guinean, supporting in part the faunal zones recognized by Roberts (1975) but modifying the boundaries. Hugueny and Lévêque (1994) provide details on endemism and on geologic and drainage history of the West African region. In southern Africa, Bell-Cross and Minshull (1988) consider the fauna south of the Zambesi increasingly limited by cold and to consist of invaders from the Zambesi and Nile systems.

South America

Lowe-McConnell (1987) attributes to Gery (1969, original not seen) eight fish faunal regions in South America, including (1) Guyanan–Amazonian, (2) Orinoco–Venezuelan, (3) Paranaean, (4) Magdalenean, (5) Trans-Andean, (6) Andean, (7) Patagonian, and (8) East Brazilian coastal rivers. Lowe-McConnell (1987) notes the low numbers of species in the Andes and a fauna in southern South America "completely different" from the rest of the continent. Recall that my multivariate analysis of fish families of the world detected South America south and west of the Parana as an entirely different fish cluster; Günther (1880, p. 248) also noted the distinctiveness of southern South America from the rest of the continent (and its similarity in fishes to his "Tasmanian" and "New Zealand" subregions).

Eigenmann and Allen (1942) documented the occurrence of fish species by principal river basins for the Andes and west–central South America, establishing nine defined fish zoogeographic provinces for the region. Lundberg et al. (1986) commented on the depauperate nature (~ 150 species) of the Magdelana River basin fish fauna relative to the nearby Orinoco and Amazon basins and provided evidence of a Miocene vicariance event (mountain uplifting) "trapping" species from the Orinoco and Amazonian to form the original Magdelanean fauna (followed by apparent catastrophic extinctions).

Australasia and New Zealand

For New Zealand, McDowall (1978, pp. 206–211) gave a thorough account of patterns in distribution of freshwater species but did not formally define ichthyofaunal regions within the islands. Instead of faunal "regions" within New Zealand, McDowall and Whitaker (1975) described all freshwater taxa as having two historical origins: a Southern–Temperate Element, and an Indo-Pacific Element. Much of the distribution of freshwater fishes in New Zealand can be explained by diadromous life histories, and even the purely freshwater taxa have relatives that can tolerate brackish or marine water (McDowall, 1978).

Coates (1993) reviewed the relationships of the fish fauna of New Guinea both within the island and as related to its connectivity until recently with northern Australia and summarized previous zoogeographic works on fishes of the island. Northern and southern New Guinea may qualify as separate fish provinces on the basis of a high degree of endemism at the species level in each, in spite of mostly shared families (Coates, 1993). Coates concluded that the relatively recent age of northern New Guinea rivers and the lack of estuaries on the north coast sharply limit the numbers of freshwater fish species in that region.

The Australian proper remains the largest island of the group that broke away from Gondwana to form the Australasian region. Freshwater fishes are limited, and no fish regions have been formally proposed. However, Merrick and Schmida (1984) noted that drainage division boundaries are the most important boundaries to fish distribution within the Australia and mapped (their Fig. 1) a total of 14 "geographical drainage features" of Australia and New Guinea (including the separation of northern and southern New Guinea, as above). Merrick and Schmida summarized previous studies by various authors to suggest that Australia had relatively low relief and few inland barriers to fish dispersal until perhaps 15 mya, and a wetter climate until the Pleistocene, promoting wide dispersal of fishes within the continent (and a consequent lack of sharply demarcated fish provinces).

5.4 Regional Biogeography of Fishes

Importance of River Basins in Fish Biogeography

Because fish, as a general rule, only move from place to place by swimming, no single factor is more important in regional biogeography of freshwater fishes than drainage basin limits and affinities (Gilbert, 1980). This fact is so long entrenched (such as Meek, 1881) that most ichthyologists accept it as *the* fundamental truth in understanding a regional fauna [e.g., the arrangement of all sections in Hocutt and Wiley (1986); fundamental arrangement of ichthyofaunal consideration in Jenkins et al. (1972)].

Zoogeographers have, in general, recognized the importance of the isolation of fishes in distinct drainages. For example, Brown (1995a, p. 171) noted for fish that "isolation has preserved the products of the speciation events that occurred either within single bodies of water or when waters became fragmented and rejoined," and also that "each isolated lake or river system tends to have its own distinctive fish fauna, with a substantial number of endemic fish species." Although there clearly can be important differences among the fish species present in small tributaries within any basin (Matthews, 1982), fishes typically are more similar at any locations within a river basin than they are among locations in neighboring basins.

Many of the regional fish faunal papers in Hocutt and Wiley (1986) quantify levels of faunal similarity among adjacent pairs of river drainages or among all possible pairs of drainages within a larger region (basin). Many of those chapters use some similarity index or multivariate similarity measure to detect areas of marked change versus drainages that differ little in fish faunal affinities. In general, coastal drainages, isolated by marine waters (at present), have relatively substantial differences in faunal composition. For example, Swift et al. (1986) showed that the range boundaries for many fishes in southeastern United States (Gulf and Lower Atlantic coastal region) coincided with either the boundaries of river drainages or the Fall Line distinguishing Piedmont from Coastal Plain parts within basins. Swift et al. (1986) showed, for example, that the adjacent Alabama and Appalachicola basins (contiguous in uplands, but with some smaller coastal river drainages between them, near the Gulf) had a total of 157 versus 86 species, respectively, with a faunal similarity of only 0.51 (on a scale from 0 to 1.00). Crossing the basin divide between the Alabama River and the Tennessee River (to the north, inland) would result in a shared faunal similarity of only 0.46. Hence, both interior and coastal river basin boundaries can have large effects on fish species distributions. Matthews and Robison (1988), for example, showed marked differences even within the Ozark upland fish faunas, with a total of 14 species present in the White River basin lacking in the adjacent Arkansas River basin.

Burr and Page (1986), in all possible comparisons of included river drainages (one outlier excluded) in the Ohio River basin, found average faunal resemblances ranging from 40% to 88% (mean = 77.2%), and for all possible comparisons of river drainages in the upper Mississippi River basin, they found faunal resemblances from 62% to 93% (mean = 77.1%). Likewise, Cross et al. (1986) showed percent similarities in species composition for all possible pairs of 18 river drainages in the western Mississippi River basin of North America to have values ranging from about 20% to more than 90%, but most values were in the range of 40–70%. Thus, for much of the interior of North America, river drainages within the same major basin only share on average of about three-fourths (or less) of their species in common. River basins and river drainages indeed seem to be substantial barriers to fish movement and are appropriate as "regional" units for estimating ecological variables like potential species pools.

Hocutt et al. (1986) provided Jaccard similarity indices (ranges from 0 to 1.00) comparing faunal composition for all Middle Atlantic coastal drainages in eastern North America. For 11 comparisons of adjacent river drainages (from the Edisto to the Potomac rivers), the mean similarity of adjacent drainages was 0.697. The similarity of the two most distant drainages (Edisto versus Potomac) was only 0.224. In general, drainages that shared a common estuary had higher faunal similarities, like the Tar and Neuse rivers (0.943) and the York and Rappahannock

(0.868). However, the York and James rivers (sharing a common estuary) had a faunal similarity of only 0.667 (apparently because the James extends much further across the Piedmont and into the Appalachian highland, than does the York). Thus, the James drainage (70 species, total) has upland fish species lacking in the York drainage (49 species, total).

Hugueny and Lévêque (1994) quantified faunal similarities among 52 west African river systems that were either coastal or drained to large inland lakes. In addition to defining fish faunal regions for much of west Africa, Hugueny and Lévêque (1994) provided measures of faunistic distance (Euclidean distances, from three axes of a multivariate analysis) between paired river systems. As detected from the data of Hocutt et al. (1986) above, there was in West Africa a stronger similarity in the faunas of two adjacent coastal rivers if the rivers were linked by coastal lagoon networks—presumably allowing intersystem dispersion of some fish species via the brackish lagoons.

Overall, although faunas of river basins are clearly influenced by their extension (or not) among major elevational regions (e.g., Montaine, Piedmont, or Coastal Plain), it is obvious that in the majority of comparisons, there is a marked difference in the composition of fish faunas among river basins. To overlook fundamental basin-level differences or distinctions among fish faunas in local or comparative ecological studies of structure or function within fish assemblages could result in a failure to properly attribute some portion of observed results to the underlying zoogeographic patterns.

Finally, although it has been well known that river basins have at least moderately distinctive faunas, there has been less attention to the degree that hierarchical connectivity among smaller river systems, river drainages, or large river basins influence similarity of fish faunas in each. Matthews and Robison (in press) compared the faunal similarities among all possible pairs of nine upland river systems in the Ozark and Ouachita uplifts of Arkansas to the similarities of those same systems in actual drainage connectivity. Using a Mantel test, Matthews and Robison (in press) showed a strong correspondence between the degree of faunal similarity and the level of drainage connectivity among all river systems. In other words, the similarities and differences in the fish faunas within each of the upland drainages were concordant with the levels of connectivity of those systems within the greater Mississippi River basin. Hugueny and Lévêque (1994) also showed, for rivers of West Africa, strong correspondence between faunistic distance and actual geographic distance between drainages. Both of these studies thus reinforce the concept that although river basins tend to have unique fish faunas, there is a relationship between their geographic and faunistic differences.

Local Endemics

One of the most important features making river basins unique is the fact that many fish species occur only in one drainage or river basin. In many cases, the number of endemics is a primary factor in the degree to which a given river basin differs from others. For example, Cross et al. (1986, their Table 11.5) document the number of endemic species in major river drainages of the western Mississippi River basin of North America to range from a low of 4 to a maximum of 25 (White River basin of Arkansas and Missouri). This feature alone helps make the upper White River basin a distinct fish faunal unit within the state of Arkansas (Matthews and Robison, 1988). McCallister et al. (1986) depicted the distribution of endemic species on equal-sized zoogeographic grids for all of North America. Their map makes clear that major centers of endemism correspond to centers of overall species richness and that basins with a long history of relatively stable, nonglaciated environments (e.g., the Appalichian and Interior highlands, the Edwards Plateau of Texas, and the Pacific Northwest south of the Pleistocene glaciation) dominate with respect to the number of endemics. Other locations with substantial endemism in North America include isolated springs or relict environments in the formerly pluvial Great Basin (Hubbs et al., 1974; McCallister et al., 1986) and Trans-Pecos region of west Texas (Echelle and Echelle, 1992). In those latter locations, differentiation of cyprinodontids and cyprinids in isolated habitats as a result of the drying of Pleistocene lakes (a major vicariant event in the Great Basin) or former river systems in the Trans-Pecos has resulted in a high proportion of endemic species. However, it should be noted that the correspondence between total species density (total species per geographic grid) and the number of endemics is not complete (McCallister et al., 1986, their Figs. 2.2a and 2.2b), in that there are grids with a high species density in the southeastern United States along the Gulf Coast, downstream of the centers of endemism. This high species density appears to result from (1) some encroachment of brackish-water taxa into freshwater river systems and (2) the enhanced number of species occurring in river systems at about the proximity of the Fall Line, which comprises a major ecotone where low-gradient coastal and higher-gradient fish taxa coexist.

At a wider geographic scale, that of the fish faunal provinces of North America, Burr and Mayden (1992, p. 26) show the percentage of endemic native species per province to range from 69%, 63%, and 50% in the Colorado, Central Mexico, and Rio Grande provinces, respectively, to none in the Arctic Archipelago, Yukon–Mackenzie, and Hudson Bay provinces. The complete lack of endemics in the three northernmost provinces corresponds to their history of strong glaciation in the Pleistocene and the fact that most of their fish taxa only exist today by virtue of postglacial colonization from provinces to the south (with the exception of Beringian refuge colonists, as detailed in Section 5.3). In contrast, the provinces with the high percentages of endemics share the traits of a long history of isolation as major river basins or as basins with endorheic drainages. The high endemism in the Central Mexico province is also increased by the existence of a species flock of atherinids in lakes within the region (Barbour and Chernoff, 1984; Echelle and Echelle, 1984).

Hugueny (1990) depicted range size of fishes in rivers of West Africa in terms of the number of river basins occupied by a given species. His summary (his Fig. 4) showed that a very large number of species (more than 130) occurred in only 1 river drainage, that about 50 species occurred in 2 river basins, and that relatively few species occurred in as many as half (10) of the 20 river basins being considered. This suggests that the range size for individual species in rivers draining tropical West Africa is on average quite small, with many localized endemics.

Vicariance Biogeography and "Historical Ecology"

Wiley (1983) described two fundamental approaches to biogeography, including "descriptive biogeography" (documentation of ranges of taxa and summarization of patterns into regions) and "interpretive biogeography" (attempts to synthesize basic descriptive biogeography into explanations or hypotheses). Wiley (1983) further subdivided interpretive biogeography into (1) ecological biogeography (study of dispersal of organisms and mechanisms that relate to this dispersion) and (2) historical biogeography [study of spatial and temporal distribution of organisms as explained by past historical (evolutionary and geologic) vicariance events]. It is this latter approach that has been much used in recent years to form hypotheses about the ways that organisms came to be distributed across the face of the Earth both at the intercontinental and intracontinental geographic scales (Grande, 1990). Both geographic scales have been studied by historical zoogeographers [e.g., Parenti (1981, pp. 538–540), intercontinental; Mayden (1987) and Gorman (1992), intracontinental or regional]. Historical biogeography requires information on (1) faunal distributions, (2) phylogenetic relationships among the species of the group(s) under study, and (3) a geologic history of the area.

Vicariance or "historical" biogeography owes its origins to articles by Croizat and his supporters (e.g., Croizat et al., 1974) and other early workers [summarized in Parenti (1981) and Keast (1991)]. The approach combines phylogenetic hypotheses of the study organisms with information on geologic or geographic changes in an area to ascertain "generalized tracks" (Croizat et al., 1974) involving multiple taxa and presumably defining distribution patterns of the ancestral biota. Croizat et al. (1974) specifically rejected the "centers of origin" and dispersal hypothesis as the primary mechanism in the zoogeographic history of most groups and proposed that a generalized track "estimates an ancestral biota that, because of changing geography, has become subdivided into descendant biotas in localized areas." They accepted that dispersal does take place, but explicitly suggested that vicariance events resulting in allopatric speciation from a common ancestor best explained, on a global basis, the distribution of modern biotas. In other words, Croizat et al. (1974) advocated the importance of geological events fragmenting an ancestral biota, followed by local differentiation in isolation, producing descendant sister taxa (usually within more than one group) on either side of the line of fragmentation. When two species, demonstrated to be sister species on a totality of biological evidence do occur together in overlapping ranges, it is generally concluded that the sympatry is due to secondary dispersal back into an area following a vicariance event that promoted allopatric speciation.

Rosen (1978) expanded the vicariance biogeography concept operationally as follows: "To seek evidence of these historical connections, cladograms of geographic areas, representing sequences of disruptive geologic, climatic, or geographic events, may be compared with biological cladograms, representing sequences of allopatric speciation events in relation to those geographic areas"; Rosen illustrated these principles with examples from (1) fishes from the Caribbean region and (2) the worldwide post-Pangaean breakup and movement of continents. Rosen (1978) made clear that phylogenetic reconstructions of taxa or estimates for zoogeographic history areas could only be done with a minimum of three entities—that any two items would be uninformative.

Debate followed among supporters of "vicariance" or historical biogeography (Nelson and Platnick, 1980) and others who doubted that vicariance models accounted for biogeographic history of various regions (e.g., Briggs, 1984) and favored a center of origins explanation (Briggs, 1987). Chernov (1982) cautioned that historical biogeographic explanations from phylogenies based on morphological relationships could be confounded by local morphological variation (in space or across time) that was exogeneously mediated and not related to the actual pattern of vicariance. However, Chernov's comments applied to quantifications of mensurative or enumeration characters, rather than to shared unique traits (synapomorphies) on which most phylogenies depend. Acceptance of a vicariance/historical biogeographic approach among ichthyologists as typified by Parenti (1981) in her combined phylogenetic and biogeographic analysis of cyprinodontoform fishes, worldwide. The vicariance biogeography approach was reviewed by Wiley (1988a, 1988b), summarizing that the technique relies on the congruence between phylogenetic and biogeographic patterns for monophyletic groups, and the larger patterns to be derived from groups that share these patterns.

Grande (1990) recently urged the inclusion of fossils as additional information on area relationships, suggesting that fossil biotas can provide "time control" (i.e., aid in detection of patterns obscured by Recent distributions of organisms).Grande (1990) also succinctly summarized the requisite four steps in application of vicariance biogeography (which I quote or paraphrase) as:

1. Obtain a resolved estimate of phylogenetic relationships among a minimum of three taxa within a group.
2. Translate the biological group relationship (phylogeny) into a pattern of area relationship. (This indicates which geographic areas are most related to each other.)
3. Look for a repeating pattern of area relationships. [Repeated patterns across dissimilar taxonomic groups (e.g., animals and plants) can indi-

cate the likelihood of a general vicariance event that divided a variety of groups such that sister species evolved by allopatric speciation after the vicariance.]

4. Seek evidence for a nonbiological (geological) event that is logically likely to explain the biological patterns (i.e., this will be *the* vicariance event).

Mayden (1985, 1988b) reviewed the eastern North American freshwater fish fauna in the context of vicariance biogeography, emphasizing sister-species relationships of taxa relative to their distributions among major river basins or highlands regions. Among the Interior Highlands and Appalachian Highlands of North America, Mayden (1985, 1988b) detected numerous generalized tracks connecting the Ozark and Ouachita highlands, and connecting those highlands to the highlands east of the Mississippi River. Mayden concluded that the generalized tracks for fishes and aquatic organisms were consistent with the geologic hypotheses of a continuous pre-Pleistocene highland west of the Mississippi River and the influence of Pleistocene glaciation on modern distributions of a number of fishes. The primary vicariance events proposed by Mayden (1985) to separate the three highlands were the division of the Ozark and Ouachita highlands by upstream erosion of the Arkansas River, and a "northern vicariance" separating eastern from western highlands as advancing glaciers created habitats unsuitable for upland fishes in the central Mississippi River lowlands.

Mayden (1987) expanded the vicariance or historical biogeographic paradigm to include its incorporation with known ecological traits of extant fish species and to advocate a perspective emphasizing historical "vicariance" events in interpreting present-day compositions of fish assemblages. Mayden (1987) suggested that such an approach comprised a highly desirable program of "historical ecology" that posed past known geologic/vicariant events as potentially accounting for as much (or more) of the composition of modern assemblages as do "extant" ecological explanations such as competition or other biotic interactions. More recently, Gorman (1992) used a historical biogeography or "historical ecology" hypotheses to account for patterns in occurrence of fishes among streams of the Ozark uplands. Mayden (1992b) reinforced the concept that historical ecology can shed light on formation of local fish faunas and support testable ecological hypotheses about assemblage members. Brooks and McClennan (1993) demonstrated the utility of phylogenetic and geologic reconstruction to explain the extant distributions of freshwater stingrays and their parasites in South America, and they outlined parsimony analysis as an efficient method for reconstructing patterns in distribution of fishes in local assemblages and posed "historical ecology" as a research program (Brooks and McLennan, 1992). The majority of fish zoogeographers appear at present to support the vicariance/allopatric speciation theory as a model that can explain much of the variation in fish faunas and to consider generalized tracks as convincing evidence of the biogeographic history of a region.

Some faunas, like the cichlid fishes in the African rift lakes, have been the subject of much debate over allopatric versus sympatric speciation [summarized in Meyer (1993)]. Meyer (1993) favored the view that the explosive speciation of cichlid species in Lake Victoria could relate to within-lake microallopatric speciation, when a period of drying ~ 14,000 years ago might have left ponds or rivers at the margin of the lake as refugia where formerly panmictic species could have rapidly undergone cladogenesis. It should be noted, however, that recent geologic evidence suggests that some of the most dramatic episodes of speciation (e.g., the cichlid flock in Lake Victoria) may have resulted from rapid sympatric speciation. Johnson et al. (1996) present evidence that Lake Victoria was completely dry no more than about 12,400 years ago, that opportunities for isolated small lakes in the basin were lacking, and that the hundreds of extant (or recently extinct) species of cichlids therein are the result of very rapid evolution. It seems possible that both allopatric and sympatric speciation may play roles in the formation of fish faunas and that a view allowing both "vicariance" and "dispersal" biogeography to influence fish faunas is most conservative.

What does all this mean to a fish ecologist studying one or more local freshwater assemblages? It can be particularly important where a combination of organismal phylogenies and geographic area cladograms can define generalized tracks and identify a likely vicariance event. Strong inference could be gained into the design or interpretation of comparative ecological studies or experiments. Knowing the likelihood that suites of species [even across phyla, (e.g., fish, crayfish, snails, etc.)] have occurred together in similar environments for a long period of geological or evolutionary time could have a large influence on expectations for their sharing common traits (e.g., physiological tolerances) or having affected each other's evolutionary history through coevolution (Chapter 9). Take, for example, a hypothetical scenario in which an ecological experiment is planned with two co-occurring fish species (A and B) somewhere in the Interior Highland of midwestern North America. If it is known that the sister species to Species A occurs in the Appalachian Highland, east of the Mississippi River, with other taxa in one resolved zoogeographic track, whereas the sister to Species B is, say, in the Edwards Plateau of west–central Texas, it can be assumed that the two species (A and B) occur together at present as a result of two different vicariance events and that they do not share a lengthy, potentially coevolved heritage. In the example above, we would expect them to have coexisted since no more than 1–2 mya at most [i.e., the approximate date of Pleistocene vicariance postulated by Mayden (1985)].

Alternatively, if both species A and B had sister species to the east of the Mississippi River, they *both* could have arrived in the Ozarks via the Pleistocene vicariance and potentially share a much longer heritage of mutual co-occurrence of ancestral forms. This also means that, in general, it would be desirable to test extant occurrences of fish species in assemblages against historical hypotheses, to ascertain if there is any need to invoke recent adaptive explanations for observed distributions or occurrences of ecological traits. To this extent, it is species that show exceptions to generalized "historical" patterns that may be most informative about ecology of a group. Information, if it can be estimated, on historical influences in modern assemblages should be of immense value to ecologists seeking appropriate organisms (with *either* very different *or* potentially common heritage) in planning experiments or comparative studies. As summarized by Mayden (1992b), "explanatory statistical comparisons that treat species and their attributes as independent variables do not control for structural integrity in the variation and distribution that may be accounted for by their phylogenetic histories." Thus, where possible, it seems increasingly incumbent on ecologists to be aware of and seek the most recent information on phylogenies for groups with which they are working and to incorporate such historical information into their studies.

Directional Affinities of Species in a Fauna

Even for taxonomic groups for which resolved phylogenies are lacking [for summaries of North American groups with or without resolved phylogenies see Burr and Mayden (1992)] or independent of detailed analysis of river basin affinities, an awareness of zoogeographic affinities of local species with respect to their global range patterns may help ecologists to make predictions or form hypotheses about ecological traits of species in local assemblages. One way to estimate the influence of zoogeographic range patterns of species to the composition of regional or local fish faunas, hence local assemblages is to quantify the relationships of local species with regard to the direction (on the compass) with which their global range is affiliated. Figure 5.11 illustrates such a quantification for 168 native fishes in the state of Arkansas [excluding species for which a "direction" of affinity could not be determined (i.e., if Arkansas was virtually in the center of their range) and also excluding endemics found only within the state].

From the dot distribution maps in the "fish atlas" (Lee et al., 1980) and from a few additions, such as *Notropis suttkusi* (Humphries and Cashner, 1994), I examined the total range of each species that occurs within Arkansas. Outward from the map of the state (taken as the center on the "compass"), I drew a line, estimated by eye, along the compass vector intersecting the center of the range of the species. Although clearly subjective, I attempted to draw this vector in a way to combine the greatest area of distribution of the species outside Arkansas with its greatest density in occurrences. Figure 5.11 includes all of the species' affinity vectors in a "clock" diagram. Although obviously imperfect in detail and fraught with errors in my ability to judge a species' "center" of range, this approach nevertheless seems instructive relative to judgments about the affinity of fishes in Arkansas. Note that this approach does not attempt to place a value

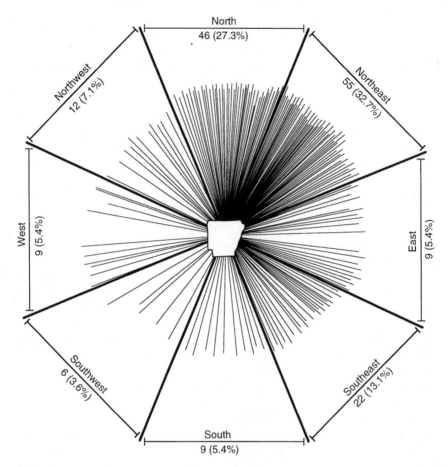

Figure 5.11. Directional zoogeographic affinities of 168 native freshwater fish in the state of Arkansas (USA).

judgment on the known location of the origin of any species (and, for most, this is imperfectly known) but merely indicates where, outside the region of interest (Arkansas), the greatest number of conspecifics of a given species would apparently be found.

From this analysis, it is possible to broadly consider the preponderance of fish species with range affinities to the north, northeast, east, southeast, and so forth. Black (1940, pp. 217–223) discussed in detail the distributions and zoogeographic affinities of Arkansas' fish species with the Great Plains and with the Ohio valley, emphasizing this latter connection, as is apparent in the results of my analysis (below). Pflieger's (1971) classic analysis of fishes in Missouri also takes a somewhat similar approach, but descriptively for each species, evaluating its direction of origin or affinity. Additionally, Pflieger (1971, pp. 293–308) describes in detail many of the pre-Pleistocene, Pleistocene, and Recent events that have helped shaped the fish fauna of Missouri and the faunal affinities of fishes in various parts of the state; much of this summary would also relate to Arkansas. More recently, from the perspective of historical biogeography, based on distribution patterns of sister species or nearest relatives, Mayden (1985, 1987, 1992b) reviewed the distributional relationships of the upland fishes of Arkansas to the Appalachian Highlands to the east, defining the Central Highlands biogeographic track, linking the Ozark, Ouachita, and Appalachian highlands as an evolutionary unit.

For the entire fish fauna of Arkansas, Figure 5.11 (divided into eight 45° compass or clock directions) suggests that 46 species (27%) of the state's species have distributional affinity to the north (including the upper Mississippi–Missouri river basin), 55 species (33%) are allied with northeast (eastern Mississippi River–Ohio River basin tributaries), and 22 species (13%) of the state's species have range or distributional affinities to the southeast [including the rich Gulf Coastal fauna (Swift et al., 1986)]. Fewer species (12, = 7%) have affinities to the northwest (e.g., Kansas, central Great Plains), and relatively few have affinities to the west (9 species), southwest (6 species), or south (9 species) (Oklahoma, Texas, and Louisiana drainages, respectively). This is not surprising in light of the less species-rich faunas of those latter regions (Connor and Suttkus, 1986; Cross et al., 1986; Robison, 1986) and the great richness of fish species in the upper Mississippi and Ohio river basins (Robison, 1986; Burr and Page, 1986). However, with respect to the ecological traits of species inhabiting streams in Arkansas, this analysis might make it possible to predict that a substantial number of species with northern affinities might, for example, tend to be cold tolerant and that relatively few species associated with the hotter, more arid West and Southwest [some, at least, with high heat tolerance (Matthews, 1987a)] occur in the state. Thus, it might be possible to make predictions about reactions of the fishes of the state to harsh drought years, to global warming, and to similar very broad trends. Similarly, knowing that the majority of fishes in a given regional species pool came, say, from the north, might suggest hypotheses about life-history patterns for fishes in those streams (e.g., with respect to factors like age at first reproduction, trade-offs between egg size, numbers, and numbers of spawnings, etc.).

Finally, the clock diagram in Figure 5.11 shows a surprisingly small number of species with affinity to the east (9, = 5%) in light of Black's (1940) assessment of a strong relationship of the upland Ozark fish fauna (north Arkansas) to the upland fish fauna of the Appalachian Mountains to the east of the Mississippi River. However, although numerous upland-limited species do join the Ozarks and Appalachians in faunal similarity (e.g., *Cyprinella galactura, Notropis tele-*

scopus, and *Fundulus catenatus*), it is clear that they do not make up a preponderance of the fishes in Arkansas as a whole. If the goal is to understand the overall fish fauna of this political unit (or for any other political or naturally bounded region), it may be more practical, for ecological purposes, to consider the directional affinities of all species—rather than to focus on a few species [or sister-species pairs (Mayden, 1985, 1987, 1992b)] that connect two regions with zoogeographic "tracks" of evolutionary affinity. Probably the best approach to understanding the regional zoogeographic affinities of any place on Earth is to take both approaches: look for evolutionary tracks *sensu* Mayden (1987, 1992b), and for preponderance of distributional affinity for all of the species. In this way, the ecologists is aware of (1) the probable vicariance events or routes of dispersal that ties a region of interest to other recent geologically–faunally related regions and (2) the totality of directional affinity of the entire fauna, which can suggest ecological capabilities, limits, or opportunities in tolerance, life history, or other extant properties of interest.

A similar analysis for the fish fauna of the Piney Creek watershed, Izard County, Arkansas, is presented in Figure 5.12. Piney Creek is a clear-water upland stream with warm but not "hot" water, in which I have made many collections over more than 20 years and from which the fauna is very well known (Matthews and Harp, 1974; Matthews, 1986c). Figure 5.12 suggests that the fishes of this north Arkansas stream have, much more than for the state as a whole, a majority of species (22, = 51% of the stream's fauna) with distributional affinity to the northeast (i.e., the eastern Mississippi and Ohio river basins). Eight (19%) of the species in Piney Creek have affinities to the east [i.e., evolutionary connectivity of Ozark uplands with nearest relatives in the Appalachian highland, cf. [Mayden, (1987)]. Fewer Piney Creek fishes area allied with the southeastern fauna, and only three species exhibit affinity with the fauna to the northwest, west, or southwest. Obviously, the occurrence of a given species in Piney Creek (no Arkansas endemics occur here) does depend on (1) the geographic origin of the species and (2) the avenues of approach for dispersal into the Ozarks. For most Piney Creek fishes, the most likely path of dispersal is along the White River, which, in turn, is a tributary to the Arkansas River near its juncture with the Mississippi River, offering a present-day avenue of dispersal from the east. Furthermore, there is evidence that during the Pleistocene, the Mississippi River flowed along the eastern escarpment of the Ozark Mountains, in the location of the present-day Black River, which would dramatically shorten the prehistoric connection of Piney Creek, via White River–Old Mississippi River, to faunas of the east or northeast.

From an awareness that most of the Piney Creek fishes have affinities to the northeast, north, or uplands to the east, it should be possible to predict that fishes in Piney Creek, as a whole, might have substantial tolerance for cold conditions but only modest tolerance for hot or intermittent conditions (which are more common in habitats to the south or southwest). This is precisely what Matthews

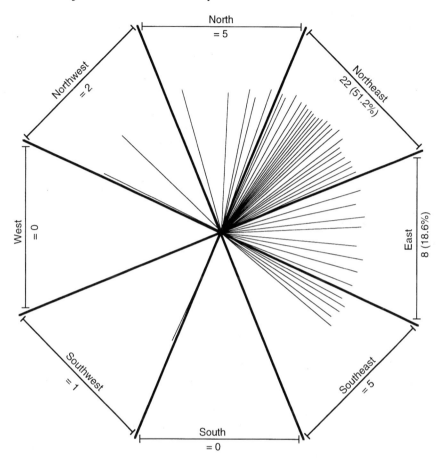

Figure 5.12. Directional zoogeographic affinities of 43 native fish species known from Piney Creek, Arkansas (USA).

(1987a) demonstrated in laboratory temperature and oxygen tests for some of the minnow species that occur in Piney Creek (but not with individuals actually collected in Piney Creek).

Obviously, such conclusions from these "compass-direction" analyses (Figs. 5.11 and 5.12) ignore many known differences in physical tolerances among species that share common distributional directions outside Arkansas (e.g., Smale and Rabeni, 1995a, 1995b), and some of the suggestions above ignore the well-known fact that for at least some species (Feminella and Matthews, 1984), there is substantial variation in physicochemical tolerances among local populations. Hence, the directional analysis approach suggested above should not be regarded as a final answer about the ecology of the fishes in a local or regional species pool, but should be regarded as a potential tool for use in the framing of testable hypotheses about the species or about local assemblages as a whole.

"Ecoregions"

There has been a strong interest, particularly among regulatory agencies in North America, in using "ecoregions" to define boundaries of geographic regions for management standards relative to freshwater fishes (Hughes and Larsen, 1988; Larsen et al., 1988; Hughes et al., 1994). In North America, these ecoregions have largely been based on the map of Omernik (1986) entitled "Ecoregions of the Conterminous United States." Ecoregions as depicted by Omernik are "defined as mapped regions of relative homogeneity in land surface form, soil, potential natural vegetation, and general land use" (Hughes et al., 1994). There is an alternative set of ecoregions for North America by Bailey (1995) that attempts to capture a hierarchy of "ecosystems of regional extent." Bailey (1989) has also provided a hierarchical map of ecoregions of the world.

However, in all cases, the "ecoregions" are based on terresrial or climatic phenomena, without regard for aquatic systems. They are also very large in size relative to the scales at which aquatic system differ. For example, within the state of Oklahoma, the "Central Oklahoma–Texas Plains" ecoregion (Omernik, 1986) would include streams ranging from extremely clear-water, spring-fed, riffle–pool creeks of the Arbuckle Mountains (Stewart et al., 1992) to very turbid, low-gradient, mud-bottomed creeks in the flat agriculture lands of northern Oklahoma (Matthews, personal observation). The fish species occurring in these two extremes of habitat are almost completely different (Matthews, personal observations), and any attempt to manage or regulate streams from these two kinds of habitats (merely because they are in one "ecoregion") in a similar manner would be fruitless.

There is a well-known problem with the use of ecoregions at a larger scale (Hughes et al., 1994). Although some studies have purported to support the use of ecoregions as having well-defined differences among fishes (e.g., Rohm et al., 1987), it is clear from comparison of boundaries of "ecoregions" to boundaries of fish faunal regions (based on actual collections of fishes from a large number of sites) that there is very little correspondence between the two (Hughes et al., 1994). Although I agree with Hughes et al. (1994) that it would be desirable to have hierarchical sets of known fish regions from which to pose ecological questions or management expectations, it clear that the ecoregions of Omernik do not, in many states or larger parts of the United States, provide an adequate framework for any detailed expectations about local fish assemblages. Throughout North America, and much of the world, the data do exist (in museums, other well-documented collections) to allow formulation of realistic "fish faunal regions" (e.g., Pflieger 1971; Matthews and Robison, 1988) that can be of direct use in knowing what kinds of fish assemblages to expect in pristine local habitats. Lyons (1989b), for example, showed that a combination of habitat characteristics (including water temperature, stream gradient, substrate composition and shoreline vegetation) more precisely classified Wisconsin streams than did the ecoregions of Omernik. Obviously, there are regional patterns that influence local fish assemblages, as suggested below. However, it would be better in most cases to consider those regions on the basis of river basin boundaries than on the totality of terrestrial features, as seems to be the case with "ecoregions."

Regional Versus Local Species Richness

At the finest scale of "zoogeography" is the assessment of linkages between regional zoogeography and the structure of local assemblages. There have been many recent papers focused on local versus regional influences on local fish assemblage composition (e.g., Minns, 1989; Kelso and Minns, 1996), with some substantial part of fish species richness accounted for by regional factors (e.g., geology, drainage patterns, postglacial colonization). However, Oberdorff et al. (in press) have recently showed only modest historical effects (e.g., glaciation) on species richness for fish in Western Europe and North America. Recent "macroecological" approaches (Brown, 1995a) have emphasized that local patterns in animal assemblages may be influenced strongly by larger-scale "regional" or "continental" phenomena. One of the most important of the postulates from this perspective is that the richness of species in local assemblages can be influenced strongly by the richness of the regional species pool available as members of the local assemblages. For this postulate to be true, it is a requisite that some defined area or region would contain species that it is reasonable to assume would comprise a source pool for the local assemblages within the region. For terrestrial animals, the defining of a regional species pool can be difficult (e.g., it may be difficult to know geographic limits to the "region"). For all of the reasons cited above, it may be reasonable for fishes to assume that individual river basins (or some discrete portions within river basins) comprise a geographic region containing a logical source pool of species that could occur in local assemblages.

For a variety of terrestrial animals, there is a history of recent studies suggesting a positive relationship between regional and local species richness. Given that there is indeed a positive relationship of local to regional richness, there is then interest in the pattern of this relationship: Is it linear, suggesting a lack of "saturation" of species, or does the local richness reach an asymptote, beyond which there is no continued increase in local assemblage richness in spite of the greater richness of the regional-species pool?

One of the first to define these relationships was Cornell (1985a, 1985b), who defined three models for the increase in numbers of local herbivorous insect species relative to the regional richness of the group. Per Cornell (1985a), the two extremes were a "pool-enrichment" model (Model I), in which the relationship of local- to regional-species numbers is linear and without an asymptote, and a "local-saturation" model (Model III) lacking a relationship between local- and regional-species numbers and having the local-species numbers fixed by local saturation (at a relatively low proportion of the regional-species pool). Cornell's (1985a) intermediate model was a "compromise model" (Model II) in which the number of local species initially increases with increases in regional-species pool size, but reaches an asymptote (fixed by local saturation of species). This model was simplified to two types of models in Cornell (1985b), Ricklefs (1987), and Cornell and Lawton (1992) with an "unsaturated" linear model of "regional enrichment" or "proportional sampling" contrasted with an asymptotic "local-saturation" model having an initial rise but reaching an asymptote in local richness (Fig. 5.13A). As a minor variation on this scheme, Lawton (1990) named a Model I in which *every* species in the regional pool is in every local assemblage (very unlikely), a Model II (proportional sampling), and Model III (local saturation). These regional–local models for local assemblages are highly important for at least two reasons (Ricklefs, 1987). First, if a biological community or assemblage is saturated, then "new species could not join the community without the compensatory disappearance of others" (Ricklefs, 1987). Second, "if local conditions determined local diversity, variation in regional diversity should have little influence on local diversity" (Ricklefs, 1987). Thus, a careful sorting out of the two predominant models [Model I = proportional sampling; Model II = local saturation, following Cornell and Lawton (1992)] should have important implications for the fundamental effects of regional zoogeographic properties on the composition of local assemblages.

How did the data fit the models? Ricklefs (1987) reviewed two previous studies of songbirds by himself and collaborators, finding a better fit to the proportional sampling than the local-saturation model. However, Terborgh and Faaborgh (1980) reported saturated bird communities in the West Indies. For the specialized cynipine gall wasps on oak trees, Cornell (1985a) found a high correlation of local-to-regional species richness, no asymptote (i.e., the proportional-sampling model), and concluded that "excess niche space is available on host oaks for additional cynipine species." Cornell (1985b) suggested that this unsaturated pattern for cynipid wasps might be characteristic for herbivores, which have in general "weak interspecific interactions." Lawton (1990) reviewed many of his previous studies of insects and mites on bracken fern, also finding unsaturated local communities relative to regional species pool richness. Hawkins and Compton (1992) found no evidence for saturated communities among different kinds of African fig wasps. Cornell and Lawton (1992) reviewed all available studies to that date and found a preponderance of Model I (unsaturated, or proportional sampling) communities among a variety of animal taxa.

For fishes, there have been a limited number of assessments of local versus regional-species richness. Tonn et al. (1990) found an upper limit to local-species

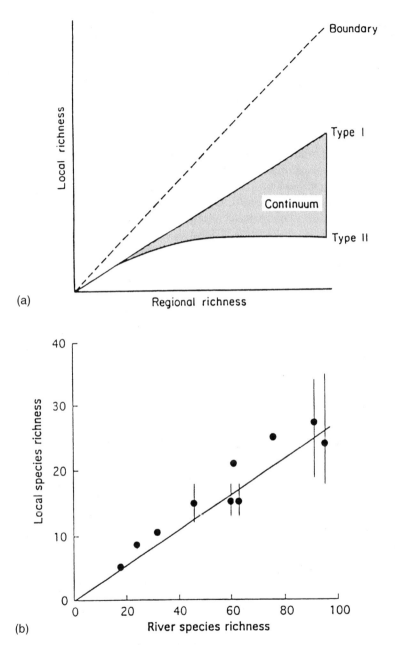

Figure 5.13. (A) Model relationships between regional and local species richness. [From Cornell and Lawton (1992)]. (B) Actual local-species richness versus regional-species (river basin) richness. [From Hugueny and Paugy (1995), The American Naturalist. Copyright © 1995 by The University of Chicago.]

richness in small lakes in Finland and Wisconsin and suggested that "ecological saturation of local assemblages predominated." However, Cornell and Lawton (1992) pointed out that this conclusion was based on a sample size of only two regions, so it is not possible to be sure which model (proportional sampling of regional species versus local saturation of species) is really supported by Tonn et al. (1990). For a total of 9679 lakes in Ontario, divided in more than 80 small "tertiary" watersheds, Minns (1989) showed an overall increase in mean species richness per lake with an increase in tertiary watershed species richness (i.e., an increase in local with regional richness) up to a regional richness of about 30 species, then the local richness (overall) exhibited a "leveling off" (Fig. 5.14). Minns (1989) cited Ricklefs (1987) to assume that "this indicates local saturation of species and thus that local factors are more important than regional factors in determining species richness in these lakes." For lakes pooled among all three primary watersheds (Great Lakes–St. Lawrence, Hudson–James bays, and Nelson River), Minns' Figure 3 does suggest an asymptote, consistent with a "local saturation" model. However, inspection of Minn's (1989) Figure 3 also suggests that this could vary among the three primary watershed areas. Lines I fitted by eye to data points in Figure 5.14 suggest that there is (1) a sharp increase in local

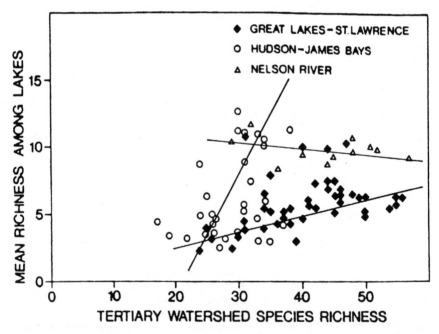

Figure 5.14. Mean species richness among lakes by major watershed versus watershed richness in regions of Canada. [From Minns (1989), with lines depicting slopes for three individual regions drawn by eye by Matthews.]

with regional richness, without an asymptote for lakes and tertiary watersheds in the Hudson–James bays primary watershed, (2) an increase and "leveling" for lakes in the Great Lakes–St. Lawrence primary watershed, and (3) a *negative* and approximately linear relationship between local within-lake fish species richness and tertiary watershed richness for the Nelson River primary watershed. Thus, a pattern detected over a large scale (among primary watersheds) might be dissected into highly differing patterns within individual primary watersheds (large river basins). Overall, there may be no single, simple answer to the question, "Do fish in lakes exist in locally-saturated assemblages?"

Hugueny and Paugy (1995) tested local- versus regional-species richness for 47 samples in 10 African river basins, as "regions" (Fig. 5.13B). They found a marked increase in local-species richness with river basin richness (Fig. 5.13B). Hugueny and Paugy (1995) were able to rule out "spurious correlations," such as distance from mouth of river and local physical variables, and concluded that the linear relationship between richness of local samples and basin faunas was "real." Their conclusion was that riverine fish communities may, indeed, be unsaturated and noted that only one other published paper (Tonn and Magnuson, 1990) had tested such relationships for fishes.

My work with Henry Robison provides an interesting extension of these questions (Matthews and Robison, in Press), in that we have found a relatively small influence of regional-species richness on local assemblage richness for 404 sites in the Ozark and Ouachita uplift in Arkansas (USA). For the upland portions of nine river basins draining this highland, we documented the regional-species pool as an enumeration of all known species that were present (Robison and Buchanan, 1988). Within each of the nine river basins, we found a very wide variance in numbers of local species, resulting in a weak overall relationship between local- and regional-species numbers. We presume that much of this wide variation in local assemblage richness is related to the very wide array of locally available habitats, coupled with the likelihood that many species in the "regional" (river basin) pool do not find appropriate habitat at all stream sites in the region. For example, large-bodied, riverine species would not be expected in small headwater creeks. Hence, local habitat variability and suitability [as noted by Cornell (1985a)] clearly must be taken into account in comparing local- to regional-species richness for fishes.

We found, if all fish taxa are included, a weak relationship of average local-species richness to regional (basin) richness with a slope of only 0.05. (Matthews and Robison, in press). When we limited analyses to minnows (Cyprinidae) alone or darters (Percidae) alone, there was no significant relationship between local and regional richness, suggesting that within these two most speciose and small-bodied groups (which would not be ruled out of local assemblages by body size versus habitat size, for example) there must be local physical phenomena or biotic interactions overriding any regional richness effect in determining the numbers of local species. Thus, our results (Matthews and Robison, in press) conflict with virtually all previous findings for local versus regional richness. All previous studies have shown a positive relationship (increase) of local-species richness to regional richness—the question has been whether the patterns were asymptotic (suggesting local saturation of species) or not. In contrast, we found little or no increase in local-species richness with increased regional richness, more reminiscent of the original "flat-curve," local-saturation model of Cornell (1985a). We, therefore, suggest that local stream fish assemblages in the Interior Highland of North America may indeed be saturated, with a relatively low percentage of regional species present in any stream reach, and with that saturation likely mediated as much by the presence or absence of appropriate habitat as by any interactions among species. In other words, our results suggest that "saturation" may exist, but in terms of saturation of appropriate habitat types rather than by saturation of any totality of interspecific interactions within sites. Clearly, this all needs much more investigation in a wide range of kinds of freshwater regions.

Zoogeography of Individual Species: Range Sizes and Body Sizes

The zoogeography of individual species, or sizes of species' ranges within a taxon, may relate to numerous mechanistic hypotheses such as capability for dispersal, physiological constraints of climate (latitude), interactions with other species, or success of generalist taxa during Pleistocene climatic stress (Brown, 1995a, pp. 112–116). Taylor and Gotelli (1994) found that body size, range size, and latitude were all correlated for 27 species in the North American minnow genus *Cyprinella*. However, with latitude removed as an effect, western and eastern clades within the genus differed in their response. The western clade showed no remaining effect between body size and range size that could be explained by speciation and isolation events in the American Southwest (where ranges of individual species have been restricted by recent increases in aridity and isolation of river basins). In contrast, within the eastern clade of *Cyprinella*, there was a phylogenetic effect, with less derived species having larger ranges and more derived species having smaller range sizes. Their conclusion was that "macroecological patterns (e.g., in body and range size) are sensitive to phylogeny and speciation history" and suggested that macroecological patterns may be most useful in comparing species that share a common climatic history within a clade. However, overall, there was support (after statistically controlling for effects of latitude, longitude, and phylogeny) for the fundamental postulate that body size and range size were positively correlated within this genus (Taylor and Gotelli, 1994). Hence, at the interface between zoogeography and local patterns, it is necessary to keep in mind that evolved traits of individual species can affect results at all geographic scales.

5.5 Merging Concepts of Geological, Zoogeographic, and Basin-Level Effects in Ecological Time

Knowing fundamental evolutionary history and derivation of fish lineages at the continental scale and knowing major and minor basin-level historical differences is all very interesting, but what does it mean to an ecologist wanting to understand the occurrence of modern fishes in extant assemblages? The answer is, in part, hierarchical. Assume a fish ecologist samples a stream in the mid-Mississippi Valley and finds a gar, a minnow, and a darter species in the assemblage. She knows the gar is of a very old lineage and that it may have been present in that habitat (or at least at that location) essentially since the Cretaceous. She knows that the minnow is much less likely to have a long evolutionary history in that particular location, as the evolution of cyprinids in North America is much more recent. Further, many minnow species have range limits that coincide with boundaries of large river basins (or fish faunal units) or with lesser drainages (e.g., individual river drainages) within basins. In many cases (e.g., known information on transfers of fishes across the Atlantic Divide), she may know approximately how long ago given minnow species gained access to a particular drainage, and there is increasing information from biomolecular studies (e.g., Richardson and Gold, personal communication) to help pinpoint dates of entry of particular species into given drainages. The darter, she knows, is probably of geologically relatively recent occurrence at the location, either through dispersal or evolution in place. So—knowing the probable differences in the length of time a taxon has been in one place—what can this help her predict about ecological relationships among these three taxa?

First, perhaps "priority effects" (e.g., Morin, 1984) are operational at geological as well as immediate ecological time scales. Barring anthropogenic disturbance, perhaps over very long periods of time local assemblages will be dominated by taxa that have been there longest. They will have had more time to adapt to a given habitat, even changing with the habitat as it changes climatologically, and could be resistant to invaders from outside the protoassemblage. Note that an early arriving species could inhibit entry of another taxon into the stream *either* through competition (full niches) *or* predation (entry of the naive taxon, if smaller, could be long inhibited by the presence of predators). This could all raise the issue of evolution of assemblages by gradual or punctuated processes, analogous to arguments about evolution of taxa (Futuyama, 1986, pp. 401–402). See also Roberts' (1975, p. 264) argument for "faunal succession" in African streams.

What else can she predict from knowledge of the timing of availability of taxa or of their evolution/dispersal into an area? In areas where no fishes existed until recently (e.g., glaciated areas, with all taxa via dispersal from refugia in last 14,000 years), there are probably few ancient interconnectivies among taxa (although note that such connectivities could have been present, maintained, or even evolved while fishes were dwelling in more southern refugia outside the glaciated areas).

On a finer scale, the ecologist can make other better guesses about the complexity of her local assemblage (e.g., presence–absence of taxa, or numbers of taxa) by knowing on which side of a ridge line or to which major or minor basin the tributary belongs. For example, headwaters of upland streams of the White and Arkansas river basins are closely interdigitated near the communities of Deer, Swain, and Nail, Arkansas. However, if our ecologist is working in a south-draining creek of the Arkansas River basin, she has 14 fewer fish species with the potential to be present, because these taxa do not cross the divide between the White River basin to the north and the Arkansas River basin to the south (Matthews and Robison, 1988). Thus, conclusions about the richness of the local fauna, the potential for interspecific interactions, or even phenomena such as "assembly rules" (Diamond, 1975) would, for these fishes, be strongly dependent on the overriding role of "basin affinity." Perhaps the overview is that the continental scale and long-term geological timetable-related phenomena make "big" differences in the ecological interpretations that are possible for a local assemblage, that major basin or faunal province-level differences play a more recent and more specific role in ecology of fishes in local assemblages, and very specific "minor-basin" zoogeographic phenomena (Matthews and Robison, 1988; Gorman, 1992) play major but very precise roles in fine-tuning our expectations about composition or interspecific interactions in a local stream reach. For experimental ecologists taking a largely reductionist approach, zoogeography raises cautionary notes about the kinds of experiment predictions (or even experiments) that are logically appropriate. It would seem that the best and most logical experiments in fish ecology would be based on a large-scale view of "ichthyology" of a region (in both space and geological time) and that the length of time for potential contact between taxa should be of paramount importance in designing experiments.

Interactive Ecology

There is recent evidence that North American land mammals do not exist in old, integrated communities—species moved independently as climates changed—not as species in communities that remained together over long periods of time (Graham et al., 1996). What are some of the common interspecific or intertaxon ecological or behavioral phenomena in North American freshwater fishes that we might view differently if we knew the length of contact of the taxa, or knew that individual fish species, like land mammals, had moved about in an individualistic manner over evolutionary time? How might we interpret each of the following interactions from a perspective of "who occurred with whom, and how long ago"?

1. Nest mutualisms or parasitisms. Many minnows nest on nests built by other minnows (e.g., *Notropis* or *Phoxinus* spawning in nests of *Nocomis*). There are also uses of sunfish nests (e.g., *Lepomis auritus*) by shiners. Centrarchids and minnows are known from North America since about 50 mya. Thus, the internesting behavior has the opportunity to be an old trait, evolved in ancestral species or genera, and perhaps not "invented" in the present species. There are also reports of darters (e.g., *Etheostoma spectabile*) spawning in smallmouth bass nests. Darters may be more recent arrivals in North American streams; thus, such nest associations might be more opportunistic and less integrated into the behavior.
2. Minnows follow suckers and feed on items suckers stirred up from the bottom. Juvenile smallmouth bass will also follow suckers. Suckers, centrarchids and minnows are old in North America; thus, interactions could be very old.
3. Bass protect larvae of minnows by driving away intermediate-sized fish that might prey on the larvae (e.g., sunfish) (Harvey, 1991a, 1991b).
4. Multitrophic level interactions that include centrarchids, algivorous minnows (*Campostoma*) and attached algae in streams.

All of these interactions seem to be among taxa that have, at a minimum, been present on the North American continent for a very long time; thus on a broad scale, it is logical that fine-tuning of ecological interactions has been possible both through phylogenetic and recent modifications of taxa. Other interactions among taxa known to have reached North America more recently would lack this rich history of opportunities for fine-tuning ecological relationships over evolutionary time.

6

Physical Factors Within Drainages as Related to Fish Assemblages

6.1. Introduction

Local fish assemblages are influenced by physical or ecosystem factors within individual drainages (lake basins, rivers, small streams) or local sites (stream reaches, parts of a lake). This chapter examines the structure of fish assemblages relative to (1) drainage area, (2) local-site dimensions (e.g., width, habitat volume, discharge, depth, and pool development), (3) habitat heterogeneity, (4) physical structure, (5) biotic productivity, (6) zonation in lakes or longitudinal patterns in streams, and (7) the landscape in or surrounding the water. This is a mixed array of factors sharing in common that all (1) characterize external environments of fishes on a scale at which an individual fish, in its daily or seasonal activities, could "sample" the variable and (2) are mostly abiotic or "noninteractive" variables (which the fish generally do not control, in contrast to biological-interactive variables dependent on or involving the fishes). However (Chapter 11), some fishes strongly influence their own external environment or that of others (e.g., if algivorous fishes alter the physical structure of the environment for other fishes (Power et al., 1985, 1988; Gelwick and Matthews, 1992; Gelwick, 1995; Flecker, 1996), so strict categorization of factors as "extrinsic" is difficult.

6.2 Area Effects at "Drainage" Level

Drainage Area and "Insular" Effects

The influence of habitat size on species richness has been tested for many animal communities, with emphasis both on actual and conceptual islands (Simberloff, 1974). Species–area theory was derived for oceanic islands (MacArthur and Wilson, 1967; Diamond, 1975; Rosenzweig, 1995), but the concept of a positive relationship between habitat size and number of coexisting species applies to most vertebrate and many invertebrate groups (Bengtson and Enckell, 1983; Ricklefs and Schluter, 1993; Brown, 1995a; Rosenzweig, 1995). Mechanisms underlying species–area relationships were summarized as an equilibrium between colonization and extinction processes (MacArthur and Wilson, 1967), with islands of greater size having more complex habitats, lower extinction rates, and, therefore, an equilibrium at a greater total number of species. Conversely, the degree of isolation from species source pools would result in slower rates of colonization and equilibrium at a lesser number of species.

Originally, species–area curves were applied to oceanic islands and focused on the slope of a line describing the increase in number of species as a function island size. With data in log-log format, a species–area curve often is linear, with a slope of z that describes the increase in number of species with area (MacArthur and Wilson, 1967). Although the biological meaning of slopes (z) of species–area curves has been questioned (Abbott, 1983), Preston (1962) suggested z equaled 0.25 or greater in most instances of true insularity, a value subsequently supported by much empirical evidence for terrestrial animals on actual islands. Diamond and May (1976, pp. 162–186) reported $z = 0.20 - 0.35$ for animals and plant taxa of true insular habitats. The finding of a species–area relationship can be strongly scale dependent (Palmer and White, 1994), and the z values (slope) for a set of habitats are strongly dependent on the range of size of habitat units—with surveys across smaller habitats generally yielding larger absolute values of z (Martin, 1981).

The concept of "island" was soon expanded to include many kinds of isolated habitats [e.g., tops of mountains with isolated cold, high-altitude, often relictual, floras and faunas (Bengtson and Enckell, 1983; Lomolino et al., 1989)] or habitat fragments. Rosenzweig (1995) noted that z values for islands are often about 0.25–0.33, whereas mainland z values for continguous, increasingly large areas range from about 0.13 to 0.18. However, Rosenzweig (1995, p. 277) noted the importance of distinguishing between z values for samples within terrestrial provinces (large areas containing ranges of animals), between terrestrial provinces, and for true islands).

That some kinds of fish habitats might represent functional islands seemed likely, in that fish are restricted not only to water but often to particular kinds of habitats. A wide range of fish habitats could function as "islands," including the littoral zone around actual islands, coral reefs, natural lakes, individuals rivers draining into the ocean, smaller watersheds (creeks) confluent to a large river, or even individual pools or riffles within a stream. Barbour and Brown (1974) found $z = 0.15$ for 70 lakes worldwide and 0.16 for 14 lakes in North America. Using the results of Barbour and Brown (1974), Grande (1984, pp. 179–181) made a fascinating comparison of species richness versus area for the extinct Green River Lakes (Wyoming, USA) fossil fish fauna, finding that species richness in one extinct lake fauna was almost exactly as predicted by Barbour and Brown's regression line but that two of the fossil faunas were much lower in richness than would be expected by the species–area regression for modern faunas. Grande attributed these differences in relative richness to environmental stability, with the deeper fossil lake having the highest relative richness, and the two shallow fossil lakes presumably having fluctuating temperatures that resulted in a lower than expected number of species.

Magnuson (1976) noted parallels between lakes (for fishes) and oceanic islands for terrestrial vertebrates. Magnuson (1976) hypothesized that "drainage lakes" (connected to natural streams) versus "seepage lakes" (lacking connection to natural streams) and distance from other lakes would influence the rate of colonization of the lake by new fish species. The result would be an equilibrium at a higher number of species for lakes connected to a ready source of colonists or located closer to other lakes. Magnuson pointed out that larger lakes with more complex habitats should have slower rates of local extinction, hence more species. Magnuson (1976) noted that "both the number of species for a given area and the slopes of the species area curves are strikingly similar for fishes and other organisms" (with z approximating a theoretical "island" value). Magnuson (1976) also postulated that human influences increase immigration rate (e.g., stocking), reduce lake isolation (by habitat modifications), and reduce the extinction rate, and surmised that anthropogenic changes will accelerate changes in lake faunas from evolutionary to ecological time scales (e.g., within our lifetimes).

For 12 lakes in central New York, Browne (1981) found a species–area slope of $z = 0.24$, with area positively related to fish species richness within one lake district with relatively homogeneous environmental conditions. For 2931 lakes in Ontario, Matuszek and Beggs (1988) found a z value of 0.20 for the mean number of species (within an area size class) and area. Lake area predicted fish species richness better than any of 18 other physical and chemical variables they evaluated, accounting for 18% of the total variance in species number. (Matuszek and Beggs also found a strong positive slope for the maximum number of species versus lake area.) In a separate analysis of 9679 Ontario lakes, Minns (1989) found species richness significantly related to surface area (and five other variables, including the depth; Section 6.3). Marshall and Ryan (1987) found lake area to be the best single predictor of species diversity in 75 Canadian boreal forest lakes. Taylor (1996) showed a strong and significant positive relationship between riffle area and benthic fish species richness in upland streams in Oklahoma.

There have been tests of species–area curves for entire streams or stream drainage areas. These have included coastal rivers isolated from each other by seawater, as well as small clear-water tributaries of large, turbid mainstreams (which are inappropriate habitat for many small-stream fish species). Hugueny (1989) reviewed eight previous studies for rivers, mostly in Africa. In most of these studies, authors used drainage area as a relative measure of habitat area. For 39 rivers in West Africa, Hugueny (1989) reported $z = 0.32$, compared to $z = 0.43$, for a geographically wider sample of African rivers (Welcomme, 1979) and a value of $z = 0.55$ for South American rivers (Welcomme, 1979). For European rivers, Hugueney reported z values ranging from 0.19 to 0.24 by three previous authors (Welcomme, 1979; Daget and Economidis, 19795; Daget, 1968) and suggested that species richness increased more rapidly with surface area in tropical rivers than in temperate rivers. The z values cited by Hugueny, with comments on regional patterns and causes of species–area relationships for fishes, are also published by Welcomme (1985). For 47 of the largest river systems, worldwide, Welcomme (1985, p. 94) reported a strong species–area slope with a z value of 0.48.

For watersheds in North America ranging in size from creeks to entire river basins, Cross (1985) showed a positive relationship between drainage area and number of species, but this relationship was different in depauperate drainage basins. For example, Cross (1985) showed (his Fig. 4) that rivers within the Colorado Basin (which has, basinwide, only 32 native species) had far fewer species than "expected" on the basis of area alone, with a much flatter slope within that basin relative to most of the rest of the continent (reinforcing the conclusion in Chapter 5 that drainage basin influences, in a major way, the structure of local fish assemblages). For midwestern North American streams, Karr et al. (1986) found a general increase in species number both with watershed area and stream order (Section 6.5). Swift et al. (1986, their Fig. 7.7 and Table 7.2) showed a positive general relationship between drainage area and number of fish species, for total species and for the combined total of cyprinid, catostomid, and percid species, for major and minor drainages in the southeastern United States, and within each of three zoogeographic subregions. For 72 sites with increasingly large drainages in the Embarras River, Illinois (USA), Fausch et al. (1984) showed a positive relationship between species richness and watershed area (although only one of their axes was on a log-scale so no z value can be determined). At a much larger, global scale, Oberdorff et al. (1995) found area of river basin to be the most important single variable influencing species richness.

The original species–area curves were for islands varying by several orders of magnitude, but these relationships can break down in small areas. H. W. Robison and I (unpublished data) found a low z value ($z = 0.11$) for the number of species versus drainage area in 19 creeks throughout Arkansas. For clear, upland creeks or small rivers within the Interior Highlands, z was 0.06 for 17 Ozark Mountain drainages, and slightly negative for 11 Ouachita Mountain drainages. However, when somewhat larger streams (medium-sized rivers with drainage areas as large as 4000 km^2) were included, a significant positive slope existed between species number and area of the watershed, with an overall z value of 0.30 for 26 upland watersheds in Arkansas (Matthews and Robison, 1997).

There is at least other one exception for fish to the "larger area = more species" generalization—sites on eight rivers in France (Oberdorff et al., 1993). For species richness per site relative to upstream drainage area, Oberdorff et al. found a peak at catchment areas between about 1000 and 10,000 km^2 and a decline at sites with drainages greater than about 10,000 km^2. However, in their analyses, the number of species per site was considered, rather than the total number of species in the entire basin. Furthermore, the analysis by Oberdorff et al. is nested (i.e., with downstream sites having larger drainage areas that included the smaller drainage areas of upstream sites). Thus, the analysis of Oberdorff et al. does not test for "insular" effects, and tests instead whether an increase in upstream area equates to more species at a given place in a system.

To summarize, in spite of the exceptions noted above, species richness in many aquatic systems is positively related to area. For lakes, area has usually been considered as the actual surface area; for whole streams, total drainage area has typically been considered. The positive relationship between area and species richness seems to fail if the range of sizes of systems being compared is too small, and it is clear that comparing species richness for a small stream in a highly species-rich versus a depauperate river basin on the basis of area alone would be invalid. Overall, however, it appears that, all other variables being equal, the actual area of a system is a good first clue to the total species richness that might be expected to be present. It is instructive to next consider the effects of local habitat size on stream fish assemblages.

6.3 Local Habitat Size

Local habitat size factors that influence fish assemblages include one-dimensional measurements like width and depth, two- or multi-dimensional constructs like local pool area or volume, and abstractions that include time as a dimension, such as annual discharge (= volume of water passing a point in a stream per unit time). In some cases the variable has been related only to increases in sizes or densities of a particular population; in most the species richness of whole assemblages has been considered. Minckley (1984) showed that habitat size was related to species richness for stable, but not for unstable habitats in the Cuatro Ciénegas basin, Mexico.

Stream Width

The width of a stream at any particular place can correlate not only with total numbers of species but also the presence–absence of particular species. The distribution of some closely related species pairs in freshwater streams is related to stream width (e.g., Robison and Buchanan, 1988). For example, in medium-sized upland watersheds of the Ozark Mountains, the riffle-dwelling darters *Etheostoma caeruleum* and *Etheostoma spectabile* are partially segregated by stream size (W. J. Matthews, unpublished data, Piney Creek, Arkansas), with the former in wider and the latter in narrower stream riffles. However, this relationship also is related to stream order (Section 6.5) (i.e., wider riffles tend to occur further downstream). A similar pattern is true for the sculpins, *Cottus hypselurus* (larger streams) and *Cottus carolinae* (smaller streams) in Piney Creek and elsewhere in the Ozark Mountains. However, Taylor and Lienesch (1996) showed that two closely related minnow species (*Lythrurus*) had nonoverlapping distributions in upper versus lower parts of streams of the Ouachita Mountains (Oklahoma–Arkansas, USA), but that the differences in the occurrence of the two species did *not* correspond to local stream width (and was, instead, best explained by a complex combination of seven physical variables). Anderson (1985) found that sculpins in larger streams grew faster and had higher fecundities and lower densities than sculpins in small streams. This paradox suggested to Anderson that greater predation pressure in large streams lowered sculpin density.

Some studies have tested species richness in local assemblages as a function of stream width. In part, this relates to the larger issue of "longitudinal zonation" of fishes within drainages (Section 6.5) because width is generally related to position in a stream continuum. However, the width of a stream at a given point is highly variable within any stream order. In the broadest sense, there exists in many parts of North America a "big river" fish fauna that is unique relative to the fishes of small rivers and creeks, with taxa such as carpsuckers (*Carpiodes* spp.), buffalo fishes (*Ictiobus* spp.), blue suckers (*Cycleptus elongatus*), catfish (*Ictalurus* and *Pylodictis* spp.), gars (*Lepisosteus, Atractosteus*), or paddlefish (*Polyodon spathula*), and minnows like *Notropis girardi, Notropis bairdi, Macrhybopsis meeki, Notropis potteri, Hybognathus placitus*, or *Macrhybopsis storeriana* particularly common. Big river faunal regions were identified in Missouri (Pflieger, 1971) and Arkansas (Matthews and Robison, 1988) (Fig. 6.1). Oklahoma has a distinctive large-stream "ecoregion" (Murray, 1996). Species adapted for an environment characterized by "persistent high turbidity, wide seasonal fluctuations in flow and temperature, and an unstable sand-silt substrate" (Pflieger and Grace, 1987) are characteristic of fish faunas in many large rivers of the United States (e.g., C. L. Hubbs, 1941; Cross and Moss, 1987).

Matthews and Robison (1988) found a significant overall correlation between stream width and multivariate axis scores based on fish species abundances for 101 drainage units in Arkansas (although width was not as strong a predictor of axis scores for a drainage unit as were turbidity, local substrate, and geology). In a comparison of 48 sites throughout the upper Red River basin in Oklahoma, Taylor et al. (1993) found maximum stream width highly correlated with ordination scores for the fish assemblages.

M. L. Smith and Miller (1986) showed a positive relationship between local habitat width and number of species per collection in the Rio Grande basin (United States and Mexico), but their Figure 13.6 also showed a difference in the slope of the species versus stream width curves for connected tributaries of the Rio Grande versus localities in endorheic drainages that lacked actual connection to the Rio Grande proper. Historical factors at work in this relationship probably included more speciation in isolated systems.

For local reaches of a small Ozark stream, Gelwick (1990) found that the first

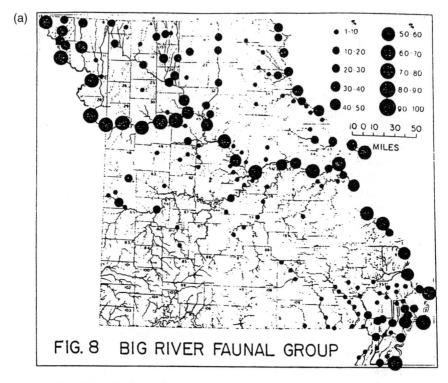

Figure 6.1. "Big River" faunal groups in (A) Missouri [from Pflieger (1971)] and (B) Arkansas [from Matthews and Robison (1988) Copeia 1988(2), p. 367, Fig. 9: *Fish faunal regions of Arkansas, based on DCA Axis 1 as described in text*].

axis of a multivariate analysis of fish assemblage structure in pools was strongly related to stream size (width) but that no such relationship existed for riffles. Matthews and Gelwick (1990) found only a nonsignificant trend in the increase in numbers of species versus stream width for 57 sites in central Oklahoma. In this latter study, mostly of low-gradient streams with harshly fluctuating environmental conditions, it is likely that physicochemical stress (e.g., high temperature, low oxygen) sets upper limits on numbers of species that can exist.

Stream width typically is confounded with other factors. In Conowingo Creek, Maryland–Pennsylvania (USA), Hocutt and Stauffer (1975) found the number of species to be correlated with both width ($r = 0.61$) and depth ($r = 0.60$), but more strongly related to stream gradient ($r = -0.90$). Similarly, Paller (1994) found fish assemblages in South Carolina (USA) coastal streams strongly correlated with both stream width ($r = 0.69$) and depth ($r = 0.78$), as well as to stream cross-sectional area ($r = 0.78$). This appears to be the only case in which fish assemblage structure has been tested against cross-sectional area of the stream.

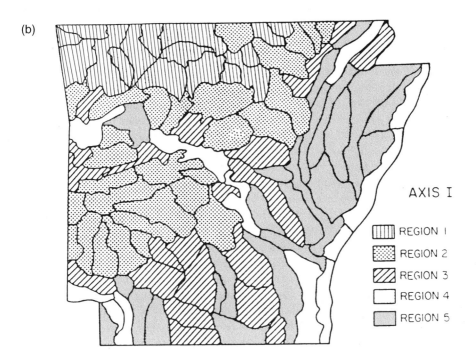

Figure 6.1. Continued.

Stream Depth and "Pool Development"

Depth has important influence on the fish assemblages in streams. Gerking (1949) showed correlation between fish biomass and volume of deep water in pools of small streams. Deep water is related to environmental stability (e.g., damping temperature variation) and allows vertical separation of microhabitats of fish species (e.g., Baker and Ross, 1981; Gorman, 1988a, 1988b). Some individual species, such as blue sucker (*Cycleptus elongatus*), are typically found in deep parts of larger rivers in the North American Midwest. Power (1987) showed that avian predators (herons, kingfishes, etc.) restrict large fish to deeper habitats. Harvey and Stewart (1991) reported a strong correlation between depth and size of the largest fish in pools, and that minnows survived longer in deeper pools.

Sheldon (1968) found in a New York creek that depth was the single most important factor related to the number of species present in a stream reach, accounting for 70% of the variance in species richness (although position in the watershed also was important), and that depth accounted for 66% of the variance in species diversity. In a multivariate analysis, Taylor et al. (1993) found mean depth strongly correlated with a canonical correspondence axis which accounted for 16% of the variance in fish species abundances. In a small east Texas (USA) stream, Evans and Noble (1979) found fish diversity highly correlated with depth of stream (and less with longitudinal position in the watershed). Tramer and Rogers (1973) found no relationship between fish species diversity and mean depth of the stream for creeks near and in Toledo, Ohio (USA), but attributed this lack of pattern to local pollution effects. Shetter et al. (1946) reported an increase in numbers of large trout after stream pools were deepened artificially.

Schlosser (1987) suggested a general model of fish assemblage properties relative to depth and "pool development" for small warmwater streams in the midwestern United States. Schlosser (1987) postulated a gradient in fish community attributes from habitats that were shallow and temporally variable with little habitat heterogeneity, to those that were deep, more temporally stable in environmental properties, with high habitat heterogeneity. Species richness was predicted to increase from the former to the latter, but that "species density" (number of species per 100 m^2) and total fish density were predicted to peak midway between the two extremes of stream habitat. Schlosser (1987) suggested the shallow "colonizing" parts of a stream to be dominated by small bodied, short-lived species (e.g., minnows) or the young/small individuals of larger species (e.g., centrarchids). At the other extreme, deep areas (greater pool development) would be characterized by large-bodied species or individuals, but potentially show a reduction in "species density" because of predation from large centrarchids (adult sunfish, black bass) in deeper pools. Schlosser (1987) also pointed out that shallow areas would have harsh winter conditions (freezing, oxygen depletion) and be dominated (in warm weather) by individuals that had colonized the habitat since the most recent excessive winter conditions or had been produced by species with high rates of reproduction or extended spawning seasons.

Shallow stream areas also can be extremely hot during summer, comprising a very harsh habitat for fishes (Matthews, 1987a; Matthews et al., 1982c) (see Chapter 7). Schlosser suggested that the intensity of competition would vary annually in shallow habitats, that trophic structure of the shallow-water assemblages would be simple, and that competition could potentially limit summer growth, thus the maximum size that individuals could attain before onset of winter. Overwinter survival can be related to body size attained by young-of-year before onset of cold temperatures.

Schlosser (1987) predicted that at the other extreme, in well-developed, deep pools, there would be greater habitat heterogeneity, with deeper water being a refuge from harsh winter conditions. I found immense aggregations of minnows in deep pools in the upper Roanoke River, Virginia, during extended winter cold weather, and C. M. Taylor (personal communication) found similar aggregations in the Red River basin, Oklahoma. Schlosser (1987) predicted that in deeper pools, colonization and recruitment would have less effect on assemblage structure, intraspecific interactions between size classes would be more important, and trophic structure of the assemblage would be complex. He predicted that in deeper pools, competition among large individuals (e.g., sunfish) would be important, and that small fish would have to "choose" between predation threat in deep pools and the potentially harsher physical conditions or competition from other small fishes in shallow areas.

These conflicting factors may work differently in summer and winter. In summer, centrarchids or other predators in pools may be very active (e.g., Harvey et al., 1988), hence a substantial threat to small fishes, whereas in winter, the centrarchids appear to be less active and less threatening as predators, allowing more small fishes to coexist with piscivores in deep pools (i.e., accepting some level of predation pressure to avoid harsh winter conditions). Schlosser's (1987) model may be most applicable to small warm-water streams sufficiently far north that winter cold is a significant threat to fishes, and it may need to be modified to apply fully to streams in hotter climates where winter cold is rarely a problem, but summer heat is.

Depth of Lakes

Whereas streams generally vary in depth from a few centimeters to perhaps 100 m (in deep parts of the largest rivers), the range in depth of lakes is orders of magnitude greater. It is difficult to equate maximum depth (alone) of lakes with fish assemblage structure, because depth in lakes is confounded by variables like area, geologic origin (e.g., tectonic fault block lakes), and other geographic factors. For 8852 Ontario (Canada) lakes, Minns (1989) found a significant relationship of species richness to mean depth (although lake area was a better predictor of number of species). However, Minns (1989) showed that mean depth was only one of six variables that contributed significantly to species richness in a stepwise multiple-regression analysis of fish species richness.

To really compare fish assemblages in lakes of similar area but differing depths would require data on lakes from approximately the same geographic region, but differing in relative depth, mean depth, or volume development (Cole, 1994). One variable that should be examined is the epilimnion to hypolimnion (E/H) ratio that has been used by limnologists. In lakes with a large, deep hypolimnion (small E/H ratio), the greater volume of water in the hypolimnion relates to more total hypolimnetic oxygen available after summer stratification seals off the water column into "upper" and "lower" lakes. Therefore, lakes with large hypolimnetic volume should have oxygen available in the hypolimnion longer into the period of summer stratification and be less likely to go anoxic. Such lakes would offer more of the water column as habitat for fishes during summer stratification and, thus, might have richer or more diverse faunas.

Estimates of fish production or complexity of fish assemblages relative to depth (often combined with other variables) do exist. Rawson (1952) found a trend for lower average standing crops in deeper lakes, whereas lakes less than about 18 m mean depth ranged from very high to low density of fishes. The depth of lakes relative to the density of fishes is incorporated in the "morphoedaphic index" (MEI), originated by Ryder (1965) and reviewed in Ryder (1982). The MEI is calculated as the total dissolved solids (mg/L) divided by the mean depth of the lake (in m). The MEI has (Ryder, 1982) had uneven success across a wide range of applications, but Ryder emphasized that MEI used as intended was useful in predicting fish density. Ryder (1982) noted that the most important factors in fish yield in lakes are "global" (e.g., temperature), but that within relatively homogeneous lake districts, a combination of nutrients and mean depth is a good predictors. Schlesinger and Regier (1982) concluded that the MEI was a better predictor of fish production within than among globally different lakes (e.g., those differing markedly in temperature). Jenkins (1982) found the MEI useful as a predictor of sport fish harvest or fish standing crops across a wide range of reservoir types. Rigler (1982) reviewed attempts to improve or modify the MEI for local (e.g., latitudinal) variation. Schneider and Haedrich (1989) summarized several mathematical problems with the MEI related to the intercorrelation of variables and conversion to metric units and made recommendations for improving the predictive limits of the model.

Effects of the MEI are species-specific. Marshall and Ryan (1987) found that the peak in abundance for lake trout, burbot, and lake whitefish was in lakes deeper than 6.3 m, but that peak abundance of walleye, northern pike, white sucker, and yellow perch was in lakes shallower than 6.3 m mean depth. Marshall and Ryan (1987) found that the MEI could distinguish two types of fish assemblages, with northern pike, white sucker, walleye, and lake whitefish in small, shallow, transparent lakes (high MEI), and lake trout plus lake whitefish or white sucker in large, deep, clear lakes (low MEI).

The depth of water characterizing the available habitats at a given location within a lake is a different matter than maximum or mean depth of the whole lake. There is very strong difference in most lakes among the fishes that occur in (1) shallow, inshore, or littoral waters, (2) the open-water, limnetic zone offshore, and (3) the deep benthic (profundal) parts of a lake. Differences include species, dominant families, or size of individuals (e.g., more cyprinids or centrarchids in the littoral zone of large lakes, and more clupeids in limnetic waters). Many fish species, however, occur both in littoral and limnetic zones or migrate between those habitats (Hall et al., 1979; Hubbs, 1984; Matthews, 1986a). Within the inshore zone, depth can have a particular effect on species that build nests, with excessive depth or steeply sloping lakebed prohibiting nesting for some species. Furthermore, the composition of fish assemblages in the littoral zone of large lakes may be ephemeral or change seasonally (e.g., Gelwick and Matthews, 1990) or spatially (Hinch et al., 1994).

Volume

Volume is more abstract than a direct measurement of width or depth, in that it is a combination of several measurements. Volume provides living space for fishes, whereas area influences surface phenomena such as gas exchange or number of terrestrial insects falling to a pool surface per unit time. Angermeier and Schlosser (1989) sampled discrete pools and riffles in Illinois, Minnesota, and Panama and found that species richness generally correlated with site volume. Habitat volume predicted species richness better than habitat area did. Their plots were not log-log, so no value for z was given, but their Figure 1 shows a distinct increase in total species with increase in log volume for pools and riffles, separately. In a small Oklahoma creek, Taylor (1997) showed that pool volume (for 30 small pools) was significantly related to species richness and to the abundance of three species (*Lepomis megalotis, Semotilus atromaculatus,* and *Lepomis cyanellus*) in isolated pools. However, for pools connected by flow, the distance from a large-stream source of species as well as pool volumes influenced abundances for some species (Taylor, 1997).

For 24 stream reaches in the Ouachita uplands of Arkansas Matthews (unpublished data) found a positive slope between pool "size" (based on width and length of pools) and numbers of species, with a slope (z) = 0.53 (Fig. 6.2). In 7 surveys of 14 individual pools in Brier Creek, Oklahoma (Matthews et al., 1994), there was a positive but nonsignificant (linear regression) $z = 0.104$ for the relationship of log mean species and log pool volume (Fig. 6.3A). However, raw numbers of species plotted against pool volume for the 14 Brier Creek pools (Fig. 6.3B) suggested a curvilinear relationship with the greatest number of species, on average, in pools of intermediate volume, and a slight decrease in numbers of species in pools with the greatest volume. This would coincide

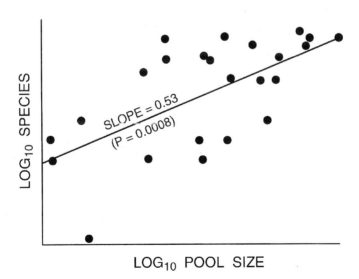

Figure 6.2. Log of number of species versus log of pool size, for 24 locations in streams of the Ouachita Mountains, Arkansas (USA).

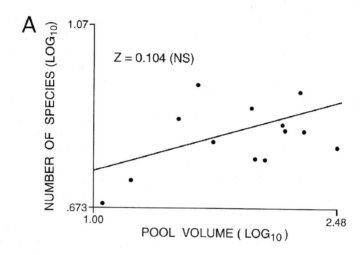

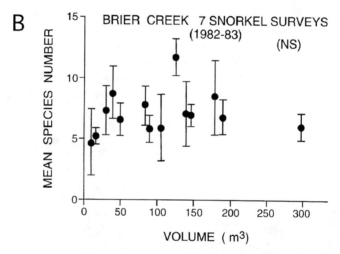

Figure 6.3. Number of species versus volume of pools in Brier Creek, Oklahoma.

with Schlosser's (1987, his Fig. 3.4) prediction that habitats with greatest pool development might show a slight decrease in species density in large or deep pools if predators limit the number of species. We also found fewer individual prey fish in pools with high numbers of piscivorous *Micropterus* basses (Matthews et al., 1994).

Discharge

Discharge is a more complex, essentially four-dimensional measure of habitat size (consisting of a volume of water passing a given point per unit of time—

often annual total or daily mean). Aadland (1993) quantified the relationship between discharge and changes in the amount of habitat available for fishes that required various current speeds. Moyle and Baltz (1985) showed that the habitat available to individual species at different flows depended on species-specific habitat requirements, and many regulatory agencies depend on applied uses of "instream flow" analyses, which are beyond the scope of this book. M. L. Smith and Miller (1986) found a positive relationship between number of fish species per drainage basin and the "basin discharge" (in m^3/year). Further, a line fitted by eye to their Figure 13.5 showed that the slope of species versus discharge was much steeper for tributaries connected by flow to the Rio Grande as compared to disconnected tributaries in endorheic basins in Mexico. Hughes and Omernik (1983) showed that annual discharge in a stream reach is highly correlated with basin area upstream of that point.

Three important studies address the question of discharge (or variability in discharge) and the structure of temperate-stream fish assemblages (Horwitz, 1978; Schlosser, 1985; Poff and Allan, 1995), and Power et al. (1995a) recently incorporated hydrologic regime in models of fish assemblage structure in tropical streams. For 15 North American rivers, Horwitz (1978) compared the magnitude of change in fish assemblages from headwaters to lower mainstems. Horwitz found headwater diversity lowest in rivers with variable headwaters [similar to the patterns suggested by Hubbs (1995) for differences between "constant" spring-fed streams versus environmentally variable headwaters]. Horwitz also found that the highest diversity in fish assemblages was in lower mainstems with low variability in discharge. Overall, Horwitz (1978) showed a strong negative correlation between fish diversity and variability in stream discharge and concluded that these findings supported either of two hypotheses relating fish assemblage structure to flow variability: (1) a "competition": trophic structure hypothesis (in which fluctuating habitats have generalist species, and vice versa) and (2) an "extermination" hypothesis (in which rates of local extermination of species by drought are greatest in variable headwaters, coupled with slow recolonization from downstream).

Schlosser (1985) showed that between-year differences in discharge strong affected composition of fish assemblages in a reach of Jordan Creek, Illinois (USA). Jordan Creek is a low-gradient, agricultural–woodland stream of the midwestern United States. Schlosser found juvenile darters (Percidae) and suckers (Catostomidae) little affected by interyear differences in discharge, but increased numbers of juvenile minnows (Cyprinidae) and sunfish (Centrarchidae) in a year with low, stable discharge. With stable flow and increased abundance of juveniles, Schlosser (1985) found greater species richness in individual habitat patches. He also pointed out that variable flow would have differential effects on different life stages within species. That discharge can affect life stages of a species differently is well known. Larimore (1975) showed that floods wash out young but not adult smallmouth bass. Harvey (1987) found larvae to be killed by spates. Matthews et al. (1994) showed more displacement of juvenile than adult sunfish by floods.

Multivariate analyses by Poff and Allan (1995) for 34 stream sites in the northern United States showed that hydrologic data separated sites dominated by two ecologically different groups of fishes. One type of fish assemblage was associated with highly variable discharge (daily variation and frequency of spates); the other assemblage type was in streams with stable and predictable discharge. Their findings agreed with Horwitz (1978) that generalist species should dominate streams with variable discharge because of uncertainty of any particular resource, whereas specialists should be common in streams with predictable discharge. Poff and Allan (1995) suggested that functional as well as taxonomic properties of stream fish assemblages are strongly related to hydrologic variability and that large streams with unpredictable flow or widely fluctuating flow might function more like the "colonizing" headwaters streams postulated by Schlosser (1987).

Power et al. (1995a) suggested that for tropical rivers, the major mortalities of fishes related to extremes of hydrologic variability (flood versus dry seasons) might actually enhance persistence of ecological communities and/or increase biodiversity. The fluctuating environment would play a "keystone" effect, preventing the dominance of any particular species. Models by Power et al. (1995a) coupled hydrologic variability with food-chain dynamics for a tropical floodplain river system (Fly River, southern New Guinea). Their models are based on Lotka–Volterra equations for food webs involving detritus, vegetation, herbivores, and predators. The model allowed the river to inundate the broad floodplain or not during wet and dry seasons. Fish assemblages were most stable in simulations permitting inundation of the floodplain, whereas simulations with low flow restricted to the river channel failed to sustain the predators. These initial models by Power et al. (1995a) have the potential for linking discharge dynamics and fish assemblage structure in other habitats—including north temperate rivers—where floodplain–channel interactions are important.

Sadly, as pointed out by Power et al. (1995a) and by John McPhee (1987), in his wonderful essay on attempts to "control nature" in the Atchafalaya River basin, the largest formerly "floodplain" river in North America (the Mississippi) is now a channelized and levee-bound remnant of its former self, so we may never know how much of the dynamics of fish assemblages in this system *did* depend on hydraulic variability. See Bennett (1958, pp. 163–164) for an account of the dispersal (and mortality of) fishes moving onto the floodplain of rivers in Illinois in the late 1800s, before most "control" of Midwestern rivers. Bennett (1958, p. 164) concluded that, "We now suspect that the phenomenon of fluctuating water levels . . . may have been *highly favorable to the well-being of the populations of fishes* . . . (as) natural predation and water level fluctuations *prevented excessive competition among the coexisting species and allowed excellent survival of game fish*" (italics mine). This was published several years before ecologists accepted, in general, that fluctuating conditions or episodic mortality could prevent domination of an assemblage by any one or a few species.

6.4 Habitat Structure, Cover, Complexity, and Productivity

Fish use actual physical structure such as rock, submerged wood, macrophytes, algae, and so forth, as shelter from predators or swift currents, as foraging sites (Benke et al., 1985), and as spawning sites. A similar factor is "cover," which includes not only the physical structure in the water but also overhead protection that allows a fish to hide from above-water observation (Heggenes and Traaen, 1988) by predators. Examples include logs lying across a stream or shoreline vegetation protruding over the water. Shade produced by overhead or floating cover influences the visibility of predators and prey and offers advantages to fish that hover out of direct sunlight (Helfman, 1981). More abstract is the concept of "complexity" of habitat, often measured statistically in terms of habitat "diversity" or "heterogeneity" (i.e., the degree to which a habitat differs from being homogeneous) at a scale or grain detectable by the fishes. Finally, primary productivity in a habitat can influence the density of individual species or local-species richness.

Structure or Cover

Underwater structure falls into two general categories: "hard" or relatively persistent objects, like large wood debris or rock, and "soft" or less persistent objects, like macrophytes, algae mats, or leaf packs (Fig. 6.4A). Coral reefs may represent the closest integration of activities or structure of fish communities with physical environmental features (Smith and Tyler, 1972). Many reef fishes are highly structure dependent, with holes in the reef important in their life history and habitats. In inshore marine waters, kelp beds provide the structure on which suites of fish species depend. In freshwater, woody debris or rock structure (e.g., in riffles) affords shelter from flow (Schlosser and Toth, 1984) or predators and many riffle-dwelling fishes use or depend on physical structure of the stream bottom (however, just how much interstitial spaces are used and in what ways largely remain to be quantified). Matthews and Hill (1979a) found that a small physical structure in the water—even as small as sticks or leaf litter—attracted small fishes in the sand-bottomed Canadian River of central Oklahoma. In lakes, rooted macrophytes provide structure (Savino and Stein, 1982, 1989) that is important in resource partitioning, foraging, or sheltering of fishes (Werner et al., 1981).

Woody Debris

In streams, downed treetops, trunks, or rootwads provide substantial and important shelter or cover for fishes (e.g., Fansch and Northcote, 1992). A question that remains is whether a relatively permanent structure under water (e.g., debris dams, rootwads, or downed treetops, with their intricate physical and hydraulic

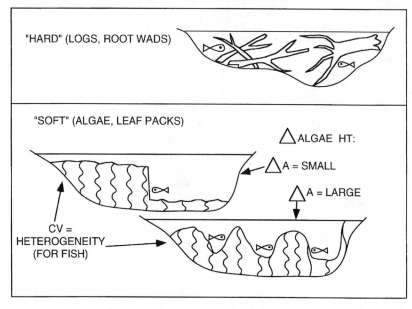

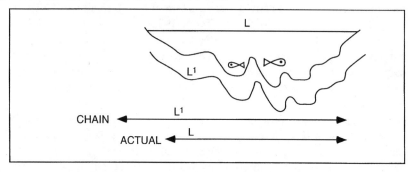

Figure 6.4. Heterogeneity in a typical stream for structure above the stream bed (A) and for variation in the profile of the stream bed (B).

structure) favor predators (e.g., smallmouth bass) more than prey (e.g., minnows and small sunfish). Angermeier and Karr (1984) indicated that although fish were always more abundant, overall, in manipulated stream reach with enhanced woody debris, large fish preferred large woody debris (LWD) but some small fish avoided such microhabitats. Propst and Stefferud (1994) found Chihuahua chub (*Gila nigrescens*) typically associated with tree rootwads in a desert stream (New Mexico and Mexico). Rootwads provided shelter from strong currents and protection from predators for juvenile salmonids (Shrivell, 1990) in a British Columbia stream. Quinn and Peterson (1996) reported that overwinter survival of juvenile coho salmon (*Oncorhynchus kisutch*) was strongly correlated with the amount of large woody debris.

In very turbid cypress swamp streams in northeast Texas (J. Killgore, J. J. Hoover, and W. Matthews, unpublished data), woody debris, stumps, and "rootwad" structure were related to the presence of some minnow species. Rock bass (*Ambloplites* spp.) in Ozark streams use shelter in submerged tops or roots of trees. By snorkeling, it is possible to find as many as 10–15 rock bass motionless and almost invisible among the roots or branches, positioned to attack prey that approach unaware of the predators. Woody debris also influences whole fish assemblages. Flebbe and Dolloff (1995) showed that several trout species in the Appalachian Mountains were virtually always found in stream sections that had large amounts of large woody debris. Logjams, (e.g., at bridges or at natural constrictions in streams) also provide structure for fishes. In upland streams in northeastern Oklahoma, large logjams or debris jams are common, consisting of large and small logs and associated debris, lodged at bridge pilings or natural constrictions in the channel. These logjams invariably harbor large sunfishes (*Lepomis* spp.), smallmouth bass (*Micropterus dolomieui*), and a large number of minnows.

In addition to comprising shelter from currents or predators, the sheltered structure of woody debris affords feeding or spawning sites for fishes. Many species like bluegill (*Lepomis macrochirus*) feed by picking invertebrates associated with aufwuchs on underwater woody surfaces. Algivorous *Campostoma anomalum* (Matthews et al., 1986) and loricariid catfishes [Panama (Power, 1984b)] often scrape attached algae from wood as well as stony surfaces. Benke et al. (1985) showed that in the Saltilla River, Georgia, a sand–mud-bottomed stream of the southeastern United States, as much as 60% of all invertebrate biomass and 16% of invertebrate production came from large woody debris snags, and that about 78% of the invertebrate biomass in drift (major source of food for water-column fishes) came from snags. Benke et al. (1985) estimated that at least 60% of the prey of three sunfish (*Lepomis*) species and of pirate perch (*Aphredoderus savanus*) originated on snags.

Many fish spawn on or in woody debris. In Pennington Creek, Oklahoma, I observed blacktail shiners (*Cyprinella venusta*) spawning in crevices of a submerged log, and in the upper Kiamichi River, Oklahoma, I collected a submerged log with developing eggs in a crevice, which hatched out in the lab to be *Cyprinella whipplei*. The use of woody structure for spawning is not limited to minnows. For example, some catfishes (Ictaluridae) in North America may spawn in hollow, submerged logs (Pflieger, 1975).

Large woody debris can indirectly affect habitat for fish by changing hydraulic scour (e.g., Fausch and Northcote, 1992), resulting in a deeper pool with an intricate physical structure and complex patterns of currents, often with more protection from strong flows than are found in the open channel of the river. Shetter et al. (1946) installed 24 log deflectors in a Michigan trout stream so that hydraulic scour deepened pools an average of about 15 cm and raised the number of "good" pools from 9 to 29 in a stream reach. They reported an increase in numbers and size of brook trout (*Salvenlinus fontinalis*) after the increase in pool number and size.

Rock Structure in Streams

Natural rock in streams also plays roles similar to that of wood, with large instream boulders providing physical/hydraulic habitat for smallmouth bass, large sunfish, and minnows. Much like a fallen tree, a boulder breaking loose falling into a sand- or gravel-bottomed stream results in a scour pool—often 1.5–2 m or more deep—which will be inhabited by large sunfishes, basses, and minnows. Rock of various sizes and complexity offers habitat for fishes, and Fuselier and Edds (1995) showed that artificial rock riffles attracted a suite of fishes similar to those in natural riffles in the Neosho River, Kansas (USA).

"Soft" Structure: Macrophytes, Algae, Leaf Packs

Macrophytes in streams also provide habitat for fishes. Small sunfish, minnows, or juvenile bass use water willow (*Justicia americana*) for shelter or foraging. Darter species may differ in habitat use according to their affinity for aquatic plants (McCormick and Aspinwall, 1983). Savino and Stein (1982, 1989), Schaefer et al., (1994), and Gotceitas and Colgan (1989) showed that macrophytes in streams or lakes could change the behaviors of fishes or affect predator–prey interactions.

In some streams of relatively low structural complexity (e.g., slow-flowing, riffle-pool reaches of low-gradient streams), attached or floating algae can play an important role in the underwater architecture of a stream and offer shelter or foraging space for fishes. Sunfish (*Lepomis* spp.), *Micropterus* basses, and minnows often use columns and masses of algae in Brier Creek, Oklahoma, as cover or shelter. Small green sunfish (*Lepomis cyanellus*), for example, have an affinity for algae columns (as well as rocks, etc.) and often shelter near or within algae columns when threatened. Gelwick (1995) showed, in artificial streams, the importance of algal structure in the use of microhabitats by small fishes and their escape from predation. Mosquitofish (*Gambusia affinis*) use algae as shelter, but they often occur over the algae (i.e., just at the surface, over a floating mat of algae), gaining protection from piscivores below. Casterlin and Reynolds (1977) showed that algae influenced habitat selection by *Gambusia affinis*. Craig (1996) and Fulling (1993) have showed that the local abundance of *Gambusia geiseri* in microhabitats in the South Concho River, Texas, was strongly related to the type of aquatic vegetation (algae, aquatic moss, watercress).

Leaf packs stranded in shallows (e.g., lower end of pool, in shallows) or accumulations of leaves and small debris in depositional areas (e.g., bottoms of deeper pools) also comprise an important structure for some fishes. Some species are often found within such mixed packs of organic detritus, including pirate perch (*Aphredoderus sayanus*), tadpole madtom (*Noturus gyrinus*), or stippled darters (*Etheostoma punctulatum*).

Cover

Slightly different from the notion of structure is that of "cover" in streams, which implies some behavioral use of objects such that fish have a decreased chance of being detected (e.g., by terrestrial or avian predators). Sechnick et al. (1986) showed that both juvenile and adult smallmouth bass, *Micropterus dolomieui*, selected habitats with cover that also provided low light levels and protection from strong currents. Angermeier and Karr (1984) indicated that fish appeared more attracted to woody debris because of camouflage than because of enhanced feeding or protection from fast currents. However, Wilzbach (1985) found that food availability was more important than cover in habitat use of adult cutthroat trout (*Salmo clarki*) in laboratory streams; thus, results for one species or system may not readily extrapolate to another. Heggenes and Traaen (1988) found that Atlantic salmon had a strong preference for overhead cover, but that brook trout, brown trout, and lake trout were little influenced by cover.

One particular form of "cover" that is important to fishes is shade at the water's surface or canopy cover above the water level. Matthews and Hill (1979a and b) found that shade was important in habitat use by minnows of a wide, shallow river of the Great Plains, with more individuals occurring in shaded than in fully sun-exposed microsites. A canopy that reduces sunlight to stream pools reduces primary productivity of attached algae and results in a lower density of algivorous catfish in Panamanian streams (Power, 1983). In lakes, Helfman (1981b) showed that fish hovered in shade because it lowered their visibility to potential predators while simultaneously making it easier for the fish to see objects that were approaching, and that numerous fish species were attracted to shaded space under objects (Helfman, 1979b). For small fishes, sun flecks or dappling due to a partially open canopy may increase the general visual heterogeneity of a pool and make it more difficult for a predator to see the prey fish against a changing and heterogeneous background of sun and shade. Finally, cover can benefit fish in shallow water where solar radiation is intense. W. L. Minckley (personal communication) has found the minnow *Tiaroga cobitis* in a desert stream taking refuge from direct sunlight within tufts of algae.

Habitat "Heterogeneity"

There have been many approaches to characterizing habitat "diversity" or "heterogeneity" in aquatic systems. Two general approaches are (1) to use measures to summarize proportional diversity of different kinds of habitat in a stream reach or in an individual pool or riffle and (2) to use heterogeneity measures that express some function of the number of changes in a feature (e.g., depth) within a reach of a stream. The former approach is characterized by use of formulas such as those from "information theory" [e.g., Shannon's index or Levins' index (Levins, 1968)] to describe quantitatively the proportions of different kinds of defined microhabitats exist in a reach. An example is that of Gorman and Karr (1978) who measured depth at a large number of points within a stream and used different categories of depth as values for p in the indices. Note that such an approach ignores the distribution within a reach of the categorical features under consideration [i.e., one could have all "deep" and all "shallow" values separated spatially in two halves of a pool, and one would obtain the same index value as if all deep and shallow points were completely intermingled (i.e., alternated) point by point within the pool].

The latter approach (heterogenity) employs field estimates or statistical measures to describe point-to-point change or variation in a stream. One simple approach is to merely use the differences in depth (or height of structure, etc.) between adjacent points on transects to characterize how much change there is in the feature within the pool. For example, in Brier Creek, Oklahoma, F. P. Gelwick and I (in press) measured the height of algae at successive 1.0-m points on cross-pool transects. For a measure of "heterogeneity," we merely determined the difference in algae height from point to point and determined the average of such "delta values" for each pool of interest (Fig. 6.4A). For rigid structural features such as a rocky substrate, a variety of field approaches has been taken, but one simple one (conceptually, if not in the field) is to lay a chain across the stream from bank to bank, ensuring that it is in contact with the substrate at all points. If the substrate were perfectly flat, the length of chain required to cross the bottom would be the same as the length to stretch across the stream [i.e., the ratio of actual to ideal length would be 1.0, for no heterogeneity of stream bottom structure (Fig. 6.4B)]. As irregularity of the substrate increases (e.g., presence of cobble, boulders, woody debris, etc.), the length of chain needed to cross the stream on the substrate becomes increasingly long relative to the minimum "ideal" distance (straight line) across the stream.

Heterogeneity of habitat in streams or lakes has focused most on nonuniformity of physical structures, but this concept needs to include heterogeneity of flows or current speeds in microhabitats as well. This hydraulic heterogeneity could be viewed across a range of spatial scales from that of variation in flow pattern within a single pool (Fig. 6.5A) to variation in current speeds or flows within long reaches of a stream system (Fig. 6.5B). In this broader view, hydraulic heterogeneity may be greater in transitions zones (e.g., the Fall Line in streams of the eastern United States, where waterfalls, cascades, etc.) make highly variable current speeds conditions for fishes where the Piedmont falls to the coastal plain. Such regions in a stream system can properly be considered an ecotone, and it is often in such reaches that the numbers of fish species are highest. In a multivariate analysis of fish assemblages among 10 sites times months for a year in the upper

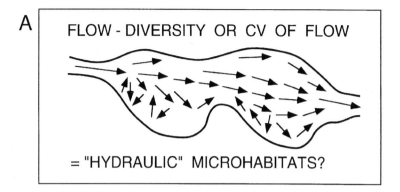

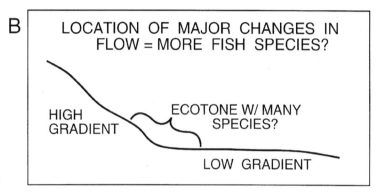

Figure 6.5. Heterogeneity in (A) current speeds within one typical pool in one stream and in (B) gradients and current speeds in a longitudinal profile of a typical river system.

Red River basin, the change in assemblage structure corresponded to the change in environmental conditions (Taylor et al., 1996, their Fig. 6).

I compared fish species richness to structural heterogeneity of local habitat, for 14 pools in Brier Creek, Oklahoma, in which I surveyed the fish assemblages 7 times in a year by snorkeling (Matthews et al., 1994). Goldstein (1978), Power and Matthews (1983) and many other authors have validated underwater observations as an excellent census technique in clear water, if the fauna is well-known. The mean number of species per pool was compared to "heterogeneity," as estimated subjectively by B. C. Harvey, then a graduate student who was also making underwater observations of fishes in the same Brier Creek pools. Harvey subjectively ranked the 14 pools for "heterogeneity" on a scale from 1 (low heterogeneity) to 5 (highly heterogeneous) based on his impression of physical complexity underwater in all pools (and without knowing how I would use his estimates). I then plotted mean species number versus Harvey's estimates of habitat heterogeneity for all pools (Fig. 6.6), resulting in an interesting positive trend (although nonsignificant for this small data set; $r = 0.396$) for more species

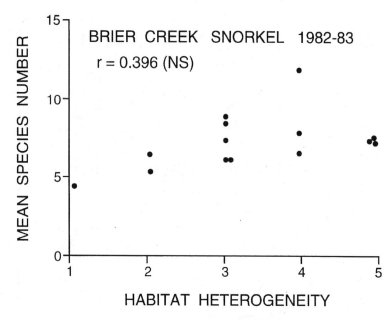

Figure 6.6. Number of species versus habitat heterogeneity in pools of Brier Creek, Oklahoma.

in pools classified as having higher heterogeneity. More attention needs to be given to evaluating, quantifying, and understanding the role of habitat "heterogeneity" and complexity of local fish assemblages.

More statistically complex measures to detect homogeneity versus heterogeneity in systems have been developed in recent years. Autocorrelation and geostatistical measures have gained increasing use to depict patterns in structural variation in streams or other aquatic habitats (Isaacs and Srivastava, 1989; Rossi et al., 1992). When structural features at any point are similar to surrounding points, there will be substantial autocorrelation in the system, and a correllogram can be used to depict the degree to which the structural features vary across space within the stream reach. However, use of autocorrelation or geostatistical approaches require large numbers of randomly or gridded points in order to achieve validity; thus, field workers are only recently attempting to use these approaches rigorously in fish ecology (cf. Gelwick, 1995).

Using information theory approaches (i.e., diversity measures), McNeely (1986) found fish species diversity in Spring Creek (Oklahoma) correlated with general habitat diversity (more so than with longitudinal position in the watershed). Gorman and Karr (1978) found a correlation between diversity of the habitat and diversity of fish communities, at the scale of "within reach" complexity of depth, flow, and so forth for streams in midwestern United States and in Panama. Using geostatistics, Gelwick (1995) quantified patterns in underwater algal structure in pools grazed and not grazed by the algivorous *Campostoma anomalum*, and testing the habitat use and survival of *Notropis boops* and juvenile *Lepomis cyanellus* in experimental streams having or not having well-developed algae columns, combined with the presence or absence of a predator (*Micropterus salmoides*).

Heterogeneity at Increasing Scales: Connectivity of Habitat Units

In the discussion above, habitat complexity was addressed at the scale of heterogeneity within habitat units like individual pools or riffles. However, there is also need to understand, in both streams and lakes, the consequences of heterogeneity at a next higher level [i.e., in the patterns in distribution of habitat units (like riffles, pools, particular kinds of beach zones, etc.)]. Schlosser (1995) has also addressed this issue as an aspect of "landscape ecology" (Section 6.6). For example, in a hypothetical "typical" stream, shallow riffles occur at a relatively uniform distance apart and can be predicted statistically to occur at distances equal to about five to seven stream widths (Leopold and Langbein, 1966). Although many streams may approximate the theoretical riffle to pool distances and ratios, real streams, particularly in uplands, often are rock bound at least in part, providing an endless array of sequences of pool and riffle units of all sizes.

With regard to heterogeneity at the scale of multiple habitat units, it is well known that different species typically inhabit pools and riffles and that the temporal dynamics of pool and riffle assemblages may differ (Gelwick, 1990). What has been little studied is the effect on local fish assemblage, of having varying patterns in sizes of pools, riffles, or even small pools within riffles [but see Schlosser (1995)]. For example, consider in a small upland creek having, in sequence, five small pools (SP) (perhaps 5–10 m long and no more than 0.5 m deep) interspersed by small riffles (SR), then five larger, long, deep pools (LP) (50–100 m long each, with depths exceeding 2 m), separated by larger, wider, deeper riffles (LR). Such a sequence is possible in streams with a geologically complex habitat. This sequence for pools and riffles (Fig. 6.7A) would produce,

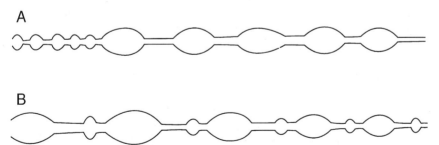

Figure 6.7. Hypothetical variation in distribution patterns of the same number of large and small pools in the total landscape of a portion of a watershed: (A) five small pools, then five large pools; (B) large and small pools interspersed.

for fishes, a long reach of stream where only small, shallow pools were available as habitat, then a sequence of much larger pools. Furthermore, crossing the large riffles (LR) with faster flow might be more difficult than crossing small riffles for small-bodied fish (but the opposite could be true for large-bodied fish). Added to this complexity, in the real world, might be additional small "pools" (of relatively quiet microhabitats) within the "riffles." If they exist, fish could use these "pools within riffles" as resting points for crossing the large riffle proper to move from one pool to the next.

The sequence above, with all small pools and then all large pools (and their related riffles), would be very different than a sequence of alternative large and small pools and riffles for fish requiring a diversity of habitat, as shown in Figure 6.7B, in which, within a relatively short reach of stream, a fish could find all four kinds of habitats ranging from large pools to small riffles. Little research has been done on the effects of habitat heterogeneity at this scale, although D. M. Lonzarich (personal communication) has begun a study of such effects on stream fishes in the Ouachita Mountains, Arkansas. Taylor (1997) has shown that the sequence of stream habitat units and potential for colonization from downstream influence fish assemblages in individual pools in Cucumber Creek, Oklahoma.

For pool-dwelling fish, the physical characteristics of flowing riffles or rapids may have a major impact on their ability to move among pools. Variables such as length of the riffle, current speeds in the riffle, or depth of the thalweg (deepest point on stream cross section) may be important considerations relative to movement of fish across the riffles. In general, small fishes (many minnows) seem better able than deep-bodied fishes (bass, sunfish) to move among pools if the potential difficulty in the riffle is shallow depth. However, where riffles are slightly deeper but more swift, large-bodied fish may be better able than small individuals to move upstream against currents. To the degree that a riffle is an impediment to fish movement for either reason above, riffles of greater length should have more effect on fish than shorter riffles.

There is relatively little testing of these ideas. For a sequence of 14 pools and 13 riffles in Brier Creek, Oklahoma, where I frequently enumerated fish in pools by snorkel survey in 1982–1983 (Matthews et al., 1994), I found that the number of species in a given pool showed a nonsignificant ($r = -0.342$) trend to be greater for pools that had the least isolation as measured by length of riffles separating it from other pools. Hypothetically, shorter riffles could increase the rates of colonization into a pool from nearby pools, similar to the manner in which Magnuson (1976) conceptualized lakes connected to other lakes by natural drainage features would have a higher colonization rate (hence, a higher number of species at equilibrium). Also, assuming that shallow riffles (least thalweg depth) would comprise more substantial barriers to fish movement, I tested species richness per pool versus average thalweg depths in riffles leading into the pool. The relationship between thalweg depth and number of species per pool was, however, actually negative and nonsignificant ($r = -0.24$). In other words, pools with deep, adjacent, riffle thalwegs actually had slightly fewer species.

Primary Productivity in Local Habitats

The density of individual populations and total species richness may increase with increased productivity in a habitat. Diamond (1975) provided a conceptual link between primary productivity and potential species richness by his model of "hot spots" (areas of high productivity) within islands. According to Diamond's model, landscapes were heterogeneous in productivity. Even within one island, some areas will be more productive than others, with some portions of the island lacking sufficient productivity to satisfy the usual needs of a given species (of birds, in this case). However, there would be smaller local areas with unusually high productivity (i.e., "hot spots") to which the local distribution of a given species might shrink during any period of short supply or ecological "crunch" [per Wiens (1977)].

Diamond's (1975) model addressed a single species and the way it might react to a period of low-resource availability. This model also was extended to apply to levels of competition that might ensue in the "hot spots" during stress periods of low resource supply. For our purposes, it seems appropriate to assume that there will be variable productivity both within and among lakes or streams. The areas (or whole streams/lakes) with highest productivity might logically comprise refugia to which many species could retreat and maintain viable populations during harsh periods.

Jones and Hoyer (1982) found a strong correlation between sport fish yield and summer chlorophyll-a values (commonly used as a surrogate for primary productivity) for 25 artificial and natural lakes in Missouri. Murphy et al. (1986) evaluated the density of fishes in cold-water Alaskan streams and found that logged areas with an open canopy and more periphyton production also had more coho salmon fry than old-growth reaches where canopy cover was greater. Johnson et al. (1986) also showed that removal of the forest canopy by logging resulted in increased solar radiation, more primary production, and more or larger salmonid fry than in reaches with dense canopy. Rader and Richardson (1994) showed approximately the same species richness, but a much lower density of fishes per unit area in low-productivity than in medium- or high-productivity parts of the Florida Everglades. In one warm-water system (Tyner Creek, northeast Oklahoma), I found, in snorkel surveys of individual pools, that the algivorous *Campostoma anomalum* was very abundant in pools with open canopy and direct sunlight, where there was a dense "felt" of algae attached to stones, but that the species was absent in pools further downstream where the creek entered a dense forest [and streambed stones were not even "slick" to the touch (i.e., lacking periphyton)] (Matthews et al., 1987).

In a review of fish production in rivers Mann and Penczak (1986, p. 239) noted that numerous authors showed geographic variation in production of species, related to "basic productivity" of streams, and that sucrose-enrichment experiments by Warren et al. (1964) supported these results. However, they cited a study by LeCren showing no correlation between salmonid production and "stream productivity" and a study by Mortensen (1977, original not seen) showing that, in many streams, fish production may be less than would be predicted from the general productivity level. Gascon and Leggett (1977) showed differences in resource segregation among species at two ends of a Canadian lake that differed in productivity, with more overlap where resources were abundant.

6.5 Zonation of Fishes in Lakes and Streams

Individual physical factors described in previous sections are correlated with distributions of individual fish species or composition of local assemblages. However, many of these phenomena, like width, depth, discharge, or habitat complexity, are interrelated. In most studies above, statistical approaches were used to detect individual variables that best predicted fish assemblages structure—with mixed results. An alternative approach is to seek patterns in fish assemblages correlated with or explained by higher-order descriptions of systems, such as spatial zonation within lakes or longitudinal zonation in streams. In this approach, individual measurements (e.g., of dimensions) are subsumed in overarching, abstract variables, like "stream order," "link number," "downstream link," or other measures that relate to large-scale patterns in systems. There is a long history of such evaluations, dating to at least Shelford (1911) for streams, but with fewer tests of broad patterns for lake fish assemblages.

Habitat Zonation in Lakes

Zonation in lakes is well defined in limnology (e.g., Cole, 1994). Terminology relates to depth of light penetration (photic versus aphotic zone; littoral zone = portion of the lake in which light penetrates to the bottom, thus rooted macrophytes can grow), or to onshore–offshore physical and biological phenomena (littoral versus limnetic), or depth zonation (limnetic versus profundal). Additionally, it has long been recognized in limnology that summer-stratified lakes can have a distinct epilimnion and hypolimnion that relates to differences in temperature, oxygen, and other physical variables (Chapter 4). All of these variables strongly influence the distribution of primary productivity, phytoplankton, zooplankton, or other biotic components and can be thought of as boundaries in a lake (Brandt and Wadley, 1981). Therefore, it is possible to borrow directly from limnology to describe zonation of habitats in lakes. However, to describe zonation of fish distribution or habitat use in lakes, it is necessary to expand from purely limnological terminology to emphasize zonation of potential effects on fishes (e.g., Fig.

6.8). For example, Pearse (1920) showed that in summer in a Wisconsin lake, there was distinct vertical zonation of fishes, with ciscoes (*Coregonus* sp.) deeper than 40 m, few fish at depths of 20–40 m, and centrarchids, yellow perch (*Perca flavescens*), and pike (*Esox lucius*) within 10 m of the surface. Keast and Harker (1977) showed distinct zonation of fish and benthic invertebrates in a northern lake.

Two distinctions need to be made. First, with regard to the use of different depth zones of a lake, fish in the upper portion of a stratified lake (epilimnion) often are exposed in summer to sufficiently high temperatures that their optima are exceeded, or such that actual direct thermal harm to fishes could result in the epilimnion. Thus, it may be best to think of a lake or reservoir (at least at low latitudes) in terms of an upper zone that potentially is hotter than optimal for most fishes. At the other extreme, fish in the hypolimnion may experience hypoxia or anoxia. As a result, few fish are found in this zone (Matthews et al., 1985). Pearse (1920) showed that oxygen depletion in deep waters changed the distribution of yellow perch during the summer. For stratified lakes or reservoirs in warm climates, there may be a three-way division of the water column (Fig. 6.8) into (1) upper—too hot, (2) lower—too little oxygen, and (3) a refuge zone in some portion in the lower epilimnion and extending downward into the hypolimnion to the depth of oxygen depletion, where conditions are either beneficial (or at least as little harm) to fishes as is available within the lake. Coutant (1985) found that striped bass (*Morone saxatilis*) responded strongly to these

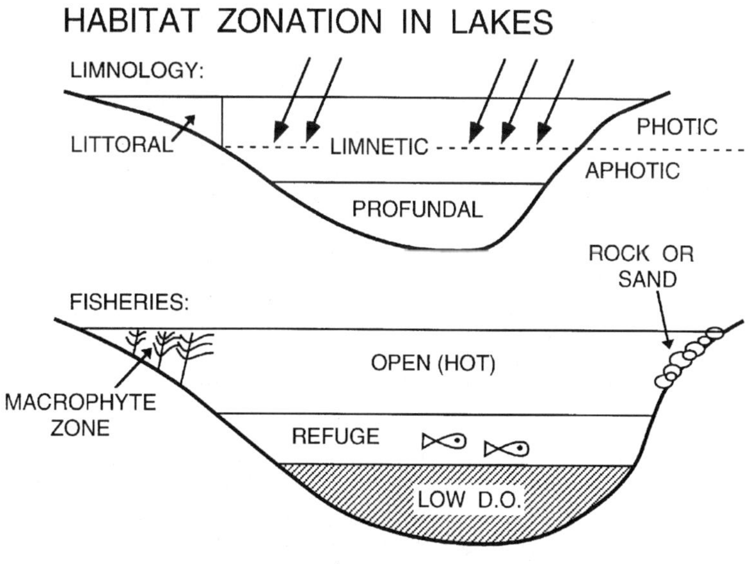

HABITAT ZONATION IN LAKES

Figure 6.8. Hypothetical habitat zonation in lakes, emphasizing limnological terminology (top) and fisheries zones (bottom).

conditions in man-made reservoirs and "sandwiched" themselves between hot and anoxic conditions. Thus, although these zones often corresponded roughly to the epilimnion, metalimnion, and hypolimnion of classical limnology (Hutchison, 1957b; Wetzel, 1983; Cole, 1994), fish ecologists may need to think of vertical zonation of fishes in direct terms. Brandt and Wadley (1981) described the thermocline in Lake Michigan as a thermal "front," comprising an ecotone and separating fish species or life-stage distributions according to temperature. Brandt and Wadley (1981) also showed that the "thermal fronts" in Lake Michigan had three-dimensional structure, as the intersection of the thermocline with the lake bottom created substantial horizontal and vertical local variation in temperatures to which fish responded. Thus, temperature variation may result in much more complex zonation of lakes for fish than a mere vertical gradient. Neill and Magnuson (1974) and Magnuson et al. (1979) also described fish distribution with respect to thermal gradients in lakes.

From the perspective of whole-lake limnology, it may make only modest difference (e.g., to phytoplankton or zooplankton) whether the littoral inshore zone is or is not dominated by macrophytes, or if nonvegetated substrates consist of sand, mud, rock or other structure. However, probably nothing is more critical to zonation of fish in lakes than the existence of different kinds of littoral structure, substrate, or vegetation in shallow waters. Shirley and Andrews (1977) showed survival and growth of young large mouth bass to be strongly influenced by presence of inundated vegetation in the littoral zone of an Oklahoma reservoir. Many studies (e.g., Gelwick and Matthews, 1990; Hinch and Collins, 1994) show that differences in such characteristics results in different kinds of local fish assemblage in the littoral zone of lakes or reservoirs. For example, Chapman and Mackay (1984) showed that northern pile (*Esox lucius*) preferred shallow, vegetated areas (although modifying habitat use in response to meterological conditions). Venugopal and Winfield (1993) reviewed the importance of macrophyte zones to fishes in lakes and ponds, suggesting that the macrophytes have greater concentrations of food items (Mittlebach, 1981a) and offer important shelter from predators (Savino and Stein, 1982; Werner et al., 1983a and b). Hence, a lake with a macrophyte zone will have different available zones for fishes than a lake (like many fluctuating reservoirs) devoid of rooted aquatic plants. Thus, there may be a first layer of zonation in lakes that is vertical, and a second that is horizontal, depending on presence–absence of rooted macrophytes and substrate types in nonvegetated areas.

The orientation of wind produces contrasts between inshore habitats that are subject to nearly as much wave shock as in the oceanic intertidal, versus backwater habitats that virtually never have substantial wave action. Small fish species, in particular, show sharp differences between wave shock and protected habitats. For example, in Lake Texoma (Oklahoma–Texas), there is almost complete turnover in fish species from open, wave-swept sites to wind-wave protected sites in a nearby cove (Matthews, unpublished data) (see Fig. 4.9).

All of the above suggests that it may be more difficult to envision unidirectional zones in lakes, in a fashion analogous to longitudinal zones in streams. Figure 6.9 suggests a hierarchy of use of "zones" in a lake in terms of increasingly detailed choices faced by individual fish. Initially, many fish species (largely due to evolutionary history and morphology) "choose" to be surface- or bottom-oriented. Given that fish are bottom-oriented, they must cope with adequate or inadequate oxygen, foods, and so forth, and if conditions warrant, give up use of benthic habitat to occupy midwater "refugia." In contrast, surface-oriented fishes or those of the upper water column "choose," in general, to live offshore versus nearshore or in the littoral. In the littoral, the increasingly detailed choices involve the use of macrophytes or not and, if nonvegetated habitats are chosen, the kinds of substrates or other physical structure of choice. In all of the habitats, or zones, short-term phenomena like oxygen depletion, episodic turbidity, or shortage of food items could cause a fish to move to another zone. However, this scheme of broad to narrow habitat choice can provide a framework for thinking about "zonation" of fish within the habitats of a lake. Other details not depicted in Figure 6.9 include details about where, in a lake, a given habitat is with respect to wind, or main channel versus cove (in reservoirs).

Coves as Special Habitats?

The feature of wind-wave versus "not" also underscores (at least for reservoirs) the pervasive difference between habitats in open-channel areas versus those within cove habitats. In run-of-the-river reservoirs, main-channel habitats, especially up-reservoir, have sufficient inflow and water currents to bring nutrients or silt from upstream and, at least at times (high inflow events), to potentially wash abiotic and biological materials (e.g., zooplankton or larval fish) further down-reservoir. Cove habitats (as contrasted to open-water, main-channel habitats) have received relatively little study by limnologists or fish ecologists, with the exception of one comprehensive model for a cove in Lake Texoma, Oklahoma–Texas, by Patten (1975).

Coves differ from main-channel habitats for fish in several important ways. First, they are important locations of local input of flow, nutrients, and allochthonous or terrestrial materials from local or even intermittent creeks; thus, they may have a high potential for primary production. Second, water exchange between cove and main-channel habitats may be very slow; thus, these habitats may remain minimally altered by main-river–channel inflows. Third, even a slight increase in lake level may result in flooding of large areas of terrestrial habitat, providing huge areas with standing or decomposing annual vegetation— hence, the high availability of small invertebrates, particulate matter, and other items that may offer feeding opportunities for larval or juvenile fishes. Zonation of lakes or reservoirs by "main channel" versus "cove" or backwater habitat may be critical to an understanding of fish ecology for the total system. As an example,

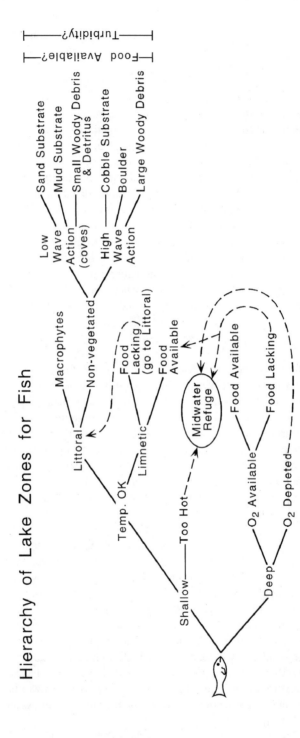

Figure 6.9. Hierarchical schematic of decision making in habitat selection by an individual fish among typical lake zones.

Figure 6.10 summarizes differences I found in weekly sampling of larval fishes with surface and midwater trawling in a cove in Lake Texoma and the open-water surface and deeper habitat of the main channel near the cove. In Figure 6.10, the large type indicates large numbers of a given taxon, and the smaller type indicates the presence of a given taxon, but at lower densities. There were differences in relative distributions of the common larvae with respect to cove versus main channel, and with depth in the open main channel. Additionally, sharp differences existed in relative abundance of larvae relative to wind exposure in the littoral, with *Lepomis* larvae abundant in rocky beach areas, shad larvae in intermediately exposed shallows, and a silversides (*Menidia beryllina*) most abundant far up in the wind-protected cove (Fig. 6.10).

In coves of reservoirs where the lower portions of creeks have been impounded, there may be distinct local habitats (Fig. 6.11) and episodes of local high turbidity. For example, the impoundment of Lake Texoma drowned the lower portion of Hickory Creek. Where the "creek" enters the "reservoir," a local delta has grown during the last two decades. During periods of high inflow from the creek (heavy rainfall), much silt is carried into the lake. Where the creek waters spread laterally and lose competency, the silt load is dropped, building up a substantial "levee" on either side of and extending lakeward to the mouth of the creek (Fig. 6.11), which is becoming stabilized by terrestrial vegetation. Where such levees are produced, there can be a complex array of local habitats, ranging from areas with strong creek inflow, to shallow backbay habitats that provide protected nursery habitats for fishes. Seining in these shallows suggests that a large amount

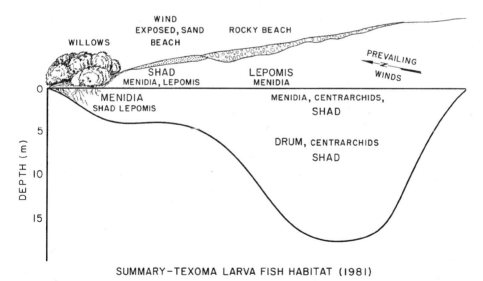

Figure 6.10. Distribution of common larval fish taxa among zones in Lake Texoma, Oklahoma.

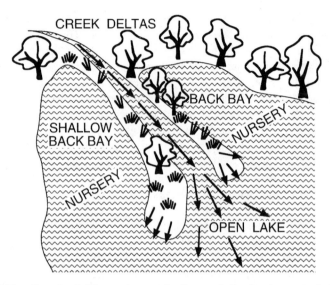

Figure 6.11. Growth of delta at the mouth of a creek flowing into a typical shallow, protected cove of a reservoir.

of small woody debris and detritus is present, providing habitat for large numbers of larval or juvenile shad (*Dorosoma*), *Menidia,* and crappie (*Pomoxis*). In that such species are some of the most important forage fishes in many reservoirs, these complex creek–delta–lake transition zones may deserve more attention than they have received.

In addition to producing local physical habitat variability, the inflow of creeks to lakes or reservoirs can produce local high turbidity. High turbidity near the mouth of Hickory Creek and similar areas within Lake Texoma is common, where the sharp increase in turidity and visibility probably results in altered behaviors of fishes or in their foraging capabilities. Thus, it seems that in addition to the very-large-scale zonation of lakes with respect to fishes, it is also necessary to consider localized but sharply defined zonation in local or unique habitats (e.g., creek mouths and coves).

Longitudinal "Zonation" or "Continuum" of Stream Fishes?

Most river systems on Earth arise from small headwater streams, and increase longitudinally in size, discharge, and a large number of related physical and biological parameters. In most streams, there is a progressive increase in numbers of fish species downstream (Horowitz, 1978), and numerous other species-specific (e.g., growth rates, abundance) or assemblage-specific (e.g., diversity, biomass) variables change trenchantly from headwaters to lower mainstreams. For example, Schlosser (1990) showed longitudinal zonation in body size and in life-history parameters for fishes in streams of North America, and Schlosser (1987) showed the population density to be higher in small stream segments than in large stream segments (because of many small individuals in shallow headwaters). Oberdorff et al. (1993) found longitudinal variation in the relative abundance of different trophic groups of fishes in eight French rivers, with more invertivores near headwaters and more omnivores and piscivores in midreach and downstream locations. Penczak and Mann (1990) also showed increases in mean body weights of fishes at higher stream orders and lower population densities of fish in the Pilica River, Poland. Jennifer Thompson and E. Marsh-Matthews (unpublished data) have found longitudinal differences in schedules of reproduction in two minnows species in Piney Creek, Arkansas (USA), with greater variability in reproductive readiness of females from the environmentally unpredictable headwaters. My samples in Piney Creek, Arkansas, show distinct clustering of upstream versus downstream sites, on the basis of the abundance of species (Fig. 6.12).

Exceptions do exist to typical longitudinal patterns of increases in species or related variables upstream or downstream, particularly where local factors such as habitat diversity override longitudinal patterns (McNeely, 1986). Additionally, fish species richness may peak, then actually decline in the lowermost reaches of some systems, as suggested by the River Cotinuum Concept (Vannote et al., 1980) and shown for eight rivers in France by Oberdorff et al. (1993).

Not all stream fish zonation is "longitudinal." Winemiller and Leslie (1992)

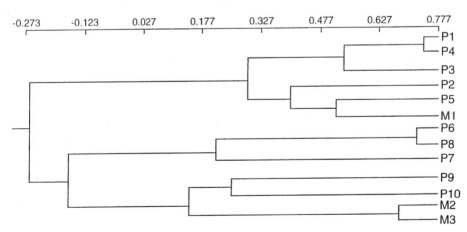

Figure 6.12. Dendrogram of similarities of fish assemblages at 12 stations in Piney Creek, Arkansas.

separated fishes in a tropical river system in Costa Rica into four major faunal zones that included "creeks," "rivers," "lagoons," and "ocean," with most fish occurring primarily in one zone and high species turnover between zones. Most of the variation among zones in fish assemblages and in trophic categories of fishes corresponded with a longitudinal gradient of stream order, but lagoons were ecologically more like lentic than lotic waters (Winemiller and Leslie, 1992). In large rivers, such as the large, lower Orinoco River system, Venezuela, fish are distinctively divisible into species predominantly in the wide, deep main river, in tributary creeks, or in secondary backwater channels ("caños") with slow-moving water (E. Marsh-Matthews, personal communication and unpublished data with J. Lundberg and others). Ibarra and Stewart (1989) found general longitudinal patterns in fish assemblages of beaches of the Napo River, Ecuador, but that the very strong influence of clear-water–blackwater gradients was a more prevalent influence in fish community structure within lowland parts of the system. Galacatos et al. (1996) found that fish assemblages in lagoons of the Napo River system differed between two distinct altitudinal regions (below 220 m elevation and above 235 m elevation). However, Galacatos et al. (1996) found less distinction between clear-water and blackwater lagoon systems than had been found by Ibarra and Stewart (1989) for riverine habitats.

In spite of such exceptions to downstream patterns, many studies, beginning with Shelford (1911), have addressed the phenomenon of longitudinal "zonation" of stream fishes and fish assemblages (Hawkes, 1975). Eigenmann (1920) described the longitudinal and vertical zonation of fishes and the upper elevational limits of individual species in the Magdalena Basin, Columbia. The discussion of stream fish zonation has been dominated in Europe by the "fish zones" concept (Huet, 1959), and in North America by contrasts between "addition" (Burton and Odum, 1945; Harrell et al., 1967; Sheldon, 1968; Jenkins and Freeman, 1972; Lotrich, 1973; Evans and Noble, 1979; Morin and Naiman, 1990) versus "replacement" (Tramer and Rogers, 1973; Gard and Flitner, 1974; Horwitz, 1978; Maurakis et al., 1987) of species from headwaters to lower mainstreams (Matthews, 1986 and references therein). Welcomme (1985) indicated that the addition of species rather than replacement was the prevailing pattern in tropical streams lacking geographic discontinuities, but that so many large tropical rivers have major interruptions to fish movement (e.g., large waterfalls, Jubb, 1977; Bell-Cross and Minshull, 1988; Moyle and Senanayake, 1984) that it is difficult to find one longitudinal pattern that really applies as a broad generalization. Discontinuities are probably the "key" to species additions versus replacements in many systems (C. M. Taylor, personal communication); thus, any broad tests of these concepts should attempt to match river systems with similar degrees of interruption by waterfalls or other geological features, making passage difficult by fish.

Stream fish occur in discrete stream zones or as parts of distinct assemblages in parts of Europe (Huet, 1959; Penczak et al., 1991; Kirchhofer, 1995), southern

Asia (Hutchinson, 1939; Edds, 1993), or western North America (Gard and Flitner, 1974; Rahel and Hubert, 1991) where high-elevation streams support a cold-water fish assemblage that changes abruptly to warm-water taxa as streams leave the mountains. However, for small and large stream systems outside montane regions, transitions from one fish "zone" to another may be so gradual that it is better to ignore "zonation" and consider the fish assemblages to change in modest increments at increasing distances downstream. Furthermore, within such a gradual transition pattern, it is obvious that a repeated occurrence of local microhabitat (e.g., pools of certain sizes) will likely result in a repeated occurrence of a characteristic assemblage. Rahel and Hubert (1991) showed that a combination of large-scale zonation (relative to temperature) and "within-zone" addition of species downstream might best describe longitudinal distributions of fish species in rivers basins with both montane and low-gradient regions. For an upland river at the border of the Great Plains, Echelle and Schnell (1976) showed distinct longitudinal faunal assemblages, albeit with some overlap.

Downstream changes in fishes can be abrupt, with the addition or replacement of species coinciding with changes in physical habitat, particularly where rivers are highly modified (Platania, 1991). Most studies show that longitudinal increases in numbers of fish species coincide with increased habitat complexity downstream in river systems (Maurakis et al., 1987), although Rahel and Hubert (1991) showed a lack of increased habitat diversity downstream for a very long river gradient in the Great Plains. In Australia, Pusey et al. (1995) showed an "orderly change" in species richness relative to longitudinal position in the Mulgrave River but no such effect in the South Johnstone River because of local barriers (waterfall and gorge) to fish movement. However, for two congeneric minnow species in the Ouachita Mountains (Oklahoma–Arkansas, USA), Taylor and Lienesch (1996) a nonoverlapping, parapatric distribution related to longitudinal differences in the streams, but best described by a complex discriminant function that included seven significant variables: algae, macrophytes, cobble, conductivity, elevation, gradient, and pH. By its complexity of variables, this study exemplifies the challenges of attempting to predict fish species presence, or composition of assemblages, on the basis of any single variable, whether direct measurement or more abstract (like stream order).

The postulate that ecosystems change in a continuum downstream is well established in stream ecology, as summarized for temperate streams by the River Continuum Concept (RCC) (Vannote et al., 1980) and modified for some regions but generally validated (Allan, 1995). However, Welcomme (1985) noted the difficulty of transposing RCC and related concepts to tropical systems. The original RCC dealt only minimally with fish, but it seems to be a logical hypothesis that fishes change gradually overall along some downstream continuum if streams lack abrupt thermal or geological transitions. See DeCamps and Naiman (1989) for a figure overlaying the European "fish zones" of Huet (1959) on a downstream pattern of the RCC for a large hypothetical river. It would almost seem unrealistic to expect the river ecosystem, with the exception of fish, to change gradually along a continuum, yet have fish make saltatory transitions from one "zone" to the next. The fish should match the ecosystem. Zalewski and Naiman (1985) considered stream fish communities, worldwide, to exist more in a "fish community continuum," regulated by "a continuum of abiotic and biotic factors." Blachuta and Witkowski (1990) noted clinal variation in variables such as number of species, diversity, density, biomass, growth rate, production, and fecundity of fishes in a number of previous European studies. In the Nysa Klodzka River, Poland, Blachuta and Witkowski (1990) found changes in density and biomass of several species to increase or decrease significantly with stream order, and higher growth rate of all fish species in high order streams, but that changes were gradual. They reported that most species (of 17 in the system) had a "rather wide distribution" along the longitudinal profile of the river, corresponding to "gradual change in abiotic and biotic factors." Thus, for this European stream, there seemed to be little evidence of fish "zones," and a continuum of longitudinal change in the fish assemblage seemed to prevail. Based on a review of much of the European literature and their own findings in the Nysa Klodzka River, Blachuta and Witkowski (1990) concluded, "The division of rivers into zones or regions of occurrence of particular species or whole communities is of limited use, usually only for small, geographically or climatically limited areas. In practice, field data often do not support such theoretically distinguished zones."

Before testing the hypothesis of a longitudinal continuum for stream fishes, note that in North America there are at least three broadly differing kinds of streams with regard to their potential for "zonation" or transitions of fish species in a downstream direction. The first major kind of stream is that indicated above; that is, streams with a sharp transition from montaine to lowland conditions, with concomitant increases in summer temperatures, and an obvious transition from cold-water to warm-water fishes (Gard and Flitner, 1974). However, throughout much of the world "warm-water" river systems exist that lack cold, high-altitude headwaters. These low-altitude, generally warm-water streams are of two distinctly different types which can influence the distribution of fishes.

Hubbs (1995) described differences in environmental conditions and fishes in streams with headwaters that are or are not spring-fed. Matthews and Styron (1981) also noted differences in physicochemical stress to fish in spring-fed and non-spring-fed headwater streams. Streams that arise from large cold or cool springs have a relatively stable, cool (at least not hot) temperature year-round (Stewart et al., 1992; Hubbs, 1995). In such headsprings and spring runs, and sometimes for many kilometers downstream, there will be a cool-water zone that eventually grades into warmer water and mimics, within a short spatial scale, the kind of cool-water zone found over much longer stream reaches at high elevations. These cool-spring headwaters support, even in hot climates, a local assemblage of cool-water, clear-water species such as some darters, minnows like *Phoxinus erythrogaster,* and locally endemic gambusias (Hubbs, 1995).

For example, the South Fork of the Concho River arises in hot, semiarid west Texas (USA) from large, cool springs. The headwaters support an assemblage (now augmented by introductions) of species like *Gambusia geiseri* that are present only in the cool headwaters and rapidly decline in abundance a few kilometers downstream where stream temperatures are warmer (E. Marsh-Matthews, personal communication). Marsh-Matthews and her students at Angelo State University studied distributions of fishes throughout the South Concho River for a decade. After some large floods, *Gambusia geiseri* were found further downstream, where the congener *Gambusia affinis* was usually more abundant. Over time, the segregation of *geiseri* and *affinis* was reestablished as the cool-warm gradient was reestablished (although there is evidence that some *G. geiseri* now remain further downstream). The local boundary between the two species appears fuzzy and not entirely temperature-driven, but in the large picture of the Concho River, the two species clearly exhibit a cool-headspring distribution versus a downstream, warm-water pattern (E. Marsh-Matthews, personal communication). As further evidence that temperature (or at least spring-water flow) produces zonation of the two *Gambusia* species, Hubbs and Hettler (1958) reported that during the severe drought of 1951–1954, the headsprings of the South Concho River failed, and *Gambusia affinis* temporarily replaced *Gambusia geiseri* as the dominant in the headwaters.

Although streams arising from large, cold headsprings often have a very distinct headwater fauna (Hubbs, 1995), more rivers in eastern North America arise from the coalesence of small headcreeks, some spring-fed and some not, and commonly augmented downstream by spring seeps [e.g., middle reaches of Brier Creek, Marshall County, Oklahoma (Power and Matthews, 1983)]. Most warm-water stream ecosystems in North America seem to fit the River Continuum Concept of gradual change relatively well. Therefore, it is logical to predict gradual longitudinal changes in distributions of fish species in streams of eastern North America, outside high-elevation, cold-water habitats, resulting in a continuum of fish assemblage changes instead of abrupt transitions from one fauna to the next, in "zones." Such a pattern for fishes would coincide with those demonstrated (RCC) for virtually every other aspect of stream ecosystems.

The abundance of some individual species is clearly related to longitudinal position in a watershed. For example, Matthews et al. (1978) found *Luxilus pilsbryi* the most common minnow at sites low in the Piney Creek (north Arkansas) watershed, but that one of three other minnow species (*Cyprinella galactura, Notropis telescopus,* or *Notropis nubilis*) was consistently the most abundant species at upstream sites. In Piney Creek, there is also a sharp difference in local abundance of the species pairs *Cottus hypselurus–C. carolinae* and *Etheostoma caeruleum–E. spectabile* (Fig. 6.13), with the former most abundant downstream and the latter more abundant upstream within each pair. Taylor and Lienesch (1996) showed a similar dichotomy in distribution of *Notropis snelsoni* and *Lythrurus umbratilis* in streams of eastern Oklahoma and western Arkansas.

302 / Patterns in Freshwater Fish Ecology

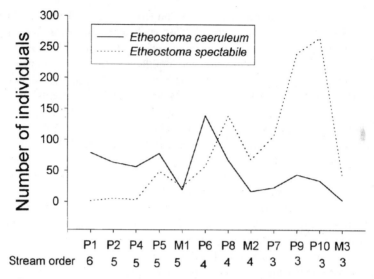

Figure 6.13. Abundance of *Etheostoma caeruleum* and *Etheostoma spectabile* at 12 stations on Piney Creek, Arkansas, from headwaters (M3) to downstream (P1), relative to Horton–Strahler stream order.

Because of the obvious differences in fish assemblages from headwaters to the lower mainstream of most drainages, numerous systems have been suggested to characterize "longitudinal zonation" of stream fishes.

The earliest formal suggestion of longitudinal zonation for stream fishes in North America was that of Shelford (1911), who likened patterns in change in fish species from mainstream to headwaters to changes that occur in geologic succession of the stream, with the extreme headwaters comprising the youngest, and the lower mainstreams containing the most mature of the fish assemblages. According to Shelford's concept of "physiographic succession," the stream at any fixed point, and its fishes, would undergo a series of transitions from being the extreme headwaters to becoming mature "downstream," as the upper reaches of the stream eroded headward. Although this is probably not in a detailed description of how stream fish assemblages work, recent studies by S. T. Ross and collaborators show that as the geological "nickpoint" moves headward during periods of increased downcutting, there is an upstream movement of structures like riffles, with concomittant upstream movement of associated darter species. In the study by Ross, changes in the Mississippi River mainstream in this century have allowed reconstruction of the phenomenon of nickpoint movement from old aerial photos compared to those of today. Trautman (1942, 1957) incorporated the concept of stream gradient as predictive of fish distribution from headwaters to lower mainstreams in North America.

Huet (1959) summarized his and other authors' descriptions of fish zones in

Europe, including, from headwaters to downstream, the (1) trout zone, (2) grayling zone, (3) barbel zone, and (4) bream zone. In Huet's scheme, there also is a direct link between elevation and cross-sectional shape of streams in the different zones, ranging from steep-sided, V-shaped streams in the "trout zone" to very wide, shallow streams in the "bream zone," all suggesting the interplay of basic stream geomorphology with resident fishes. Huet's system worked well in much of Europe, where there are relatively clear transitions (and a simple fish fauna) from cold-water upland to warm-water lowland fish species and habitats. Kirchhofer (1995) found, for Swiss streams, that species-rich communities occurred only below certain elevations, following the fish zones of Huet (1959), but indicated that fish species richness is dependent mostly on two physical factors: stream gradient and width. Hutchinson (1939) also established a system of four longitudinal fish zones (I. headwater streams, II. large streams, III. rapid turbid rivers, and IV. slow rivers, lacustrine swamps and their channels and lakes) for streams in southern Asia and compared this system to classifications of European streams by earlier workers. Hawkes (1975) summarized riverine fish zonation, indicating that "most workers report a zonal distribution in which there is a more-or-less sharp border line between successive zones," and suggested stepwise changes from zone to zone (his Fig. 14.11, p. 339). Hawkes (1975) also provides a detailed review of numerous other early schemes of zonation in European rivers according to their local fish or invertebrate faunas. Day et al. (1986) suggested three major zones for South African rivers, including headwaters, middle reach, and mature lower reach.

Balon and Stewart (1983) dismissed early efforts at defining fish zones in Europe as "too simplistic" to be ecologically significant. Balon and Stewart (1983) also suggested that Huet (1949, 1954, in French, originals not seen) added little to the earlier European concepts of fish zones, and that Hynes (1970) and Hawkes (1975) had overlooked some of the objections to fish zones (most of which were published in European languages; see references in Balon and Stewart, 1983).

Balon and Stewart (1983), to amplify the "evolutionary and hydrological" arguments relative to fish zones, described four river zones (Fig. 6.14), each with a distinctive fish fauna, in the Luongo River of the Zaire system, Africa, which lacks typical patterns of stream gradients except in headwaters and extreme lower reaches. However, Balon and Stewart specifically anticipated that the use of fish zones might be "rebuffed in modern studies." Their headwater zone collection in the Luonga River contained only five species, three of which existed only in that zone (i.e., there was species replacement, relative to arguments in North America). One of the species was an epibenthic algae-scraper, *Kneria paucusquamata*, represented by specimens 11–60 mm SL, with most about 35 mm. Localities 2 and 4 river mainstream, in a central floodplain zone, contained 25 species. A floodplain zone tributary (Location 3) had 15 species, of which 8 were in common with mainstream floodplain collections. A site in the downstream "exit zone,"

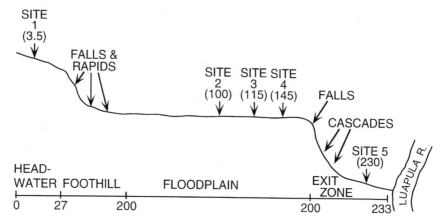

Figure 6.14. Schematic of Luongo River system, Africa. [Modified from Balon and Stewart (1983) with kind permission of Kluwer Academic Publishers.]

from the last cascades to confluence with the larger Luapula River, (Location 5) had 27 species, 14 of which occurred only in that zone. In steep headwater and exit zones, 20% and 22% of the species were benthic, but in low-gradient middle reaches, only 12% were benthic.

Balon and Stewart considered the Luongo River to be fragmented into isolated (fish) sections by "numerous ecological and physical barriers" (Fig. 6.14). They commented in detail on the river basin affinities that contributed over geologic time to the composition of the Luongo River (e.g., contribution of faunally richer Zaire system), but concluded that the connection to the Zaire system has been overshadowed by the extant series of physical and ecological barriers—high gradient stream reaches and swamps—that have led to speciation (by vicariance). Balon and Stewart (1983) concluded that the boundaries of four river zones in the Luongo, based on extant hydrography (high-gradient reaches), "clearly separate distinct fish taxocenes." They noted that the two high-gradient headwaters and "exit" zones were faunally related and that two mainstem stations in the low-gradient midreach were most similar. They point out that previous schemes for identification of fish zones applied only in rivers with a "normal" gradient: steepest in headwaters and becoming progressively less downstream. Their overall conclusion for fish communities (and invertebrates) was that, "distribution along fluvial gradients is strongly in favour of distinct zones with optimal conditions inhabited by one taxocene fading into and overlapping with another optimum for the next."

It appears that Balon and Stewart (1983) favored a stream fish model in which largely distinct zones existed, with minimal overlap of assemblages of one zone into another (although allowing ecotonal regions with high diversity from both adjacent zones). It should be very interesting to further test the generality of nonoverlapping or distinct zones with fishes in rivers elsewhere of about same length as the Luongo. Finally, Balon and Stewart (1983) suggested that formal "hydrographic zones" can be refined through identification of "homeostatic" fish taxocenes. Presumably, this means that one can initially identify physiographic or geologic features (e.g., stream gradients) that are *likely* to define river zones, but that determination of consistent fish taxocenes (assemblages) can help clarify the zones.

A Test of the "Continuum" Versus "Zonation" Hypotheses for Stream Fishes

To test the hypothesis that fishes in warm-water North American streams change along a longitudinal "continuum" instead of by discrete units of change (yielding "zonation"), I arbitrarily selected a stream system with which I am relatively familiar, and for which fish distributions are well known (Robison and Beadles, 1974; Robison and Buchanan, 1988). The system selected for analysis consists of the mainstream of the Strawberry, Black, and White rivers of northern and eastern Arkansas, beginning at the extreme headwaters and ending at the confluence of this system with the much larger Arkansas–Mississippi rivers (Fig. 6.15). The approximate distance from the headwaters of the system to its confluence with the Arkansas River is 310 km.

The Strawberry River arises as numerous small headwaters in Izard County, within the Ozark uplands, and makes a relatively gradual transition to low-

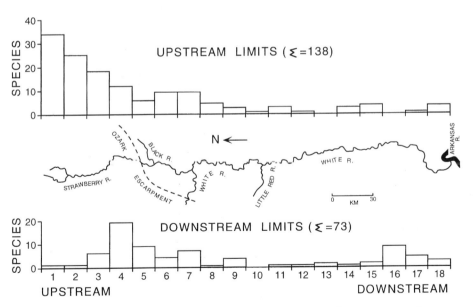

Figure 6.15. Frequency distribution of numbers of species meeting upstream or downstream limits to distribution in the Strawberry–White river system in east–central Arkansas, USA.

gradient conditions at the edge of the Ozark uplift before flowing into the Black River in lowlands of the Mississippi Alluvial Plain. The Black River then flows southward near the upland–lowland boundary to join the larger White River, which then flows southward entirely through lowlands to join the Arkansas River near the confluence of the latter with the Mississippi River. There are no dams in this system from the extreme headwaters of Strawberry River to the confluence of White River with Arkansas River, and, although there are some cool springs in the Strawberry River headwaters, the entire system is "warm water."

To test the hypothesis for this large river system that changes in the fish fauna are gradual and along a continuum, rather than exhibiting faunal "zonation," I plotted the distribution limits (upstream or downstream, or both) for all fishes known to occur in this system from detailed "dot maps" in Robison and Buchanan (1988). Only occurrence records on the actual river mainstreams were considered (i.e., tributaries were ignored, to emphasize the downstream changes within the mainstem). The maps in Robison and Buchanan (1988) include the extensive collections throughout the Strawberry River by H. W. Robison, made over a period of many years (Robison and Harp, 1971; Robison and Beadles, 1974; Robison, 1979; and subsequent unpublished collections in the system by Robison).

I divided the system into approximately equidistant segments (Fig. 6.15) from the headwaters of Strawberry River to the confluence of White River with Arkansas River (ignoring minor differences in segment length that would result from more meandering in some segments). The numbers of distributional limits (DL) of species within segments showed a strong tendency for many species to reach their upstream limit in the system in the uppermost segments, at some point within the Ozark upland. A total of 89 species (65% of all species in the system) reached their upstream limit within the upland, whereas none of the 14 more downstream (low-gradient habitat) segments had more than 10 upstream DLs.

A total of 33 species scored in Segment 1 penetrated to or very nearly to the extreme headwaters of the system. Another 25 and 18 species in Segments 2 and 3, respectively, were not found in the most upstream segment, but occurred well into the Ozark region (Fig. 6.15). There were surprisingly few upstream limits to species in Segments 4 and 5, immediately above and below the transition from uplands to lowlands. Neither the juncture of Black River with the larger White River (Segment 7) nor the influx of the Little Red River to the White River (Segment 10) was marked by any particularly great numbers of upstream species limits. Thus, it appeared that the lowland (~ 250 km) portion of this river system was characterized by broadly overlapping species distributions with only gradual changes in species composition, whereas in the upland, there was abrupt loss of species in an upstream direction (Fig. 6.15). (Note that the transition from many to few upstream species limits per segment mimics the rapid changes in elevation (high-gradient streams) in the upland and the much more gradual changes in elevation within the lowland segments.

There were downstream distribution limits for 73 species in the system (with the remaining species showing no decline in occurrence in the lowermost White River, and occurring further downstream in the Arkansas or Mississippi river mainstreams). Two segments (Segment 4 and Segment 16) appeared to be moderate locations of the downstream terminus of species distributions in the system. Segment 4 comprised the lower Strawberry River, which included the transition from upland to lowland across the Ozark Escarpment (although no sharp break exists in Strawberry River). In Segment 4, a total of 19 typical Ozark species find their downstream limit, most of which are well known to be "clear-water" species. Downstream from the departure of the system from the uplands, there is a very long reach (~ 200 km) of mainstreams with few downstream limits of species. Apparently, most fish species of the lowland portion of the drainage are able to use the habitat of these rivers very broadly. No obvious environmental correlate exists, but in Segment 16, 10 species reach their downstream limits of known occurrence within this system (although most of these also occur elsewhere in the Mississippi Alluvial Plain; that is, this is not actually the limit of their geographic range).

Overall, considering the frequency distributions of both upstream and downstream species limits in the Strawberry–Black–White river system, it appeared (Fig. 6.15) that there is, indeed, a distinct upland region where many species find upstream or downstream distributional limits. However, considering the entire system, there are no other marked transitions from one "fish zone" to another. The locations of gains or losses of species upstream and downstream in the system may well mimic the changes in stream ecosystem conditions predicted by the River Continuum (Vannote et al., 1980). Although these phenomena have not been tested rigorously in comparison, it would seem highly desirable to simultaneously examine rates of downstream change in "ecosystem" variables (kinds and amounts of primary production; particulate organic matter fractions; functional groups of aquatic invertebrates; processing of carbon and movement of carbon downstream; nutrient spiraling lengths) and composition of the local fish assemblages for a large river system.

Stream Order and Longitudinal Distribution of Fishes

Since its introduction to fish ecology by Kuehne (1962), the Horton–Strahler system (Horton, 1945; Strahler, 1957) of stream order has been much used as a framework for variation in fish assemblages from headwaters to lower mainstreams (e.g., Zalewski and Naiman, 1985), but its utility has been called into question by Hughes and Omernik (1981, 1983) who suggested that measures of stream discharge predict patterns in streams better. Hughes and Gammon (1987) showed that longitudinal changes in fish assemblages in the Willamette River, Oregon (USA), did not match stream order. However, many important ecosystem patterns in streams (e.g., carbon dynamics) are related to or coincide with stream order (Naiman et al., 1987). Other features related to stream habitats have also been shown to be related to stream order, including macroinvertebrate assemblages and stream metabolism (Ross and Wallace, 1983) and even species of bottomland woody vegetation (Hupp, 1986).

Naiman et al. (1987) found significant linear increases in the numbers of freshwater fish species with stream order, and that channels of fourth order or larger (> 16 m) typically offered an increased overwintering habitat in a cold-climate stream where severe ice conditions precluded the use of streams less than third order by fish throughout the year. Rozas and Odum (1987) offered the interesting finding that more fishes were collected at sites on lower stream orders in a tidal marsh creek than in lower mainstreams and related this to the presence of more submerged aquatic vegetation (offering fish refugia) at low stream orders.

Edds (1993) found that in the Kali–Gandaki River, Nepal, which has one of the steepest overall river gradients on Earth, fish assemblages were completely different in headwaters in the Himalayas versus downstream in the Gangetic Plain, with "longitudinal additions and replacements." Edds summarized that suites of geographic, water quality, and stream hydraulic parameters had strong correlates with the fish assemblage structure as revealed by multivariate analyses. Stream order was only one of seven "geographic" factors that correlated well with fish assemblage structure in this river system (Edds, 1993). He noted that although stream order was highly correlated with local fish assemblage scores on multivariate axes, "it (stream order) alone does not sufficiently explain a high proportion of the variance in fish assemblage composition in this river." Edds suggested that the most important influence on longitudinal distribution of fishes in this river was a combination of physicochemical variables, mainly of "stream hydraulics."

Many authors have reported a strong correspondence in number of species or other aspects of fish community structure with stream order [e.g., Kuehne (1962), Harrell et al. (1967), Beecher et al. (1988), and Paller (1994)]. Mundy and Boschung (1981) found a strong correspondence of the abundance of 28 species to stream order in the Tombigbee River system of Alabama and Mississippi, but they cautioned that, "stream order offers only a gross approximation of continuously varying limiting factors and stream order is not applicable to fishes in every drainage system (for watershed management)." Lotrich (1973) summarized that "stream order in most cases equals a biological unit," and that the addition of species proceeded by "discrete units of stream order." Fausch et al. (1984) found a strong correspondence between the numbers of fish species and stream order in the Embarras River, Illinois, and in Red River (of the North), Minnesota. Bramblett and Fausch (1991a) demonstrated that, although the total faunal richness was low, there nevertheless was a distinct increase in the numbers of fish species with stream order in harsh-climate tributaries of the Arkansas River in Colorado. In one small upland creek, Gelwick (1990) found that longitudinal distribution of pool-dwelling, but not riffle-dwelling, fishes corresponded well with stream order.

Matthews (1986b) reviewed essentially all of the literature to that date on longitudinal zonation of stream fishes in eastern North America and tested quantitative and qualitative differences in assemblages through the lengths of stream mainstems. Matthews (1986b) showed that although fish assemblages were roughly distributed along stream order gradients, there was no greater difference in local fish assemblages between than within stream orders, suggesting that, at a minimum, changes in stream order do not necessarily result in abrupt changes or "breaks" in fish faunas. Matthews (1986b) attributed this lack of change in fish assemblages with stream order to the strong influence of local physical environments (e.g., in individual reaches, pools, or riffles) that are often similar among stream orders. Evans and Noble (1979) earlier found a similar result for Big Sandy Creek, Texas, in which fish communities changed gradually along a continuum, in patterns not related specifically to stream order. Likewise, Stewart et al. (1992) found that species assemblages differed as much within as among stream orders and suggested that local habitat structure combined with physicochemical tolerances of individual species accounted for much of the longitudinal variation in fish species in a stream of the Arbuckle Mountains, Oklahoma.

Penczak and Mann (1990) concurred that changes in fish faunas with stream order occur only if stream order actually reflects stream size and gradient, diversity of habitat, and "other physical phenomena." Penczak and Mann (1990) found, in the mainstem Pilica River, Poland, that stream order, distance from source, and drainage basin area were all highly correlated. They found that the number of species, number of phytophillous spawners, and fish density were correlated with stream order but that the diversity of local assemblages was not. Some of these stream order relationships broke down when the entire Pilica basin (including tributaries as well as mainstream sites) was included, but the number of species was highly correlated with stream order for the entire basin.

Morin and Naiman (1990) found that the number of fish species increased with stream order in streams in Quebec, but that biomass of fishes peaked at fourth order and was lower in their (relatively few) samples in sixth order streams. They concluded that this pattern in fish biomass coincided with patterns in efficiency of carbon processing in streams. Morin and Naiman (1990) found that stream order and stream habitat each (analyzed separately) accounted for 61% of the variation in total fish biomass. However, they concluded that the fit of fish parameters to stream order was a function of the correlation between stream order and the geomorphic and habitat variables that dominated their study region. Overall, Morin and Naiman (1990) concluded that there was little evidence that stream order directly influenced the organization of fish assemblages and that the relationship of local fish assemblages to stream order was due instead to the "shifting influence of geomorphology and hydrology along the stream–river continuum," with an emphasis on links between patterns in fishes and efficiency of carbon processing within the stream.

Ross et al. (1990) suggested that examination of individual species in response to longitudinal positions in streams can enhance the understanding of longitudinal effects on whole assemblages. They found that both geographic (stream order, distance from creek confluence with river) and local microhabitat suitability were important in predicting the abundance of one local darter species (*Etheostoma rubrum*). Neither geographic nor microhabitat factors alone could account for the distribution of this "stenotypic" species. Otherwise "suitable" microhabitat (swift, shallow water, and coarse, consolidated substrate) located too far upstream (where desiccation dominated in summers) or too far downstream (where floods caused greater disturbance to substrates) had few individuals, whereas the "suitable" microhabitat within midreach orders of this stream had large numbers of individuals.

Potential Problems with the Horton–Strahler System of Stream Order for Fish Ecologists

What are some problems related to Horton stream order and fish distribution?

1. It works in general, but it is going too far to suggest that the distribution of fish species is "controlled" by stream order. It is clear that within any given stream order, there may be a diverse array of habitats types within stream reaches of 200 m or so. For example, even in small headwater streams, large pools can exist wherever rock-bound stream channels dictate such structure.

2. First-order streams are difficult to identify on maps or to adequately incorporate in the scheme, as has been noted by numerous authors. In 1983, I commented in a letter to a colleague that

   Order 1 streams are really very different biologically and physically, even within watersheds, depending on the source of water. An Order 1 stream that derives permanent flow from springs is as biologically different from an intermittent order 1 stream as day is from night. It is time that we quit considering all first-order streams as harsh environments, although many are.

   Spring-fed first-order streams are perennial, with comparatively stable environmental conditions (Hubbs, 1995), whereas runoff first-order streams may demonstrate high variability in flow and be intermittent or ephemeral, comprising a harsh environment for fishes. The extreme of this condition may be reached in arid-land streams, where many "first-order" streams consist only of isolated pools much of the year (e.g., Fausch and Bramblett, 1991). At higher elevations, snow-melt streams may also have flashy first-order tributaries, and in cold climates, headwater, first-order streams may freeze solid in most winters, eliminating fish except by colonization from downstream. Overall, there are probably more environmental, biological, and ichthyofaunal differences among different kinds of first-order streams than among stream reaches in lower orders; hence, knowing that a stream is "fourth order" may tell us something about expected fish assemblages, but "first order" is only informative if we know what kind of a first-order stream it is.

3. In some regions, e.g., "ridge-and-valley" provinces, streams follow long river valleys, attaining small "run" tributaries from the facing mountain slopes. Because the smaller tributaries are not sufficiently large to cause an increase in stream order (by the Horton–Strahler system), a very large creek in a ridge-and-valley geographic region might appear no larger than third or fourth order (i.e., one to two orders lower on the Horton–Strahler scale than a similarly sized stream would be in a dendritic watershed). This increases the potential importance of using more detailed measures like the D-link that take into account the size of adjacent stream reaches.

4. Local variation in flow or other physical variables can override general longitudinal or stream order patterns. For example, McNeely (1986) showed that distribution of fish species in a spring-fed Ozark upland stream did not correspond nearly so well with stream order as with local vagaries in flow or habitat size, as this gravel-bottomed stream had great variation in surface water availability at various points downstream.

5. Finally, fish obviously cannot respond directly to stream order, area of drainage, or any of the surrogates used by humans to describe streams, and stream order is strong correlated with other factors like stream size. Fish respond to local features of habitat in their behavioral responses and daily activities. Thus, regardless of overarching patterns of downstream transitions of streams from one suite of conditions to another, in most stream systems there are repetitions of habitat types (or "patches") interspersed through long reaches of stream. Fish species respond to these local patches of habitat, not to stream classification schemes of humans. For example, it is possible throughout much of the upstream, midreach, and even lower mainstream of a large creek in eastern North America to find pools with small- to moderate-sized substrate, modest flow, and depths of about 1 m or more. Wherever such pools exist, regardless of position in the watershed, longear sunfish (*Lepomis megalotis*) live and build nests. It is only as this particular kind of habitat ceases to be available (e.g., much further downstream in wide muddy big river habitat) that longear sunfish may decline in numbers. In general,

fish species live where there is sufficient local availability of habitat types that correspond to their morphologically and physiologically mediated ecological and behavioral requirements.

Alternatives to Stream Order?

Hughes and Omernik (1983) showed a strong correlation among watershed area, stream order, and mean annual discharge, and recommended using the two "measurable" variables (upstream drainage area and discharge) to best characterize a particular place on a stream. Gorman (1986) first suggested the importance for local fish assemblages of the size stream to which a given stream reach is tributary, with small streams that directly join large rivers having different assemblages (due to migration) than small streams that increase gradually in size downstream. Fowler and Harp (1974) also showed that a creek flowing directly into a large river had elevated species richness at the most downstream site because of migration of large-river species into the lower parts of the small stream. Osborne and Wiley (1992) showed a positive relationship between local numbers of fish species and four stream-size variables, including drainage area, stream order, and two measures of linkage of stream reaches. They showed that the "downstream link" (D-link), which included local stream size and the size of the stream at the next confluence downstream, best explained variation in fish species richness. Osborne and Wiley (1992) showed that D-link (defined for any place in a stream network as the link magnitude of the stream below the next downstream confluence) was, of several common approaches to describing stream size, the only measure sensitive to the overall position of a stream segment within a drainage basin. They concluded that it is not only the size of a particular tributary but also the location of that tributary relative to larger stream reaches downstream. Their findings suggested that small tributary streams joining a large river will have a greater source of colonists (in the event that the small triburary suffers local extirpations) than will other small tributaries that join only slightly larger streams. Osborne and Wiley (1992) suggested that it is necessary to know the position of a given stream within the overall basin drainage network in order to adequately predict its potential for species richness.

6.6 Landscape Ecology: Looking Laterally, Instead of Just Up and Down the Stream?

From a mostly "longitudinal" view of streams, fish ecologists are moving more to a "landscape" perspective. Stream ecologists have been increasingly aware since Hynes (1975) that streams do not operate independently but are deeply intertwined with the landscape of the whole watershed (Welcomme, 1988; Gregory et al., 1991; Richards et al., 1996). Likewise, natural lakes typically reflect intimately the geology of the region or a common history of formation processes

(Wetzel, 1983). Ward (1989) summarized that stream ecosystems actually interact in four dimensions: upstream and downstream, laterally, vertically (with groundwater), and time. Bretschko (1995) showed that instream patch and landscape dynamics are most important to fish in headwaters, whereas land-water ecotones were more important in downstream reaches. The ecology of fishes in streams and lakes also is very clearly related to landscape phenomena like inputs or availability of nutrients (Moyle, 1946; Wright, 1976; Vitousek, 1977; Randall et al., 1995), sources or quality of water (Matthews et al., 1992; Hubbs, 1995), geology (Matthews and Robison, 1988; Nelson et al., 1992), sources of carbon supply (Adams et al., 1983), availability of terrestrial insects as food (Lotrich, 1973), woody debris and its effects on fishes (Benke et al., 1985), effects of silt on stream fishes (Berkman and Rabeni, 1987; Waters, 1995), water withdrawals and land use (Cross and Moss, 1987), hierarchy of stream habitats (Hawkins et al., 1993), or interruptions of natural stream continuity by impoundments (Ward and Stanford, 1983; 1995a, 1995b).

"Landscape ecology" has been used in at least two ways with regard to fishes. From a broad perspective, the "landscape" includes all of the features of environmental patches and their interactions outside the water, as related to ecological phenomena within the water (where fish live). From a more narrow view, landscape ecology has also been applied to the spatial relationships or placement of distinct habitat patches within a lake or stream. Magnuson (1991) emphasized the landscape connectivity of lakes and their basins. Kneib (1994) provided a succinct overview of the interaction of "landscape" ecology with ecology of fishes (regarding feeding, in this case), emphasizing the effects of scale on studies of fish responses, and the concept of source habitats (producing individuals in excess of mere replacement rate, thus providing colonists) and sink habitats (that absorb colonists, with mortality exceeding local natality). Schlosser (1991) considered that land-use practices had pervasive effects on fish assemblages, and landscape ecology had potential for improved views of fishes relative to the dynamics of their landscapes. Schlosser suggested that landscape ecology focuses on structural relationships among parts of the landscape; interactions among those parts of the landscape by movement of water, nutrients, organic matter and animal species; and anthropogenic or natural changes in the relationships of parts of the landscape over time. Schlosser (1991) emphasized the potential of landscape phenomena to affect life-history requirements of fishes, population size structure, fish–habitat associations, nutrient and energy fluxes, and top-down versus bottom-up regulation of fish assemblages, with emphasis on land–water links increasing heterogeneity of aquatic habitats.

Schlosser (1995) noted the need for more research addressing general views of the ways large-scale, landscape disturbances (deforestation, agriculture, grazing) influence fish assemblages at small scales. Two studies with potential to capture a landscape view of land use that might influence fish communities are by Montgomery et al. (1995, 1996). Both of these articles indicate the presence of large woody debris (LWD) influences spacing of pools in stream channels, with logjams and LWD loading resulting in more pools, or pools in channels where they might not otherwise occur. Such an alteration of pool spacing or location by change in woody input appears to provide a direct link between land use (e.g., clear-cutting) and potential impacts on fish assemblages.

Schlosser (1995) emphasized the importance of landscape-level influences on trophic processes in streams, on population dynamics, and on survival of harsh environmental conditions. Three of the important landscape feature within streams addressed by Schlosser (1995) were (1) "habitat complementation"; that is, the pattern in which discrete habitat patches, each fulfilling different needs of a fish species, exist in a stream (similar to the different pool size sequences depicted in Section 6.3) and (2) "habitat supplementation"; that is, the potential for substitution of one kind of resource for another to meet needs of fishes, and (3) the ways stream habitat patches are arranged in "neighborhoods," providing fish ready access to needed resources, without barriers that comprise high-risk or high-cost (energetic) transitions from one habitat type to another.

Bart (1989) showed that most fish in an Ozark stream used a variety of habitat patches, and that few fish met all their requirements within a single kind of habitat. Freeman and Grossman (1993) showed that linkages between habitats used by a given fish species changed seasonally. In summer, but not in other seasons, rosyside dace (*Clinostomus funduloides*) used habitat patches having slack currents adjacent to areas with fast currents (Freeman and Grossman, 1993). Many drift-feeding minnow species appear to similarly combine the two kinds of habitat use—hovering just out of strong currents, but darting into the current to obtain drifting items (Matthews, personal observations).

A final note seems appropriate regarding landscape–habitat dynamics for stream fishes. Much of the theory reviewed above is obviously most relevant to fish species that migrate or that move actively over substantial space within their lifetime. This indeed applies to many species of stream fishes. However, some stream fishes (e.g., darters, sculpins) may remain within very short segments of streams throughout much of their lives (Hill and Grossman, 1987), even completing the life cycle within a single riffle (Scalet, 1973). Additionally, different fish species have quite different energetic costs of using flowing-water environments (Facey and Grossman, 1992; Hill and Grossman, 1993). It seems necessary, in considering landscape effects on stream fishes, to take into account such differences among species in potential to use or move among habitat patches, as a direct consequence of their normal levels of movement or energetic costs of movement, across daily, seasonal, or lifetime spans of time. Body size and related home range size (Minns, 1995) may be critical to the interaction of a fish with its landscape, as may be details of movements of individuals among population units (Fausch and Young, 1995). Sheldon and Meffe (1995) also show that density of source populations can affect movement of fish among habitat reaches. Schlosser (1995) also noted the need to understand ontogenetic changes in the ways fish species use landscapes. Thus, the "landscape" for an orangebelly darter living its entire life in one riffle may be very different from that of a diadromous fish species moving thousands of miles in its life cycle, or even for a stream minnow or sucker species that migrates many meters or even kilometers, daily (e.g., Matheney and Rabeni, 1995). Thus, landscape ecology, while providing a robust overview of the ways anthropogenic or other effects influence physiognomy of streams or lakes, can only be most useful when the life histories and requirements of the individual fish species are well understood (e.g., Goulding, 1980). The hierarchical ecosystem framework of Imhof et al. (1996), linking physical habitat formation to requirements of life stages of individual fish species and the "fit" of species to physical habitat features, may be a step in this integrated direction. Unfortunately, few fish species are as well known as the brown trout (*Salmo trutta*) that Imhof et al. (1996) use as an example; obviously, for such a framework to be used for fish assemblages, much more research is needed into the basic life history and individualistic ecological attributes of *all* fish species that are involved. Otherwise, efforts to "improve" or "save" the habitat template for one species (quite possibly a game fish) may ruin it for many others (quite possibly the small or poorly known species that lack the attention of the general public). If large-scale "landscape ecology" approaches are to be used to "improve" habitat for fishes, it is of paramount importance that *all* species that contribute to the biodiversity of a system be considered—not just game fish!

6.7 Multivariate Analyses of Fish Distributions, and Influence of Environmental Variables

Although some individual environmental variables (like temperature, oxygen, or salinity) clearly influence distribution of freshwater fishes, there is much interest in using multivariate approaches to characterize differences in fish assemblages among sites at both small and large scales, and to relate those differences to suites of environmental factors. One of the first multivariate analyses of local fish assemblages was by Smith and Powell (1971), who used factor analysis to characterize fish assemblages in Brier Creek, Oklahoma. At about the same time, at a much larger scale, Smith and Fisher (1970) used factor analysis to relate distributions of fishes in Kansas to environmental variables. Stevenson et al. (1974) also used factor analysis to describe the distribution of fish faunas in western Oklahoma relative to suites of environmental variables. Echelle and Schnell (1976) reported distinct associations of species in the Kiamichi River basin, Oklahoma, related to environmental variables. In addition to ordination techniques like factor analyses, clustering approaches (many options in cluster analyses reviewed in detail in Sneath and Sokal, 1973, and Gauch, 1982) commonly have been used to group fish samples on the basis of taxonomic composition or abundance of species. As examples, Ross et al. (1985) used cluster analysis to assess spatial and temporal patterns in fish assemblages in two midwestern (United States) streams, and Hughes et al. (1987) used a similarity measure followed by clustering to produce a dendrogram depicting similarities and differences among fish samples in Oregon streams. Discriminant functions analysis has also been used to assess variation among local stream fish assemblages, e.g., Grady et al. (1983).

Although factor analyses were commonly used as a tool in the 1970s to compare fish distributions to environmental patterns, the incorporation of both species abundances and environmental variables into one analysis based on assumptions of linear relationships later came into question, and some authors began using a "two-step" approach to the problem of "fishes" versus "environment". Such an approach first classifies sample sites on the basis of environmental characteristics, then separately overlays in some manner the distribution of fish species or abundances.

For example, Matthews and Hill (1980) carried out a principal components analysis (PCA; Sneath and Sokal, 1973) of microhabitat-scale fish samples from the South Canadian River, Oklahoma, on the basis of 10 environmental variables measured within each sample site (= first step), and then depicted the relative abundance of each of the common fish species in each microhabitat (second step), with the samples arrayed on PCA axis. In such analyses, patterns in abundance of each species can be viewed as a function of the known individual variables that are correlated with each principal components axis. Another example of this approach was by Matthews (1985c), in which 101 sample sites in midwestern streams were classified by PCA of local (in-stream) environmental variables, and abundance or presence/absence of common species were then overlain on the samples in principal components axis space. Again, this two-step approach allowed determination of abundance or occurrence of individual species relative to the combinations of individual variables underlying each of the PCA axes.

More recently, other multivariate approaches have come into favor for analyzing similarities of fish assemblages among sites, or defining composition of fish assemblages relative to environmental variables. Polar coordinates analysis (Gauch, 1982) has been used successfully to classify temporally variable fish assemblages (e.g., Gelwick and Matthews, 1990). Detrended correspondence analysis (DCA), a method of indirect gradient analysis (Gauch, 1982), has frequently been used to determine similarities and differences among local or regional fish faunas. Hughes and Gammon (1987) and Hughes et al. (1987) used DCA to classify fishes in Oregon into regional groups, and Matthews and Robison (1988) used DCA to classify the fishes of Arkansas into "faunal regions" on the basis of the abundance of more than 150 species among more than 100 "drainage units". Gelwick (1990) assessed year-around distribution patterns of fishes in Battle Branch, a small upland stream in northeast Oklahoma, by DCA. DCA was used by Cashner et al. (1994) to classify fish samples from Louisiana bayous on the basis of abundances of species, and by Taylor et al. (1996) to examine spatial and temporal differences in fish assemblages at 10 sites sampled monthly for a year in the upper Red River basin, Oklahoma. Hubbs et al. (1997) recently used DCA to assess comparative levels of change in local fish assemblages at many sites in East Texas, over a span of 33 years.

Detrended correspondence analysis of fish samples includes no data on environmental conditions, and results of DCA can only be compared to environmental factors by additional steps. For example, to relate fish faunal composition to environmental conditions, Matthews et al. (1992b) combined separate multivariate analyses of fish abundance in Arkansas (by DCA) and 14 water quality variables (by PCA). Matthews et al. (1992b) found strong correlation between the first PCA axis of water quality variables and the first DCA axis of abundances of fish species, suggesting that the water quality and distributions of fishes were strongly linked, statewide.

At present, there is a growing trend to use a "direct gradient" analysis, such as canonical correspondence analysis (CCA, ter Braak, 1986) to map abundances of individual fish species or depict suites of species in the multivariate space described by environmental factors. In this way, the occurrences or abundances of species can be examined with respect to similarities and differences among species, and with respect to the ways the fishes are related individual environmental variables or suites of variables. As examples, Pyron and Taylor (1993) carried out a CCA of fish distributions among small-stream, oxbow lake and swamp habitats in southeastern Oklahoma, and Taylor et al. (1993) used CCA to describe distributions and associations of fishes in the upper Red River basin (Oklahoma) with respect to the suite of environmental variables measured at each site. Lyons (1996) used CCA to show that local stream sites in Wisconsin were distributed along gradients of both fish species and environmental conditions. Polivka (1997) used CCA analysis to map the occurrence/abundance of common fish species in the South Canadian River, Oklahoma, relative to a suite of physical characteristics in small microhabitat samples. Rodriguez and Lewis (1997) used CCA to summarize structure of fish assemblages in a wide variety of aquatic habitats in the Orinoco River basin, and found by this approach that a small number of environmental variables accounted for most of the differences among local fish assemblages.

These indirect and direct gradient analyses reveal much, at least in an exploratory fashion, about the overall relationships between distributions of fishes and the suites of underlying environmental conditions. However they also may have great potential for hypothesis testing when used appropriately. Although the visual outputs, typically scatterplots, of multivariate ordination or gradient analysis programs often are judged by eye in an "exploratory" manner, it is possible to use the axis scores for fish samples as original characters in further statistical analyses of properties of distribution patterns or to compare to environmental factors, as was done by Pusey et al. (1995). This approach—that is, using scores from multivariate analyses as raw data for further analyses—is just beginning to be used in a productive manner by fish ecologists, but it seems that using such scores as a *beginning* of analyses, rather than the "end", will be profitable.

7

Disturbance, Harsh Environments, and Physicochemical Tolerance

The rapidity of dispersal of a type (of fish) depends entirely on its facility to accommodate itself to a variety of physical conditions, and on the degree of vitality by which it is enabled to survive more or less sudden changes under unfavourable conditions. . . .

—Günther (1880, p. 213)

In the deserts we see in most dramatic operation the principle that fresh-water fishes, whether we think in terms of individuals or of species, or present or of past, are inseparably bound to the springs and streams and lakes which they inhabit. When any waters completely dry up, as they so often do in arid regions, the fish life perishes. When canyon creeks are transformed by cloudbursts into raging torrents that no living thing can resist . . . , the entire fish life is stranded on the normally dry flats.

—Carl Hubbs (1940)

7.1 Introduction

Disturbance, stressful environments, and the ability of fish to cope with physicochemical challenges all affect local fish assemblage composition or dynamics. Abiotic disturbance and stress, or abilities of fish to survive under harsh conditions, can explain the presence of some fish species or life stages in particular kinds of habitat. The phenomena addressed in this chapter are "abiotic," although as abiotic factors change (e.g., thermal stress), some of the "biotic" variables (e.g., competition or predation; considered in Chapters 9–11) may also change. The phenomena considered in this chapter under "disturbance" or "stress" are conceptually and operationally more difficult to measure than are simple measures of structural features like dimensions of a habitat. Furthermore, to identify phenomena as a disturbance or as stressful to local fishes involves measuring baseline properties of the environment over a sufficiently long time to gain an estimate of actual system variance. Finally, understanding response of fish to stress can require laboratory tests of ability of fish to survive, or accomplish basic life/reproductive functions or behaviors under controlled conditions. For example, Breitburg and Moher (1994) have argued that disturbance may affect fish species differently, as a function of their mobility. However, before turning to those details, it is appropriate to first consider the general background on contrasts between abiotic and biotic explanations for assemblage structure, including fishes, and some general concepts about "disturbance."

Abiotic Versus Biotic Regulation of Fish Assemblages

There has been much debate about regulation of communities by biotic interactions versus abiotic limitations (Connell, 1975, 1980; Wiens, 1977; Diamond, 1978; Schoener, 1982). Biotic and abiotic factors both influence community structure, with their relative importance relating to environmental harshness of particular systems (e.g., Peckarsky, 1983; Power et al., 1988). Several, unpredictable "stochastic" events can strongly and rapidly affect community structure. In addition, perenially or continually harsh/stressful conditions in a system (e.g., very hot habitats or those with chronic low oxygen concentrations) may exclude, in ecological or evolutionary time, those species unable to withstand their impact, leaving only relatively hardy species in whole systems (e.g., entire watersheds).

To understand the impact of abiotic factors, we need to know (1) the degree to which severe short- or long-term events impact stream or lake fish assemblages *within* generations, in the context of known long-term baseline variation in individual systems under "normal" conditions, or (2) the degree to which nonhardy species are excluded by stress at a multigeneration time scale.

Fish assemblages show evidence of both biotic and abiotic regulation, and it is likely that neither provides a total explanation for success of a fish assemblage in a habitat. Biotic factors that influence fish assemblage structure include competition (Echelle and Schnell, 1976; Page and Schemske, 1978), resource partitioning (Zaret and Rand, 1971; Werner and Hall, 1976; Mendelson, 1975; Baker and Ross, 1981; Paine et al., 1982; Wynes and Wissing, 1982; Ross, 1986), predator–prey interactions (Moyle and Li, 1979; Zaret, 1979; Fraser and Cerri, 1982; Fraser et al., 1995), scale of food availability (Power, 1984b), morphological adaptations (Gatz, 1979a and b, 1981; Welcomme, 1985), or "ecomorphology" (Douglas and Matthews, 1992). In contrast, Starrett (1951), Smith and Powell (1971), Kushlan (1976), Gorman and Karr (1978), Harrell (1978), Horowitz (1978), Grossman et al. (1982), Capone and Kushlan (1991), and Poff and Allan (1995) have shown that abiotic limitations, variability, or stress can have a strong influence on the structure of stream fish assemblages. Abiotic factors can directly limit fishes in local assemblages through adverse physical effects, or they can make responses like seasonal migration necessary for fishes in a system (Zalewski et al., 1990). I have found both orderly and predictable biotic relationships among stream fishes (Matthews et al., 1982b; Surat et al., 1982; Power and Matthews, 1983; Matthews et al., 1988; Douglas and Matthews, 1992) and strong influences of abiotic stress or physicochemical changes on their distribution or dynamics in habitats (Matthews, 1986d, 1987a; Matthews and Maness, 1979; Matthews and Hill, 1979a, 1979b, 1980; Matthews and Styron, 1981; Matthews et al., 1982b). B. C. Harvey suggests (personal communication) that it would be advisable to distinguish between situations (1) in which abiotic and biotic factors are consistently more important in different places (e.g., each could dominate dynamics in two different streams) and (2) situations in which abiotic and biotic variables sequentially are important in the same location (e.g., over a time line from abiotic control in a flood, to biotic control during long periods without disturbance). Dudgeon (1993) showed that predation by fish could have negative effects on benthic macroinvertebrates, but that these biotic interactions were suppressed by disturbances produced during spates. Where physical harshness excludes most species [e.g., in hot, headwaters reaches characterized by low oxygen (Matthews, 1987a)], the species that are present may enjoy some freedom from interspecific competition (albeit individuals may experience extreme intraspecific competition). Additionally, in some situations, heat or oxygen stress might exert abiotic control on an assemblage at the same time that competition between species would also be important (e.g., in a shrinking pool during drought). Thus, local physical harshness is a distinct filter to assemblage composition in harsh places (Smith and Powell, 1971), but this filter may not remove competition or predation as important variables in the success or survival of an individual fish.

7.2 Definition and Time Scales of Disturbance

What Is "Disturbance"?

Disturbance in aquatic communities has been defined in various ways (Resh et al., 1988; Poff, 1992; Karr, 1994). Some concepts of disturbance include the criterion that an event results in a change in structure or functioning of the biotic assemblage. By this criterion, no event is a disturbance unless it results in a demonstrated change in the organisms in the system, no matter how physically dramatic it may seem. Disturbance and "perturbation" are often used interchangeably, both implying that an extrinsic factor or event changes some property in an assemblage. An obvious example might be a turbid-water inflow from a river to a reservoir, causing changes in light penetration, movements or habitat use of fish, feeding by fish and other aquatic organisms, subsequent growth, and so on. However, by the definition of "disturbance" above, the perceived event (muddy water) is only a disturbance if it changes some property of the biotic community in a detectable way.

Poff (1992) has argued that the "2 standard deviation" or similar approaches lessens the ability to predict disturbances because of its "purely statistical criteria." Poff suggested some actual physical benchmark for disturbance, such as the point at which the stream bed begins to move during a spate. Poff (1992) also argued that viewing disturbances as predictable at various time scales allowed assessment of evolution of stream organisms for such episodes.

Stanford and Ward (1983, p. 266) suggested that, "any stochastic event which forces normal system environmental conditions substantially away from the mean is a convenient definition of disturbance." Resh et al. (1988) formalized the concept that an event could be defined as a disturbance on the basis of the magnitude or amplitude of physical factors that are presumed important to the local organisms, without the requirement of a demonstrated effect on organisms. One suggestion of Resh et al. (1988) is to view any departure from normal factors (e.g., stream flow) as a "disturbance" if the variable reaches a magnitude that exceeds the long-term mean + 2 standard deviations. This definition also is based not on year-round averages, which can ignore normal seasonality, but on monthly means or some other shorter time scale that recognizes seasons. For example, a bank-full flood, by this definition, might be a "disturbance" in late summer in streams of the North American Midwest (when flows are usually low), but a flow of similar magnitude might *not* be classified as a disturbance during late spring, when repeated rainstorm spates are "normal" in long-term records for the region. It might also be possible to predict potential disturbance events related to high stream flow well in advance. For example, McCabe (1995) showed a very strong statistical correlation between winter air-pressure anomalies over the northern Pacific Ocean and the American West, and annual stream flows related to melt of snowpack. Given an early prediction from such meteorological data, ecologists might be able to predict stream flows for the region after spring warming and their effects on fishes.

Use of a "+ 2 standard deviations" rule to define disturbance may be useful from the perspective of identifying "disturbance" from high flows (i.e., floods greater than usual for the system and time of year). However, at the other extreme of low-flow or "no-flow" periods, an absolute criterion focused on discharge may not be amenable to a "− 2 S.D." rule, in that two standard deviations below mean flow for small, flashy streams can be less than "zero." The critical feature that needs to be included to detect disturbance in small or low-gradient streams that are subject to summer intermittency (e.g., Capone and Kushlan, 1991) is probably the length of time (days) of "no flow" (i.e., the length of time that fish are isolated in individual pools). Obviously there is also the problem of loss of water from the isolated pools, in a sequence that I suggest for "phases" of drought later in this chapter. It does not seem possible to use long-term flow or discharge records (e.g., from stream gauges) to adequately predict the local impact of the cessation of flow and desiccation of pools. However, it might be possible to use flow records for small streams to estimate the average number of "no-flow" days that is typical for the system, thus to use a "more than 2 standard deviations" longer period of no-flow to consider a "disturbance." At least for individual fishes in local assemblages, I suggest that at any time surface flow connecting pools ceases, those fishes are subject to a disturbance from their normal suite of biological or behavioral options (e.g., the ability to migrate among pools to escape predators, to forage, to reproduce, and so forth), even if they possess the absolute tolerance to endure the conditions until rains next establish flows. The difference between no-flow and at least minimum-flow can be critically important to stream fishes. Travnichek et al. (1995) showed that the establishment of a minimum-flow protection below a dam on the Tallapoosa River, Alabama, more than doubled species richness downstream of the dam, with a marked increase in fluviatile species. Becker et al. (1982) showed the negative effect of dewatering of streams on salmon eggs and larvae in redds. In hot climates, cessation of flow is often closely related to increased temperature or decreased oxygen concentrations to levels stressful to fish.

Using the suggestion of Resh et al. (1988) to base "disturbance" on physical measurements might have the advantages of (1) giving investigators in disparate sites a common currency for discussion of event magnitude and (2) providing a threshold that investigators might use to initiate studies of "disturbance" events. However, at the very longest, evolutionary time scales for disturbance, it is impossible in most cases to calculate "averages." Such is the case with respect to the disturbance of habitats from tectonic activity, as suggested to be important in the history of extinctions of fishes in the Magdalena River basin, Columbia (Lundberg et al., 1986). Finally, it is not possible in defining "disturbance" to ignore the schedules of disturbance relative to the ability of organisms to recolonize. Rapid recolonization may make even a sharp disturbance only transitory in its effect on a system [e.g., Rosser and Pearson (1995) and Johnson and Vaughn (1995), both of which showed highly resilient assemblages of stream invertebrates that rapidly recolonized artificially disturbed patches of substrate].

Theoretical Implications of Disturbance

Disturbance, in the broadest sense, is suggested to interrupt, simplify, or cause changes in communities. The intermediate disturbance hypothesis was extended to aquatic communities by Ward and Stanford (1983), who suggested that moderately disturbed aquatic systems would have the highest richness of species. Karr and Freemark (1985) considered that interactions of different disturbances can have an important influence on stream fish assemblages, particularly in light of individualistic responses of species to the disturbances. Poff and Ward (1990) suggested that physicochemical variability interacted with complexity or stability of substrates to influence the ability of stream systems to recover from disturbances. Detenbeck et al. (1992) reviewed a large number of case histories for recovery of stream fish communities from disturbance and found that "press" (chronic) disturbances altered stream fish assemblages, with recovery times of 5 to more than 50 years after the disturbance was abated. In contrast, Detenbeck et al. (1992) suggested that "pulse" (short-term) disturbances had very different trajectories of effect on stream fishes, depending on a variety of site-specific or disturbance-specific factors. In general, they found warm-water families more resilient to disturbance than cold-water (salmonid) fishes. They included the availability of refugia and distance from source populations for colonists as having substantial impact on recolonization times. Note that "refugia" for various fishes could consist of very small spaces in the streambed (e.g., Lancaster and Hildrew, 1993) or larger spaces [e.g., deep pools providing refugia for recolonization of fishes into shallow headwaters (Schlosser 1987b)].

There is substantial empirical evidence to support theoretical constructs about disturbance effects in animal assemblages. Nash (1988) found higher fish species richness and diversity in Norwegian coastal sites with less fluctuation in oxygen values, which they equated to lower disturbance. Death (1995) showed that disturbance or low stability of sites led to similar invertebrate communities, but that at stable sites, invertebrate communities diverged into site-specific configurations. Death and Winterbourn (1995) showed that invertebrate diversity was maximal at stream sites of intermediate stability (with respect to temperature and discharge) but that highest diversity was in sites with a low intensity of disturbance and high habitat patchiness or heterogeneity, suggesting synergism of variation in flow regime and habitat heterogeneity. Wootton et al. (1996) showed that loss of flood disturbance in riverine food webs could result in dominance of predator-resistant insects and divert energy from fishes (i.e., that flood disturbance helps to maintain complexity in food webs).

Time Scales of Disturbance

Regardless of the definition of disturbance (physical deviation from average versus detection of biotic response), it is apparent that in fresh waters, disturbances occur over temporal scales ranging from minutes to centuries, and at spatial scales from individual microhabitats to whole river basins. Bayley and Li (1992, p. 255) provide a comprehensive model for stream fishes, suggesting the importance of floods or low flow, and of their predictability. Fish are probably well adapted for frequent or regular disturbance events but less able to deal with uncommon events. Further, there may be substantial differences in the degree to which an event "disturbs" a fish assemblage, depending on whether the event is predictable or unpredictable in the annual reproductive cycle of the species. For example, a flood during the period of nesting by sunfish (*Lepomis*) or the presence of minnow larvae may destroy a cohort (Larimore, 1975), whereas a flood of equal magnitude in winter might have minimal effect on subadults or adults. Finally, it should be recognized that disturbances or events take place at a wide range of time scales from minutes to millenia, and that even very modest events (by standards of human observers) could change dynamics in fish assemblages. Perhaps the "2 standard deviation" rule could be applied with respect to present changes relative to recent magnitude of effects, even on short time scales, like minutes (e.g., if a "calm" lake becomes windy).

At the shortest temporal scale, there are events that vary from minute to minute that can alter behavior or success/survival of individual fish. Such events may be trivial from the perspective of the whole assemblage but potentially significant to individuals. A brief increase in wind speed can induce ripples on the surface of a calm lake or stream pool or set streamside vegetation in motion, altering the underwater patterns of light/dark or shade (Helfman, 1981b). The phenomenon of "flickering" of light underwater, which is critical to perception of underwater objects, is highly influenced by wave action at the surface (McFarland and Loew, 1983). Thus, wind might alter the balance in visual acuity or detectability of predators or prey and change the outcome of a predation attempt. Similarly, rain, clouds, or other ephemeral phenomena events can change the influx of light, on which the ability of fish to find prey depends (Closs, 1994). Thus, very brief phenomena, like the passing of a cloud, could change activity of fishes. Such events collectively influence a fish assemblage, but they are so commonplace that they are rarely considered "disturbances."

At the scale of daily events, many disturbances are more noteworthy, with an increased potential to alter conditions in a fish assemblage. Day-to-day phenomena are often weather related, occurring with some degree of predictability over a cycle of days, like the succession of springtime storms in the midwestern United States as meteorological fronts arrive in sequence from the west.

In three different summers, I made, with D. R. Edds or P. Lienesch, unpublished data) daily early-morning collections of fish by seining at one site on the north (downwind) shore of Lake Texoma (Oklahoma–Texas). The lake commonly varied, daily, from calm to having waves of 25–35 cm. On days with waves greater than about 20 cm in height, fewer inland silversides, *Menidia beryllina* (a highly abundant species in the lake), were present in the littoral zone, but those that were present had significantly greater stomach contents than on calm days (perhaps due to eating items suspended from the sand substrate by the wave action). Littoral zone fishes of this reservoir seem able to cope with these daily wind–wave action disturbances, and at a population level, there is probably little effect. However, it is apparent that such daily disturbances markedly alter habitat use, foraging sites, and resource acquisition by *Menidia* and other small fishes of the littoral zone. Daily differences in winds also change the distribution or aggregation of larval fish in the open waters of Lake Texoma, as shown in Chapter 4 (Fig. 4.8).

Other potential disturbances in freshwaters are of greater magnitude and occur less frequently than day-to-day events. Examples include major stage rises in small streams, or major windstorm or inflow events in river–reservoir systems. Such events can be reliably predicted to occur at some time in the appropriate season of any year, but it is not possible to predict more than a few days in advance just when the events will occur. Brier Creek, Oklahoma, exhibits stage rises sufficient to disturb fish nesting or kill small larval fishes several times each spring in a typical year (Harvey, 1987). Modest stage rises due to springtime storms often occur no more than days to a few weeks apart, with each successive rise having the potential to disrupt normal breeding or feeding activities within the fish community. Most of these stage rises seem to be normal events for this creek, with relatively minor effects on the local fish assemblage (Matthews et al., 1994). For example, sunfish will renest if nests are lost in spates. Recent observations of fish in 14 pools of Brier Creek by snorkeling showed no more changes in fish distribution between surveys if a stage rise occurred than if one did not (Matthews and E. Marsh-Matthews, unpublished data). Thus, even relatively dramatic events to a human observer (e.g., a major spring flood) may have little effect on the local fish assemblage other than to temporarily disrupt activities or displace individuals, so long as the event is within the range of "normal" (*sensu* Resh et al., 1988), for which local fishes are adapted.

At longer time scales, there is a strong probability of Brier Creek having a major and more physically eventful stage rise [i.e., a half-bankfull (or greater) flood sometime in any year]. Such events occur at some time in most, but not all, years. These events are much greater disturbances to fish communities, from the perspective of their food base, but, again, may have little effect on the actual structure of the fish community in a stream reach (Matthews et al., 1994). These "annual" floods are sufficiently powerful to move or overturn most of the rocks in the gravel–cobble streambed, scouring most of the attached algae (Power and Stewart, 1987; Gelwick and Matthews, 1992) from the substrate and removing a high proportion of the benthic invertebrate community (Matthews and Gelwick, personal observation). Dramatic, system-resetting flood events may displace relatively few adult fishes (Matthews et al., 1994), but they do alter, for some weeks or more, the productivity (Power and Stewart, 1987) and food base for the stream, with likely effects on the feeding success of many species.

Another seasonal or annual event of great potential importance to fishes (and the entire reservoir ecosystem) is the major inflow of highly turbid water from Red River into Lake Texoma that typically happens once or twice a year (Matthews, 1984). Heavy rains in the upper basin result in distinct plumes of very turbid water entering the reservoir from Red River. Often, there is a definite front between lake water with a secchi depth of 1.0–1.5 m, and highly turbid inflowing water masses with secchi depth as little as 15 cm (Matthews, 1984; Dirnberger, 1983; Dirnberger and Threlkeld, 1986). The inflow is sometimes a surface overflow, but, on some occasions, light-meter readings have indicated that the turbid river water moves as a "throughflow" (Dirnberger and Threkeld, 1986). These turbid water inflows radically alter the distribution of zooplankton and small fish (Matthews, 1984; Dirnberger 1983) and the movement and feeding activity of fishes, including predators like striped bass, channel catfish, and blue catfish.

Another category of disturbance that is important in aquatic systems consists of very regular, largely predictable, events related to annual weather patterns, such as tropical wet–dry seasons, snow-melt floods in montane rivers, or the monsoon in deserts of the American Southwest. Such events are a regular pattern within a year, and there may actually be a "disturbance" to fish assemblages if the annual pattern changes or is delayed. For example, a late rainy season in the tropics can result in continued desiccation of isolated pools and massive death of fishes (Lowe-McConnell, 1987). Snow-melt floods in mountain streams are an annual event to which reproduction of invertebrates and fish may be timed, even for many miles outside the mountains downstream in plains rivers. However, in years with an unusually deep snowpack, flooding may provide channel formation and maintenance and prevent stabilization of islands or sandbars that actually isolate parts of the normal river–floodplain system (Ward and Stanford, 1995a, 1995b). Such floods may cause harm to individual fish (e.g., by destruction of microhabitats), but likely benefit fish assemblages in the long term by keeping the needed spawning and nursery habitats available and maintaining the opportunity for food-web linkages. Thus, the "severity" of a disturbance must be examined not only in the context of the immediate killing of some individuals, but in the potential long-term benefits of maintenance of the physical structure that is the template for the whole ecosystem.

Finally, there are very unusual disturbance events in aquatic systems so rare that there are many generations of fish between episodes. The effects of such rare floods may persist for decades (Friedman et al., 1996) or longer [e.g., formation of the major terrestrial benches in the floodplains of rivers (Hefley, 1937)]. Such events were two "100-year floods," slightly more than a year apart in Brier Creek, Oklahoma, and Piney Creek, Arkansas, where I have long-term records of assemblage structure; drought throughout the southwestern United States in the early 1950s (Hubbs and Hettier, 1958); floods in the Mississippi River basin in 1927 and 1994; and extreme high water in Lake Texoma in 1957, 1981, 1982, and 1990. These events can be either physically dramatic, with extensive destruction and rearrangement of physical habitats (e.g., maximum floods) or relatively "nonphysical," like the great increase in the expanse of water in a flooded reservoir with concomitant inundation of large amounts of usually terrestrial habitat. Such rare events have the potential to modify local fish assemblages for years, as did the drought of the 1950s (Hubbs and Hettler, 1958) or a flood in the Sierra Nevada (Strange et al., 1992). Alternatively, the assemblage may recover relatively rapidly to a former condition (Matthews, 1986c).

7.3 Floods

Rain pumped snakes from their holes—and rain was so much rain it began—to leak up and bear on its back—the froth of rain that came—to cover the rain that came before. Rain with rain on its back goes—where its load needs to go. . . .

—From "Record Flood," in the collection entitled *"Flood"* (1982), by William Matthews
(a poet who lives in New York)

. . . in the absence of scouring floods, the food web beneath the fish collapses.

—Wotton et al. (1996)

Effects of Floods on Streams

An important dichotomy exists between nonerosive (laterally expansive) and erosive floods and their effects on stream fishes. Where channel morphometry, such as canyon walls, exposed rock substrates, or deep incision, constrain the stream, floods tend to be erosive. In low-gradient terrain with wide, gently sloping floodplains, floods tend to have less erosive force. Laterally expansive flooding in natural river systems results in an intimate relationship of fishes with the floodplain in both tropical (Goulding, 1980; Welcomme, 1985; Power et al., 1995a, 1995b; Wootton et al., 1996) and temperate (Ross and Baker, 1983) river systems. Expansive flooding cycles influence the composition of local fish assemblages in the tropics, as resident fishes are joined transiently by migrating species (Lowe-McConnell, 1975).

Nonerosive floods typically inundate vast amounts of terrestrial habitat, resulting in large inputs of nutrients. Fish in temperate streams may move into flooded areas to feed, with some species showing improved reproductive output in flood years (Ross and Baker, 1983). Moses (1987) showed that the annual catch of fish in the Cross River, Nigeria, depended on flooding of the previous year and the degree to which fish could use allochthonous materials in their diet. However, Rutherford et al. (1995) found that the growth of fishes in the constrained (levees) lower Mississippi River of North America was influenced more by primary and secondary production within the river channels than by allochthonous inputs from floods. Kwak (1988) found that fish occupied floodplains in direct proportion to the river stage of the Kankakee River, Illinois, and that these floodplains were important nursery habitats for fishes. Lateral floodplain pools associated with the Parana River, South America, are important habitats for juvenile fishes, and the return of juveniles to the main river coincides with a complex schedule of flooding (Bonetto, 1975). Goulding (1980) is the classic study of movement of fish into flooded tropical forest, where they feed actively on plant material during the rainy season.

Erosive floods are characterized by fast-moving, turbulent water with power to entrain and move substrates and scour benthic algae or macroinvertebrates (Grimm and Fisher, 1989). Erosive floods can be accompanied by the dramatic destruction of the physical habitat in the stream channel and the riparian zone. Erosive floods can have pervasive effects on the physiognomy of a stream system [e.g., forming and maintaining heterogeneity of structure and habitat patches (Ward and Stanford, 1995a, 1995b)]. Erosive floods also can change the dynamics of ecological linkages within stream ecosystems (Fisher et al., 1982).

Immediate effects of floods on individual fish may largely depend on size or life stage or on habitat complexity. Larval and young fish may be washed out by floods, but adult fish, particularly in complex habitats, may avoid displacement by floods. Heggenes (1988b) and Matthews et al. (1994) suggest that larger fishes are not washed away by tropical flooding. Pearsons et al. (1992) showed by snorkeling before and after floods that fish assemblages in stream reaches with heterogeneous, hydraulically complex habitats had greater assemblage similarity before and after floods than did reaches with hydraulically simple structure. Lobon-Cervia (1996) concluded that complex habitats provided complex hydraulic conditions even during severe floods, which facilitated the survival of a very severe flood by salmonids in a river in Spain. In general, native fishes should be reasonably well adapted for the local flood regime, and there is, indeed, some evidence that non-native fishes may suffer more changes during severe floods (Meffe, 1984; Strange et al., 1992).

If stream reaches are scoured severely or the stream channel is displaced laterally, the fish fauna may initially be reduced. Over time, however, the diversity of physical structure at may increase (e.g., by reestablishment of riffles and pools) and the complexity of the fish assemblage may concomitantly increase. Trautman (1957) argued that ultimately the stream gradient is a major factor controlling fish assemblages and that, given time between disturbances, the fish assemblage will reflect the basic physical characteristics of the reach. In Piney Creek, Arkansas, I found that even in severely scoured reaches, the fundamental riffle–pool structure had begun reforming in the stream by 4 months after a great flood (Matthews, 1986c).

Most studies of erosive flood effects on fish have been at individual fixed sites (e.g., Matthews, 1986c; Erman et al., 1988) with focus on changes in fish assemblages within each site. However, Reeves et al. (1995) suggested a model for disturbance effects on stream fishes that focused on broader patterns of asynchronous events throughout a basin. In the model of Reeves et al. (1995), various streams or stream reaches will be at different stages of recovery from disturbance (landslides, debris flows, floods), creating a mosaic of habitat patches. The model of Reeves et al. (1995) might apply broadly to flood effects if local flooding at different times and places (different tributaries in a basin) can additively produce a heterogeneous landscape for fishes at the basin scale—promoting habitat and biological diversity. However, Reeves et al. (1995) note that good information on the degree to which fish populations act as metapopulations distributed among parts of a basin is largely lacking.

Floods may also help maintain the complexity of food webs. Erosive floods, by periodically creating short-term disturbances, may prevent dominance of food webs by some limited number of species. Wootton et al. (1996) and Stanford and Ward (1983) suggest that aquatic invertebrates, as well as the predatory fishes in food webs, may depend on occasional floods. The difference between the two approaches is that Stanford and Ward (1983) leave "intermediate disturbance" as a black box, whereas Wootton et al. (1996) show mechanisms of effects of floods in food webs. They suggest that floods, by moving rocks and crushing interstitial invertebrates, function to remove taxa that are poor food for fishes. Junk et al. (1989) emphasized that the "flood pulse" in streams ranges from unpredictable (to which organisms have little chance to adapt) to highly predictable (affording organisms the possibility to not only survive floods but to directly benefit from flood inputs). Thus, floods can be both "good" and "bad" for fishes.

Some of the immediate effects of erosive floods on fish, or in stream ecosystems include the following:

1. Scour stream bottoms and remove silt or organic detritus. Floods can remove accumulations of detrital particulate material or silt, deepening pools by removal of sediment or detritus that has accumulated in depositional zones during base-flow conditions. One flood in Brier Creek, Oklahoma, deepened a stream pool by more than a meter (E. Marsh-Matthews, personal communication) making it nonwadable. Gammon and Reidy (1981) attributed an episode of severe hypoxia in the Wabash and Big Vermillion rivers, Indiana (USA), to a lack of floods, which allowed buildup of detritus and subsequent oxygen depletion by organic oxygen demand. Because many fish are strongly tied to the substrate (e.g., by nest building or deposit of eggs, or by algivory), modification of substrate by floods can have a strong effect on fish assemblages, decreasing the proportion of fine material or detritus or exposing clean cobble and bedrock. Fine materials and detritus can smother nests or eggs and interfere with growth of attached algae; thus, flood-induced changes in "fines" could have direct effects on reproduction or feeding of fish. Berkman and Rabeni (1987) showed that silt affected the composition of fish assemblages, interfered with feeding of algivorous fish, and lowered the reproductive success of some spawning guilds (e.g., lithophils).

2. Remove algae. Floods grind or strip away both filamentous and adnate periphyton (Power and Stewart, 1987; Grimm and Fisher, 1986) or bury algae in sediments (Allan, 1995, p. 100). Catastrophic losses of periphyton occur in floods sufficient to move the streambed (Uehlinger et al., 1996). Regrowth of algae can be rapid (e.g., within weeks; Fisher et al., 1982; Power and Stewart, 1987) or postflood recovery can be quite long [e.g., if a nutrient is limiting after the flood (Grimm and Fisher, 1986)]. Fisher et al. (1982) showed virtually complete removal of standard crops of algae by a desert flash flood, followed by recovery in as little as 2 weeks. Biggs (1995) showed that streams sites with frequent flood disturbances supported less annual production of periphyton, altering the entire "habitat template" for organisms associated with algae. Many stream fish eat attached algae (Matthews et al., 1987) or invertebrates living in algae (Power, 1990a); thus, loss of algae during floods could have a direct impact on short-term feeding opportunities for fish.

3. Remove macrophytes. Henry et al. (1994) showed that floods in the Rhone River destroyed macrophytes and that most recolonized only very slowly. Particularly in shallow or slow-moving streams, stems

of macrophytes provide shelter and foraging opportunities (attached epiphytes/aufwuchs/biofilms) for small fishes, and predatory fish lurk near weedbeds. Thus, macrophyte removal by floods drastically alters the underwater landscape available to fish.

4. Wash out invertebrates. Floods that move the substrate strip gravel- or cobble-bottomed streams of invertebrates (Fisher et al., 1982), but they may recolonize rapidly, particularly for taxa with flying adults (Fisher et al., 1982). Stock and Schlosser (1991) showed loss of more than 90% of the benthic insects in a stream following a flash flood, and that after 2 months, densities were only 62% that of preflood. Fish densities were concomitantly lowered. Scouring floods can reduce the benthic macroinvertebrate fauna so drastically that it takes months to reach high densities (Flecker and Feifarek, 1994). Miller and Golladay (1996) showed 90% reduction of invertebrates by spates in two Oklahoma creeks. Some benthic invertebrates in flash-flood-prone streams are adapted for the flood regime by behavioral avoidance of strong currents and rapid reproduction (Gray, 1981). Loss of macroinvertebrates can be equated with loss of feeding opportunity for many benthic or drift-feeding fish (although most pool-dwelling fishes can also eat terrestrial insects that fall to the water's surface).

5. Remove spawning gravel. Erosive flooding may scour and remove patches of gravel used by some fish for spawning in streams where such substrates are in short supply, limiting annual production of some salmonids (Kondolf et al., 1991). These gravels are normally found where low shear stress favors deposition, but high-flow years can remove these patches of spawning substrate.

6. Destroy nests. Floods can destroy nests or eggs by mechanical grinding in the substrate as the bed load moves (Erman et al., 1988), by depositing silt onto nests, washout of larvae from nests, or destroying nesting habitats (Lobon-Cervia, 1996).

7. Wash out or kill free-swimming larval fish. Ottaway and Clarke (1981) showed that young salmonids were vulnerable to being washed downstream by high currents. Harvey (1987) showed that minnow and sunfish larvae were washed out of nursery pools and killed by spates.

8. Increase turbidity. Increased turbidity during spates can interfere with sight feeding by fish, changing their foraging regime (Crowd, 1989; Barrett et al., 1992; Gregory, 1994), or with their use of visual references in streams (Larimore, 1975).

9. Fill existing pools. Pools as much as 3 m deep have been completely filled by gravel deposition in a single flood in the gravel-bottomed Baron Fork River in eastern Oklahoma, forcing movement of resident

smallmouth bass (*Micropterus dolomieu*) and other large-bodied fishes (Matthews, personal observations).
10. Maintain channel. Erosive floods remove encroaching vegetation (that could stabilize midstream bars, forming islands or braiding), prevent lateral silt or sandbars from blocking backwaters (important nurseries for riverine fishes), and scour and deepen stream channels.
11. Remove or deposit large woody debris. Floods can undercut stream banks, allowing riparian trees to fall into and lodge in the stream channel, which provides additional complexity to instream habitats. Alternatively, floods can refloat existing woody debris, stranding it in terrestrial floodplain habitats some distance from the stream, thus lowering habitat complexity or stream retention.
12. Crowd fish into refugia, or alter their behaviors. As described below, I have often found fish using refugia at stream edges during flooding, and small tributaries may also serve as refuges for mainstem fish during floods. Fitzsimons and Nishimoto (1995) found that reproductive behaviors of stream fishes in Hawaii were interrupted by flooding and took as much as a year to return to normal.
13. Strand fishes in temporary pools. Chapman and Kramer (1991) showed that floods could cause a loss of 75% or more of the fishes in stream pools if they became trapped in isolated pools formed during high water and subsequently desiccating.
14. Stimulate upstream movement of fishes. Dalquist (1957) reported that the atherinid *Labidesthes sicculus* moved long distances upstream in the Red River basin in Oklahoma and Texas during prolonged flooding in 1957.

Potential Long-Term Effects of Floods on Stream Ecosystems and Fish Assemblages

The effects described above occur during or immediately after a flood. Many immediate negative effects of erosive floods on fish may be ameliorated by vagility of the fish, by their ability to find small hydraulic refugia associated with banks or perhaps streambed, or by rapid reproduction after floods. However, other effects of great, erosive floods might only become apparent long after a stream returns to base flow. In particular, if a flood changes the basic habitat template, or the balance of materials moving into or through a stream, the fish assemblage will adjust over the long term to reflect these altered conditions. Some of the following mechanisms might cause longer-term or delayed changes in a stream ecosystem, thus in the fish assemblage.

1. Scouring of the stream bottom can cause changes in microhabitats available to fishes. Riffle or shallow channel areas may be swept clean

of sand, leaving more interstitial space under and between cobbles available to fish (and invertebrates). This could cause marked changes in fish microhabitat use, in nest sites (e.g., for *Nocomis*), in protection for fish eggs, or in standing crop and availability of invertebrate prey. It is not possible to predict whether alteration of interstitial space will increase or decrease the abundance of various fish species, but it is plausible that increased interstitial space might increase the capacity of stream reaches for fish biomass or numbers of individuals. Changes in microstructure of habitats might alter predator–prey relationships among small benthic fishes and their macroinvertebrate prey, or between larger piscivorous fishes and small fishes that occupy the riffles (Crowder and Cooper, 1979, 1982).

2. Alteration of the riparian zone could change primary production in a stream system and ultimately affect fish populations. Although riparian trees and shrubs are resistant to flooding (Stromberg et al., 1993), major erosive floods can remove trees from stream banks and alter canopy. A decreased canopy allows more sunlight to reach the streambed, and periphyton may increase, followed by increases in invertebrate or vertebrate grazers. Floods can strip underbrush and accumulated leaf litter from terrestrial areas, or result in altered hydrology or input of nutrients via runoff. A flood also can deposit such thick layers of sand (Stromberg et al., 1993; Matthews, personal observation) that there will be significantly reduced local growth of streamside vegetation (e.g., grasses) and input of allochthonous materials could be changed.

3. Floods erode earthern banks, which may remain a source of increased turbidity or silt. Silt and turbid water can directly affect production of invertebrates, reproductive success of fish (e.g., siltation of nests), predator–prey interactions, and growth or production of fish biomass.

4. The phenomena above could affect standing crop or taxonomic composition of the aquatic invertebrates in a watershed for a long period of time. Altered food availability can cause changes in altered feeding ecology or trophic interactions of fishes in streams. Elwood and Waters (1969) showed that although immediate effects of a severe flood were negligible, there was a drastic delayed impact on trout populations due to decreased habitat and food supply.

5. Major floods are important events forming or maintaining the physiognomy of not only the streambed but also the terraces and terrestrial–riparian plants and habitats associated with rivers (Hefley, 1937). Friedman et al. (1996) demonstrated that a single major flood in Plum Creek, Colorado, left effects on the morphology of the stream channel and adjacent bottomlands that were evident after more than 30 years.

Overall, the prediction of long-term or delayed effects of a flood on fish assemblages is tenuous, but the literature suggests that the flood-related phenomena could result in changes in macrohabitats and microhabitats, productivity of streams, food availability to fishes, and shelter. Such changes might affect composition of the assemblage and interactions among fishes for years.

Field Studies of Flood Effects on Fishes

Erosive floods can reduce populations of fish, in some cases lowering densities of adults (Paloumpis 1958; Seegrist and Gard, 1972; Collins et al., 1981) as well as larvae (Harvey, 1987). Some floods wash out small fish and have little impact on adults (John, 1964; Hoopes, 1975; Rinne, 1975), or the majority of individuals may remain in place during floods (Gerking, 1950). Kushlan (1976) and Schlosser (1982) showed that the timing of floods influence their impact on fish assemblages. There is evidence that both timing of floods and the kinds of stream habitats that are affected can influence flood impacts on fishes. Thomas (1970) showed that floods interfered with reproduction of four darter (*Percina*) species in an Illinois (USA) river. Pearsons et al. (1992) also showed that the timing of floods is critical, with species vulnerable to floods at their reproductive times of year. Salmonids that spawned early in the year were affected more by spring floods, whereas suckers (Catostomidae) and minnows (Cyprinidae) were more negatively affected by summer floods (Pearsons et al., 1992). Erman et al. (1988) found that greatly increased shear stress on the streambed during flooding caused mechanical grinding or crushing of benthic fishes or their eggs and found, at long-term study sites, a reduction in sculpins and salmonids in the next year as the result of winter floods. Starrett (1951) hypothesized that floods were an asset or a liability to different minnow species, depending on their timing within the year. Harrell (1978) showed that a flood altered species richness, diversity, and species associations in habitats of the Devils River, Texas. However, he found that the species numerically dominant before the flood also dominated the postflood system and hypothesized that successful species were well adapted to the flood-prone environment through ecological plasticity.

Most studies of flood effects on fish have been in environments that flood rather often. If fish of such environments are well adapted for floods or swift currents, as Smith and Powell (1971), Harrell (1978), Goulding (1980), and Constantz (1981) suggest, long-term effects of floods on structure of the fish assemblages and/or ecological interactions might be minimal. Relatively few studies (e.g., Hoopes, 1975) exist on flood effects on fish in streams that rarely flood. A major flood in systems that are usually stable could represent the temporally rare "crunch" hypothesized by Wiens (1977) to strongly influence community structure or interactions among species.

Habitat Use by Stream Fish During and After Floods

Quantification of fish behavior during erosive floods is limited (Jowett and Richardson, 1989), largely because of logistical problems with field work during flooding. Meffe (1984) showed in the laboratory that native fish in flash-flood-prone Arizona streams have behavioral responses that facilitate survival, such as remaining close to structures during increased flow. Minckley and Meffe (1987) noted crowding of fishes into stream margins during flood. Larimore (1975) experimentally found that *Micropterus dolomieui* fry use tactile and visual orientation to maintain its position in floodwaters. Paloumpis (1958) showed field evidence that during erosive mainstem floods, fish move into small tributary refugia. Scrimgeour and Winterbourn (1987) speculated that during floods, fish seek refugia, moving away from gravel–cobble areas into backwater pools and river margins, where they are safer from destruction from bed-load movements. Heggenes (1988a) documented use of low-velocity refuges by trout during flood conditions. Matheney and Rabeni (1995) found that northern hogsuckers (*Hypentelium nigricans*) in Missouri streams moved into low-velocity flooded riparian areas during flooding, thus remaining in approximately the same stream segment as before flooding.

Lancaster and Hildrew (1993) described areas of streambed with low hydraulic stress in spite of increased flow, comprising refuges for invertebrates (and, presumably, small benthic fishes). Both Erman et al. (1988) and Lancaster and Hildrew (1993) suggested that shear forces on streambeds resulted not only from complexity of the streambed habitat but also from the constraint of stream reaches by bank morphology or (in winter) by streamside snow.

I documented habitat use by stream fishes in the Roanoke River, Virginia, during and immediately after severe floods in 1977 and 1978. On 27 March 1978, when the Roanoke River in Salem, Virginia, was in severe flood, I measured current speeds of 150–166 cm/s in the main channel near shore; mid-channel was not wadable because of strong currents. The river was about 50 m wide; the water was 7°C. I seined downstream in the edge of the swift main channel, by having a student on the bank anchor a 4.6-m, 0.5-cm mesh seine while I let the current carry me and the seine downstream to land at the bank. No fish were found in the swift main channel, and it seemed very unlikely that water-column fishes could have held their position against the current. However, in nearby low-velocity eddies associated with irregularities of the bank, we took substantial numbers of cyprinids and darters, some from under mats of floating debris. At one site I collected 11 *Luxilus cerasinus*, 9 *Luxilus albeolus*, 12 *Pimephales notatus*, 16 *Lythrurus ardens*, 2 *Nocomis leptocephalus*, 1 *Percina rex*, and 1 *Etheostoma flabellare*. The darters *P. rex* and *E. flabellare* rarely use backwater habitats under normal flow conditions (Matthews, personal observation in more than 100 collections in the Roanoke River drainage, 1977–1979). This collection suggests that during peak flood-current speeds in the main channel, adult fish take advantage of small recessed areas and eddies just off the main current, from which they could disperse back into typical habitats after flow slackens.

On the same day in the Roanoke River, in a small backwater (current speed = 32 cm/s) in what was normally a grass-covered terrestrial slope, floodwaters had apparently concentrated a large number of fish. In two to three seine hauls in this backwater, we collected 17 *Luxilus cerasinus,* 13 *L. albeolus,* 57 *Nocomis leptocephalus,* 9 *Lythrurus ardens,* 44 *Campostoma anomalum,* 1 *Pimephales notatus,* 1 *Rhinichthyes atratulus,* 6 *Hypentelium roanokense* or *H. nigricans,* 1 *Moxostoma cervinum,* 1 *Moxostoma* sp., 1 *Moxostoma rhothoecum,* and 1 *Lepomis* sp. Several of these species, including the *Hypentelium, Moxostoma,* and large numbers of *E. flabellare,* are normally found in substantial flow in riffles and runs of main channels. This collection also suggested that many adult fish could use channel-edge refugia and remain in a given reach of river even in a major flood. Although I have not directly observed most of these species in strong flow (e.g., in artificial streams), Matthews (1985) showed that *E. flabellare* resists displacement by direct behavioral (posturing) mechanisms. Webb et al. (1996) reviewed a variety of behaviors used by benthic fish to increase their streamlining and avoid the necessity for active swimming to hold position.

Another set of observations in the Roanoke River suggested that during floods, benthic fish may leave or be forced from their usual habitats but remain in generally the same reach of river. From August 1977 through July 1978, I made seven collections in large riffles of the main channel in the Roanoke River at a site where the river was 30–50 m wide, with a large cobble riffle spanning the entire channel. The focus of this series of collections was on taking the darters *Etheostoma flabellare, E. podostemone,* and *Percina roanoka* for studies of resource use (Matthews et al., 1982b). In most warm weather collections, we took substantial numbers of all three species in swift mainstream riffles, including 23 April 1978 (Table 7.1). On that date, measured current speeds in the riffles ranged from 11 to 88 cm/s. The largest Roanoke River flood in 6 years crested on 29 April. Next, we were able to make collections on 2 May while the river remained high, with riffles 30–50 cm deep and very swift. In a total of 15 kicksets

Table 7.1. *Numbers of Three Darter Species Collected in One Large Riffle of the Roanoke River, in Warm Weather Collections 1977–1978.*

Date	*Etheostoma flabellare*	*E. podostemone*	*Percina roanoka*
30 August 1977	Several	7	17
8 April 1978	70	1	1
23 April 1987	33	7	14
2 May 1978 (main channel)	0	0	0
2 May 1978 (stream edges)	7	2	10
18 May 1978	18	6	13
7 July 1978	16	13	2

in large rubble of the main channel (which normally was an ideal habitat for these darters), we took only one fish, a *Campostoma anomalum* about 100 mm Standard Length (SL). Current speeds were 85–129 cm/s in the mainstream riffles at this time, which would not of itself exclude these darters (Matthews et al., 1982b), but currents in the main stream during peak flow 3 days earlier must have been substantially faster. On 2 May, in slow riffles and pool habitats (8–27 cm/s) near shore, we collected 7 *E. flabellare*, 2 *E. podostemone*, and 10 *Percina roanoka*, which more usually were found in the mainstream riffles. Other fishes taken at that time near shore in slow current, or in other habitat away from the swift main channel, were 15 *C. anomalum*, 2 *Clinostomus funduloides*, 24 *Nocomis leptocephalus*, 3 *Phoxinus oreas*, 15 *L. albeolus*, 17 *L. cerasinus*, 5 *L. ardens*, 1 *Pimephales notatus*, 4 *H. nigricans*, 1 *Moxostoma erythrurum*, and 1 *M. rhothoecum*. By 18 May, darters had returned to the swift main-channel riffles, where they were also present in July (Table 7.1). Thus, during erosive floods, main-channel river fishes, particularly benthic species, may abandon (or be swept from) their typical habitats in mid-channel but survive in substantial numbers in slow-current refugia. However, some species, through unknown behaviors, seem able to remain in main-channel habitats in some floods. On 27 October 1977, at Glenvar, Virginia, the Roanoke River was turbid and flooding after 2 days of rain. In 13 kicksets in swift riffles, 9 *Percina roanoka* were taken at current speeds exceeding 125 cm/s. Obviously, swift currents and turbulent waters of even highly erosive floods do not have the same effects on all species—and the capabilities of only a few species to maintain position in floods have been measured in controlled conditions.

A Major Flood event in a Typically "Benign" Warm-Water Stream: Effects on the Fish Fauna

Discharge in streams of the interior highlands generally fluctuates less than in stream of prairies or deserts in North America. Their spring-fed origin typically results in a stable base flow (Pflieger, 1971). Severe scouring floods are infrequent, occurring on average only once every 2.5 years. Thus, an individual fish has a reasonable chance of experiencing a flood once in its lifetime, but floods are not nearly so regular a feature in Ozark streams as they are in prairie streams to the west. Benign periods of relatively stable flow are of sufficient length (years) in Ozark streams that biotic interactions could be important in community interactions.

Piney Creek, Izard County, Arkansas (USA) is a medium-sized Ozark mountain stream, with permanent flow in all but the extreme headwaters, benign physicochemical conditions, and a diverse fish fauna of 48 known species (Matthews and Harp, 1974; Matthews, 1986c; Matthews et al., 1988). In its physiognomy, size, and fauna, it is an excellent model of streams in uplands of eastern North America. The most severe flood in this century in Piney Creek was in December

1982. At 12 sites (see Fig. 4.11) in the watershed, I had sampled fish in the previous August, and I followed recovery of the fish assemblages during the 8 months after the flood (Matthews, 1986c). This flood, from approximately 28–35 cm of rainfall throughout the drainage in 48 h, resulted in a stage rise of as much as 11 m in the normally wadeable creek. Whether the 1982 flood resulted in ephemeral or persistent change in the structure of the fish assemblage in Piney Creek is of interest in the context of historical data on its fishes of the stream as an opportunity to measure the importance of such an extreme event within a relatively stable stream that lacks annual floods.

In July, December, and April 1972–1973, I had made fish collections by seining at all fixed locations (Matthews and Harp, 1974). In 1982–1983, I repeated collections at the same locations on, as nearly as possible, the same calendar dates as in 1972–1973 (Matthews, 1986c). The goal of the resampling initially was to evaluate changes in the composition of the fish fauna, species distributions, and their ecological associations across the 10-year hiatus. During the decade from 1972 to 1982, there had been little apparent change in land use or cultural activities in this rural watershed and rainfall records (EarthInfo) showed no major floods.

In early December 1982, following the fish collections of August 1982 and just prior to planned winter collections, the most severe flood known for this watershed took place with more than 40 cm of rain in the watershed in a 2-day period. Trees as large as 1 m in diameter were uprooted by the flood. Massive piles of dead trees were deposited at the high-water mark or in curves of the streams. Flood currents stripped riparian underbrush and leaf litter, and stream banks were severely eroded. Massive amounts of sand were scoured from streambeds and either redeposited within the stream, filling pools, or deposited in piles as much as 1 m deep, and as far as 100 m from the stream banks.

Within the channel, sandy riffle or channel reaches were scoured, leaving only coarse gravel or cobble. Shale and cobble were piled up by the flood, forming new bars in the stream. After the flood, aquatic invertebrates were almost completely absent. Despite the tremendous flow, the basic pattern of pools and riffles remained or was rapidly reestablished (by April) at most stations. Stations with large substrate or steep rock walls were not as altered as much as were stations with sand or small gravel substrates. However, the general picture of Piney Creek in the immediate postflood environment was one of a drastic alteration of physical habitats. In the context of the typically benign environment of Piney Creek, this flood clearly was a major disturbance (*sensu* Resh et al., 1988) event with potentially strong impact on the biota.

Results for fish sampling in 1972–1973 are detailed in Matthews (1973) and Matthews and Harp (1974). The entire 1972–1973 sampling, including a collection with rotenone, produced a total of 44 known fish species in the watershed. The 1982–1983 collections produced only one species of fish not previously taken in the watershed (*Erimystax dissimilis,* one specimen). Rank correlation of the 12 most abundant nonbenthic species in summer 1972 versus summer 1982 collections was highly significant.

The flood caused no drastic loss of fish from the watershed as a whole. In January 1983, I took 5217 individuals, compared to 5911 individuals in the preflood August collections (Table 1 of Matthews, 1986c). Immediate postflood collections in January 1983 included 36 species of fish compared to 38 in August. Diversity of the collections (all stations pooled; Shannon–Wiener index) was little changed after the flood (2.68 and 2.53, respectively). Thus, pooled data suggested that the immediate effect of the 1982 flood was not a wholesale change of the fish fauna of the watershed. However, some individual collecting stations and species abundances showed marked changes.

Table 1 of Matthews (1986c) indicates the number of fish species collected per station before the flood (August 1982) and in the first and last sampling in the year after the flood (January 1983 and August 1983). At seven of the locations, the change in total number of species collected was minimal after the flood. However, at five locations, there was a marked decrease in numbers of species collected immediately after the flood (losses in richness ranging from 7 to 10 species). Overall, the mean number of species per station before (19.1) and after (15.8) the flood was significantly different. The five stations with greatest reduction in numbers of species all showed severe scour, channel alteration, or reduction of habitat complexity. It was obvious on the basis of comparisons within years and across years that the period including the December 1982 flood resulted in short-term change in relative abundance of the common fish species of the watershed, and that the January 1983 fish collection was unique in comparison to collections at all other sampling periods.

To summarize the tangible immediate (i.e., 6 weeks postflood) effects of the flood on the fish assemblage, there was little change in total numbers of fish or fish species in the watershed or in fish faunal diversity. However, there were major changes in the fish fauna of several of the fixed sampling localities, and there was more change in the relative abundance of species during the flood period than in the decade prior to the flood. The flood altered distributions of some species, drastically altered the presence of fish at some localities, and changed some fish species associations (Matthews, 1986c).

Although there were noteworthy changes in local fish assemblages in Piney Creek immediately after the flood, the local assemblages and the entire creek fauna had recovered by August 1983 (8 months postflood) so that they were indistinguishable from those of the previous August (Matthews, 1986c). For the creek overall, faunal composition was highly similar in August 1982 and 1983, and the rank correlation of common species was significant (Matthews, 1986c). There was excellent reproduction of most common species in the summer after the flood, and no common species was lost from the system. Two relatively "rare" species (locally), not taken in the year after the flood (*Phoxinus erythrogaster* and *Erimystax dissimilis*) were again found in Piney Creek in samples taken a decade later in 1994–1995 (W. Matthews, E. Marsh-Matthews, and C. M. Taylor, unpublished data). With the exception of one headwater station that was very severely scoured, all other sites had recovered to nearly as many or more species in August 1983 than had been present in August 1982. Across all 12 sites, the mean number of species in August 1983 was 20.3, compared to 19.1 in August 1982.

Finally, it could be possible for a major flood, with rearrangement of habitat patches, to change distributional associations among fishes in a watershed. However, for Piney Creek, distributional associations among the 15 most common species were compared across 36 collections (12 sites × 3 samples) for 1972–1973 and 1982–1983. Product moment correlation was used within each decade to produce a similarity matrix for all possible pairs of the 15 species, based on their abundance at each sampling site. Figure 7.1 shows two resulting dendrograms when each of the decadal similarity matrices was clustered by UPGMA (Sneath and Sokal, 1973). A Mantel test comparing the two similarity matrices showed a highly significant ($p = 0.0004$) concordance of species associations in the samples taken a decade apart. Therefore, the distributional associations of the common species within Piney Creek were essentially unchanged even through the 1982 flood at least temporarily displaced some species from local assemblages.

Thus, although the immediate effects of the flood of December 1982 were physically dramatic and there were marked changes in local fish assemblages at some sites, by 8 months after the flood and following successful reproduction it was not possible to detect any major negative change in the fish assemblages attributable to the flood, and fish species associations, overall, remained unchanged between the 1970s and 1980s. Sampling of all 12 sites year-round in 1994–1995 (12 years postflood, with no such major floods since 1982) revealed that fish assemblages were much like their local counterparts in the early 1970s and 1980s, and ongoing multivariate analyses showing the degree of changes in local assemblages throughout the three decades suggest no more changes associated with the flood year than during any other intervals in this long-term data set (Matthews and Marsh-Matthews, unpublished data). Overall, it appears that even highly erosive floods can occur without causing lasting change in stream fish faunas [but see Strange et al. (1992) for a differing example, in which one major flood caused apparently persistent changes in the fish assemblages in a California stream].

It would be a highly desirable to follow a truly great flood to assess the time scale, persistence, and mechanisms of flood effects on fishes. Unfortunately, the chances to follow a severe flood, in a system with preflood records, are rare. If such an opportunity arises, a team might focus on (1) assessing the time course of flood-related changes in the fish assemblage during and immediately after a flood, (2) detecting and evaluating longer-term (e.g., 3 years) or delayed effects of the flood on fish assemblages or interactions of species, and (3) continuing

340 / *Patterns in Freshwater Fish Ecology*

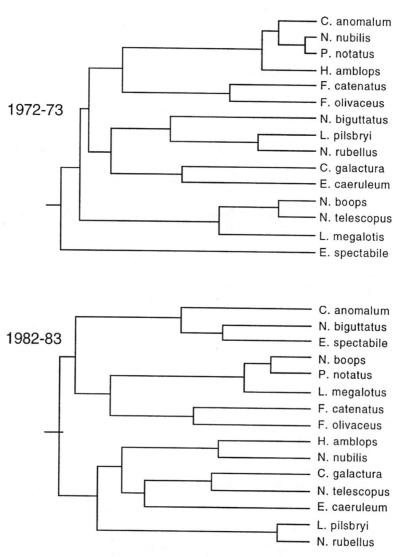

Figure 7.1. UPGMA clustering of common fish species in Piney Creek, Arkansas, based on their abundance at 12 permanent field sites collected 3 times in a year, in the 1970s and the 1980s.

long-term (in terms of fish generations) study of physical phenomena, ecosystem processes, standing crops of algae and invertebrates, and structure of fish assemblages in a watershed. We have carried out a limited part of these activities on Piney Creek, Arkansas, and for those variables, it would be desirable to know if the effects are similar in other regions or stream types.

7.4 Drought

Water scarcity is chronic in many parts of the world, and droughts have become worse in places like the southwestern United States, where growing populations demand more and more water withdrawal from streams (Tarboton, 1995). Bonacci (1993) noted that "drought events are becoming increasingly frequent worldwide." Hubbs (1990) described reduction of spring and stream flows as "a universal arid land fish faunal insult." Hubbs noted that reduced flow not only decreases available water but also results in increased thermal oscillations, and that "in many instances the thermal effects of decreased flow may exceed the direct impacts." Reduced flows, whether from local human impacts, drought, or interaction of the two, are devastating to fishes. Anthropogenic modification of stream channels may act synergistically to make drought worse for fish. Bret Harvey (personal communication) notes that in disturbed, aggraded stream channels, subsurface flow is important and substantial discharge is necessary to maintain the surface flow on which most fish depend. Thus, anthropogenic disturbance to stream channels can produce drought-like effects even without changes in precipitation (Harvey, personal communication). Additionally, drought conditions can exacerbate diurnal fluctuations in stream flow that results naturally in small forested streams due to evapotranspiration (Kobayashi et al., 1990).

Although droughts may decimate fish populations in streams, it is clear that over evolutionary or geologic time, fish species that persist in harsh environments must have had the capacity for some populations to persist during even extended periods of drying. Sections of streams with deep pools often allow some individuals to survive droughts (Griswold et al., 1982). Fish populations of a species, distributed across different watersheds, might function as metapopulations, with survivors in one watershed "rescuing" those in others by dispersal after drought ends (Reeves et al., 1995).

Drought is a natural phenomenon in North America (Trenberth et al., 1988). Some drought periods can be very long, relative to generation times of fishes. Tarboton (1995) used tree-ring reconstructed stream flow for a period of 442 years for the Colorado River in Arizona (USA) to detect a critical period in the years 1579–1600 when there were three droughts in rapid succession, and a "drought of historic record" from 1943 to 1964. Tarboton (1995) considered as a best estimate the following return times of three levels of drought severity for the Colorado River in Arizona.

1. Drought of historic record: 50–100 years
2. Severe drought of 1579–1600: 400–700 years
3. Most severe drought (1579–1600), but with years rearranged in order of decreasing stream flow (as if no relief in flow had occurred in the 22 years): 2000–10,000 years or more

Stine (1994) used dating of relict tree stumps to show two periods of extremely severe and prolonged drought in California from approximately A.D. 900–1100 and from A.D. 1200–1350, with a drought period matching the first in temperate South America. Thus, it is possible that temperate fishes, worldwide, have had during their evolution periods of "epic drought" (Stine, 1994) that could have been a substantial selective pressure. Stine (1994) cautioned that global warming might initiate another period of severe drought, which could have devastating consequences to fish biodiversity (Matthews and Zimmerman, 1990).

Soule (1993a) compared overall drought patterns for the United States across a 90-year period of record. For the contiguous 48 states as a whole, Soule found no indication of increased drought conditions. However, at a regional scale, he detected significant changes in moisture conditions in the interior of the continent from north–central Rocky Mountains to northern plains. In that region, there was a transition from wetter to dryer conditions from the early to middle part of the period of record, persisting to the present (Soule, 1993a). Desiccation in the streams of the Great Plains, thus, may relate not only to increased anthropogenic withdrawal of water (Cross and Moss, 1987) but also to true meterological changes. Soule (1993b) also detected 20 individual drought events in southwest Utah (USA) (each with 6 or more months with unusually dry conditions) during a 94-year period, with droughts common in both warm and cold seasons. Soule (1993b) noted that, "it is well established that droughts in the North American interior (especially the northcentral Great Plains) are more persistent than in other regions."

It is apparent from estimated drought return times (Tarboton, 1995), the frequency of drought in some regions (Soule, 1993b) and the propensity in some parts of North America for periods of increased aridity (Soule, 1993a), that native fishes of arid lands have probably already passed through an "evolutionary filter" with adaptation of life history or physiological tolerances for drought conditions. However, one requirement for continued existence through drought periods may be downstream refugia from which colonists return in wetter periods. We have no way of knowing, but in the long perhistoric drought there may have been widespread destruction of riverine fish faunas, followed by years of recolonization when pluvial conditions returned. Such a "fall and rise" of a fish fauna would appear, to an observer, to include major changes in distributions of fishes across the region. The widespread construction of reservoirs in semiarid regions could have impacts on the potential for fish to migrate downstream during dry periods and return upstream during wet periods. The normal route of migration downstream may now be blocked by a "lake" containing many more piscivores than would naturally exist in a river. As a result, small fishes moving down into a reservoir might not survive the period of drought because of greater predation pressure than before the reservoir was built. Winston et al. (1991) have identified at least one case in which a mainstem reservoir has apparently caused elimination of upstream fish species by blocking the ability of fish to refuge during drought.

Droughts do not just affect the obvious surface water occupied by fishes. There is a growing appreciation in stream ecology that the hyporheic zone is an important part of the ecosystem, as habitat for invertebrates and as the location of important chemical processes (Bretschko and Klemens, 1986; Stanford and Ward, 1993; Valett et al., 1993; White, 1993). Stanley and Boulton (1995) showed in Sycamore Creek, Arizona, that the shallow hyporheic zone remained well oxygenated early in a drought, but that deep hyporheic and hyporheic zones lateral to actual stream channel became very hypoxic. At this time, the hyporheic became a nitrate sink rather than a source. As this desert stream is typically nitrogen limited (Grimm, 1988), there could be substantial consequences to the algivorous fish *Agosia chrysogaster* that is common in the stream. Stanley and Boulton (1995) suggested that during prolonged drought, the reducing, nitrogen sink conditions would likely spread through the creek.

Drought has potentially severe effects on stream fish by killing fish directly and, additionally, by destruction of (1) invertebrates (Griswold et al., 1982), (2) algae, and (3) other normal components (e.g., habitat or export of materials) in a stream system (Cuffney and Wallace, 1989). Boulton and Stanley (1995) also found that the hyporheic invertebrate fauna was gradually altered by the disturbance effect of prolonged drying. In that hyporheic invertebrates can emerge to be foods for fishes, changes in this source of prey could be important to a stream fish assemblage that might reinvade after rewetting of the system. Drought can also have pervasive effects on stream ecosystems [e.g., by reducing downstream transport of particulate organic matter (Cuffney and Wallace, 1989)], which can potentially interact through food webs to affect fish assemblages.

Ladle and Bass (1981) documented that in a small intermittent stream, water temperatures approximated those of the air after stream flow ceased in late summer. There were also marked changes in the invertebrate fauna as dry conditions prevailed (Ladle and Bass, 1981), with some taxa like amphipods nearly eliminated, and others (mayflies) increasing. Smock et al. (1994) found that the benthic macroinvertebrate community of a small coastal stream was "initially devastated" by drought and that streambed drying caused "loss of all active aquatic individuals." Miller and Golladay (1996) showed that summer drought and dewatering of riffles in Oklahoma streams truncated reestablishment of taxa after spring spates, and that periodic drying altered the composition of the benthic invertebrate assemblage. Such changes would have major effects on foods available to fishes during drought.

Attached algae, on which many fish and invertebrates depend, can produce drought-resistant propagules, but prolonged drought and hot temperatures appear to result in effectively killing most of the periphyton. Hence, even rewatering does not immediately provide all of the components of a functioning stream ecosystem. In contrast, a flood, no matter how physically powerful, leaves aquatic components "wet" and destructive of components of the ecosystem come more from actual grinding and washout. When the flood peak subsides, the stream is in many ways left intact [although components like long filamentous algae are obviously reset, and must regrow—often in weeks (Power and Stewart, 1987)]. Drying of the streambed usually does not act alone but is most common during summer, thus accompanied by high temperatures and or oxygen depletion, which interacts with the loss of habitat to be severe for stream fishes. Specific effects of temperature and oxygen stresses are presented in a later section of this chapter. There can, however, be drought—prolonged lack of rain and low stream flow in winter, which may have consequences for stream fish by reduction of habitat and crowding.

Known Changes in Fish Assemblages During and After Drought

James (1934) chronicled the effects of a disastrous drought in 1934 throughout the Midwest and West of North America. Although based on general and anecdotal reports from the game and fish departments of the states, this report aptly summarized the widespread destruction of fish during the drought. James (1934) noted that drought may harm fish more from "obscure and indirect" losses than from absolute drying of streambeds. Thirteen of 17 states reporting described serious loss of fish from drought. Many of the losses were from outright drying of streams or lakes, but reporting agencies also attributed losses to indirect causes, including oxygen depletion, excessive concentrations of fish, decay of vegetation and "stagnation," and loss of natural foods. Two states (James, 1934) explicitly blamed at least some of the fish loss to crowding, causing fish to become "prey to predators" (Arkansas) and "forage fish are being . . . forced to deeper water where they are subjected to other predatory fish" (Indiana). It is obvious that more than half a century ago, fishery biologists were sensitive to the need to consider an array of indirect effects of drought on fish assemblages.

Wickliff (1945) reviewed "some effects of drouths and floods on stream fish" in Ohio. Because this publication is not easy to obtain, I quote liberally. Wickliff noted that drought of summer and early fall resulted in part from evapotranspiration of plants, producing dry riffles, low or dry pools, and high water temperatures. He also showed that at such low conditions, there was little or no cover for fishes near shore. Wickliff described the early loss of riffle habitat, but that, subsequently, typical riffle species "survived for months in cool, deep or shaded pools." He included a photograph of a dry pool "strewn with the skeletons of young suckers, minnows and darters." [A similar situation was found in Colorado River (Texas) drainage by E. Marsh-Matthews and Selma Glascock (personal communication) who made daily counts of dead fishes at the margin of shrinking pools.] Wickliff found no mass downstream movement of young fishes during low flow, but found that fish remained in place in the stream and "each major disconnected pool seems to retain more or less its population of fish, plus the riffle species," which can be crowded yet "survive for weeks or months if excessive decomposition does not develop, or if the pool does not dry up, or if predators do not eat the fish." Wickliff (1945) summarized that "drouth period causes an abnormal concentration of live fish in a very limited water area" and that "species in headwaters are able to survive for limited periods of time under the most unfavorable living conditions," which describes very well my observations in Brier Creek, Oklahoma, and elsewhere. His final conclusion was very simple: "The first requirement for the production of stream fish and other aquatic life is an adequate supply of suitable water for twelve months of each year" (Wickliff, 1945).

Starrett (1950a) showed that differences in the tolerance of low oxygen and crowding coincided with differential survival of several midwestern (USA) minnow species during drought conditions. Starrett (1951) suggested that low-water periods reduced space for fish and success of spawning in the Des Moines River, Iowa. Paloumpis (1958) considered streams to be unstable environments for fish because of water-level fluctuations, with intermittent streams comprising the extreme. He indicated that flood and drought were common events in Squaw Creek, Iowa, but from 1953 to 1956, the stream suffered from extreme drought conditions. Paloumpis (1958) found pools during drought to be important "havens" for fish and that some species (e.g., *Cyprinella lutrensis*) survived crowding in stream pools better than did others, like *Notropis atherinoides*. This finding parallels the results of Matthews and Maness (1979) that the latter was less tolerant of heat stress than the former. Drought was thus confirmed by Paloumpis' field studies as a selective force, favoring some species over others. He concluded that fish in this stream depended entirely on the limited habitats that remained during drought, but that "these stream havens were not always safe and the fish in them were subjected to the dangers of concentration, predation, and suffocation." Paloumpis (1958) also reported that during the winter in drought years, oxygen was depleted in the isolated pools and fish suffocated or died in pools when ice froze to the bottom. Paloumpis was obviously impressed that, "Most of the species have been successful in maintaining themselves regardless of the drastic change which occur in habitat. The fish population changes seem to be rather small compared to observed habitat changes." These early studies seem to provide excellent empirical evidence of the strong impact of harsh physical conditions on fishes and the obvious sorting of regional species pools into local assemblages by stress factors in areas like the North American Midwest.

Hubbs and Hettler (1958) found noteworthy changes in stream fish distributions after a severe drought in the 1950s in Texas (USA). The reduction of stream volume in the Guadalupe River from 1951 to 1954 apparently resulted in loss of *Percina sciera* from a long reach of that stream. Drought conditions also caused the replacement of a spring-head fish species (*Gambusia geiseri*) near the large headsprings of the South Concho River by typical central Texas warmwater species, including *Notropis amabilis* and *Cyprinella venusta* (Hubbs and Hettler, 1958). However, the gambusia later became and remains to this date highly abundant in the spring-fed reach of river, after drought conditions ended

(E. Marsh-Matthews, personal communication). Fish distributions in other streams have been influenced by drought. For example, Closs and Lake (1996) showed that low-flow periods, combined with high summer temperatures, mediated longitudinal distribution of an introduced salmonid and a native galaxiid in an Australian river.

Larimore et al. (1959) showed rapid reestablishment of fish and invertebrates once a small stream was rewatered. Drought that prevailed throughout much of the southwestern United States (1953–1954) (Hubbs and Hettler, 1958) also affected streams in Illinois. Larimore et al. found both fish and invertebrates exposed to "desiccation, stagnation and predation." Effects in isolated pools were described as worst during decay of autumnal leaf inputs, in winter under ice, and as a result of extreme temperature fluctuations. However, Larimore et al. reported that most species withstood drought in refugia and rapidly colonized rewetted stream reaches when flow resumed. Within 2 weeks of resumed flow, they found 21 of 29 species back in usual habitats, and that 25 species had returned to previous habitats by the end of the first summer after the drought. Larimore et al. also noted the ability of fish to move upstream across very shallow riffles. Adult fish reentering the stream appeared to reproduce rapidly, with the populations subsequently dominated by young-of-year individuals. Larimore et al. (1959) surmised that rapid recovery of both fish and invertebrates in this small stream was due to a combination of "versitility of stream organisms, adaptations and movements" and their "life cycles." They also noted the importance of the kinds of pool substrate during drought (i.e., that clay-bottomed pools remained as deep areas, but gravel pools drained rapidly and dried up).

Griswold et al. (1982) demonstrated the value of deep pools as refugia for fishes during droughts and but showed that recolonization of dewatered streams by immigrants from river mainstreams is an important part of the reestablishment of fish populations after drought. Fish assemblages in the dewatered lower Little Auglaize River (Ohio) took up to a year to return to normal richness via arrival of colonists from the main Auglaize River. Farther upstream in the Little Auglaize River, where some deep pools remained during the drought, fish assemblages remained depressed in richness even after a year, because a low-head dam blocked upstream migration of fishes from the main Auglaize River. This case history amplifies the important long-term negative effects that drought combined with anthropogenic barriers to migration can have on local fish assemblages in streams.

Drought or low-flow periods can have substantial effects on stream invertebrates, hence on fishes. Cowx et al. (1984) also showed long-term effects of drought on invertebrates in a stream in Wales, with changes in invertebrate community structure in the year after a drought, and the loss of a year-class of salmon because of high temperatures during the drought. Boulton et al. (1992) showed that desert-stream invertebrates were more negatively affected by drying than by flooding and suggested that shrinking of stream pools intensified biotic interactions that resulted in declines of the invertebrates. Meyerhoff and Lind

(1987) showed that the invertebrate assemblages in pools of McKittrick Creek, in arid western Texas (USA), isolated by lack of flow diverged markedly from one another, but that assemblages in pools connected by flow remained quite similar. Closs and Lake (1994) demonstrated for an intermittent Australian stream that nonflowing periods acted as a "filter," determining the composition of the invertebrate assemblages.

Paloumpis (1957) described the devastation of local fish assemblages in ponds during drought conditions and the subsequent interaction of winter freezing on ponds with lowered water level during the drought—killing many fishes as they were trapped in ice or sections of ponds froze to the bottom. Deacon (1961) summarized the effects of drought on fish populations of the Marais des Cygnes and Neosho rivers, Kansas (USA), noting the reduction in some species during the drought and expansion of populations in subsequent years. Deacon (1961) suggested the suppression of year-classes of channel catfish in Kansas rivers for 5 consecutive years during the drought in early 1950s, and that the now threatened Neosho madtom (*Noturus placidus*) declined in numbers during drought years and only recovered to previous levels after 3 years of resumed flow. Deacon also found that two species of darters (*Etheostoma flabellare* and *Etheostoma spectabile*) were either decimated by drought in some stream segments or forced from small tributaries into the mainstem river. In contrast, Deacon found that spotted bass (*Micropterus punctulatus*) apparently were favored by and increased in numbers during low-flow years. Overall, Deacon concluded that most fishes that were common in the two Kansas rivers in years before the 5-year drought were again found by the last year of his study, 3 years after the end of the drought—suggesting that even severe drought may have modest long-term effects on local fish assemblages in parts of the world where the fish species are able to cope with stress through a combination of tolerance and migrations to refuge habitats.

Canton et al. (1984) reviewed the effects of late-summer reduction in stream flow on fish, emphasizing their crowding in shallow, hot, stagnant pools. Canton et al., (1984) found sharp reductions in catostomid and salmonid populations, and their complete absence from some stream reaches in a low-flow year in an upland Colorado (USA) stream. Titus and Mosegaard (1992) showed that drought interfered with spawning migrations of brown trout (*Salmo trutta*) and caused mortality of fry in small coastal Baltic stream. Rutledge et al. (1990) showed changes in population genetics of small-stream fish after summer drought.

Bayley and Osborne (1993) viewed widespread stream desiccation in Illinois in a 1988 drought as an experiment in "natural rehabilitation" of small streams in a river basin. They found no significant differences before and after the drought in species richness or biomass, either in desiccated or perennial stream reaches. Basinwide, they reported a 17% lower biomass relative to either before or after the drought. However, desiccated areas had recovered within a year, with the level of recovery related to the distance of a given site from permanent water that retained a full complement of species as potential colonists. This finding of rapid recolonization of an Illinois stream was in agreement with that of Larimore et al. (1959) for the same region.

Fish of desert streams may be markedly adapted for survival in hot, desiccating habitats. Minckley and Barber (1971) described rather remarkable behaviors by longfin dace (*Agosia chrysogaster*) in streams of the Sonoran Desert in Arizona (USA). During drought conditions, when evapotranspiration removed virtually all surface water from the streambed in Sycamore Creek, they found *Agosia* alive under damp algae mats, surviving the day and then moving around during the night in the few millimeters of water available at night and early morning. Temperatures under the alga mats were measured at 23.0–25.4°C in full sun when the temperature of nearby dry sand surface reached 50°C! They also noted the high vagility of this species, permitting rapid recolonization of individuals many kilometers from permanent water and rapidly reproducing to repopulate rewetted areas, and the ability of individuals to concentrate in available shaded areas, thus avoiding much warmer stream channels. Obviously, some species are quite tolerant of drought, so long as some water is available.

Tropical fishes may also be well adapted to resist drought. Roberts (1972) quotes Beebe (1945) as reporting "an astonishing variety of fishes from a small, all but dried up mud-hole" in Venezuela. A total of 34 species was recovered from "malodorous mud and decayed vegetation covered by damp slime (but no free water) in what was left of a drying pool" that had been "almost unswimmable" slime for weeks. Roberts (1972) did, however, cite old reports of the massive mortality of fish in dry years in the Amazon or Congo basins when even large rivers were reduced to highly stagnant pools and curious reports that "immense quantities of fish" were killed by drought and heat from related forest fires.

Fewer studies have addressed effects of drought on lake or pond fishes. However, Schwartz (1988) reported changes in fish assemblages in seven freshwater ponds of the North Carolina outer banks before and after three consecutive drought years. Changes were moderate, with two out of the original five total species having been replaced over the drought interval. Marshall (1988) showed that drought years with low river inflow to Lake Kariba, Africa, resulted in less availability of nutrients to the lake and subsequent reduction in zooplanktivorous sardines (*Limnothrissa miodon*).

Short stream reaches [e.g., headwaters of Brier Creek, Oklahoma (Matthews, 1987a)] can rapidly approach predrought fish assemblage structure, if there are nearby refuges (deep pools, etc.) from which colonists can return. Huge numbers of fish can crowd into refuge pools, and the timing of the next rain becomes critical. On one occasion, I marked more than 900 individual fish in two small (~ 30 m and 50 m in length) remaining pools in a 2-km reach of upper Brier Creek that was otherwise dry. If rains had come in time, many of these fish might have recolonized the creek. In this case, drought continued, and virtually all of the fish that I marked died as the pools dried up.

Phases of Effects of a Drought

To summarize effects of droughts on stream fishes, envision the stages that a local fish assemblage in a typical medium-sized stream might go through during two scenarios: (1) a brief, harsh drought of 3 months with no rain and unusually high heat and (2) an extended drought exceeding a decade (with rainfall, but sufficiently little to be considered a severe drought like the one apparent in southwestern United States in the 1500s. In the first scenario, assume that before the 3-month period, there had been typical base flow in the stream. The stream has consisted of pools interspersed by riffles and has a fauna divisible between pool- and riffle-dwelling subsets. Riffle fishes like darters (Percidae), madtom catfishes (Ictaluridae), and sculpins (Cottidae) have had an abundant supply of food consisting mostly of insect larvae associated with stony substrates. Pool fishes like minnows (Cyprinidae), topminnows (Cyprinodontidae) and sunfishes (Centrarchidae) have also had an abundant food source in drift of invertebrates from riffle habitats and from terrestrial insects falling onto the water. Riffle fishes have enjoyed minimal threat of predation, because most piscivores occupy pools. Within pools there is adequate habitat volume and complexity for prey (minnows, topminnows, small sunfish) to spatially avoid piscivores by occupying shallow water (Harvey, 1987) or by emigrating from pools containing predators by crossing riffles (Power et al., 1985; Power, 1987). Water remains relatively clear, seasonally normal in temperature, and well oxygenated.

In the first stages of the drought (Phase 1), flow drops below the base level but surface flow continues across riffles (Fig. 7.2). Water levels in pools drop, but fish can still cross riffles and also gain access to shallow areas in pools (where large piscivores cannot go). However, the decreased depth in riffles and pools results in increased exposure to avian predators (Power, 1987) for fish in shallow or edge microhabitats, and decreased aeration of water inflowing from riffles lessens the capacity of the pools to be oxygen saturated. Lessened current speed in pools promotes the growth of longer filaments of green algae like *Spirogyra* and *Rhizoclonium*, and dense columns or floating mats appear. These may supersaturate pools with oxygen during the afternoon, but their respiration during hours of darkness creates partial anoxia in early morning hours. The decreased volume of water and inflow from riffles results in less input of drift into pools, so pool fishes begin to suffer density-dependent competition for foods that are available. Summer heating increases, so that metabolic demands increase as feeding opportunities decrease.

At the transition from Phase 1 of drought to Phase 2, surface flow ceases across the most shallow of riffles, beginning to isolate pools physically (Fig. 7.2). During this phase, subsurface flow through gravel bars continues to move water downstream between pools, but fish can no longer move from pool to pool. Some riffle fishes have migrated into pools; others were trapped in small pools within riffles, which are now drying and those individuals die. As Phase 2

350 / *Patterns in Freshwater Fish Ecology*

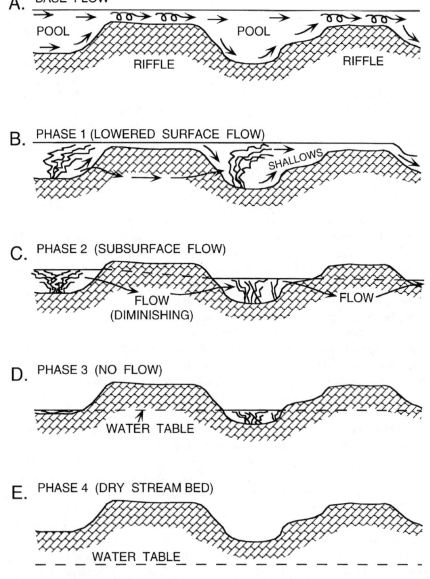

Figure 7.2. Hypothetical stages of drought, as defined in text, for a typical small gravel-bedded stream with riffle–pool structure.

increases in duration, surface flow ceases across all riffles, although there is still some movement of water through gravel bars, with biological filtration of water by virtue of bacteria in the gravel beds. Virtually all fish are now sealed into the pools in which they must remain until the drought is broken, and predation intensifies as water levels drop, pools shrink, and small fishes lose access to shallows that existed in pools edges. There is now no detectable current in the pools, and algae mats and fill much of the water column. These algal structures provide some shelter for small fishes and some foraging opportunities, but the increased algae load in the now-stagnant pool results in periods of worsening anoxia during darkness. Oxygen levels become critical for more vulnerable fish species or life stages and direct mortality from hypoxia begins. Other individuals, weakened by low oxygen and high daytime temperatures, as well as a lack of food sources, become diseased or are more vulnerable to predators because of generally poor condition. Of the small fishes, those best adapted for aquatic surface respiration (Lewis, 1970; Jobling, 1994) are in the best physical condition (such as *Gambusia* and topminnows of the genus *Fundulus*). Predators that are tolerant of heat and low oxygen (e.g., some sunfish (*Lepomis*) species or large-mouth bass adults (*Micropterus salmoides*)] do quite well in the pool, with abundant feeding opportunities. However, juvenile sunfish and bass themselves become prey to adults.

In Phase 3 of the drought, the water table drops so low that there is no longer any water moving through riffles, and pools themselves are left perched in clay or rock pans above the water table. Fish are now fully trapped in shrinking pools, with no input of water or removal of nutrients. Each pool system behaves like a drying pond, with survival of a fish dependent on avoiding the intense predation, finding at least minimal food to match metabolic needs and avoiding outright death from high temperature and low oxygen. The pools are now strongly stratified vertically, with as much as 5°C difference from upper to lower waters during the heat of the day, so fish seeking cooler water are further crowded within the water column. However, oxygenation now depends almost wholly on photosynthesis; hence, oxygen is most available high in the water column in association with floating algae. The dense algae is becoming scenescent [common in stream algae in a matter of 2–3 weeks, Power et al., 1985; Gelwick and Matthews, 1992)] and dying, contributing an oxygen-demanding, decaying mass of detritus on the pool bottom. As the pool shrinks, fishes become increasingly easy targets for wading or diving birds, and watersnakes (e.g., *Nerodia*) and mammals (e.g., raccoons) begin to take a daily toll, particularly on weakened individuals. At this stage of the drought, some individuals would survive if rains came, but only the most tolerant of individuals and of species now remain [see Matthews (1987a) for an actual account of midwestern U.S. species that survived longest in such a scenario].

The last stage of drought (Phase 4) includes complete desiccation of all stream pools in the reach. All fish die. All water-dependent invertebrates die, although some, like burrowing crayfish, may be able to move deep into the hyporheic and find enough moisture to survive for a long period of time. Most invertebrates lacking drought-resistant stages will not reappear until recolonization from flying adults or upstream movements from the nearest permanent water. Algal reproductive stages desiccate and eventually become nonviable if exposed too long to drying or heating. The stream reach, from the perspective of fishes, is dead. Phase 4 persists until rewatering of the stream reach at some point in the future.

Now contrast this extreme event scenario with the likely response of stream fishes during a prolonged period of lower than average rainfall—a drought by statistical assessment, but likely with more than one acute episode like the one described above. The question of impact on fish assemblages, or distribution within a basin, becomes one of refugia and the frequency in time of the severe episodes of drying. Additionally, total lessening of habitat space (assuming years of lower than average rainfall) is worst in headwater segments of the drainage. Former headwaters may be dry for years, and the drainage network shrinks overall. The headwaters are now further downstream than formerly, with small-volume streams flowing through physically larger spaces (the larger channels occupied by large streams during wetter times). As a result, the new headwaters consist more of isolated, large pools, and a more substantial rain is required to cause flow across riffles. Further downstream, there is a smaller deep mainstem than in wetter periods, and large-bodied main-channel fishes (e.g., large ictalurids, ictiobine catostomids, sturgeons, gars, etc.) have less available habitat. Over time, the fish fauna of the drainage adjusts to favor small-stream species, which now are located many kilometers downstream of their original locations.

Physicochemical conditions are harsher overall, with less flow to oxygenate water, less flushing of sediments from stream pools, more "fines" filling interstitial spaces in riffles and in cobble-bottomed pools, and general homogenization of the formerly well-defined riffle and pool structure (that was maintained by occasional high-flow events). Lacking flood events (there is not only less rainfall, but with soils in the basin very dry, it would require a large rain to cause flooding), there is also less chance for flushing of non-native fishes from the system, and physicochemically tolerant, aggressive species (like red shiners, *Cyprinella lutrensis*) increase in abundance to the detriment of native fishes. After 20 years of drought, the original fauna, with a mixture of tolerant and nontolerant species and with a wide array of body sizes and life histories, has been largely replaced by species that are tolerant of stressful physical conditions and able to reproduce even under adverse conditions. Biodiversity decreases, and the local assemblages in the drainage have fewer species and more dominance of a few particularly hardy taxa.

In contrast, even very frequent flood events would probably not have this much effect on the fish fauna of a basin in 20 years. To sum it all up, "fish need water" (and stream fish need flowing water, at that). However, in the final analysis, it may be best to consider drought effects on any stream or lake fish fauna in the context of frequency of "disturbance" as suggested early in this chapter. In some regions, drought (or lengthy dry periods) is a frequent event (such as southwestern United States and Mexico, or tropical dry seasons), and the fish that do exist in those areas must already be selected for drought resistance [like the fauna of the stream described by Paloumpis (1958)]. The worst effects of drought could be expected in places where drying of flow is unusual (i.e., occurring in a region that normally has perennial stream flow where drought would be a much less frequent disturbance). There, the local assemblages have had little selection to be heat or drought tolerant, and sensitive species may experience catastrophic losses.

7.5 Physicochemical Stress

Tolerance of Individual Species for Physicochemical Stress in the Environment

Biotic factors (morphology, competition, predator–prey interactions, and resource partitioning) and habitat complexity influence distributions of animals in benign environments, but in harsh environments, the ability of organisms to tolerate or avoid physicochemical stress may be more critical to their success. At the extremes, there are two trenchantly different kinds of aquatic habitats (cf. "harsh–benign" hypothesis of Peckarsky (1983): (1) "benign" habitats where neither physicochemical extremes nor rates of change cause stress for fishes and (2) "harsh" habitats, where absolute extremes, diel fluctuations, or unpredictable schedules of change cause potentially stressful or lethal physical or chemical conditions. Some habitats are typically benign, but with occasional years when unusual drought, flood, heat, or other variables impact fish assemblages.

Primary physicochemical stressors for fish include thermal extremes, oxygen minima, sharp salinity gradients, extreme acidity or alkalinity, and, commonly, their interactions. In many freshwater environments, it is the combination of high (or low) temperature with reduced oxygen that can be lethal to fishes or that can impair reproduction (e.g., Coutant and Benson, 1990). Castleberry and Cech (1986) showed that a combination of stress from increasing temperature and low oxygen enabled a metabolically more efficient invading fish species (*Gila orcutta*) to displace the native *Gila bicolor mohavensis* in an environmentally fluctuating Mojave River in California (USA). Stressors can influence the use of space by fish both zoogeographically and locally, with some physicochemical challenges acting on fish at subcontinental scales (e.g., effects of aridity) and others at the scale of meters or centimeters (e.g., thermal gradients where differing water sources meet, cold seeps, etc.). Physicochemical stressors can be a continual feature of some habitats, or linked to ephemeral disturbances in others. In some habitats, it is the fluctuation of temperature (or oxygen or other variable) that may cause stress in fish; thus, there have been attempts to adequately estimate thermal fluctuation in habitats from a limited number of observations (Stoneman and Jones, 1996). The kinds of physicochemical stresses typical in harsh streams are not considered here in physiological detail [which is well described in physiological ecology texts, like Schmidt-Nielsen (1975) and Jobling (1994)] but from a view of the potential responses of fishes to these stressors—hence their influence on local assemblage composition.

Effects of High Temperature

There is a voluminous literature spanning more than half a century (Fry et al., 1942; Fry, 1947; Hart, 1947; Hutchison, 1975) on the biological effects of temperature on fishes or its effects on distribution of fishes in streams and lakes. Eaton et al. (1995) used a very large, continentwide database to relate "upper extreme habitat temperature" (UEHT) to distributions of 12 fish species, concluding that high temperatures broadly limited distributions of lentic and lotic fishes. One of the most pervasive zoogeographic patterns, worldwide, is that of "coldwater" versus "warm-water" fish species, with a sharp sorting of species along altitudinal gradients in montane regions (e.g., Rahel and Hubert, 1991; Edds, 1993). Salmonid distribution in North America (G. Power, 1990) and worldwide is strongly linked to temperature. Clearly, there is a strong evolutionary and phylogenetic component to such broad differentiation of fish taxa. Here, I focus not on such deep evolutionary patterns but on responses (mostly of warm-water species) to temperatures so high or so low that either absolute values or fluctuations are prohibitive to some species in extant (daily) time. This apparently plays a strong role in sorting broad subcontinental fish faunas into regional faunas (Matthews, 1987a).

There also can be strong local thermal sorting of fish species. At 4 P.M. on 6 July 1996, during 3 days of record heat in the region, I observed minnows active in the littoral zone of Lake Texoma, at 34°C. Many of the species in Lake Texoma [e.g., striped bass (Coutant, 1985)] either actively avoid or cannot tolerate such temperatures. The result is an obvious littoral to limnetic gradient in fish-species distributions in the lake (that would be more sharply defined than during less extreme thermal conditions). The sorting of fishes along regional or local thermal gradients can be direct, relative to survival, or indirect (e.g., if competitive or predator–prey interactions change between coexisting species at different temperatures) (Baltz et al., 1982; Reeves et al., 1987; De Staso and Rahel, 1994). De Staso and Rahel (1994) make a convincing argument that differential outcomes of competition at high versus low temperatures could influence distributions of cutthroat trout (*Onchrhynchus clarki*) and brook trout (*Salveninus fontinalis*) in streams of western North America.

Thermal refugia may be important, allowing thermally sensitive fish to exist in some systems where they might otherwise be unable to survive. For streams in the Midwest and in California (USA), Peterson and Rabeni (1996) and K. R. Matthews and Berg (1997) showed that springs or spring seeps could be critical thermal refugia, allowing trout to survive or providing thermally stable habitats for warm-water stream fishes.

Temperature is pervasive in its effects on aquatic organisms (Hutchison, 1975). Small fish have internal body temperatures that approximate external water temperatures, thus profound changes in physiology accompany environmental temperature changes (Crawshaw, 1979), and death occurs quickly outside tolerance limits. Direct death of freshwater fishes from thermal stress has been documented in natural habitats (Bailey, 1955; Tramer, 1978). During severe heat and drought in 1980, we found heat death of *Etheostoma spectabile* in Brier Creek, Oklahoma, at temperatures matching their tolerance limits in the laboratory (Matthews et al., 1982c). Physiological effects of high or low temperature on fishes are well known (Hutchison, 1976; Schmidt-Nielsen, 1975; Jobling, 1994). Acclimation and seasonal acclimitization to thermal changes are important in the ecology of fishes (Hutchison, 1976) although caution is needed in extrapolating laboratory studies to field habitats (Kleckner and Sidell, 1985). Larimore and Duever (1968) showed that temperature was an important factor influencing swimming ability of juvenile smallmouth bass (*Micropterus dolomieu*) and their resistance to washout in floods. Locally, thermal gradients can also influence the activities of fishes. Persson (1986) showed that increasing temperatures increased the search rate and the capture rate and shortened prey handling times for roach (*Rutilus rutilus*) and perch (*Perca fluviatilis*) in laboratory tests, and that the vertical distribution of these species in lakes matched the optima of about 17–19°C found in the laboratory. Nevermann and Wurtsbaugh (1994) showed that juvenile *Cottus extensus* migrated vertically on a diel basis to occupy temperatures optimizing digestion rates and growth.

Many authors have measured the limits of tolerance of fish for high temperature (e.g., Black, 1953) using a wide variety of methods (Hutchison, 1976). Eggs or larvae are typically less tolerant of high temperatures than are adults (Hubbs and Bryan, 1973). At extreme high temperatures, proteins denature, limiting all fishes to temperatures below about 44°C. Record high temperature for thermal tolerance of any fish species was reported for the sheepshead minnow (*Cyprinodon variegatus*), with a critical thermal maximum (CTM) (Hutchison, 1961, 1976) of 45.1°C for individuals acclimated at 37–42°C (Bennett and Beitinger, 1996). Very few fish can tolerate exposure above 40°C, and for many North American freshwater fishes, the CTM is from about 32°C to 38°C (Coutant and Talmadge, 1977; Matthews, 1987a; Smale and Rabeni, 1995a).

One potentially important factor that needs more investigation for freshwater fish is the presence and/or role of "heat shock proteins" (hsp) (Marx, 1983; Fader et al., 1994; Coleman et al., 1995). Heat shock proteins, first discovered in fruit flies more than 30 years ago (Ritossa, 1962, 1964), are now known to occur in virtually all animals (Marx, 1983; Fader et al., 1994). Animals exposed to high temperatures typically turn off most "normal" protein synthesis and produce a different suite of proteins (hsps) that appear to protect existing proteins and membranes from heat stress (Coleman et al., 1995) and increase the resistance or recovery of cells (hence, whole organisms) to heat stress (Marx, 1983; Fader et al., 1994). They may play a role in the phenomenon of "heat hardening" reported for fishes and other ectotherms in the 1970s (Hutchison and Maness, 1979; Maness and Hutchison, 1980), whereby fish once exposed to near-lethal temperature subsequently exhibited short-term increases in thermal tolerance.

Although many animals produce hsps in response to elevated heat (Sato et al., 1990) or other stressors (Jobling, 1994), it now appears that fish among numerous families have normal background levels of hsp, year-round. Fader et al. (1994) found that *Pimephales promelas, Salmo trutta, Amieurus natalis,* and *Ambloplites rupestris* in natural habitats all had a strong seasonal effect in Hsp 70 heat shock protein concentrations, with a marked increase in their production in the spring.

Thus, for fish it would appear that some levels of heat shock proteins protection is available, normally, year around. The ecological and evolutionary context of heat shock proteins is just now beginning to be appreciated (Coleman et al., 1995). It seems logical that stream fishes, which can be subject to dramatic temperature fluctuations even in winter, would benefit from maintaining at least a moderate level of heat-resistant proteins year-around. For example, I have observed a 10 C increase in water temperature in Brier Creek, Oklahoma, on a sunny and unusually warm day in mid-winter, which approximates the magnitude of thermal flux on the hottest summer days in that system. Keith Gido (personal communication) recorded a decrease in stream temperature of 10 C when a summer hailstorm cooled a southwestern USA river. Continual existence of some level of hsp in fish in fluctuating environments would be similar to the finding that some desert ants "presynthesize" hsps at relatively low temperatures so that they are available as a buffer against extreme heat shock when they leave the nest to forage [Gehring and Wehner, 1995, original not seen, cited in Coleman et al. (1995)].

Coleman et al. (1995) suggest that production of hsps is costly to organisms and that if natural selection has actually optimized the cost/benefit ratio, hsps should be found most in species or developmental stages with the most likelihood for exposure to high temperatures. Clearly, this could equate to a pattern among freshwater fishes for more or less hsps. White et al. (1994) found among six closely related *Poeciliopsis* species a pattern in occurrence of two kinds of hsps (Hsp30 and Hsp70) consistent with the exposure of these fish to heat in natural habitats. Comparative surveys of heat shock proteins among related species of fish in different thermal environments is needed. A possibility would be to survey for hsps the species that have been compared for thermal tolerance across a gradient from thermally harsh to benign habitats (e.g., Matthews, 1987a; Smale and Rabeni 1995a). Species successful in harsh, prairie streams and having high thermal tolerance, like *Cyprinella lutrensis* or *Notropis girardi*, might be predicted to produce more hsp or to produce "anticipatory" hsp at lower temperatures than would species that are limited to thermally benign upland streams. Such a mechanism could correlate directly with field and laboratory observations of different tolerance levels within families (Matthews, 1987a). Coleman et al. (1995) suggest that research should focus not only on immediate survival but also on the role of heat shock proteins in the long-term fitness of organisms exposed to high heat events.

Events of Cold Temperature

Fish may die during the winter from cold torpor, or from lack of oxygen in ice-covered stream pools or lakes, or long periods without feeding, but these effects have rarely been quantified. Lyons (1997) showed that in Wisconsin (USA) streams, winter starvation was not likely to limit smallmouth bass (*Micropterus dolomieu*) populations, although there was more than 30% starvation mortality in one study stream, and starvation worsened with the length of winter cold.

Low-temperature extremes have a less distinct threshold than does hyperthermia for causing the death of fish. Many freshwater fish species become torpid near 0°C, so that detecting death is difficult. I have observed minnows to lose equilibrium and appear "dead" as experimental temperatures approached zero, then to revive with no apparent harm after warming.

Many warm-water fishes of lower latitudes are limited by absolute winter temperatures. Some clupeids of warm waters (e.g., threadfin shad, *Dorosoma cepedianum*) die at lake temperatures below about 6°C. However, there are many habitats in temperate parts of the world where water temperatures do not go below 6°C in the winter, and for fish where winter cold is not a stress factor, the lowered metabolic demands due to cooler temperatures might result in benefits (e.g., less need to feed, thereby less exposure to predators). As a result, there is a distinct sorting of fishes from south to north in North America into latitudinal groups, but this level of sorting rarely has much effect on the local distribution of fishes within those bands of latitude. Exotic warm-climate fishes may be limited in their northward invasions by lack of tolerance to cold (Shafland and Pestrak, 1982). Winterkill is a major problem for fishes in many lakes in cold environments, but although low temperatures may cause stress, the death under ice is most often related to low oxygen concentrations. Deacon (1961) and Paloumpis (1957) suggested that crowding of fish (by drought) combined with winter cold could harm fish populations.

Oxygen

Dissolved oxygen is critical to fish survival, although many species can tolerate brief periods of anoxia. Most fish avoid hypoxic waters, but some are known to enter potentially lethal low-oxygen conditions in order to forage if prey are lacking in oxygenated water (Rahel and Nutzman, 1994), and direct breathing of air is common in many tropical fishes (Kramer, 1983). Fishes that are physiologically or behaviorally capable of functioning in low-oxygen environments might enjoy some benefits from reduced predator threat or interspecific competition. For example, central mudminnows (*Umbra lima*) that function adequately in hypoxic environments appear to largely escape from piscivorous fishes by occupying lakes with low oxygen (e.g., Tonn and Magnuson, 1982). Gammon and Reidy (1981) found that during an unusual event of hypoxia in a long (35-mile) river reach, many fish died, but individuals of some species were concentrated at the mouths of oxygenated tributary creeks, which were apparently important refugia.

Winterkill from low dissolved oxygen in lakes is well known to fish biologists (Cooper and Washburn, 1946), and oxygen is long been understood to be a directive factor in habitat selection in the water column of lakes (Pearse, 1920). More recent studies have shown hypoxia to limit the use of specific habitats by some northern fishes (Suthers and Gee, 1986; Tonn and Magnuson, 1982). However, in the past, anoxia or low-oxygen conditions were considered less a pervasive agent than temperature in influencing distribution of stream and lake fishes. However, in Great Plains streams, oxygen concentrations become low enough to stress or kill fish, and oxygen concentrations in a local stream reach can vary widely among microhabitats and with time of day (see Fig. 4.3, Chapter 4).

For lakes, Nurnberg (1995a) formulated an "anoxic factor" (AF) relating duration of anoxia per area of sediment to whole-lake surface area. Nurnberg (1995b) showed that the anoxia factor, considered at various times of year, could explain as much as 75% of the variation in fish species richness among 52 Ontario (Canada) lakes. Likewise, Matthews (1987a) showed that hypoxia within habitats, and tolerance of species of minnows for low oxygen, has significant impact on distribution of fish species in streams. Smale and Rabeni (1995a, 1995b) showed, in Missouri upland and upland border streams, that oxygen tolerance influenced fish distribution much more than did temperature tolerance. Coble (1982) found that the number of fish species and several indices of sport fish abundance were higher at sites throughout the Wisconsin River where the average summer-dissolved oxygen exceeded 5 ppm. Castleberry and Cech (1992) showed that for five California freshwater fishes, tolerance for hypoxia was more important than thermal tolerance with respect to survival in their habitats, which had been culturally altered to produce increased temperatures and lowered oxygen levels. Castleberry and Cech also note that dissolved oxygen was lowest near the bottom of Upper Klamath Lake, hence most likely to impact bottom-dwelling catostomids. Hence, there is a strong suggestion from field and laboratory studies that oxygen conditions and adaptations of fishes for hypoxic episodes can have a strong sorting effect on composition of natural local fish assemblages.

Low oxygen often acts in conjunction with thermal stress. Stefan et al. (1995) found that a model combining oxygen and temperature conditions performed well in predicting fish presence across more than 3000 lakes in Minnesota (USA). Interestingly, Kalikhman et al. (1992) found a strong correspondence of fish (and zooplankton) in Lake Kenneret, Israel, to sharp gradients of oxygen and temperature (not absolute values), suggesting that patterns in distribution of these two physicochemical variables can be more intricate than a mere matching of "fish" with "absolute" values of "best conditions."

In the past, low oxygen seemed more often to have been considered a result of man-made pollution than a natural challenge for fishes, and half a century of sewage control and so forth has been based in North America on maintaining minimum oxygen conditions for fishes. In general, values below 4 ppm have been considered "harmful," "unacceptable," and so forth. Such values are no doubt necessary for adequate protection of all life stages of many fish species. However, fishes native to harsh environments often tolerate much lower oxygen values for substantial periods of time.

Many examples of physical or physiological adaptation for low-oxygen conditions are well known. Physiological adaptations in oxygen affinity and blood oxygen loading enable Sacramento blackfish (*Orthodon microlepidotus*) to occupy very-low-oxygen environments (Cech et al., 1979). Jobling (1994) provides a detailed chapter on physiological and morphological adaptation of fishes to low oxygen and direct use of atmospheric oxygen. Many freshwater fishes have accessory organs (lungs; intricate swim bladders; oral, opercular, or alimentary diverticula) to permit direct breathing of atmospheric oxygen, at least as a supplement to gill respiration. Other fishes, lacking special respiratory structures, have general morphology, allowing respiration at the aquatic surface ["aquatic surface respiration" (ASR) (Jobling, 1994)] [e.g., killifishes (Lewis, 1970) and guppies (Jobling, 1994)]. The immediate behavioral responses [e.g., feeding, growth, and mortality (Weber and Kramer, 1983; Kramer and Braun, 1983)] and evolutionary implications (Kramer, 1983) associated with air-breathing in fishes have been described in detail. Kramer (1983) provided an overview of costs and benefits of air-breathing, which included the risk of predation balanced against the limitations placed on fish by oxygen demand in hypoxic environments, suggesting the advantages of bimodality in air-breathing and aquatic respiration.

Chronic low oxygen of some habitats either favors air-breathing fishes or precludes those lacking this capability. However, oxygen deficiency may occur in some habitats where hypoxia is not usual. Headwaters of small streams may become anoxic or reach levels below which most fish cannot survive, with early-morning oxygen values in isolated pools at 1–2 ppm (Matthews, 1987a). In such habitats, there are species that lack accessory organs for air-breathing but are able to survive for some periods of time (e.g., perhaps until midday, photosynthesis increases oxygen values to "normal" levels). In my laboratory, oxygen-tolerance tests with species of minnows from the Great Plains have shown not only a marked variation in tolerance of low oxygen among individuals but also that many individuals can survive for many hours, sometimes more than a day, at oxygen levels about 1 ppm. This capability could be a distinct advantage in many small midwestern streams, and the species of these streams are overall much more tolerant of low oxygen than are congeneric species from perennial upland streams of the Interior Highland (Matthews, 1987a). Oxygen relative to winterkill also provides a strong sorting of small-lake fish assemblages in northern North America (Tonn and Magnuson, 1982; Rahel, 1984). Small lakes that became anoxic in winter and that lacked accessibility of a refuge stream for fishes formed a distinct subset within the suites of lakes in Wisconsin studied by Tonn and Magnuson (1982) and Rahel (1984).

Finally, evidence is emerging that some warm, subtropical river systems can have extensive areas with very low oxygen. Recent studies of the Achafayala River basin in Louisiana (USA) by faculty and students at Louisiana State University (ASIH abstracts, 1996) show that extensive reaches of this river basin (hundreds of km^2) consist of flooded backwaters and swamps throughout which oxygen values are chronically low. Brunet and Sabo (1996) showed strong evidence that this low oxygen limits reproduction by altering egg size, hence potentially eliminating many fish species from this part of the river basin. Cashner et al. (1994), in related work, found a paucity of sunfish and bass (*Lepomis*, *Micropterus*) in a Louisiana bayou where low oxygen values were often found. Perhaps the "abnormal egg at low oxygen" hypothesis is more widespread than is generally known, and potentially limiting to fishes in low-oxygen environments. Thus, we have evidence from at least three extremely different kinds of systems that low oxygen can play a strong role in sorting regional fishes into local assemblages. This phenomenon needs more study as a major factor in fish distribution.

pH and Acidification, Natural "Stained" Water

In some Great Plains streams, the pH fluctuates widely during diel or seasonal cycles, with photosynthesis raising the pH to 9.5–10.0 or higher (Oklahoma State Dept. of Health 1977, USGS 1978). In poorly buffered streams in areas like much of the tropics or the southern United States, acid-stained streams with high concentrations of humic acids may have a pH as low as 4.0 under natural conditions. However, in much of the world, anthropogenic acidification of streams and lakes is a major challenge for fishes.

Fish are relatively poor regulators of internal pH; thus, changes in the pH of their environment can alter enzyme activities or electrolyte composition of body fluids, producing severe stress (Packer and Dunson, 1970; Eddy, 1976). Low pH interferes with oxygen uptake, and pH outside a range of ~ 4.0–10.0 can kill fish (Doudoroff and Katz, 1950; Fromm, 1980). The well-known Bohr shift decrease in capacity of blood to hold oxygen at a lower pH in vertebrates is exacerbated as the "Root effect" in fishes (Schmidt-Nielsen, 1975) such that at a low pH some fishes are unable to saturate blood at the gills with oxygen, regardless of how high the oxygen concentration of water may be. Low pH also changes acid–base regulation at the gills, mucous secretion, and gill structure [reviewed in McDonald (1983)]. Fishes also change behaviors at low pH, reducing activity or feeding (Jones et al., 1985a and b, 1987). However, some freshwater fishes are unusually well adapted, via altered morphology and physiology of the gills, to live in alkaline waters as high as a normal pH of 10.0 (Laurent et al., 1995). In some environments occupied by fishes, such as aquatic weed beds, the pH can be markedly elevated in the afternoon hours by the effects of photosynthesis. However, Sefafy and Harrell (1993) showed that fish do not avoid a pH even as high as 10.0 if there is concomitant supersaturation of oxygen (as can occur in weed beds).

One of the most widespread environmental problems in the Northern Hemisphere in this century is acid rain and its effect on aquatic systems (Haines, 1981). As a result of acidification of systems, fish have suffered a wide range of lethal and sublethal effects of low pH as reviewed by Fromm (1980) and Haines (1981). Rahel and Magnuson (1983) showed for 138 Wisconsin lakes ranging from pH 4.0 to 9.2 that values below 6.2 limited many minnows and darters, but emphasized that lakes can be variously affected by acidification because of their biogeographic history and the composition of the fish assemblage. Rahel (1986) showed for 100 lakes in northern Wisconsin that low alkalinity and concomitant low pH limited the occurrence of many minnow and darter species. Local extinctions of species related to increased mortality of eggs after acidification have been shown (Hulsman et al., 1983). Kelso and Lipsit (1988) showed that the young-of-year fishes in northern lakes were sensitive to changes in pH and suggested that changes in the young-of-year assemblages associated with pH changes could be used to predict changes in the whole-lake fish assemblages. Townsend et al. (1983) showed for streams in England that locations with a low pH had fewer individuals, species, or evenness in relative abundances for fishes and invertebrates, and Muniz et al. (1984) suggested that even relatively minor improvements in pH (e.g., 0.2 units) could result in marked enhancements of lake fish populations in Norway.

Intraspecific Variation

Some investigators have shown differences in tolerance or preference (considered separately) among populations or subspecies of a fish species (Huets, 1947; Hubbs and Armstrong, 1962; Hubbs, 1964; Otto, 1973; Hall et al., 1978; Matthews and Styron, 1981; Guest, 1985), and two articles demonstrate apparently inherited tolerance within different strains of a species (Robinson et al., 1976; Swarts et al., 1978). Other intraspecific studies have produced contradictory results. Hart (1952), McCauley (1958), Hutchison (1961), and Spellerberg (1973) found intraspecific differences in thermal tolerance of some species of fish or amphibians, but not of others. Others found little difference in temperature selection among populations despite physicochemical differences in their environments, suggesting that thermal responses are genetically fixed in some species (Reynolds and Casterlin, 1979; Beitinger and Fitzpatrick, 1979; Winkler, 1979). However, life stage can be critical to thermal tolerance: Elliot et al. (1994) and Elliot and Elliott (1995) found no difference in thermal limits for feeding or survival among adults from three populations of bullhead (*Cottus gobio*) and stone loach (*Noemacheilus barbatulus*) in England, but they did find significant differences between adults and juveniles. The literature is thus contradictory and provides no clear indication of the prevalence of differences in tolerance/selectivity strategies among populations, nor of its importance in speciation.

Feminella and Matthews (1984) and Matthews (1986d) showed conflicting results in studies of intraspecific variation in thermal tolerance of a darter species (*Etheostoma spectabile*) and a minnow species (*Cyprinella lutrensis*), respectively. The darter species exhibited dramatic differences in thermal tolerance among four populations that were within 100 km of each other, but occupied sharply differing thermal environments (ranging from a constant 17–19°C to very fluctuating thermal conditions). The minnow species, in contrast, showed no significant difference in thermal tolerance across a 1200-km north–south gradient of river basins from north Kansas to south Texas (USA). However, in the harshest environments, there was a suggestion of more heterogeneity in thermal tolerance within populations (i.e., means did not differ, but there was wider variance in CTMs among individuals at harsh stream locations).

Therefore, even using the same techniques in the same laboratory, we find one example of a taxon apparently well adapted for local variation in stress, and one in which tolerance was invariant (but high) throughout a large region. It would be possible to argue that virtually all populations of red shiners across the Kansas–Texas gradient are exposed at least at times to very hot stream temperatures, but United States Geological Survey (USGS) water temperature records show that streams in the southern part of the study area are indeed hotter (as much as 39°C or greater in some mainstreams!).

Finally, intraspecific variation in tolerance/selectivity could result from the summation of "between-phenotype" differences among individuals of a population (Roughgarden, 1972). However, our impression from our past studies of various fish taxa is that fish populations are not generally separable into differentially adapted subunits with regard to either tolerance or selectivity. In thermal tolerance tests we have found very low variation among individuals [standard deviation ~ 0.2–0.4°C (Matthews and Maness, 1979) at many sites, but see Matthews (1986d)], whereas in pH and oxygen tolerance tests, distributions, although wider, are unimodal. In our tests of fish populations in temperature, oxygen, pH, and salinity gradients in laboratory chambers, we have found biomodality of selection to be very rare (Matthews and Hill, 1979a); instead, most populations exhibit sharp, unimodal selection peaks. The distribution of a fish population in our gradient chambers has typically resulted from the movement of most individuals within a similar section of the total gradient, not from the separation of the population into subgroups that each select and remain in different parts of the gradient. The between-phenotype component of fish tolerance or selectivity is small. One final note of interest (see Chapter 9 on competition) is evidence that competitive ability between closely related species could switch, depending on temperature or other environmental factors (Reeves et al., 1987).

Sublethal or Indirect Effects of Stressors

Much of the discussion above has implied that a stress factor like temperature or oxygen influences the composition of local fish assemblages by directly killing some life stage—egg, larva, juvenile, or adult. However, sublethal effects of low oxygen, high temperature, or combinations of stressors may weaken or otherwise contribute to mortality without reaching lethal limits directly. Further, within a given lake or stream, the distribution of a stress factor may strongly influence the habitat used by a fish species as it avoids extremes.

As an example, consider landlocked and introduced populations of striped bass, *Morone saxatilis,* that exist in more than 100 reservoirs of the southern United States. Massive die-offs of adult striped bass have been reported from a substantial number of these reservoirs (Matthews, 1985b), and Coutant (1985) has formed a detailed temperature–oxygen "squeeze" hypothesis to account for habitat use and/or mortality of this species in some reservoirs. Coutant (1985) suggests that because many reservoirs are strongly stratified in mid to late summer, with anoxic hypolimnion and very warm epilimnion, the striped bass are restricted to a relatively narrow zone just above the thermocline or to isolated inflowing refugia (if they exist). Matthews et al. (1985) showed in Lake Texoma (Oklahoma–Texas, USA) that striped bass and virtually all open-water fish species were restricted by late summer to a few vertical meters of the water column, just above the epilimnion–hypolimnion boundary at a depth of about 13 m (Fig. 7.3). At this time in Lake Texoma, the hypolimnion was completely anoxic, and surface waters were at or near 30°C (Matthews et al., 1985), far in excess of the usual tolerance of striped bass (Coutant, 1985). The net effect of the temperature–oxygen squeeze described by Coutant (1985) and confirmed by Matthews et al. (1985) in a different system is to crowd large numbers of fish into a very limited portion of the total available aquatic habitat. As a result, feeding opportunities become limited or nonexistent, potential for disease transmission is increased, and if the fish remain too long in temperature–oxygen refugia, they apparently are weakened in general. If the temperature–oxygen condition becomes too severe, massive die-offs, particularly for large fish, occur in some of the reservoirs of the United States (Matthews, 1985b).

This example is extreme and involves a widely introduced species existing outside its natural habitat. However, the message is the same in many systems. It may be a mistake to look for limitation of fishes across environmental gradients only as a result of direct lethal limits of physicochemical variables. Physicochemical stressors may function most often through indirect pathways—weakening fish, hence harming the ability to forage, commit energy to gametes or nest

364 / Patterns in Freshwater Fish Ecology

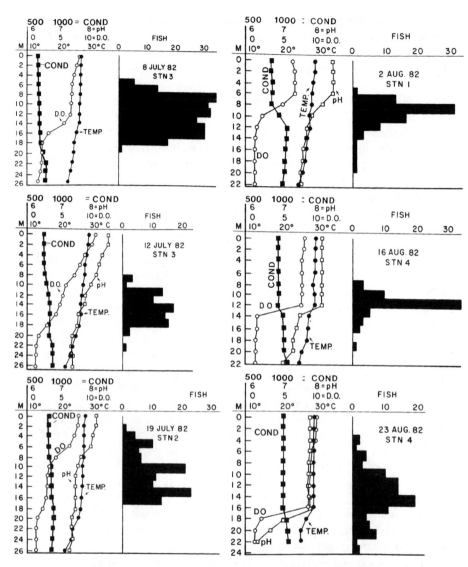

Figure 7.3. Vertical physicochemical profiles and depth distribution of fish in Lake Texoma, Oklahoma–Texas, in summer 1984. [From Matthews et al. (1985).]

construction, or maintain vigilance against predators; causing potential prey (like invertebrates) to be lacking; or generally interfering with the life processes vital to the production of the next generation of fish. In such a situation, a species will not exist in the local assemblage, except perhaps by emigration from other habitats.

"Harsh" Conditions—Detriment or Benefit to Individual Species?

As suggested above, physicochemical conditions that would typically be considered harsh (e.g., high temperature, low oxygen, excessive salinity, exaggerated pH) for freshwater fishes may, in fact, act as a filter (Smith and Powell, 1971; Tonn, 1990) that sorts the fishes in a stream or lake system into distinctly distributed assemblages. In a stream reach with harsh conditions in headwaters (e.g., Brier Creek, above), only a limited subset of the regionally available species will be able to survive upstream. In many streams, the upstream limit to distribution of fishes is temperature or oxygen (e.g., if the stream lacks spring-fed headwaters). Matthews and Styron (1981) provide an example of only the most tolerant species inhabiting extreme headwaters of Roanoke River, Virginia, tributaries, where intermittent pools become very hot in summer. Hubbs (1995) provided a clear account of species limited in the opposite direction—that fishes adapted to relatively constant temperature environments in large Texas (USA) springs are unable to extend very far downstream, where thermal fluctuations are greater.

Physicochemical factors other than oxygen and temperature can stratify fishes longitudinally in a watershed. One example is that of a creek near Sulfur, Oklahoma, in which strong concentrations of hydrogen sulfide emerging at a springhead apparently set limits to the upstream distribution of fishes, with even the very hardy *Gambusia affinis* dropping out well downstream from the springhead. Salinity is well known to structure the distribution of fishes between freshwater streams and estuaries. Cashner et al. (1994) and many other authors have reported strong shifts in composition of fish assemblages in streams entering brackish water whenever wind, high tides, and so forth drive salt wedges further upstream than normal.

An Example of Salinity Structuring a "Freshwater" River Assemblage

The upper Red River and its tributaries in Oklahoma and Texas (USA) are unusual in that the headwaters have salt concentrations approximating that of seawater. The source is natural: Springs arising through deep Permian salt deposits carry high concentrations of sodium chloride, and further downstream, an influx of streams from gypsum areas add sulfates to the stream. The salt load lessens further downstream as a result of dilution from "fresh" tributaries. In the headwaters of the main Red River and its tributaries, where salinity concentrations are highest, only two species of fish are found: Red River pupfish (*Cyprinodon rubrofluviatilis*) and plains killifish (*Fundulus zebrinus*). Further downstream, additional species are added, until more than 70 species have been described from Lake Texoma (a mainstream impoundment of Red River) and its tributary creeks. Echelle et al. (1972a) provide a very comprehensive account of fish distribution throughout the upper Red River system, with a detailed analysis of fish assemblages that are controlled by the salinity gradient. Echelle et al. identified suites of species that comprised distinct assemblages associated with various ranges of salinities. This is an important example of salinity limiting species distributions in a major way in a river system not influenced by marine habitats. However, the most important contribution of Echelle et al. (1972a) is in the suggestion that the fish species that are very abundant in but limited to the saline headwaters (Red River pupfish, plains killifish) are able to maintain a high level of success not only because they are tolerant of salinity, temperature, and other stressors, but because these harsh conditions preclude invasion of other, potentially more competitive, fish species from further downstream.

For example, the red shiner, *Cyprinella lutrensis,* is very abundant in most of the middle Red River below the high-salinity reaches. Where red shiners have invaded into other systems (e.g., Arizona, Nevada, Colorado) where they were not native, they have had devastating impacts on native species (Douglas et al., 1994). It is logical that if red shiners were able to penetrate upstream into the high-salinity reaches, they, along with other Red River cyprinids that are excluded by the salt, could have serious consequences for the pupfish or killifish. Anecdotally, Larry Cofer (Regional Biologist for Oklahoma Department of Wildlife Conservation, personal communication) has observed red shiners, Red River pupfish, and other fish species in aquaria. Cofer indicates that with all species present, red shiners are highly aggressive and the pupfish and others species act as submissives. However, with red shiners removed, the pupfish then become aggressive and dominant.

There is good evidence, therefore, that in the Red River system it is elevated salinity levels that (1) maintain large-scale zonation of fish species in the river and (2) that this salt-driven zonation is a likely factor in the existence of unique fish assemblages in the saline reaches. The U.S. Army Corps of Engineers has proposed a huge project designed to sharply reduce the salinity in the Red River, to bring its chloride concentration to levels acceptable for agricultural and municipal uses. (The plan involves a variety of engineering devices to divert or stop highly saline waters at source springs, capture the salt in evaporation lakes or similar devices, or reinject salt water in deep wells.) There is a strong suggestion that the Red River Chloride Control Project, if carried to fruition, will have a devastating effect on the native fish assemblages of the upper river, where they are now maintained by the natural salinity gradients, and it is, at present, unclear if the project will be completed.

Physicochemical Tolerances and Selectivity of Stream Fishes as Related to Ranges and Local Distributions

Echelle et al. (1972a and b), Kushlan (1976), Harrell (1978), Moyle and Li (1979), Matthews and Styron (1981), Matthews (1987a), and Smale and Rabeni (1995a, 1995b) demonstrated the influence of physicochemical factors on geographic distributions or on local habitat use by stream fishes. Smale and Rabeni (1995a) found that distributions of Missouri fish species coincided with oxygen, but not temperature tolerance. Matthews and Hill (1979a, 1979b, 1980) demonstrated the influence of physicochemical factors on habitat use by stream fish in a harsh, unpredictable environment, and Matthews and Maness (1979) showed a direct relationship between thermal tolerance (based on CTMs) or oxygen tolerance and the relative success of four minnow species in that stream system. Because of their common link as ectotherms, studies of reptiles, amphibians, and fishes are all of interest in understanding fish distributions.

Relationships exist between latitudinal distributions of some ectotherms and their physiological tolerances (Moore, 1949; Snyder and Weathers, 1975). Hart (1952), Brett (1956), Ferguson (1958), Hagen (1964), John (1964), Lowe et al. (1967), Cross and Cavin (1971), McFarlane et al. (1976), Gee et al. (1978), and Kowalski et al. (1978) have showed relationships between fish tolerance or preferences in laboratory tests and conditions in their typical habitats or their ability to live in harsh environments. However, Brown and Feldmeth (1971) found thermal tolerances of closely related pupfish unrelated to thermal differences in their environments [although Hirshfield et al. (1980) subsequently reported differences among some of those populations], and Hutchison (1961), Licht et al. (1966a,b), and Dawson (1967) reported interspecific thermal tolerance/preference differences among ectotherms that correspond more to phylogenetic relationships than to physicochemical characteristics of their habitats.

The research in Matthews (1987a) was designed to test hypotheses about the importance of absolute tolerance versus microhabitat selection as ways fish could survive in harsh environments. "Selectivity" was measured as the sharpness of the distribution of a species in artificial environmental gradients of oxygen and temperature, and tolerance was estimated in tests of critical thermal maximum (CTM) and length of survival in hypoxic conditions.

Matthews (1987a) argued that tolerance sets ultimate limits on space available to ectotherms. Fish are often active near their limits of tolerance (Reynolds and Casterlin, 1979; Matthews and Zimmerman, 1990), so minor differences in their tolerances might be a strong selective advantage. However, selectivity of a species for "good" physicochemical microhabitats could influence their use of space, hence the assemblages of species in particular habitat patches. It is well known that fish select certain parts of artificial temperature or oxygen gradients (e.g., Shelford and Allee, 1913; Doudoroff, 1938). Most fish avoid harmful temperatures when offered a choice in laboratory gradients (e.g., Stauffer et al., 1984), and laboratory studies may relate directly to distribution of fish in the wild (Stauffer et al., 1976). For example, Bulkley et al. (1981) and Black and Bulkley (1985) showed that thermal selectivity could explain avoidance of artificially cold tailwaters below dams in the Colorado River basin. Kellogg and Gift (1983) showed, for three fish species, a final thermal selectivity that was similar to optimal temperatures for growth. I evaluated temperature and oxygen tolerance and selectivity of upland minnow species versus species from prairie streams, habitats that contrast sharply in physicochemical extremes or rates of change (Matthews, 1987a), and found that species from prairie systems were, indeed, much more tolerant of physicochemical stress than were species of upland streams.

Matthews (1987a) demonstrated that minnow species successful in harsh prairie streams of the Great Plains had a greater tolerance for temperature or oxygen stress than did congeners in "benign" environments of upland streams. Although temperature is important with respect to the distribution of freshwater fishes, it may not be the primary determinant of fish distribution in or among harsh streams. In Brier Creek, Oklahoma, the intermittent headwaters are harsh, with daily thermal fluctuation of 10°C or more in midsummer, and oxygen concentrations in isolated pools approaching 1 ppm in early morning hours. Further downstream in perennially flowing reaches, temperature fluctuates less (1–2°C, daily) and oxygen is usually adequate for fish. There was no significant relationship between upstream limits of distribution of fishes in Brier Creek and thermal tolerance of the species (Matthews, 1987a). Apparently, Brier Creek is so hot (Matthews et al., 1982c) that all resident species have high heat tolerance. However, there was a clear relationship among oxygen tolerance, the upstream limits of distribution of common species in Brier Creek, and their ability to successfully colonize the harsh headwaters in a year after desiccation and rewatering (Matthews, 1987a).

Hypothetical Tolerance and Selectivity Strategies

Levins (1968, 1969) suggested that (1) animals exposed to strong environmental variation within generations may exhibit a broad tolerance to diverse conditions through physiological flexibility and (2) animals with strong habitat selectivity may show decreased limits of tolerance. Matthews (1987a) hypothesized that fish in harsh environments might (1) have wide limits of tolerance but relatively nonselective within physicochemical gradients or (2) have no enhanced physiological tolerance but have acutely selectivity for advantageous physicochemical habitat patches (Fig. 7.4). Alternatively, fish in benign environments, lacking an evolutionary history of exposure to stressful conditions, might have neither wide limits of tolerance nor acute physicochemical selectivity (Fig. 7.4).

Matthews (1987a) reported a direct relationship between thermal tolerance and distribution of minnow species across the harsh–benign geographic gradient at the prairie–upland boundary. However, the prairie minnow species showed no greater thermal selectivity than did upland congeners. Prairie minnow species also had a much greater tolerance of hypoxia than did species from uplands, where oxygen depletion is rarely if ever a problem. In contrast to the lack of increased thermal selectivity, the prairie species showed more acute avoidance of low-oxygen conditions in laboratory gradients than did the upland species, suggesting that minnow species successful in the harsh prairie environment could combine tolerance with acute selectivity of microhabitat to withstand periods of low oxygen (Matthews, 1987a).

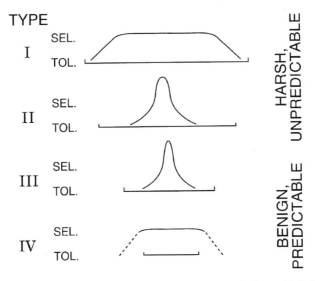

Figure 7.4. Hypothetical "types" of strategies of fish for dealing with habitat use in harsh versus benign environments.

Matthews (1987a) also tested ideas about tolerance versus selectivity to the phenomenon of longitudinal zonation of fishes in small streams. Typically, intermittent headwaters are environmentally harsh and fluctuating relative to large tributaries and mainstreams (Starrett, 1950a; Neel, 1951; Metcalf, 1959; Whiteside and McNatt, 1972). Headwaters are characterized by a limited number of species which have been presumed highly tolerant of harsh conditions (Thompson and Hunt, 1930; Burton and Odum, 1945; Starrett, 1950a), but there had been few tests comparing headwater to mainstem species. Matthews and Styron (1981) found significantly greater temperature, oxygen, and pH tolerance of a Roanoke River (Virginia) headwaters minnow species than of three closely related mainstream minnows. They also showed that within one common species (fantail darter, *Etheostoma flabellare*) individuals from intermittent headwaters were more tolerant of low oxygen than were individuals of the mainstream. Matthews (unpublished data) also compared oxygen tolerances of three minnow species that occupied a spring run originating at a deoxygenated head pool. One species, which was consistently found well upstream into the low-oxygen zone showed 91% survival after 24 h at an oxygen concentration of 0.8 ppm, but two other species that were common downstream but not in the springhead had only 57% and 29% survival in low-oxygen tests.

Matthews (1987a) also compared the longitudinal distribution of all of the common fish species in Brier Creek, Oklahoma, to their tolerance of low oxygen and high temperatures, and to their selectivity in physicochemical gradients. Brier Creek headwaters are extremely harsh, fluctuating 10°C daily in summer, with hypoxia in isolated pools. Further downstream the stream flows continually in most years, and neither daily fluctuations nor extremes are as potentially stressful as they are in the headwaters. (However, in extremely hot years, all of Brier Creek clearly is a harsh system.) All species in Brier Creek had high-temperature tolerances (relative to fish species in more benign, upland streams) (i.e., differences in thermal tolerance offered no explanation for longitudinal differences among species). Temperature and oxygen selectivity in gradient chamber tests showed no relationship to interspecific distributions in this stream. However, species common in and capable of colonizing headwaters after drought all showed enhanced tolerance for oxygen-deficient conditions. In this case, one very simple factor, oxygen tolerance, appeared to coincide with differences in the pattern of species assemblages in this small, environmentally harsh system.

Influence of Phylogeny on Physicochemical Tolerance?

An unanswered question about the success of species in harsh environments is whether they invaded because their phylogenetic history conferred tolerance to extreme physicochemical conditions, or, if species evolved tolerance for harsh conditions by selection after immigrating to those environments (or as the environment became harsh where they already existed). Within a genetically plastic species, gradual adjustment of absolute tolerance or physicochemical selectivity in populations encountering harsh environments could facilitate dispersal of a species into different kinds of habitats. Zimmerman and Richmond (1981) showed genetic modification of minnows in a fluctuating thermal environment less than 40 years after anthropogenic modifications that caused changes in the thermal regime. Some laboratory studies (e.g., Winkler, 1979), as well as the field observation of the abrupt border of the ranges of many fish species (e.g., at the edges of uplands) suggest that the "fixed-tolerance" model may explain extant distributions of some species. The alternate extreme is that within limits of zoogeographic chance and biotic pressures, populations of a species might colonize a wide array of habitats and adapt to local physical and chemical conditions (Huey and Slatkin, 1976). For numerous amphibians and some fish species, this latter explanation seems possible.

To test the hypothesis that high oxygen or temperature tolerance is an old evolutionary event involving whole clades, rather than a recent ecological phenomenon, it might be possible to map physicochemical tolerance of species onto a group cladogram. Block and Finnerty (1994) mapped a physiological trait (endothermy) for one suborder of fishes onto their phylogeny. Unfortunately, there are not sufficient data in the literature for rigorous mapping of tolerances of species onto cladograms for all members of any freshwater clade.

However, two published data sets are suitable to partially compare tolerance to phylogeny for two groups of North American fishes. Smale and Rabeni (1995a, 1995b) measured oxygen and temperature tolerance for 35 and 34 species, respectively, in seven families for fish in Missouri (USA). Distributions of fishes in streams with varying levels of oxygen stress closely matched the oxygen tolerance of species (Smale and Rabeni, 1995b). There was some evidence in Smale and Rabeni's (1995a) data that families differed in tolerances (i.e., consistent with a hypothesis that tolerance was a family trait). I divided the species in Table 2 (oxygen tolerance) and Table 3 (critical thermal maximum) of Smale and Rabeni (1995a) into two groups, representing the more and less "tolerant" half of the species. With respect to thermal tolerance, all species tested by Smale and Rabeni (1995a) in Ictaluridae and in Cyprinodontidae (three species each) were in the "more tolerant" group. Of the Centrarchidae, four of six species were "more tolerant." In contrast, only 1 of 4 darters (Percidae) were "more tolerant," and 7 and 17 minnow species were in the tolerant group. Of the species tested by Smale and Rabeni for oxygen tolerance, five of six centrarchids, all three cyprinodontids, and two of three ictalurids were "more tolerant." Of the 17 minnows, only 6 were in the "more tolerant" group for anoxia. Thus, Centrarchidae, Ictaluridae, and Cyprinodontidae had more high-tolerance members than did Cyprinidae. Obviously, if more species within any one family could be tested, the impression might change, but it appears that general tolerance could be a family-level trait for some groups.

However, examining levels of temperature and oxygen tolerance within the most speciose family (Cyprinidae) of Smale and Rabeni's study suggested little relationship between evolutionary relatedness and physicochemical tolerances. For this test, I combined the cyprinid "shiner" and "chub" clades of Coburn and Cavender (1992) for minnows of eastern and central North America (ignoring the "western minnow clade" that remains poorly resolved and has little bearing on questions for midwestern North America. The minnows tested by Smale and Rabeni (1995a) are from nine genera across the shiner and chub clades (Fig. 7.5), with one (*Notemigonus*) in a very distant Old World minnow clade (Coburn and Cavender, 1992). Six of 11 species (55%) in the shiner clade were "more tolerant," compared to only 1 of 4 (25%) in the chub clade (Fig. 7.5), suggesting a moderately interesting difference. However, within the shiner clade (Fig. 7.5), the trait of "more tolerant" for high temperature was distributed across four of the five genera, with no tolerance pattern reflecting evolution within the clade. Instead, with this limited suite of species, it appeared that "more tolerant" was mixed within some of the genera and appeared to have arisen independently at least three times (even assuming that *Cyprinella* and *Pimephales* might have higher tolerances because of their similar evolutionary origins). One potentially confounding factor in this comparison and in those below is that laboratory tolerances were generally measured for individuals of a species from only one

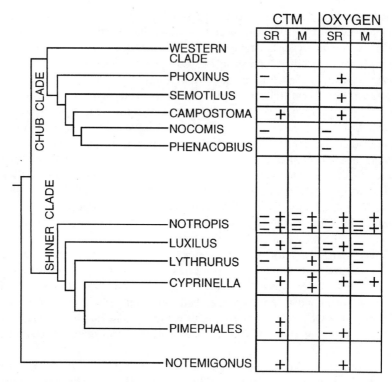

Figure 7.5. Phylogeny of minnows of the midwestern United States [simplified from Coburn and Cavender (1992)] versus critical thermal maximum (CTM) and oxygen tolerance for 11 species. SR = data from Smale and Rabeni (1995b); M = data from Matthews (1987a).

or a few locations; thus, any local variation in tolerance might be lost in the analyses I show.

With regard to oxygen tolerance, there was virtually no pattern between the chub and shiner clades and none within those clades, individually (Fig. 7.5). Tolerance for anoxia existed in seven genera that were widely dispersed throughout the cladogram (Fig. 7.5). This finding was corroborated by data from Matthews (1987a) for minnows in upland versus prairie streams in Oklahoma and Arkansas. Where the Great Plains of North America meet the Interior Highland, there is a rich array of minnow species both in harsh streams of the plains and in physicochemically "benign" streams of the uplands (Ross et al., 1985), with many minnow species having a sharp distributional boundary at the juncture of the upland and the plains (Matthews, 1987a). We can ask to what extent the more tolerant, "prairie stream" minnow represent evolutionary clades using the same cladograms from Coburn and Cavender (1992). I obtained CTM and oxygen-tolerance data for 11 minnow species, divided between upland and prairie species. Matthews (1987a) confirmed that the prairie group had, overall, a greater tolerance for both factors than did the species of upland streams. Since the original work, the genus *Notropis* has been divided so that the species are now in four genera of the shiner clade (Fig. 7.5). As in the results for Smale and Rabeni (1995a), there is little evidence of a phylogenetic influence on temperature or oxygen tolerance of these species. "High" thermal and oxygen tolerance was found for species in two or three of the four genera but not for the *Luxilus* that are embedded within the clade (Fig. 7.5). Within the species remaining in *Notropis,* there are both tolerant and intolerant species. Again, future addition of more species of these clades in a survey of physiological tolerances might change the results, but, for now, I argue that in my data and those of Smale and Rabeni (1995a), there is little evidence of the evolution of physicochemical tolerances at the level of genera or larger clades of cyprinids. The alternative hypothesis that the ability to tolerate harsh conditions arose on numerous occasions within the family seems more plausible. This would support the hypothesis that species have individually attained increased tolerance for harsh conditions as they invaded harsh environments or remained in regions as environments became increasingly harsh (e.g., during transition of the Great Plains from pluvial to xeric conditions).

Finally, there is evidence that in another family common in the midwestern United States (sunfish family, Centrarchidae), there may be tolerances related to position of species in major clades. Figure 7.6 depicts four *Lepomis* and two *Micropterus* whose thermal and oxygen tolerances were tested by Smale and Rabeni. (No *Ambloplites* or *Pomoxis* were tested, but these genera are included for completeness of the abbreviated cladogram, which is modified from Wainwright and Lauder (1992). Details in positions of the four sunfish species (Fig. 7.6) within a complete phylogeny of the genus may require further modification, but the overview for six species tested by Smale and Rabeni can be summarized that the species in the sunfish clade (*Lepomis*) appear to have as a group relatively high tolerance of temperature and oxygen stress. This finding would coincide with the known existence of all of these species in relatively harsh environments (Matthews, 1987a).

The results of all these comparisons of tolerance to phylogeny suggested that tolerance for thermal and oxygen stress may be a recent adaptation by some cyprinids and that some families are more tolerant of harsh conditions than are others. However, there is no reason to suppose that enhanced tolerance for physical stress should be entirely due to either "phylogenetic" or "recent adaptation." Families (or higher taxa) with long histories of success in harsh environments (Chapter 5) seem likely to have had members with the capability to survive environmental crunch periods, and there are clearly global differences in distributions of families (Berra, 1981) that would suggest some families to be more tolerant than others for hot conditions or low oxygen. Additionally, within families, subfamilies, genera, or species, there is no reason to preclude changes

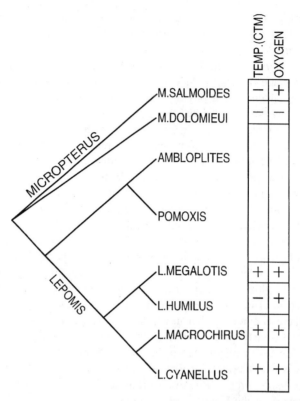

Figure 7.6. Phylogeny of centrarchids of the midwestern United States versus critical thermal maximum and oxygen tolerance. [Data from Rabeni and Smale (1995b).]

in physiological or behavioral traits that enhance fitness in harsh environments. The challenge would seem to be in deducing what part of the tolerance of an array of species is attributable to the history of the group across all levels of increasingly fine taxonomic distinction.

Fish in Microrefugia from Physicochemical Stress: When to Stay and When to Go?

The ability of fish to either remain in place and survive physicochemical stress (high temperature, low oxygen, etc.) or to immigrate to seek a better refuge can raise the evolutionary question of when (i.e., at what point in deterioration of habitat) to leave one habitat patch to seek a better one, in a habitat matrix that is full of potential hazard. I consider this the problem of "when to stay and when to go?" In this problem, particularly in harsh, variable habitats, an individual fish is assumed to have imperfect or no knowledge of conditions in microhabitats at some distance from the one it occupies.

Consider the following scenario, based on actual observations (Matthews, 1977) of fishes in the main-channel and adjacent creek mouth habitats in the South Canadian River of central Oklahoma. The shallow, sand-bottomed river has very harsh and potentially lethal conditions for fishes in the main channel for many days in summer. In the shallow, main-channel water column, I have measured temperature as high as 36°C during August days [and values to 40°C have been found in similar rivers (Matthews and Zimmerman, 1990)]. However, within the general matrix of a very hot exposed river channel, there are microhabitats like shaded pools, creek mouths, or slightly deeper pools adjacent to main flow in the riverbed. These habitats may be a few degrees cooler (Fig. 7.7), but, apparently from crowding of fishes into these refugia and because of decomposition of depositional materials, they also may have lower oxygen values than do main-channel microhabitats. Within a kilometer of river and creek where I studied microhabitat use by minnows (Matthews, 1977), there were numerous (possibly a dozen or more) "safe" microhabitats within the larger matrix of the riverbed.

Therefore, during the course of a hot summer day, an individual fish might be in a particular refuge pool, with temperatures significantly cooler (biologically) than the shallow channel, but with limited availability of oxygen. Particularly in shaded pools and with lack of algae, there may be a minimal production of oxygen, but respiration of the many small fishes crowded into the pools can lead to deterioration of water quality. By early afternoon, even the refuge pools become hot (because of the exchange of water with the main channel) and water quality (oxygen, decreasing pH values) may begin to produce acute stress in fish. The individual fish is faced with the dilemma of (1) remaining in the refuge and surviving until conditions improve or (2) immigrating to try to find a better pool.

The costs of staying in the initial pool are directly related to effects of declining water quality on the physiology of the fish, combined with the worsening of conditions for immigration if the main channel becomes hotter or if the fish undertakes the migration in an increasingly weakened condition. The costs of leaving to seek a new refuge include the certain penalty of having to cross hotter water than is in the refuge, and the possibilities of either (1) death enroute or (2) finding no other refuge better than the initial one. The problems for immigrating (Fig. 7.7) may be summarized to include not only the hot corridors but also the energetic cost of travel, with greater metabolic demands (Q-10) at higher temperatures, the probability of death en route, and the impossibility of knowing anything useful about either the route or the conditions in the other possible refuge pools. It may be that in contrast to the Ideal Free Distribution (which suggests animals should leave a patch when its quality is as low as the average quality elsewhere) that fish in the "deteriorating refuge" game will win, on average, if they remain in the refuge until the water quality deteriorates to some point less than "average," so long as they have a fundamental tolerance for at least brief episodes of poor conditions. The "game" is made more complex not only by individual differences in absolute tolerances of fish species or life stages

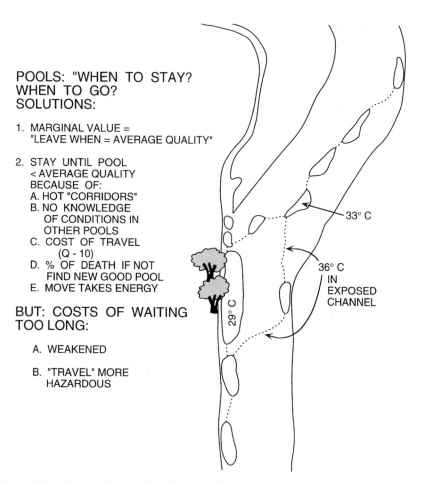

Figure 7.7. Hypothetical problem that must be "solved" by fish faced with degrading water quality in "refugia" embedded within an inhospitable environmental matrix. Habitat depicted is that of the main-channel South Canadian River and a tributary (Pond Creek) in central Oklahoma, during midsummer.

but also by differences in sublethal stress on the ability to swim or function and by possible differences in competitive abilities of fish species under different stress scenarios. However, I suspect that in the crowded refuge pools, competition is not much of an issue and that survival of direct physicochemical stress and predation (if a larger fish is present) is much more a concern for a fish than is any interference with feeding. It seems likely that the autecology of individual species with respect to tolerances of stress or ability to rapidly migrate among refugia would be of greatest importance in this scenario.

7.6 Disturbance and Physicochemical Effects Combined

Integrating Disturbance and Physicochemical Stress into Distribution of Fishes in a River/Lake System

Peterson and Black (1988) provided a conceptual view of the way physical stress can interact with the density of organisms to increase the effects on mortality of organisms. How can disturbance (floods, drought) and physicochemical stress be integrated into the picture of which fish species exist in local assemblages? Assume a hypothetical temperate-zone, large river system with a mainstem several hundred kilometers long, natural backwater lakes, and a couple of artificial mainstem impoundments. Assume that one major tributary of this river originates in low mountains (but not high, cold mountains), and another originates on a large elevated plain or steppe grassland. Place the system with a rainfall gradient increasing in a downstream direction from an average of 30–120 cm per year. Assume very hot summers and cold winters but no montane snowpack. This particular river basin does not exist in any one place, but it captures much of the variation in physicochemical and disturbance factors that are common in stream/lake systems of North America.

Also, assume that the fish fauna of this river basin has already been selected over evolutionary time by global, continental, and major subcontinental factors, and that regional physical factors like area of tributaries, "size" of habitat, and the like have been taken into account. The primary potential stressors in the basin are high temperature, low oxygen, and high salinity in one tributary. By excluding high, cold mountains from the hypothetical river, truly cold-water fishes (e.g., salmonids) are omitted (Rahel and Hubert, 1991). Hypothetically, all of the fish species in the basin are "warm-water" species with the potential to migrate anywhere in the system.

Now, how do disturbance and physicochemical stress sort the regional (basin-wide) fish fauna into subbasin faunas, and eventually into local assemblages of particular species (where biotic factors like competition, predation, and the like can ultimately "fine-tune" the community)? The two factors (disturbance and physicochemical stress) are not easily separated. Of the two major disturbance factors (flood and drought), floods are probably less linked to physiological challenges for fishes. As described early in this chapter, floods exert negative influences more through (1) direct hydraulic stress on adult fishes, restricting them to a limited subset of habitat, (2) flushing out and killing of larvae, (3) destruction of nests, eggs, or interstitial microhabitats by moving and grinding substrates, (4) destruction of substrate-dwelling invertebrates and/or algae that provide food, or (5) removal of physical structure like woody debris that provide shelter and production of food on their surfaces. There are also potential benefits to high-energy floods, in that they may remove scenescent materials (aging algae, detritus) from stream systems (reducing oxygen demand), potentially making more space available to juvenile fishes [an old concept by Starrett (1951) for midwest streams, which has never directly been tested].

Floods are most likely to remove or restrict fishes in headwaters of our hypothetical river, where small to medium-sized tributaries can flood with erosive force as a result of local storm events. Whether rock bound (in canyons) or bound by high earthern banks (as in many midwestern streams), streams in the upper reaches may reach very high, channel-forming velocities before the stream flows over the bank. Fish in reaches that normally flood (canyon areas or arroyos) may be adapted behaviorally to survive floods. Meffe (1984) showed that native fishes in canyon streams moved to and remained near hydraulic safe sites along rock faces, whereas introduced fishes lacked this capacity. Thus, in the upper reaches of the hypothetical river system occasional erosive floods (every few years) may be sufficient to remove or maintain, at a low level, populations of fish species that are not adapted to survive flood effects—either by behaviors of adults during floods or by life-history traits that facilitate renesting after loss of a cohort (e.g., Harvey, 1987).

Downstream in middle and lower parts of the basin, streams are wider, somewhat deeper, but with much less potential for direct removal of fishes by floods. Downstream, where banks are less elevated relative to the width of the stream, (1) it takes much greater, widespread rain to effect a major flood and (2) even a major flood will relatively rapidly spill over the bank, minimizing scouring of the channel and maximizing potential use of the floodplain by adult (Ross and Baker, 1983) or juvenile (Finger and Stewart, 1987) fishes as feeding or nursery habitat. Another factor in flood effects in headwaters versus lower mainstem is likely in the effects on woody debris. In upstream reaches, rapid stage rises may refloat large woody debris that has accumulated in streambeds and deposit it in the terrestrial riparian zone during the falling hydrograph (Brown and Matthews, 1995; A. V. Brown, personal observation). However, lower in the river system, the hydrograph may rise more slowly, but the flood peak may be of greater magnitude (Matthews, 1986c) and the stream may have greater power; thus, floods downstream are likely to remove woody debris from the streambed by flotation and stranding. However, erosion of stream banks bordered by riparian forests can result in major inputs of whole trees to the river channel, which form large snags that may remain in place for years. These large woody snags provide important habitat and foraging opportunities for larger river fishes as well as small-bodied taxa. Bilby and Ward (1989) showed a positive relationship between stream order and size of woody debris in the stream. Downstream in the largest parts of rivers (where channels are wide), few taxa are probably washed out by floods, and some taxa may be favored by large, nonscouring floods that bring wood debris or riparian-zone organic matter or nutrients into the stream system (Junk et al., 1989). Laterally expansive floods provide opportunities for "flood-exploitative" species to produce strong year-classes or enhance production (Ross and Baker, 1983). Hence, there is some sorting of the regional fish fauna by floods, with some unable to persist in upper reaches where scouring floods are common; however, virtually none of the fauna are excluded by floods in downstream reaches.

Drought and physicochemical stresses are inextricably linked. In extended drought and "no-flow" events, fish become restricted to isolated pools where water temperatures become very high and oxygen depletion or reduction is common. Additionally, drought in both temperate and tropical streams likely crowds many fish species into close proximity (Zaret and Rand, 1971) and likely increases the potential for a variety of predator–prey and competitive interspecific interactions (see Chapters 9 and 11). Drought is seasonal in much of the world, more predictably so in some regions (tropics) than others. However, for non-spring-fed small streams in the temperate zone, a cycle of dry–wet season may, in harsh climates, be just as important and potentially stressful, as is better documented in the tropics.

Further downstream in some middle and lower sections of the hypothetical river, drought may have less effect on fish. However, of the two major tributaries, the one that originates in the prairie may experience substantial late-summer drying. This tributary is typically very wide, shallow, and sandy, much like the South Canadian River of central Oklahoma. Hefley (1937) described the actual South Canadian River as "one of the harshest environments on Earth," because of the unstable sand substrates, and the propensity of the river to shrink from bank-full width to a series of barely connected, hot pools in summer. Matthews (1977) documented fish distributions during extreme drought in the South Canadian River in summer 1976 and found that most fish were crowded into very limited amounts of habitat space (shaded pools) avoiding the very hot, exposed main channel almost completely. Thus, dry, hot conditions can limit fish species distributions in the hypothetical system by creating harsh, intermittent pool conditions in headwaters, where both oxygen and temperature are a distinct challenge for fishes, and by making mainstems too hot for many species even relatively far downstream. Overall, there will be a filtering of the regional fauna from almost unlimited use of habitat downstream (at least regarding physical or flow challenges) to a much smaller suite of flood-, heat-, or drought-tolerant species in the upper reaches. In an imaginary north temperate river like the one I have described, it would be highly probable to have 90 or 100 species in lower portions of the river basin, including the main channel and its direct tributaries, yet only a dozen or fewer species able to survive upstream where physicochemical conditions and schedules of disturbance are significant obstacles to survival and successful reproduction of fishes. From this perspective, it is clear that to understand local fish assemblages, at least in "harsh" regions of the world, it is necessary to understand the ways that physical stressors limit distributions of individual fish species within river basins or among isolated refuge habitats.

8

Morphology, Habitat Use, and Life History

> *What the humming-birds are in our avifauna, the darters are among our fresh-water fishes. Minute, agile, beautiful, delighting in the clear, swift waters of rocky streams . . . they do not seem to be* dwarfed *so much as* concentrated *fishes—each carrying in its little body all the activity, spirit, grace, complexity of detail and perfection of finish to be found in a perch or a wall-eyed pike. . . . They have taken refuge from their enemies in the rocky highlands where the free waters play in ceaseless torrents, and there they have wrested from stubborn nature a meagre living. Although diminished in size by their continual struggle with the elements, they have developed an activity and hardihood, a vigor of life and glow of high color. . . .*
>
> —Stephen A. Forbes (1880b)

> *. . . the study of Himalayan fishes is also full of charm. It is for the reader to strive for and enjoy their beauty of structure and the remarkable way in which they perform their life functions.*
>
> —Sunder L. Hora (1952)

8.1 Introduction

Previous chapters have focused on progressively smaller scales in space and time, from global–continental–deep evolutionary phenomena to regional or local, extant phenomena, and have addressed factors like disturbance and stress that filter species for local habitats. This chapter addresses the morphology of individual species and the ways physical abilities of fishes interact with the variation in their habitat (both structural and hydraulic) to place species in particular microhabitats in a stream or lake. This chapter also addresses ways that morphology of fishes and environmental conditions in their habitats influence feeding, patch choice, movement, and migration, as well as reproduction, growth, mortality, and related life-history features. This chapter includes topics that seem superficially quite different, but all address the ways that the unique ecology of a given species (autecology), independent of other species, influences its presence or abundance in a local assemblage.

A central question of this chapter is that of how a single fish species would live, use habitat, or reproduce, it is did so independent of any other species. This is the Gleasonian view of local assemblages—it is the individualistic capabilities and limitations of each species that causes it to be present at a given place and time, with the assemblage in a stream reach or local lake habitat consisting of aggregation of species that all have traits (recent adaptations or features acquired in the history of a clade) appropriate for that habitat. In this view, it is the traits of individual species, or the "fine-tuning" of those traits for local conditions, that influence accumulation of species in local assemblages. It follows that geographic variation in morphological or ecological traits is important, with intraspecific variance being of interest, instead of merely bothersome noise.

In this chapter I consider the following:

1. Constraints of the liquid medium on morphology, based on incompressibility of water, and fluid drag.
2. Morphological relationships (based on descriptive assessment of individual characters, functional morphology, and multivariate analyses of multiple anatomical characters) of fishes to the use of physical habitat and feeding.
3. Life-history traits of fishes as influenced or constrained by local physical environments, by predators, and by phylogenetic or recent adaptive features.
4. Hydrodynamics of microhabitats in a local reach or patch of a stream or lake, and ways those hydrodynamic and structural properties affect habitat or patch use due to (a) energy costs in the habitat, (b) availability of foods, and (c) reproduction and survival of young in the habitat.
5. Migrations or movements among habitat patches, as related to seasonal or daily movements, home range use, or longer "migrations," and the functional basis for these habitat changes.

Chapter 8 addresses autecological phenomena that can influence the ways individual species occur in particular habitats. Each of these topics has a massive literature, which cannot be reviewed here in detail. I am most interested here in the ways that autecological variables influence composition of local assemblages (i.e., how habitats and traits of fish match to produce, at least in part, the fish assemblages in particular habitats).

8.2 Water as a Medium: Fluid Drag

Unlike air, water is incompressible for all practical purposes and has a high viscosity. As a result, a fish moving through water faces the problem that the medium offers substantial resistance to movement and streamlining is requisite for any relatively rapid movements in water. Webb (1975) provides the most comprehensive review of fluid mechanics, hydrodynamics, streamlining, and drag in fishes, and Aleev (1963), Alexander (1967), and Gosline (1971) give good accounts of ways fish deal with drag, swimming, or resistance to currents morphologically and behaviorally. Aleev (1963) is a detailed mathematical treatment of fish shapes and positions of fins relative to hydrodynamics. Vogel (1981) is a thorough review of the physics of fluid dynamics and the ways that organisms in general deal with "life in moving fluids." Weihs (1989) described mathematically the efficiency of transfer of momentum from muscles of a fish to the water mass, and the ways various nodes of swimming and placement of muscles affect hydromechanical efficiency of fishes. Studies of swimming biomechanics also need to take into account the mechanical limitations of the structures in use. For example, Lauder (1982) showed the relationship between the ability of bone to be deformed without breaking and the strain imposed on bone during "fast-start" performance in sunfish (which often feed by burst speed to overtake prey).

From the perspective of fluid dynamics, fish, birds, and airplanes all are similar: those that move rapidly have design constraints for movement through fluid that set limits to variation in shapes. There are, of course, many fish with highly unique shapes, very divergent from any notion of streamlining. However, typically, it is only those fish (or birds or aircraft) that will move very slowly through the fluid medium that succeed in any substantial departure from a streamlined body plan. [This restraint by viscosity and drag of the medium may be a central reason that fish and birds have diverged less from a primary body plan than have, for example, mammals—whose rate of movement through their fluid medium (air) is rarely a constraint on body form, except in bats.] Restraints to movement or holding position of a fish include drag, which tends to push or pull backward on the body of a fish, and the Bernoulli effect, which tends to lift the body of a stationary benthic fish off the substrate. The Bernoulli effect (lift above an airfoil or hydrofoil because of reduced pressure above the structure) occurs in water as well as in air, whenever a hydrofoil effect of any part of a fish results in the more rapid movement of water molecules across the upper surface than the lower surface of the structure (Webb, 1975, p. 26). However, if the structure (e.g., a rayed fin) is appressed onto the substrate (e.g., stream bottom) such that no water can flow under it, there is no Bernoulli lift (because there is no water pressure underneath the structure).

Drag is of two basic types: friction and pressure (Alexander, 1967; Webb, 1975). Friction drag results from shear or viscosity forces produced because water molecules (or polymers; see Chapter 4) in immediate contact with the body of a swimming fish tend to move with the fish, forming a boundary layer, but water molecules at increasing distances from that layer of molecules move less with the body of the fish, out to some distance from the fish at which the water is essentially motionless (Fig. 8.1). This results in division of fluid around the body of a fish into two layers (the boundary-layer concept; Webb, 1975, p. 12): the inner (boundary layer) of which moves with the fish, creating high viscous forces, and the outer flow layer, where viscous forces are negligible.

The resistance between the boundary-layer water adhering to or moving with the body of the fish and slower-moving water molecules farther away creates friction drag. Because this drag depends on molecules adhering to the surface of the fish, an increase in surface area of the fish (relative to its volume) results in increased friction drag. Thus, a very-long-bodied or attenuate fish with a high surface-to-volume ratio would have a relatively high friction drag, as does a fish whose body is laterally flattened (Alexander, 1967, p. 26).

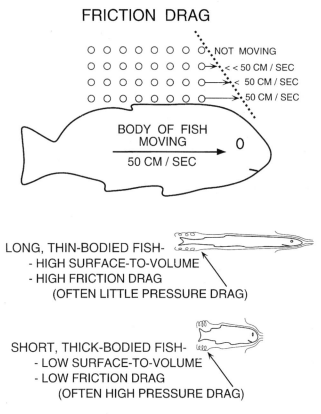

Figure 8.1. Drag forces encountered by a fish moving forward at a speed of 50 cm/s, through motionless water, and hypothetical differences in friction versus pressure drag encountered by fish of extremes of body shape.

The phenomena producing friction drag are the same for a stationary benthic fish holding position against currents as for a fish swimming forward through a nonmoving water column, except that for a stationary fish, it is the molecules adjacent to the fish that do not move, whereas those further from the body of the fish are moving increasingly faster, up to the velocity in the habitat. The microvelocity gradient is of the same magnitude, merely reversed, and the result is friction drag that tends to pull the fish downstream.

Pressure drag results when turbulence establishes eddies along and behind a body in a moving fluid and there is a separation of the laminar boundary-layer flow from the body (Webb, 1975). In ideal laminar fluid flow over a body, there is no turbulence. However, in the real world, there is no totally laminar flow at dimensions the size of a fish, and two phenomena are most important with respect to pressure drag. The further posteriorly that laminar flow over the body is maintained, the further toward the rear separation of the boundary layer from the "fuselage" occurs, hence the further rearward is the onset of eddies and turbulence (Webb, 1975). Changeover from laminar to turbulent flow occurs at a critical Reynolds number (a function of inertial force divided by viscous force), with laminar flow in the boundary layer at low Reynolds numbers (< about 5×10^5), and with turbulent flow at high Reynolds Numbers (> about 5×10^5). See the nomogram by Webb (1975, p. 15) for a convenient estimate of Reynolds number and changeover to turbulence for fish a given size and swimming speed. Also, see Statzner and Holm (1989) for very detailed measurements of drag forces and adaptation of macroinvertebrate shape to flow—much of which may apply to hydrodynamics of small fishes.

The "ideal" streamlined body, with maximum depth approximately one-third of the distance from the leading edge to the tip of tail, results in smooth flow over the body of the fish and defers posteriorly (e.g., to the caudal peduncle) the turbulence that causes pressure drag. Second, propulsive energy is lost when a swimming fish creates an eddy, and efficient use of fins for propulsion is lost within a turbulent wake. As a result, fishes that swim at high speeds typically have a thin caudal peduncle, which minimizes the depth of the turbulent wake band, and a tall caudal fin, which places the propulsive surfaces above and below the turbulent wake and allows those portions of the fin to provide efficient propulsive power (Aleev, 1963).

8.3 Morphology

> *The river fish are bulky in the middle because they move with their head and tail: the lake and tank fish are similar . . . but are characterized by a relatively smaller head; the spring and pool fish . . . are extremely deep behind the head; the fishes of the torrents are traditionally well known by the possession of . . . greatly flattened body on account of their habit of crawling with the chest, and a relatively reduced anterior part of the body.*
>
> —Translation by S. A. Hora (1952) of a Sanskrit passage written between 600 B.C. and A.D. 500.

Body Regions and Relation of Their Morphology to Ecology

Detailed descriptions of correspondence of morphology with ecological capabilities of fishes are in Alexander (1967) and Gosline (1971). Details on structure versus hydrodynamics is in Webb (1975), in which phenomena like cross-section shape of body, elongation, and streamlining are related to hydrodynamic performance of a fish. The body of a fish has three primary regions, discernable mostly from external morphology, that are critical to understanding its ecology: (1) the "body" (hull or fuselage, i.e., lacking fins), (2) the fins, and (3) the head and trophic structures including the pharyngeal jaw apparatus (PJA). The bony skeleton is also critical, as it affects the abilities of fishes to move in different ways or attain flexure of the body. In addition to understanding each of these body regions relative to the influence of individual morphological traits on ecology, it is these body regions that have been used most in multivariate morphological analyses ("ecomorphology"). Thus, understanding of either functional or multivariate morphology of fishes depends on a basic knowledge of descriptive morphology.

Approaches to Morphological Studies of Fish Ecology

Three approaches exist to linking morphology of a fish to its ecology: (1) descriptive morphology, (2) "functional" morphology, and (3) "ecomorphology." In the following sections, I review general ideas from descriptive and functional morphology as they relate to swimming, feeding, and other life activities, then survey studies that have used a multivariate ecomorphological approach.

Descriptive morphology, which certainly is not new (Hora, 1952), infers the role of a morphological feature of a fish or of its general body shape from observations of the fish in field or laboratory, or from quantification of relationships such as mouth size and prey size (e.g., Matthews et al., 1982b). "Functional morphology" differs from "descriptive morphology" in that functional morphology typically includes tests or quantitative measurements of capacities of the structures to perform in particular ways, as typified by Lauder and Wainwright (1992) and Wainwright and Lauder (1992) for oral structures, or by Aleev (1963) or Webb (1975) for body and fin shape relative to hydrodynamics of swimming. Westneat (1995) pointed out that a lack of functional or biomechanical information allows only a correlation between morphological and ecological characteristics of a fish, and that biomechanical analysis "enables one to advance a much stronger argument for causality in terms of morphological and functional transformations affecting changes in ecology." Functional morphology is often referred to as "biomechanics" when muscle–bone levers or strengths are assessed or if muscle–nerve firing is measured (Westneat, 1995), or to "hydromechanics" when measurements are made of the interaction of the structures of a fish with flowing water, with precise measurement of drag forces, Bernoulli effects, and similar problems. Where the tools of descriptive morphology are observation or intuition, the tools of functional morphology include techniques like high-speed cinematography of living fishes performing motor functions; electromyography, providing measurements of action potentials in nerves associated with muscle movements; measurements of pressures associated with suctions, grinding/biting teeth, or jaw closure; kinematic profiles to partition of head–jaw movement into levers with measurements of forces produced by muscle action; or measurement of forces on a fish in flowing water. Functional morphologists often focus on one structure or an isolated group of structures in animals, although Liem (1980) cautioned that a change in morphology of one structural region might not be "adaptive" for the whole animal in the context of its environment if the change under consideration actually made some other portion of the body mechanics less efficient in the environment. Statzner and Holm (1989) pointed out that five different factors simultaneously influenced morphological adaptation of invertebrates to flow, and that it was physically impossible to optimize all of these factors in one body plan. Aerts (1992) reviewed the potential application of biomechanics to many problems in fish biology or ecology. There is also a growing demand for rigorous mapping of functional morphological traits onto phylogenies in order to determine evolutionary origins of functional capabilities, rather than assuming adaptation to the recent environment. Lauder et al. (1995) have recently highlighted the link between functional morphology, comparative physiology, and phylogenetic hypotheses, arguing that congruence of these approaches make an understanding of evolution of function within a clade possible.

"Ecomorphology" takes a holistic approach, summarizing large numbers of morphological traits (usually by multivariate analyses) and linking multiple features of morphology to specific ecological features for a species or assemblage of fishes (e.g., Douglas and Matthews, 1992). Motta et al. (1995) review the many uses of the team "ecomorphology" and the philosophy of the approach, and distinguish ecomorphology from "functional anatomy," in that the former attempts to compare morphology by "actual resource use among individuals, populations, species and higher taxa, or communities." Norton et al. (1995) discussed the role of ecomorphology within the field of comparative biology for fishes. Many studies of "ecomorphology" have assumed a "form–function" link based on descriptive studies, but Norton et al. (1995) note the desirability of performance tests to establish explicit cause-and-effect between morphological characters and known capabilities or actions of the structure, and Felley (1984), Douglas and Matthews (1992), and Wood and Bain (1995) suggest the need to determine if suites of morphological characters used in "ecomorphology" of fishes do, indeed, match their ecological traits.

Generalizations from Descriptive Morphology

Classic "descriptive" morphology of fish is based on perceived correlations between morphological features and use of habitat (e.g., Page, 1983, p. 174) or foods (Keast and Webb, 1966). This approach is largely univariate or may use ratios of direct measurements like body width versus depth (Page, 1983). Many generalizations have come from such studies—for example, sunfish (*Lepomis*), with slab-sided bodies, live in a vertical structure such as stems of aquatic plants, whereas catfish (e.g., *Amieurus*) have flattened bodies and live in holes in mud banks, or under logs and stones. Anatomical studies spanning several centuries have described the fundamentals of fish morphology summarized in any ichthyology text. However, a few references to detailed descriptions may be useful in understanding the anatomy–ecology of fishes. For skulls of fishes, the primary general reference remains Gregory's (1933) "fish skulls," although details on some species have been questioned by more recent authors. Harrington (1955) provides the most detailed account of bones of the cranium of a cyprinid at various levels of dissection and a synonomy of bones of the teleost skull. For girdle elements of fishes as related to their ecology, Lundberg and Marsh (1976) showed that suckers (Catostomidae) had divergent patterns in the evolution of pectoral fin rays, with simplification of anterior rays and shortening and increasing widths ("foreshortening") of fin-ray segments, which increased flexibility, strength, and resistance to buckling in species that use the pectoral fins as hydrofoils, props, or holdfasts on the substrate in swift water. Lundberg and Marsh (1976) also described parallel evolution of position-holding pectoral fin structures in Asian fishes of swift torrents and in the evolution of other major groups. Gosline (1977) illustrated the potential for and limitations to movements of fishes dictated by structure of the pectoral girdle.

An explicit early "descriptive" treatment of individual morphological characters versus ecology of stream fishes was by Carl Hubbs (1941) who related body form to hydrodynamics of the habitat. Hubbs (1941) noted slimmer bodies (within and between species) of fishes in fast-flowing habitats, and features such as reduction of scales or size of eyes in fish of swift, turbid habitats. Hora (1952) summarized much of his earlier work on "organs of attachment" (modification of ventral fins to form a suction disk, depressed body form, rugiosity of ventral surface) of torrent fishes in the Himalayas that permit their existence in rapid mountain streams. More recently, descriptive morphology (body shape) was linked to relative foraging behavior and efficiency of centrarchids in different microhabitats of small lakes (Werner, 1977).

Many other general descriptions relate body shapes or trophic morphology of fishes to their ecology or habitats. Keast and Webb (1966) compared the diverse body forms of fishes in a north temperate lake (see figures in Chapter 1) to habitats of the species, and the morphology of the head and mouth to food use. Keast (1978a) summarized the differences in body form for fishes in northern lakes relative to their swimming capabilities. Lowe-McConnell's (1987) account of diverse body shapes in African and South American tropical fishes suggested that these fishes have morphology appropriate for hydrodynamics of the habitat and for their behavior within microhabitats (e.g., the upturned shape of the arawana). Welcomme (1985, p. 133) depicted adaptations of African fishes for swift streams, including oral suckers, stiffened barbels, stout pectoral spines, and elongate body form. Page (1983) provided a comprehensive review of morphology versus habitats of the North American darters. The question remains unanswered in many cases—whether apparent adaptations of morphology for microhabitats arose within that species and habitat, or if a morphological feature that facilitates use of a given microhabitat is a product of phylogeny (i.e., present in extant species because of origin an ancestor). The answer to this question should become clearer with continued mapping of ecological traits on phylogenies of fishes (Mayden, 1992b).

Swimming or Holding Position Against Fluid Drag

To understand the role of morphology in movement or habitat use of fish, it is necessary to first consider swimming and hydrodynamics of the body relative to physical principles of drag forces. Swimming speed depends on (1) the power of propulsion and efficiency of transformation of power into motion and (2) drag, which resists forward motion (Webb, 1975). Station holding by a benthic fish depends not on propulsive power but on streamlining of the body or modifying and strengthening structures of attachment (Aleev, 1963; Hynes, 1970) and maintaining body postures that help to resist drag (Matthews, 1985a). Energy is used in the effort required to hold fins immobile against drag forces or changes in postures as flow rates (and drag) increase (Matthews, 1985a; Webb et al., 1996).

Few water-column fish swim all the time, and few benthic fishes live exposed to direct currents all of the time. However, to function in the water column or to live in flowing benthic environments, a fish must be capable of forward motion or position holding in currents of the habitat for at least brief periods of time. Even though benthic fishes may spend most of the time out of direct currents (e.g., sheltering behind cobbles or sand ridges), it is, nevertheless, the ability to be exposed to direct current (i.e., resist drag) that may set the upper limit on current speeds in the microhabitats occupied by different species (e.g., Matthews, 1985a).

A swimming fish must overcome drag on the body and minimize turbulence on the tail. The aspect ratio of the caudal fin of the fish equals the vertical height squared, divided by the area of the fin surface (Pauly, 1989). (A perfectly square caudal fin would have an aspect ratio of 1.0.) Alexander (1967, p. 22) and Webb (1975, p. 27) express the aspect ratio in a more usual form borrowed from aerodynamics, as AR = height (or span)/width (or chord). The two formulations give identical values *if* a fin is a perfect rectangle; Pauly's formulation takes into account the irregularities in shape of the fin of a fish, which are much greater than for aircraft wings. Regardless of the formulation, a large aspect ratio indicates a "tall" fin that places much of the propulsive surface outside the band of turbulence (wake) produced by the "fuselage" of the fish (Aleev, 1963). Second, the caudal fin itself is a hydrofoil, which is cross section (as viewed from above) can have a streamlined shape. For example, the caudal fin of a fast-swimming tuna (marine) is near the ideal streamlined in cross section and smooth, whereas the caudal fin of a black bass (*Micropterus*) is rather flattened in cross section from anterior to posterior, and thus not very streamlined (Alexander, 1967, p. 24).

With regard to drag on the body of a fish, there is a trade-off between being shortened or round-bodied versus being either very elongate or slab-sided (Alexander, 1967; Webb, 1975). An elongate or slab-sided body has a high surface-to-volume ratio, hence increased friction drag relative to amount of musculature or internal organs. However, although packaging muscle, gonads, and internal organs in a shortened or rounded body decreases the surface-to-volume ratio, hence friction drag, it increases the tendency for anterior separation of the boundary layer from the surface, and the onset of turbulence further forward on the body of the fish, increasing the pressure drag. The compromise for minimum drag from both sources (friction and pressure) exists if the body is smooth, has a gradually tapering and elongate caudal peduncle, has greatest width or depth of body about one-third of the distance toward the tail, and has a "fineness ratio" (body length/maximum diameter of body) of about 4.5 (Webb, 1975, p. 21). For fish of different size but identical shape, there also is a tendency for flow to be more turbulent over large than small individuals [i.e., Reynolds numbers are larger (Webb, 1975, p. 15)] swimming at the same speed or exposed (benthic fish) to the same currents. Thus, diminution of body size [e.g., in darters, *Etheostoma* (Page, 1983)] may offer an advantage in resisting drag due to currents, which may be important for species that spend time motionless on the substrate, holding position against flowing water. Page and Swofford (1984) reviewed the morphology of darters (Percidae) relative to habitat use and considered that diminution of body size was an important evolutionary advance that led to habitat specialization.

The external surface of a fish is not smooth, and small irregularities (spicules, spines, etc.) can increase drag. Roughness of the body surface inside the nonmoving boundary layer has little effect on friction or drag, but, because the boundary layer becomes thinner at higher velocities, small irregularities of surface may substantially increase drag at higher current speeds (Webb, 1975).

Functional Morphology in Swimming

Basic swimming of fish consists at the two extremes of (1) undulation of the body or (2) complete tail swimming (Gosline, 1971). Some taxa or life stages also "row" through the water using paired fins, or use paired-fin swimming for gradual movements while minimally setting up turbulence or vibrations in the water, or maintain position in the water column by "fanning" the pectoral fins (Alexander, 1967). Small pickerels (*Esox americanus*), for example, swim mostly by movements of the pectoral fin as they approach a prey item. Webb (1975, pp. 89–90) depicted and described in detail the categories that have been used to classify swimming modes of fishes.

Undulation or "anguilliform" swimming (as evident in swimming of eels, lampreys, most moderately-swimming teleosts) consists of producing a series of flexures in a posterior direction along the body (Gosline, 1971, p. 15; Weihs, 1989). In eels and very elongate fishes, the flexure begins at the head; in less flexible taxa, the undulation may occur more in the midbody and caudal peduncle. Each flexure produces a backward pressure on the water mass, with successive lateral forces canceling each other out, thus thrusting the body of the fish forward (Gosline, 1971).

Tail swimming has the same general principle as undulation (i.e., that some body surface pushes backward against the water mass, thrusting the body of the fish forward). In fishes that mostly swim by thrusts of caudal peduncle and tail [i.e., carangiform swimming in Moyle and Cech (1988)], the bulk of the swimming musculature is further forward in the body of the fish, with the fore part of the body comprising a steady fulcrum and the contraction of more anterior musculature translated to the caudal peduncle and base of the tail via tendons [e.g., in tunas (Gosline, 1971, p. 15)]. Gosline (1971) pointed out that flexibility of the caudal fin also is important, because a completely rigid tail (like a wooden paddle) would, in swinging forward for a new stroke, tend to cancel out the forward thrust of the body of the fish that had been produced by the previous rearward stroke. Thus, the caudal fin of a typical tail-swimming fish, when viewed from above, tends to be rigid during the rearward power portion of a swimming stroke, but flexible and draging the tips of the fin rays as it moves forward in preparation for the next power stroke (Gosline, 1971, p. 16). The consequence of all this is that a shift from undulatory swimming to tail swimming required a variety of morphological modifications. Undulation is more efficient for slow swimming, with more total propelling surface in contact with water and sharing of contractions for swimming among many smaller muscle masses along the body of the fish. Tail swimming may be less efficient but can produce higher speeds and can also give a much larger "burst speed," allowing a predator to accelerate from zero to a high velocity in a very short period of time—making possible the ambush tactics of many fish [e.g., largemouth bass (*Micropterus salmoides*)].

Modification of the body for tail swimming involves more than external shape of body and location of muscle masses. A body being pushed from the posterior has to have some way of remaining relatively rigid against the thrusting force (Gosline, 1971). A notochord, which is mostly an elongate, gelatinous structure made firm by a cylindric sheath and an external, elastic membrane (Romer, 1962, p. 145), is appropriate for undulatory swimming, but this flexibility would be inappropriate for tail swimming and the structure would be too weak to resist compression when propelled from behind (Gosline, 1971). The bony vertebral column of teleosts, with the cushioning intervertebral joints retained from the notochord (Gosline, 1971, p. 18) is sufficiently flexible to allow motion from side to side, but the bone prevents longitudinal compression of the body during thrusting by the tail. In general, the number of vertebrae have been reduced as tail swimming increased, whereas undulation requires more joints; thus, taxa like eels that retain undulation may have hundreds of vertebrae, whereas advanced teleosts typically have 24 or fewer vertebrae (Gosline, 1971).

Body and Caudal Fin Shape Relative to Hydrodynamics

Overall body shape is the single greatest factor in hydrodynamics of a species. Principles of fluid dynamics apply to movement of water over the body of a fish just as they do for airfoils (Webb, 1975; Vogel, 1981), but an obvious difference is that whereas the body of an airplane generally does not change shape, the flexible body and fins of a fish result in a constantly changing relationship of surfaces to water moving past those surfaces. Webb (1975, p. 67) points out that differences between rigid and flexible bodies may make theoretical calculation or "dead" measurement of drag problematic if applied to living fishes. There also may be slightly different problems for a swimming fish (which is vigorously changing shape every second) moving through relatively stationary water and benthic fish, largely immobile, holding position on the substrate and exposed to flowing water. However, in both cases, drag produced by movement of fluid across the body of the fish produces forces that impede forward motion of the swimming fish or cause a downstream pull on a stationary benthic fish. The problem becomes that of reducing drag on the body (and fins) of a fish so that it requires less energy to swim or maintain position, allowing more available energy to be used in other life functions.

Lagler et al. (1962) described general body shapes of fishes, including the following: (1) fusiform—maximum depth of body located 36% of the distance back from tip of snout to end of tail; (2) attenuated—elongate, with less capability to contain muscles than in a fusiform fish but creating less turbulence; (3) laterally compressed—with the body flattened side to side and elevated dorsoventrally; (4) truncated—shortened from snout to tail, often producing a deep and short-bodied fish. In general, the shape of a fish is a trade-off between elongation of the body, with increased surface-to-volume ratio (less "volume" to contain musculature, internal organs, reproductive products, etc.) and thickening of the body, producing lower surface-to-volume ratio and allowing greater volume to contain muscle, organs, or gonads. Both designs (elongation versus thickening of body) have costs and benefits relative to drag forces (Alexander, 1967). The shapes may also relate to maneuverability. Werner (1977) noted that a shortened body may facilitate rapid turning by bluegill sunfish.

For several darter species, Paine et al. (1982) found no relationship of body form (as measured by caudal peduncle length, pectoral fin length, or flatness/compression of body) and use of microhabitats. In contrast, Matthews (1985a) found differences in the occurrence of two darter species and of two life stages of one species in the Roanoke River, Virginia, that related to differences in fineness ratios, position of pectoral and pelvic fins, bulkiness of the head, and general body and head shape (more or less cylindrical). Paine (1986) suggested that allometric differences during development of three darter species, producing ontogenetic changes in body form, might relate to "common, perhaps ancestral" developmental restrictions (comprising an early suggestion to examine phylogeny as well as extant ecology in such comparisons).

Axial Fins: Role in Propulsion or Controlling Movements

Fins have two skeletal components: the pterygiophores of the endoskeleton and the lepidotrichs or external components. Lepidotrichs consist of soft rays or stiff spines. Soft rays each comprise, in cross section, two halves of a hollow cylinder, surrounding a central membrane (Lagler et al., 1962; Gosline, 1971). Soft rays are jointed and may branch distally. Spines lack segmentation, lack the double structure in formation, originate differently in embryogenesis, and do not branch. In teleosts, the entire fin can be raised and lowered by muscle action, and soft rays can be swung side to side independently by contraction of individual muscles (Gosline, 1971). Joints in the soft rays provide flexibility, whereas the central membrane gives elasticity to the structure.

Alexander (1967) described the ways fins of fishes increase stability of a swimming fish, controlling pitch (vertical deflection), yaw (sidewise motion) and roll (tendency to roll on the long axis). Posteriorly placed stabilizing fins can be compared to feathers on an arrow. Alexander likened the correction of movement of a fish through water by use of its fins to the correction in flight of an arrow produced by posterior feathers. If the body of a fish in forward motion is deflected from its most efficient position with the long axis parallel to direction of motion, fins (or feathers on an arrow) positioned posteriorly efficiently deflect the rear portion of the body so that the whole body resumes its parallel orientation to direction of motion. However, fins placed further forward on the body of a fish lose their utility in producing hydrodynamic stability with regard to orientation of the body axis, much as might be the case if one attempted to shoot an arrow "backward" with feathers at the advancing end of the object—resulting in magnification instead of correction of any departure from body orientation (Alexander, 1967). Posterior location of fins facilitates their potential as stabilizers to control pitch or yaw. In general, stabilizers must be posterior to the center of gravity of a fish to be effective.

Aleev (1963, pp. 219–224) described in detail four "zones" from anterior to posterior on the body of a fish in which fins were most efficient in different hydrodynamic roles. These zones were, from anterior to posterior as follows:

Zone I: Zone of anterior rudders and bearing surfaces

Zone II: Zone of keels

Zone III: Zone of stabilizers

Zone IV: Zone of posterior rudders and locomotor organs

Within this general framework, Aleev (1963) illustrated the positions and postulated roles of fins for a large number of representative genera of fishes.

The caudal, dorsal, and anal fins (all in line with vertical axis of the body) typically provide locomotion, serve as stabilizers, or function as keels (to prevent sidewise slippage due to centrifugal force when a fish swims in an arc). In fish with the dorsal and anal fins near or anterior to the center of gravity, they may be most functional as keels [as described for bluegill sunfish, *Lepomis macrochirus*, by Werner (1977, p. 554)]. In advanced teleosts, the posterior end of an elongate dorsal and/or anal fin can serve as a rudder to prevent yaw, and in a rearward position, the dorsal and anal fin tips can work in concert with the caudal fin to produce braking action for the fish [i.e., with the caudal fin curved in one direction and the dorsal and anal fins curved in the opposite direction (Breder, 1926; Gosline, 1971)].

In advanced teleosts, the dorsal fin includes a spinous, anterior portion that is folded down when the fish is in forward motion, but which is erected as a keel or potentially as an anterior rudder (Gosline, 1971) when the fish is turning (Alexander, 1967). The rigidity afforded by spines enhances the function as a stiff keel. The posterior portion of the dorsal fin of advanced teleosts, consisting of soft rays, acts as a stabilizer, aids in generating thrust, or helps with braking forward motion (Alexander, 1967).

The caudal fin of teleosts generally consists of an upper and a lower lobe (Gosline, 1971). Teleosts can move the upper and lower lobes simultaneously for a powerful thrust or they can move each lobe separately for fine control of body position. Fish that occupy high-density habitats, like sunfish in weed beds, have caudal fins with great flexibility and fine control of fin rays, allowing finely executed corrections of position. Because the individual caudal fin rays can swing independently from side to side, the fish can accomplish vertical undulations within the caudal fin, similar to the horizontal undulations in the body of a swimming fish. In this manner, thrust can be exerted upward or downward on the water mass, lowering or raising the posterior part of the body (Gosline, 1971, p. 35). Open-water, continuously swimming taxa (e.g., *Morone*) generally have less capacity for fine adjustments of tail movements. In teleosts that have evolved squared or rounded tails, like *Cyprinodon* or *Fundulus,* the primary thrust is near the middle of the caudal fin, and the longest rays occur at that point. Accordingly, the central fin ray supports are strongest, and the central hypural plates are fused with the last vertebral centrum [the "urostyle" (Gosline, 1971, p. 39)].

Paired Fins: Control of Motion in Swimming or Their Use to Hold Position in Flowing Water

Most teleosts have two sets of paired fins: the pectoral and the pelvic fins. Pectoral fins have a basic function as rudders, helping change the angle of the body by pushing the fish up or down as it moves through the water. In lower teleosts, pectorals tend to be more ventral, with a horizontal base (Gosline, 1971; Alexander, 1967). In more advanced teleosts, pectorals are usually higher up on the body, near the midline, and have a vertical base. This higher position on the body makes "fanning" movements possible, which allow the fish to remain stationary or even, in some, to swim backward, and the possibility of thrust swimming, by thrusting the pectoral fins rapidly backward (as do darters of the genus *Etheostoma*).

Pectoral fins can be used as hydrofoils to hold benthic fish to the substrate (Hubbs, 1941; Jones, 1975; Lundberg and Marsh, 1976) or for grasping the substrate with fin-ray tips, as in North American darters (Matthews, 1985a), or sculpins (Webb et al., 1996). Aleev (1969, p. 141) illustrated some of the more extreme modifications of pectoral fins for position holding by benthic fishes. Use of the pectoral fin for support on the substrate [as in flatfishes (Marsh, 1977)], or as a hydrofoil results in catostomids in "foreshortening" of basal fin-ray segments, and in less branching of the fin rays (Lundberg and Marsh, 1976). Jenkins and Burkhead (1994) suggested that for northern hogsuckers (*Hypenelium nigricans*), a relatively large, benthic species that occurs in swift rivers, both the fins and a concavity of the head could serve as hydrofoils to press the fish onto the substrate. When a benthic fish rests propped up on its pectorals, the leading edge of the fin is appressed to the substrate, allowing the water passing over the fin to help force the fin downward into the substrate [although in comparisons I made for two darter species, the existence of pectoral fin as a hydrofoil did not significantly change the critical current speed (Matthews, 1985a)]. However, in suckers (e.g., Lundberg and Marsh, 1976) the pectoral fins are more ventral, and a large part of the surface of the fin can be appressed on the stream bottom, whereas in darters (e.g., Matthews, 1985a), the pectorals are higher up on the body, with less of the total fin area actually in contact with the substrate. Having more of the fin in contact with the substrate can lessen or prevent flow of water under the fin, lessening or negating the tendency of a fish to lift off the substrate due to the Bernoulli effect.

Kessler et al. (1995) considered that a benthic sculpin and four benthic darter species had differences in habitat use in streams at high flow that corresponded to differences in morphology, with two species having "robust bodies and . . . large pectoral fins which allow them to withstand currents on smaller smoother substrates." They also noted that the laterally compressed body of another darter species "is well-suited to hiding under rocks."

The pelvic fins of less advanced teleosts are posterior, whereas, in advanced teleosts (e.g., perciforms), they are more anteriorly near to but below the pectoral fins. In advanced teleosts, the pelvics can be extended to deflect moving water and force the body of the fish downward. Additionally, for water-column fishes like sunfish (*Lepomis*), extending the pelvic fins can counteract the tendency of the body of a fish to rise when pectorals are extended as a brake [i.e., giving the fish a "four-point" braking system (including the two extended pectorals and the two extended pelvics)], and the ability to rapidly stop without the head rising or falling in the water column (Gosline, 1971, p. 42). Pelvic fins also can be important structures of benthic attachment in swift waters. Hora [in Hynes (1970)] depicted the modified pelvic fins by swift-water, benthic fishes in Asia (e.g., hillstream loaches), with the fins serving as suction devices for attachment to rocks.

Additional Morphological Features of Benthic Fishes

The previous sections suggest that benthic fishes have many of the same hydrodynamic challenges as do water-column fishes but may meet them in different ways. Hynes (1970) noted the following general adaptations of benthic fishes:

1. Gill openings lateral, not vento-lateral (where they would be in contact with the substrate)
2. Enlarged ventral lobe of caudal fin (helps to keep the head down, near the substrate, to avoid being swept downstream by currents)
3. Inner rays of pectoral fins act to pump water from under the body of the fish, which help to keep the body in contact with the substrate.
4. Flat, naked belly for benthic fish in sandy streams
5. Laterally placed fins and dorsal eyes
6. Pectoral fins are spread as hydrofoils, pressing fish down on substrate
7. Reduction in swim bladder
8. Ventral suckers and friction pads, by virtue of thickened anterior pectoral rays, and movement of pelvic fin forward as a friction pad (e.g., *Cottus* and *Etheostoma*).
9. Thickened upper and lower lip of a ventral mouth (e.g., the African cyprinid *Garra*).

Gosline (1994) reviewed the various uses of paired fins by benthic scorpaeniform fishes (including some in fresh water, like sculpins), with uses of the pectoral or pelvic fins including defense (displays and warning), locomotion (clambering, hopping, "walking"), perching (vertical perch or horizontal prop, by pelvic fins), digging (with pelvic fins), and as a suction disk (with fleshy pelvic fins). Aleev (1969) described the "walking" on pectoral fins by some benthic fishes. No single species of fish possesses all of the traits above, but numerous taxa have one or more of these traits coinciding with a primarily benthic existence. It would be interesting to broadly map these presumably adaptive traits on phylogenetic hypotheses for clades of fishes that include benthic forms (e.g., minnows, darters, etc.) to determine whether these traits are indeed derived, how many times they may have arisen independently in various lineages, or if they are primitive in some clades.

Shapes of Fishes in Nonflowing Habitats

In lakes, there is typically less water movement than in streams, and fish may have fewer current-related or hydrodynamic challenges (except in wave-swept shorelines). However, body or head shapes of fish in lentic waters may relate to their ability to live at or near the surface, at midwater, or on the substrate. Shape of head and position of dorsal fin are related to vertical position in the water column, with flattened heads and posteriorly placed dorsal fins (e.g., some *Fundulus*) representing adaptations to surface habitats in nonflowing waters. Note that morphological features of some fishes facilitate aquatic surface respiration (ASR). Lewis (1970) showed a strong relationship between the head shape and the angle at which various fishes could remain suspended near the surface, using the oxygenated microlayer. Species with a dorsally flattened head maintained position at the surface longer, using less energy to do so than fish with more "normal" heads, which could only breathe at the aquatic surface with the body at an acute angle, presumably tiring more rapidly than those that could remain almost horizontal while using the oxygenated microlayer.

For fishes in pools of tropical streams, Welcomme (1985) described three communities that "can readily be distinguished within the water column," including "a pelagic community which tends to consist of small species, silvery in color with upward facing mouths, a mid-water community of larger silvery fishes streamlined with terminal mouths, and a bottom living community of drab coloured fishes with dorsally humped profiles and ventrally positioned mouths." Note that this latter description (dorsally humped profile, ventral mouth) would also fit some of the large humped minnows or suckers of large rivers of western North America, so these traits in benthic fish are not particularly aligned with lentic waters.

Trophic Morphology of Fishes

> The explanation of certain structural conditions about the mouth, throat and gills, has proceeded so far as to make it very likely that ... definite correspondences between structure and food will be made out, which will enable us to tell with considerable accuracy and detail what the food of an unknown fish must be, by a mere inspection of the fish itself. ...
>
> —S. A. Forbes (1880b)

Shape and hydrodynamics have a strong influence not only on where a fish lives, but also what it eats; thus, its role in a local assemblage is strongly influenced by morphology of the head, oral teeth, or pharyngeal jaw apparatus (including pharyngeal "teeth"). There is tremendous divergence among fishes in morphology of feeding structures, like placement of mouth, size of mouth, strength of oral apparatus, or mechanical attachment and lever arrangement of upper and lower jaws. These factors all influence modes of attacking or handling prey, ability to retain prey, and propensity to crush, chew, or otherwise fragment prey items. Most fish actively ingest prey by use of jaws, and only a few (like parasitic lampreys that rasp through body wall of prey and consume body fluids) deviate from this pattern. Among the fish that consume prey by jaw action, some bite frontally or sidewise, many suck in prey by lowering the floor of the buccal cavity to create strong negative pressure, and some protrude jaws to pick prey from substrates (Alexander, 1967). The oral structure of fishes probably has a greater range of variation than any part structure of the body, but evolutionary patterns can be described.

A detailed survey of fish skulls, jaws, and pharyngeal apparatus is beyond the scope of this book, and references like Gregory (1933, skulls), Romer (1962, jaws), or Sublette et al. (1990, pharyngeal teeth) provide details of nomenclature and relationships among bones or comparison among taxa. Gosline (1971, pp. 50–75) and Alexander (1967, pp. 87–108) provide excellent summaries of the evolutionary patterns in structural and mechanical functions of oral jaws of fishes. The composition of local fish assemblages depends at least in part on a matching of the possible biomechanical properties of the fishes with the foods available in a particular habitat.

Primitive bony fish (e.g., *Amia*) typically use the entire mouth to seize and process prey, and teeth were generally present throughout the mouth in some (Gosline, 1971, his Fig. 16). Typically, advanced fishes are more specialized (e.g., with anterior teeth for biting, with posterior oral teeth or a pharyngeal jaw apparatus suited for mastication or grinding of prey, and with general reduction in teeth in various parts of the upper or lower oral surface). Gosline (1971) depicts the evolution of fish jaws as they became modified from a relatively fixed upper jaw and simple lower jaw condition to a more efficient apparatus consisting of highly mobile or protrusable jaws.

In the extinct paleoniscoids, the premaxilla and maxilla were sutured to the cheek and immobile (Alexander, 1967). There was an evolutionary transition to allow movement of the maxilla in a typical preteleost, like *Amia,* and further mobility of upper jaw structures in advanced teleosts. In *Amia,* a heavily toothed premaxillary is fused to the cranium (Romer and Parsons, 1977, their Fig. 171; Gosline, 1971). Behind the premaxillary is the maxilla, whose anterior end is wedged behind the premaxillary, partly free to move, and whose posterior end is hinged to the mandible. As the mandible of *Amia* moves downward to open the mouth, the maxilla is pulled forward by connective tissue and also moves forward, forming part of the biting gape (Gosline, 1971). In protacanthopterygians like salmonids, the anterior end of the maxilla is wedged between a partially movable premaxillary and the cranium (Gosline, 1971). Both the maxilla and premaxillary are included in the gape, and with the increased mobility of upper and lower oral structures, the straining of prey or suction feeding can be enhanced, because the mouth can form a more rounded oral opening than is possible in the mouth of *Amia* (Gosline, 1971). This allows the most typical teleost feeding mode (i.e., extending jaws around or near prey, and exerting negative pressure to pull prey into the mouth), and the partial mobility of the premaxillary allows specialization of oral movements for sucking, picking at prey, or other fine-tuned motions (Gosline, 1971). Alexander (1967) noted that a rounded, tube-like mouth opening is most efficient for sucking in prey, but that it may lose some efficiency in that the fish must be directly facing the prey item.

In more advanced teleosts, the premaxillary extends in front of the maxilla (Fig. 8.2), excluding the latter from the active gape of the fish (Gosline, 1971, his Fig. 17.D1). There may be an advantage (e.g., in strength) to having the upper part of the gape consist of one continuous, movable bone. In highly advanced fishes like perciforms, the premaxillary is often protrusable (Gosline, 1971; Alexander, 1967); that is, it can be rapidly extended forward from the

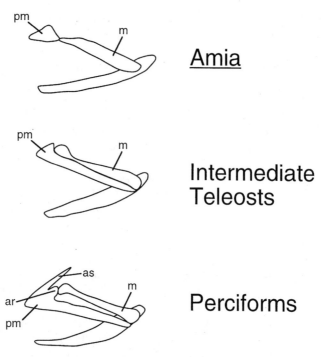

Figure 8.2. Jaw structure of *Amia,* typical intermediate teleosts, and typical perciforms. pm = premaxillary, m = maxillary, ar = articular process of premaxillary, as = ascending process of premaxillary. [Modified from Gosline (1971)].

maxilla, adding overall to the velocity of the upper jaw (Nyberg, 1971) and allowing the fish to modify the size or shape of the mouth opening to fit the food item. Alternatively, Alexander (1967) argues that protrusable jaws may offer more advantage to fish that feed by taking benthic food, in that the protrusability allows the fish to capture items on the bottom without having to hold the body at a sharp angle to the substrate. Dutta and Chen (1983) describe the protrusion of the jaw in the perciform *Micropterus salmoides* (largemouth bass), with emphasis on all bones involved in jaw movement and the associated rostral cartilage and ligaments, and on the role of the palatine in enhancing movement of the maxilla. Additionally, during advances in teleost evolution, the lower-jaw support changed from direct articulation with the skull to hinge on an intervening hyomandibular bone, which allows the mouth to be opened wider (hyostylic jaw suspension Romer and Parsons (1977)].

There is a direct relationship between the relative size and strength of bones of the jaw in fishes and use of food items. For example, the South American *Colossoma,* which feeds by crushing large seeds (Goulding, 1980), have massive jaws. At the other extreme, many fish species that feed by scraping attached algae from stones or other hard surfaces have a lower jaw modified by a ridge of cartilage (e.g., the North American *Campostoma*). *Campostoma* uses the cartilaginous ridge in several different behavioral modes for feeding (Matthews et al., 1986).

Pharyngeal "teeth": In many teleosts the fifth pharyngeal arch has lost its respiratory function and is modified with "teeth" for grinding or tearing of prey items, as a pharyngeal jaw apparatus (PJA). In cyprinids, for example, the pharyngeal teeth vary from relatively flattened to sharp, hooked elements that protrude across the pharynx and interdigiate like the fingers of two human hands loosely joined. At other extremes, the pharyngeal teeth of some catostomids are "comb-like," with many flattened teeth that help retain small prey items in the oropharyngeal cavity (Eastman, 1977). Pharyngeal teeth of some cyprinds, catostomids, centrarchids, and many other taxa are flatted and "molariform" for crushing hard prey (e.g., snails) or grinding either animal or vegetable materials. Greenwood (1984, his Fig. 4) illustrated the wide diversity of pharyngeal teeth in haplochromine cichlids of the African Great Lakes, and the occurrence of teeth in the mouth (his Fig. 5), both as related to trophic ecology of the fishes. One of the best surveys of pharyngeal teeth across orders and families of North American fishes is in Sublette et al. (1990), providing electron micrographs of the pharyngeal arches of many species in New Mexico (USA).

Descriptive Studies of Morphology Relative to Food Use

There are many descriptive studies of trophic structures of fishes and their relationship to food use. Turner (1921) noted a correlation between mouth position and protrusion of the jaw with the use of large and active prey by darters (*Etheostoma*). Matthews et al. (1982) showed a relationship between relative gape width and size of prey for three darters (*Etheostoma* and *Percina*) in the Roanoke River, Virginia. Paine et al. (1982) considered that the diets of four darter species were closely related to morphology, on the basis of three measures of mouth structure: mouth height, mouth position, and snout protrusability. Eastman (1977) described the pharyngeal cavity and teeth and their trophic function in suckers (Catostomidae). Schmitz and Baker (1969) detailed the anatomy and putative functions of the pharynx, gizzard, and intestine of the largely planktivorous shads of the genus *Dorosoma*. Miller (1983) showed the comparative food use of two riverine darter species related to the overall size of the body (length). McComas and Drenner (1982) showed that differences in oral structures between two atherinid species led to one replacing the other in a reservoir, because the species *Menidia beryllina* with a funnel-shaped oral opening was more efficient at capturing zooplankton than was *Labidesthes sicculus*, which has a scissor-like mouth structure.

Pauly (1989) described a link between food consumption rates of fish species as related to shape of their caudal fin (aspect ratio), hence their swimming abilities. Pauly (1989) suggested for marine fish a variety of estimates of "shape-related indices of fish activity" which have not been tested for freshwater fishes.

Gosline (1977) focused on the way the pectoral girdle influences the ability to move the head relative to the vertebral axis of a fish, and the expansion of the pectoral girdle to permit food to pass into the esophagus [which is very important in the functional limitations of oral versus pharyngeal structures relative to size food ingested in the functional analysis of Wainwright (1987) later in this chapter]. Gosline (1985) suggested that teeth on the third basibranchial or the parasphenoid make it possible for fish to retain a grip on hard-shelled, active prey items that might otherwise escape before they could be crushed in the oral cavity. Stevenson (1992) showed a relationship between food use and divergence in trophic morphology for four members of the species flock of *Cyprinodon* in Lake Chichancanab, Yucatan. Humphries (1993) showed for three atherinid species that the species having the most protrusable jaws feed most at the surface. Blaber et al. (1994) reviewed the oral dentition and diet of 13 species of ariid catfish (freshwater and marine), with detailed figures depicting arrangements of palatine and vomerine tooth patterns as related to their diets. Figure 3 of Blaber et al. shows a wide diversity of tooth patterns ranging from extreme simplification to proliferation of teeth in various pad configurations. Luczkovich et al. (1995) showed that two estuarine species showed ontogenetic changes in mouth size and dentition that affected kinds and size of prey they selected. Wainwright and Richard (1995) demonstrated that ontogenetic changes in mouth size were more important than changes in body length in the transition of centrarchids from juvenile diets of zooplankton to adult diets of fish or large invertebrates. Wainwright and Richard noted that many studies of predator–prey size have overlooked the most important dimensions in prey size in evaluations of predator versus prey size, and that the most important dimension of prey size is the cross-sectional diameter of the prey item (not length of the prey). Wainwright and Richard also showed the very important fact that some predatory centrarchids may have a smaller pharyngeal than oral gape (i.e., if they are gape-limited, it could be by size of the pharyngeal opening, not the oral opening, at least for hard-bodied prey.

Functional Analyses

A recent example of a functional morphological analysis of jaws is by Wainwright and Richards (1995), who measured lever distances and calculated ratios of effective lever distances involved in opening versus closing the jaw. A ratio of the effective lever distances provided a good discrimination of species that use ram–suction feeding versus biting or manipulation of prey [following the definitions of Liem (1980) citing by Wainwright and Richards]. For both freshwater and marine fishes the biomechanical assessment of jaw capabilities corresponded fully with their primary feeding mode.

Functional analyses of jaws or feeding morphology of fishes date at least to the 1920s {[protractile mouths of sunfish and catostomids [Jones, 1925; Edwards, 1926, originals not seen, cited by Dutta (1979)]}, with increased emphasis on functional mechanisms by the 1960s [e.g., Shaeffer and Rosen, 1961; Liem, 1967; Alexander, 1966; originals not seen, cited by Dutta (1979)]. Use of cinematography and electromyography to assess functional mechanics of vertebrate feeding began in the 1960s [references in Dutta (1979)]. Dutta (1979) provided a detailed description of functional mechanics of anabantid fishes, relative to jaw movements for feeding and for gulping of air, by dissection and motion filming from which he obtained estimates of movements of the bones of jaw and hyoid region, and explained differences in movements of three species by the limitations set by physical structures of the jaws.

Liem (1980) combined electromyographic and kinematic profiles to detail feeding mechanics of cichlids, with emphasis on fine details of various actions to consume prey during different feeding nodes. He illustrated that even a very highly specialized species, presumably adapted morphologically for scraping attached algae, had at least eight distinct feeding modes, each with its detailed neuromuscular repertoire. Liem (1980, p. 312) offered a detailed schematic of the manner in which all parts of the head, jaw, pectoral girdle, and hyoid operated in concert to allow or facilitate feeding patterns, and emphasized that functional morphological analysis of a single, isolated action could result in faulty interpretation of overall activities of a species. Likewise, Lauder (1983) showed that specialization of *Lepomis microlophus* (redear sunfish) for crushing snails involved morphological increases in molariform pharyngeal teeth, increased cross section of pharyngeal muscles, and a radical "evolutionary transformation in neuromuscular pattern" as measured by patterns of firing of the muscles in the feeding action. The derived "crushing pattern" of muscle activity with most pharyngeal muscles contracting simultaneously (to crush a hard-bodied snail) is so stereotyped in *L. microlophus* that it is used even when the species consumes soft prey. Thus, the analyses of Liem (1980) for an algae-scraping cichlid and that of Lauder (1983) suggest different degrees of evolutionary neuromuscular canalization for two specialized fish species. Lauder (1983) suggested that such differences in degree of specialization could account for the "radically different patterns of trophic diversification in centrarchid and cichlid fishes."

Lauder (1985) reviewed functional feeding morphology of lower vertebrate clades, including salamanders, lungfish, and teleosts, concluding that suction feeding involving the hyoid as well as the jaws is the basic mode, with prey capture by filter feeding or by biting with the jaws a derived or secondary specialization. Liem (1986) used x-ray cinematography and electromyograms to describe highly complex and stereotyped chewing and swallowing mechanisms for surfperches (Embiotocidae; mostly shallow marine but with a freshwater species; trophic generalists) that included a "figure-8" motion of the pharyngeal jaws, with both shearing and crushing phases (with shearing produced by movement of upper over lower pharyngeal jaws), and indicated the strong similarity of this mechanism to that in cichlids and labrids. This motor pattern, made possible by a specialized muscle "sling" suspending the pharyngeal jaw apparatus, appears to allow the use of a wide variety of kinds of food. Wainwright (1989) used electromyography to determine species-specific patterns in pharyngeal jaw activity in four species of marine haemuliids, in which he found patterns (1) highly conserved interspecifically but (2) capable of modulation intraspecifically, depending on prey types. Wainwright (1989) concluded that this pattern was evident across a wide range of lower vertebrates.

In addition to the measurement of structural motor patterns, Wainwright (1987) argued for the importance of performance testing in fish functional morphology. For a mollusk-crushing marine species, *Lachnolaimus maximum* (Caribbean hogfish), Wainwright (1987) found (based on calculations from dissections of muscle mass) that the maximum size snail an individual could ingest was limited by its ability to crush prey ("force limited) rather than by its gape width. Westneat (1994, 1995) has taken performance analysis to a sophisticated level, using principles from mechanical engineering to make complex assessments of force vectors associated with opening and closing levers or jaws of labrids to determine kinematic transmission of force associated with jaw opening and closing and hyoid depression, as well as maxillary rotation and gape angle. Westneat (1994) showed that species feeding on evasive prey emphasized the velocity of jaw action, whereas species eating benthic prey emphasized force transmission in the trophic systems. Westneat (1995) provided a detailed summary of methods and offered advice on techniques for mapping biomechanical character states on a phylogenetic hypothesis and noted that information on biomechanical limitations may have important implications for the ordering of character-states in a phylogeny—thus helping select among possible polarizations of character states that affect interpretation of cladograms themselves. For labrids, Westneat (1995) showed the correspondence among evolutionary relationships of species within clades, the kinds of prey they use, and the possession of high transmission of jaw force, concluding that use of evasive prey and high kinematic transmission of force evolved independently numerous times with the lineage.

Two other noteworthy examples of mapping biomechanical or functional traits onto a phylogeny are in Lauder and Wainwright (1992) and Wainwright and Lauder (1992). Lauder and Wainwright (1992) examined, broadly across fish clades from chondrichthyes to the neoteleosts, the presence and function of the pharyngeal jaw apparatus, finding that acquisition of a given muscle does not necessarily equate to its playing the same functional role in all taxa in which it is present. Wainwright and Lauder (1992) mapped feeding biomechanical traits of the centrarchids onto a phylogeny for the family, pointing out the importance of comparing both kinds of information in order to understand if the functional and phylogenetic history of the group is congruent. For mandibular jaw and suction feeding, Wainwright and Lauder (1992) showed that although motor patterns associated with suction feeding were conserved, morphology, feeding performance, and diet were unique in each of four species tested (representing different genera). For the pharyngeal jaw apparatus, Wainwright and Lauder

(1992) showed the ability to crush snails was present in two parts of the sunfish clade (i.e., the capability arose twice) or else was lost by species embedded in the "crushing" clade but not possessing the trait. The fact that green sunfish (*Lepomis cyanellus*) have the capability to crush snails, but do not predominantly do so, was interpreted by Wainwright and Lauder (1992) to indicate that the acquisition of a biomechanical capability may not be congruent with the application of that trait in actual behaviors. It seems clear that no detailed functional analysis of a fish is complete with careful "natural history" or behavioral observation to determine if its physical capabilities are indeed used in the natural environment.

Functional morphology of fishes in not limited to the interpretation of capabilities or actions of structures independent of the activity of the fish. For example, Wiehs (1980) showed that forward motion of the body of a largemouth bass increased by 60% the distance in front of the bass from which an item can be captured by suction feeding. Wiehs (1980) provided a detailed review of the general physical and hydrodynamic principles of suction feeding in fishes.

"Other" Morphological Features Relative to Ecology

Moore (1950) described increases in cutaneous sense organs to compensate for reduction of eyes in cyprinids adapted to turbid streams in North America, and Moore (1956) showed the enlarged cephalic portion of the lateralis system in *Lepomis humilis,* a sunfish most common in turbid or silty waters, relative to the more usual condition for other *Lepomis* species. Page (1977) described comparative morphology of the lateralis system for all major groups of North American darters (Percidae), with emphasis on reductions in the lateralis during evolutionary advances in subgenera of *Etheostoma* and *Ammocrypta*. Bauchot et al. (1977) calculated an index of "encephalization" comparing brain and body size across a large number of fish orders. Huber and Rylander (1992) analyzed in detail the anatomical features of the eyes and optic nerve of minnow species from clear versus turbid waters, finding that clear-water taxa (where sight presumably is useful) had twice as many optic nerve fibers as did species of turbid waters. The structure of the retina of species of haplochromine cichlids in Lake Victoria were analyzed morphologically by van der Meer et al. (1995), who concluded that structural differences among species coincided with differences in their visual threshold, which promoted, via "optical differentiation," observed differences in feeding behavior and food use by those fishes. Ecology and structure of fish gills are reviewed by Hughes (1972) and Hughes (1995).

Ecomorphology

Although ecomorphology is included in this chapter along with "autecological," single-species phenomena, it is important to note that ecomorphology of a single species has only rarely been the focus of studies {[but the approach is useful to compare populations [e.g., geographic variation in body form relative to habitat, as in Matthews (1987b), for red shiner minnows throughout their range]}. Most ecomorphological studies have assessed the *relative* morphology of members of a local assemblage in comparison to each other, or to make comparisons among higher taxonomic groups (Douglas, 1987; Wood and Bain, 1995) or assemblages (Strauss, 1987; Winemiller, 1991a and b). In practice, this has most often, for fishes, involved a dozen or more species and a dozen or more morphological characters, all reduced into phenetic ecological morphospace on a small number (one to three) of axes.

Measurement of overall shape has often relied on intuitive characters describing the position of fins on the body, depth or width of body, or location of maximum diameter, sizes of fins, or distances between points on the head and jaws presumed to be related to trophic ecology or aspects of microhabitat use. Alternatively, there has been a growing trend in the last decade to use a geometric "truss" system, first introduced to ichthyology by McCune (1981) to describe the outline of a fish as a series of triangles or polygons obtained from positions of homologous points.

Multivariate ecomorphological studies only advanced substantially after the availability of high-speed computing which allowed multicharacter approaches (e.g., factor analysis, principal components analysis, cluster analysis) to incorporate a large number of physical features of an organism into an "overall" comparison of similarities and differences among taxa. Multivariate assessment of morphology as related to ecology originated for bird, rodent, lizard, insect, and bat communities (Findley, 1973, 1976; Ricklefs and Cox, 1977; Smart, 1978; Sheldon, 1980; Ricklefs et al., 1981) and was initially applied to stream fish communities by Gatz (1979a, 1979b). Gatz used a large number of measurements (56 characters) for 44 species in a factor analysis to summarize niche overlap as Euclidean distance among species pairs, for fish in three different drainages. The first axis accounted for 31% of the morphological variance and largely separated species among families. Patterns in morphological distances were similar across drainages and different from nonrandom, indicating that, morphologically, nonrandom, structured assemblages of species coexisted (Gatz, 1987a). Gatz (1979b) described the morphological characters in detail, with the rationale for inclusion of each and its putative ecological significance.

Since the work by Gatz, multivariate ecomorphology has been applied in many stream fish assemblages. Felley (1984) compared morphological traits of cyprinid species to their microhabitat use in 138 field collections in Oklahoma, Louisiana, and Mississippi (USA). Using suites of morphological traits taken from Gatz (1979b), Felley (1984) found no substantiated relationships to microhabitat variables like current speed, depth, food habits, vegetation use, or substrate type. However, using a principal components analysis of the morphological variables to select covarying suites of species a posteriori, Felley (1984) found significant relationship of four characters related to fin height or location and scale size to depth of fish in the water column, and another suite of characters including intestine length, pigmentation of the peritoneum, and size of cerebellum that was related to food habits (based on detritus in the gut).

In the same year, Mahon (1984) demonstrated by principal components analysis of 12 morphological characters a relationship between morphology of stream fishes and a longitudinal gradient in current velocities for streams in Poland and Ontario. Mahon's first factor axis distinguished benthic fish of fast waters (like *Cottus bairdi, Noturus flavus, Etheostoma flabellare,* and *Hypentelium nigricans*) from species of slow-moving waters (like *Notemigonus chrysoleucas* and *Lepomis gibbosus*), and his second axis separated small, active swimmers (e.g., *Notropis heterolepis* and *Notropis rubellus*) from larger, slower-swimming species (like *Ambloplites rupestris, Amieurus nebulosus,* and *Noturus flavus*). Mahon's results suggested that multivariate analyses of body or fin shape, and position of structures on the head could separate fishes along axes that intuitively were reasonable relative to general known habitats or activity levels of the fishes. However, it remained to be determined as of 1984 if there was a rigorous predictive capacity of multivariate ecomorphology and details of microhabitat or food use by coexisting fish species.

Watson and Balon (1984) used 15 characters similar to those of Mahon (1984) to carry out a principal components analysis of ecomorphology of fish assemblages in five streams in northern Borneo. Watson and Balon (1984) identified four factors that separated species morphologically and (presumably) ecologically that corresponded most strongly to (1) caudal fin aspect ratio, lateral compression, eye position, and pectoral fin aspect ratio, (2) head length, mouth width and height, body depth, and pectoral and caudal fin areas, (3) body size and peduncle compression and length, and (4) mouth orientation and ventral flattening. Watson and Balon (1984) interpreted these four axes as corresponding to surface, pelagic, benthic, and substratum-dwelling species and suggested that all five fish assemblages shared similar morphological structuring. They reported a decrease in ecological similarity among the dominant species at higher elevations and current velocities, independent of species richness. They found that, in comparison to assemblages with fewer species in Poland and Ontario, the total niche space was greater in streams of Borneo where species richness was higher.

Page and Swofford (1984) scored 66 morphological variables for 78 darter species (Percidae) and found strong correlations between several suites of meristic, measurement, or "miscellaneous" morphological variables and the typical habitats for darter species. They concluded that there is much evidence of morphological convergence of relatively unrelated species that occupy similar habitats and that habitat has been "a prime influence in the evolution of darter morphology." However, all evidence for the linkage between multivariate morphology and the ecology of fishes remained correlative as of 1984.

Moyle and Senanayake (1984) examined the degree of morphological, diet, and habitat specialization of species in assemblages in tropical rain-forest streams of Sri Lanka, using 14 ecomorphological characters from Gatz (1979b). They found all of the species were separable from each other on the basis either of shape or trophic variables, and, like Gatz (1979a), Moyle and Senanayake's results show that fish separate morphologically among families (i.e., they are more similar within than among families, confirming again that families do comprise recognizable, phenetic morphological groupings if viewed independently of any information on phylogenies). They also concluded that the morphological segregation of these fishes coincided with microhabitat and diet specialization.

Two papers in Matthews and Heins (1987) were based on multivariate analyses of ecomorphology of stream fishes (Strauss, 1987; Douglas, 1987). Strauss (1987) tested the morphological structure of fish assemblages at five localities in North America and two in Paraguay to detect patterns in body form within those assemblages and to determine if patterns differed in tropical versus temperate habitats. Strauss used a truss system (Strauss and Bookstein, 1982) augmented by measurements of body width, resulting in 56 measurements summarized by a principal components analysis. All assemblages, including those in North America and Paraguay, overlapped strongly in multivariate space, suggesting that patterns in morphology were fundamentally similar although no species were shared between the North and South American sites. Mean nearest-neighbor distances in morphospace were not significantly different among any of the assemblages, but morphological variance was greater in species-rich assemblages. Overall, Strauss (1987) supported the hypothesis that as richness of species in assemblages increases, it does so by adding species to the "edges" of morphospace rather than increasing the "packing" of species within existing morphospace. Strauss (1987) also showed that at this global level, the phylogenetic history of species in the assemblage had more effect than did extant ecological constraints in determining morphological structure of an assemblage.

Douglas (1987) discussed strengths and weaknesses of ecomorphological approaches for studies of fish ecology, and discussed methodologies. Douglas compared members of the minnow genus *Notropis* (as previously used, including species that have now been placed into separate genera) and the sunfish genus *Lepomis* to evaluate morphological diversity and the packing of species in morphospace for these two genera, which differ widely in numbers of species. Within the two clades, minnows differed much less in body size than did the sunfishes. However, each was determined by Douglas to fill approximately the same amount of morphospace, but minnows (by virtue of having many more species) were much more packed in morphospace than were sunfish. Douglas (1987) concluded that sunfishes should be much less constrained ecologically than are *Notropis* minnows by congeneric taxa (i.e., that a given sunfish species should have more surrounding morphospace into which to expand via plasticity of phenotype) and that *Notropis* should tend to be niche specialists, whereas *Lepomis* species should be niche generalists.

Winemiller (1991) compared the dispersal of "core" species (those numerically dominant in the assemblage) in morphospace defined by principal components analysis of 30 morphological characters taken from Gatz (1979b) for five sites ranging geographically from Alaska to the Neotropics and Africa. Winemiller (1991b) specifically advised against use of a "truss" system (e.g., Strauss, 1987), arguing that the use of a truss to define shape in a sagittal plane might overlook ecologically important details of body form. Winemiller also used nearest-neighbor distance and distance to the assemblage centroid in morphospace to compare species packing and niche diversification among the tested assemblages. Winemiller (1981) found evidence of both morphological divergence with phylogenetic divergence and ecomorphological convergence of distantly related taxa. Winemiller concluded that (1) the ecomorphological patterns indicated that interspecific competition had a strong influence on fish assemblages, (2) assemblages with more species had more ecomorphological diversity, and (3) morphological convergence among distantly related taxa was common. Winemiller (1992a) expanded this analysis to conclude that there is more morphological, taxonomic, and ecological diversification in tropical than in temperate faunas, based on the assumption that morphology does reflect ecological traits.

Winemiller et al. (1995) used a canonical correspondence analysis to compare multivariate morphological and dietary data sets for riverine cichlid assemblages in Costa Rica, Venezuela, and Zambia, and further divided the morphological data into attributes associated with the head and those associated with body shape. All three cichlid faunas overlapped considerably in morphospace and trophic space. However, there were more trophically specialized cichlids in the Costa Rican fauna, and rates of morphological divergence (relative to phylogenetic differences) were faster in Costa Rica than in the other two regions. Using multivariate morphological estimates of divergence relative to known phylogenetic divergences might allow similar identification of rapidly evolving clades for other fish taxa.

Winston (1995) used an ecomorphological assessment, based on 56 characters in Strauss (1987, based on a truss system), of similarities and differences among all possible pairs of 27 species of minnows (Cyprinidae) to test potential explanations of the co-occurrence of species in the Red River basin of Oklahoma and Arkansas. Winston found that morphologically very similar species pairs occurred together in 219 assemblages less than would be expected by chance, ruled out phylogenetic effects as causing the pattern, and accepted the hypothesis that competition between morphologically very similar pairs was sufficient to maintain complementary distributions of such species.

Does morphology, as revealed by multivariate analysis of many characters, faithfully or adequately reflect the ecology of a species such that ecomorphology can be a surrogate for field evaluation of niche of a species? Felley (1984) tested the relationship of morphology to ecology for a large number of minnows (above), but used only rather generalized measures of habitat. Wikramanayake (1990) used a suite of 16 measurement or qualitative morphological characters and field-determined microhabitat and food use to ask if there was a strong relationship between morphology and ecology for fishes in streams of Sri Lanka. Wikramanayake carried out, separately, principal factor analysis of (1) ten microhabitat variables and (2) ten morphological variables putitatively linked to habitat, then compared the results with canonical correlation. Similarly, he subjected nine morphological variables and, separately, five dietary categories for each species to principal factors analysis and tested for correlation of results. The first ecological and morphological axes were highly correlated, explaining 65% of the variance and demonstrating that the array of morphological measurements adequately captured microhabitat differences among the species. Diet was strongly related to a second trophic morphology axis that consisted of mouth orientation, presence of barbels, and gut length. Thus, both diet and microhabitat of these stream fishes, which are considered highly specialized as a result of interspecific competition (Moyle and Senanayake, 1984; Wikramanayake and Moyle, 1989), could be adequately evaluated by an ecomorphological approach.

However, most studies have simply assumed that measurements of body features should perform well as a surrogate for ecological performance measured in the field. Douglas and Matthews (1992) statistically tested the relationship between ecomorphology of fish species and their microhabitat and food use in the Roanoke River, Virginia. For 17 species of fish from the Roanoke River and its tributaries, Douglas and Matthews (1992) used Mantel tests (Mantel, 1967) to compare 3 separate data sets, including (1) a morphological data set, based on 34 measurement characters, (2) a data set on foods of wild-caught individuals, and (3) a data set on microhabitats. If morphology is to be useful as a surrogate for ecological information, then a classification of species on the basis of morphology should concur with a classification of the species on the basis of ecological traits.

In an analysis of all 17 species, which included 5 families, Douglas and Matthews found the results relatively uninformative, in that ecomorphological measurements largely grouped species by family (in multivariate cluster analyses), and those groupings matched the grouping of the species with respect to food but not microhabitat use. In essence, the ecomorphological approach merely confirmed that families may be important determinants of general trophic ecology (i.e., food use within families tends to be broadly overlapping, but less so among families, at least in temperate streams). Douglas and Matthews (1992) concluded that the ecomorphological approach should be used with caution if it is applied across families [even though this has been done commonly (e.g., Gatz 1979a, 1979b; Douglas, 1987; Strauss, 1987)], because it may merely serve to underscore unsurprising differences among families. Furthermore, Douglas and Matthews (1992) showed that both phylogenetic history and food use were related to body shape for these 17 species, with a stronger correlation of body shape to taxonomic grouping than to foods used. Motta et al. (1995) also showed (for estuarine fishes) that multivariate analyses of morphology and foods did not correspond well.

However, when analysis of Roanoke River fishes was limited to eight species of minnows (Cyprinidae), the results were much more ecologically informative: multivariate clusters of species based on morphology and (separately) on habitat use exhibited statistical congruence (Mantel test), indicating that it would be possible with some confidence to estimate microhabitat use by those fish species from an array of morphological measurements. In that a microhabitat has been considered a more important determinant of minnow community structure than feeding (Mendelson, 1975), one might conclude that for minnows the ecomorphological approach "worked" and could be most useful at a within-family level. At any rate, it would appear, at least for this group, that morphological measurements had the potential to capture important aspects of autecology of the individual species (i.e., it reflected their differences in microhabitat use). Note that this does not rule out that some portion of their microhabitat use was influenced by other species (e.g., interactive "community" control; Chapter 9), but it does suggest that the morphology of a species strongly influences the kinds of microhabitats it uses.

Wood and Bain (1995) also tested the link between morphology and ecology of stream fishes in systems in Alabama. For 15 species, a total of 21 morphological measurements of shape were determined from a truss, followed by sheared principal components analysis. Microhabitats were determined from capture in prepositioned electrical grids within the streams. Morphological shape components provided the wide separation of species within the minnow and darter families, but not within the sunfish family. Three sunfish species were strongly overlapping within morphospace. For the minnows (Cyprinidae) and darters (Percidae) there was a significant regression between principal components axes based on ecomorphology and microhabitat, suggesting a strong link between the two classifications of species. For sunfish (Centrarchidae), there was no significant relationship between morphology and microhabitat. Therefore, the results for the two families with the greatest intrafamilial morphological variation were consistent with the results of Douglas and Matthews (1992) (i.e., that ecomorphological measurements can reflect microhabitats by at least some common families of stream fishes).

Ecomorphology has been used to assess morphological features of fishes other than external shapes or sizes. Reinthal (1990a) used 24 neurocranial measurements in a principal components analysis to compare 11 species of cichlids from Lake Malawi and found that foraging behavior and diet were related to a suite of characters located in the vomerine region. For 51 species of North American minnows, Huber and Rylander (1992) used principal components analysis to compare brain morphology (13 characters, including eye size) to turbidity of their usual habitat. Huber and Rylander found evidence that turbidity of habitat was related to brain structure, but that there also was a substantial phylogenetic component to brain structure. For 189 cichlid species from African Great Lakes, van Staaden et al. (1994/95) used 23 brain-eye characters in a PCA to show substantial similarities in brain evolution related to vision in species flocks in different lakes, which suggests that similarities in feeding or activity patterns might also be high.

Overview

Multivariate analyses of ecomorphology have been carried out for a substantial number of fish assemblages since 1979. Initially, most studies were based on the assumption that careful measurement of a large number of morphological traits should allow useful predictions of ecological traits (mostly feeding or microhabitat use) of the fishes in an assemblage. On the basis of this assumption, numerous studies quantified niche diversification, species packing in morphospace, or similar patterns to test niche theory or strengths of potential competition among species. However, the most recent studies, which have statistically compared suites of species on the basis of matrices of ecological and morphological characters, have not suggested a strong correspondence between "ecomorphology" and diets, except, broadly, among families. The best judgment at this point seems to be that ecomorphological analysis of large suites of morphological characters may be more useful as a surrogate for microhabitat than for trophic performance of a species in an assemblage. Furthermore, multivariate ecomorphology may perform well among families if broad generalizations are acceptable, but if predictions of details of ecology are needed, the technique may be most useful for detecting fine-scale differences within families. Application of multivariate ecomorphological techniques across diverse taxa is very likely to first separate species merely by family groups (Gatz, 1979a; Douglas and Matthews, 1992), which seems minimally informative. However, to the extent that multiple axes allow the separation of family groupings into phenetic morphospace, broad views like that of Winemiller (1991a) may provide a robust comparative view of assemblages at very large scales.

Whether the ecomorphology of a species sets limits or opens possibilities remains to be seen, but one might predict that the morphology is the fundamental template of microhabitat use (and much behavior) that can be further modified by the presence of other species in the assemblage. What remains to be seen is the degree to which various suites of morphological features of fishes are fixed by phylogenetic history versus recently or even locally adapted. As more phylogenetic hypotheses become available from systematic research, this question can be resolved, but for now, it is difficult to estimate what portion of the ecology of a species is "old" versus "new" evolutionarily.

Intraspecific Variation in Morphology

Within species, there are two levels of morphological variation: differences between populations, and differing morphs within a single location. There can be ecologically important differences in morphological traits among populations of a single species. Toline and Baker (1993) showed differences in body shape among 18 populations of *Phoxinus eos*) by truss analysis and principal components analysis (PCA), showing a major axis in shape from deep-bodied to shallow-bodied populations, with the deep-bodied forms better at capturing evasive prey. Likewise, Foster and Baker (1995) found interpopulation differences in *Gasterosteus aculeatus* (three-spine stickleback) that ranged from deep-bodied to slender populations, with each foraging most effectively in particular kinds of habitat, and the deep-bodied populations were effective cannibals of young in nests. Schluter (1993), although considering deep-bodied and slender three-spine sticklebacks to represent separate species, showed that differences in their body shape resulted from character displacement where the species were sympatric. The two forms were differentially efficient at feeding in different habitats, and Schluter concluded that "adaptation to one habitat has occurred at the expense of feeding rate in the other."

There are also well-known examples of two or more ecologically important morphs of a species (species delimitation sometimes questioned) in a single location. Hindar and Jonsson (1982) showed food and habitat segregation between dwarf and normal-sized Arctic char (*Salvelinus alpinus*) in a lake in Norway. Four morphs of arctic charr are present in Thingvallavatn, Iceland with differences in head morphology that have a genetic basis and are related to feeding habits (Skulason et al., 1989; Skulason et al., 1993). Robinson et al. (1993) have showed the divergence of two trophic morphs of pumpkinseed (*Lepomis gibbosus*) in a lake lacking other sunfish, with the pumpkinseeds filling the niche usually occupied by bluegill sunfish (*Lepomis macrochirus*), providing a view of potential evolutionary consequences of release from competitors.

Summary: Morphological Approaches and Fish Ecology

Morphology has been studied for three reasons: (1) systematics and ecology of taxa; (2) understanding functional capabilities, and the resulting ecology, behavior, or physiology of individual species; and (3) understanding the ecological patterns of coexisting species. Comparative morphology has been the very core of studies of classification, function, and evolution of organisms since Linneaus, Cuvier, and Darwin (Lurie, 1960). Modern fish systematics continues to rely on morphological comparisons of species (e.g., Lundberg, 1992; Coburn and Cavender, 1992; Stiassny et al., 1996), along with allozyme (Cashner et al., 1992; Echelle and Echelle, 1992), cytogenetic (Amemiya et al., 1992), and a wide array of developmental, physiological, and other evidence (Smith, 1992). Studies of the functional or biomechanical capabilities of individual species have allowed the understanding of optimization of feeding or habitat use (Werner, 1977). Both biomechanics (Westneat, 1994) and comparative ecomorphology (Winemiller, 1991a) have allowed detailed conclusions about the ways species fit into assemblages.

For each of the three basic purposes for morphological studies, the different approaches to morphology (descriptive, functional, and ecomorphological) are variously useful. Descriptive morphology of body parts of individual species has been the foundation for forming testable hypotheses about the ecology of individual species (e.g., Lundberg and Marsh, 1976). Uncritical acceptance of form–function adaptation can overlook "engineering" or phylogenetically constrained traits that influence ecology of a species (Jaksic, 1981). Studies of biomechanics can test hypotheses about form–function to ask if a structure is, indeed, suitable for its supposed task and can provide mechanistic explanations for patterns revealed by ecomorphology. Comparison of structures of individual species to those of outgroups can reveal phylogenetically determined patterns within clades (Jaksic, 1981).

Broad studies from an ecomorphological perspective may be the only practical way to extrapolate functional morphology to large groups of species forming assemblages or faunas—in that detailed biomechanical studies of all functional parts of all species in a complex assemblage is probably beyond the scope of most laboratories. However, when comparative biomechanical studies can test the assumptions inherent in multivariate ecomorphological studies, the latter approach becomes much stronger. Without any direct tests of capabilities of particular structures, the multivariate ecomorphological approach provides only a hypothesis about ecological patterns of organisms in assemblages, relying on the assumption (as did Cuvier) that form reveals function. The marriage of descriptive, functional, and ecomorphological studies of fishes (e.g., Norton et al., 1995) to include performance tests of morphological characters and accommodation of phylogenetic causalities (Norton, 1995) should be the most productive approach to problems of autecological responses of individual species or assembly of species under a given suite of environmental conditions. Finally, to the extent that intraspecific variation in morphology exists (Matthews, 1987b; Toline and Baker, 1993), multivariate ecomorphology at the species level will be confounded by variance. However, ecomorphological comparisons of individual populations could help clarify rates of evolution of species or conditions under which the ecology of a species can be modified.

8.4 Hydraulics, Morphology, Microhabitat, and Food Use

The morphology of a species sets limits on its capacity to use habitats or foods in general. However, how does morphology, combined with hydraulic characteristics of a habitat and availability of foods, translate into the actual use of resources? The hydrodynamic architecture of habitats, including all tiny "microhydrodynamic" variables (small eddies, turbulence, microzones), influences, at a scale of centimeters or millimeters, the particular space in the water column occupied by a given fish or the food available to it. Current speeds, often modified by the presence of structure (Gleason and Berra, 1993), have often been considered a critical aspect of microhabitat selection by stream fishes (e.g., Baker and Ross, 1981; Matthews et al., 1982b; Wynes and Wissing, 1982; Felley and Felley, 1987; Fisher and Pearson, 1987; Gorman, 1987; Ross et al., 1987).

The most common structure in streams and lakes consists of large woody debris or rock ranging in size from sand particles to bedrock (Chapters 4 and 6), which produces a complex hydraulic variation in the environment, and provides refuge from currents. For example, Gleason and Berra (1993) found that female *Luxilus cornutus* and *L. chrysocephalus* hovered downstream from nest sites in a microhabitat where woody structure provided protection from currents. In wide, shallow, sand-bottomed streams of the North American Great Plains, sand ripples create small dead zones with little or no current. Flow over unstable sand also produces a larger sand-ridge habitat behind which current speeds may be slower. Some minnow species are commonly associated with such a sand-ridge habitat. However, the species that are found in flowing sand-ridge habitat may move to nonflowing habitats in cold water and can be found in large numbers in backwater pools in winter (Matthews, 1977; Matthews and Hill, 1979a).

Substrates larger than sand provide complex hydraulic structure. Gravel and cobble bottoms set up complex flow patterns and have associated microlayers and dead zones (Hynes, 1970; Statzner and Higler, 1986). Schlosser and Toth (1984) provide an example of two darter species with different morphology using gravel–cobble habitat differentially, with the stouter, more robust species in microhabitats exposed to flow, and the more slender species with a more flexible vertebral column living downstream or beneath stones. Gravel in flowing water also provides important substrate for feeding for lampreys, salmonids, minnows, suckers or darters, and attendant water-column fishes. Many suckers (Catostomidae), like northern hogsuckers (*Hypentelium nigricans*), actively overturn gravel by "rooting" in a haphazard pattern across the bottoms of streams, ingesting invertebrates that are exposed by this activity [and putting many invertebrates into the water column to be consumed by an accompanying entourage of omnivorous–insectivorous minnows; Reighard, 1920, pp. 20–21; Matthews, personal observation). Some darters (Percidae) of the genus *Percina* use the snout to flip over individual stones, which they then inspect for macroinvertebrates.

Flowing water brings food, a benefit, but it imposes an energetic cost on fishes (Facey and Grossman, 1990) that depends on their streamlining and swimming capabilities to use this subsidized food delivery system. Many articles link feeding of fish in streams to drift of invertebrates (e.g., Reisen, 1972). Amounts of drifting food can influence longitudinal distribution of fish in rivers (Shannon et al., 1996) or influence the size of local territories used by drift-feeding individuals (Grant and Noakes, 1987). Drift is greatest at night or peaks in early evening, apparently to minimize loss to visual predators (Hynes, 1970; Allan, 1978; Skinner, 1985; Kohler, 1985), but different invertebrate taxa drift at different schedules (Cowell and Carew, 1976). However, there is a correspondence between peaks in drift and in feeding by some fish (Elliott, 1970; Cadwallader, 1975). Drift can consist of benthic macroinvertebrates that actively abandon the substrate to drift downstream when food availability is low (Kohler, 1985), that drift to escape predation, that are incidentally entrained by flowing water, or that enter the drift by numerous other mechanisms (Brittain and Eikeland, 1988). Drift appears to be a valuable subsidy for numerous fish species (Brittain and Eikeland, 1988), delivering food to feeding stations for fishes like salmonids and minnows.

For example, some species of stream fish (*Notropis boops*, *Notropis rubellus*) are seen most often just out of mainstream currents, adjacent to, but not within, the fastest water. From this position, they can dart into swift water to capture drift items or they can eat particles that drop out of the current in eddies. Trout take foraging positions (Fausch and White, 1986) that maximize food gain relative to energy costs of holding position, and feeding of many stream fishes is influenced by drift of invertebrates (Cadwallader, 1975b). In Pennington Creek, Oklahoma, students sampled every 2 h for 24 h and compared the patterns of food in water-column drift, in benthic samples, and in stomachs of blacktail shiners (*Cyprinella venusta*). Macroinvertebrates in stomachs of fish were more similar quantitatively to those in the drift (percent similarity, PSI, = 87.3%) than to those in the streambed (PSI = 59.4%), and there was a strong correlation ($r = 0.86$) between amount of drift and number of food items per fish stomach at 2-h intervals (Matthews and A. A. Echelle, unpublished data).

Schaefer (1995) evaluated the ability of longear sunfish (*Lepomis megalotis*) and bluegills (*Lepomis macrochirus*) to occupy and feed in flowing-water environments. Longears are more common and abundant in Oklahoma streams than are bluegills, which are more abundant in lentic waters. Bluegills more nearly approximated a theoretically ideal streamlined shape (location of deepest part of body nearly one-third of body length distant from snout), and laboratory tests showed that there was less drag on the body of longears than bluegills (Schaefer, 1995). Artificial stream tests showed that longears were better than bluegills at feeding on drifting food items (Schaefer, 1995). These results suggested a direct link among morphology, streamlining, and differential ability to acquire food in a flowing environment that could result in the differential success of the two species in streams.

Bioenergetics influence the behavior of fish in flowing water. Metabolic costs of position holding or swimming can change with season and temperature (Facey and Grossman, 1990). Facey and Grossman (1990) showed that four fish species differed in their ability to passively hold position in flowing water, with increased metabolic costs when changeover to swimming was required in order to hold position. The critical swimming velocity (measuring ability of fish to swim against currents) was related to the morphology of two salmonid species (Hawkins and Quinn, 1996). Matthews (1985b) showed differences in critical current speeds for benthic position holding between two darter species that reflected their relative streamlining. Simonson and Swenson (1990) showed that maximum feeding benefits versus energy expenditures was at current velocities between 8 and 13 cm/s for young smallmouth bass (*Micropterus dolomieui*).

Much of the work in bioenergetics presently focuses on modeling designed to predict or assess growth of fish populations (Hansen et al., 1993). Bioenergetic efficiency in microhabitat patches can even be affected by shade or color background, with more movement or expenditure of energy in microhabitats lacking protective background conditions (e.g., MacCrimmon and Robbins, 1981). From the perspective of understanding assemblages of fishes in particular habitats, bioenergetics help link morphology to hydrodynamic efficiency to provide an understanding of choices that an individual or groups of individual fishes make on a day-to-day basis. A case in point involves the winter ecology of the abundant stoneroller minnow (*Campostoma anomalum*) in the Baron Fork of the Illinois River, Oklahoma.

Stonerollers are large algivorous minnows, often reaching 150 mm total length (TL) as adults and occurring in incredibly large schools of thousands of individuals (Matthews et al., 1987) that forage actively on gravel bottoms of flowing pools (e.g., 15 cm/s or more) throughout the day during warm weather. They are the most conspicuous fish in the Baron Fork in summer, moving in huge schools with their flanks "flashing" in the sunlight as they turn head and body sidewise to feed on attached algae. They feed actively in the flowing main channel through autumn, but at the onset of cold weather [e.g., water temperature below ~ 10°C (check field notes)], they largely abandon the main channel and move in large numbers into nonflowing backwaters where spring seeps keep water 1–2°C warmer than the mainstream. In such backwaters, we have regularly found stonerollers in numbers estimated in many thousands of individuals per school, behaving in a very "non-stoneroller" fashion, hanging suspended in the water column and appearing nearly inactive (Matthews and F. Gelwick, personal observation). During these times, even on brightly sunlight days, we have found few or no stonerollers in the main channel, and it appears (although we have not investigated this in detail) that they have taken the energetic tactic of abandoning feeding opportunities (more periphyton is in the main channel than in the sand-bottomed backwaters) in favor of conserving energy by near-quiescence.

In contrast, in nearby small streams, Tyner Creek, large numbers of stonerollers remain at least partly active, feeding in the mainstream pools throughout the winter. However, in small creeks, current speeds are negligible in the pools; thus, fish may have a different cost–benefit balance in remaining active in these small habitats. It would appear that this is an example of a single species finding two different solutions to the problem of energetic balance, even though they are in habitats only a few kilometers apart and there is likely much gene flow between the sites (Tyner Creek is a direct tributary to Baron Fork). The comparative benefits of the two tactics (winter-active versus winter-quiescent) remain to be investigated, but it would be of much interest to know which results in the greater energy reserves (lipids, etc.) for production of eggs in the spring.

Larval Fish Success Relative to Hydraulics of Habitat

Flow patterns in streams (and lake zones) have strong impact on the success of larval fishes. Some species, like darters (Percidae), may spawn in flowing riffles, where eggs develop in interstitial spaces among or under stones, but the very young and minimally developed larval subsequently drift downstream to available pool (depositional) environments where they feed and develop morphologically (Paine, 1984). Coon (1987) emphasized the variation among darter species in the survival of young during a flash flood, depending on differences in nest location and in microhabitats used by larvae. However, the bowl-shaped nests of most fish in stream pools may offer some protection from washout (at least at base flow) for eggs and larvae that remain in the nest. One of my students in summer at the OU Biological Station demonstrated that there were significant differences in depth and shape of nests of longear sunfish in different flow environments. Once larvae (e.g., minnows) leave the nests in typical small streams, they occur in "clouds" of individuals, suspended in the water column in zones with apparently little flow. In snorkeling Brier Creek, Oklahoma, B. C. Harvey (personal communication) was able to directly observe locations of larval minnow "clouds" in stream pools. My snorkeling observations agree with his— that postlarval fish typically occur as a very distinct patch of individuals while they are at small postlarval size. As minnows grow, they disperse more, but I still typically observe very small juvenile minnows (e.g., *Notropis boops*) in dense schools associated with algae columns or in beds of *Chara*. Changes in discharge and flow velocities have profound impacts on larval fishes in streams. Harvey (1987) showed not only a near-complete washout of larval fishes from a 1-km reach of Brier Creek, Oklahoma, on two occasions, but that the greatest flux of individuals downstream came early on the rise of the hydrograph. Long before the spate peaked, virtually all small larvae had been washed out of the reach and killed (based on the examination of individuals collected in his drift nets) (Harvey, 1987). Therefore, there may be a fine division between flows that are beneficial, bringing increased downstream transport of food items or clearing stream bottoms of accumulations of detritus, and the slightly greater discharge that can wash out and kill a cohort of larvae. Even though most species can renest, loss of a cohort represents a substantial cost to a population.

In larger rivers, the role of flow in creating or destroying nesting sites and/or nurseries for larval and juvenile fish may be even more important than in small streams. In large rivers, the main channel is often an inhospitable habitat for nesting or for development of larvae to juveniles, because of depth, unstable substrates, and strong currents. Accordingly, much reproduction in large rivers is associated with backwaters (Finger and Stewart, 1987), side channels, or other available microhabitats away from the main channel. Under normal (nonregulated) conditions, a river forms such backwaters during channel-forming floods, and periods of moderately elevated flow may help to keep the backwaters connected to the main channel so that they are available for adults to use as reproductive microhabitat.

Rivers form backwaters into which they predictably (or irregularly) overflow (Welcomme, 1979, 1985; Power et al., 1995a and c). Anthropogenic changes in flow regimes (e.g., those found downstream of many dams) may (1) prevent the very high flows that help to form and maintain desirable backwater habitats and (2) create a daily or other very frequent stage rise with undesirable flushing flow on a schedule that does not permit available backwaters to provide good nursery habitat. For example, the Colorado River downstream from Glen Canyon Dam was formerly characterized by substantial warm backwaters which numerous fish species used as spawning or nursery sites. Now, the dam operation results in the release of water too cold for spawning, and an essentially daily flushing of those formerly quiescent backwaters on a schedule too frequent to permit successful rearing of larvae that might be produced (Minckley, 1991). Minckley (1991) also points out that modifications of river systems by dams, changing temperature, and flow regime may invite proliferation of non-native, introduced species, which provide the final variable forcing native fishes to extinction. This extreme situation underscores the impacts that dams or flow modification can have on riverine fishes and the desirability of maintaining natural temperatures and flows regimes in streams.

Habitat Selection, Patch Choice, Effect on Feeding or Reproduction

To what extent do animals "learn" and to what extent is behavior "innate"? "Innate" behavior, in terms of modern evolutionary ecology has come to refer to behavior directed by inherited gene complexes that have developed through decisionmaking by individuals over evolutionary time. With respect to habitat or patch choice, foraging sites, optimal foraging, and similar concepts, the evolutionary perspective lessens the role of the individual's minute-by-minute choices and emphasizes that animals will make the best moves to accrue energy, survive, reproduce, and enhance their enclusive fitness. In other words, individual "choice" has been subsumed in recent decades under the concept that animals, including fish, "play the odds" and over the course of a lifetime will take prey items, forage in particular places, or choose habitats that confer greatest fitness. There is also interest in variance in behavior; that is, not all individuals in a population choose habitats or foods in the same way [e.g., individual behavior of trouts, Bryan and Larkin (1972); individual behavior of bluegill (Gotceitas and Colgan, 1988)].

For nearly 30 years, ecologists have considered habitat use and food searching as a function of optimal foraging (Schoener, 1971; Pyke, 1978; Werner, 1974, for fish). Foraging optimally, fish use microhabitats and foods that maximize energy return for energy invested, thus gaining the greatest net benefits for growth, maintenance, and reproduction. With respect to habitat use or patch choice, two related concepts include the Ideal Free Distribution (Fretwell and Lucas, 1970;

Fretwell, 1972) and the Marginal Value Theorem (Charnov, 1976). The Ideal Free Distribution suggests that animals in unevenly profitable habitats, with "perfect knowledge" of their options, "free" to move unrestrained among habitats, and with inconsequential travel time among patches, will first occupy the "best" patches of habitat (i.e., those that return the most benefit for each individual, on average). Then, if population density or average number of animals per unit area increase beyond a given point, some individuals will spill over into previously "next best" habitat patches, specifically at the point at which net profit or benefit to individuals becomes equalized between the two habitat patches. Additional population increases or densities per unit area will result in a trickle-down effect, into successively less profitable (originally) habitat patches, as the numbers of individuals per patch and productivity of the patches [or other commodity (e.g., shelter, mates, etc.)] reach a balance at which all individuals in the population enjoy equivalent net gains. The original Ideal Free Distribution models have been subsequently modified to incorporate factors like incomplete knowledge of the total environment, unequal competitive ability of individuals, competitive interference, predators, and travel time between discrete patches (Åström, 1994; Kennedy and Gray 1993).

One important modification of ideas in optimal foraging and Ideal Free Distribution is consideration of the situation in many real-world environments that patches of useful habitat or foraging sites are not continuous and occur as discrete units embedded within a largely nonusable background matrix. For example, for water-column-dwelling stream fishes, individual pools may be the functional habitat patches (Matthews et al., 1994), with the intervening riffles largely inhospitable habitats through which a pool fish must pass (and incur an energetic cost—particularly if moving upstream).

A classic example is that of Pyke (1978) who studied foraging by hummingbirds as they moved between inflorescences of wild plants, obtaining nectar from individual flowers. A major question with regard to habitat use or foraging in such discrete habitats is that of "when should an individual leave." The Marginal Value Theorem of Charnov (1976) can be summarized by the postulate that an animal will arithmetically maximize its gain if it departs a particular foraging patch to go to another when the net gain in the original patch declines (because of the withdrawal of resources or other factors) to a level that is equal to the average within the total environment available to the animal. In other words, if an individual remains to forage in a patch when its value drops to less than the environmentwide average, it inevitably "loses." However, one variable not yet incorporated into this simplified version of the model is that of travel time to another more profitable patch (e.g., Åström, 1994), and the level of certainty of finding another suitable patch in an efficient manner (i.e., without perfect knowledge of the landscape), the individual must incur a cost of searching for the next patch in addition to the energy needed to move to it. Finally, it should be realized that some movement among habitat patches is necessary as "sampling," in order for an individual to improve its imperfect knowledge about other habitat patches (including, perhaps, the location of predators).

One of the problems in attempting to understand patch use, habitat selection, or foraging in stream fishes is that of asking if we can directly apply any of the ideas from optimal foraging theory. Most optimal foraging seems to imply that the individual animals are "always" foraging (i.e., actively seeking food throughout the day). Most optimal foraging and related theory was developed with respect to endotherms (birds, mammals) or invertebrates with very high energy demands (e.g., bumblebees). This is not true for many fishes, which, as ectotherms, have substantially lowered metabolic demands at colder temperatures. Although some fish species (e.g., *Campostoma anomalum*) appear to forage actively and consistently throughout the day (in warm weather), others give the impression at times that they are not foraging and that they spend considerable time resting, suspended in the water column, and relatively inactive (cf. Mendelson, 1975). This may be particularly true for large-bodied, piscivorous fishes that take meals at only irregular intervals. Additional evidence for "active" versus "inactive" periods for fishes comes from Brett's (1971) finding that salmonids in lakes move between warmer foraging and colder resting or "digesting–growing" habitats in related to vertical thermal stratification in the water column. Thus, for fishes, patch choice may not be driven by food or active foraging for prey during much of their time. This may all be just a matter of degree (i.e., that fish may engage in optimal foraging just as much as endotherms), but it would seem that they also have the option to not forage; thus, for fishes, there may need to be "optimal habitat theory" to include important variables like resting places (with shelter, lack of currents) within habitat patches that provide safe or low-energy-cost sites.

8.5 Reproduction and Life History

Much as the environment sets limits on morphological configurations that are efficient in a particular habitat, environmental constraints may also set limits on life-history traits that are successful or enhance fitness of a species in a particular situation. Only those fishes whose hydrodynamic, trophic, and reproductive traits all fit the available microhabitats will be resident in a given location. To the extent that not all traits "work" in a single location, a fish may be a transient, and its migrations among sites may reflect requirements at various stages in the life cycle of the species. Furthermore, life-history traits that are advantageous in one kind of environment (e.g., seasonally fluctuating rivers) may also be advantageous in new habitats (e.g., man-made reservoir), as shown for barb (*Barbus anoplus*) that had early maturity and high reproductive efforts (Cambray and Bruton, 1984).

Life History of Fishes

"Life history" has two usages in fish ecology. The original, broader meaning encompasses virtually all of the "natural history" of a particular fish species, including where and how a species lives, what it eats, its reproductive traits, parasites, diseases, and other aspects of biology (e.g., Mayden and Walsh, 1984). The life-history publications on individual fish species during the last 25 years at the Illinois Natural History Survey typify the best of this approach (e.g., Page and Smith, 1971). Such studies are indispensible for understanding the basic biology or planning management of a species, and they provide the raw data from which many of the syntheses of life history have been derived. One of the most comprehensive "life histories" of an individual species is for *Etheostoma blennioides,* by W. E. Fahy (1954), which includes details on distributional, trophic, and reproductive ecology of this species. Highly detailed and quantitative information on early life-history stages of individual fish species also can be found in volumes like Wallus et al. (1990).

In recent decades, "life history" has taken on a more narrow perspective, focused mostly on the suites of reproductive or "life-table" characteristics (such as egg size, number of offspring, age at first reproduction, sexual size dimorphism, survivorship, and frequency of reproduction) or in the evolutionary responses of a species expressed in sexual selection, mating strategies, and parental care (Heins and Matthews, 1987), all as modified by predation, productivity, or other extrinsic factors. These variables or suites of variables hypothetically should be adapted to maximize the reproductive success of a species in a particular environment. However, the traits possessed by a particular population of a species could have its origin in the evolutionary past or represent fine-tuning of that population (Leggett and Carscadden, 1978) for the extant environmental conditions.

This chapter focuses more on the narrow definition of life history, asking how or in what ways those traits maximize the reproductive success of a species in particular physical habitats. The approach will be to consider the major components that comprise the life-history strategy of a species (including intraspecific variation). Life-history theory focuses at different levels of strategic responses, considering suites of life-history characteristics as adaptations that maximize reproductive success under different circumstances, within constraints imposed by fundamental morphology or phylogeny. There have been a variety of attempts to correlate suites of these characters with the environment of the fish with respect to habitat types (e.g., Heins, 1991; Winemiller and Rose, 1992) or placement of a location on an $r-K$ continuum (e.g., Paine, 1990), but these approaches have typically included a large number of species, making it impractical to incorporate intraspecific variation.

This chapter first reviews various syntheses of life-history strategy in fishes, and then considers individual traits that are components of the life-history strategy of a species, beginning with the attainment of mature reproductive status. Intraspecific variation in these traits is considered, as is the influence of the environment on each. The component traits included are as follows:

1. Age at first reproduction and reproductive life span
2. The reproductive season and regulating conditions like temperature, photoperiod, or food availability
3. Prespawning actions, like nest-building or establishment of territories.
4. Finding or gaining access to mates; mate choice; mating systems
5. Actual spawning, to include clutch factors like number of eggs, of what size, and semelparity versus iteroparity (within and among years), and physical constraints to clutch size, like body size
6. Parental care; guarding
7. Early life history and survival of larvae
8. Growth, survival, and recruitment to the reproductive population; mortality patterns

Some phenomena, like sexual dimorphism, costs of reproduction, and evolutionary trade-offs can affect more than one stage of the scheme above. Additionally, evolutionary constraints to life-history strategies can be detected through a broad comparative view across many taxa. Emphasis will also be placed on intraspecific differences among habitats, interspecific differences in reproductive mode within a single habitat, and on the genetic basis of life-history traits. All of these traits and strategies will, in keeping with the theme of this book, be considered from the fundamental view of why certain fish occur together in local assemblages (i.e., what works in a given habitat, and what does not). Thus, this section will address applicability of these traits for conditions (hydraulics, microhabitats or environmental stability) of local environments and the ways that different traits may be beneficial in differing environments.

Overview of Life-History Patterns

Reproductive diversity in fishes is enormous (Breder and Rosen, 1966), with noteworthy interspecific (and in some cases intraspecific) variation in most traits. Some of the fundamental traits relative to reproductive success of a species include size at maturity, number of eggs, size of eggs, reproductive season length or timing, and longevity (Wootton, 1984), as well as the number of clutches per female per year, and the variation in those traits among populations or individuals, or within an individual. Egg production combined with longevity provides the familiar l and m_x on which life tables and lifetime reproductive expectation or output are based, but these have rarely been measured for fish in the wild. For live-bearing fishes, fecundity is determined by the number of embryos, interbrood interval, and length of breeding season (Hubbs and Moser, 1985).

Notable efforts to document or understand life-history diversity include Breder and Rosen (1966), Balon (1975, 1984), Page (1985) and Winemiller and Rose (1992). Breder and Rosen (1966) is a comprehensive catalog describing reproductive modes of marine and freshwater fish groups, reviewing virtually everything known about fish reproduction to that date. The organization of Breder and Rosen (1966) is by phylogenetic groups, with detailed descriptions of breeding behavior and life-history traits for individual species. Then follows [in Breder and Rosen (1966, pp. 620–675)] charts summarizing the reproductive modes of most freshwater and marine families, worldwide, with respect to secondary sex characteristics, mating system, breeding sites, kinds of eggs, parental care, and migration. However, although highly useful, Breder and Rosen (1966) is largely a listing of traits of species and families, without attempting a synthesis. Detenbeck et al. (1992) also compiled for a large number of North American freshwater fishes fundamental data on life-history characteristics. Miller (1996) summarized costs and benefits of small body size, and its implications for life-history.

Balon (1975, 1984) provided a classification of evolution of reproductive modes in fishes, detailed transitions in life stages of fishes, and described general size relationships among gametes, larval sizes, and adult body size. Wootton (1984) compiled data on size at maturity, life span, months of spawning, fecundity, egg diameter, time to hatching, and migration for 162 freshwater Canadian species, showing correlations of the various parameters and classifying the species into 10 clusters having similar traits. The strongest correlations were between body size and life span, body size and egg size, and spawning month and egg size (due to autumn-spawning salmonids with large eggs). Wootton (1984, p. 10) concluded that in a given geographic region, only a certain number of types of reproductive strategies are possible, in that environmental factors, morphology, and phylogeny constrain the kinds of strategies that can evolve. Paine (1990) showed differences between darters (Percidae) with small and large body size, with the latter exhibiting the combined traits of fast growth, maturation at large body size, larger clutches, shorter spawning seasons, and longer reproductive life spans.

Page (1985) provided a comprehensive summary of life-history traits of North American darters (percidae), with emphasis on reproductive behaviors and egg-laying strategies. Page classified darters by egg laying traits including: "broadcasting", "stranding", "burying", "attaching", "clumping", and "clustering". His classifications of egg-laying and reproductive strategies were used as characters (along with morphology and other features of behavior) to test established phylogenies of darters. Correlations of general life-history traits to specialized reproductive traits for these diminutive fishes included: production of larger but fewer eggs, early reproduction, semelparity (repeated spawnings), shortened life-spans, and increased population density (Page, 1985).

Winemiller (1989, 1992), to tie life-history traits to habitat, suggested that the original r versus K dichotomy of reproductive strategies (MacArthur and Wilson, 1967; Pianka, 1970) failed to capture some important variance in fish reproduction and suggested three basic patterns in freshwater fish life history, including "equilibrium," "opportunistic," and "seasonal" modes. Winemiller and Rose (1992) reviewed 16 life-history traits for 216 North American fish species of 57 families by multivariate analyses. They showed fundamental differences between marine and freshwater species in modal values for numerous traits, like clutch size, egg size, spawning season, and degree of parental care, but patterns for the marine and freshwater species, in separate principal components analyses (PCA) were very similar. For freshwater fishes, a PCA for 12 traits showed a continuum from fishes having late maturation or maturation at large size, large clutches, small eggs, fewer spawning bouts, and little parental care, to small or early-maturing species with small clutches, large eggs, longer spawning season, and more parental care, mimicking, at a broader taxonomic scale, the results of Paine (1990). A second PCA reduced to 5 traits for 82 species showed two major gradients, with the first placing taxa on a continuum from small size, small clutches, and more spawning bouts to those with larger size, larger clutches, and fewer spawning bouts. The second axis identified species on a continuum from small egg size and little parental care to the opposite traits. Winemiller and Rose (1992) found a strong effect of phylogeny in overall patterns, with some higher taxa showing particular suites of the traits, but some speciose freshwater taxa, like the cypriniforms and perciforms, included species that were distributed widely through multivariate space, suggesting differentiation within major groups of fishes. Bart and Page (1992) also showed strong differentiation in reproductive traits for species in the family Percidae, with respect to body-size-related traits. Winemiller and Rose (1992) also showed a strong relationship between suites of life-history traits and the habitats occupied by the taxa (their Fig. 5a).

From a review of previous life-history classifications (e.g., Mahon, 1984; Wootton, 1984; Winemiller, 1989; Paine, 1990; and others) and their findings for North American freshwater and marine species, Winemiller (1992a) and Winemiller and Rose (1992) hypothesized a triad of life-history strategies termed "periodic," "opportunistic," and "equilibrium." According to their classification, periodic fishes delay maturation to a size sufficient for producing large clutches, often with synchronous spawning and promoting survival of adults during adverse environmental conditions (e.g., tropical dry seasons). Egg size is small in periodic fishes, but growth of larvae and young-of-year is rapid. Periodic spawners also tend to be highly migratory, moving to best places at the right time within a relatively predictable seasonal environment.

Opportunistic strategy fishes typically are of small body size, with early maturation, frequent reproductive bouts over a long spawning season, rapid larval growth, and high population turnover rates, with a resulting high intrinsic rate of population increase, and populations of these small fishes are often present at high density in spite of high adult mortality (Winemiller and Rose, 1992). This strategy is appropriate for disturbed habitats, but differs from the classic "*r*-strategy" by having small instead of large clutches (Winemiller and Rose, 1992).

Winemiller and Rose (1992) note that their "equilibrium" strategy matches those traits considered "*K*-strategy" by Pianka (1970) and other authors, with adaptation for life in resource-limited environments. These traits include large eggs, parental care, and production of small clutches that develop into individuals being at an advanced stage when they assume living independent of the parent(s). However, Winemiller and Rose showed that the equilibrium-strategy species tended to be small bodied, in contrast to the predictions from *r* versus *K* theory that *K*-selected life histories would prevail in large-bodied taxa.

Finally, Winemiller and Rose (1992) note many examples of taxa that are intermediate between the extremes of this three-strategy system for viewing fish life history. They note, for example, that sunfish (*Lepomis*) are periodic or seasonal spawners but have moderately large clutches and guard nests. I would also point out that sunfish can produce numerous clutches in a season, and often renest if a spate destroys original nests (personal observation, Brier Creek, Oklahoma). Berra (1987) showed that Australian grayling (*Prototroctes maraena*) could readily repopulate streams in a harsh and unpredictable environment by virtue of its having an amphidromous life cycle, with high fecundity and potential early spawning. This combination of traits seems appropriate for its habitat, but it is also intermediate within the scheme of Winemiller and Rose (1992).

To summarize, there are numerous schemes for describing and categorizing overall life-history strategies for freshwater fishes. Breder and Rosen (1966) provide the most comprehensive descriptive summary, worldwide. Balon (1984) and others have provided various schemes for classification of reproductive traits, and Winemiller and Rose (1992) provide the most detailed and integrated assessment of broad life-history patterns to date. In the context of these large overviews, I now turn attention to the components of the life history of a fish species within the context of its life cycle, beginning with attainment of reproductive maturity and its environment.

Attempts to understand why particular suites or combinations of suites of reproductive characters work under certain conditions has been the subject of theoretical life-history models such as optimality, trade-offs, bet-hedging, and evolutionary stable strategies (ESS) (Gross, 1984) that consider not only benefits but also costs of reproductive investment, constraints on reproduction, and trade-offs between current reproductive effort and future reproductive value [as summarized by Pianka and Parker (1975)]. These approaches have proved useful, but most studies focus on subsets of characters. However, Tuomi et al. (1983) argued that it was not correct to assume a priori that reproductive effort had costs in survival and future reproduction, and more attention should be paid to whole organisms and their total life history and the elimination of unfit strategies instead of seeking optimization of traits. However, to understand either optimization of traits or elimination of disadvantageous traits (Tuomi et al., 1983), these complex suites of characters need to be examined in the context of each stage in life history of a fish species in order to assess how variation and trade-offs associated with that stage influence life history as a whole.

Age at First Reproduction and Reproductive Life Span

As suggested above (e.g., Winemiller and Rose, 1992), a delay in maturation until attainment of larger body size generally confers on a fish a capacity for larger egg volume, and large body size at maturity is often related to longevity or lifetime reproductive span. Many studies show intraspecific differences between populations in life-history traits, including age at first reproduction, which can be related to thermal effects like winter mortality (Conover, 1992) or to food availability (e.g., Jonsson and Sandlund, 1979). Baltz and Moyle (1984) showed that populations of tule perch (*Hysterocarpus traski*) had marked differences in age at first reproduction, which related to longevity and size of broods. Nigro and Ney (1982) showed that a population of alewife (*Alosa pseudoharengus*) stocked in a lake south of its normal range had earlier maturation, combined with shorter life span and high growth and fecundity relative to populations in the north. Garrett (1981) showed that members of a pupfish population in a stable environment matured at a larger size and lived longer than did those in a fluctuating environment, but had smaller ovaries and produced fewer, larger eggs.

Additionally, within a population, some individuals may mature earlier or at smaller size than others, often as a result of the existence of alternative reproductive strategies (below) or in response to predation pressure (Belk, 1995). Within chinook salmon (*Oncorhynchus tshawytscha*), there are two life-history types, with one maturing earlier and at a smaller size, possibly mediated by temperature and photoperiod factors that influence potential for growth (Taylor, 1990).

Stearns (1983) showed marked variation in age at maturation in *Gambusia* that had been introduced to various bodies of water in Hawaii, as well as marked plasticity in age or size at maturation, and Hubbs (1996) has shown large differences in interbrood interval among populations of several *Gambusia* species. Stearns and Crandall (1984) provided a mathematical and empirical argument that plasticity in timing of maturity is adaptive, allowing fishes to optimize their fitness in differing demographic situations. Thorpe (1994, and references therein) argued that age-at-maturity differences among salmonid populations suggested multiple solutions as evolutionary stable strategies (ESS) within species. Thorpe (1994) speculated that lipid storage in excess of some critical level initiates maturation in some salmonids. Baylis et al. (1993) showed that differences in age or size of smallmouth bass at first reproduction alternated between generations in a northern lake, but that these differences (young versus old) at onset of spawning were mediated by the environment (i.e., were not a genetic polymorphism). Reznick and Braun (1987) showed that differences in timing of maturation and size at maturation for male *Gambusia affinis* was related to the presence of mature males in the population and to timing of fat storage. Immature males delayed maturation in summer in the presence of existing mature males but stored fat which allowed them to overwinter after autumnal maturation. Males of livebearing fishes generally are thought to cease growth upon maturation, but Snelson (1982, 1984) has reported continued growth of male sailfin mollies (*Poecilia latipinna*) subsequent to attainment of sexual maturity.

It is well known that latitudinal differences exist in life-history traits of fish species, both in marine (e.g., *Menidia* on the North American east coast) and fresh water. Many northern populations are slower growing, later in onset of reproduction, and potentially longer-lived than conspecific populations farther south in warmer temperatures (Carlander, 1969, 1977). Mills (1988) showed a range in age at first reproduction from 1 to 13 years of age for minnows (*Phoxinus phoxinus*) from south to north across their range, with concomitant trade-offs in growth and clutch size. These differences can often be attributed to physiological phenomena such as food conversion schedules at various temperatures and to length of temperature-mediated growing seasons at different latitudes. Latitudinal differences among populations is not limited to large-bodied or long-lived fishes: Matthews (1987b) found that male red shiners (*Cyprinella lutrensis*), which are generally small bodied and short-lived, nonetheless had minimum adult body size that was smaller in the southern relative to the northern parts of its range.

Factors Regulating Length or Timing of the Reproductive Season

Most freshwater fish species have seasonal reproduction, with peaks in reproduction in spring or summer (Wootton, 1984; Settles and Hoyt, 1978), but darters (Percidae) commonly spawn during cold months [e.g., November to April (Hubbs et al., 1968; Marsh, 1980)], and many salmonids spawn in autumn. Heins and Machado (1993) reported spawning of *Etheostoma whipplei* from February to May, which they considered typical for many darters of southeastern United States, and Knight and Ross (1992) showed the bayou darter (*Etheostoma rubrum*) to be a spring to summer spawner. For other species of temperate freshwater fish, the period of spawning can be very brief [e.g., brown bullheads (*Amieurus nebulosus*) in Michigan (USA) mostly reproducing during no more than 15 days each year (Blumer, 1985)]. Within a single stream, species can differ markedly in seasonality of spawning (DeHaven et al., 1992).

Environmental temperatures both initiate and cause cessation of reproduction in many species, but temperature cues are often confounded with photoperiod. DeHaven et al. (1992) found that in one stream, the spawning of *Cottus bairdi* and *Rhinichthys cataractae* were regulated both by water temperature and photoperiod. Hatching time and attainment of various stages of embryonic development are highly temperature dependent (Hubbs et al., 1969). High environmental temperatures appeared to be responsible for termination of breeding activities in

Menidia beryllina (Hubbs and Bailey 1977) and in many darter (Percidae) species (Marsh, 1980; Hubbs, 1985). Lehtinen and Echelle (1979) showed that bigeye shiners (*Notropis boops*) in Brier Creek, Oklahoma, terminated spawning and exhibited ovarian regression earlier in a year with drought conditions and high temperature than in a cooler year. Lehtinen and Echelle (1979) also showed latitudinal differences in length of the reproductive season for *Notropis boops*, with a protracted spawning from April to August in Oklahoma. I have collected red shiners (*Cyprinella lutrensis*) as small as 14 mm in the South Canadian River of central Oklahoma during the month of January, strongly suggesting an even greater extension of this species into mid or late autumn spawning.

Temperatures at which fertilized eggs can develop at least partly set limits to spawning seasons of fishes. A large number of studies by Clark Hubbs (e.g., Hubbs, 1961a; 1961b; 1965) and colleagues have shown a correspondence between environmental temperatures and hatching success, with species spawning at cooler temperatures having eggs more tolerant of cold water, and vice versa, as well as locally distinctive variation among populations of the same species (Hubbs, 1961b, 1965; Hubbs and Armstrong, 1962; Hubbs and Strawn, 1963; Echelle et al., 1972b; Wilson and Hubbs, 1972). Wilson and Hubbs (1972) suggested that summer-breeding fishes may be more tolerant of temperature fluctuations than are winter-breeding fishes. Hubbs and Burnside (1972) also showed that within a single species, developmental sequences during ontogeny could be shifted by environmental temperatures, instead of being canalized into a fixed sequence of development of morphological traits. Thus, environmental temperature, and a matching of spawning time of a species with the developmental capabilities of its larvae may be a critical feature of seasonality. The alternative hypothesis relative to differences in spawning seasons among species is that by staggering spawning times among species, there will be less interspecific competition for food by early-feeding larvae or juveniles. Both of these competing hypotheses (developmental temperature ranges versus feeding advantages) might be operative in natural assemblages, but to my knowledge, no comparison has been made relative to fitness of the progeny.

Single or Repeated Spawnings?

Many fishes spawn more than once in a season [e.g., *Cyprinella* species (Gale and Gale, 1977, Gale and Buynak, 1978)]. Murphy (1968) expanded from the review of iteroparity (more than one reproductive bout in a lifetime) by Cole (1954) to argue that repeated reproductions (within or among years) dominated when juvenile mortality was high or variable. Since that time, a huge literature has developed regarding trade-offs between maximizing immediate reproductive output (at the cost of decreased chance for future reproductions) and "bet-hedging," or holding back some reproductive effort at the present in order to increase the potential to survive to reproduce again at some future time. The winning solution is the one that maximizes lifetime reproductive output of viable offspring, in the context of a particular environment. Thus, in environments in which adult survivorship is likely to be low, it would seem best to reproduce early, and as much as possible. The only problem with this approach is that in risky or unpredictable environments, the production of all possible offspring in one "big bang" might have the effect of placing young into the environment at just such a time as a catastrophe would destroy them all.

Thus, in Brier Creek, Oklahoma, a system prone to unpredictable, erosive floods that kill eggs or larvae (Harvey, 1987), a fish placing all its reproductive output into a single lifetime bout would run a high risk of losing 100% of that output and be genetically dead. The best solution may be to parcel out the reproductive effort over time, "betting" that adult survivorship to make additional cohorts will offset reduced reproductive output early on. This problem is obviously made more difficult by the fact that adult sunfish in Brier Creek may be vulnerable to unpredictable drought periods that lower pool volumes, create oxygen stress, and make adult sunfish vulnerable to avian or terrestrial predators, and also by unpredictable food shortages under such circumstances.

Additionally, if solutions to such questions are, in reality, a result of evolution in different systems (Partridge and Harvey, 1988), how much has anthropogenic tampering with stream or lake systems (exacerbating schedules of flood or drought, etc.) changed those systems from being the kinds of systems in which the fish evolved? Partridge and Harvey (1988) called for more empirical work testing life-history theory in the field and argued that allocation of reproductive effort evolved in response to different environments within the constraints of genetic variance and evolutionary history. However, I wonder how much the present-day environments in which we find fish resemble the environments in which they evolved, even in the same stream or lake system. Sunfish in Brier Creek, which now floods and often goes intermittent, may have ancestors that a century ago lived (and had for many generations) in a much more constantly flowing system, which would make the rules of the game under which the species would have evolved iteroparity versus semelparity, or bet-hedging within a single season quite different. How much has anthropogenic effect uncoupled mating systems or clutch parameters from the systems in which they evolved?

The degree of iteroparity is not fixed within a species and may differ widely among populations. Legett and Carscadden (1978) showed a strong latitudinal gradient in repeated spawning in American shad (*Alosa sapidissima*).

In what is probably the most thorough study of reproduction and life history in any species to date, Raffetto et al. (1990) showed that male smallmouth bass (*Micropterus dolomieu*) surprisingly spawned on average only one time in their entire lifetime! In a closed-lake population, size at age 3 seemed to determine when an individual would spawn, and the opportunity to breed may be the most important event in the entire life of an individual in this lake (Raffetto et al.,

1990), with the trait of "being able to breed" (at all) being selected more than reproductive success among breeders.

Prespawning Phase of Reproduction

Lampreys, some minnows, and salmonids build or use nests (or redds, for salmonids) in gravel (Reighard, 1908), which must be of an appropriate size for manipulation by the fishes. Various lampreys and minnows use their mouth to move gravel to build nests. Salmonids clean redds by fin action. Successful nests of smallmouth bass (*Micropterus dolomieu*) in an Ohio stream were those placed where they were protected from increased current speeds during a flood (Winemiller and Taylor, 1982). Many other species select spawning sites in crevices in wood or stone (Gale and Buynak, 1978; Wallace and Ramsey, 1981) or beneath stones (Page and Mayden, 1979). Some stream fishes are known to spawn in nests constructed by other species (Robison and Buchanan, 1988; Vives, 1990), and the association can be obligatory for the "guest" species [e.g., yellowfin shiners, *Notropis lutipinnis* (Wallin, 1992)]. Wallin (1992) also suggested that the "host" species that constructed the nest can benefit, in that the presence of eggs of the "guest" species in the nest might dilute the effects of egg predators.

That there is a cost to males as a result of reproductive activities, including energy-consuming actions like nest-building and nest-guarding, was confirmed by Jennings and Philipp (1992), who showed that high levels of male reproductive investment were related to less somatic growth for populations of longear sunfish (*Lepomis megalotis*).

Mating Systems or Mate Selection

Mating Patterns

Fish are well known for their tremendous variation in numbers of individuals involved in matings, ranging from some cichlids that pair for life to open-water broadcast spawners in which hundreds or thousands of individuals may all release gametes without any discrete pairings of individuals. Most fish species show intermediate matching of males to females, with many species (like darters) typically having a temporary pairing for fertilization of ova (Winn, 1958), or spawning in small groups of three or more individuals with more than one male accompanying a single female in spawning (e.g., Reighard, 1920; Burr and Morris, 1977). Many minnow species spawn in groups, often in gravel nests of other species (Robison and Buchanan, 1988; Gleason and Berra, 1993).

Sexual Selection

The ability of a member of one sex to identify "best" partners among individuals of the other sex should be selected for in a species, because appropriate choices for a mate confers advantages on the chooser's genes (or may help the chooser avoid a "bad" male, one with parasites, risky behavior, etc.). Examples exist across many taxa of fishes in which male–male competition, female mate choice, or (more rarely) male mate choice (Pyron, 1996b) influences the outcome of which male mates with which female fish. For many fish species, female choice of mates seems to dominate reproduction, and females have been shown to chose males on the basis of sperm availability or immediate prior spawning history (Nakatsuru and Kramer, 1982), freedom from parasites or apparent "good genes" conferring parasite resistance (Milinski and Bakker, 1990; Houde and Torio, 1992), presence of eggs already in the next or existence of morphological "egg mimics" (Knapp and Sargent, 1989), male color patterns (Endler, 1983; Hunde and Hankes, 1997), size of males relative to the choosing females (Downhower et al., 1987), absolute male size (Hughes, 1985), size of males relative to the ability to guard nests (Sabat, 1994a), male color intensity versus risk of predation (Breden and Stoner, 1987; Stoner and Breden, 1988), or complex mosaics of male traits (Kodric-Brown, 1993; McLennan, 1993). Female preference for male coloration may be population-specific (Bakker, 1993; Hunde and Hankes, 1997) and under either environmental or genetic control (Bakker, 1993). Endler and Houde (1995) showed that female guppies (*Poecilia reticulata*) from different localities in Trinidad chose males from their own populations over those from other localities, and that populations differed in the criteria used by females for chosing males. Female choice in *P. reticulata* appears highly complex, depending on a combination of male dominance, courtship intensity, and coloration, resulting in successful mating; thus, both male competition and female choice are active in this species. Additionally, for at least one darter species (*Etheostoma spectabile*) in which males have bright nuptial coloration, there was no evidence that females selected most colorful males, suggesting that other less obvious factors, including male behavior, were more important in their obtaining matings (Pyron, 1995).

Male–male competition is also a common mode of achieving fertilizations among freshwater fishes, and life-history theory generally predicts that a male can increase his fitness (number of successful offspring produced in a season) by increasing the total number of matings (Loiselle, 1982). Theoretically, male–male competition can result in evolution of males larger than females (Parker, 1992). This competition can consist of active male–male combat, as in salmonids, or many others compete for females by construction of breeding territories or nest sites [sunfish (Keenleyside, 1972); pupfish (Echelle, 1973)] combined with active attempts to coax females into spawning and aggressive displacement of intruding males (Keenleyside, 1972). In this situation, female choice can be active, but aggressive defense of the nest against intruders (especially other parental males), comprising male–male competition, enhances the chances of the resident male to fertilize eggs.

Male mate choice is less common among fishes but does occur, because a male correctly recognizing the relative value of females for producing offspring can most efficiently expend his sperm. Male pupfish (*Cyprinodon macularius*) aggressively displaced small females but actively selected larger individuals, presumably because mating with ripe or large females can maximize the number of zygotes fertilized per unit male effort in courtship (Loiselle, 1982). McLennan (1985) showed that male sticklebacks could discriminate between interspawning and gravid, nuptially colored females, and increased mate selectivity as their own parental investment increased (as they had more clutches of eggs already in the nest). However, Pyron (1996b) showed, in an empirical field test of a theoretical model of male mate choice, that male orangethroat darters (*Etheostoma spectabile*) did not choose larger females, either because of a high male-to-female ratio (7:1) in the environment or because there was little variation among females in the number of eggs deposited per spawning bout.

In addition, in closely related species that overlap within habitats, it may be advantageous to accurately identify the status of a potential mating partner as conspecific. Ridgeway and McPhail (1984) showed that sympatric sticklebacks, both male and female, correctly selected conspecific mates when offered a choice.

Winemiller (1992b) noted that the propensity for sexual selection was related to overall reproductive strategy and that species with "periodic" reproduction typified by high fecundity and small offspring produced in pulses were unlikely to have selection for conspicuous males. Kodric-Brown and Hohmann (1990) demonstrated that sexual selection by male–male competition and female choice in pupfish (*Cyprinodon pecosensis*) resulted in stabilizing selection for a host of independent morphological characters, including meristic variables apparently unrelated to reproduction.

In summary, there have been many studies during the past two decades that directly tested mate choice in freshwater fishes, but only a few have explicitly tested differences in levels or in mechanisms of mate preference among populations of a single species, making it difficult to know the degree to which local environments might alter the context of mate choice. Pyron (1996b) has suggested that mate choice might differ locally with population features such as sex ratios. In the context of the ways life-history strategies operate at the local level, it would be additionally interesting to know if or how much the mating system for a given species would change in the context of the total composition of local fish assemblages. To date, it would seem that only predators have been considered as interspecific agents of changes in mate choice (e.g., Breden and Stoner, 1987; Stoner and Breden, 1988), but it would be interesting to know if the total suite of species in a local assemblages alters the expressed system of mate choice of a focal species.

Alternative Life Histories

In some salmonid (Gross, 1984; Montgomery et al., 1987) or sunfishes species (Keenleyside, 1972; Gross, 1979, 1984; Jennings and Phillip, 1992) and in at least one minnow species [*Pternotropis hubbsi*, bluehead shiner (Fletcher and Burr, 1992)], there are alternative life-history patterns. In these systems, some males delay reproduction until they attain larger body size, becoming nest-defending males (in sunfish or bluehead shiner) or "hooknoses" (in salmon) who obtain most of their fertilizations through male–male fighting for females. Other males mature at a smaller size, do not take on the appearance (color or morphology) of "full" males, and gain matings through sneaking or cuckoldry. Gross (1985) demonstrated disruptive selection favoring two sizes of coho salmon (*Oncorhynchus kisutch*), with intermediate-sized males at a competitive disadvantage. Jennings and Philipp (1992) demonstrated that cuckolder male longear sunfish (*Lepomis megalotis*) sacrifice growth rates for gonadal production and suggested that early investment in reproductive materials at the expense of potential stunting of growth could be adaptive in harsh environments (e.g., headwater streams where this species is common), in that early maturation would allow rapid reproductive output at a younger age. Gross's (1984) review indicated that alternative male mating strategies occur in at least 12 families of freshwater or shallow marine fishes.

Gross (1984) reviewed these alternative systems for bluegill sunfish (*Lepomis macrochirus*) and Pacific salmon (*Oncorhynchus* spp.) in detail and argued that their persistence in populations is as an evolutionary stable strategy (ESS). For sunfish, there was theoretical and behavioral support for equal fitness for males adopting either strategy (Gross, 1984). Furthermore, the optimal density of cuckolders in a given situation was environmentally mediated—with the advantage being to single cuckolders per nest if cover (e.g., weed beds) was dense, but more cuckolders favored if cover was sparse.

For Pacific salmon, sneaking "jacks" could have equal fitness to "hooknoses," with density-dependent decreases in profitability to being a jack. Because profitability of the sneaking alternative was density dependent, Gross (1984) concluded that in both cases, the presence of alternative male mating system was a mixed evolutionary stable strategy (mESS). This is in contrast to situations in which adopting the alternative strategy is at some apparent cost to the individual, and this does not represent an evolutionary stable strategy [as in dwarf versus normal-sized arctic charr, in which both males and females have a dwarfed and normal morph (Jonsson and Hindar, 1982)]. Montgomery et al. (1987) note also that sneaker "parr" individuals of Atlantic salmon (*Salmo salar*) may eat some of the eggs they have just fertilized via sneaking behavior, presumably regaining some of the energy expended in spawning.

Spawning and Clutch Parameters

Morphology sets limits on some clutch parameters of fish. In any species, fundamental morphological configuration of bone, muscle, and connective tissue, combined with basic body shape (Foster et al., 1992), limits maximum size of the body cavity, which may, in turn, set an upper limit on the volume of reproductive materials that can be produced by females at any one time [clutch volume (Foster et al., 1992)]. Wootton (1984) showed a strong correlation between body length and total egg volume for Canadian freshwater fishes. However, Baker et al. (1995) showed that other morphological factors like body armoring or pelvic configurations, in sticklebacks, may not alter clutch volume and suggested that total clutch volume was also related to foods, predator pressure, or behavioral traits of a species. Hubbs et al. (1968) argued that it may be the area of alimentary tract absorptive surface (facilitating nutrient transfer to ova) that limits potential egg number in females. In general, larger female fish produce more eggs (Lack, 1954; Hubbs et al. 1968), but, given either a body cavity of a maximum volume to accommodate ova (i.e., body cavity minus requisite space for internal organs) or a limiting alimentary absorptive surface, an individual female can either produce fewer larger eggs or more smaller eggs (Hubbs, 1958; Hubbs et al. 1968; Williams, 1959, 1966). Rubenstein (1981) showed that Everglades pigmy sunfish (*Elassoma evergladei*) had reproductive effort that increased under high-density conditions, with increased ovary size and more eggs, but at the sacrifice of slower growth rate.

Although morphology at the species level probably sets maximum potential clutch volume, populations (Hubbs and Delco, 1960; Bagenal, 1971; Marsh, 1984; Heins and Baker, 1987) or individuals within a population (Marsh, 1984; Heins and Baker, 1987) can differ in the ways the volume is filled with eggs. Additionally, the constraint of body-cavity volume fixes only the upper limit on egg mass at any one time. The number of times the egg mass space is filled and then emptied by spawning is not constrained by morphology, but by energetics and environmental cues like as temperature, day length, or length of spawning season (Hubbs and Strawn, 1957; Marsh, 1980; Bye, 1984). Many darters have multiple clutches, with some individuals spawning repeatedly in a year (Marsh, 1984). Some populations of darters may breed nearly year-round in thermally stable environments (E. Marsh-Matthews, personal communication). Many freshwater fishes are known to spawn multiple clutches, sometimes as often as daily per female (Weddle and Burr, 1991; Mire and Millett, 1994). To the extent that species have multiple clutches, they are freed from the constraint of body-cavity size on total fecundity in a season or a lifetime (Paine, 1990).

Egg Size Versus Egg Number

Marsh (1986) showed that the total number of eggs that could be produced was constrained by egg size relative to size of female; thus, females producing larger eggs were limited to producing fewer eggs. Big eggs generally produce larger larvae at hatching, which may confer on those individuals better resistance to starvation (Wilson and Hubbs, 1972; Marsh, 1986) or other environmental challenges. Marsh (1986) and Wootton (1992) noted the potential advantage of producing large eggs at the expense of numbers of eggs, in that egg size and size at hatching are positively correlated for many fishes. The same is true for larger neonates in live-bearing fishes—with larger offspring better able to cope with the vissicitudes of early life (e.g., finding foods and escaping predation). The only study of which I am aware in which egg size and hatchling size were not correlated was for pupfish (Mire and Millett, 1994).

For orangethroat darters (*Etheostoma spectabile*) from Texas streams Marsh (1986) showed that larger eggs resulted in larger size of larvae at hatching. When larvae were fed (in the laboratory, without predators present), the large hatchlings survived no better than did small individuals. However, when food was withheld, to simulate low-food availability in the wild, the larger larvae (from larger eggs) survived longest. Marsh (1986) calculated that a 0.100-mg increase in the mass of the egg could result in a 0.23-mm increased body length at hatching, that would have an increased time to starvation (if food was lacking) of 3 to 7 days, depending on temperature. Most importantly, even in the absence of food, the larvae that hatched at larger initial size continued to grow. Larger larvae could, therefore, be favored in the field under low-food conditions, with growth continuing which could allow those individuals to reach sizes where predation (or washout from floods, etc.) might be less of a threat. How often such size differences at hatching might make a difference in the wild is problematic, but populations of *Etheostoma spectabile* do exist in harsh stream systems (e.g., Brier Creek, Oklahoma), where episodes of low resource availability seem possible. In addition, it would be desirable to know how many species in naturally occurring assemblages exhibit similar life-history traits of egg size to larval size to enhanced survival in harsh conditions. In certain environments, it seems possible that species with larvae capable of withstanding uncertainties of resources would have an advantage, linking life history and environmental challenges in the local habitat to composition of the assemblage.

Egg size can depend on time the clutch is produced within the reproductive season (Williams, 1967). Heins and Machado (1993) reported a decrease in the mass of mature oocytes in one but not the other habitat studied, between February and March, for *Etheostoma whipplei*. Marsh (1984) found that most female *Etheostoma spectabile* produced smaller eggs later in the season (at higher water temperatures). Hubbs et al. (1968) suggested that production of smaller eggs, having a higher surface-to-volume ratio, might be appropriate at higher temperatures when oxygen availability to eggs in nests could be lower (because of lower oxygen concentrations at saturation at higher temperatures). Food availability can also influence the condition of the female producing eggs, altering strategies, or change the best approach to egg size, depending on foods available to larvae in the environment (Kamler, 1992).

Intraspecific Variation in Clutch Parameters

There can be substantial variation in egg or clutch parameters among populations of a single species. Guill and Heins (1996) showed dramatic variation

(300%) in clutch size for banded darters (*Etheostoma zonale*). Marsh (1984) showed great variation in egg size among populations of orangethroat darters (*Etheostoma spectabile*), with much of the variance being among drainage basins. Heins and Baker (1987) showed the variation in egg size among populations of blacktail shiners (*Cyprinella venusta*) from different coastal river systems, and Heins (1991) showed variation in egg size with stream discharge for *Notropis longirostris*. Healey and Heard (1984) showed strong intraspecific variation in fecundity of chinook salmon (*Oncorhynchus tshawytscha*) and argued that this species had, in general, sacrificed high fecundity for increased body size, which has survival value for these fish in the wild.

Individual Variation

Evidence from some wild populations suggests that although larger females have more eggs, they do not necessarily have larger eggs (Lack, 1954; Mathur, 1973; Hubbs et al., 1968; Marsh, 1984; Mire and Millett, 1994). For 16 populations of *Cyprinella venusta* (blacktail shiner, Cyprinidae) Heins and Baker (1987) showed that egg size was not related to female size within populations but that egg size was correlated with female size across all populations. Therefore, populations with larger females had larger eggs. Heins and Baker (1987) showed larger egg size on average in stream systems with a higher annual runoff. The extent to which these differences are genetic versus environmentally induced is not known. Meffe (1990) also pointed out that the size of neonates in live-bearing *Gambusia affinis* was only partly accounted for by environmental factors, season, size of mother, or clutch size, leaving a high proportion of the variation in neonate size as unpredictable "noise" not under maternal control.

Effect of Reproductive Status on Morphology and Hydrodynamics of Individuals

Reproductive readiness in females (i.e., possession of an abdomen swollen with eggs) can alter the morphology of an individual relative to hydrodynamics or efficiency of swimming in a habitat. Referring to ideal values for streamlining (Section 8.2), it is obvious that a highly gravid female, or a pregnant female live-bearer, will have a substantially altered maximum body diameter, or rearward displacement of maximum diameter, which should alter the dynamics of pressure drag on its body. I know of no tests in the literature but would hypothesize that gravid/pregnant females in many species would be less able to swim efficiently or to hold position on the substrate (for benthic species) exposed to currents. Thus, reproductive status of an individual might alter its hydrodynamics and thus alter the hydraulic microhabitats that it could most efficiently use.

Parental Care and Guarding

Blumer (1979) defined parental investment as "nongametic contributions that directly or indirectly contribute to the survival and reproductive success of the offspring" and defined 15 forms of parental care in bony fishes, including guarding, nest-building and related activities, fanning, internal gestation, removal of dead or diseased eggs, oral brooding, retrieval of eggs or fry that fall out of the nest or stray, cleaning eggs, external egg carrying, egg burying, moving eggs by mouth from one location to another, coiling of the body around eggs to protect them, ectodermal feeding, brood pouch, and splashing water on eggs deposited out of the water (e.g., those exposed by low tides). Parental care could also include details of effort in egg-laying, differing even within closely related species (Page, 1985). Parental investment theory would generally predict greater risk-taking when broods are larger. However, Fitzgerald and Lachance (1993) found some sticklebacks to not follow this pattern (i.e., taking no greater risks as a function of brood size) and suggested that environmental characteristics rather than brood characteristics would influence parental investment. Winkelman (1996) found the opposite response for dollar sunfish (*Lepomis marginatus*), in which males returned more quickly after disturbance (bird predator model) to guard nests containing offspring than nests that were empty. Thus, both environment and status of reproductive cycle may influence risk-taking or parental investment.

Blumer (1979) reviewed male parental care, which is common in freshwater fishes, concluding that major factors in the maintenance of this trait include relatedness of the male to eggs or larvae in his care (in his nest) and the effect of parental care on the male's future reproduction potential. Blumer concluded that the cost of male care-giving was least in situations in which males could guard existing clutches or broods, yet simultaneously solicit additional eggs (e.g., for centrarchids or sticklebacks seeking to add fertilized zygotes to an active nest). Males may also increase parental investment when environmental conditions increase threats to nests (e.g., Fitzgerald and Caza, 1993). Ridgeway and Friesen (1992) showed interyear differences in length of male parental care for smallmouth bass (*Micropterus dolomieu*) related to variation in water temperature. In years with slower spring warming, smallmouth bass showed more variance in the date of initiating parental care, whereas rapid environmental warming tended to synchronize deposition of eggs and onset of parental care by males. They also showed a strong correlation between male size and the degree of parental care given.

Male parental care can be costly in terms of future or residual reproductive potential. Some species, such as river bullhead (*Cottus gobio*) reduce food intake, lose weight, and exhibit peak mortality of males associated with guarding in the breeding season (Marconato et al., 1993). In rock bass (*Ambloplites rupestris*), Sabat (1994b) showed for 191 brood-guarding males a marked decrease in survival of individuals to the next reproductive season that was related directly to relative loss of body mass during nest guarding.

Constantz (1985) described for the tessellated darter (*Etheostoma olmstedi*) an unusual system of male care ("allopaternal"), in which the parental male might abandon a nest site in order to attract females to new sites, but in which the abandoned eggs might be "adopted" and cared for by nonterritorial males. Thus, the larger, wandering males maximize the number of fertilizations, and the smaller "adopting" males may also benefit, because females may be stimulated to spawn by the presence of at least a few viable eggs in a potential spawning site.

Both sexes of a monogamous pair of fish may be involved in parental behavior or care (e.g., Itzkowitz and Nyby, 1982), but there often is division of labor between the male and female (Itzkowitz, 1984; Yanagisawa, 1986). Male:female body size may relate to parental care (Pyron, 1996a). Dupuis and Keenleyside (1982) showed that levels of predation risk can increase the division of labor between sexes, dividing between roles in defense and care for eggs in a nest. However, in some species (e.g., brown bullhead, *Amieurus nebulosus*), some nests are attended by both sexes, but some are attended only by the male. The presence of both parents was shown by Blumer (1985b) to be most successful, and he suggested that males benefitted more by staying on a given nest than by leaving, if the male was aided by the female. In some cichlids, females are the persistent guarder of nests, with males coming and going to visit other females, controlling several breeding territories simultaneously (Yanagisawa, 1987).

Early Life History: Survival of Eggs and Growth of Larvae

Bond (1996, p. 473) reviewed the tremendous growth of published information on "early life history" (eggs and larvae) of fishes and summarized schemes for terminology describing various stages in the life span of a fish species from spawning to attainment of adulthood in the next generation. Terminology related to egg and larval development in Bond (1996) adapted from Kendall et al. (1984) focuses on morphological benchmarks in the egg and flexion of the notochord and metamorphosis to full complement of fins in larvae–postlarvae. In addition, "annual" fish have an additional complexity of sequences, including up to three periods of diapause (Cunningham and Balon, 1986).

At each stage in development, a growing embryo or larva faces different environmental challenges, and natural mortality is high in early life stages (Dickie et al., 1987). Egg viability or survival can be closely linked to differences in local environments, such as variation in oxygen concentrations (Deacon et al., 1995). Paine (1984) and Paine and Balon (1984) describe highly differing early ontogenetic sequences for various darter (Percidae) species, and relate microhabiat, drifting of larvae, feeding, and overall strategies for survival to the differences in timing of development of vitelline circulation or oral structures in these species. Differences in morphology, use of resources, and survival are closely linked in these species.

Habitat conditions and predator density influence the survival of nests or larvae. The quantity and quality of habitat in a stream may change rapidly relative to the potential to allow maintenance of nests or eggs. Nests built in midchannel flow are rapidly destroyed in spates, but nests in hydraulically protected areas may survive. Spates may roll rocks in riffles, crushing eggs deposited in interstitial crevices in erosional zones. Jennings and Philipp (1994) showed that for sunfish nests in Jordan Creek, Illinois, eggs and larvae suffered highest mortality from biotic interactions like predation in years with stable flow. In years with variable flow, most losses of early life stages were to floods or nest desertion associated with floods.

In lakes, there is lower probability of disruption of spawning by flowing water, but storm-driven waves can destroy shallow-water nests [e.g., those of sunfishes (Popiel et al., 1996)]. Other catastrophic destruction of eggs or larvae is common. Predators can cause significant mortality of eggs or larvae in nests (particularly when non-native, introduced fishes are the predators) (Marsh and Langhorst, 1988). Hubbs (1996) showed that the larger size decreased susceptibility of neonate *Gambusia* to predation. In Brier Creek, Oklahoma, we commonly observe that if a guarding male longear sunfish leaves the nest, smaller sunfish and minnows swarm to the nest and consume large numbers of eggs in a matter of seconds. Temperature fluctuations can also kill larvae. Hubbs (1964) showed that populations of greenthroat darters (*Etheostoma lepidum*) from variable thermal environments had greater tolerance for temperature fluctuations than did those from constant-temperature habitats.

Finding Food as Free-Living Larvae

As free-living larvae out of the nest, the greatest challenge to a larval fish is the finding of food for the critical "first feeding" (before yolk-sac reserves are exhausted) as well as outgrowing predator threats. Unger and Lewis (1991) found that larval *Xenomelanaris venezuelae* in Lake Valencia, Venezuela, survived only when zooplankton (rotifer) densities exceeded about 100 per liter. Letcher and Bengston (1993) showed that growth of larval and early juvenile *Menidia beryllina* depended on feeding levels, temperature, and age or size of the larva. Shute et al. (1982) showed that growth of darter larvae (*Etheostoma perlongum*) can be very rapid, with females attaining adult size 5 months from hatching. However, the entire life history of this species seems accelerated, with early reproduction and mortality.

From a theoretical perspective, fish "decide" during the egg production phase (above) between many small and few large ova, which can result in size differences among larvae at hatching (Marsh, 1986). Paine and Balon (1986) found that a darter species with larger eggs and more yolk volume had a longer embryonic period and produced better-formed larvae upon hatching than did a related species with smaller eggs. Larger, better developed larvae are generally better able to search for or consume prey items (typically zooplankton) and also may better escape predators. Many fish larvae starve because they are unable to forage in an energetically profitable manner (i.e., the energetic cost of finding food items may be greater than the return on the effort at swimming to locate them). Kashuba and Matthews (1984), for example, found by histological examination that a substantial percentage of shad (*Dorosoma* spp.) larvae in Lake Texoma (Oklahoma–Texas) were in advanced stages of starvation on any given day. The larger size at hatching may also allow larvae to persist longer without starving if food availability is low.

Winemiller and Rose (1993) demonstrated by a simulation model that various trade-offs between egg (thus, larval) size and number conferred the greatest advantages in different kinds of environments. When prey were random, clumped, or scarce, the strategy of fewer but larger larvae conferred the most survival, but if prey was abundant, the strategy of many small larvae produced the greatest number of survivors from a cohort. However, the scale of clumping of prey within the environment also affected the outcome of their simulations, suggesting that if prey were patchy at large spatial scales, then investing in large numbers of small eggs was the best evolutionary strategy.

Growth, Recruitment, and Mortality

After advancing from the larval to the juvenile stage, growth is essential to eventual recruitment as a mature adult. There are, as noted above, fishes in which parallel adult or mating systems emerge. However, for most species, juvenile growth is necessary to escape predation threat and to attain a minimum size for maturation, so rapid growth is favored. Growth is influenced by temperature (Keast 1985a), food, and numerous environmental variables (Weatherley, 1972). In some fish species, growth can be dramatically different, even within a single cohort. In some centrarchids like *Pomoxis* and *Micropterus,* individuals within the same cohort may convert from zooplankton to fish as food at different schedules, resulting in bimodal size structure. Some individual largemouth bass (*Micropterus salmoides*) can be double or more the size of smaller members of the same cohort after a year's growth (W. Shelton, personal communication). In an experimental test of various biological and social factors, Koebele (1985) showed that food acquisition is the primary mechanism promoting divergent growth in juvenile cichlids. However, Jobling (1985) showed for Arctic char (*Salvelinus alpinus*) that social interactions that included aggression affected feeding and resulted in growth suppression in some individuals. DeAngelis et al. (1993) showed by simulation modeling that once an individual fish gains an advantage in growth rate at one time, this advantage results in a tendency to grow faster in subsequent days, leading to substantial divergence in sizes of fishes. Because winter mortality or condition of young bass or bluegills (Cargnelli and Gross, 1997) is size dependent, attainment of substantial size in the first season of growth can be critical (Wismer et al., 1985).

In many species, growth differs markedly among populations (e.g., Hansson, 1985). Cowx (1990) showed for both roach (*Rutilus rutilus*) and dace (*Leuciscus leuciscus*) populations in English rivers that growth of each species varied within a river system, that faster growth allowed one or the other species to differentially dominate the local assemblages, and that reproductive effort was positively related with growth rate. Cowx also showed that mortality curves differed within and among rivers for dace and for roach. Victor and Brothers (1982) showed that daily and annual growth of fallfish (*Semotilus corporalis*) in different populations was strongly linked to conditions in the local habitat, including stream size, and was also density dependent. Osenberg et al. (1988) showed density-dependent growth of young of two species of *Lepomis* in lakes in the northern United States, but that size-specific growth rates depended most on the influence of being in individual lakes (rather than on differences between years). However, the review by Pagel et al. (1991) showed that across many fish taxa, there was no pervasive pattern between population density and body size in natural assemblages, and Carl (1983) showed a lack of density-dependent growth in two salmonid species in the northern United States.

After adulthood is reached, high levels of parental investment in a given reproductive period can decrease growth or increase mortality in fishes. Jennings and Philipp (1992) reported a trade-off of increased reproductive output at the expense of somatic growth for male longear sunfish (*Lepomis megalotis*). In that body size confers some safety from piscivorous fishes or from harsh changes in environmental conditions (Schlosser, 1987a), sacrifice of somatic growth can be related to increased mortality risk. Numerous analyses of overall life-history strategy (e.g., Winemiller and Rose, 1992) link large body size to later maturity and longer life span, and the notion of large-bodied species living longer than small species is well established in the fisheries literature (Carlander, 1969; 1977). In general, rapid growth, even after achieving sexual maturity, may allow a fish to outgrow the risk of predation from increasingly large piscivores, to win in intraspecific competitions, or gain territories or microhabitats that confer a fitness advantage. If adults reach a size at which they are too large for most predators, mortality may remain low until some eventual period of stress (e.g., spawning runs (Sigler et al., 1985)]. Thus, growth may be linked directly to length of life within a given environment. However, across a latitudinal gradient, it is typical to have both slow-growing, longer-lived populations, and faster-growing, shorter-lived ones.

Evolutionary and Environmental Constraints on Reproductive Traits

Morphological constraints (e.g., on clutch volume) can obviously relate to recently acquired traits of a species or to ancestral traits acquired by a clade in deep evolutionary time. Wootton (1992) reviewed a hierarchy of constraints that influence life-history traits of a fish species, including physical, allometric growth, physiological, demographic, and genetic factors. Wootton (1992) considered that physicochemical limitations in the environment defined the outermost boundary of reproductive capabilities of a species, and that body size (as total length) was a good overall predictor of total volume of eggs that can be produced. Wootton next argued that allometric growth of fishes (slower at older ages) imposes constraints on the evolution of fish life-history patterns, in that there can be no continued linear increases in fecundity throughout life because the body volume available to eggs reaches an asymptote. Physiological constraints on fecundity and egg size relationships relate to requirements for energy and oxygen that limit some life-history options to fish, and an integrative constraint across all life-history traits is that they must not result in demographic patterns or cycles that result in the extinction of local populations (Wootton, 1992). Finally, Wootton (1992) pointed out that genetic constraints exist on life-history pattern—that is, that the gene pool inherited as a result of selection may constrain a particular species from taking opportunistic advantage of some changes in environmental conditions (but, interestingly, did not comment specifically on the role of phylogeny or common traits within evolutionary clades). In summary, Wootton (1992) described a suite of major constraints on life-history patterns of fishes, within which still remain a wide array of expressed patterns.

Bart and Page (1992) reviewed and contrasted constraints in North America percids on the basis of morphology (body size) and phylogeny, showing that evolutionary trends in body size within clades dominated traits relating to longevity, time of maturation, fecundity, and iteroparity, whereas there was possible adaptive "fine-tuning" of egg size or reproductive behavior within clades. For example, Bart and Page (1992) suggested, from a phylogenetic analysis, that egg-attaching evolved independently three times as a reproductive behavior with the genus *Etheostoma*. Thus, there is the suggestion that some life-history traits constraints (e.g., those dependent on body size), obtain from common ancestry in clades, whereas other traits (e.g., behavioral) have developed as more recent adaptations. In addition, Winemiller and Rose (1992) reported substantial influence of phylogeny on broad suites of life-history traits across 57 fish families (as noted above).

Intraspecific Differences in Reproductive Traits

As seen for many of the individual traits above, there can be wide intraspecific variation in life-history strategy among populations of a species, often related to temperature or predation pressure in the environment (e.g., Picard et al., 1993). Meffe (1990, 1991) has shown that mean offspring size and clutch sizes changed for *Gambusia affinis* populations from thermally altered environments. Dorsey's (1990) work with *Gambusia geiseri* from a relatively constant thermal environment agreed with the suggestions of Meffe (1987, 1990) that *Gambusia* in relatively constant environments might have less variability in size of offspring than do those in fluctuating environments. Weeks and Gaggiotti (1993) found differences in sizes of neonate *Poeciliopsis* at birth in two populations and suggested that cannibalism might lead to an advantage for larger offspring.

Winemiller (1993) showed that for three live-bearing fishes in Costa Rica, there were substantial seasonal differences in female-size, brood-size relationships, with larger broods in wet season. He suggested for one species that this difference was due to enhanced feeding, but for the other two, the difference could be due to enhanced potential for survival and growth of neonates.

Interspecific Differences in Life-History Traits

Many cases demonstrate that either closely or distantly related fishes have substantially different life-history strategies within the same kind of habitat (Schloemer, 1947; Williams and Bond, 1983; Cambray, 1994). Why do fish species occupying the same environment have quite different life-history traits or reproductive schedules? If optimality or fine-tuning for local environments were the only force regulating selection of life-history strategies, it is logical that all resident species in a given location might converge to a common strategy. However, even within the "common garden" of a single site, different species use different solutions to the problem of survival and reproduction. Part of the answer may lie in the spreading out of reproductive timing of species so that the young do not all saturate the environment simultaneously (Kramer, 1978). However, there is a trade-off between (1) saturation of the habitat with larvae of many species simultaneously and (2) staggering of reproductive peaks among species. The latter may help to ameliorate potential competition among larvae of various species; the former may, however, sufficiently saturate the environment that predators on the larvae are not sufficiently dense to eat as high a proportion of the predators as they would if each species produced young-of-year at a different time. Many freshwater fishes are voracious predators on eggs or free-swimming larvae. Furthermore, predators may have evolved to spawn earlier then their prey species, providing an adaptive advantage for feeding by young of the piscivorous species (Keast, 1985a). Given that the environment has a strong influence on the reproductive traits or tactics that work in a particular kind of environment, why do not all species in a particular assemblages (sharing a common environment) converge to a single "best" reproductive strategy?

Life History Versus Habitat Features

The question of why a given suite of fishes occurs in a particular stream reach or portion of a lake is tied to reproduction and life history of each species in the context of available habitat or environmental conditions at that location. Reproduction of individuals provides the raw material for the next generation, upon which mortality will exert pressure to produce the empirical survival and reproductive success in the new generation. The interplay of mortality on a given generation, beginning (arbitrarily) at the point of fertilized ova and ending with the death of the oldest individual, will produce the "life history" for a population within a species. The life history of a given species is not fixed but can vary in space or time. There is abundant evidence that life-history characteristics can vary among populations of a species (Heins and Baker, 1987; Marsh, 1984, 1986; Garrett, 1982; Hubbs, 1996), and also that life history features like mortality curves can vary within a population across time ("good" versus "bad" years). Berven and Gill (1983) showed strong evidence that life-history variation in wood frogs was related to differences in environments from northern Canada to temperate lowlands in the eastern United States.

Effects of Predators on Life History

Suites of life-history traits are influenced not only by the abiotic environment or environmental variability but also by biotic interactions like predation (Reznick and Miles, 1989). In a series of articles, Reznick, Endler, and others have showed a variety of effects of the potential predators (*Crenicichla alta* and *Rivulus harti*) on life-history patterns for guppies (*Poecilia reticulata*) in streams in Trinidad [although Mattingly and Butler (1994) have questioned the effectiveness of at least one of these putative predators in the wild]. In these streams, there are natural segments that have high, medium, and low levels of predator threat, comprising a natural laboratory in which life-history traits of a single species can be compared in similar environments differing mostly in predator threat.

Reznick and Endler (1982) showed that populations of guppies (*Poecilia reticulata*) in which the greatest predator threat was to adults increased reproductive investment by (1) devoting a higher percentage of body weight to developing offspring, (2) having shorter interbrood intervals, (3) producing more, smaller young, and (4) beginning reproduction at smaller body size, and Reznick (1982) showed that these traits persisted for two generations in the laboratory, hence were under genetic control. Reznick and Endler (1982) argued that their findings "make a strong case for the direct role of predators in molding guppy life history patterns," and the genetic findings of Reznick (1982) would suggest that populations have evolved differentially in light of predator pressures.

From subsequent introduction experiments, Reznick and Bryga (1987) showed that some traits expressed in the high-predation (*Crenicichla*) environment were altered when they were introduced to a modest-predation (*Rivulus*) environment, with changes in maturation size and a shift to larger, fewer offspring; they argued that guppies were typified by marked plasticity in life-history traits. Reznick (1989) showed that the initial patterns in life-history traits between *Crenicichla* and *Rivulus* sites (high and medium predator threat, respectively) were persistent across additional field sites, and between seasons, although in the rainy season, some differences were less than in the dry season. Reznick et al. (1990) summarized an 11-year experiment in which a natural population of guppies was exposed to a change from predation on adults to predation on juveniles. The resulting changes in life history of this population after 30–60 generations paralleled the changes observed earlier in natural streams and was in accord with predictions of life-history theory, with later maturation, lower reproduction effort, and fewer, smaller offspring when adult mortality is low.

Strauss (1990) reanalyzed much of Reznick and Endler's data and affirmed that high- and moderate-predation sites differed in life-history traits, but that moderate- and low-predation sites were indistinguishable in a multivariate analysis of traits. Strauss (1990) also suggested the alternative hypothesis that the life-history traits of the Trinidadian guppies might differ among populations in response to suites of local environmental factors, to which extremes the putative predators also were attracted. In this interpretation, both the predators and the guppies responded individualistically to local environmental differences, but the predators do not drive life-history differences among guppy populations. However, Rodd and Reznick (1991) pointed out that if there was an overriding environmental influence on guppy life-history patterns, it was not obvious in any geographical pattern among the study localities.

Genetics or Heritability of Life-History Traits

Reznick (1981) provided one of the first views of the genetics of life-history traits in fishes, and Zimmerman (1987) reviewed broad relationships between genetic structure of populations and general life-history traits. Vrïgenhoek et al. (1987) found different levels of genetic divergence related to life history traits in *Poeciliopsis* and Arctic char. Substantial questions remain as to whether most life-history traits are evolutionarily adaptations selected for in a given environment, or if they are an epigenetic manifestation of the environmental setting. The question of genetic modification of life-history traits in individual populations can best be examined only in "common garden" experiments. Because many species of fish are too long-lived for multigeneration experiments to be practical, attention has been focused on short-generation-time taxa like *Gambusia* or on raising a single generation from fertilization to adulthood in controlled, common environments. Consider the difficulty of working even with a relatively short-lived minnow from a North American stream. To ask a simple question like, "Are differences in egg size genetically fixed in different populations?", it would be necessary to (1) bring adults from wild populations to a common-garden facility (with replicated tanks or holding chambers within each population), (2) allow maturation of ova and sperm in those adults in the common-garden environment, (3) obtain fertilized ova (probably by stripping and artificial fertilization; possibly by natural reproduction by the adults); (4) keep the embryos alive through the larval stage; (5) raise this first laboratory generation to adulthood in the common-garden facility, holding all possible features of their environments truly "common," including food, water chemistry, and so forth; (6) bring them into reproductive condition to obtain mature ova for counting, measuring, and so forth. This may all be conceptually simple, but it could be a logistical nightmare.

Even a successful common-garden experiment cannot tease apart the link between genetic basis of a given life history and the environment in which that suite of characteristics is employed. Even though two populations might have genetically based differences in a life-history trait that would produce phenotypically different results in the first laboratory generation, those same genetic systems placed in the wild, with differences in factors like temperature, might operate differently—due to the interplay (interaction) of the genetic systems and environmental factors. To even begin to tease apart the genetic versus genetic × environment interactions would require repeated common-garden trials under differing suites of environmental conditions (i.e., much larger physical facilities or many more years of work). The question of heritability of a trait in a population is best decided by tests in which individual females and/or males are chosen by the investigator on the basis of "largest eggs," "most active sperm," and so on and are mated in an intentional manner to discern if the trait in the parent(s) also exists in the next generation (i.e., its "heritability"). The bottom line is that we have very little information on genetic adaptation of life-history traits in individual populations, and most empirically perceived life-history information is based on field-caught individuals, in which separation of "nature and nurture" is actually impossible.

Different Life-History Strategies May Work Within a Single Place

Returning to the question of differences in life history among species in a single environment, consider two species of sunfish (*Lepomis*), both of which are abundant and successful (for at least the last 25 years) in Brier Creek, Oklahoma. Green sunfish (*Lepomis cyanellus*) and longear sunfish (*Lepomis megalotis*) are both common throughout the stream. Smith and Powell (1971) reported substantial numbers of each in summer 1969, and our collections since that time (Matthews et al., 1988; unpublished data through 1994) reveal that they consistently are successful in the system. What they do there seems to work, but they appear to have quite different life-history traits. Both construct nests (males) in gravel or sand bottoms in pools of the stream. Longears are conspicuous, colonial nesters, often with five to six nests in close proximity in shallow water (but always relatively near to escape space in deeper water), which the colorful males very actively defend against smaller egg-eating sunfish individuals or minnows. In the 14-pool reach of Brier Creek which we have studied for the last decade, it is common to observe several dozen longear nests at any time in the summer. In contrast, green sunfish have far fewer nests in this reach of Brier Creek, typically placed less conspicuously and slightly deeper, and often in coarser gravel than the nests of longers.

Although there clearly were more longear than green sunfish nests in this reach of Brier Creek during the 1980s, snorkeling surveys (Matthews et al., 1994) of this reach showed that there were far more young-of-year or small juvenile green sunfish than longears (Fig. 8.3), and they were present over a much longer period

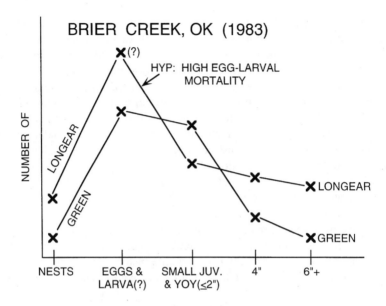

Figure 8.3. Hypothetical differences in lifetime reproductive strategy and survival of longear sunfish (*Lepomis megalotis*) and green sunfish (*Lepomis cyanellus*) in Brier Creek, Oklahoma. Relative values approximated from snorkel surveys by W. Matthews in 1982–1983.

of the year. In fact, in this reach of Brier Creek, small (apparently young-of-year) green sunfish were abundant as early as March (not shown) and remain abundant in September; that is, large numbers of small green sunfish were present throughout the warm season of the year, apparently being produced at a steady rate from the relatively few observed nests. [This assumes no substantial immigration of green sunfish young from outside this reach of the creek, but my visual inspection of reaches of Brier Creek upstream and downstream from the main study reach provided no indication of a "green sunfish factory" at any nearby part of the stream.] At least in the year of most intensive study (by snorkel surveys, 1983), longears appear to have had a much more compressed period of successful production of young (Fig. 8.3), with large numbers of young-of-year and small juveniles in June. There was a major flood between the 10 June and

24 June collections, but there was little evidence that this factor alone limited the production of young by either species—in fact, the adults of these species were displaced very little by the flood: most remained within original pools (Matthews et al., 1994).

An interesting difference in life history of the two species in Brier Creek is the apparent difference in survivorship curves (assuming that size of individuals is a surrogate for survivorship from one age class to the next; Fig. 8.3). Although there are many more small green sunfish than longears, the number of individuals becomes virtually equal by the time they reach 5–6 in. in total length, and indeed is even slightly reversed, with more large (≥ 4 in. total length (TL) longears than green sunfish by September (Fig. 8.3). Although the above is based on only 1 year of intensive observation, it is my impression that this pattern holds in most years in this stream. Clearly, both species are successful in this environment. However, green sunfish appear to have fewer nests, but to be more successful at producing large numbers of small young than are longears. Apparent mortality of these small green sunfish must be high, however, suggesting a survivorship curve more like that of a "Type III" of Pearl (1928), as summarized in Pianka (1974). For longears, it appears much more likely that an individual growing to certain minimal size will survive to reach maturity and reproduce, reminiscent of a Type I or Type II survivorship curve. Determining mechanisms causing the two species to differ in life history and survivorship in Brier Creek would be of interest. However, for now it remains empirical evidence that both life-history approaches apparently work successfully, side by side in the same reach of a stream. Perhaps there is no single "best" solution to blending life history with the environment, and to understand alternative solutions in terms of mortality, it would be necessary to also know the remaining reproductive potential of an individual at each age for each species, so that a life-table approach to estimating strategies maximizing lifetime reproductive output (Pitcher and Hart, 1982, p. 94) might be possible.

Finally, instead of focusing on differences in a few species or populations in life-history traits, what would the patterns be if one were to summarize (multivariate analysis?) the life-history traits for *all* of the species in a given assemblage? It should be possible, with a combination of literature review and field research, to construct the kind of comparative life-history-trait table produced by Paine (1990), Bart and Page (1992), or Winemiller and Rose (1992) for all of the species in a place. For a given locality, would multivariate analyses reveal separable groups of species with divergent traits that all "work" in that habitat, or would there be a continuum of species [much like that in the multivariate analyses of Winemiller and Rose 1992)] on a gradient from one extreme to another. In the case of the latter outcome, would those species nearest to the center of the gradient be most successful (i.e., most abundant, or most stable over time) in that location? How would patterns in multivariate analyses of life-history traits appear for more than one locality? In a question analogous to that often asked in ecomorphological analyses of assemblages (Section 8.2), would species-rich assemblages show "packing" of species within multivariate life-history space, or would the size of life-history space occupied on multivariate axes expand outward at the edges as more species were added to the assemblage?

8.6 Movement and Migration

Movements and home ranges of fishes have been very well studied, with dozens of books and hundreds of publications addressing patterns for individual species. A sample of fishes at a site on a stream or lake provides a snapshot of those present at a particular date and time. However, home range or homing behavior, local movement, and migration all can influence the composition of an assemblage. More is known about movement and migrations of salmonids (Hasler, 1966; Hasler et al. 1983) than any other freshwater fishes, but a large number of species exhibit (1) home range or homing affinities, (2) daily or seasonal movement patterns, or (3) longer-distance migrations. Most of these patterns are truly "autecological," with little apparent influence from other fish species. Highly predictable seasonal movements include phenomena like long-range migrations of tropical fish in response to seasonal floods (Lowe-McConnell, 1987; Goulding, 1980). However, in temperate freshwaters, upstream migrations also occur [e.g., by catostomids and cyprinids prior to spawning (Hall, 1972; Matheney and Rabeni, 1995)]. Not all movement or migration of fish in freshwaters is regulated by oriented swimming. In one of the most unusual of movement mechanisms, fish associated with floating islands of vegetation (broken loose from swamps and set adrift during storms) in Lake Malawi, Africa, can be transported to various parts of the lake over a period of months (Oliver and McKaye, 1982). Similar phenomena have been reported in the Amazon by Junk (1970, 1973).

Home Pools or Limited Home Ranges

Minns (1995) reviewed the literature on home range size for temperate freshwater fishes and found that (1) larger individuals had larger home ranges and (2) home ranges were larger in lakes than in streams. Various freshwater fishes may have home ranges varying in size from small to quite large. Mundahl and Ingersol (1983) showed that > 96% of *Etheostoma nigrum* (johnny darter) and > 87% of *Etheostoma flabellare* (fantail darter) marked in a small stream moved very little from the point of initial capture. Hill and Grossman (1987) showed that average moves between captures by a sculpin species and two minnow species in an upland stream of North Carolina (USA) were only 12–20 m in length. Grossman and Freeman (1987) showed that fishes in an Appalachian river might move as little as 30 m during a lifetime. Scalet (1973) showed that marked orangebelly darters (*Etheostoma radiosum*) in an Oklahoma stream might occupy the same riffle area for long periods of time and that desiccation of a stream reach caused marked *E. radiosum* to move substantially downstream in an intermittent Oklahoma creek in midsummer (F. P. Gelwick, unpublished data). Conversely, radio-tagged northern hogsuckers (*Hypentelium nigricans*) moved an average daily distance of 425 m in summer and 276 m in winter, with an average calculated home range calculated of 812 and 426 m of stream in winter–spring and summer–fall, respectively, in Missouri (USA) streams (Matheney and Rabeni, 1995).

Many centrarchids exhibit home pool behavior or homing behavior (Gerking, 1950, 1953, 1959; Larimore, 1952; Fajen, 1962; Lewis and Flickinger, 1967). Gerking (1959) and Funk (1955) suggested that a substantial part of a centrarchid population remained in home areas, although a "wandering" stage allowed juvenile dispersal. Hasler and Wisby (1958) found that green sunfish (*Lepomis cyanellus*) and largemouth bass (*Micropterus salmoides*) released in ponds returned to their original area, although sunfish moved more precisely than bass back to the initial location. Berra and Gunning (1972) showed displaced longear sunfish (*Lepomis megalotis*) to return to home pools. Smallmouth bass released as much as 11 km away showed a strong tendency to return to the initial capture site, with almost 80% of the recaptured individuals at or en route to the capture site (Pflug and Pauley, 1983). A large flood in Brier Creek, Oklahoma, appeared to redistribute adult sunfish or bass only minimally among pools (Matthews et al., 1994).

Home ranges and movements of resident salmonids in lakes and streams also have been well studied. Walch and Bergersen (1982) estimated home ranges of lake trout (*Salvenius namaycush*) in Colorado (USA) mountain lakes to be 199, 162, 173, and 62 hectares in spring, summer, fall, and winter, respectively. Native brown trout (*Salmo trutta*) in a river in Wales rarely moved than 50 m; most recorded movements were less than 15 m, and they had a tendency for homing (Harcup et al., 1984). As suggested for centrarchids (above), both static and mobile individuals were detected. Hesthagen (1990) reported home ranges of 40–50 m^2 for two salmonid species in a Norwegian stream. Shirvell (1994) also reported only very short movements of salmonids in a coastal stream in British Columbia (Canada). Thus, for resident (nondiadromous) salmonid populations, much of the evidence has pointed to a relatively sedentary existence with limited movement of adults. However, in contrast to many of the earlier studies, Riley et al. (1992), Gowan et al. (1994), and Gowan and Fausch (1996) detected much greater movement of resident stream salmonids, with brook trout in Colorado streams commonly moving as much as 2000 m between captures. They also showed that these fish will opportunistically move into stream reaches that have been enhanced with cover or deeper pools (Riley et al., 1992). This group has suggested that the high degree of movement is an adaptive response of the species to the highly heterogeneous conditions in Colorado mountain streams. Young (1996) has also reported larger home ranges for stream salmonids than predicted by earlier studies [e.g., home ranges averaging more than 200 m of stream length for cutthroat trout (*Oncorhynchus clarki*). Therefore, the degree to which "restricted movement" or home reach affinity actually characterizes resident salmonid populations, worldwide, remains in question.

Short-Term Movements

Movements of freshwater fishes (at the scale of hours to a few days) can influence the composition of a local assemblage. Deacon (1961) showed distinct differences between sedentary and mobile fish species in rivers in Kansas (USA). One of the first large studies of fish movements in temperate streams was by Thompson (1933), based on tagging approximately 7000 fish in streams of Illinois (USA) over a 3-year period. From recaptures, Thompson calculated that some species moved as much as 2 km per day, and others ranged in total movement from about 1 to 15 km per month.

Mendelson (1975) documented frequent movement of sand shiners and other minnow species from pool to pool in a shallow, sand-bottomed northern stream. Movements of various species from outside a pollutant "kill zone" in an Arkansas creek resulted in relatively complete recolonization of the affected area in 2–3 months (Olmsted and Cloutman, 1974). In Brier Creek, Oklahoma, Matthews et al. (1994) found in monthly snorkel surveys a substantial movement of minnows and other small fishes from pool to pool, and in that same stream, Power et al. (1985) and Power (1987) documented rapid (within hours) emigration of stoneroller minnows (*Campostoma anomalum*) when predatory largemouth bass (*Micropterus salmoides*) were introduced to a pool.

In lakes, some species show daily vertical migration (Appenzeller and Leggett, 1995), or onshore and offshore, like golden shiners in small northern lakes (Hall et al., 1979) and numerous species in Lake Texoma (Hubbs, 1984; Matthews, 1986a), apparently to feed or avoid predators. Brett (1971) showed diel vertical migration of salmonids among thermal zones in lakes. Starrett (1950a) summarized hourly and diel movements of fish in Iowa (USA) streams, with distinct day–night patterns in habitat use by some cyprinids. Southern duskystripe shiners (*Luxilus pilsbryi*) move into or near swift currents during the day to feed, and return to slackwater areas near the channel edge at night (personal observations, October 1974, Piney Creek, Arkansas). Matheney and Rabeni (1995) also found that northern hogsuckers (*Hypentelium nigricans*) moved into feeding habitat during the day and into backwaters at night. Inshore movements of fish at night have been shown in a European river (Copp and Jurajda, 1993), but the effect differed among types of stream banks. Fish also exhibit diel movements from spring runs into river mainstems in some systems (personal observations, Roanoke River, Virginia).

Some short-distance movements occur in response to changes in flow or discharge. Ross and Baker (1983) showed that "flood-exploiting" fishes in Mississippi streams migrated out onto inundated floodplains to feed and benefitted in improved physical condition or growth. Starrett (1951) noted that fish in shallow, sandy mainstream rivers of the Great Plains would move into expanded "edge" habitat as waters rose, but as waters dropped, they left pools. Thus, they could expand into "new" habitat to feed, but avoided stranding. As described in Chapter 7, I have found movement of midchannel riffle fish species to stream edges during floods in the Roanoke River, Virginia, and in Brier Creek, Oklahoma.

Daily or short-term changes in physicochemical gradients can influence movement of fish among habitats. In large springs of west Texas, endemic headpool *Gambusia* species invade further downstream during high flow, which displaces thermal gradients (C. Hubbs and E. Marsh-Matthews, personal communication). Brown (1971) showed that desert pupfish (*Cyprinodon* spp.) track minute-by-minute the moving local microgradients of warm versus excessively hot water, in order to maximize their feeding areas. Matthews (1977) and Matthews and Hill (1979a) showed that red shiners (*Cyprinella lutrensis*) had distinct preferences in thermal, oxygen, and other physicochemical gradients in the laboratory that corresponded to their movement among microhabitats in field sites along a large Great Plains stream (South Canadian River, Oklahoma).

Rheotaxis

The tendency of fishes to move upstream, or to move into a current, is known as rheotaxis. Rheotaxis can differ among populations of a species and has been suggested to allow riverine populations to maintain positions in streams (Kaya, 1991). Large aggregations of fish are common downstream of impediments to upstream movement. For example, in the Little River near Wright City, Oklahoma (before increases in pollution; personal observations), there was a particularly diverse fish assemblage just below an old low-water bridge that allowed flow of water but blocked upstream movement of fish at normal discharge. In July 1976, we collected 28 species at this site, more than at any of 8 other sites in a 2-day field trip.

Large numbers of fishes also accumulate below mainstem dams. In summer 1976, Clark Hubbs and I (as teaching assistant) took a reservoir fish ecology class into the draft tubes below the (inactive) turbines of Denison Dam (which forms Lake Texoma). In the cavern-like interior of the draft tubes, in water about 40 cm deep, was an incredibly dense aggregation of large shad, gars, buffalo, carpsuckers, and numerous other species that had apparently moved upriver until encountering the dam. Probably more than mere rheotaxis was involved, as predatory or scavenging fishes apparently feed on fish injured by the dam.

Spawning Movements

Some fish species, like some of the darters, make short movements to appropriate microhabitats for spawning, but remain within the same general reach of the stream (Winn, 1958; Ingersoll et al., 1984). However, longer rheotactic movements are commonly associated with spawning. Even some darters exhibit upstream migration prior to spawning [e.g., *Etheostoma spectabile* (Ingersoll et al., 1984)]. *Luxilus cornutus* displays rheotactic behaviors at the onset of spring warming and lengthening of days prior to the spawning season (Dodson and Young, 1977). Indirect evidence also suggests upstream bias in the location of large adult minnows (e.g., spawning-sized individuals) like *Luxilus pilsbryi* in Piney Creek, Arkansas (Matthews et al., 1978) and more young-of-year or juveniles at downstream sites. In Piney Creek, larger fish apparently move upstream before spawning, and small fish move downstream to use quiet-water locations near edges of large habitats low in the watershed as nursery areas.

Sucker (Catostomidae) "runs" for spawning are well known in larger North American creeks and rivers (e.g., the genus *Moxostoma* migrating upstream in large numbers to spawning areas in gravel riffles). Reynolds (1983) detected extensive upstream migration of as much as 1000 km by golden perch (*Macquaria ambigua*) in Australia, attributing the upstream movement as an adaptation to offset the downstream floating of the buoyant eggs of this species. In western North American, some of the large, endangered minnows (squawfish, humpback chub) and suckers (razorback sucker) move to known spawning areas within the Colorado River system [e.g., lower Little Colorado River (Minckley, 1991)].

Less dramatic migrations for spawning are movements of fish in North American rivers onto floodplains. Although such movements are now largely impeded by levee construction, a century or more ago agencies like the Illinois wildlife department invested huge amounts of energy in "saving" fish from stranding in backwaters along the Mississippi and other large rivers (Bennett, 1958). Places still remain in the Mississippi basin where riverine fishes move into vast flooded forests to spawn. There, larvae develop with access to rich food sources (Finger and Stewart, 1987). Near the White River, Arkansas, wooded terrain still flooded during the springtime, and species like freshwater drum (*Aplodinotus grunniens*) move into these areas to spawn (personal observation). Many fish make similar movements to spawn in shallow coves and flooded forests or fields at Lake Texoma during high water in the spring (personal observation).

Nonspawning Seasonal Movements and Migrations

Fish also make many seasonal movements not associated with spawning, and presumably to seek acceptable or optimal physical habitats. Fuselier and Edds (1994) found that riffle-dwelling madtoms moved to microhabitats with slower currents in the winter, and Baltz et al. (1991) showed that rainbow trout (*Onchrhynchus mykiss*) occupied different microhabitats in summer and winter (which was also related to ontogenetic differences in habitat use). Thompson (1933) described the movement of fish to deep pools in Illinois streams in winter, thus using only a small part of the available habitat. Thompson also reported a regular upstream movement of fish in spring and downstream movement in summer or fall. Thompson (1933) interpreted these latter movements as driven by the tendency of a particular species to occur in a stream of a given size [as summarized in Thompson and Hunt (1930)]. Prewinter movements of fish during autumn have been reported in streams in Canada (Derksen, 1989; Brown and Mackay, 1995) and Norway (Hesthagen, 1988). Lowe-McConnell (1991) reported "spectacular upriver dispersal" of small fishes like characins during falling water levels in tributaries rivers of the Amazon, but the cause or purpose of these migrations is not known.

Many fish species migrate vertically on a seasonal basis to avoid anoxia or high surface temperatures in lakes (Chapter 4). In Lake Texoma and Keystone Lake, Oklahoma, adult striped bass (*Morone saxatilis*) migrate long distances to avoid hot uplake areas and spend most of the summer in the deeper (cooler) main body of the lakes near the dams (e.g., Matthews et al., 1989). Vertical movement of fish in lakes in New Zealand was caused by seasonal changes in oxygen in the water column (Rowe, 1994).

Diadromy

In contrast to fishes that live their lives wholly in either freshwater or the ocean, or that occur at various estuarine salt concentrations, a small but important number of taxa are "diadromous" (McDowall, 1987). McDowall (1987) defined diadromy as "a specialized migratory phenomenon of fishes involving regular, seasonal, more or less obligatory migrations between fresh and marine waters." These movements he classified into three categories: "anadromy" (moving into fresh water to spawn); "catadromy" (moving into the ocean to spawn); and amphidromy (involving migrations of immature individuals). McDowall estimated that less than 1% of the world's fishes are diadromous and that, of those, more than half are anadromous. McDowall (1987) reviewed in detail the prevalence of diadromy in freshwater systems of various zoogeographic regions of the world. In general, anadromy dominates in cool to cold temperate regions (and also occurs in the tropics), and catadromy is more prevalent in tropics but also occurs in cool temperate regions (McDowall, 1988). McDowall (1988) reviewed in detail the biology and ecology of diadromous fishes.

Although relatively few taxa may exhibit diadromy, some of them are among the most important fishes commercially or ecologically as members of assemblages. A large number of salmonids are anadromous, and the long-distance migrations and very precise return of spawning adults to home streams are legendary (Hasler, 1966). Literally thousands of studies have documented in detail the spawning runs of salmonids, worldwide, and the biology of many of these taxa is known in great detail. For many salmonid species, there can be both resident (fresh water) and anadromous populations (Castonguay et al., 1982), or even individuals within populations (exhibiting the alternative life histories described in Section 8.4).

In addition to the commercially and ecologically important salmonids, many species that penetrate far into fresh water, like some sturgeons, are diadromous. In the freshwater streams and lakes of the interior of large continents like North America, South America, Africa, and Asia, where fish faunas are highly diverse, only a small percentage of species are diadromous, and local assemblages are influenced little by those species. Additionally, in many rivers of the world, high dams have blocked migratory routes of diadromous fishes, and some, like eels (*Anguilla rostrata*), are rapidly disappearing as part of the local fish assemblages. Because of damming of North American rivers, *Anguilla* has become very scarce in the interior. In all of my field sampling since 1972 in midwestern North America I have found the species abundant only in one river system (in North Texas), and the old adult individuals that now exist in midwestern rivers are not likely to be replaced by new immigrants from the ocean.

In contrast to the aforementioned continents, other regions dominated by cold-water rivers (e.g., northern Europe and northern Asia) and with depauperate fish faunas (Moyle and Herbold, 1987) may have local assemblages largely dominated by diadromous salmonids. Oceanic islands, like Hawaii, have streams dominated by small diadromous fishes (Fitzsimmons and Nishimoto, 1995). Many native stream fishes in Australia and New Zealand are diadromous, and many of their present freshwater forms were derived from diadromous taxa (McDowall, 1978). Therefore, the structure of freshwater fish assemblages in much of the world may be considerably influenced by the migrations of species to or from the ocean, in addition to the presence of purely freshwater forms.

9

Interactive Factors: Competition, Mixed-Species Benefits and Coevolution

9.1 Interspecific Competition and Resource Partitioning

> *What groups crowd upon each other in the struggle for subsistence? Do closely allied species (of fishes), living side by side, ever compete for food? What relation, if any, do specific and generic differences bear to differences of food? ... Prominent peculiarities ... will probably be found merely to extend a little the capacities of the species, or to enable it to take those slight advantages of its competitors when the struggle for existence comes to the death grapple. ... That a full understanding of the* competitions *among the fishes of a stream or lake is necessary ... is evident at once.*
>
> Forbes (1878)

> *... so the existence of an animal may be decided by the presence or absence of some structural modification adapted to carry it safely through a single brief period of unusual scarcity or of extraordinary competition.*
>
> —Forbes (1888b).

Background

To understand the many studies of competition and niche segregation (or "resource partitioning") among fishes, it is necessary to place them in the context of historical thinking in ecology about "the niche," competition, and community structure. Stimulated by the empirical or conceptual work of Grinnel (1917), Elton (1927), Hutchinson (1957a), MacArthur (1958), and others, generations of ecologists studied community structure from the perspective that animals have "niches," that typical communities are relatively stable equilibrium (or at least predictable) entities, and that over the course of evolutionary or ecological time, coevolution results in the members of a community segregating along important resource axes, permitting similar (but not ecologically identical) species to coexist. Competition between species was generally accepted as a (if not "the") driving force in the structure of communities.

Grinnel (1917, 1924) first used "niche" to indicate the habitat in which a species lived. Elton's (1927) niche reflected the trophic role of animals in their environment. As late as mid-century, different sections of the premier text in general ecology in North America (Allee et al., 1949) referred to the "habitat niche" and the "food niche," discussing the two as separate concepts.

Beginning in the 1920s, theoretical treatments like the Lotka–Volterra equations, based on saturation of the environment at K, or stable carrying capacity for a population, were developed to describe idealized views of change in population numbers over time. Modification of the Lotka–Volterra equations to include competition coefficients allowed modeling of the effects of a given species on the population numbers of another.

Gause (1934) concluded from experiments with microorganisms that species with requirements that were too similar could not exist indefinitely in a community because one would outcompete the other for resources, with the latter becoming locally extinct. This prediction was codified by Hardin (1960) as the "competitive exclusion principle" and became the dominant paradigm for organization of communities.

Hutchinson (1957a) formalized earlier niche concepts, along with thinking from Volterra and Gause, into consideration of the "niche space" of organisms, providing concepts like the "fundamental niche" versus the "realized niche" and describing the total niche of a species as an "n-dimensional hypervolume," having an infinite number of niche axes (along which animals presumably could segregate, ameliorating competition). However, by 1949, Allee et al. (1949, p.11) had noted that, "Competition is avoided, at least in part, by the evolution of space and time separations (for plants, presumably, as footnote mentioned), or by some combination of these." Allee et al. (1949) further noted that, "Important as competition may be, it can readily be overstressed," but also that, "Competition is a potent factor in animal life . . . ." A succinct account of the history of niche concepts is in Pianka's *Evolutionary Ecology* (1978, pp. 237–245). Pianka (1974) shifted thinking from two-species competitive effects in communities to mathematical treatment of the total "diffuse" competition experienced by a given species by virtue of its interactions with all other potentially competing species in a community.

To this point in time, many ecologists considered competition between/among species as a major organizing factor in communities (see articles in May, 1976). However, other views began to emerge, emphasizing that competition might not be pervasive under all conditions and that other variables (abiotic stress, "crunch" periods of scarce resources, levels of disturbance, and predators) also played substantial roles in the structure of communities. There was a general shift away from the competition pagadigm and its corollary, resource partitioning, as ecologists like Wiens (1977) raised the issues that resource uncertainty made it difficult to assume equilibrium populations and that an empirical finding of equilibrium communities regulated by smoothly operating levels of competition balanced by careful "partitioning" of the available resources was difficult to find. Wiens (1977) emphasized that interspecific competition might only be a major factor in community structure during "crunch" periods of resource shortages. Weatherley (1972) had indicated earlier (for fish) that severe interspecific competition for food was probably discontinuous, "restricted to more or less acute convergence of requirements for a resource for limited periods." Forbes' 1888 quotation introducing this chapter suggests that a century ago he recognized a link between competition intensity and a shifting resource base. Wiens (1984) emphasized the positioning of communities on a gradient from equilibrium to nonequilibrium, with competition most important only near the equilibrium end of the spectrum.

Ecologists found that keystone predators (Paine, 1969) could ameliorate competition in communities, promoting coexistence of similar species, and that disturbance ["intermediate disturbance hypotheses" (Connell, 1978)] also could prevent dominance of one species over another, promoting species richness or diversity (Rogers, 1993) in communities. Seifert (1984) also argued that facilitation may be more important than competition in the structure of some communities. Another troublesome problem was that of demonstrating ongoing competitive interactions in presumably competition-driven communities. Low levels of overlap in resource use observed in a field study of extant species could result either from ongoing "interactive segregation" (Nilsson, 1967) in ecological time or from separation of species into "niches" that were shaped by competition in the past that is now unobservable.

The difficulty of accepting competition as the most important extant factor regulating communities was emphasized by Connell's (1980) admonition of the unproven "ghost of competition past" (i.e., that a lack of observable competition in communities could be due to past coevolutionary divergence in resource use). Connell (1980) concluded that "the notion of coevolutionary shaping of competitor's niches has little support at present" and that "it is more likely that they (coexisting species) diverged (in resource use) as they evolved separately." Finally, neutral or "null" models (Connor and Simberloff, 1979, 1984) that suggested random structure in many communities—hence no need to explain any "structure" by competition or any other interactive factor—further eroded confidence that there was strong deterministic structuring of animal communities. [For a full exposure of both sides of the argument, see Connor and Simberloff (1984) and Gilpin and Diamond (1984) with their "rejoinders".] Finally, in continued emphasis on effects of physical factors on individual species of fish (e.g., Hutchison, 1976), animal ecologists touched on the long-standing controversy in plant ecology over the contrasting "Gleasonian" (1917, 1926) and "Clementsian" (1916) ideas about interactively structured communities versus autecologically driven aggregations of species in time and place (McIntosh, 1995).

Niche Segregation or "Resource Partitioning"

Extension of the concepts of Gause (1934) suggested that for two similar species to coexist, they should segregate along at least one resource axis ("partition resources"), ameliorating competition to a level at which both could persist. However, long before Gause's work, Pearse (1916, p. 282), noted, for fish in Wisconsin lakes, that, "when a particular sort of food is abundant, a number of different fishes may feed upon it, but if it becomes scarce the fishes do not all turn to the same diet for a second choice. *This specificity of food enables different species to live together in the same habitats.*" MacArthur (1958) provided the classic field example of segregation of closely related species, describing empirically the distribution of five sympatric species of warblers (*Dendroica*) with emphasis on height and position in trees (i.e., resource partitioning) [although MacArthur (1968) himself also pointed out that in some circumstances an increase in one member of a direct-competition pair could benefit the other (Stone and Roberts, 1991).] Schoener's (1974) review of all studies of resource partitioning to that date concluded that three major resources axes (food, space, and time) could account for sufficient ecological segregation to permit species to coexist.

There is a huge literature on niche segregation or resource partitioning among members of stream or lake fish communities. Ross's (1986) review of resource partitioning in fishes pointed out that at least some of the perceived effect could be attributed to ways studies were done. Ross showed that there was more resource partitioning on average (defined as separation along at least one resource use axis) between species within genera than for species pairs from more distantly related taxa (i.e., across families or orders).

Most studies of resource partitioning among fishes were based on quantification and comparison of niche axes in the wild. In many studies, ecologists went to the field [e.g., with darters in the Roanoke River, Virginia (Matthews et al., 1982b)] and measured habitat, food, or time use by groups of related fishes, then determined how much or little they overlapped in space or diet (cf. MacArthur's warblers; Pianka's lizards). Measures like Levin's or Pianka's indices were used to estimate niche breadth or overlap, asking how one species affects niche breadth of another (Douglas et al., 1994), or how much they overlap. MacArthur (1972) estimated theoretically how much species could overlap and coexist, but there has been little test of these theoretical limits for fish. For fish, most authors have based judgments on the magnitude of overlap values, with little comparison to MacArthur's "how much overlap can occur"?

Segregation of species along various resource axes for fish assemblages has been shown in many studies. Important corollaries to any empirical finding of ecological segregation of species are several questions: (1) Is the segregation caused by ongoing competition (i.e., "interactive segregation" of Nilsson); by phylogenetic differences among species; or by past competition that caused divergence within a group so that there is no ongoing competition? (2) Does the segregation facilitate or permit coexistence of species? With regard to the first question, Winemiller (1989) presented a modern view as, "contemporary patterns of niche partitioning could be founded in genetically-based morphological and behavioral traits that evolved in response to past environmental conditions . . . (sensu Connell 1980) . . . (which) contrasts with . . . (an) explanation that contemporary patterns of resource subdivision are adaptive in the context of present day biotic interactions." Winemiller (1989) noted that interspecific competition remained the most invoked "ongoing" factor in studies of resource partitioning, but that numerous authors had emphasized the influence of predators in habitat selection (e.g., Fraser and Cerri, 1982; Werner et al., 1983a; Mittlebach, 1984; Power, 1984a; Power et al., 1985; Greenberg, 1991) by freshwater fishes (see Chapter 10, on piscivore effects).

Finally, it should be emphasized that the empirical finding of a given level of overlap between two (or more) species on a resource axis cannot, lacking other information, be interpreted to indicate that competition is or is not occurring. At a minimum, information on the level of resource availability can influence the interpretation. Consider in Figure 9.1 that the magnitude of overlap on a resource axis is shown relative to resource availability. At moderate levels of resource availability, species may diverge on this resource spectrum, with each specializing on that part for which it is best (optimally) adapted. However, *at either extreme* of resource availability (scarce versus superabundant), it is possible to empirically find high overlap in field studies. However, to the right of Figure 9.1, the resource is so abundant that all species may use it opportunistically, whereas to the left

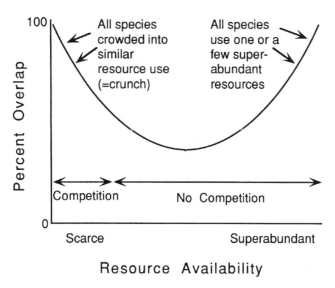

Figure 9.1. Hypothetical overlap on resource use axes as a function of resource availability in a local habitat.

in the figure, all species may converge to using almost identical resources [e.g., if essentially nothing else is available (i.e., during the "crunch" periods of Wiens (1977)].

As an example from field studies, *Luxilus pilsbryi* are opportunistic (e.g., feeding activity on insect pupae during hatches or emergences) (Matthews et al., 1978). If other minnow taxa do likewise, several species collected at the same time might all be engorged with the same kind of pupae (and this actually makes a very efficient use of an ephemeral resource in the system). A niche overlap index calculated for food use would be very high, although competition would not be going on in any biologically meaningful way. Alternatively, for fish confined to shrinking pools of an ephemeral stream (Matthews, 1987a), examination of fish stomachs would likely reveal similar contents. In this case, a high overlap value could be equated correctly to severe competition for remaining items. Although the two situations are obviously very different, there are many published studies of resource partitioning and competition between fish species in which high overlap values are presented with scant information on resource availability or environmental conditions.

Competition Among Fishes: Historical Perspective

Until the 1970s, fish ecologists and fishery managers alike largely accepted competition as a major organizing force in natural or artificial fish communities, and a wide range of empirical phenomena (e.g., replacements of one species by another in space or time) in fish communities was attributed to interspecific competition. More than a century ago, the influential ecologists S. A. Forbes (1878) pointed out, "That a full understanding of the *competitions* among the fishes of a stream or lake is necessary to anything better than guess-work in fish-culture ... is evident at once."

Clemens et al. (1924) indicated that in Lake Nipigon (Canada) the ling (burbot, *Lota lota*) was a competitor of the cisco (*Coregonus*) for food (and also its predator), and also a competitor of lake trout (*Salvelinus namaycush*) and pike perch (walleye, *Stizostedion vitreum*). They suggested that "in view of the large numbers of ling in the lake, there is little doubt that this competition is serious and that reduction in numbers of ling would mean increase in numbers of lake trout." Clemens et al. also noted that the "common sucker" was an important competitor (for food) of the sturgeon but that it "does not commonly inhabit the deeper water, and so is not as serious a competitor of the common whitefish as is the northern sucker. ..." This latter statement suggests that by 1924, fishery biologists recognized that segregation into different habitats in a lake could ameliorate competition between species, but another comment reminds us that this article was published before any suggestions of competitive exclusion in the ecological literature: "There appears to be no fundamental difference in the food of the two species (of stickleback) occurring in Lake Nipigon." This comment a few decades later would have stimulated some discussion about "resource partitioning" or lack thereof!

Several other comments in Clemens et al. (1924) suggest that they were acutely (if erroneously) alert to the potential for noncommercial species (e.g., suckers) to compete with commercially important species:

> The northern sucker, the common sucker, and the ling are the most serious in view of their occurrence in large numbers. The first two consume very large quantities of bottom food stuffs and thus enter into *direct competition* [italics mine] with two of the most important commercial fish in the lake, namely, sturgeon and whitefish, while the ling competes with the lake trout for food, directly and indirectly. *A great reduction in the numbers of these fish* [italics mine] would seem to be desirable in the interests of the conservation of the staple food materials of species of commercial importance.

Hubbs and Eschmeyer [1938; quoted in Allee et al. (1949, p. 659)] suggested that total biomass of fish in a lake would remain relatively constant over time, but that fluctuations might occur in the weights of "competing species."

Starrett (1950b) noted that Des Moines River fishes tended to be either specialized or generalized feeders, that generalists could modify feeding habits and "thus avoid serious competition," but that specialists might be food limited. Starrett suggested, however, that "severe interspecific competition for food was not thought to exist among the minnows during the course of this investigation" and that "the abundance of dipterous larvae, other compensating foods, and diversified feeding habits of minnows reduced the competition to a minimum." Northcote (1954) found that two species of sculpins (*Cottus asper* and *C. rhotheus*) used similar habitats and foods in northern lakes. He concluded that competition could exist, but also noted the impossibility of assessing actual competitive relationships between the species without data on food availability. Citing Gause, Northcote suggested that one "would be expected to replace the other," but attributed the continued coexistence of the two species to "diversity of habitat" within the littoral zone of lakes, where they occurred.

Larkin (1956) cited numerous fisheries papers supporting the concept that competition does exist among fish populations, even if it has been poorly demonstrated scientifically, and noted the widely accepted concept of "carrying capacity" for lakes (Moyle, 1949). Larkin (1956, pp. 337–338) noted "two schools of thought" on the control of fish populations, contrasting density-dependent (i.e., competition) factors (e.g., Nicholson, 1933, 1954) with abiotic (climate) factors (Andrewartha and Birch, 1954). Note that Larkin (1956) was generally very skeptical of competition and concluded that it was likely of little importance in freshwater fish communities. Larkin (1956) also cited Lagler (1944) as concluding that "studies demonstrating exact effects of competition . . . are few in our (fish management) literature" and lamenting that fish managers are confronted with "a vastly great unknown" of competitive effects in fisheries systems. Larkin suggested that, in many cases, interspecific competition would be "as unpredictable as the physical environment, but with little chance of being an important controlling factor," and that "in some aquatic habitats . . . climatic controls would seem to outweigh biological factors (e.g., competition) in controlling populations (Larkin, 1956, p, 338). However, Larkin (1956) also indicated (p. 338) an appreciation of the "interplay of biological factors" (such as competition) with physical factors and "higher order" biological interactions like predation and parasitism as regulating fish populations in fresh water. This appears to be a precursor to the skepticism about competition that was soon to develop in ecology in general and to the modern view that many other variables, as well as competition, can influence fish communities.

Overall, Larkin (1956) believed that, "Freshwater environments offer comparatively few opportunities for specialization in fishes . . . many species share . . . habitat type, a flexibility of feeding habits and in general share many resources . . . with several other species of fish." He believed that growth and reproductive potential would "permit fish populations to tide over unfavorable periods of competition" and appeared to give little credence to competition as a factor in population control. However, Larkin (1956) noted that competition might function as a "subordinate factor" causing fish to be lost from other causes and thus having indirect effect on population structure. Larkin (1956) noted little literature directly addressing interspecific competition in freshwater fish populations and an "absence of reviews on the subject" concerning freshwater fish, perhaps because "it is not important in freshwater fish relationships."

These comments are somewhat surprising for 1956, as many fisheries managers in that era appear to have not questioned that competition between species was important. Larkin explored the notion that two populations might "compete for a niche" and that "competition" should be enlarged in meaning to encompass any and all mechanisms (including life-history variables: longivity and fertility) by which one population might be favored over another in a particular environment. Larkin (1956) assumed that a predator might influence the outcome of a competitive interaction among fish populations; that is, "a predator . . . may find one prey species more easy to catch, and by its predation may determine the course of competitive relationship." Larkin cited the work of Foerster and Ricker (1941) indicting that predation is "the" factor controlling competition between "undesirable" species and sockeye salmon. A modern analog to this idea is in Mittlebach (1984, 1986) in which predators can influence the degree of competition between two species.

Larkin (1956) pointed out "a host of factors in natural environments which influence competitive relationships and their outcome" and attributed a listing of those ideas to Crombie (1947). Also in Larkin's interpretation of Crombie (1947) is the indication (Larkin, 1956, p. 330) that "the organism itself, the physical environment or the biological environment may all change so much during the course of competition between two populations, that it is unlikely that the relationship will be positively resolved in favor of one species or another." This can be read as a forerunner of ideas articulated three decades later by John Wiens (Wiens, 1977) in which competition changed during "crunch" times in variable environments, or that of Grossman et al. (1982), suggesting that conditions in some streams vary too much to permit highly structured, equilibrium assemblages.

Larkin (1956, p. 331) also argued (citing earlier works) that there is little demarcation of fish into ecological zones; that a wide range of habitats is characteristic of freshwater fish, with many taxa occurring both in lakes and streams; and that there is little specialization in feeding habits of freshwater fishes. Larkin was skeptical about attempts to classify food use by fish species (p. 332), cited strong overlap in food use among species, and indicated that because flexibility in diet is the "rule," there are few clear cases of multispecies demand on a "readily defined mutual (food) supply." Larkin (1956) cited Starrett's (1950b) study of Des Moines River fishes to indicate that fish "may change diet 'rather than enter into severe competition.'" Although Larkin indicated that this made it difficult to establish competitive relations between fishes for food, his conclusions could also suggest the possibility for diet shifts or partitioning of resources if potential for competition in a fish assemblages increased at any particular time. Larkin (1956) was equally skeptical of discrete zonation of fishes in lakes in any kind of ordered habitat segregation, and cited earlier workers' evidence (p. 333) of much overlap in habitat preferences and seasonal changes in distribution breaking down any spatial partitioning of the environment. Basically, Larkin (1956) eschewed the notion that species might eat similar foods but segregate by habitat in any structured fashion (that might ameliorate competition). Note the contrast between this opinion and that of Werner et al. (1977) or Werner (1984) for centrarchids, also the disagreement with Keast's (1977) concepts of morphology versus ecology of Lake Opinicon fishes and their distribution among habitats.

Keast (1977) emphasized that all life stages of bluegill (*Lepomis macrochirus*) in small northern lakes ate essentially a similar diet, but diverged in habitat use with life stage, *contra* Larkin (1956). Larkin's underlying argument for the lack of freshwater fish specialization was based on the premise (p. 333) that most of the diversity in physical structure of freshwater habitats is short-lived in geological time, concluding that "specialization in such impermanent environments would be untenable."

Despite Larkin's view, work by Zaret and Rand (1971), Mendelson (1975), and much of the work by Werner and collaborators (late 1970s–1980s), as detailed in subsequent sections, appeared to confirm the prevalence of competitively based resource partitioning (*sensu* Schoener, 1974) in fish communities, and most fish ecologists appeared to agree. However, Grossman et al. (1982) postulated in a hotly debated paper (e.g., Yant et al., 1984) that abiotic factors were so pervasive in streams that fish assemblages had little opportunity to function like highly structured or "deterministic" systems, and that they were, instead, "stochastic" in structure and function. The views of Grossman et al. (1982, 1985, 1990) suggest lack of stability in fish assemblages, but others (e.g., Matthews et al., 1988; Matthews, 1990) have suggested that more order exists in stream fish assemblages.

Is Competition Important in Fish Communities?

In spite of the change in view of many ecologists in the 1970s and 1980s away from a competition paradigm, there is substantial evidence to suggest that intraspecific and interspecific competition are important in fish communities, operating at various scales of space and time. The prevailing view among fish ecologists now probably coincides with that of "general ecologists" that competition is important at times or under certain conditions, and that one important mission of community ecology is to determine how, where, and when competition compares in importance relative to other factors regulating community structure and function. Years of environmental "calm" (with lack of disturbances) can permit populations to reach densities or assemblages to become saturated to a point at which competition becomes important. Many fish ecologists would now agree that competition can be important, but only along with a variety of other factors like the physical and chemical template, distribution of habitat units or "patchiness," abiotic disturbance, predation, and bottom-up control, and varying in importance along an equilibrium–nonequilibrium community continuum (Wiens, 1984).

Lines of evidence putatively showing competition between fish species have included the following:

A. Adjacent but nonoverlapping ranges of similar species
B. Differences in resources used where two species are naturally allopatric versus sympatric ("natural experiment")
C. Co-occurrence of species at field sites within their ranges less than if they were distributed at random
D. Changes in abundance or resource use of native species in the wild after introductions of exotic fishes to lake or stream systems
E. Controlled field or laboratory experiments imposing a potential competitor on a target species and looking for niche shifts.

Interactions between two species have been evaluated in many empirical field studies that show overlap (or lack thereof) of distributions or resources used by the two species. For example, it is common to find longitudinally in a watershed, or within given zones of a lake, that certain species overlap slightly with others. Clearly, this is weak evidence for competition, because, lacking other evidence, it would be equally likely (among numerous other alternative explanations) that the two species do not overlap in space because of dissimilar requirements for some physical feature of the habitat. Many ecologists now are skeptical of concluding that competition is ongoing or matters in interactions between species on the basis of field observations only. Only manipulative experiments (e.g., Edlund and Magnhagen, 1981) showing changes in one species' resource, growth, survival, and so forth, in the presence of a second will convince most ecologists that competition is ongoing. Furthermore, most fish ecologists now want to know not only that competition can be demonstrated in a one given experimental situation but also "how often," "under what conditions," and similar questions before accepting that competition drives the outcome of a situation.

Two examples—one in which competition is probable and one in which competition does not likely mediate the observed interactions—illustrate the weakness of assuming that observed segregation is due to competition. In four years (1981–1984) of year-round gill netting in Lake Texoma, a large reservoir on the Red River (Oklahoma–Texas, USA), D. R. Edds and I found that two closely related catfish species segregated strongly in habitat use. Blue catfish (*Ictalurus furcatus*) was most abundant in deep water 1 km offshore, whereas channel catfish (*Ictalurus punctatus*) were much more common in coves and shallower habitats. Numbers taken offshore and in a shallow cove respectively were as follows: blue catfish, 318 versus 96; channel catfish 19 versus 208. Although this level of habitat separation would have been attributed to competition and "resource partitioning" if viewed uncritically, it is likely that in this artificial impoundment, each species is merely using the habitat most like the habitat in which it naturally occurred in preimpoundment conditions, with little competition between the species in Lake Texoma. Blue catfish are a species of large, deep midwestern river mainstreams (Pflieger, 1975), whereas channel catfish in the prairie region are common in smaller streams and even shallow creeks (personal observations).

In Piney Creek, Izard County, Arkansas (USA), two similar darter species (of the subgenus *Oligolepis*) *Etheostoma caeruleum* and *Etheostoma spectabile* are both common, but *E. caeruleum* is more abundant downstream and *E. spectabile* is much more common upstream (see Fig. 6.13). This apparent segregation could be explained by many variables other than competition, including factors like an affinity for smaller gravel, more gentle currents, and so forth in upstream reaches by *spectabile* (e.g., due to its body size or hydrodynamics).

Slightly stronger evidence that the two darter species might compete comes from the fact that midway down the creek [at Station 6 of Matthews et al. (1988)], where both species are common, *E. caeruleum* occupies parts of the riffles with large cobble and *E. spectabile* tends to be more common near riffle edges where current speeds are less and cobble is smaller (personal observations). This suggests that where they coexist, each diverges to a habitat most like that where it is common, either upstream or downstream of Station 6. However, none of this is sufficient to "prove" that ongoing competition between the taxa results in the observed microhabitat segregation: Where they coexist, they each (independently) may merely be segregating to the kinds of microhabitats in which they are favored by hydrodynamic factors, feeding efficiency, or other unknown factors. Note also that even within a darter species [e.g., *Etheostoma flabellare*, Roanoke River, Virginia (Matthews, 1985a)] juveniles and adults may separate across gradients of current speed.

There is substantial evidence for some competitive control in species occurrence from a broad distribution study in which Winston (1995) examined the distribution of 27 species of minnows (Cyprinidae) among 219 field sites in streams throughout the Red River drainage in Oklahoma and Arkansas. Winston (1995) evaluated the co-occurrence of morphologically similar species pairs and found that the most morphologically similar pairs co-occurred at sites less than would be expected from random placement of species across the collection sites. Additionally, Winston ruled out a phylogenetic basis for this finding (i.e., that lowered co-occurrence did not coincide with evolutionary relationships among the taxa and thus may be related to extant or recent ecological interactions).

Additional tests of competition are in the large number of "experiments" in which species have been stocked or accidentally introduced [e.g., Jester (1971) in Elephant Butte Reservoir, New Mexico; and McCarraher et al. (1971), who listed 19 species that have been stocked into Lake McConaughy Reservoir, Nebraska] into a lake or stream, or gains access accidentally. For example, in the 1950s, the planktivorous atherinid *Menidia beryllina* (inland silverside) gained access to Lake Texoma, where another member of the family (with superficially similar body form), *Labidesthes sicculus* (brook silverside), was very abundant. Within a few years, the former had completely replaced the latter (so completely that in 15 years of collecting in the reservoir I have never captured a *Labidesthes*, although they remain in some nearby tributary streams). Lacking any evidence of gross environmental changes in the reservoir, it seems logical to conclude that one taxon outcompeted the other, and detailed work by McComas and Drenner (1982) on the relative ability of the two species to capture zooplankton suggest morphological and functional mechanisms for the replacement. *Menidia* is significantly better than *Labidesthes* at capturing open-water zooplankters (McComas and Drenner, 1982) because of differences in the morphology of the mouth. *Labidesthes* has a "scissor" or "snapping" oral structure, whereas *Menidia* is better able to form a "tube" mouth with improved suction feeding on zooplankton (Fig. 9.2).

Empirical Evidence for Competition in Fish Assemblages

Suggestions of one species displacing or replacing another come from many examples in fishery management literature (e.g., Jester, 1971). However, in many cases it may be logical that competition could drive the replacements, but in most, the actual cause or nature of the competition is not established (e.g., competition for food or space; eating of larvae or young; aggressive displacement from suitable habitats, etc.). Declines of a preexisting species, following stocking

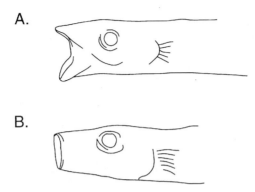

Figure 9.2. Shape of oral opening (mouth open) for (A) *Labidesthes sicculus* and (B) *Menidia beryllina*. With the mouth open, *Menidia* has a tube-like oral structure, in contrast to snapping mouth of *Labidesthes*. [Redrawn from McComas and Drenner (1982).]

or natural increases of another species, has often been assumed to result from competition. However, other factors often may be involved. For example, in Lake Texoma until the mid-1970s, white bass (*Morone chrysops*) were highly abundant. The winter of 1976–1977 was unusually cold, resulting in the loss of much of the clupeid forage (especially the introduced threadfin shad, *Dorosoma petenense*), and the numbers of white bass dropped sharply. By 1979, huge surfacing schools of the introduced striped bass (*Morone saxatilis*) were common in Lake Texoma, with this species having largely displaced white bass. However, striped bass did not fully replace white bass in this lake, and in recent years, white bass have increased in numbers. However, for years, the numbers of white bass remained very low relative to striped bass (Matthews, unpublished gill netting data, 1980–1987). There is some evidence to suggest that competition between striped bass and white bass of similar size could have caused the shift in abundance of the two *Morone* species. For juveniles < 150 mm Total Length (TL), there was substantial overlap in use of habitat and very strong overlap (0.88 0.97) in their diets throughout several years (Matthews et al., 1992a). However, these two species may overlap strongly in habitat and diet and yet be successful in Lake Texoma during most years. Was it the scarcity of forage (*Dorosoma*) late in the 1970s acting to exacerbate competition that mediated the shift in dominance from white bass to striped bass in the reservoir? The data to answer the question do not exist, but this is a clear demonstration of the need to consider factors other than competition, or the degree to which environmental factors can influence the outcome of potential competition between species.

Experimental Evidence of Competition Between Fish Species

There is a growing body of direct experimental evidence on competition between freshwater fish species in natural or laboratory settings. Most experiments have been of short duration (e.g., asking "What happens to species A if we add species B?") and typically focusing on immediate changes in habitat or food use [e.g., small tank experiments in Matthews et al. (1992a)]. In a few cases (Schluter, 1994), tests have lasted several months to ask if the presence of one species influences survival, growth, or condition of the other, but virtually no experiments (at least in fresh waters) have addressed these issues over a sufficiently long time to incorporate reproductive effects. Failing to find short-term habitat shifts says little about what might happen in field environments over many years of interaction between two species.

Some of the earliest experiments on competition between fish species were in Europe (Svardson, 1949). Allee et al. (1949, p. 660) quote an account by Meek (1930) of plaice being transplanted into habitats in which they competed with blennies, and also that predation on the blennies by adult cod influenced the outcome of the competition between plaice and blennies. Svardsen (1949) showed (in what would now be considered an oversimplified experiment without replication) that *Salmo trutta* fry survived much less well than did *Salvelinus* when they were housed together under controlled feeding conditions. Although the experiment was simplistic, Svardson specified that "there must be two kinds of food competition, i.e., interspecific and intraspecific competition pressures" and discussed contrasting effects of intraspecific competition promoting variance in size within a population and intense interspecific competition (which resulted in apparent death of individuals in his experiments). Svardson noted that "some sort of standard test for measure of competition must be proposed. . . ."

Numerous more recent investigators have tested interspecific competition by fishes in field or laboratory experiments. Fausch (1984) showed in experimental streams that juvenile salmonids would compete for positions that were most energetically profitable (value of drift minus cost of position holding), and that these differences in positions were related to growth rates of individual fish. Salmonids rapidly establish positional heirarchies from which dominant individuals can be identified. Subsequently, Fausch and White (1986) used laboratory stream experiments based on position dominance hierarchies to evaluate competitive interactions among three salmonid species, including one introduced taxon. The introduced coho salmon (*Oncorhynchus kisutch*) was the competitive dominant of two native species, displacing them from the energetically most profitable positions in the stream. These same experiments demonstrated that brook trout could dominate brown trout with respect to most beneficial feeding positions. Moreover, when experiments were reversed, competitive release of subdominants allowed them to take up energetically more beneficial positions. However, because results of the brook–brown trout competition differed from that expected on the basis of field studies, Fausch and White (1986) suggested that competitive relationships between species could change with differing physical conditions. Frank Rahel and Fausch are now working on interspecific experiments with western North America salmonids based on hierarchical dominance to predict potential changes in species' distributions in the event of global warming.

For two darter species, Greenberg (1988) showed that when one was removed, the other expanded its use of habitat to include that previously occupied by the removed taxon. For benthic riffle fishes, Taylor (1996) experimentally showed asymmetric competition of banded sculpins (*Cottus carolinae*) on orangethroat darters (*Etheostoma spectabile*), consistent with field-survey data showing negative association of the two species at natural stream sites. One relatively complex experiment (Bergman, 1990) showed that "two-species" competitive systems do not operate in a biological vacuum. In Bergman's experiments, perch (*Perca fluviatilis*) were expected (on the basis of earlier studies) to shift toward a more benthic diet in the presence of roach (*Rutilus rutilus*), which is a more effective planktivore. However, the expected diet shift did not occur, which Bergman attributed to the presence of a third species (ruffe, *Gymnocephalus cernua*) which is an even better benthivore than perch. This finding underscores the difficulty of extrapolating "two-species" laboratory experiments to field situations, unless the ecology of other common taxa in the system is also known.

Schluter (1994) showed experimentally that competition was most intense between species of the *Gasterosteus aculeatus* complex (three-spine sticklebacks) that were most similar ecologically, and that evolution would favor adaptive radiation of morphs away from that of similar, competing morphs.

In field studies in the Colorado River basin of western North America, Douglas et al. (1994) showed that competition from the introduced red shiner (*Cyprinella lutrensis*) displaced native spikedace (*Meda fulgida*) into inferior habitats. Specifically, where spikedace were syntopic with the invading red shiner, they used habitats with faster currents than they normally used without red shiners present. Such a displacement into energetically costly habitats, to which spikedace are not adapted, would seem likely to be harmful to the maintenance of a population (e.g., by leaving less energy for reproduction or other maintenance activities). Douglas et al. (1994) were unable to positively identify the mechanism of displacement of spikedace by red shiners, but suggested that aggressiveness of the invading species played a role. In artificial streams at our laboratory, we have noted that the presence of red shiners (that exhibit more rapid "agitated" movement than other species of minnows) often results in a general increase in levels of activity and movement among all members of the assemblage (F. P. Gelwick and J. Stewart, personal communication).

Fausch (1989) examined wide-scale distribution patterns for four salmonids in western North America to argue that interspecific competition based on behavioral dominance partially explained patterns in occurrence. Fausch hypothesized that each species could be superior to and outcompete others in local habitats for which it was physiologically best adapted. In another field-based analysis of habitat use in sympatry and allopatry, Gatz et al. (1987) concluded that there was strong evidence of asymmetric interspecific competition, with brown trout (*Salmo trutta*) displacing rainbow trout (*Oncorhynchus mykiss*) from preferred habitats.

Field studies continue to be important, but there is increasing use of manipulative field or laboratory experiments to estimate strength of competition. In studies based on field observations or collections, substantial overlap in food use led to the conclusion that competition was likely or at least possible in several studies (Gaudreault et al., 1986; Thonney and Gibson, 1989; Parrish and Margraf, 1990), and this conclusion was supported by negative changes in the putatively impacted population in some cases. Alternatively, Persson (1983a) found a low diet overlap between two species, but concluded that roach were having a negative effect on perch by having caused a trophic niche contraction of the latter. In several other field studies, there was direct evidence of a diet shift in one species when a competing species was present (Johnson, 1981; Brabrand, 1985; Magnan, 1988).

Among experimental studies, one showed an example of competitive release, with one species rapidly expanding its feeding habits when the competitor was absent (Hume and Northcote, 1985). Other experimental studies demonstrated shifts or displacement in habitat (Hearn and Kynard, 1986; Hindar et al., 1988; Donald and Alger, 1993; Beall et al., 1989) or diet (Paszkowski, 1986; Persson, 1987; Reeves et al., 1987; Crowl and Boxrucker, 1988; Hindar et al., 1988) in the presence of a competitor among taxa ranging from salmonids to minnows to perciforms. In several examples, such as Hindar et al. (1988), competition was asymmetrical, with the subordinate species negatively affected in sympatry and no effect on the dominant. Reeves et al. (1987) showed an important interaction between temperature and competition (habitat displacement) effects. In laboratory trials, redside shiner (*Richardsonius balteatus*) and steelhead trout were differentially able to displace each other from preferred microhabitats, depending on temperature: Trout dominated in cool water, and the shiner dominated in warm water (Reeves et al., 1987). Only three of the studies seeking evidence about competition found none. Ensign et al. (1989) found no diet shift by potentially competing trouts, in allopatry or in sympatry. Baltz and Moyle (1984) found little competition for habitat in a field study of Sacramento suckers (*Catostomus occidentalis*) and rainbow trout. Likewise, Moyle (1977) reported only very weak evidence that competition (by predation) by sculpins could negatively influence salmonids in streams of western North America.

Resource Partitioning in Fish Assemblages: Through 1970

Although there are numerous accounts of comparative use of foods by coexisting fishes dating to the last century or the early 1900s [e.g., Forbes (1878, 1880a, 1880b, 1883, 1888a, 1888b), on foods of Illinois fishes; Turner (1921), on foods of Ohio darters), one of the earliest accounts that can actually be considered a study of niche segregation or resource partitioning by fishes is that of Hartley

(1948) for feeding relationships of species in the River Cam and a small tributary brook in Great Britain. In his introduction, Hartley cited 14 earlier references as having examined food relationships among fish species from "one water" and indicated that although some of these articles did not explicitly specify "competition," it was implicit in their content. Hartley (1948) was concerned with both interspecific and intraspecific competition by fishes; that is, "Interspecific competition for supplies may be fully as rigorous as intraspecific competition in its limiting effects upon multiplication, growth and well-being (of fish)."

Hartley concluded that many species had wide and variable food habits, that most generalized feeders were facultative, and that "between no two species is there complete identity of feeding habitat, but that there is much *general competition* [italics mine] between all the fish of the community for certain staple foods." Hartley considered that "none of the species present have clearly defined individual roles to play" and that the community overall was a "loosely organized assemblage in which the members are distinguished by no more than the varying proportions in which they draw upon the constituents of a common stock (of food items)." However, Hartley also concluded that "the degree of general competition is such that few of the members of the fish fauna will be wholly unaffected by an enforced change in the diet of some other species." Hartley thus appears to have considered competition important among the 11 species in these streams, but that partitioning of resources (he did not actually use the term) was by proportional differences in use of foods rather than not qualitative differences in diet. In the language of a few years later, Hartley would have found a high dietary overlap (although he did not calculate overlap indices). Another early reference to resource partitioning (although not so named) in fishes was that of Larkin (1956), who indicated that "two species may be said to occupy *partially separate niches*" if they coexist but compete for (some?) foods.

Gee and Northcote (1963) carried out one of the first experimental studies of interspecific resource segregation (microhabitat use) by freshwater fish, explicitly asking if leopard dace (*Rhinichthys falcatus*) and longnose dace (*Rhinichthys cataractae*) defied Gause's (1934) suggestions. They found that the two species overlapped in use of foods in the field but were segregated by depth of water. In an artificial stream tank, Gee and Northcote (1963) showed that recently emerged fry had similar habitat requirements, but suggested that at the time of the emergence of both species, food was superabundant in the Fraser River drainage due to flooding of terrestrial areas. They concluded that the subsequent divergence of adults to distinct habitats likely allowed overall survival of both species in the drainage "without entering into severe competition for those resources of the environment essential for their existence."

Some of the next important contributions to studies of resource partitioning in fish assemblages were by Allen Keast, of Queen's University, Canada. Keast (1965) published on "Resource subdivision amongst cohabiting fish species in a bay, Lake Opinicon, Ontario," and a related article the next year (Keast, 1966) addressed "Trophic interrelationships in the fish fauna of a small stream." In both articles, however, Keast [like Hartley (1948) focused only on differences in one resource axis—foods]. In the lake article, Keast (1965) sampled not only the fishes of the lake, but documented plankton and benthos available as foods for fishes. His diet diagrams show marked month-to-month fluctuations in proportion use of foods by various species of minnows (*Notropis, Pimephales*), topminnows (*Fundulus*), and perciforms (*Perca, Pomoxis, Ambloplites*, and *Lepomis* spp.). The occurrence of some invertebrate food items in the diet corresponded with numerical peaks in their availability, and he concluded that "clearly . . . fish feed to the greatest extent on a resource when it is at its numerical peak." When a food item was superabundant, Keast found that many species simultaneously used the item, but that most species maintained a "certain degree of selectivity." Keast also noted high variability in foods used by a species at different locations within the lake and their versatility in feeding and rapid adaptation to alternate resources (e.g., in sunfishes).

Keast (1965) did not calculate numerical indices of overlap but summarized that the fishes in Lake Opinicon were "partly separated ecologically" by food preference and that interspecific competition was diminished by the fact that greatest overlap in use of food items came when those items were most abundant (hence, not in short supply). Keast found that Lake Opinicon fish species could be separated not only as trophic specialists or generalists but also by position in the water column at which they fed and by the general propensity of a given species to be a "deep water" or "shallow water" form (p. 128). He also noted that age groups of species are only "in limited competition" and that habitat differences between age classes could ameliorate competition. He concluded that "many mechanisms are operative" (p. 129) that tend to direct species into "different ways of life," including food specialization, use of alternative resources, minor habitat differences, and behavioral or structural specializations.

Keast (1966) addressed resource use by fishes in a small Canadian stream, and, as in the lake article, compared food use to food availability. As in the lake, he noted specialized versus generalized feeders, and the capability of species to use alternate foods. Food overlap was greatest during periods of greatest prey abundance, suggesting, as in the lake, that competition between species might not be intense.

Nilsson (1967) applied the concept of "interactive segregation" [which he attributed to Brian (1956)] to freshwater fish to indicate magnification of ecological differences (segregation into different niches) because of active, ongoing competition (or predation). Nilsson suggested that the capacity of fish species to interactively segregate into different niches depended on plasticity in ecological requirements [which was emphasized by Larkin (1956) to be a major feature of fish assemblages].

As evidence of interactive segregation among fish species, Nilsson (1967) cited numerous examples of situations in which the introduction of a competitor resulted in changes in the niche of a preexisting species. He cautioned, however, against "snapshot" views of resource use, as the resource being used at a given time might not be the resource for which two fish species previously competed, even in the recent past. [As a recent example, Wilson and Williams (1992) showed that environmental conditions in which esocids evolved were much different than those in habitats they now occupy.] Nilsson (1967) cited Johanes and Larkin (1961) as an example in which an introduced dace and a resident salmonid population competed intensely for amphipods, which later declined in number so that "an observer (in later years) would hardly suspect that amphipods had been the most important item of competition." Nilsson (1967) pointed out numerous examples of species differing in resource use (diverging) when sympatric, relative to the use of food or other resource by individual species when allopatric. He also cited evidence that "interactive segregation" was not restricted to closely related fish species and that it might be a seasonal process.

In general, examples used by Nilsson pointed to more intense segregation into different feeding niches during periods of resource scarcity (cf. Zaret and Rand, 1971; Wiens, 1977). Nilsson (1967) suggested that in extreme cases, there is complete displacement of one species by another, resulting in local extirpation or in a retreating to marginal habitats [see also Douglas et al. (1994)] by the "defeated species." Nilsson (1967) suggested five possible mechanisms for interactive segregation in fishes: (1) exploitation, (2) territoriality, (3) "food fighting," (4) predation, or (5) "other interference." As an example of direct exploitation, Nilsson cited Johannes and Larkin's (1961) report that the redside shiner was quicker than rainbow trout to notice prey and begin feeding, and cited the many experiments of Ivlev (1961) as evidence that fishes would change their food electivity when housed with various competing species. Similar results were obtained years later by Werner and Hall (1979) who showed *Lepomis* switching prey use in later summer, depending on which other species were present. Northcote (1954, in Nilsson) showed that minor differences in mouth size could allow one species to better exploit large prey than did another.

The principles of "interactive segregation" were summarized by Nilsson (1967) as follows:

1. Interactive segregation can occur between taxonomically distant species, but may be most severe between closely related species if competition is at work.

2. Total or partial coexistence may be apparent, but in each case, the species will be segregated into different food niches or habitats.

3. Total displacement or exclusion of one or more species may occur as a result of interactive segregation.

4. Segregation is governed by behavior . . . .
5. The degree of competition varies with the food supply . . . pronounced when resources are limited and less pronounced or nonexistent when resources are superabundant.
6. A given species with "a broad ecological potency" [quotation marks mine] is forced by interactive segregation to take a smaller but more specific share of the resources . . . however, the resources should be more and more completely exploited if more species are present to exploit them in their own specific ways.

Nilsson (1967) also expanded on this last point and the view of Larkin (1956) that most temperate fish have generalized food habits, to suggest that the available niches of a body of water are easily filled by even a small number of fish species. This agrees with the later indication by Herbold and Moyle (1986) that "empty niches" are not the rule in fish communities.

Empirical Evidence of Resource Partitioning: Early 1970s

The next major addition to the dialogue on resource partitioning in fish assemblages was Zaret and Rand (1971), examining food overlaps of fishes in wet versus dry seasons in a tropical Panamanian stream. Zaret and Rand (1971) suggested that observation of changes in niche overlap during periods of varying environmental extremes (wet versus dry) would relate the degree of overlap to ongoing changes in conditions. They included information on habitat, food type, and time of day (day versus night feeding) and concluded that overlap was greatly reduced in the dry season (a period of limited habitat, but when foods were still available) by interspecific divergence in habitat and food use. In the wet season, they found less segregation of species (i.e., they converged, at least in part, to common resource use during benign times). Zaret and Rand used Morisita's (1959) index to estimate food overlaps and assumed (without comment) that any overlap > 0.60 between species (range $= 0$ to 1) was "significant." On this basis, they reported five overlapping pairs in the dry season, compared to eight significant overlaps in the wet season. They concluded that there was a clear change in this stream from distinct food niches in the dry season to widely overlapping diets in the wet season, with the greatest niche separation when lower levels of food were available. They concluded also that there was increased food competition in the dry-season pools and that their study supported the principle of competitive exclusion (Hardin, 1960).

Moyle (1973) described ecological segregation (food and habitat use) among three common minnow species (*Notropis volucellus, Luxilus cornutus, Pimephales notatus*) in a small northern (USA) lake. Ecological segregation of these species reflected differences in feeding habits, location in the water column, and activity patterns. Moyle concluded that temperate freshwater fish species were "strikingly" segregated from similar coexisting species, that the mechanisms for this segregation were behavioral, and that the "interactive segregation" of Nilsson (1967) was operative.

In a comprehensive study of resource use of temperate stream fish, Mendelson (1975) examined spatial and trophic relationships of four species of insectivore–omnivore cyprinids [all then assigned to the genus *Notropis*, but one now considered a *Cyprinella* (Mayden, 1989)] in a small stream in Wisconsin, and availability and microdistribution of invertebrate prey. Mendelson concluded that throughout the year, these species were segregated more by microhabitat than by food use and that interspecific differences in food habits were due primarily to microhabitats occupied. He found substantial overlap of the species in the genera of invertebrates eaten and in the relative proportions of the taxa consumed (i.e., little separation of species by trophic segregation). However, Mendelson (1975) concluded that the observed stable subdivision of microhabitats would act to reduce competition in these coexisting fishes. He explicitly noted (p. 227) that these fishes "are spatially separated and hence can prey on the same kinds of animals without competing." Importantly, Mendelson pointed out (p. 227) that morphological and behavioral characteristics permitting the species to coexist (through spatial partitioning) are "preadaptations allowing these species to coexist wherever they are found together but not demanding a coevolutionary origin," and that they had not necessarily been "molded by mutual interaction into a functional unit." These statements suggest that Mendelson believed the four common minnow species in this creek differentially occupied microhabitats on the basis of individualistically acquired characteristics (i.e., that these species did not reflect a coevolved community) [cf. Connell, (1980, "Ghost of Competition Past").

Two other articles in 1975 addressed competition or resource overlap of species in stream assemblages. Cadwallader (1975a) found for four native New Zealand stream fishes and one introduced species (brown trout, *Salmo trutta*), that all species used the same food organisms, but in varying proportions. For most combinations of native species, he noted dissimilar foods and feeding locations, with no significant concordance (Kendall's rank correlation) in foods used by most species pairs. The two most abundant fish in the river (*Galaxis vulgaris* and *Gobiomorphus breviceps*) took foods of the same size, but there were differences in feeding mechanisms, timing, and locality. However, between the introduced brown trout and the native species food, similarity was high, and for galaxiids, essentially identical foods and microhabitats were used. Cadwallader attributed observed losses of number of galaxiids in the presence of trout to direct competition for food between the taxa (along with aggression and microhabitat overlap). He emphasized the competition of small trout and galaxiids, in that they used the same foods, feeding manner, and microhabitats.

In three salmonid streams in Wales, Jones (1975) found partial segregation of dominant fishes across different microhabitats (riffles, runs, pools) and that those differences were similar across the three streams. Jones considered it probable that the observed "partial segregation" was due to adaptations of individual species for particular environments, as well as to interspecific interactions. He also noted the difficulty of knowing if a species used a particular habitat due to "choice because they have inherited an adaptation to that habitat" or "as a result of competition, with the most successful . . . taking the most favorable habitat and leaving the remainder to the weaker. . . ." Jones' last sentence agrees with the opinion of most workers today: "It is probable that the interactions involving several fish species are complex and that experimental study will be required to elucidate the interrelationships."

Experimental Work—Werner, Hall, and Colleagues

The most influential work on resource use and niche relationships of freshwater fishes in the 1970s and early 1980s was research by E. E. Werner, D. J. Hall, and associates at Michigan State University. Werner incorporated optimality theory into studies of niche relationships among fishes. Initially, Werner and Hall (1974) showed that bluegill sunfish (*Lepomis macrochirus*) exhibited optimal foraging (maximum return for effort spent in foraging) with respect to prey size. Werner (1974) expanded this to examine relationships among fish size, prey size, and handling time for bluegill (*Lepomis macrochirus*) and green sunfish (*Lepomis cyanellus*), and again found that fish took prey that, based on their morphology, yielded optimal gain per foraging effort.

In 1977, Werner published a paper in *American Naturalist* detailing comparative foraging efficiency of bluegill, green sunfish, and largemouth bass (*Micropterus salmoides*) in relation to morphology and food size. The ultimate product of Werner (1977) was a "cost curve" for each species describing the energetic cost of obtaining prey of varying sizes. Werner concluded from empirical measurement of prey sizes and from the cost curves that bluegills and bass should be able to coexist in the same habitat because of differences in optimal food size, but that green sunfish (intermediate in body morphology and food optima) should not be able to "pack" into the food axis because of too much similarity to the other two species. He reasoned (p. 571) that if green sunfish were to coexist in lakes with largemouth bass and bluegills, it should be by segregation along a habitat niche axis.

Werner et al. (1977) field-tested this postulate by underwater observation in two small Michigan lakes. As predicted, bass and bluegill were similar in their distribution, with wide range but occurring more in deeper areas, whereas green sunfish, as predicted (Werner, 1977), were strongly limited to very shallow inshore waters where only a few small bass and bluegills were found (Werner et al., 1977). Werner et al. (1977) also noted that differences in the vegetation associated with the species further contributed to niche differentiation. The body of work thus indicated that it was possible to begin with observed differences in body form, move to theoretical and laboratory analyses of optimal prey relationships, and translate these into empirical differences in foraging habitat among three closely related species (family Centrarcidae) in natural lakes. Overall, Werner et al. (1977) concluded that only one species pair in the natural lakes (largemouth bass and bluegill) was segregated by food size and that the other pairs of common species showed niche complementarity by separation into different habitats.

Hall and Werner (1977) found seasonal differences in these relationships, with less habitat segregation in spring when food was abundant, and retreat of species to more narrowly partitioned use of habitat throughout summer when habitat separation was apparent. Finally, Werner et al. (1978) concluded that in Florida as well as Michigan lakes, the littoral zone was saturated with species (i.e., that competition could structure habitat use). Although these studies had tremendous influence on thinking about niche relationships within fish communities, most of the work was with fish in stable environmental conditions (small vegetated lakes), although one of the lakes in Florida had fluctuating water levels; thus applicability to harsh, unstable systems (e.g., Matthews and Hill, 1980) remained to be seen.

A series of experiments by Werner and Hall (1976, 1977, 1979) tested for niche shifts and competition among sunfish species. In small experimental ponds, Werner and Hall (1976) examined niche relationships among bluegill, green sunfish, and pumpkinseed (*Lepomis gibbosus*). In natural lakes, the pumpkinseed, like the bluegill, occur more offshore than green sunfish, but pumpkinseeds are deeper in the water column, near the bottom. Although their results were controversial because of lack of a control for density effects (Maiorana, 1977), they found that all species stocked alone used similar foods, but diverged in food use when stocked together, consistent with expectations from interspecific competition. Additionally, species stocked alone showed "competitive release," growing more rapidly than when stocked with other species. They concluded that "considerable ecological segregation occurs when congeners are present," because competitors were thought to change the profitability of a given patch type for less competitive species.

Werner and Hall (1977) showed that when green sunfish and bluegills were confined experimentally to vegetated parts of ponds, there was asymmetric competition favoring the more aggressive green sunfish, which also had a larger mouth, ate larger prey in the vegetation, and consumed a wider spectrum of foods in general. They suggested that the open water column, with only small foods available, offered a "competitive refuge" for bluegills, which ate zooplankton more efficiently than did green sunfish. Werner and Hall (1977) suggested that, "Segregation on the habitat dimension is probably the most important means of niche separation in freshwater fishes." Werner and Hall (1979), for the same sunfish species, showed that the addition of a competing species that is more effective in a given habitat could change the profitability of that habitat for a target species and cause the latter to abandon what had been a preferred habitat, or that species would change habitats during a season as initially profitable habitats became less so.

Other Studies of Resource Partitioning: Late 1970s

At about the same time as Werner and Hall's work, other research in lakes and streams addressed comparative food or habitat use of fishes from the perspective of competition. Adamson and Wissing (1977) noted that "little is known of . . . partitioning of food among cohabiting darter species." They examined food and feeding periodicity of three darter species in riffles of a small Ohio (USA) stream, finding that two species (*Etheostoma caeruleum* and *E. flabellare*) had strong similarity in diet, eating a fairly wide array of aquatic insect larvae, but that their diet was quite different from that of a third, *Etheostoma zonale*, that fed almost exclusively on chironomid and simuliid larvae and at a different schedule than the other species. Adamson and Wissing (1977) gave no overlap values, but their figures suggest that the former two species actually showed little niche segregation on a food or time axis, and they all coexisted in the same riffles. This seems to be an early suggestion that fishes may not partition resources in streams as clearly as they do in lakes.

In a Canadian lake with a nutrient-mediated production gradient, Gascon and Leggett (1977) found in the less productive part of the lake much more intensive competition for foods by six small fish species which "strongly influenced the patterns of food utilization." In the more productive end of the lake, all fish species had a tendency to similarly eat abundant chironomid larvae or other benthic foods. Gascon and Leggett (1977) noted that these findings agreed with earlier workers (Nilsson, 1960, 1967; Zaret and Rand, 1971) showing divergence to separate foods when/where food was scarce, and convergence to feeding in common when or where food was abundant. Baker-Dittus (1978), for three estuarine species of killifish (*Fundulus*), also found that diet overlap increased when foods were abundant, whereas the species segregated in food use during periods of food scarcity. Baker-Dittus (1978) considered that both food partitioning and seasonal variation in species' abundances in the habitat reduced interactions among the species. Likewise, Desselle et al. (1978) reported interspecific diet segregation in four species of sunfish in the Lake Pontchartrain, Louisiana (USA), estuary, with greater niche segregation in summer than winter, and diet partitioning in this brackish-water habitat similar to that seen for centrarchids in freshwaters in Werner's articles.

Conceptual bases for segregation of species in communities were unclear in some works of the 1970s. Schoenherr (1979) indicated strong "niche separation" among introduced fishes in a non-natural habitat, which he interpreted as "illustrating rapid selective forces in action." However, this article could be interpreted to support the idea that fish found together in a habitat where they could not possibly be coevolved might exhibit as much niche segregation as fish that evolved together in a natural habitat—suggesting that in each case, fish might merely be diverging into different habitats that reflect innate differences in morphology or other traits that preadapt them for particular microhabitats. Schoenherr also noted, as did Harrell (1978) in Devils River, Texas, and Matthews and Hill (1980) in Canadian River, Oklahoma, that modification of habitat by flooding markedly changed the microhabitat relationships among fish species.

George and Hadley (1979) attributed ecological segregation of young-of-year rock bass (*Ambloplites rupestris*) and smallmouth bass (*Micropterus dolomieu*) in the Niagra River, New York, to both food and habitat axes, with the greatest overlap in foods early in the year (when foods were abundant?), and divergence to very low diet overlap later in the summer. This agrees with the pattern shown by Werner and Hall (1979) in which food use by sunfishes (their Experiment IV) overlapped strongly early in the summer (all species feeding in vegetation) but diverged to species-specific diets later in the summer. However, George and Hadley attributed early overlap and later divergence in diet to ontogenetic factors: Small young-of-year smallmouth bass and rock bass all were limited (by gape?) to eating zooplankton and small invertebrates, with divergence as they assumed foods more like that of adults as they grew during the summer. Note two possible interpretations of any empirical finding of high overlap values: (1) fish overlap when foods are abundant, when they can use similar foods with no competition or (2) if fish overlap strongly, it is at this time that competition might be greatest. Clearly, finding of given overlap value, say 80%, could be interpreted either way if evidence on food availability is lacking [as it is in most field studies; but see Mendelson, (1975)]. It is also interesting to note that George and Hadley (1979) introduced their article by stating that the goal was to "examine the resource partitioning" by the two species—implying a priori that "resource partitioning" did indeed exist.

Niche partitioning has most often been considered to include "space," "diet," and "time" (Schoener, 1974), but other approaches have been suggested. For percid communities, Kerr and Ryder (1977) proposed a description of niche space in terms of metabolic scope for activity. Engel and Magnuson (1976) related seasonal changes in habitat partitioning to temperature differences; Magnuson et al. (1979) proposed the "thermal niche" hypothesis to describe distributions of fishes in lakes. MacLean and Magnuson (1977) evaluated partitioning of space in percid communities as a function of temperature, but they also considered food and time of feeding, and competition important in regulating percid communities. Although arguing that effects of competition could not be clearly separated from effects of predation (without manipulative experiments) on percid community structure, it is obvious that MacLean and Magnuson considered resource partitioning to play an important role in interactions of fishes in northern percid communities (with *Perca* or *Stizostedion*).

In 1979 and 1980, two reviews of resource partitioning by fishes (neither restricted to fresh water) appeared (Sale, 1979; Fishelson, 1980). Sale (1979) indicated that it was the "usual" situation to find that the requirements of each species differ some from those of coexisting species, and that the available resources are "divided up or partitioned" among the species present (p. 323).

Sale emphasized that the greater the overlap in resource use, the greater the possibility of competition, but that high overlap does not of itself mean that there is strong competition. Sale suggested that even if resources are scarce, habitat segregation can effectively ameliorate competition, as the species will be searching for food in different places.

Sale also noted that temporal partitioning of feeding activities reduces potential for competition in freshwater communities, although less than in reef fishes. Sale pointed out that most studies of "resource partitioning" appear to have been undertaken with uncritical acceptance of the premises that (1) communities are regulated by competition and (2) resources are usually limiting. Sale reviewed the difficulty of assuming competition always exists and also pointed out problems with "natural experiments" in which niche differences in sympatric and allopatric populations in natural settings is taken as evidence for competition (pointing out that natural "replicates" rarely differ with respect to only the primary factor being considered). Sale advocated manipulative experiments [as did MacLean and Magnuson (1977)] focused on niche shifts as tools for detecting extant competition between species, but indicated that (as of 1979) such an approach rarely had been taken to study fish communities. In fact, Sale considered that only in the work of Werner and associates was there convincing demonstration of partitioning of habitats by sympatric species to ameliorate interspecific competition.

Fishelson's (1980) review indicated that "ability to coexist and to share common food and space resources in a given aquatic habitat optimizes exploitation of various niches and produces interspecific behavioral optima that decrease competition." He believed that some such strategies were innate and others learned. Like most others, Fishelson (1980) considered space and food the two primary variables determining kinds of fish assemblages. Fishelson (1980) emphasized for both marine and freshwater fishes, the importance of underwater shelter (e.g., habitat space within plants in the shallow littoral zone of lakes) where juveniles can hide. He also reviewed interspecific differences in breeding and in general habitat use by freshwater fishes in lakes in Israel and Africa, emphasizing changes in shelter or habitat type during ontogeny of a species. Fishelson (1980) provided references to several articles discussing food partitioning in tropical freshwater habitats, including the extensive information in Fryer and Iles (1972) (who suggested little "partitioning") and Lowe-McConnell (1975). With regard to food partitioning, Fishelson pointed to the separate strategies of (1) living in the same space and eating different foods or (2) using similar foods but doing so in different places (e.g., herbivores, separated topographically).

Resource Partitioning: 1980 to the Present

The 1980s saw a plethora of articles on interspecific resource partitioning, niche overlaps, or related ideas in freshwater fish communities. Matthews and Hill (1980) found that small-bodied species of a Great Plains (USA) river were spatially segregated at times, but that these relationships were transitory and no stable patterns were found. Habitat relationships in this shallow, river with environmental extremes appeared driven more by physical conditions than by any regularity of interspecific habitat partitioning; by autumn, the most abundant species were crowded into relatively similar habitats. Harrell (1978) had found similar instability in habitat use by fish species in the Devils River, Texas, with a single flood markedly changing habitat associations.

Helfman (1981a) demonstrated interspecific differentiation in resource use by freshwater fishes, clearly showing the separability of species (even within a given family) into diurnal and noctural foragers. Such separation of feeding activity could lessen interactions between similar species. Baker and Ross (1981), a benchmark paper on resource segregation by stream fishes, focused on eight cyprinids species in a low-gradient stream in Mississippi (USA). Six of these species showed habitat segregation by the vertical position in the water column or association with aquatic plants. Two species showed essentially no habitat segregation, but one (*Notropis buccatus*) fed at night, resulting in niche separation from the other (*Notropis longirostris*). In a subsequent article (Ross et al., 1987) based on the same stream and species assemblage, relative abundance of common species remained similar across 9 years, but there were observed changes in 22% of the species–habitat associations over time. However, species varied least on habitat axes (in a multivariate analysis) on which they showed most interspecific segregation. It appears that in spite of spatial (upstream versus downstream) and temporal changers in microhabitat use, there remained substantial habitat segregation among the species across years.

For three darter species in Little Miami River, Ohio, Wynes and Wissing (1982) found little segregation on the basis of prey numbers, modest segregation (of two species) by prey size, and low food overlaps based on biomass. The darters also showed low overlap in microhabitat use, and Wynes and Wissing (1982) considered that they coexisted because partitioning of food and space reduced competition. [Caution is needed in evaluating magnitude of overlaps, as the answer can depend strongly on how the overlap values were calculated. See also Matthews et al., 1992a, concerning diet overlaps based on ennumeration versus biomass of prey items.]

Two particularly important articles in the dialogue on niche separation in stream fishes are those of Angermeier (1982, 1987). These articles are of interest because Angermeier's assessments of diets (Angermeier, 1982) and habitat use (Angermeier, 1987) showed there was little consistent segregation of common species by either food or microhabitat, yet the taxa included (five minnow species, three centrarchids, and a madtom catfish, *Noturus*) are a common combination of species for midwestern (USA) streams. Therefore, in spite of prevailing paradigms in fish ecology suggesting that related species "should" partition habitat, it is clear that these species do coexist successfully in many streams without strong resource partitioning.

Likewise, for darters of the upper Roanoke River, Virginia, Matthews et al. (1982b) found only "partial ecological segregation" where three species were sympatric, with generally high overlaps in diet and microhabitat values. These species were all common in the Roanoke River and have been so since at least a century ago (Jordan, 1889). Apparently, "resource partitioning" in the sense of classical competition-based community ecology is not a requisite for successful coexistence of species, given that the habitat is sufficiently large. However, in very small tributaries of the Roanoke River, only one of the three darter species (*Etheostoma flabellare*) was common (Matthews et al., 1982b; Matthews, personal observation). Matthews et al. (1982b) presented evidence on food availability that supported the hypothesis that all three darter species coexisted where foods were more abundant (mainstreams), but that *E. flabellare* occurred alone in small tributaries due to less food and its inherent competitive advantage in taking a wide range of prey.

With respect to levels of overlap on resources, the magnificently diverse, endemic cichlid faunas of the rift lakes of Africa have been particularly challenging to ecologists. Earlier studies of these fishes (Fryer and Iles, 1972) suggested, as did Greenwood (1974), that within trophic groups it was common to find numerous species completely overlapping in food and habitat use. Witte (1984) reviewed the various opinions of earlier workers and noted that Fryer and Iles considered these fishes to be "important exceptions to Gause's principle of competitive exclusion" and that their high degree of overlap "emphatically refutes" the concept of competitive exclusion. Fryer and Illes (1972, p. 287) noted that "the situation among the Mbuna (rock-dwelling cichlids) seems not to agree with this maxim (Gause's), and very strongly suggests that food preferences can become not only very similar but even identical, and yet the species concerned, of which there may be not two but several, can and do co-exist." They further noted that "requirements are so similar that inter-specific competition differs in no way from intra-specific. From the point of view of feeding it appears that two (or more) species behave exactly as do individuals of a single species." They considered that the species coexisted "in flagrant disregard of Gause's contention" (Fryer and Illes, 1972, p. 287), and that in Lake Malawi, "three remarkable assemblages of cichlid fishes (algae grazers, zooplanktivores, and piscivores) . . . in each case defy Gause's hypothesis." Witte (1984) cites Greenwood (1981) as stating that the explosive evolution in this species flock "has led to the coexistence in one niche of two or more species (often close relatives) with what appear to be identical demands on that niche. In other words, this appears to be an apparent negation of the competitive exclusion principle."

Witte (1984) argued a different point of view from his studies of Lake Victoria cichlids, indicating that "each case of apparent total interspecific overlap that has been studied in detail reveals niche segregation." van Oijen (1982) also examined the piscivorous haplochromine cichlids in one part of Lake Victoria and found that, within the group, each was specialized for somewhat different foods, concluding that his findings "contradicts the prevailing notion that many haplochromine cichlids in Lake Victoria feed on the same food items in a similar way." However, Witte (1984) noted that "whether current competition for resources plays a role in segregation of species is as yet unknown because it has not been shown that the fishes would expand their niches in the absence of other species." Witte concluded that ecological relationships among cichlid species flocks in lakes Victoria, Malawi, and Tanganyika may be relatively similar, and that within each lake there is more niche segregation of the species than had previously been supposed, and that segregation could play an important role in intralacustrine speciation processes.

Surat et al. (1982), assessing food and habitat segregation of three minnow species in the upper Roanoke River, found that one species (*Lythrurus ardens*) was segregated from two others (*Luxilus albeolus* and *Luxilus cerasinus*) on the basis of vertical spacing in the water column and in kinds of foods eaten. The two *Luxilus* species were nearly alike, segregating only partly in the water column (Fig. 9.3). Hinting at an idea (phylogenetic "baggage") that has become one of the important organizing themes in fish ecology as well as systematics (Mayden, 1992b), Surat et al. noted that "overlaps in resource use among the three species pairs suggested congruence of *phylogenetic and ecological similarities*" and that "general ecological relationships (among the three species) reflect similarities and differences in their morphology which *are congruent with their taxonomic relationships at the subgeneric level*" [now generic level, with reassignment of cyprinid genera (Mayden, 1989)]. In other words, Surat et al. suggested that the levels of ecological differences empirically found between species were consistent

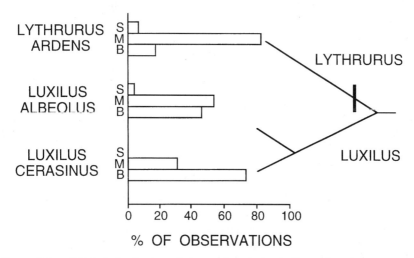

Figure 9.3. Vertical distribution of three minnows in the Roanoke River, Virginia, overlaid with a simple cladogram of species-level relationships. S = near surface; M = midwater; B = near bottom.

with a "null" hypothesis that taxonomically more similar species might also be more similar in ecology, behavior, and so forth, merely because of inheritance of pleisiomorphic (ancestral) characteristics. In modern interpretation, it might be argued that "there is nothing (causal) to explain" with respect to the observed niche "partitioning" by these three Roanoke River minnows, if the differences can be explained by ancestral relationships.

In a major review of resource partitioning of fishes, Ross (1986) showed that more distantly related species pairs segregated more on resource axes, and Winemiller (1989) indicated for tropical fishes that "due to genetic and morphological similarities, closely related species frequently exhibit relatively less ecological segregation than more distant taxa." Likewise, Dudgeon (1987, p. 359) noted that for Asian stream fishes with relatively modified body form, microhabitat use would result from inherited morphology, and that "phylogeny rather than competition determines niche occupation." However, this does *not* mean that any inherited characteristics would not be *operational* in the extant community (i.e., this could lend support to the argument that communities do not have to consist of coevolved species in order to exhibit segregation of species on axes of resource use) (cf. Connell, 1980). Obviously, if species in an assemblage differ ecologically for any reason, whether phylogenetic history, physiological requirements, and so forth, it may yet facilitate their coexistence.

Within a clade, the phylogenetically most similar species may not be the morphologically most similar, and resource partitioning can relate more to morphological than phylogenetic similarity. Two articles by Greenfield et al. (1983a, 1983b) evaluating resource use by morphologically similar and dissimilar species-pairs of *Gambusia* in Belize illustrate this concept. Of the three species examined in these two articles, *Gambusia luna* and *Gambusia sexradiata* are most closely related phylogenetically, but *G. luna* evolved in rapidly flowing mountain streams (Greenfield et al., 1983b) and is very streamlined. The second species-pair consisting of *G. sexradiata* and *G. puncticulata yucatana* are less related phylogenetically but are morphologically very similar (Greenfield et al., 1983a). Reading of both articles yields the conclusion that it is morphological, not phylogenetic similarity that determines resource use relationships between the species-pairs. The morphologically divergent pair (*G. luna* and *G. sexradiata*) occupied very different habitats in sympatry, with substantial segregation in diets as a result of the spatial segregation. the morphologically similar pair (*G. sexradiata* and *G. p. yucatana*) showed only minor separation in microhabitat and feeding behavior, which Greenfield et al. (1983a) suggested still "may allow coexistence." However, based on fecundity data, Greenfield et al. (1983a) suggested that the "subtle differences" in food use between these morphologically similar species might not be enough to "avoid the negative effects of competition." They concluded on the basis of all information that when two species were morphologically distinctive, there might be substantial segregation by habitat differences, but that morphologically similar species might segregate more by minor differences in feeding than by habitat segregation.

Hlohowskyj and White (1983) examined a trio of darter species in two Ohio streams and found that although there was only partial trophic segregation, all species relied heavily on chironomids, which were superabundant. Hlohowskyj and White concluded that in spite of relatively high overlap values, there was probably little competition for food because all species used a very abundant resource. However, whereas one stream contained only two of the darter species (*Etheostoma blennioides* and *E. caeruleum*), the presence of the third species (*Etheostoma flabellare*) in another stream resulted in apparent displacement of the former two species from shallow-water habitats (i.e., there was evidence of a spatial niche shift in this "natural experiment").

By the early 1980s, most authors addressing comparative resource use by fish species did not attribute segregation on resource axes directly to competition as the only causal factor. Fish ecologists, like those working with other taxa, were emphasizing that low overlap values could mean (1) two species were each diverging to that part of the resource they used most efficiently (cf. Zaret and Rand, 1971) because of competitive pressure from the other or (2) two species were innately different, perhaps due to morphology, behavior, or phylogeny, and showed little similarity in use of a resource simply because each was adapted (possibly because of past competition) for using different parts of the resource (i.e., resulting in no ongoing competition). Low overlap values had often been interpreted to mean that two species "avoid competition via resource partitioning," but they may be so morphologically or behaviorally different that there is no competition to avoid.

Schlosser (1982), although not explicitly addressing resource partitioning, concluded from a comprehensive study of distribution of fishes in a small stream (Jordan Creek, Illinois, USA) that, at least in upper reaches and in riffles, variables like recolonization, physical conditions, and temporal variation in reproductive success were more important than competitive exclusion (or predation) in determining the organization of the fish community. Miller (1983) found minor differences in prey size for two *Percina* species in the Tombigbee River, Mississippi (USA) but concluded that there was no evidence that the segregation was due to competition. Two sculpins, *Cottus bairdi* and *C. girardi*, in a Virginia (USA) creek differed sharply in mean occupied current speeds and preferred substrate size (Matheson and Brooks, 1983), and there also was segregation by sex and age class and seasonal variation in the relationships. Matheson and Brooks (1983) concluded the competition for space "is minimized" between these two sculpins by the specializations above. They also concluded that "interactive segregation" likely was ongoing, on the basis of literature comparisons to habitats of *Cottus* in allopatry and on the basis that where present, *C. girardi* eliminated *C. bairdi* from beds of aquatic plants or from (for juveniles) slow, silty areas near shore.

These results are likely correct, but this seem to be only a rather weak "natural experiment."

Moyle and Senanayke (1984) quantified levels of resource partitioning among 20 abundant fish species in tropical streams of Sri Lanka. Their "ecological key" to the fishes (their page 216) demonstrates that virtually all species could be differentiated ecologically. These fishes were highly specialized with respect to body shape or feeding structures, well separated on either habitat or food niche axes, with 88% of all possible species pairs slowing an overlap < 0.33 on either diet or water depth axes. Overall, these fishes exhibited traits predicted for equilibrium assemblages, including "a remarkable degree of ecological segregation" that coincided with differences in their morphology. Moyle and Senanayke concluded that "the morphological diversity of the fishes reflected their ecological diversity, implying coevolution."

Schut et al. (1984) described habitats for eight herbivorous species of *Barbus* in Sri Lanka [in terms of "haunts" (i.e., "a part of a waterbody where fish of a particular species are typically found ... characterized by local depth, current velocity, presence of objects, etc.")]. They statistically found three associations of species that used similar habitats, but within each habitat association, individual species used partly differing foods (e.g., differences in kinds of algae eaten). However, they explicitly indicated (p. 176) that it was impossible to know if the observed separations were due to competition, either in the present or in the past (i.e., "divergent evolution as a result of former character displacement").

Schlosser and Toth (1984) quantified microhabitat and food use relationships of two benthic species (*Etheostoma caeruleum*, rainbow darter, and *Etheostoma flabellare*, fantail darter). Both are common on rocky substrates of steams throughout the midwestern United States. Schlosser and Toth (1984) examined food use of the two species relative to availability of food in the natural streams, documented microhabitat use both spatially and in terms of behaviors in the stream, and conducted single- and mixed-species experiments to seek evidence of habitat shifts in sympatry. They found that overlaps were high in kinds and size foods used regardless of food availability. However, they found that the two species used substantially different microhabitats (at the fine scale of use of tops and exposed surfaces of stones by *E. caeruleum*, whereas *E. flabellare* occupied interstitial spaces) and that the differences in microhabitat use could be attributed to differences in morphology. [Paine et al. (1982) found similar partitioning of microhabitat by these species, as *E. caeruleum* "took prey from rock surfaces" and *E. flabellare* "took prey from between and beneath rocks".] *Etheostoma flabellare* was much better at flexing the body to move into small interstitial spaces, whereas *E. caeruleum* had larger scales and a more rigid body, making access to interstitial spaces difficult (Schlosser and Toth, 1984). Finally, Schlosser and Toth ruled out interspecific interactions (e.g., competition) as driving the perceived differences, in that the species did not change in microhabitat selection whether either or both were present in experiments. Although the field studies encompassed only the warm parts of the year (and darters are active during winter), the study seems to be a good model for combining field sampling, detailed behavioral observations (to produce estimates of mechanisms), and experimental tests for niche shifts in order to determine magnitude and mechanisms for ecological relationships between species.

As Schlosser and Toth (1984) did for warm-water stream fishes, Dolloff and Reeves (1990) used laboratory manipulations and field observations to test for interactive segregation versus innate habitat selection (based on species-specific behavior or morphology, independent of the other species present) for two coldwater salmonids. In the field, these species differed in depth of water occupied, with coho salmon (*Oncorhynchus kisutch*) in deeper water than Dolly Varden (*Salvelinus malma*) (although confounded by variation among age groups). In large artificial stream channels, the species retained relative differences in microhabitat depth, regardless of the presence of the other species. Dolloff and Reeves (1990) concluded from the lack of habitat shifts in experiments that microhabitat selection was innately controlled (rather than resulting from interactive segregation). However, they did attribute (p. 2305) these innate differences to "a consequence of a long evolutionary history of coexistence" of the two species. This latter argument appears to assume sufficient coexistence for coevolution to have taken place (i.e., this article suggests the strong influence of "competition past") (Connell, 1980).

Many additional articles on resource partitioning appeared 1980–1994, as summarized below, that quantified "resource partitioning," "niche segregation," "interactive segregation," or similar ideas in fish assemblages of streams and lakes. Of 51 other articles published in 1980–1994 (Table 9.1), 9 were on food use (only), 19 were on microhabitat, and 23 addressed both food and microhabitat (and some included activity time). Across all of these studies, there was a trend from research in the early 1980s consisting almost exclusively of empirical data from field-caught (or observed) fish, to later articles that often included experimental or manipulative components. Almost all articles seeking evidence of segregation on a resource axis or axes found it, but others like Grossman and Freeman (1987, Appalachian upland stream) and Schlosser (1987b, low-gradient midwestern creek) reported essentially no separation of fish species. Note that whereas Grossman and Freeman (1987) found no consistent microhabitat partitioning among stream fishes, Gorman (1987) described very fine details of habitat segregation in a fish community in a physically similar kind of stream. Much of the discussion is subjective: Some authors considered numerical overlap values such as 0.70 as "considerable overlap," whereas others considered that value "partial segregation." However, earlier articles often accepted uncritically the notion that "fish partitioned habitat or food, thus competition was avoided," whereas later articles asked more critically just what the meaningful interpretation

Table 9.1. Summary of Published Articles 1980–1994 on Habitat and/or Food Segregation of Fish Species

Author, Year	F[a]	FE	LE	Segregate?	Comments
Habitat Only					
Finger, 1982	x		x	Yes	Due to competition
Englert & Sehgers, 1983	x			Some	
Moyle & Baltz, 1985	x			Little	
Mahon & Portt, 1985	x			Yes	
Felly & Felly, 1986	x			Yes	
Allan, 1986			x	Yes	Niche shift lessens competition
Grossman & Freeman, 1987	x			Moderate	Not due to competition
Grossman et al., 1987	x			Yes	Due to predators and evolution
Johansson, 1987		x	x	Yes	Interactive segregation
Schlosser, 1987b	x		x	Not much	Predators increase/overlap in experimental streams
Greenberg, 1988	x	x		Yes	Temperature & flow related
Gorman, 1987	x			Yes	Competition & coevolution
Gorman, 1988a	x			Yes	Competition & coevolution
Gorman, 1988b			x	Yes	Predation influenced
Brown & Moyle, 1991	x	x		Yes	Due to predators
Freeman & Grossman, 1992	x			No	No resource partitioning
Kessler & Thorp, 1993	x			Yes	Lessens competition
Greenberg & Stiles, 1993	x	x		Yes	
Dolloff & Reeves, 1990	x		x	Yes	
Trophic Only					
Gillen & Hart, 1980	x			Yes	Lessens competition
Elrod et al., 1981	x			Some	
Martin, 1984	x			Little	
Angermeier, 1985	x	x		Yes	Re. resource depression
Scrimgeour & Winterbourn, 1987	x			Yes	Lessens competition
Greger & Deacon, 1988	x			Not much	Harsh environment
Cowx, 1989	x			Yes	Not due to competition
Sturmbauer et al., 1992	x			Yes	Aufwuchs—cichlids

Study		Description	Notes
Habitat and Food			
Laughlin and Werner, 1980	x	Adults segregate, not juveniles, for food & habitat	Predator driven
Nilsson & Northcote, 1981	x	Food & habitat	Niche shift due to competition
Magnin & FitzGerald, 1982	x	Food & habitat	Due to competition
Paine et al., 1982	x	Food & habitat	Morphology & competition
Greenfield et al., 1983a	x	Food & habitat	Due to competition & morphology
Greenfield et al., 1983b	x	Food & habitat	
Welton et al., 1983	x	Food & habitat	Lessens competition
Walsh & FitzGerald, 1984	x	More segregation for food than habitat	
Mittlebach, 1984	x	Food & habitat	Predator-driven
Todd & Stewart, 1985	x	Food & habitat	
Moyle & Vondracek, 1985	x	Food & habitat	Partly due to competition
Vøllestad, 1985	x	Food & habitat.	
Kraft & Kitchell, 1986	x	Food more important than habitat	Due to catchability of prey
McNeely, 1987	x	More by food than habitat	Avoids competition
Dudgeon, 1987	x	No	Not due to competition
Winemiller, 1989	x	Food & habitat	Avoids diffuse competition, predators influence habitat.
Delbeek & Williams, 1988	x	Habitat more important than food	Due to innate differences, lessens competition
Winemiller, 1989	x	Food more segregated than habitat	Due to diffuse competition
Wikramanayake & Moyle, 1989	x	Food & habitat	Food shift in natural experiment, lessens competition, coevolved community
Goldschmidt et al., 1990	x	Food & habitat	No niche shift, not due to competition
Winemiller, 1991a	x	Food & habitat	Not due to competition but due to evolution in distinct niches
Greenberg, 1991	x	Strongest segregation for habitat, but also for food/time	Due to phylogeny, predators, body sizes
Naesje et al., 1991	x	Food & habitat	Due to morphology

[a] F = field study; FE = field experiment; LE = laboratory experiment (including in outdoor artificial streams).

was of observed differences along one or more niche axes. It is instructive to ask how much "partitioning" or "segregation" on niche axes was reported in these 51 articles.

Of 19 articles (Table 9.1) dealing with microhabitat (but not food use), all but 3 included field observations. Three included manipulative experiments and one was a "natural experiment" due to an introduction. The articles ranged from studies of as few as two closely related species (e.g., two darter species; Greenberg, 1988) to whole assemblages of 20 or more species (e.g., Felley and Felley, 1986; Schlosser, 1987b). Benthic fishes like darters (Percidae) and sculpins (Cottidae) as well as water-column minnows (Cyprinidae) dominated much of this work.

Six of the 19 articles included a laboratory or a "seminatural" manipulative experiment, suggesting that many authors by the 1980s wanted to directly assess the effects of one fish species on microhabitat use by others (e.g., Finger, 1982; Allan, 1986; Greenberg, 1988; Gorman, 1988a), the effect of microhabitat manipulation (Greenberg and Stiles, 1993), or the effects of other fishes (e.g., predators) on interactions among potential prey species (Schlosser, 1987a; Brown and Moyle, 1991). Of these 19 studies, four showed little or no microhabitat segregation (Moyle and Baltz, 1985; Grossman and Freeman, 1987; Schlosser, 1987a; Freeman and Grossman, 1992). It also was clear that authors were not willing to consider "resource partitioning" in a vacuum, and at least six of the articles showed or suggested that predators could play an important role in changing or regulating the magnitude of microhabit overlap among their potential prey species (e.g., Schlosser, 1987; Brown and Moyle, 1991). (See also the work by Mittlebach and others in Chapter 11, on prey refuging in response to predators.)

Of nine articles published in 1980–1992 addressing comparative diets or trophic partitioning, only two included an experimental component, that of Angermeier (1985), testing effects of altered structure on feeding, and Delbeek and Williams (1988), in laboratory trails of feeding behavior. All others were based on foods of field-caught fish. Some authors implied that competition was a primary cause of community structure and coexistence. Sturmbauer et al. (1992) indicated for aufwuchs-eating cichlids that, "These sympatric species are able to coexist on account of their slightly different feeding habits and diets." Sturmbauer et al. also postulated that "partitioning of resources by trophic specialization seems to be one driving force of speciation," linking ongoing species differences to evolution of species [as more recently did Grant (1994)]. However, others, like Delbeek and Williams (1988) found that "competition for food is thought not to occur" because of abundant prey and morphological specialization that led to differing food usage among species.

Martin (1984) found little evidence of diet segregation, as all species of darters he examined were highly opportunistic, often using similar foods. All other authors reported some substantial interspecific segregation in food type or size, at least at certain times of year. A common pattern was that of the use of similar foods early in the growing season when invertebrates were more abundant, with divergence to separate foods as resources dwindled to moderate levels late in the year. Only Scrimgeour and Winterbourn (1987) emphasized the importance of diel segregation in the use of foods, but indicated that this factor resulted in lowered potential for competition by two New Zealand benthic fishes.

Of 23 articles since 1980 that considered both food and habitat segregation, most found evidence for substantial differences among species on one or both axes. All 23 articles included field measurement of resource use; 7 also included a field "experiment" (typically measuring resource use by a given species in presence or absence of another); only 4 articles reported laboratory experiments. Four of the articles reported niche shifts in the presence of a putative competitor. Seven articles from 1980 to 1989 interpreted the niche segregation as being due to competition; five attributed the difference to "innate," morphological, or phylogenetic differences among species; three attributed segregation of species to effects of predators. Eight of the 23 articles indicated that the existence of niche segregation resulted in less competition, avoidance of competition, or similar conclusions. Thus, many of the articles indicated that the observed level of segregation did lessen competition among species, but several attributed the niche differences to innate interspecific differences (e.g., morphological or phylogenetic) than to extant "interactive segregation."

The article by Goldschmidt et al. (1990) is typical, examining food and habitat relationships among 14 zooplanktivorous cichlids in Lake Victoria. Goldschmidt et al. concluded that although the species were segregated in habitat and food use, they were "already ecologically segregated" (p. 352) by morphology, and finding no change in use of space by some species when others were or were not present, they further concluded that ongoing interactive segregation was not operative. Goldschmidt et al. (1990, p. 353) also presented a heuristic flowchart characterizing relative and hierarchical importance of factors (substrate, horizontal segregation, vertical segregation, and, finally, food) in segregation of these species (Fig. 9.4).

Greenberg (1991) quantified resource segregation among a group of 13 benthic species in Little Tennessee River (USA) and found that all species were segregated along niche axes of habitat (most important), food, or activity time. However, Greenberg (1991) pointed out clearly that although competition among some species no doubt occurs (cf. Greenberg, 1988), all of the observed segregation in habitat and food use among these species could equally well be explained by alternative hypotheses focused on phylogenetic affinities, body size, and avoidance of predation.

Conclusions: Competition and Resource Partitioning in Fish Assemblages

Most published studies show that species in fish assemblages differ along axes of food, habitat, or activity time (Ross, 1986). However, there are cases in which examination of diet and microhabitat show little or no segregation, particularly

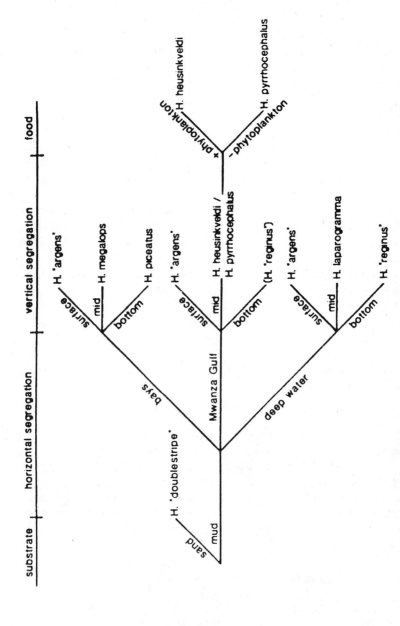

Figure 9.4. Idealized scheme of ecological segregation of zooplanktivorous cichlids in Lake Victoria, Africa. [From Goldschmidt et al. (1990)].

in tropical streams [Dudgeon, 1987, and three additional references cited therein, but see Moyle and Senanyake (1984) for a markedly different view]. The literature suggests that niche overlap between/among species will vary over time, and that, at times, in most systems fish will converge to common use of temporally abundant food items. These facts are supported by several dozens of articles, but this does not mean that (1) there is highly regular, competitively mediated sharing of the resources or (2) that competition is the dominant factor imposing structure or function on fish communities.

In 1984, Schlosser and Toth (1984) suggested that elucidating factors regulating community organization would require "a combination of experimental manipulations of biotic interactions and long term studies of resource availability, trophic interactions, and population dynamics." Experimental manipulations have become a substantial part of work in fish ecology, but truly long-term studies of resource availability and interactions among fish species are mostly lacking. Additionally, Fausch (1988) has argued that most experiments on competition among fishes fail to compare interspecific to intraspecific competition.

Competition among fishes no doubt takes place. In a riffle of a medium-sized upland stream of the midwestern United States, there might be two to six (or more) species of benthic, largely insectivorous darters, sculpins, and madtom catfishes. Each species might exhibit modest differences in mean size of food items consumed, proportions of various items eaten, or mean current speed at which individuals occur. However, simple underwater observation by snorkeling (even when one is careful to not disturb the fish) will reveal much movement of fishes within the riffle. Thus, within the various microhabitats of a single riffle, an individual of almost any species might occur at any given time, even if the *mean* current speed, substrate size, and so forth for that species differs statistically from that of other taxa. In other words, despite modest documentable differences in average microhabitat use by species, a given food item (i.e., a chironomid, baetid, or simuliid larva) might have about as good a chance of being eaten (as it, too, moves about in the riffle) by one species as another, so long as the invertebrate will fit within the gape of whichever fish first encounters it. This view of fishes and their prey in riffles thus suggests a very dynamic movement of all the players, with consumption of the prey items including a very strong component of which fish encounters which prey. From the perspective of functional processes in the riffle, I tend to be more impressed with the old concepts of Hartley (1948) and others who emphasized the substantial *overlap* of species, and suggested only modest differences in food use among coexisting fish species, instead of the work of many others (including myself: Matthews et al., 1982b; Surat et al., 1982) who have documented moderate to substantial *differences* in foods used by benthic fishes as though those differences permitted coexistence.

The issue seems to be that, given a finite number of food items in the riffle under consideration, coupled with a finite rate of replacement of those items from the drift, there is, by definition, competition among actively foraging fish species for the prey items. The question is whether or not that competition is sufficiently severe or frequent to impose structure on the composition of the riffle community.

Competition for food in riffle habitats would seem likely to be relatively intense, in that space is mostly useful to fish only in two dimensions; that is, the benthic species do not function efficiently off the substrate within the water column. For stream fishes occupying pools, the option exists for vertical segregation, as shown for many minnow species (Mendelson, 1975; Surat et al., 1982), but within particular zones of a stream pool, competition for food appears to have the potential to be intense, intraspecifically and interspecifically. Consider the typical aggregation of minnows that can be observed by snorkeling near the head of a pool; that is, in the area just below a riffle in which individuals can remain largely out of the strongest currents, yet dart into the main flow to grasp items being delivered by the drift. Here, many individuals of taxa like *Notropis rubellus, Notropis boops*, and other minnow species that are efficient (relatively streamlined, with large eyes) at feeding in the head-of-pool currents occur. The impression gleaned from snorkeling is that this relatively dense crowd of feeding minnows is highly active and likely quite efficient at catching prey items that come downstream through or from the riffle immediately above the pool.

A single invertebrate drifting into a stream riffle may face conditions somewhat like a "lottery" (*sensu* Sale, but with modified meaning—here the invertebrate enters the lottery to determine if it lands in a safe site, near a given fish species, or near another). The invertebrate drifting downstream from the riffle into the pool is likely observed by many pairs of cyprinid eyes and runs a substantial risk of being eaten by one of the predators. The species to obtain that particular invertebrate may depend largely on chance, out of the several fish species that often occur in the upper end of the pool. The fact that each fish species may exhibit, on average, modest differences in mean current speed, depth in water column, and the like (Moyle and Vondracek, 1985) may create a bias overall toward a drift item being eaten by one species or another, but in the complex system of midwestern (USA) stream pools, with perhaps as many as a dozen minnow or "minnow-like" species present, there seems to be substantial likelihood that a single prey item could be the subject of rather intense and immediate competition among individuals, both of the same or similar species.

Additionally, the competition for the food item coming across the riffle in the drift does not end at the head of the pool, because even if a pool-head individual successfully catches and eats it, the item is thereby removed from its potential as food for a different species of fish that occupies space near the middle of the pool. This, at least is midwestern streams, could as likely be a juvenile or adult sunfish (*Lepomis*) as another minnow species. Many midwestern streams are more "sunfish streams" than "minnow streams," with as many as five *Lepomis* species living simultaneously in a single stream pool (i.e., longear, green sunfish, redear, bluegill, and orangespotted sunfish, all a common combination in larger pools of low-gradient streams in Oklahoma.

The main point is that it would be possible (in all likelihood) for an ecologist to collect fish in this hypothetical pool–riffle and open stomachs and find that there were moderate differences in the kinds and proportions of foods contained, on average, in each species. This finding would have been interpreted in many past studies to suggest that "resource partitioning" existed that "ameliorated competition" among species and "facilitated coexistence." However, the species described above, living in a riffle–pool sequence in this hypothetical midwestern stream may, indeed, have contained in their stomachs those food items that they were most successful at catching, relative to other fishes in the pool that were competing for individual food items, but almost any individual fish in the pool (at least of the omnivore–insectivores, and many "piscivores") would have eaten the invertebrate food item, given the chance and if not preempted by another individual (of the same or another species).

A question related to competition among stream fishes is that of the density of fish versus available prey items, which leads to another possible hypothesis. Strong overlap in food (or habitat) use by stream fishes may be possible because there is sufficient overwinter mortality ("winter as predator") to keep populations (or total density of fishes) in a reach of stream sufficiently low that there is enough food for all individuals during the active feeding and growing season. Overwinter attrition may be most important in keeping fish densities low (1) in moderate to large streams and (2) in spring and early summer *before* the seasonal influx of juveniles into the competitive arena. Clearly, this too will depend on (1) harshness of the previous winter [fish overwintering on fat stores in winter refugia (deep pools?) may be critically affected by a few weeks' difference in warming and resumption of activity] and (2) timing of spring or summer floods that can interfere with nesting and reproduction, kill larvae (Harvey, 1987), or even eliminate a year-class [smallmouth bass (Larimore, 1975)]. Size of stream should be important in this consideration because a large stream (e.g., the Roanoke River, Virginia, with large riffles 50–100 m long, and long, deep pools) seems less likely to be saturated by fish density than are smaller riffles or pools of a lesser tributary.

I have found the greatest densities of riffle fishes in small tributaries (e.g., up to dozens of *Etheostoma spectabile* per square meter in riffles of small creeks in Missouri, as compared to one to three per square meter in large riffles like those in the Roanoke River). On average, there might be no greater fish density in one system than another, but if the fishes do become concentrated in a part of the habitat (for whatever reason), they can clearly create local patches of very high density in small streams. I suggest that as a result, competition is more likely to be an important factor in more of the year in the smaller streams. In larger streams like the Roanoke River, reductions of fish densities by winter stress, combined with production of large numbers of invertebrates in riffles, could provide sufficient food to minimize or eliminate competition as a major factor among the individual adults that survive the winter. Additionally, it may be difficult for fish of various species to congregate at high densities in riffle (rapids) habitats of larger streams, where current speeds and turbulence are much greater than in riffles of small creeks.

Remaining Questions About Competition and Resource Partitioning in Fish Assemblages

Many questions about competition and resource partitioning remain to be answered in stream and lake fish assemblages. For example:

1. How much, how often, and under what conditions does competition have an important influence on composition of fish assemblages?
2. Can "competition–assay" protocols be developed that will allow investigators to go to streams or lakes at various times or under varying conditions to "measure" competition strength in the field or in seminatural enclosures, and so forth?
3. What would long-term, multigenerational studies of competition and resource partitioning in freshwater systems reveal?
4. How critical are particular species in the overall competitive arena or in resource sharing within an assemblage?
5. Are the most abundant species also the most important with respect to structure of resource use in a fish assemblage, or do the rare species also have the potential to alter structure of resource sharing?
6. How much or how often does the presence of predators (cf. Mittlebach, 1984) and/or refuge sites alter resource use or sharing within a fish assemblage?
7. To what extent do factors like primary or secondary productivity affect competition and resource sharing by fishes, and how important is the role of top-down versus bottom-up control of intensity of competition or resource sharing in communities?

Particularly with respect to questions 4 and 5, there has been some experimental work on resource partitioning in laboratory assemblages, but little work addressing larger suites of species (or the whole fish community) from the perspective of the importance of individual species. One profitable approach might be to set up replicated experimental systems (Gelwick and Matthews, 1993) with the entire suite of species of interest, then selectively remove individual species and assess the changes that deletion of each makes on the function of the remaining suite. Will the absence of any one species have a limited effect on resource use by the remaining taxa? Or will certain species be "strong interactors" (cf. Paine, 1980) whose deletion will markedly influence resource use by the rest of the assemblage? Finally, superimposition of predators on such experiments could provide direct answers to question 6.

At a broader level of questions regarding overall structure of fish assemblages, do the fish that occur together in the wild represent interactively organized communities, or are they assemblages of species, each acting individually, that exist simultaneously in space and time because of the individualistic requirements of each species (cf. the Gleasonian–Clementsian debate in plant ecology)?

It would also be interesting to test levels of interspecific resource "partitioning" on a large scale in artificial assemblages, as Moyle et al. (1985) did in Salton Sea or as could be done with data from large impoundments, most of which have artificial assemblages of fishes. If fish in these exhibit niche "segregation," it has to be because of innate differences or extand competition, not because of any coadaptation at the community level through evolutionary time. Finally, how much of the observed resource use in fish assemblages can be accounted for by phylogeny? To what extent can the observed segregation of fish species on resource axes be accounted for by evolutionary differences in relatedness among the species (e.g., Rinne, 1992)?

9.2 Intraspecific Competition, Density Effects, and Resource Partitioning

Intraspecific competition and density-dependent limitations influence resource use by fish in assemblages. Intraspecific competition can result in (1) large individuals denying or displacing small individuals from use of resources, (2) large numbers of small individuals collectively sequestering more of the available resources than can be obtained by large individuals, or (3) individuals of approximately equal size, or independent of size, interfering with each other in a density-dependent fashion. In theory, density-dependent effects can lead to greater dispersal of individuals among habitat patches [Ideal Free Distribution (Fretwell and Lucas, 1970)]. Conversely, interactions between individuals in a population, such as threat, aggressive defense of locations, and similar phenomena, can result in some individuals influencing other individuals in a manner less dependent on total population density.

Most fish have indeterminate growth, which results [particularly for long-lived taxa at northern latitudes (Keast, 1977)] in individuals of a species exhibiting a wide range of body sizes, including both juveniles and sexually mature adults, in any lake or stream. The term "adult" is somewhat difficult to define, because some species, like red shiners (*Cyprinella lutrensis*), can mature at very small body size [e.g., 29 mm (Hubbs and Ortenburger, 1929)] yet grow several times larger after reaching sexual maturity. Continued growth after attaining maturity results for many fish species in "size-structured" populations in which behavior or resource use changes markedly with ontogeny (Werner and Gilliam, 1984). Keast (1977) notes that juvenile sunfish often are much more abundant in northern lakes than are large adults, and this is true farther south (e.g., in Brier Creek, Oklahoma). However, because many fishes (after completing larval or early juvenile zooplanktivorous stages) have the potential to eat the same kinds of foods as larger adults, there is a prolonged period in which competitive or density-dependent effects could result in a negative interaction between adults and young.

From the perspective of distribution in space of similar-sized fishes in a population, there should be two conflicting factors: (1) benefits of group living versus (2) interindividual competition for resources (food, mates, sheltered habitat, breeding sites, etc.). The benefits of group living are well documented for many animals, including schooling or shoaling fishes (Shaw, 1962, 1970, 1978; Pitcher, 1986), with advantages in two general categories including enhanced feeding and better protection from predation (by predator detection or escape). Therefore, many fishes live in monospecific aggregations of some kind. However, competition from conspecifics, particularly of similar size or life stage, should be stronger than competition from heterospecifics. The balance of these two factors result in conflicting tendencies of conspecifics to aggregate versus disperse.

Figure 9.5 depicts the likelihood of relative levels of competition versus relatedness of any two individuals. I assume that animals of the same species, particularly of similar size and from the same genetic cohort (left side of figure) are most likely to use similar resources, and thus have the greatest potential to compete.

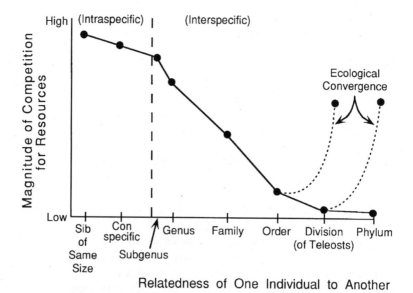

Figure 9.5. Hypothetical magnitude of competition (y axis) between any two individual animals as a function of their relatedness (x axis).

Intraspecific competition might be less intense between two individuals of different body (or mouth) size if one has access to a range of resources not available to the other (e.g., by gape limitation). To understand potential effects of intraspecific competition within an assemblage, it is also necessary to consider there relative effects of one large individual on many small conspecifics, or of many small individuals on a single, large individual. [It seems unrealistic to evaluate only the direct effects of one very large versus one very small member of a size-structured population, when, in nature, it is typical to have hundreds or thousands of small individuals for each large adult, and the collective effects of many small individuals may be important. For example, F. P. Gelwick and W. Matthews (unpublished data) showed that although a large *Campostoma* eats more periphyton per unit time than a small individual, the total amount of attached algae removed per square meter of streambed is greater for the juvenile than the adult component of the population (111 versus 58 $g/m^2/h$), because of the large number of juveniles present].

Beyond the level of intraspecific competition, species within subgenera or genera often are similar in body form or physiology; thus, the potential for competition should be high. At the level of genus, as typically used for freshwater fishes (e.g., Mayden, 1989), many fish are substantially alike in body form, and ecological traits (e.g., members of the former subgenus, now genus *Cyprinella*, are all water-column-dwelling minnows of moderate size). If two individuals are similar only at the level of family (Fig. 9.5), the average level of competition may be less, because some species within families may be quite different in ecological requirements. For example, in the minnow family, Cyprinidae, there are species that are very dissimilar, ranging from herbivorous *Campostoma* to insectivorous *Lythrurus* and *Clinostomus*.

Beyond the level of family, likelihood of competition should be lower, except in the case of ecological convergence. For example, Taylor (1996) showed asymmetric competition between a sculpin (*Cottus carolinae*; Order Scorpaeniformes) and a darter (*Etheostoma spectabile*; Order Perciformes). These rather distantly related taxa appear to have converged in adaptation for life in swift, rocky riffles of upland streams. Such examples aside, in most cases two individual fish, selected at random and related at less than the family level, would be less likely to compete for resources than would more closely related individuals, and sharpest competition should be intraspecific—between individuals of the same cohort.

Density-Dependent Effects

Some of the relationships in a fish assemblage might suggest the importance of density-dependent effects (Karlstrom, 1977) in ecology of freshwater species. Juvenile salmonids at high density may show more territoriality and become limited by supply of foods in drift (Crisp, 1993). Regardless of causal mechanisms such as competition for particular resources, it is possible to ask whether total density of a species influences its distribution among habitats. The Fretwell–Lucas Model (Fretwell and Lucas, 1970), or Ideal Free Distribution (IFD), predicts that across a patchy landscape [e.g., pools in Brier Creek, Oklahoma; Matthews et al., 1994)], individuals of a species will occupy the highest quality patch(es) until the density of individuals reduces the benefit per individual to the level equal that afforded individuals that spill over into the previously "next-best" habitat patch, and so on.

Persson (1983c, and references cited therein) and Keast (1977) predicted that increased intraspecific competition would result in expanded niche space for a fish species. Fraser and Sise (1980) tested the IFD model for dispersal of fishes in consecutive pools of a small New York (USA) stream and found that small blacknose dace (*Rhinichthys atratulus*) and small creek chubs (*Semotilus atromaculatus*), but not piscivorous adult creek chubs, were distributed as predicted by the Ideal Free Distribution. Fraser and Sise (1980) reasoned that under an ideal distribution, the dispersal of individuals among pools would be greater as population sizes increased. To test this, they calculated an index of "evenness" (actual versus hypothetical maximum diversity of distribution of fish across all pools) and found a slight but positive slope for both minnow species. Apparently, in this small stream, the piscivores tended to avoid each other, whereas the small minnows would aggregate until certain (unspecified) density levels were reached.

In Brier Creek, Oklahoma, Matthews et al. (1994) documented distribution of fish among 14 pools 8 times in less than 2 years. These date allow a test of IFD relative to changing densities of fish in that reach (~ 1 km) of stream. At higher total population density, was there greater dispersion of individuals across habitat patches (pools)? Brier Creek is a good location for a comparison to Fraser and Sise (1980), in that Brier Creek (in contrast to their New York creek) is a relatively harsh, fluctuating stream, with two particularly severe floods during the year of study described below. If intraspecific density does not reach a level at which competition challenges individuals, then in the predator-rich environment of Brier Creek (Power and Matthews, 1983), we might expect little tendency for individuals to disperse across more widely among pools.

For eight common taxa [a large piscivore, including large juveniles and adults of two black bass species, *Micropterus* spp.; two sunfish, *Lepomis megalotis* and *Lepomis cyanellus*, with the more abundant *L. cyanellus* separated into "small" [<75 mm Total Length (TL)] and "large" individuals as two "ecospecies," as in Matthews et al. (1994); a drift-feeding minnow, *Notropis boops*; a generalized omnivore minnow, *Cyprinella lutrensis*; a upper-water-column topminnow, *Fundulus notatus*; and the benthic algivore *Campostoma anomalum*], there was, in general, greater dispersion of individuals among pools at higher than at lower total population densities.

The data used were counts of individuals in each of 13 pools made by snorkeling on 7 dates from November 1982 to September 1983 and once in May 1984 (Matthews et al., 1994). On each date, the dispersion of the individuals of each species among pools was estimated by Levins' index (1968) of diversity, considering each pool as a resource state. The value of Levins' index can range, in this case, from unity (if all individuals of a species are congregated in a single pool) to 13.0 (if all individuals of a species are evenly distributed among all 13 pools, regardless of total abundance).

Figure 9.6 summarizes for each of the eight taxa the relationship between total abundance in the Brier Creek reach and its Levins' index score. Each regression line in the figure was a best fit by least squares and is limited in its length to include only the range of actual values of observed population numbers. There was substantial scatter around the regression lines, slopes of individual lines were small, and the correlation of Levins' index versus total individuals was not statistically significant for any single species at $p = 0.05$ (in part because of the small sample size of $N = 8$ surveys). However, the overall pattern in the slope of these best-fit lines was that for six (and perhaps all eight) of the taxa there was a modest positive increase in the value of Levins' index at higher population levels. This outcome was not strong for any species but collectively supports the expectation from IFD that in this reach of Brier Creek populations of the common species reached or exceeded densities at which dispersal among habitat patches increased. In Figure 9.6, for the one large-bodied piscivore that was included [bass > 100 mm Total Length (TL)], the slope is steep relative to other species. Whereas most of the sunfish (at least juveniles) and minnows tend to be schooling species, the piscivorous bass disperse among pools such that only a limited

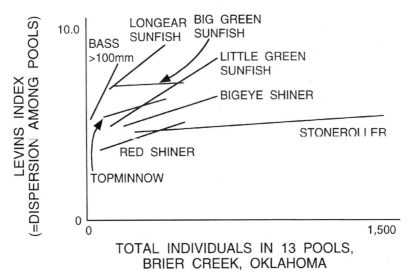

Figure 9.6. Dispersion of individuals of 8 species among 13 pools in Brier Creek, Oklahoma, as a function of total population density for each species during 7 surveys of pools by snorkeling. [Data from Matthews et al. (1994)].

number occupy a single pool (Matthews et al., 1994); thus, at higher abundance, bass show more dispersion among pools.

Clearly, dispersion of individuals among Brier Creek pools is complicated by predator threat, floods, or other non-density-dependent factors. However, environmental variation is normal for small midwestern streams, and even with such potential disturbances factored into the analysis above, the results suggested that in this stream reach, the balance shifts at higher abundances from favoring aggregation in single pools to dispersion among more pools.

Competition Between Size Classes of a Species

Intraspecific competition for resources can be between individuals of similar or of different sizes. Ontogenetic habitat use or diet shifts are common (Werner et al., 1983a, 1983b; Werner and Gilliam, 1984; Keast, 1985a, 1977), with larvae, juveniles, and adults of a species typically using different foods (Keast 1985b) and/or occupying different habitats (Matthews, 1985a). It is actually the exception when age-classes fail to show much separation in diets (e.g., Keast, 1978b) or microhabitats.

Even within an age-class, or between members of the same cohort, early differences in feeding can result in marked disparity of growth of individuals (Keast and Eadie, 1985), resulting in individuals of the same age differing markedly in size and use of resources. Svardsen (1949) contrasted the effects of intraspecific and interspecific competition for two salmonid species, concluding that "mild intraspecific competition" could produce disparity in sizes of individuals in a cohort, and was one of the first to note differential growth among individuals of a cohort.

From limited available data from diet studies, it is possible to at least illustrate the degree to which individuals of a species use similar foods at a given time and place. In Lake Texoma, 10 juvenile striped bass (*Morone saxatilis*) collected by seining at one shoreline location on 3 July 1986 showed mean percent similarity stomach contents of only 38.0% for all possible pairs of individuals. Only five kinds of foods were used. All contained chironomid larvae (but in widely differing proportions). In spite of a wider variety of prey available to these fish, they were relatively restricted in the kinds of prey they actually ate. The notion that small individual striped bass in Lake Texoma were differentially specialized in any fixed way on different food items seems unlikely. I prefer the tentative conclusion that all of the individuals had the potential to eat items in common, that the *potential* for intraspecific competition was high, and that if those items came into short supply, these individuals all might compete substantially for whatever items they could locate.

A similar comparison is possible for 10 juvenile largemouth bass (*Micropterus salmoides*), also captured by seining in the shoreline of Lake Texoma (22 July 1985). These individuals contained only three kinds of identifiable food items

Table 9.2. Percent Similarity in Stomach Contents of 10 Individual Specimens of Juvenile Largemouth Bass, Collected Together at the Same Place and Time in Lake Texoma, Oklahoma

70.7	100.0	75.0	77.8	69.0	33.3	33.3	50.6	95.2
	70.7	95.7	48.4	96.8	4.0	4.0	21.2	75.4
		75.0	77.8	69.0	33.3	33.3	50.6	95.2
			52.8	94.0	8.3	8.3	25.6	79.8
				46.8	55.6	55.6	72.8	73.0
					2.4	2.4	19.6	73.8
						100.0	82.8	28.6
							82.8	28.6
								45.8

Tabular values are all possible comparisons of pairs of individuals.

(corixiids, baetid larvae, and ostracods). Mean overlap (PSI = percent similarity index) between individuals (based on numbers of prey) was 55.4%, suggesting slightly more overlap, on average, than for the striped bass above. However, note in the triangular matrix of similarity of individuals (Table 9.2) that the mean of 55.4% is misleading. Many of the comparisons of particular pairs are high (70–100%), and numerous others are low (< 30%). Few values are intermediate, suggesting that for these fish, there may have been two dominant feeding patterns on that particular day. Whether due to chance or actual preferences cannot be detected from the data, but it is clear in Table 9.2 that a substantial number of individuals had eaten very similar foods. Schindler et al. (1997) showed high consistency in food use by individual marked largemouth bass at high population densities, suggesting individual specialization, but also reported "flexibility" in feeding by individuals.

Intraspecific Competition: Cold-Water Fishes

Numerous authors have explicitly examined intraspecific competition for cold-water fishes. Li and Brocksen (1977) found that in competition for space by trout of similar size, various individuals within a feeding hierarchy differed in their energetic gain, and that there were overall metabolic costs associated with competition. Interestingly, the individual achieving primacy (alpha individual) in the dominance hierarchy sometimes did not profit the most energetically, because it spent much time defending its position—and the beta individual (second in the hierarchy) sometimes attained the greatest energetic benefits and growth. Trout not achieving dominant status were forced into less profitable positions (e.g., holding positions in faster current speeds) and the lowest trout in the dominance hierarchy sometimes gave up trying to feed, hid in the substrate, and starved.

Symons and Heland (1978) predicted that young Atlantic salmon (*Salmo salar*) should occupy certain flow velocities preferentially. Yearlings chased smaller underyearlings from the preferred parts of a laboratory stream. However, in the artificial stream, the addition of boulders allowed both size classes to occupy similar habitats. Symons and Heland suggested that in natural streams, the ideal nursery areas for underyearlings would be shallow riffles free of large boulders, in that such habitats would not be attractive to large yearlings. Thus, underyearlings would suffer fewer negative intraspecific interactions from larger individuals in the nursery habitats.

Persson (1983a, 1983b) evaluated intraspecific competition within perch and roach populations in a shallow Swedish lake, where both were resource limited. For both species, energy intakes and growth rates were low. Persson (1983c) suggested that (1) perch age-classes could not segregate because of habitat homogeneity in the small lake, (2) roach age-classes could segregate trophically, but only by eating more low-quality foods (detritus and algae), and (3) intraspecific competition was most intense in summer when metabolic demands were high and food was of lower profitability.

Fausch and White (1981) showed intraspecific differences in salmonid habitat selection, and Fausch (1984) measured energetic profitability of positions for juveniles of salmonid species. Gotceitas and Godin (1992) also showed that dominant salmonids captured more prey than subordinates by selection of feeding sites. Fausch and White (1986) also hypothesized that slow growth (in a laboratory population in this case) of brook and brown trout was related to intraspecific competition that resulted from a lack of cover and associated visual isolation. Intraspecific dominance hierarchies were established, with dominant individuals holding positions that offered the greatest energy gain. Fausch found a strong fit between the hierarchy of linear positions and successively lower potential profits for subordinate fishes. Fausch concluded "in specific stream sections, position choice is ... constrained by formation of intraspecific hierarchies in which dominant fish hold optimal positions ...." Fausch did not explicitly refer to this as "competition," and from one perspective, it could be considered that once the dominance hierarchies are worked out within a group of trout, active competition for space or resources might be lessened—although with some individuals accepting suboptimal locations.

Fausch's (1984) model of a direct correspondence between positions and net energy gain has been criticized by Hughes and Dill (1990) as too simple, in that it leaves out a factor for the proportion of prey that an individual can actually capture relative to the rate of delivery of drifting items by the current—essentially that individuals are relatively less successful at actually catching and eating drifting prey at higher current speeds. By adding this factor, Hughes and Dill better fit their model to the positions actually selected by the solitary Arctic grayling (*Thymallus arcticus*) in instream pool trials. However, Hughes (1992) showed the Arctic grayling sorting into intraspecific spatial hierarchies in which the dominance rank of individuals matched the desirability of the position, and suggested that there might be a net energy savings to the group from maintaining stable, noncontested positions.

Brown and Brown (1993) showed for Atlantic salmon (*Salmo salar*) and rainbow trout (*Oncorhynchus mykiss*) juveniles that closely kin individuals (fullsibs) formed aggregations toward the upstream (more profitable) end of stream tanks, whereas nonkin groups tended to disperse throughout a stream channel, much like fish in Fausch's work. Additionally, there was less aggression on average in the kin than in the nonkin groups (Brown and Brown, 1993). Brown and Brown noted that the kinds of aggression also differed, with more "passive" aggression among kin and more "overtly aggressive" behaviors between nonkin. They argued that the greater tolerance of and more passive behaviors toward kin would benefit the population from reduction in (1) energy spent on territorial defense, (2) exposure to predation risk, and (3) physically risky fighting that could produce open wounds.

Paszkowski (1985) compared the impact of intraspecific and interspecific competition on foraging success of central mudminnow (*Umbra lima*) and yellow perch (*Perca flavescens*) in laboratory experiments. Increases in intraspecific densities negatively affected the feeding efficiency of yellow perch, but not of mudminnows.

In a relatively complex interpretation of the effects of intraspecific competition, Hamrin and Persson (1986) argued that an observed 2-year population oscillation of vendace (*Coregonus albula*) in a Swedish lake was caused by asymmetrical intraspecific competition between age-classes that actually favored small individuals. Their hypothesis was based on two components: (1) that small vendace had a competitive advantage in summer because of lower metabolic requirements and (2) vendace from strong year-classes would be relatively small by the next summer (at Age 1) because of greater intraspecific competition within the cohort throughout their first year. The strong cohort would also have substantial intraspecific competitive effects on the previous year's cohort, because of the large numbers of the younger fish. Thus, in the interpretation of Hamrin and Persson (1986), intraspecific competition both within and among age-classes led to the observed population dynamics of oscillating population sizes.

Whoriskey and Fitzgerald (1987) manipulated densities of adults of two stickleback (*Gasterosteus*) species in a salt-marsh environment and found a decrease in territory size in natural ponds at higher densities, but no decline in number of eggs in the nest per male (as an index of reproductive success); thus, they concluded that any effects of intraspecific competition on the populations were probably secondary to effects of unpredictable temperatures, oxygen, and water levels.

Heggenes (1988b) experimentally increased densities of adult brown trout (*Salmo trutta*) by almost an order of magnitude in a stream reach and evaluated amounts of movement and habitat choice by the original residents as well as the supplemented individuals. There were no effects of density on the amounts of movement, but the introduced individuals generally occupied less preferred habitats (shallow, less cover, fine substrate), likely from dominance (intraspecific competition) from resident trout, which were larger and also had "home" advantage. At both densities, the largest trout occupied deeper pools (i.e., the more preferred habitat).

Grossman and Boulé (1991) experimentally tested intraspecific and interspecific competition in rosyside dace (*Clinostomus funduloides*) and rainbow trout (*Oncorhynchus mykiss*) in an artificial stream. Of nine behavioral variables quantified for trout, only one showed an effect of intraspecific competition (percent time a trout spent near a conspecific). They attributed more behavioral changes by trout to intraspecific competition than to interspecific interactions, but emphasized the role of abiotic factors like floods and droughts in regulating populations of trout in Appalachian streams.

Intraspecific Competition in Warm-Water Stream Fish

Luxilus pilsbryi (southern duskystripe shiner) were collected in Piney Creek, Arkansas, at 2-h intervals from early morning feeding until midnight. Table 9.3 shows that six kinds of food items dominated the diet for 12–18 individuals, and that throughout the diel period, much the same kinds of prey were taken. There was a very high overlap overall (mean PSI = 79.3% between all possible pairs of times), suggesting that *Luxilus pilsbryi* at this particular location took quite similar food items throughout the daily feeding cycle.

One of the first experimental demonstrations of intraspecific competition among warm-water stream fishes was by Freeman and Stouder (1989), who demonstrated that small mottled sculpins (*Cottus bairdi*) in Appalachian streams were displaced from preferred deeper habitats in the presence of large individuals. They concluded from their experiments that the microhabitat segregation observed in

Table 9.3. Percentage of Individual Specimens of Luxilus pilsbryi *(N = 134) Containing Each Kind of Food Item, Collected at 2-h Intervals from One Pool in Piney Creek, Izard County, Arkansas (USA), in October 1975 by W. J. and R. S. Matthews*

Prey Type	Morning			Afternoon–night						Total Fish
	8	10	N	2	4	6	8	10	M	
Chironomid larv.	55	33	25	44	20	47	14	40	20	45
Chironomid pupae	27	80	92	89	67	53	64	73	67	91
Ephemeropterans	80	60	75	44	27	40	71	47	53	78
Trichopterans	47	60	42	17	20	33	14	40	40	46
Algae	40	47	42	44	40	47	64	33	33	58
Other vegetation	6	13	0	0	13	0	7	7	7	8
Terrestrial insect	47	40	58	67	80	40	57	53	60	75
Coleopterans	0	0	8	0	0	0	0	0	7	2
Odonatads	0	0	0	0	7	0	0	0	0	1
Simuliids	0	6	8	0	13	0	0	0	7	5
Miscellaneous	7	26	0	5	33	20	14	0	20	17

natural streams was caused more by the intraspecific competition than from any innate preference of the two size-classes for different depths.

Intraspecific Resource Partitioning?

Traditional interpretations have focused on the fact that by using different foods, often in different habitats, fishes at different life stages will not compete intraspecifically. However, instead of being driven directly by intraspecific competition, other explanations for segregation of size-classes are possible. In streams, juveniles often occupy shallower microhabitats with slower current speeds than do adults (e.g., Matthews, 1985a). For the benthic darter species *Etheostoma flabellare*, Matthews showed that segregation of juveniles from adults across current speed gradients likely had a hydrodynamic explanation. Spatial separation of juvenile from adult darters because of hydrodynamic considerations [see Statzner and Higler (1986)] does not rule out the fact that such segregation may ameliorate intraspecific competition—but does suggest that factors other than competition should be considered as potentially causing observed patterns in differences in resource use among life stages in the field. Regardless of the mechanism, however, differences commonly exist among size-classes in use of space or foods, which would result in less potential for competition than if all individuals in a population used similar resources.

Particularly at higher latitudes, where growth to sexual maturity is slower [often taking several years (Keast, 1977)], there may be present distinct age cohorts within a population (Tallman and Gee, 1982), with very abundant juveniles in potential competition with mature adults for limited food resources. Keast (1977) detailed food use by different size-classes of two centrarchids (rock bass, *Ambloplites rupestris*, and bluegill sunfish, *Lepomis macrochirus*) in Lake Opinicon, Canada.

The two species differed in the spatial relationship of juveniles to adults: Rock bass young-of-year appeared in inshore (adult) habitat soon after spawning was completed, whereas young-of-year bluegills had an "offshore" period before moving inshore. Additionally, when inshore, juveniles were mostly in weed beds, but larger individuals often occupied space near exposed rock ledges. Rock bass showed a "striking difference" in food use among year-classes (changing from plankton and larval insects to fish and crayfish as they mature), whereas bluegills showed little marked year-class differences in diet (with all age-classes eating large numbers of Cladocera, and little prey size segregation). Keast (1977) showed that for rock bass, prey size selection was important, and that even if two age-classes took the same prey items (qualitatively), they used different sizes of the prey. Bluegills "sidestepped" intraspecific competition by switching to a succession of prey as they became seasonally available.

Tallman and Gee (1982) hypothesized that, "Such age groups have evolved methods to avoid competition *as much as any species has done*." The special case of niche segregation among members of the same species (and, naturally, within the same population) raises the questions of whether observed segregation in foods or habitats among size-classes (or age cohorts) actually (1) reduces competition, (2) results from evolution to avoid intraspecific competition, as suggested by Tallman and Gee, or (3) results from morphological or physiological changes during ontogeny, but that did not evolve under pressure to avoid inter-age competition.

There is little question that, in most species, juveniles and adults use different portions of the resource spectrum. The question is whether or not this is a functional "resource partitioning" that "allows" or facilitates success of the population by lessening intraage competition, or if it is incidental to success of the population [as suggested for sticklebacks by Whoriskey and Fitzgerald (1987) because of other more important factors like stress, floods, drought, and so forth (e.g., Grossman and Boulé, 1991; Matthews, 1987a). As for many phenomena in fresh waters, the phenomenon of intraspecific resource partitioning is well studied for only a few species, such as centrarchids in lakes (Hall and Werner, 1977; Keast, 1977) or among salmonids (Haraldstad and Jonsson, 1983), and little is known about segregation of life stages or size-classes of small-bodied species of streams. [However, see Mahon and Portt (1985), who showed for four stream fish species distinct differences in size of individuals in riffles, raceways and pools, and Keast and Fox (1990), who reported intraspecific diet overlaps ranging 0.10 to 0.40 for *Fundulus* and *Notropis* in a small pond.]

In Piney Creek, Arkansas, adults and juveniles of the minnow *Luxilus pilsbryi* exhibited a differential longitudinal gradient, with more large individuals upstream and more small juveniles in downstream reaches (Matthews et al., 1978), and the largest individuals within a reach most often found in deep boulder pools (Matthews, personal observation). In the Roanoke River, juvenile and adult *Etheostoma flabellare* were segregated with respect to current speeds in microhabitats (Matthews, 1985a). Young-of-year *Etheostoma spectabile* in Brier Creek, Oklahoma, occur at the extremely shallow margins of pools over fine gravels or sand (in contrast to adults, which are mostly in flowing riffles or in mid-pool locations). Robert Hoyt and students (Floyd et al., 1984) have documented microhabitats of larval minnows (whose habitats differ from that of adults), but few quantitative studies explicitly address intraspecific niche differences among size-classes of minnows, darters, or other small fish species (Turner et al., 1994).

One notable exception is the article by Tallman and Gee (1982) on intraspecific resource partitioning by pearl dace (*Semotilus margarita*) in headwaters of small Canadian streams. From May to September, individuals of different age-classes were segregated into different habitats (although converging to common use of deep pools in winter). Monthly feeding locations and studies of focal individuals showed that age-classes were segregated by microhabitat, and overlap in diet (Morisita's index) between age groups was lower than 0.50 at least half of the time. All comparisons between Age 0 and 2+ fish were lower than 0.40, indicating dissimilar diets. Like Mendelson (1975), Tallman and Gee suggested that differences in microhabitats occupied by different age-classes were most important, and that differences in food use was related to availability in those microhabitats. Differences in environments occupied and sites of feeding "probably contributed most to partitioning of food resources." Tallman and Gee concluded that "vigorous evolutionary pressures favoring intraspecific resource partitioning should exist for headwater stream fishes" (where there are high densities of juveniles relative to adults, carrying capacity is limited, and environmental variability is high with respect to temperature, oxygen, food availability, and cover). They suggested that pearl dace were adapted to such environments by being dietary generalists but specialists in microhabitat, facilitating intraspecific resource partitioning. Moshenko and Gee (1973) likewise found different age groups of the creek chub (*Semotilus atromaculatus*) segregated by water depth and velocity in Canadian streams.

Two separate studies addressed age segregation by brown trout (*Salmo trutta*) in lakes of Norway. Haraldstad and Jonsson (1983) studied a lake in which no other fish species occurred; thus, comparisons among age-classes were unaffected by any interspecific phenomena. Haraldstad and Jonsson found distinct age/size-class segregation in Lake Myrkdalsvatnet, with younger fish mostly in the littoral zone and older fish deeper in the lake and farther offshore. They found some differences in foods of age groups, but no distinctive segregation. They concluded that habitat segregation between age groups "may be due to intraspecific competition," suggesting as evidence that young fish were restricted in habitat, whereas older (larger) fish changed habitats to follow availability of pelagic versus benthic foods. They noted chasing of small brown trout by larger individuals and that the littoral zone afforded protection (boulders, etc.) from such aggression. The alternative hypothesis that habitat segregation was due to differences in foraging profitability by different age groups was rejected, in that all sizes ate primarily small prey (zooplankton or benthos) that could readily be handled by all age groups. Haraldstad and Jonsson (1983) concluded that the regular ontogenetic advance of individuals from littoral to deeper or more pelagic habitats effectively opened up the littoral habitats to the next cohort, resulting in stable populations in the very predictable environments of these northern lakes.

Bridcut and Giller (1993) found that different age-classes of brown trout occupied different microhabitats in the Glenfinish River, County Cork, Ireland. Vøllestad and Andersen (1985) emphasized feeding of different age groups of brown trout in the same habitat (littoral zone) of Lake Selura, Norway. They found wider use of foods by young than by older fish during warm months. Age 2 and 3+ fish overlapped very strongly in use of foods at all times, but Age 1 fish overlapped little with the older fish in food use during 3 of 5 months of study. However, in July and August, all age groups fed opportunistically on abundant surface (terrestrial) insects and Trichoptera pupae. Vøllestad and Anderson concluded that the age groups of brown trout were intraspecifically segregated (at times) by "interactive segregation," with the segregation lessening in months when food was abundant.

For brown trout in an open, sea-run system (Vangsvatnet Lake, Norway), Jonsson and Gravem (1985) showed partial segregation by size and age (and sex) in warm months, both at the broad scale of streams versus the lake, and within a given zone of the lake. For example, in the littoral lake zone, Age 1 fish were more common near the bottom and Age 2 fish were more in surface waters. In the lake, smaller fish lived more in the littoral zone (which afforded structural shelter from predation), whereas older fish were more pelagic (where there were more potential predators and no physical shelter). In tributaries, they found that young fish fed on smaller items than did older fish. From their findings, they hypothesized that small and young individuals would use habitats to minimize competition or predation risk, where older fish would select habitats with better opportunities for feeding and growth. They concluded that in this lake the partial habitat segregation by age and size reduced potential intraspecific competition for foods.

Segregation within the brown trout population of Lake Vangsvatnet was partially according to life-history "morphs" (permanent freshwater residents versus returning migrants from the ocean). More complicated cases in interpretation of resource segregation occurs when distinguishable physical morphs with uncertain taxonomic status (distinct species or not?) occur in the same waters. For example, Galat and Vucinich (1983) demonstrated distinct differences in diet between two morphs of tui chub (*Gila bicolor*) in Pyramid Lake, Nevada (USA), with the morph (*pectinifer*) with more gill rakers consuming more zooplankton, and the morph (*obesa*) with fewer gill rakers eating more benthic macroinvertebrates. Here, there is distinct segregation of two forms along a resource axis, and separation is predictable from the feeding apparatus. Whether this is "intraspecific" or "interspecific" diet partitioning awaits taxonomic clarification.

Hindar and Jonsson (1982) addressed the related problem of resource segregation between two size morphs (dwarf and normal Arctic char, *Salvelius alpinus*) from the well-studied Vangsvatnet Lake, Norway. Vrijenhoek et al. (1987) provided a broader overview of the problem of various numbers of charr morphs in lakes throughout Iceland, Norway, and north Europe. In an Icelandic lake, for example, Vrijenhoek et al. reported that four completely distinct morphs (ranging from dwarfed, benthic form to large, piscivorous or mollusk-eating forms) existed without exhibiting any electrophoretic evidence of genetic subdivision. In spite of major differences in diets and habitats (i.e., very low "niche overlap"), the morphs apparently shared a common gene pool. This suggests the extreme in "intraspecific resource partitioning."

Hindar and Jonsson (1982) specifically addressed resource use by two morphs (dwarf and "normal") of charr, finding that they used different habitats during the growing season. They found [as in Vrijenhoek et al. (1987)] that the dwarf form was more benthic, whereas the normal morph was an open-water, limnetic form. They concluded that the larger morph was the superior competitor, using a wider range of habitats and foods. Because the two morphs converged in habitat and food use during periods of abundant food, Hindar and Jonsson concluded that intraspecific competition for food mediated the observed segregation in resource use. They explained differences in habitat use as due to intraspecific competition between the two morphs, as well as interspecific competition with brown trout.

Baltz and Moyle (1984) examined both interspecific and intraspecific habitat relationships of rainbow trout (*Oncorhynchus mykiss*) and Sacramento sucker (*Catostomus occidentalis*) in streams of California. They showed for both species that small individuals [< 50 mm Total Length (TL)] were in shallow water, reducing habitat overlap within species. For both suckers and trout, maximum depths occupied by young-of-year differed significantly from juveniles and adults. Mean depth for individuals differed among all size-classes of both species, and adults of both species occupied faster water velocities than did young-of-year or juveniles (Baltz and Moyle, 1984). Baltz and Moyle (1984) suggested that the observed intraspecific size segregation minimized competition.

Hamrin (1986) reviewed the evidence that intraspecific competition for food resulted in adaptive habitat partitioning by different age-classes of vendace (*Coregonus albula*) and that intense intraspecific competition was responsible for oscillations in strengths of year cohorts (cf. Hamrin and Persson, 1986). Hamrin argued that intraspecific competition for food became intense in Swedish lakes in summer, resulting in reduced growth rates in the populations. When there was well-defined thermal stratification, adult vendace were significantly more abundant in cold, well-oxygenated hypolimnetic than epilimnetic or metalimnetic waters of most lakes. Vendace were smaller on average in the warmer upper layers of the lakes (Hamrin, 1986, Fig. 3), but there appeared to be substantial overlap in sizes of individuals across the depth gradient; that is, there was no distinct segregation of size-classes by depth. Hamrin also suggested a distinct vertical migration of vendace fry into the upper layers of the lake during the night, in which the younger fish remained segregated from adults (Hamrin, 1986, Fig. 7). Hamrin found that temperature regulated the vertical distribution of adult vendace (no adults caught > 18°C) and concluded from sonar and gill net catches that adult vendace remained deeper in the lake (where water was colder) while young-of-year migrated upward through a wider range of temperatures (varying as much as 10°C in the vertical migration). Hamrin believed that the observed vertical segregation of young and adults was driven by ontogenetic differences in thermally linked metabolism, but that asymmetrical competition (for food) of the abundant young on large adults was also a contributing factor. This suggestion thus reflects exploitative competition driven by smaller, abundant individuals. This is in marked contrast to aggression-mediated intraspecific segregation, as shown in many works (e.g., Fausch, 1984) in which dominant individuals displace subordinates. For an open-water species like the vendace, there is probably little or no opportunity for interindividual aggression, dominance behavior, or similar factors to play a role, and the observed intraspecific segregation seems logically to be driven more by optimization of feeding or energy-gain opportunities for individuals.

Some studies have explicitly shown a lack of intraspecific segregation on one axis, and substantial age-related segregation on another. Keast (1978b) examined the stomachs of pumpkinseeds (*Lepomis gibbosus*) from Lake Opinocon, Canada, finding only minor differences among age-classes. However, pumpkinseeds up to Age 2 were largely found in macrophyte beds, whereas older/larger fish where common in less densely vegetated or more open habitats.

Summary: Intraspecific Competition or Resource Partitioning

In the laboratory, it is possible to demonstrate that some individuals of a species displace others, which may result in lowered growth of the subordinate individuals. In field experiments, it can be demonstrated (e.g., Freeman and Stouder, 1989) that some individuals (often larger) will displace others. In almost every study addressing intraspecific niche segregation by a freshwater fish species, some separation of adults from juveniles, or among year-classes, has been found. However, as Whorisky and Fitzgerald (1987) and Grossman and Boule (1991) have shown, it may be difficult to demonstrate that intraspecific competition has a pervasive effect on a population in terms of reproductive success, growth, habitat use, or overall fitness, particularly in streams or other harsh environments in which other factors like temperature or oxygen stress (Matthews, 1987a), floods (Harvey, 1987), droughts, or other abiotic factors may play a large role in regulating population size or outcome of fish–fish interactions. Intraspecific competition is probably important at times, especially in lakes, in determining differences between individuals in gains in energy over some periods of time or some other measure of fitness. However, the impact of intraspecific, relative to interspecific, competition as organizing factors in stream fish assemblages largely remains to be tested in field situations.

9.3 Mixed-Species Effects in Assemblages

> Many fish are gregarious.
> —Aristotle (384–322 B.C.), in *Historia Animalicum*

Background

For many years, the one paradigm has been that coexisting species segregate into distinct microhabitats, partitioning resources (Zaret and Rand, 1971; Werner et al., 1977; Baker and Ross, 1981; Ross, 1986). Taking this postulate to its logical or theoretical limits (MacArthur, 1972), the more that species segregated into distinctive microhabitats, the less they would overlap in use of space or foods. This argument would lead to the prediction that individuals of a given species would occur most often with conspecifics and not from mixed-species groups. For age- or size-structured populations (Werner and Gilliam, 1984), adults and juveniles might function as "ecological species" (Polis, 1984), and for those species, it would be expected that there would be primarily monospecific and monosized aggregations of individuals.

However, Stone and Roberts (1991) pointed out that in a community, a relatively high proportion (20–40%) of interactions among species should be beneficial or "advantageous." Their examination of field data on birds, ants, and zooplankton suggested that many potentially "competitive" interactions can actually be advantageous because of indirect interactions of the two species (potential competitors) with other species (Stone and Roberts, 1991; Boucher et al., 1982). As a result, there might logically be mutual co-occurrence of some or many species in common habitat space. Dodds (1988) pointed out that across many animal groups, field studies have skewed away from examining the importance of mutualisms. Dodd's (1988) theoretical model for communities showed a significant advantage to positive "directed" interactions (in which each species interacts reciprocally with another species).

Empirically, it is apparent that (1) in many streams (or lakes) there is a great deal of space underwater at any one time that is *not* occupied by fishes and that coexisting species could be more dispersed across microhabitats than they actually are and (2) that it is very common (see examples below) to find mixed-species "contact groups" with two or more species of related fishes occurring within a sufficiently small space (or with individuals intermingled) such that it is likely that individuals can transmit or receive information interspecifically. This leads to several questions. If they occur together in space, do fishes of one species actually respond to those of another? To what extent can different species of fishes cue on or receive useful information from each other? When fishes do exist in mixed-species schools in the wild, do they actually gain improved survival or foraging?

Potential Benefits

General benefits of schooling in fishes are well known (Shaw, 1962, 1970; Pitcher, 1986). The literature suggested benefits to animals from interspecific aggregation that include increased foraging efficiency and enhanced detection and avoidance of predators (Boucher et al., 1982). Pitcher (1986) reviewed the evidence for potential benefits to fishes of occurring in groups and suggested an array of effects within the broad categories of feeding enhancement and antipredator tactics. Pitcher (1986) pointed out that "most of the factors favouring larger group size in single-species shoals can . . . apply equally to mixed-species shoals." Like Moyle (1973) and Mendelson (1975), Pitcher suggested that in mixed-species shoals, there might be less intense competition for food while presenting a predator the impression of a large shoal. However, he also noted a possible cost of mixed-species shoaling in that some predator evasion maneuvers (during an actual attack by a predator) might be hampered by the presence of heterospecifics (with different abilities to swim, turn, etc.). Additionally, individuals appearing "odd" or unusual within a shoal may suffer increased predation risk (Theodorakis, 1989). Pitcher noted that mixed-species shoals could facilitate "genuine symbiosis" in feeding if one species flushed out or facilitated access to potential prey by a second species. Pitcher also noted that increased feeding opportunities might accrue through social observation of feeding by heterospecifics in mixed shoals.

For monospecific shoals, there is a known advantage that individuals find food faster when in larger shoals (Pitcher et al., 1982). Specific advantages regarding food-finding in heterospecific shoals could include finding food faster via enhanced prey detection (Pitcher, 1986), but the advantage would only exist if the onset of feeding on a kind of prey by one species was passed as usable information [passive information transfer (Pitcher, 1986)] to the other species. From a more direct perspective, some species in mixed groups may benefit from the flushing or mechanial extrusion of food items from cover, as when benthic feeding fishes disturb the substrate and stir up small prey into the water column for waiting omnivores or insectivores. Gorman (1988b, p. 15) reported finding *Notropis boops* to commonly swim closely above schools of benthic-foraging *Notropis nubilis*, feeding on items suspended into the water column by the latter. Gorman also reported *Nocomis buguttatus* following and feeding in tandem with the algivorous *Campostoma anomalum*. Likewise, Greenberg (1991) found that the darter *Percina evides* commonly followed the logperches *Percina caprodes* and *Percina burtoni*, feeding actively on items exposed when the logperch flipped over stones in their own unique foraging activities.

Fish in monospecific shoals are known to transfer information about predators, with receiver individuals modifying behavior in response to signals from transmitter individuals who have seen the predator (Maguarran and Higham, 1988). Fish in mixed-species groups should have increased potential for predator detection, because no two species are likely to have identical faculties for observation or detection of vibrations (the "more kinds of eyes and ears" hypothesis). This early warning would be effective only if information were transmitted rapidly (Godin et al., 1988) across the mixed-species shoal much like that shown for monospecific shoals by Maguarran and Higham (1988). Early-warning benefits in mixed groups are well known in some groups, like birds (Thompson and Thompson, 1985). However, Pitcher (1986) has also suggested a possible cost once a predator begins an actual attack, in that individuals dissimilar to the more abundant species in the shoal may actually be more subject to attack than at random, and monospecific groups may be able to maximize evasive maneuvers or predator confusion.

Evidence of Mixed-Species Phenomena

Mixed-species schooling among marine fishes is a relatively well-documented phenomenon (Barlow, 1974; Alevizon, 1976; Itzkowitz, 1977; Montgomery 1975, 1981) with most cases attributed to feeding advantages. Morse (1970) made a detailed analysis of costs and benefits of mixed-species flocking by birds. Across many animal taxa, the phenomenon of species occurring in heterospecific groups is well established. There are even documented cases of fish and invertebrates forming mixed groups [e.g., that of juvenile marine fish schooling with invertebrates (mysid shrimps)], providing each with enhanced protection (McFarland and Kotchian, 1982), and myctophids aggregating with ctenophores, sergestids, and amphipods (Auster et al., 1992). However, mixed-species schooling, shoaling, or aggregating has been evaluated less well for freshwater fish.

Awareness of mixed-species interactions among freshwater fishes is not new. Reighard (1920) reported that foraging northern hogsuckers (*Hypentelium nigricans*), which feed by rooting in gravel substrates with snout and lips, were "accompanied by ten or twelve small shiners (*Notropis*)" that "formed a little school at his sides and below him, and seemed to be waiting for fragments from his feeding." Pflieger (1975) noted that "other fishes, especially the smallmouth bass, longear sunfish, and various minnows commonly follow foraging hogsuckers to feed on the small organisms exposed by these activities," and I have seen similar behaviors of fish on numerous occasions in Piney Creek, Arkansas, with several minnows and typically a single juvenile smallmouth bass (*Micropterus dolumieu*) forming a persistent entourage that followed a hogsucker. Reighard (1920) made the intriguing comment (his page 5) that he had observed white suckers (*Catostomus commersoni*), which he considered normally very shy and easily startled, to be much less shy when accompanied by a group of logperch, which "appear to be feeding on eggs uncovered by him (the sucker) and perhaps on other crumbs from his table." Reighard (1920) suggested that because the white sucker was normally associated with logperch only in deeper water "safety and log perch have been closely linked in his experience." Although perhaps exceeding the limits of demonstrable proof, Reighard in this one article has suggested both enhanced feeding and safety as factors in these mixed-species groupings.

Case Histories: Mixed-Species Groups

Moyle (1973) included data on positive or negative associations of species in 1-m^2 plots in a small Minnesota (USA) lake. Of 21 possible combinations of three primary study species (*Notropis volucellus*, mimic shiner; *Pimephales notatus*, bluntnose minnow; *Luxilus cornutus*, common shiner) with other common species in the lake, he reported 14 significant negative associations compared to three significant positive associations at this small spatial scale. However, 6 of the 14 negative associations were between minnows and large-bodied centrarchids, which are predators of the former. The positive associations were mimic shiner–bluntnose minnow, bluntnose minnow–white sucker, and common shiner–white sucker. Moyle specifically commented on the tendency of bluntnose minnows to associate with large schools of mimic shiners (which were more abundant), but noted that the bluntnose minnows "seldom actually mingled with shiners but tended to swim in independent schools of 10–15 fish immediately behind and slightly beneath shiners." He pointed out, however, that "the association between the two species . . . was real; . . . that 17 out of 22 mimic shiner schools had bluntnose minnow schools associated with them," and that bluntnose minnow were seldom seen alone at that particular depth.

Moyle (1973) noted that, "Presumably, these fish (bluntnose minnows) were attracted by the protection and feeding advantages that a large school of fish can afford." In addition, Moyle (1973) found small common shiners (< 100 mm TL) were typically in mixed schools with mimic shiners or in small monospecific schools "on the outside edge of the mimic shiner schools." Although some of the circumstances described by Moyle suggest segregation of the mimic shiner–bluntnose minnow and common shiner–mimic shiner aggregations into monospecific subgroups, it would appear from his comments that in both cases, the two species were in sufficiently close proximity that they could accrue benefits from shared information and respond to each other. Although Moyle concluded, overall, that ecological segregation of the three minnow species in space and in food use was ecologically important, his report clearly indicates that at least at times mixed-species schooling was common in this lake.

Mendelson (1975) appears to be the first author to investigate in detail the phenomenon of multispecific schooling in closely related species (minnows). Mendelson worked with four cyprinid species of the genus *Notropis* (although one is now considered a *Cyprinella*) in a small Wisconsin (USA) stream. He noted (his page 203) that co-occurrence of minnow species was suggested by data from earlier authors, but those articles are not specific about the fact that fish actually formed mutually responsive mixed-species schools. For example, Davis and Louder (1971, cited by Mendelson) actually indicated only that "*Notropis petersoni* were usually collected with *Menidia extensa* and *Fundulus waccamensis* along the shoreline . . . ," implying that they were found in the same seine hauls. Mendelson (1975) noted that his examination of 291 collections in the University of Wisconsin museum in which at least 1 *Notropis* species was present, more than 60% contained 2 or more species.

In 20+ years of sampling fishes in streams of the midwestern United States, at the scale of whole collections of fish (often encompassing 100–200 m of a stream), I have very rarely taken only one minnow species. In observations of fishes within a visual sphere 1 m in diameter in Ozark streams, 141 of 195 sightings of such groups (72%) included more than 1 minnow species (Matthews, unpublished data). Likewise, Hatch et al. (1985), Probst and Bestgen (1991), and many other authors have commented on finding species consistently associated in particular habitats of streams.

Mendelson (1975) also noted that observations he made in aquaria "suggest that *Notropis* species are mutually responsive, individuals of one species readily following those of another. By and large, fishes of several species held in aquaria act as a single school." Mendelson conducted aquarium trials with three of his minnow species, determining the proportion of time that fish in pairs "followed, swam parallel to, or turned simultaneously with the other" over a 10-min period. In the three possible mixed-species pairings, fishes "schooled" by this definition from 46% to 72% of the time (means of all trials). In monospecific pairings, fishes schooled slightly more, from 60% to 88% of the time, but it appears from these limited tests that these species (1) would respond to a heterospecific and (2) move with them for extended periods of time. However, such trials in small aquaria (114 liters) may at least, in part, predilect fishes for mutual interactions (due to fright, crowding, etc.), and it would be interesting to conduct similar timings with focal individuals in the wild or in larger artificial streams. Nevertheless, it appeared from Mendelson's work that at least some minnow species not only occur together in the water column but also meet the criterion of mutual responsiveness needed to postulate benefits from mixed-species schooling.

Much of Mendelson's 1975 article focused on differences in morphology and microhabitat use by these species within the water column in his small creek. He noted, in particular, that "each species seems to have morphological adaptations ... sufficiently different to permit coexistence," and "in view of their rather distant relationship within the genus, (they) would seem more likely to be preadapted for coexistence rather than molded by mutual interaction into a functioning unit." Thus, although Mendelson recognized that these species often schooled together or formed "associations," his interest seems to have been more in the microscale differences in habitats each used in the wild, rather than in benefits of mixed species schooling.

Mendelson's (1975) description of the behavior of these minnows, apparently in mixed-species schools, is instructive. He noted:

> In the pools fish remain in loose aggregations facing the current. . . . Fish do not maintain ... a stationary position ... , but characteristically swim upstream a short distance, then drift backward ... Overt interactions between individual fishes of the same or different species are not common ... *If the fishes remain in species-specific schools this is not apparent from streamside observation* ... in general the aggregation is well dispersed in the water column, and directional feeding movements are rarely noted ... For long periods individual fishes, always oriented into the current, may move only short distances ... neither feeding nor interacting with their fellows.

My impression from Mendelson's description is that I have seen similar behavior by minnow species in pools of midwestern streams, but, in general, I observe greater activity, with more active feeding and intraaggregation movement than Mendelson describes. However, Mendelson's descriptions suggest (although details are a bit ambiguous) that these assemblages of minnows are of mixed species composition, and in lack of feeding activity as he noted, one might assume the mixed schooling to be related to a general propensity of these fishes to congregate (much as he noted in aquarium trials). Even if the species are only moderately active within the water column, "hovering" or awaiting feeding opportunities, the mixture of different kinds of eye, lateral line organ, and so forth, could enhance detection of either piscine or avian predators.

Mendelson's (1975) conclusion was that by existing in multispecies assemblages, these minnow species might accrue a measure of antipredator benefit, whereas species-specific spatial segregation (e.g., by vertical spacing) within the multispecific aggregations would lessen potential competition for foods. Mendelson implied that individuals in a large monospecific school of the same size might equally benefit from predator protection, but they would suffer more direct intraspecific competition for foods if all members of the school were conspecifics—with identical morphological adaptations for feeding. Thus, for stream minnows, Mendelson suggested a need to understand the balance between benefits and costs of aggregation in both mixed and monospecific groups. Moyle and Li (1979), in a review of community ecology of warm-water stream fishes, noted Mendelson's (1975) findings and further suggested that in multispecies assemblages rare species might persist in a stream by virtue of hiding within schools of more abundant heterospecifics. [However, a few rare individuals hiding within schools of a similar, more abundant species could make the rare individuals *more* conspicuous in event of an actual predator attack, cf. Pitcher (1986).]

Copes (1983) noted that young longnose dace (*Rhinichthys cataractae*) often formed mixed schools with young creek chubs (*Semotilus atromaculatus*) in streams in Wisconsin and Wyoming (USA). Allan (1986), like Mendelson (1975), pointed out that by joining mixed-species groups, fish might enjoy the benefit of being a member of a larger group (antipredation; vigilance) yet reduce the cost of intraspecific competition to less than it would be in an equally large group of conspecifics (if all were feeding in the same manner, using virtually identical morphological structures to acquire foods).

He also pointed out that rare species (with too few individuals to benefit from forming a school) might benefit from joining a more abundant taxon in shoaling. Although such rare individuals may not be well hidden in a mixed-species shoal, in that they could be conspicuous to predators (Pitcher, 1986), Allan (1986) points out that under extreme threat, they could still leave the school to hide. Allan (1986) tested three sympatric European cyprinids that often formed mixed-species shoals in streams (dace, *Leuciscus leuciscus*; European minnow, *Phoxinus phoxinus*; gudgeon, *Gobio gobio*) in an artificial stream apparatus to assess the impact of the presence of a second species on the behavior of target species as individuals or in groups. All three species exhibited shifts in habitat or in swimming behaviors in the presence of various combinations of the other two. Allan suggested, overall, that the positions and behaviors of the species in mixed aggregations reflected a balance between the tendency to avoid identical use of resources (space, food) with potential competitions, yet to remain in sufficiently close proximity to the other species to preserve the antipredation benefits of appearance of membership in a large shoal.

One additional cost to participation in mixed-species shoals (in addition to increased conspicuousness of the lesser abundant species) may be that, in the event of predator attack, some evasive maneuvers depend strongly on swimming ability and may be best performed in the company of conspecifics (Allan and Pitcher, 1986). This leads to what Allan and Pitcher (1986) considered a conflict of interest between remaining in a large mixed-species group versus separation into monospecific groups if a predator attacks. Allan and Pitcher used a model predator to simulate attacks on the three cyprinid species and in the same artificial environment as in Allan (1986). With the model predator present, the species diverged more into monospecific groups, and there were fewer mixed-species shoals. Allan and Pitcher (1986) interpreted this finding to suggest that fish segregated into monospecific schools would be more effective in high-speed predator-avoidance maneuvers and should increase predator confusion when each shoal presented a "uniform appearance." They argued that membership in mixed-species shoals, prior to an actual predator attack, provided increased foraging profitability and antipredator vigilance, but that when a predator actually attacked, these benefits were lost; thus, fish segregated into monospecific groups. I know of no investigations of this phenomenon in the wild.

Grossman and Freeman (1987), in a study of microhabitat use by Appalachian Mountain (USA) stream fishes, found that the common species could be separated into benthic and water-column guilds, but that within each guild, the microhabitats used by individual species could not be separated statistically. This is not the same as showing that taxa within a guild occurred in mixed-species shoals, but does suggest the possibility that species can occupy similar habitats. Further, they cited "unpublished data," indicating that "members of the water column guild displayed tendencies toward aggregation" (presumably multispecies).

Subsequently, Freeman and Grossman (1992) noted that rosyside dace (*Clinostomus funduloides*), other minnow species, and rainbow trout (*Oncorhynchus mykiss*) frequently foraged in mixed-species groups. Freeman and Grossman (1992) tested for competition in such mixed-species foraging groups by examining the effect of other species on arrival, departure, and feeding rates of dace in a defined area. When trout were present, dace departed a quadrat more often, but the presence of other species had no effect on dace arrival or departure rates. Dace were not more or less likely to forage in a quadrat if other other cyprinid species were present (Freeman and Grossman, 1992). They concluded that competitive cost to dace from the presence of other native minnows in a shoaling area was minimal, but that the introduced rainbow trout could lessen foraging opportunities for the dace. Combining both of the Grossman papers suggests that water-column fishes in these streams might accrue antipredator benefits from joining multispecies aggregations, yet incur minimal costs from interspecific competition.

McNeely's (1987) underwater observations of microhabitat use by fishes in a small Ozark stream of northeast Oklahoma (USA) showed that although the mean microhabitats of seven common minnow species could be separated in part, the species often formed mixed schools, intermixing vertically and horizontally within the stream. In a similar observational field study in an Ozark stream, Gorman (1987, 1988ba) found six minnow species within a pool-dwelling guild to be well segregated by a combination of microhabitat variables. However, he also documented vertical and horizontal positions of each of the minnow species within the water column when they occurred with and without heterospecifics. If a species vertically or horizontally shifted toward the habitat of a heterospecific, the interaction was considered "association" and the converse was "disassociation." Of 33 significant vertical interactions, Gorman (1988a) found about two-thirds to be associations. Of 37 significant horizontal interactions, approximately half were positive and half negative, with the overall result that taxa were often clumped into multispecies aggregations, despite the statistically detected segregation (Gorman, 1988b). Although it was possible to detect interspecific differences in mean habitats used, as depicted schematically in Gorman (1987), it is apparent that Gorman also detected much interspecific mixing. From these field observations of mixed-species effects, Gorman suggested both facilitation of feeding and enhanced antipredator effects. He emphasized that microhabitat use overall within the guild likely was a dynamic balance between segregation (lessening interspecific competition for resources) and associations (with potential benefits to mixed-group members), as influenced by changes in environmental conditions.

In a follow-up to the field observations, Gorman (1988a) tested in aquaria the vertical microhabitat use of five minnow species in monospecific versus mixed-species groups. In 18 of 22 possible two-species tests in aquaria, Gorman found significant changes in vertical habitat use, with 10 positive "associations" and 8 "dissociations." From all of Gorman's work (1987, 1988a, 1988b) it is apparent that Ozark stream minnows may segregate partially into distinct microhabitats, but that this segregation is modified by the presence of other minnow species, with slightly more positive associations (i.e., converging to common habitats when mixed-species were present) than dissociations (i.e., avoidance of each other in mixed-species pairings). Clearly, mixed-species phenomena have the potential to influence patterns of distribution of individuals in these stream communities.

Examples from Streams and Reservoirs

It is very common to capture more than one species of minnow in a single seine haul (of perhaps 5 m length), and indeed in some streams of the midwestern United States, it would be difficult to make a seine haul that produced only a single minnow species. Although not explicitly demonstrating any particular interaction or benefit between the species, such data suggest that there is a great overlap in occurrence of the species, at least in the horizontal plane. Data from my field collections in five very different systems are instructive, as detailed below.

Harrell (1978) and Matthews and Hill (1980) suggested that fishes in harsh, environmentally dynamic southwestern (USA) streams showed at least periodic strong overlap within microhabitats, and showed little long-term or persistent microhabitat segregation, as might be expected if interspecific competition caused stable resource partitioning. The South Canadian River of central Oklahoma (USA) is a wide, shallow, sand-bottomed prairie stream, subject to drastic water-level fluctuations and with extremes of high temperatures in summer (Matthews and Hill, 1979a, 1979b). On several dates within a year, I made a large number (~ 100) of very small seine hauls within discrete, randomly selected, microhabitat patches (Matthews, 1977). The stream did not exceed 1 m in depth, and there were no impediments to seining. Matthews and Hill (1980) documented differences among common minnow species in their distributions across habitat types, finding much dynamic seasonal change and little consistent use of habitats by any species. In May, August, and October, the percentage of seine hauls containing more than one species of minnows was 56%, 91%, and 75%, respectively. Overall, in a total of 212 seine hauls having at least one minnow species, 71% had two species and 46% had three or more species present. Clearly (Matthews and Hill, 1980), differences can be detected in means of habitat use among the species. However, it is also clear that in the majority of cases when one species was found, at least one other minnow species was present in the same small patch of aquatic habitat.

At a slightly larger scale of area encompassed by a seine haul, I made, over several summers, a large number of 20-m shoreline seine hauls at four fixed sites on the north shore of Lake Texoma, Oklahoma (USA), ranging from a wave-exposed, relatively rocky location to a wave-protected, mud–sand substrate. In a total of 202 such seine hauls, under a wide variety of weather conditions, 122 (60%) contained 3 or more minnow species, and 174 (86%) had at least 2 minnow species present. In three collections at Blue River and Clear Boggy River in southern Oklahoma, USA, I (with help from my class) kept records of numbers of species and individuals per seine haul. These hauls were not standardized for any particular area but were typically short hauls of the seine within a finite microhabitat. Of a total of 27 seine hauls including at least 1 minnow, 18 (67%) included 2 or more minnow species. In the upper Roanoke River, Virginia (USA), I made, in 1978, a total of 94 "kicksets," each including 2 m^2 in mainstream riffle areas, in each of which, samples were kept separate for counting of species (Matthews et al., 1982b). In 52 of the 94 (55%), I found two or more minnow species (Matthews, unpublished data).

Again, as suggested above, such mixed-species occurrences, even in very small or precise field collections, do not convey information on interactions of the species within the space, but they do demonstrate that in a high percentage of cases more than a single minnow species is found together, across very differing habitats and collecting approaches. With all samples from the studies above pooled, including a wide array of locations (a harsh southwestern river; littoral zone of a reservoir; clear upland streams; riffles of a large upland river; medium-sized streams of south Oklahoma) and a wide range of spatial scales (from a 1-m-diameter sphere to 20-m seine hauls), there is a total of 787 point "samples" containing at least 1 minnow species, in 559 of which (71%), 2 or more minnow species were present. These samples seem to substantiate at least the *possibility* for mixed-species groupings of minnows under a wide range of conditions. However, a caveat regarding mixed-species groups of fish suggested by M. Gophen (personal communication) is that it is important to differentiate between mixed-species groups that occur during "normal" conditions and those that might occur during harsh periods of time, when many species could be pressed into limited coexistence in refuge habitats. (In all of the cases of my own observations, conditions were "normal" for the stream, without obvious stress to the fishes.)

More Complex Mutualisms

In addition to potential benefits from interactions of species in feeding or general predator defense, there are examples of much more complex interspecific groupings in space in freshwater that apparently produce benefits to participants. Some focus on reproduction or young life stages. For example, Pitcher (1986) cites McKaye and Oliver (1980) in a case from Lake Malawi (Africa) in which nesting catfish defend not only their own young but also the young of two species cichlid that shoal with the catfish young. In North American streams, there are many well-documented cases of nest mutualisms (Wallin, 1989) as reviewed by Johnston and Page (1992). In many of these elaborate circumstances, visiting species lay eggs in or on gravel nests or other structures that are tended by the original (often larger) builder. Best known are the habitats of some chub (*Nocomis*) species of building large gravel nests as much as 50 cm or more in diameter and 30 cm or more tall. On these mounds of gravel spawn numerous species of stream minnows, whose eggs apparently accrue benefit from being protected in the clean gravels of the nest or from direct protection of the guarding nest owner. Wallin (1989) has suggested that some such associations may be true mutualisms, with the chub's eggs benefitting within its own nest from the continual cleaning of the nest of sediments due to the mechanical action of minnows visiting the nest.

Pflieger (1966) found that orangethroat darters (*Etheostoma spectabile*) placed their eggs within the nests of stream smallmouth bass (*Micropterus dolomieu*), thus gaining protection.

Conclusions: Mixed-Species Groups in Assemblages

In lakes or streams, it seems common to find two or more related fish species in close proximity to each other, such that information transfer or interactive benefits could accrue. If extreme interspecific segregation were "good," they have sufficient space in most systems to be more monospecifically segregated; instead, many species school or aggregate in mixed-species groups. The critical idea is that these mixed-species groups are indeed contact assemblages—that is, within a distance that they can be mutually responsive to each other.

In temperate lakes and streams, there are at least two common kinds of interspecific associations of fishes in mutually responsive aggregations. The most obvious is that of taxonomically similar species occurring together in midwater schools or shoals, as shown by Moyle (1973), Mendelson (1975), and Gorman (1988b). In such aggregations, often by abundant minnow species, fishes might benefit from the antipredator mechanisms of being in a large group, or of possible enhanced predator detection (the "diverse eyes and ears" hypothesis). The other well-documented mixed-species groupings are those often involving more distantly related taxa that provide obvious feeding advantages to at least one member of the species-pair. The best example of this phenomenon is that of smaller, omnivorous species (minnows) following larger, benthic-disturbing foragers such as hogsuckers, feeding on items loosened from the substrate by the mechanical disturbance of the larger species. In this relationship, the larger, benthic-feeding species might accrue no benefit from the trailing entourage of companion species, or the larger species might receive early warning of a threat (e.g., an avian predator) if the smaller fishes made initial detection and fled for shelter. It would be desirable to test the responsiveness of each species to the other in many putatively beneficial mixed-species groupings.

Much evidence (Ross, 1986) accumulated in the last 30 years to suggest that when related fishes coexist in a given stream reach, they should "do" things differently (resource partitioning) for reasons that may include avoidance of competition (among other factors). Nevertheless, it seems apparent that they may also "do things" alike (i.e., occupy similar microhabitats in mixed-species groupings). A balance between being "too different" and "too similar" may be optimal, as species that differ excessively in ecological traits may derive no information or benefit from each other, and if they are excessively similar (e.g., in use of foods or water-column micropositions), then heterospecifics might compete as much or nearly as much as would conspecifics (i.e., total density effects would be no less than for a monospecific school of similar size).

Optimal Numbers of Species in Mixed Groups?

Hypothetically, there might be an optimal number of fish species in mixed-species groups, related to a balance in benefits from multispecies living to detriment from interspecific competition (Fig. 9.7A). Envision a simple curve with benefits to

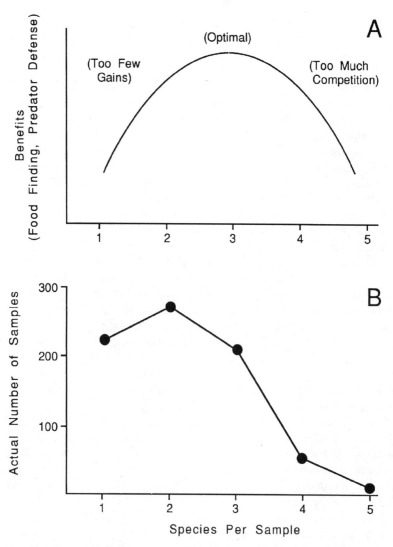

Figure 9.7. (A) Hypothetical optimal number of species present in mixed-species assemblages, to balance trade-off between benefits and costs of mixed-species existence. (B) Actual observed numbers of samples having indicated number of species per sample for surveys in streams of Oklahoma and Arkansas.

Table 9.4. *Numbers of Samples Containing Specified Numbers of Minnow Species, from Five Data Sets as Described in the Text*

No. of Species	Canadian R.	Lake Texoma	Ozarks	Roanoke	Blue-Boggy	Total
1	101	17	54	42	14	228
2	71	52	107	21	10	261
3	91	63	33	18	5	210
4	12	47	1	7	2	69
5+	0	12	0	6	1	19
					Total:	787
				Total Samples > 1 Species:		559

mixed-species grouping peaking (Fig. 9.7A) somewhere between too few species for substantial benefits to accrue (e.g., at minimum, only one species) and a maximum beyond which there is too much competition (e.g., for foods, space, etc.). Naturally, the upper bound on numbers of species in mixed-species groups for a given stream reach is fixed by the number of species in the regional faunal pool, but in every case from my field sampling, the numbers of minnow species actually observed in small samples in space is far less than the total possible "regional" species that could be present. In the hypothetical curve of Figure 9.5A, one species alone detects food or avoids predators less well than do two or three species schooling together, but at four or five species, the increased pressure of competition from too many kinds of species makes the negative interactions a dominant factor, and there could also be increases in phenomena like direct aggression between species. Thus, a limit could be reached beyond which no more species join the group, or beyond which one species joining a group might result in loss of another.

Such speculations can only be tested in detail with careful observations across space and time, and I offer the following only as a heuristic exercise. If all of my field observations and samples noted above are pooled (in spite of gross differences in technique) in Table 9.4 and converted to the curve in Figure 9.7B, the number of coexisting species peaks at two, with substantial numbers of samples with only one, or with three species. Although far from conclusive about ideal numbers of species, these relatively crude data suggest the commonness of finding two or three species together in "point" sampling.

9.4 Coevolution in Fish Assemblages?

Are patterns in resource segregation in assemblages derived more from recent adaptations of species or from the characteristics of larger parts of clades that were obtained deeper in the evolutionary past? Baltz and Moyle (1984), for example, suggested that a catostomid and a salmonid had minimized competition "by a long history of coevolution." If careful measurements are made of habitat or food use in fish assemblages, some degree of niche segregation is almost always found among coexisting species. The main question here is whether or not the resource partitioning observed in freshwater fish is a competitively driven, coevolved phenomenon, or if each species acts individualistically and the "resource partitioning" is due to innate differences that evolved in the past (either recent or distant), *sensu* Connell (1980).

Sharply divergent views on coevolution exist. Ehrlich and Raven (1964) used the term in the sense of tightly coupled species-pairs, with evolution of a given trait in one species producing subsequent evolution of some trait in the other species of the pair, which results in further modification of the trait in the first species, and so on. Tightly coupled coevolution of species-pairs may be hard to demonstrate (Farrell and Mitter, 1993). Furthermore, such coevolution of species-pairs is often attributed to situations involving parasites and hosts, or predators and prey (Connell, 1983), and less often to competitors [but see Roughgarden (1983), who argues competitors do coevolve].

Competitively driven coevolution of resource segregation by two species in the "tightly coupled" view would require that if two species used very similar resources one would begin to diverge a bit from strong overlap with another, that the second species then, in turn, would diverge a bit, then the first species diverges a bit more, and so on. At each step, each species would need to derive benefit, or, lacking adaptive value, the process should stop [which might indeed account for where (i.e., how far apart on a niche axis) two coevolving fish species would "stop"]. However, the actual mechanism by which one fish species (Species B) would benefit from "shifting right" on a resource axis following an initial "left shift" by its competitor (Species A) seems obscure. Why would not Species B remain at the original optimum on the axis, or even shift to the left (pressing its competitor further away from the original optimum)? As Connell (1983) points out, in predator–prey or host–parasite coevolutionary pairs, there is benefit to both partners in the pair. In a competitively driven system, it would seem to benefit a given species most if it does drive its competitor to extinction [unless this "competitor" actually provides some benefit (e.g., reducing predation pressure on the other species by its presence)].

An alternate view to tight coevolution of pairs is that "diffuse coevolution" could affect entire communities of species simultaneously; that is, that the community coevolves by virtue of a looser contact among species. The intricacy of connectivities within a community makes it somewhat difficult to envision mechanisms for diffuse coevolution—if one assumes that coevolution by definition involves reciprocal "steps" by all participating species. Would a change in resource use by any species be followed by adjustments in all other species in the community (or at least in a guild)? Perhaps, and perhaps not. "Tightly" and "loosely" coupled coevolution are quite different phenomena, both under a common rubric, and coevolution of communities remains controversial (Krebs, 1994).

Finally, there appears to be in the literature of fish ecology, a third, very generalized application of the term "coevolution" merely to mean that a group of fish species evolved together in roughly the same region of the Earth and same geological epoch. MacLean and Magnuson (1977, p. 1942) explicitly stated, "We assume that patterns of resource utilization and partitioning in percid communities result from coevolution driven by competition and predation as selective mechanisms rather than by chance occurrence of species with different resource utilizations." Wikramanayke and Moyle (1989) considered fish communities in streams of Sri Lanka coevolved, with competition a major structuring force. They showed that ecological overlaps were very similar between species-pairs in various streams, and that niche partitioning was constant, regardless of which other species were present in the streams.

Gorman (1987) indicated that Ozark stream minnows consisted of coadapted complexes with fine-tuned segregation and that they were the product of coevolution in the old streams of the Ozarks. However, Gorman (1988b) indicated that the segregation patterns among species are not static and emphasized the importance of the ability of species to change habitat use in response to encounters with heterospecifics, and that predation is a major factor behind interspecific interactions because it crowds minnows into limited "predator-free" space. Gorman (1988b) emphasized the importance of "positive associative interactions" and that they *were not* a result of "common microhabitat preferences." Gorman (1988a) considered multispecies interactions to influence habitat use patterns, with each species modifying its vertical distribution in response to heterospecifics, and that these minnows are "relatively flexible in their use of stream habitats," with substantial variability in habitat use in the field. The pattern of separations of species (Gorman, 1987) seemed to imply relatively consistent placement of species in horizontal and vertical space. Gorman (1988b) argued that segregation patterns are a product of species differences in habitat selection, "continuously modified by species-specific interactions over time and space," and that assemblage members have "highly flexible habitat use."

Gorman (1988a, p. 1249) stated that "co-evolution among guild members may have 'fine-tuned' the (behavioral) interactions in response to similar ecologies and in response to strong predation pressure." Gorman (1988a) invoked the "concordant distribution" of these species across the Ozarks [although Matthews (1982) shows that not all species occur together across an array of creeks]. Gorman concluded that the fine-tuned nature of the interspecific interactions and habitat segregation suggested that "only certain combinations of species are compatable" and that community assembly is not a stochastic process.

In spite of the use of the term "coevolution" in studies of fishes (above), it seems difficult to argue for dominance of tightly coevolved (in the sense of reciprocal evolutionary steps), fish assemblages in most situations for the following reasons:

1. To be a consistent factor in fish assemblages, competition-driven coevolution [as pointed out by Connell (1980)] would depend on the fish species existing together for the (unspecified) long periods of evolutionary time needed for coevolution to occur. Although some fish species are very old, it seems unlikely that species within a guild or assemblage will have evolved together in a tightly consistent fashion. Further, if the views of Wiens (1977) are correct, competition might occur at levels of biological significance only at long intervals, whereas coevolution would be more likely driven by variables that act nearly continuously (e.g., predation, parasitism). Short-term adjustments in resource use (e.g., within a few generations) in response to a competitor might occur, but might lack a genetic basis [cf. Connell's proofs (1980)]. If the interaction between species in ecological time results only in minor adjustments in use of resources, resource use might readily return to the original state if the competitor was removed.

2. Contact among species would not have been constant during evolutionary time. Gorman (1987, p. 41) suggested that numerous stream fishes in the Ozark Mountains, which are old uplifts, have had sufficient time in contact (within the region at large) to support the idea that they are coadapted, and stated that "coevolutionary adjustments have maintained the assemblage over time." I agree with Gorman's (1987) suggestion that within a given reach of a given creek, there are probably fine-tuned interactions between species—but in day-to-day extant time. With regard to "coevolution," I suggest it most likely that each species has evolved under somewhat different circumstances, partly in response to interactions with other species, but that the most likely sequence is (1) speciation events, then (2) fine adjustments in their resource use in response to other species that occur, or in response to resource levels. This is not "coevolution" in the tightly defined sense of *reciprocal* evolutionary steps between species. It is probably correct to say that in no single stream or small lake did all of the presently coexisting species evolve together. (Perhaps the best place to seek evidence of coevolution in fish assemblages is in the numerous endemic "species flocks" in lakes throughout the world (cf. Echelle and Kornfield, 1984).

3. This does not mean that species are not locally adapted. The evidence for this is mixed [e.g., *Cyprinella lutrensis* is not thermally adapted for local limits (Matthews, 1986d), but *Etheostoma spectabile* is (Feminella and Matthews, 1984)]. Consider the possibility for coevolution versus local adaptation of individual species of Ozark Mountain fishes. Several

"Ozark" species also occur in the Appalachian highlands of North America east of the Mississippi River, and there are divergent views on direction of origin and subsequent movement. However, regardless of where the species originated, taxa like *Cyprinella galactura*, *Fundulus catenatus*, and *Notropis telescopus* have widely disjunct centers of occurrence in the two highlands, separated by hundreds of kilometers of lowland and the Mississippi River delta, and occur with almost completely different coexisting small fishes east and west of the Mississippi River. Would each of these species have "coevolved twice" as part of different assemblages, or is it more likely that they show moderate local adaptations? It would be of interest to know how similar use of resources is by each of these species where they occur as members of almost wholly different assemblages (Ozark versus Appalachian).

4. There may be no completely discrete or exclusive fish communities in a typical watershed. Most species blend in distribution with others both upstream and downstream. There are few rigid breaks in stream fish assemblages [Matthews (1986b) and see Fig. 6.15] and it is likely that a species will occur with a different suites of species at different locations in a watershed. Thus, tightly coupled coevolution of species-pairs, guilds, or assemblages over long periods of time, even within a given watershed, seems unlikely. Larkin (1956) argued that lakes and creek systems are inherently short-lived on Earth (geologically) and there is not much evidence (except in the case of closed systems and old lakes—fish species flocks) that any given combination of species is now the same as those with which a given species evolved.

5. Even in extant time, there are few consistently coexisting combinations of species (i.e., with all the same species) across watersheds. For example, in 13 Ozark watersheds, some regionally common species are present in all, but others are present in part of but not all of them (Matthews, 1982). Some species abundant in one watershed are scarce or absent in another (Gorman, 1992). It is difficult, under this variable situation of coexistence at present, to accept that a particular pair of species would have coexisted long enough in the past to evolve in rigid connectivity to and under competitive pressure from a particular (unchanging) suite of other species.

6. There is evidence that resource use by a given species of fish is variable in space and time (Angermeier, 1987; Ross et al., 1987). At a given site, the same species may do very different things at a given time (seasonal, etc.), as witnessed by numerous "year-round" resource partitioning studies. From one stream to another, a given species may also exhibit different ecological traits, and even within a lake (Gascon and

Legget, 1977), there is evidence that resource sharing among the same species can differ from place to place (e.g., depending on how much food is available).

7. There is also growing evidence from experiments that what a fish does at a given time and place (concerning microhabitat or foods) will depend on what other species it is with in a given situation. This suggests that competition from or interactions with other species will influence what a species does on a given day, but not that there is a rigid, consistent, coevolved template that the community follows. Habitat (Matthews and Hill, 1980; Harrell, 1978) and food use can be very plastic, with fish operating within certain limits (perhaps set by morphology) instead of consistently operating at some fine-tuned optimum (e.g., Werner's sunfish in lakes—cost curves). Cost curves may be useful, but the presence of another species may change the practicability of operating at an optimal position on a cost curve. Comparison of the aforementioned Ozark–Appalachian species would be interesting.

Overall, it seems most reasonable to accept that fish assemblages consist of individual species adapted for certain optima and limits in tolerance of physical conditions (Matthews, 1987a) or with optimal resource use, because of evolution in the past, in places and with other species that are not the same as the places where they now occur. However, within a suite of species that exist at any given place and time, it seems likely that there can be day-to-day adjustments in feeding, microhabitat use, and so forth as competitors or predators change. McNeely (1987) noted that even if taxonomic composition of a stream fish assemblage is not regulated by competitive exclusion, there is no reason to exclude the possibility that competition or other biotic factors can influence niche relationships when populations are dense or resources are limiting. However, Robinson and Wilson (1994) have recently made the case that competition is important in evolution of fish communities and noted that "an absence of competitors allows species to expand their niche." Conversely, the presence of competitors (either a single similar species, or a large number of competitors forcing "diffuse competition" *sensu* Pianka) might be interpreted to suggest the constraining of the niche of a particular target species. In that sense, competition between species could, at least in the "loose" meaning of coevolution, influence the species that occur in assemblages.

In other words, competition is real. It exists in extant daily ecological time (Robinson and Wilson, 1994). However, competitively driven, tightly coupled coevolution of fish species does not seem well supported theoretically or documented empirically as *the* explanation for levels of niche segregation that can be observed in most extant fish assemblages at a field site. If fish in assemblages have "coevolved," it would seem most likely to be a "loose" coevolution, not requiring reciprocity between any particular species-pairs, and based only on the fact that fish within a region have probably evolved at least in part in a common environmental milieu and have had some degree of influence on each other. Clearly, when exotic species are introduced into an existing fauna, those newly arrived species will not have taken part in the thousands to millions of years of experience shared by the native fish species, and, in that sense, clearly are "not coevolved" with the native taxa.

10

Interactive Factors: Predation Effects in Fish Assemblages

> *No species can long maintain itself anywhere which cannot, in some way, find a sufficient supply of food, and also protect itself against its enemies.*
>
> —S. A. Forbes (1880b)

10.1 Predation and a Hypothetical Model

In this chapter, the effects of piscivorous fish on their prey species in fish assemblages are considered. This can include outright killing and removing prey from a population, or changing behaviors of prey because of predator threat. In Chapter 8, the consequences of predator threat on life-history parameters of prey species were described, and in Chapter 11, the effects of predation in trophic cascades or food webs are considered. Here, I examine evidence for predators having an effect on the structure of fish assemblages, or on interactions among fishes in those assemblages. Topics include the following:

1. Killing of fish by piscivory, thus individuals as living components of the system
2. Piscivore-induced differences in distribution of fishes in lakes or stream drainages at the macrohabitat level
3. Piscivore-induced changes in habitat, shelter, patch use, or other behaviors by fishes
4. Direct or indirect predator-mediated changes in competition, resource partitioning, or other interactions between prey species

Because the focus of this chapter is on the effects of piscivores on fish assemblages, I make minimal reference to the voluminous literature on behaviors of individual predator or prey fishes (learning, vigilance, response patterns, antipredator tactics), or on effects of predators on prey group phenomena like schooling behavior. Excellent reviews on these "behavior-oriented" topics already exist, including Stein (1979) on the array of behavioral responses of individual fish to predator threat; Milinski (1986) on predator risk and responses of fishes; Pitcher (1986) on shoaling behavior in response to predators; and Sih (1987) who reviewed individual antipredator responses of fishes and other aquatic organisms. Individual articles such as Magurran (1986) on predator-inspection behavior by minnows; Dugatkin and Alfieri (1992) on assessment of predator inspection by game theory; Abrahams and Colgan (1985), Pitcher et al. (1985), and Ashley et al. (1993) on hydrodynamic structure, size segregation, and changing sizes of fish shoals; and Reist (1980) on interactions of piscivores with variably threatening (spines) prey all provide entry into the more purely "behavioral" aspects of interactions between individuals of prey and piscivore species. For the purpose of assessing piscivore effects on fish assemblages and interactions in the assemblages, it is sufficient to acknowledge that there is a huge literature on individual behaviors showing that prey typically are aware of and make appropriate responses to piscivores, such as sampling of the habitat for presence of predators, vigilance, predator inspection, assessment of risk, and the like.

Hypothetical Model of Interactions of Numbers of Prey and Piscivore Species

Several "cause-and-effect" scenarios are possible with respect to the numbers of prey and piscivore species in local communities. I suggest three simple models for system regulation by piscivores or by prey species. First, assume that because of their intimately linked interactions, each has the potential to affect the success of the other. Further, assume that more species of piscivores represents a greater variety of modes of attack on prey within an assemblage, and that more species of prey in an assemblage offers piscivores a greater variety of "targets" to attack (across a greater variety of microhabitats, etc.). Thus, an increased number of species of piscivore could exert increased killing pressure on any given prey species, and an increased number of prey species could offer increased diversity of opportunity for any single piscivore species. The question to be asked from empirical field studies of assemblages is whether there is a positive or negative relationship between numbers of prey and piscivorous species, and what the potential causes of any detected relationship might be. Consideration of causality must also include timing: Do changes in number of prey species subsequently result in changes in numbers of piscivores, or vice versa?

Three models seem plausible to explain changes in numbers of prey versus piscivore species in local assemblages:

1. Increase in number of prey species causes increase in numbers of piscivorous species (prey-regulated model)
2. Increase in number of piscivorous species causes decrease in number of prey species (piscivore-regulated model A)

3. Increase in number of piscivorous species causes increase in number of prey species (piscivore-regulated model B)

The first model (prey-regulated) suggests that numbers of prey species influence a local assemblage and that changes in numbers of prey species (for any reason) are *followed* by changes in numbers of piscivore species. For example, increased numbers of prey species in a given reach of stream might make that stream reach a more attractive foraging patch for more kinds of predators, leading to increased accumulation of piscivore species. Under this model, increased diversity of prey would offer more kinds of foraging opportunity for piscivores, potentially reducing competition among the piscivores and allowing more "packing" of species (Pianka, 1978) into piscivore niche space in the assemblage.

The second model is based on piscivore regulation of the numbers of prey species, with an initial increase (for whatever reason in the numbers of piscivores in a locality being followed by a decrease in numbers of species of prey. The rationale for this decrease in prey species is that as the numbers of piscivore species increases, there is an increase in the ways in which an individual of a prey species can be attacked (from above, from below, by active cruise attack, by ambush, etc.), placing increasingly great total threat upon prey. In this model, some prey species are simply unable to cope with the multiway treats from multiple piscivore species and are locally extirpated. Hence, the numbers of prey species decreases after the numbers of piscivores increase.

In the third model, piscivores also drive the system, but indirectly by a "keystone" effect (Paine, 1969, 1995). The keystone effect depends on the assumption that prey species compete with each other and that in the absence of predation [or disturbance—"intermediate disturbance hypothesis" (Connell, 1978)], some prey species will locally eliminate others. Under this saturation model, the arrival of more kinds of piscivores in the local assemblage results in increased overall predation threat, but not sufficiently to drive any prey species to extinction. The result of more piscivore species in this model is a keystone effect in that competition between prey species is ameliorated by cropping of any highly abundant species, resulting in more prey species being able to coexist.

The first and third models depend on the importance of competition in local fish assemblages. The likelihood of competition regulating an assemblage depends on environmental stability, with more potential for competition-based models to prevail in stable systems (e.g., northern lakes) or during long intervals between disturbance (e.g., floods in streams).

Obviously, more than one of the models above (if any) could actually operate in a system. For example, in a manner analogous to the classic "hare–lynx" predator–prey oscillation (for numbers of *individuals*), the predator-dominated and prey-dominated models above could operate alternatively, with each more important in parts of a long-term cycle. Figure 10.1 suggests a simplistic view

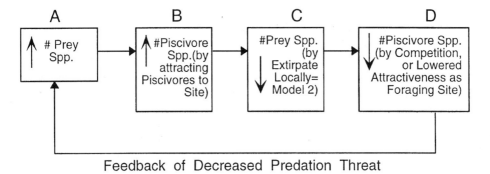

Figure 10.1. Hypothetical model for feedback between the numbers of prey fish species and piscivorous fish species in a local assemblage.

of the way that numbers of species of prey and predators could oscillate, with each controlling the other in sequence. Assume an initial increase in prey species for some unknown reason (Compartment A, Fig. 10.1). This (Model 1) causes an increase in the numbers of piscivorous species, as more generalized piscivores are attracted to the site because of more diverse foraging opportunities, and/or because the increase in the numbers of prey species lessens competition among piscivore species. The increase in numbers of piscivore species (Compartment B, Fig. 10.1) then causes a decrease in numbers of local prey species (Compartment C), via local extirpation (Model 2). The lowered numbers of prey species subsequently causes fewer different kinds of piscivore to be attracted to forage in the local area or to be able to compete there, so the numbers of piscivore species decreases (Compartment D, Fig. 10.1). This lowering in numbers of piscivore species feeds back to Compartment A (Fig. 10.1), resulting in a local increase in numbers of prey species, beginning the cycle again.

The models above all assume that it is the interactions between prey and piscivores that regulate the relative numbers of each. In actual field situations, it is obvious that uncontrolled variables such as habitat heterogeneity or complexity could be important—with greater numbers of prey and piscivore species coexisting in assemblages merely because a more complex habitat could be more attractive to each group, or because physical structure could alter the dynamics of predator effectiveness (Savino and Stein, 1982). Note also that all of the speculation above addresses only changes in assemblages in ecological time (i.e., changes that occur within or over only a few generations of fish species at a given locality). These possibilities could be extended to evolutionary time, with the numbers of prey or piscivore species in a local assemblage strongly related to the numbers of such species available in the entire stream or lake system, or within a whole drainage basin (a passive local colonization model). Additionally, there is clearly the potential for long-term coevolution of piscivore and prey species, although I have argued elsewhere (Chapter 9, following Roughgarden's ideas) that the potential for couplets of prey–piscivore species to coexist sufficiently long for true coevolution may be unlikely.

Anything is possible in models. What is the evidence from real fish assemblages? It is not possible to distinguish Model 1 from Model 3 with empirical field data. Under both models, the increased numbers of prey species and piscivore species could be observed within assemblages. To distinguish Model 1 from Model 3 would require carefully designed, long-term experiments that have not yet been performed. Regardless, is there a pattern (for North America, where I have most information) in numbers of prey versus piscivore species (which could at least minimum distinguish Model 2 from Models 1 or 3) in streams.

10.2 Empirical Studies of Piscivory

Background on Piscivory in Fish Communities

Much empirical evidence shows that piscivory can influence structure of fish communities in lakes (e.g., Miller, 1989; Robinson and Tonn, 1989). For closed systems like small lakes and ponds, there is a lengthy history in the management literature of the influence of piscivores on populations of their prey, as reviewed in many fisheries texts (e.g., Bennett, 1970). There is less evidence that piscivory (at least from native species) has strong effects on composition of fish assemblages in streams. Although the selection of contributors may have been biased by organizer interests, the symposium on "Community and Evolutionary Ecology of North American Stream Fishes" (Matthews and Heins, 1987) included only two (Fraser et al., 1987; Schlosser, 1987b) of 30 "data" papers (Fraser) with primary focus on the role of predators in stream fish assemblages. In spite of detailed accounts of feeding by some piscivores in tropical streams, Goulding (1980) does not discuss their effects on fish assemblages in those habitats.

Alexander (1979) argued that in cold-water streams (Michigan, USA), piscivory had an important impact on young trout. Alexander (1979) cited Foerster and Ricker (1941, original not seen) that the reduction of piscivorous fishes enhanced the survival of migrating young of sockeye salmon. Alexander (1979) reported an attempt to estimate the loss of young trout to piscivores (mostly avian and mammalian, but also piscivorous brown trout) along 20 miles of a Michigan river by killing a total of 240 birds (mostly mergansers, herons, kingfishers, bitterns) with shotgun and shotgunning or drowning 26 otters, mink, and racoons to obtain an estimate of trout consumption. His estimates suggested "much of the loss (of young trout) can be attributed to American mergansers, great blue herons, belted kingfishers, mink, otter, brown trout, and anglers." Alexander estimated that predation by brown trout caused as much as 58% of the mortality of brook trout between Age 0 and Age 1. Feunteun and Marion

(1994) showed grey herons (*Ardea cinerea*) to prey non-selectively on all but very small fish species, and to take 6% of the standing crop of fish in a marsh.

For warm-water streams, the review by Moyle and Li (1979) suggested that "information on predation on fishes in warmwater streams is largely lacking, and great caution has to be exercised in transferring what has been learned about coldwater streams to warmwater streams, particularly . . . with complex fish communities." Moyle and Li suggested that there may be surprisingly small numbers of piscivores per area in warm-water streams [although my more recent underwater surveys in midwestern streams suggest that a large number of piscivorous bass, *Micropterus* spp., can be present in a given stream pool (Matthews et al., 1987, 1994)]. In a companion review to the article by Moyle and Li, Campbell (1979) noted that in large rivers, "conducting experiments . . . by manipulating either predator or prey abundance is almost impossible," and that regarding information on effects of predators on prey populations, "most of the evidence for this type of change in the prey community is from lakes and ponds. . . ."

Thus, as of 1979, essentially nothing was known about the effects of piscivorous fishes on other fish species in large river systems. Since these reviews, there have been a large number of experiments demonstrating that piscivores can affect the distribution of prey species within natural stream pools (Power et al., 1985; Power, 1987; Harvey, 1991a, 1991b) or artificial streams (Fraser and Cerri, 1982; Gilliam and Fraser, 1987; Fraser et al., 1987; Schlosser, 1987a, 1988a, 1988b), and there are case histories in which stocking of a piscivore apparently altered stream fish assemblages. However, a definitive controlled experiment, with replication, testing the effects of a piscivore on structure or function of fish assemblages throughout a lengthy stream reach remains to be done. Such an experiment, with sufficient replication for statistical validity, would not be easy logistically and there could be ethical problems if an exotic piscivore was introduced outside its range. However, there may be streams in which density of a native piscivore (i.e., within a watershed) could be experimentally supplemented (or depleted), in a preplanned manner allowing detailed quantification of the response variables of habitat, food use, or densities of species that are their potential prey.

Winemiller (1991a, p. 636) cited Jackson (1961, original not seen) as indicating that intense predation by *Hydrocynus vittatus* (tigerfish) in African streams will limit the number of species in open, main river channels. However, in his own study, Winemiller (1991a) found a less apparent effect of this predator, noting that large individuals of at least six species coexisted with the tigerfish. There is good evidence that the presence of predators in streams alters or influences microhabitats used by smaller fish species (e.g., Greenberg, 1991; Power et al., 1985; Schlosser, 1988a, 1988b). Schlosser (1987a) combined information on piscivore threat with geomorphology of streams to produce his "conceptual framework for fish communities in small streams," in which greater piscivore threat in deeper, more "developed" pools resulted in exclusion of most small species or young or larger-bodies fish species.

Case Histories: Introduced Piscivores in Streams

In most cases in the literature in which the presence of predatory fishes has altered the composition of the prey fish assemblage at the scale of whole steam reaches, it has been as the result of an introduction (Moyle and Nichols, 1974; Meffe et al., 1983; Meffe, 1985; Rincón et al., 1990). Rincón et al. (1990) found that introduced pike (*Esox*) changed fish assemblages in small streams of the Esla River basin, Spain. Red shiner minnows (*Cyprinella lutrensis*) introduced to streams west of the Continental Divide in North America are purported to have had negative effects on native fishes via predation on young of the latter (Minckley, 1991, Ruppert et al., 1993). Blinn et al. (1993) reported little behavioral response of native Little Colorado spinedace (*Lepidomeda vittata*) to introduced rainbow trout (*Oncorhynchus mykiss*) and indicated that this introduced predator was now likely responsible for limiting both geographic and local habitat distributons of the native spinedace in streams in Arizona (USA). Blinn et al. (1993) attributed the low response of spinedace to a predator to its evolution in the faunally depauperate drainages of the western United States, where it would experience few kinds of predators.

Garman and Nielsen (1982) documented the effects of experimental stocking of non-native brown trout (*Salmo trutta*) on the native fish community of a Virginia (USA) creek in the Roanoke River drainage. Large brown trout ate native species, particularly the torrent sucker (*Moxostoma rhothoecum*) and caused a marked decrease in the abundance of species that were most used as prey the experimental section. My calculations from their Table 3 suggest that percent similarity in abundance per species was only 53.4% between assemblages of nongame fishes in the control section versus the section stocked with large, piscivorous brown trout. Before and after the stocking of large brown trout, the nongame fish assemblage within the experimental section showed a 65.2% similarity (although seasonal effects of comparing April to October could have had some influence). Garman and Nielsen (1982) concluded that stocking of large brown trout would likely result in declines of nongame species.

"Micropiscivory" also may have effects on stream fishes. For example, there is evidence that small-bodied, introduced *Gambusia* (mosquitofish) have by predation negatively impacted *Poeciliopsis* populations in desert streams (Meffe et al., 1983; Meffe, 1985). Mosquitofish have also had negative impacts on *Heterandria* populations by predation (Lydeard and Belk, 1993; Belk and Lydeard, 1994; Schaefer et al., 1994). Another category of "micropiscivory," that of fish predation on eggs, is known (Rahel, 1989; Paradis et al., 1996). However, Vadas (1990) predicted that the effects of predators on eggs would be small relative to the influence of numbers of ova produced. However, the fact that many stream fishes (e.g., *Nocomis* and other minnows) build some kind of gravel nest for egg protection (Johnston and Page, 1992) suggests that predation on early life stages can be considerable. In Brier Creek, Oklahoma, immediate and catastrophic loss of eggs or young larvae occurs when male longear sunfish (*Lepomis cyanellus*) temporarily depart their nest, and minnows or small sunfishes rush in to feed.

Case Histories: Piscivory in Lakes

Piscivores in lakes may be able to consume a substantial amount of the available prey fish biomass: Stewart et al. (1981) estimated that stocked salmonids could eat 20–33% of the alewife (*Alosa pseudoharengus*) biomass in a given year even in so large a body of water as Lake Michigan (USA) (Fig. 10.2). Not all "piscivory" in lakes is by "typical" predator species on prey species. Crowder (1980) noted that alewife and rainbow smelt (*Osmerus mordax*), not normally considered piscivores, consumed eggs and larvae of native species, with negative effects on some populations.

In relatively stable northern lakes, the presence of a piscivorous species can exert influence on the fish assemblage for very long periods of time [unless factors like winterkill periodically remove the piscivore (cf. Tonn and Magnuson, 1982)] during which even a small predator-driven bias against particular prey species could lead to their elimination. Tonn and Magnuson (1982) showed that differential piscivory (although mediated by the stress of low oxygen in the systems) likely caused the distinctive *Umbra*–cyprinid versus centrarchid–*Esox* fish assemblages found in small northern lakes (Wisconsin, USA). Tonn (1985) found that both predation and competition from perch regulated populations of mudminnows in lakes where both occurred. Lyons (1987) concluded that preda-

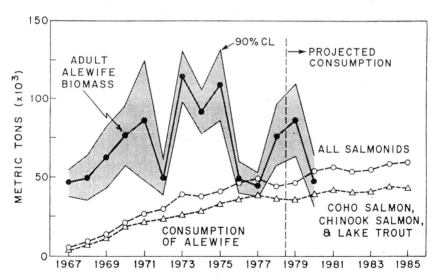

Figure 10.2. Biomass of adult alewife in Lake Michigan, compared to simulation modeling estimates of alewife consumption by three salmonid species and by "all salmonids" (including five species). [From Stewart et al. (1981).]

tion accounted for variability in size of darter populations, but not in cyprinids. Although, Lyons did suggest that predaceous yellow perch caused a minor inshore shift of cyprinids. Rahel (1984) showed distinct assemblage types in 43 Wisconsin lakes that he believed to be mediated by pH, winter anoxia, and piscivory (Fig. 10.3). Where adequate oxygen concentrations allowed centrarchids to persist, their effect as piscivores (and possibly as competitors) resulted in the exclusion of small species like cyprinids. Robinson and Tonn (1989) subsequently demonstrated similar phenomena in small lakes of Alberta (Canada), finding that piscivory by northern pike (*Esox lucius*) appeared to maintain differences between lakes with pike–yellow perch (*Perca flavescens*) and those with brook stickleback (*Culaea inconstans*) and fathead minnow (*Pimephales promelas*), and generalized that "for small boreal lakes of North America, piscivory and processes related

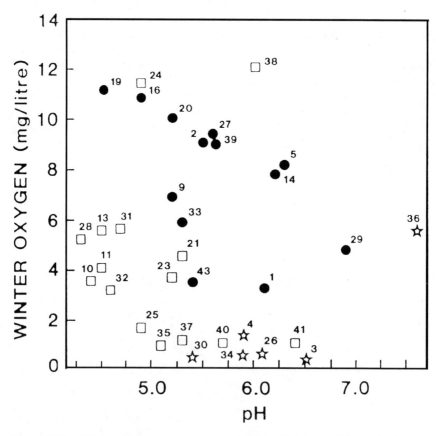

Figure 10.3. Winter oxygen concentration and pH for 34 lakes in Wisconsin. Assemblages dominated by centrarchids (solid circles), *Umbra–Perca* (open squares), or cyprinids (stars). [From Rahel (1984).]

to a small number of environmental variables ... appear to be most important in structuring fish assemblages."

The strongest evidence for the potential of predation to control lake fish assemblage structure comes from introductions of piscivores into waters to which they are not native. One dramatic example is that of Gatun Lake (Panama) in which introduced peacock bass (*Cichla ocellaris*) rapidly decimated a large proportion of the native fish stocks and caused a total change in composition of the lake's fishes (Zaret, 1979). Importantly, the introduction of this same species into the Chagres River, a triburary to Gatun Lake, had a very different effect, not resulting in local species extermination (Zaret, 1979).

Perhaps most stunning is the effect of introduced nile perch (*Lates niloticus*) in Lake Victoria, east Africa (Ogutu-Ohwayo, 1990; Kaufman, 1992; Gophen et al., 1993; Hecky, 1993; Gophen et al., 1995). Lake Victoria is one of the African rift lakes, renowed to ichthyologists and evolutionary biologists alike as the home of the most expansive fish species flock on earth [more than 400 haplochromine cichlid species (Gophen et al., 1993)]. The introduced Nile perch has, in 20 years, caused the extinction of the majority of the native, endemic cichlids (Fig. 10.4), with the result that now fewer than 100 species likely remain (M. Gophen, personal communication). From the human perspective, this episode also is a

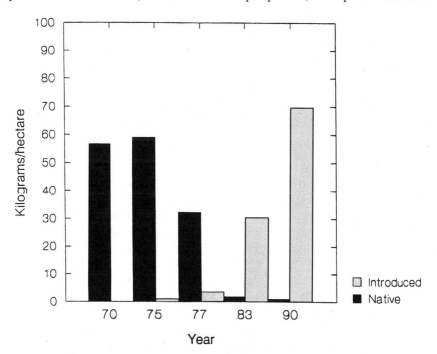

Figure 10.4. Demise of the native fishes of Lake Victoria, after introduction of nile perch. [From Kaufman (1992)].

tragedy, as the indigenous peoples have had their all-important food source altered forever, and they have changed to different fishing and fish cooking methods. The change in cooking has resulted in decimation of the surrounding forest for many square kilometers, as wood is taken for cooking fires now required to prepare the smaller fishes for edible use (M. Gophen, personal communication).

Many other examples of the strong influence of piscivores in closed bodies of water come from management case histories in small and large lakes, in which the introduction of a piscivore has changed composition of the rest of the fish assemblage (Hackney, 1979, on small pond management; McCammon and von Geldern, 1979, on large reservoirs). However, note that there is a conflict of opinion in some cases about the influence on newly introduced piscivores into reservoirs. For example, stocking and subsequent establishment of a large population of striped bass (*Morone saxatilis*) in Lake Texoma (Oklahoma–Texas) was considered potentially harmful to other piscivores (Hubbs, 1984), or innocuous to resident fish species (Harper and Namminga (1986)).

Localized Effects of Piscivores in Streams

There is no doubt that piscivores can influence the composition or behavior of the fish assemblage within individual stream pools. Power (1987) emphasized that fish in stream pools face threats from both avian (or other terrestrial) predators in shallow water versus piscine predators in deep water, resulting in a "bigger–deeper" distribution: Small fish occur more in shallows (where their small size may make them only minimally vulnerable to wading birds or diving kingfishers), whereas larger fish occur in deeper habitats (where they are relatively invulnerable to most fish predators, and escape avian predation by exceeding the depth at which birds are effective). Gorman (1988b) likewise noted that minnows in streams balance between excessively shallow habitats where they are vulnerable to birds, and deep habitats where large piscivorous fish often are present.

In bayous of Louisiana (Cashner et al., 1994), we most often saw very large numbers of small-bodied mollies and killifish in extremely shallow (a few centimeters) near-shore habitats, where piscine predators are likely ineffective. In an apparently predator-mediated system, small (neonate) eastern mosquitofish (*Gambusia holbrooki*) survived better in experiments in which adults were forced into refugia by the presence of predatory-chain pickerel (*Esox niger*) (Winkelman and Aho, 1993). Schlosser and Angemeier (1990) showed that centrarchid predators restricted smaller fishes to shallow-water refugia. Schlosser (1988a, 1988b) reported that small and large fish are segregated in shallow and deep stream habitats, respectively.

Thus, a wealth of evidence suggests that piscivores can influence the depth of habitats selected by fishes within a stream reach. In experiments in natural stream pools in Trinidad, Fraser and Gilliam (1992) recorded the occupancy of "depth zones" by prey with and without predators (*Hoplias*) present and showed that the piscivore restricted the use of the deepest areas (which the predator typically occupied) by large (but not small) *Poecilia*, and by small *Rivulus*. At twilight, when the predator became more active, all prey typically converged into shallow areas of the pools, effectively breaking down the habitat segregation by size-class that had been observed for guppies during daylight. Gilliam and Fraser (1988) summarized a preliminary experiment with guppies and *Hoplias* with similar results: Intraspecific segregation in habitat use by different size-classes of guppies during daylight broke down at twilight as the predator became more active and all sizes of guppies moved to shallow water. In experiments allowing emigration from pools, the piscivore increased the departure of prey species from pools, with emigration a more important influence than actual death to piscivory with respect to composition of pool assemblages in Trinidad.

Harvey (1991a, 1991b) showed that largemouth bass strongly changed habitat use by minnows and small sunfish within stream pools: In the absence of the bass, these fishes dominated the open waters of pools; with the addition of bass, the smaller fishes were restricted to marginal habitats at ends or fringes of pools. Power et al. (1985) showed in Brier Creek, Oklahoma, that piscivorous largemouth bass, *Micropterus salmoides*, restrict the use of habitat by *Campostoma anomalum* to shallow-pool margins and increase their emigration from pools. Power and Matthews (1983) reported a strong negative relationship between the presence of black bass (*Micropterus* spp.) and stoneroller minnows *Campostoma anomalum* among 14 pools of Brier Creek, and subsequent surveys and experimental manipulations (Power et al., 1985) supported the hypothesis that bass regulated the occurrence of this minnow species from pool to pool. From snorkel surveys of those pools, Matthews et al. (1994) also showed an overall negative relationship between the presence of bass and of all taxa of prey fishes (summed) within individual pools.

Matthews et al. (1994) also suggested that not all predators are equal, as bass were much more effective than large sunfish (*Lepomis* species) in keeping the numbers of prey low within individual pools. Harvey et al. (1988) showed in manipulative experiments, that some species of *Micropterus* basses were more effective than others at controlling the numbers or activity of small fishes in individual pools: Largemouth bass, *M. salmoides*, had strong influence on the distribution and numbers of minnows in streams pools, but smallmouth bass, *M. dolomieu*, did not. Like Harvey et al. (1988), Schlosser (1988b) found that piscivorous species differed in strength of their effect on prey species. In an artificial stream, Schlosser showed that smallmouth bass (*Micropterus dolomieu*) and adult creek chub (*Semotilus atromaculatus*), although both piscivorous, differed in their impact on habitat use by prey (brassy minnows, *Hybognathus hankinsoni*). The creek chubs consumed far less prey and resulted in only a weak shift in habitat use by the minnows (into structurally simple pools). The smallmouth bass were far more lethal and caused a strong shift of minnows to shallow refuge habitats (raceways and riffles). Schlosser reasoned that the degree of prey shift in habitat was directly related to the level of threat from each predator. Finally, Schlosser (1988a) compared his results in small artificial streams to our results in natural small (Power et al., 1985) and large (Matthews et al., 1987) streams and noted the need for future work to assess levels of piscivore influence in habitats ranging widely in pool size and complexity. As of this writing, such experiments remain to be done.

In other artificial stream experiments, Schlosser (1987b) showed that large-bodied predators (smallmouth bass, *Micropterus dolomieu*) would restrict the use of pool habitats by small fish species in patterns concordant with their microhabitat use in a natural stream (Jordan Creek, Illinois, USA). Schlosser also showed differences among prey species in vulnerability to predation, which he related to body form (e.g., soft bodied versus spines) and levels of activity (e.g., white sucker, *Catostomus commersoni*, juveniles foraged actively, potentially placing them in harm's way). Schlosser found that the degree to which prey species shifted out of their preferred (deeper/structured pool) habitats into shallow habitats in the presence of a predator was related to their differences in vulnerability to the predator. Schlosser (1987a) pointed out that the predator-avoidance by the small species resulted in their extensive overlap in the use of shallow habitats in streams; that is, perceived levels of "niche overlap" among these species may have been mediated by the predators that occupy deep pools.

This concept was amplified and generalized by Schlosser (1987a) in his model, predicting that the combination of deep pools and their resident piscivores had pervasive effects on the composition of stream fish assemblages throughout midwestern North America. Gilliam and Fraser (1988) and Fraser and Gilliam (1992) also showed that predators in tropical streams (Trinidad) could cause emigration of small-prey species from pools or force the small species into shallow "edge" habitats.

Predators in some stream environments may consume large numbers of prey without apparently influencing population numbers significantly. For example, abundant green sunfish (*Lepomis cyanellus*) and transiently cruising *Micropterus* bass spp. in the extreme upper reaches of South Concho River, Texas, near "Head of the River Spring," eat large numbers of the diminuitive largespring gambusia *Gambusia geiseri* (which is introduced at this site but is native to large spring habitats in Texas) (E. C. Marsh, personal communication). However, in spite of intense predation by the centrarchids (and also perhaps by abundant, introduced Mexican tetra, *Astyanax mexicanus*), the *Gambusia* (which are live-bearers with rapid brood production) are, by far, the numerical dominant in this reach of the river (Glasscock, 1989) occurring throughout the entire headpool and upper reaches of the river in large numbers.

Piscivore Effects Throughout Longer Stream Reaches

Whereas there is little argument that bass or other predators might lower the numbers or kinds of smaller fishes within individual stream pools, there seems to be little evidence of the ability of native piscivores to do so over longer stream reaches under natural conditions. In Baron Fork of the Illinois River, Oklahoma, I regularly found, by snorkeling, very large numbers of both smallmouth bass (sometimes dozens of individuals) and minnows (many thousands of individuals) within the same pool. Many of these minnows were central stonerollers (Matthews et al., 1987)—the same species that was regulated strongly by largemouth bass in small pools of Brier Creek. Such a co-occurrence of predator and prey species may be due in part to the difference in the species of *Micropterus* in the two systems: largemouth versus smallmouth bass. Harvey et al. (1988) showed differences in responses of stoneroller minnows to the two species, but these experiments were in Brier Creek pools only, as the reverse experiment could not be done (logistically or ethically) in the much larger Baron Fork system. In addition to differences in the species of the top predator, Baron Fork pools are at least an order of magnitude larger than typical Brier Creek pools, pools are well connected across much deeper riffles than in Brier Creek, and the density of minnows is much greater. Intuitively, it seems almost impossible for a few smallmouth bass to dominate such large pools as those of Baron Fork, and from underwater and videocamera observations, it is obvious that minnows only weakly avoid smallmouth bass, typically moving away only if bass make an obvious feeding attack, or swim closer than 1 m to a minnow school.

Differences in riffles as barriers to movement of fishes in Brier Creek versus the Baron Fork could differentially affect movement of predators (large bodied) and prey (small bodied). Whereas the shallow riffles (often only a few centimeters deep) in Brier Creek are traversed readily by *Campostoma* in most seasons, they are an apparent impediment to bass movement, especially at base flow. Therefore, a prey individual might benefit more from emigration out of a pool containing bass in Brier Creek than it would in Baron Fork, where deeper riffles provide avenues for interpool movement by both prey and predator. In general, these two examples lead to a logical postulate that abundant prey species might follow two different survival strategies in systems dominated by deep versus shallow riffles between pools. Where deeper riffles allow ready movement by predators, it may be of little advantage to the prey to escape by moving to a new pool, as the predators could readily follow. Thus, it seems possible that local populations could evolve differences in behaviors or life histories in response to the differences imposed on predator threat by physical structure.

One other solution could be in growth rate. Although Schlosser (1988b) suggested that most minnows do not grow sufficiently large to escape from smallmouth bass by gape limitation, *Campostoma* in Baron Fork grow very large [e.g., 150 mm or longer, Total Length (TL)] with robust bodies, and I suspect that rapid growth to a size exceeding *efficient* handling by the smallmouth bass could have a selective advantage. Yule and Luecke (1993) suggested that differential vulnerability of two prey species to lake trout in Flaming Gorge Reservoir (Utah–Wyoming, USA) was related to the slower growth of one than the other, with the slower-growing Utah chub (*Gila atraria*) more vulnerable to piscivores for a longer period than the faster-growing kokanee (*Oncorhynchus nerka*). Obviously, all of these scenarios place predation as a strong evolutionary force, with implications for selection of individuals of the prey species that grow rapidly or that are best at avoiding a bass (or other predator) within a given pool. However, predator swamping by large numbers of individuals seems commensurate with what we observe in the Baron Fork: *Campostoma* schools seem more focused on active grazing of algae than on long-distance avoidance of predators. Individuals do move away from bass that approach within about 1 m, and the school often exhibits a "fountain effect," parting a pathway as the bass cruises through, and circling about to a position behind the bass from which vigilance is possible. In Brier Creek, *Campostoma* exhibited a different response, rarely occurring in large numbers in pools with bass (Power and Matthews, 1983; Power et al., 1985), and rapidly emigrating from pools to which bass were introduced. Here, if riffles serve as barriers to bass movement, it may be more energetically and evolutionarily efficient for *Campostoma* to actively avoid the predators by inter-pool movement to safe nonbass pools. "Common garden" experiments assessing propensity of *Campostoma* from the two different streams to move in response to bass presence would be of interest to test this hypothesis.

Piscivore Influence on Distribution, Habitat, or Behavior or Prey Species

Winemiller and Leslie (1992) showed that in tropical lagoons, small fish species are largely restricted by predators to shallow-margin habitats. Meador and Kelso (1990) reported behaviors of fish in which species typically thought of as the "hunter" is the "hunted" [to paraphrase Gilliam (1990)]. In sluggish, marsh–stream habitats near the Gulf of Mexico, Meador and Kelso (1990) found that a local race of largemouth bass (*Micropterus salmoides*), known locally as "marsh largemouth bass," had relatively slow growth and apparently included little fish prey in their diet. In these systems, most of the small-bodied fish that could have been prey for bass were in open waters, whereas the bass were most common in highly structured macrophyte beds and edge habitats. Although experiments in such a system would obviously have been impossible, Meador and Kelso (1990) reasoned that, in these systems, the *Micropterus*, although usually a top predator in most systems, were actually only the intermediate predator. They reported abundant alligators (*Alligator missippiensis*), alligator gar (*Atractosteus spathula*), and red drum (*Sciaenops ocellatus*) adults in the system and that a high percentage of largemouth bass bore scars of apparent encounters with these large predators. They reasoned that these very large top predators restricted the *Micropterus* to the highly structured edge habitats, where they were limited in foraging opportunities but found at least partial refuge from the open-water top predators.

Sazima and Machado (1990) reviewed the interesting situation in which various species of fin-eating piranhas in Brazilian streams are only "partial predators," biting off pieces of fins of victims rather than killing them. They argued that this "unpleasant" experience, possibly reinforced numerous times in the life of a "prey" fish, had strong effects on habitat use by fish in piranha-occupied waters. Both their and Winemiller's (1989) observations suggested that the piranha *Pydocentrus notatus* exert strong habitat structuring on prey fishes, with open-water mid-pool habitats avoided by the potential prey. Sazima and Machado (1990) also noted particular defensive behaviors by potential prey, such as the forming of defensive rings while feeding.

In a study of fish movement and patch use in a small stream (New York, USA), Fraser and Sise (1980) evaluated the distribution of adults and juveniles of two minnow species (*Semotilus atromaculatus* and *Rhinichthys atratulus*) in a series of 11 pools and riffles. Larger adult *Semotilus*, which are piscivorous, were clumped in distribution among habitat patches, but small *Semotilus* and *Rhinichthys* followed the expectations of an Ideal Free Distribution (Fretwell and Lucas, 1970). Fraser and Sise (1980) suggested that predation by adult *Semotilus* could influence habitat of smaller minnows, but the effect might be weak or difficult to detect, in that "small reductions in density of young fish in a given pool due to adult predation may be only transitory except during periods of low flow when movements of the fish are restricted preventing replenishment of losses within pools." The idea that losses of small fish to predation will be compensated for by immigration into the pool is somewhat analogous to the suggestion by Allan (1982) that drifting insects arriving in a stream reach would mask any effect of insectivory by fishes, so that seeing effects of predation is improbable.

10.3 Experiments with Piscivore–Prey Systems

Experimental Evidence: Effects of Predators in Streams and Lakes

Much information on the effects of piscivores and predator threat on fish assemblages comes from experiments on stream piscivores and prey reactions by D. Fraser, J. Gilliam, I. J. Schlosser, M. E. Power, and B. C. Harvey, and on lake fishes by E. Werner, G. Mittlebach, J. Savino, R. Stein, and their colleagues. However, most experiments in North American systems have used only a limited suite of piscivores (largemouth bass, *Micropterus salmoides*; smallmouth bass, *Micropterus dolomieu*; adult creek chub, *Semotilus atromaculatus*) and prey species (mostly small creek chub; blacknose dace, *Rhinichthys atratulus*; brassy minnow, *Hybognathus hankinsoni*; fathead minnow, *Pimehales promelas*; northern redbelly dace, *Phoxinus eos*) out of the more than 900 species in fresh waters in the continent (Burr and Mayden, 1992). Therefore, the suite of piscivores in experiments has been rather limited, and it seems obvious that there must remain many naturally realistic combinations of piscivores and their prey in temperate streams that would yield further temperature insight into the ways piscine piscivore–prey systems work. I am also unaware of any experiments in which the addition of a piscivore was evaluated with respect to changes in behavior or habitat use by a whole fish assemblage.

In the first of several artificial stream experiments by Fraser and his associates, Fraser and Cerri (1982) found that small creek chubs, *Semotilus atromaculatus*, and blacknose dace, *Rhinichthys atratulus*, avoided compartments with piscivorous large *Semotilus*, regardless of the structural complexity in compartments. However, if analyzed by time of day, predator effects were significant only at night, and prey also were more likely to be in a compartment with a predator if the structure was complex. Fraser and Cerri (1982) and Cerri (1983) suggested that large, piscivorous *Semotilus* might seek shelter in daylight [e.g., from avian or terrestrial predators that were common in the area, *fide* Fraser and Sise (1980)] and thus be less active. They noted the conflicting demands faced by small fishes when habitat patches affording optimal foraging also contained piscivores. Fraser and Emmons (1984) showed in a companion experiment that dace avoided the risk of being in a habitat patch with any predators and, under most conditions, avoided one piscivorous creek chub as much as they avoided two or three. Additionally, Fraser (1983) found that availability of shelter had an important influence on dispersal of adult creek chub; that is, that numerous adults would aggregate near a shelter if it was the only one available, but disperse if multiple shelters are available. In contrast, young creek chub would not aggregate near a single shelter, probably because high densities of young might attract predators (or lower individual foraging efficiency).

Although Fraser and Cerri (1982) showed piscivorous creek chub could influence habitat use by small fishes at low light levels, Schlosser (1988a) found in an artificial stream experiment with brassy minnows (*Hybognathus hankinsoni*) as prey that creek chub were much less dangerous predators than smallmouth bass (*Micropterus dolomieu*). Smallmouth bass are more common than creek chubs as the top predator in many streams of eastern North America. (Creek chub typically are the dominant piscivore only in small, headwater streams.) Schlosser and Ebel (1989) compared the effects of creek chub on brassy minnows [from Schlosser (1988a)] with the effects of creek chub on two additional minnow species (fathead minnow, *Pimephales promelas*; northern redbelly dace, *Phoxinus eos*). Schlosser and Ebel (1989) showed that creek chubs were, overall, a relatively ineffective predator, limited by size of prey taxa, eating only (at maximum) 0.25–0.30 fish per day, and having the least effect on the largest species (brassy minnow). Creek chubs also had less effect on prey species distribution than did habitat complexity, and all three prey minnow species shifted habitat only minimally in the presence of creek chubs.

Schlosser (1988b) also tested, in experimental streams, the influence of smallmouth bass on habitat selection by hornyhead chub (*Nocomis biguttatus*), a common minnow in eastern streams, in day and night. Small *Nocomis* were more susceptible to predation than larger individuals. In the absence of bass, both large and small chub preferred deep pools in daylight (even though more insect prey was available in shallow riffles or raceways). The presence of bass resulted in the restriction of both size-classes of chubs to shallow refugia in the day. At night, in the absence of bass, larger chubs tended to remain more in deep pools, whereas small chubs shifted to shallow habitats. Furthermore, Schlosser cited evidence of threat from common terrestrial predators in shallow water both day (great blue heron, *Ardea herodias*) and night (mink, *Mustela vison*) that would cause larger hornyhead chubs to choose to remain in deeper water. Schlosser (1988b) concluded that his findings fully supported the concepts of Power (1987) with regard to shallow-water versus deep-water predator threats to fishes, and suggested the need for better understanding of the interactions of predation risk and environmental variability on structure of fish assemblages.

In considering the results of Schlosser's and Fraser/Gilliam's experiments, it is not surprizing that different piscivorous species would have different levels of effect on prey, and that factors like body size (Wahl and Stein, 1989), morphology, or behavior of different prey species (or size-classes), test arena complexity, or time of day could substantially affect the outcome of predator–prey experiments, particularly in laboratory or artificial systems. As a specific example, consider the potential difference in outcome of experiments in simple "one pool" tests versus those in which an individual predator (e.g., one bass) and its potential prey are offered access to multiple pools and riffles [as in experiments by Gelwick (1995) in large, multipool experimental streams (Gelwick and Matthews, 1993)]. This is *not* an indictment of experiments in artificial streams but is intended to suggest that investigators need to be sensitive to potential effects they can have on the outcome of experiments as they choose (particularly in complex fish assemblages) the species, sizes, and so forth of predators and prey as well as arenas and conditions used to test theory. At best, appropriately sized species are selected, and tests are conducted in sufficiently large (or complex) arenas so that results extrapolate well to the outside world. At worst, poorly chosen matches of predators, prey, and test conditions can result in experiments that are no more than mere "demonstrations" of "what these species did on a particular day in a particular system," providing little insight into the dynamic relationships that prevail in real streams or lakes.

Effects of the Environment on Outcome of Predator–Prey Contests

As suggested by Schlosser (1988b), a large number of environmental variables can strongly influence the outcome of piscivore–prey interactions. A classic example is that of Savino and Stein (1982) who showed that habitat complexity altered the outcome of largemouth bass–bluegill interactions. Largemouth bass were less successful predators on bluegills in artificial vegetation because of changes in behaviors of the prey and also because of visual blocking of prey from predators by plant stems. However, in a subsequent study with two predator and two prey species, Savino and Stein (1989) demonstrated that structure would not change the outcome of predator–prey contests unless the prey species used the protective structure effectively. Wahl and Stein (1989) showed that differences in vulnerability of juveniles of three esocid taxa to piscivorous largemouth bass disappeared when vegetation (simulated) was available. Gotceitas and Colgan (1989) showed that the effects of vegetation on piscivory by largemouth bass on bluegill was not linear: There was a threshold of cover density that adequately afforded protection to bluegill young, and the young bluegills selected habitats with more than adequate cover. As another example of environment–predation interactions, the amount of cover (vegetation in this case) changed the propensity of largemouth bass to prey on small blue tilapia or bluegills (Schramm and Zale, 1985). Not only is vegetative cover effective: Everett and Ruiz (1993) showed that coarse woody debris also provided refugia for and was used by small fishes and suggested that woody debris is a major factor in providing shelter from piscivory in many freshwater environments. [However, instream experiments (e.g., with log-drop structure enhancement) rarely seek to differentiate between foraging advantages to small fishes (cf. Benke et al., 1984) and shelter from piscivores.] Eklöv and Persson (1995) and Persson and Eklöv (1995) showed that when structural refugia were available, survival of small perch (*Perca fluviatilis*) and roach (*Rutilus rutilus*) increased under threat of predation from adult *Perca*, and that the availability of structure changed behavior of both piscivore (spent more time near the structure) and the prey (changed time in refuge or in schooling).

In one example of the effects of physical variability on predator–prey outcomes, Poulin et al. (1987) showed that changes in oxygen concentrations in environments changed the probability of capture of guppies (*Poecilia reticulata*) by the cichlid *Astronotus ocellatus*. Many authors have shown that changes in light levels influence the effectiveness of piscivores. Cerri (1983) showed, for example, that blacknose dace were more vulnerable to predatory creek chubs at low light levels than at high light levels.

Effects of Other Animals on Piscivory

Rahel and Stein (1988) described the mutual enhancement of predator effectiveness. Piscivorous fish (smallmouth bass) and crayfish (*Orconectes rusticus*) synergistically influenced habitat use and vulnerability of the benthic johnny darter (*Etheostoma nigrum*), which is common in small streams. In the presence of bass alone, the darters took refuge under stones. However, the crayfish evicted the darters from such refugia, making them more vulnerable to bass predation. Likewise, the bass increased vulnerability of the darters to crayfish, by forcing them to seek refuge under stones. Therefore, the effectiveness of the predatory fish was enhanced by presence of the crayfish. Finally, although the effects on predator–prey interactions was not addressed directly, a companion study by Rahel (1989) showed that nesting johnny darters would evict small (but not large) crayfish from shelters under stones. Therefore, would the presence of the darters in streams differentially influence the vulnerability of small versus large crayfish to larger piscine predators?

Gelwick (1995) found a "crayfish" effect in experiments in which stoneroller minnows (*Camposoma anomalum*) were in large, multipool outdoor experimental streams with largemouth bass alone, or with bass plus adult crayfish (*Orconectes* sp.). When crayfish were present, fewer minnows survived, apparently because [as in Rahel and Stein (1988)] the activities of crayfish (harassment, eviction from refugia) made the minnow more vulnerable to bass.

The effectiveness of crayfish in evicting darters from (or sequestering use of) interstitial spaces was underscored by an experiment that went awry some years ago. With the help of a class, I constructed a large number of "darter" refugia, consisting of sections of 50-mm-diameter PVC pipe, approximately 200 mm long, partially embedded in concrete to simulate spaces under large cobble in a riffle of a natural stream. In some, both ends were open; in others, only one end of the PVC allowed entry. In swift riffle habitats of the Blue River, Oklahoma, some were placed in the stream longitudinal to the current; others were placed across the current. The goal was to test combinations of openness and orientation with respect to colonization by abundant orangebelly darters (*Etheostoma radiosum*) at the site. After a few weeks, we retrieved the "darter" shelters and found almost all occupied by crayfish, with few darters successfully holding space. Apparently crayfish can be quite efficient in various systems at evicting or excluding darters from interstitial spaces, thereby potentially exposing them to greater piscine predator threat.

The following is an example of the ways complex interactions between the environmental template and fish activities can influence the outcome of piscivore–prey interactions. In the midwestern United States, streams are dominated by complex algal growth during much of the year. For example, during one rapid "visual" survey of 152 streams in Texas and Oklahoma in April 1992, I observed conspicuous standing crops (floating mats or columns of attached algae) of algae in 74 (48.7%) of the sites. Interactions among predators (largemouth bass), grazing minnows (*Campostoma anomalum*), and crayfish (*Orconectes* sp.) strongly affected such algal structure in experiments in large, outdoor artificial streams (Gelwick, 1995).

Gelwick hypothesized that the resulting differences in algal physical structure could have cascading effects on interactions among other fishes, with consequences for stream community structure and function. In the experimental streams, Gelwick tested the effects of grazed and nongrazed algal structure on habitat use and behavior of two invertivorous fishes, juvenile green sunfish (*Lepomis cyanellus*) and bigeye shiners (*Notropis boops*), and their shared predator, largemouth bass. Gelwick (1995) showed that the survival of sunfish was unaffected by minnows either in grazed or nongrazed pools. The survival of minnows in grazed pools was not affected by sunfish, but the survival of minnows in nongrazed (algae-dense) pools was higher when sunfish were present. In grazed (algae-sparse) pools, bass frequently cruised within and between pools in late afternoon and near dusk, whereas in nongrazed (algae-dense) pools, bass spent more time refuging under algae than in the open. Minnows foraged in open water under all conditions of predator and algal structure.

In grazed pools with bass, sunfish showed greater risk-aversive behavior by refuging under rocks and, thus, were less vulnerable to predation than minnows. However, in nongrazed (algae-dense) pools, sunfish foraged in the open, but closer to algal structure in the presence of bass. The dense growths of algae in nongrazed habitats may have facilitated risk-prone foraging behavior by sunfish, making them more susceptible to ambush predation by bass and thereby reducing predation on minnows in two-species assemblages. Thus, the grazing activities of crayfish and one minnow species (*Campostoma anomalum*) decreased algal structure, which, in turn, increased the mortality of another minnow species (*Notropis boops*) in a complex shared-predator situation (Gelwick, 1995). Alteration of the environmental structure by one fish species therefore changed indirectly the outcome of piscivore–prey interactions.

10.4 Theoretical Aspects of Predation

Balancing the Risk: Eat Well and Die Young(er)?

Werner et al. (1983a, 1983b) showed that when predators were absent, small bluegill (*Lepomis macrochirus*) foraged primarily in energy-efficient open areas of experimental ponds, but in the presence of a predator (largemouth bass, *Micropterus salmoides*), small bluegills would restrict foraging to areas near vegetation that provided much less energy gain. Thus, they traded off maximum net energy gain for the relative safety of near-vegetation habitat, unlike larger adults that foraged in open water even when bass were present.

Cerri and Fraser (1983) explicitly tested the hypothesis that prey fishes can balance the conflicting demands between the need to forage efficiently and the need to avoid predators, a problem in "risk balancing" (cf. Sih, 1980). Theoretically, risk-balancing predicts greater risk-taking when the potential benefits are greater. However, Cerri and Fraser (1983) found no interaction between the presence of predators and the choice of food levels. Prey avoided compartments with predators, both in low-food and high-food streams. The sequence of declining levels of patch use by prey fish was (1) high food–no predator, (2) low food–no predator, (3) high food–predator, (4) low food–predator. In general, the results suggest that predators had more influence than levels of food in determining habitat use by the prey fishes. Cerri and Fraser (1983) concluded that the small minnows were not making combined judgments about food and predators and did not balance risks. They cited similar results by Milinski and Heller (1978) for sticklebacks. Cerri and Fraser (1983) proposed as an alternative to risk-balancing a "patch choice" model based on time allocations for foraging versus attention to predators. Their overall conclusion was that minnows in streams should not risk-balance (i.e., they should not take proportionally greater risks in order to improve feeding). The one critical variable that seems to be lacking in applying the findings of Cerri and Fraser to unrestricted fish in natural streams is information on overall availability of foods in the system and on the hunger or energy needs of the prey fishes.

Magnhagen (1988) demonstrated differences between salmonid species (pink salmon, *Oncorhynchus gorbuscha*, and chum salmon, *Oncorhynchus keta*) in willingness to trade off predator avoidance for improved foraging. Additionally, there were individual differences in risk-taking, with satiated individuals foraging less than hungry individuals in predator-risky experimental habitats.

Fraser et al. (1987) essentially repeated the experiment of Fraser and Cerri (1982) in a natural (rather than an artificial) stream, to test effects of predators, structure, and light levels on habitat choice by small minnows. They found that in the dark, prey used the pool with predators significantly less. (Predators are more dangerous at low light, because the balance of detection ability may tip in favor of a motionless "ambush" predator.) Fraser et al. also found a three-way interaction of predators × light × structure: In daylight, the young avoided the pool with predators if it had supplementary structure. This implies that in daylight in low-structure habitats, small chub can co-occur with adults (who tend to hide under bright light), but that in highly structured pools, adults might remain more active or willing to attack and, hence, be a greater threat to young. Gelwick (1995) found that in highly structured pools (algal structure), small green sunfish may remain more active, "sheltering" less under stones, and thereby be more vulnerable to predation by largemouth bass.

In clear streams of the Little Missouri River drainage, of the Ouachita Mountains (southwest Arkansas, USA), headwater pools with both adult and juvenile *Semotilus* often have a high degree of structural complexity such as slab rock, boulders, or cobble, and there also can be a wide and overlapping array of fish sizes in a given pool (personal observation). The combined studies by Fraser and colleagues would seem to suggest that young *Semotilus* should not forage much (1) at night—when predators are a serious threat, or (2) in daylight in structured pools. (In the complex habitats of Ouachita Mountain streams, when *would* little creek chubs forage?) What is the size structure of creek chubs across a large number of small pools in these natural streams? Do little creek chubs predominate only in somewhat open, larger pools, where they obtain "protection" from piscivorous adults by remaining in the water column, at a substantial distance away from shelter that might hide the adults as sit-and-wait predators? Likewise, in laboratory tests, small creek chubs under bright lights foraged actively, but responded to attacks from adults by rapid escape and temporary cessation of feeding (Fraser et al., 1987). In dim light, small chubs increased use of refuges, foraged less, and avoided the piscivores (Fraser et al., 1987).

It would seem to be an interesting follow-up to Fraser et al. (1987) to use a sequence of pools in a natural stream, restrict predators within pools by fencing across riffles [see Gelwick and Matthews (1992)] with large-mesh size [as did Fraser and Cerri (1982)], and manipulate food levels and numbers of piscivores in the pools. Results might differ concerning risk-taking under varying natural conditions such as (1) low-food availability for minnows, as in immediately after a scouring flood, (2) temperature—with greater demand to forage as temperatures rise—at least up to some maximal threshold, (3) size ratio of actual predators to prey, and (4) relative burst speed or swimming speeds of predators or prey. Cerri (1983) makes clear that light levels during the experiment could be important (cloudy days, sunny days, dense canopy cover?).

Theoretical Prey Behavior Under Predator Threat

Werner and Hall (1988) summarized the costs to small bluegill (*Lepomis macrochirus*) of foraging in open water with bass predators (40–80 times as much mortality) versus the potential benefits of obtaining energetically more profitable foods in open-water microhabitats. They interpreted this as a growth rate–predation risk trade-off for bluegill in shifting ontogenetically between "safe" vegetated littoral zone habitats and open waters of ponds or small lakes. Werner and Hall (1988) empirically observed two shifts in habitat use by bluegill during ontogeny (*into* vegetation as larvae reach about 12 mm, and *out* of the vegetation to open water as they reach a size of lowered vulnerability to bass) as consistent with predictions of a model of minimizing mortality relative to growth rates for each size class of fish.

Gilliam and Fraser (1988) summarized previous work and provided a conceptual overview of the ways that levels of predator threat might influence behaviors of individual fish or evolution of predator-avoidance traits. In the model, they emphasized individual differences in responses of prey, and the fact that a given prey fish species might respond very differently depending upon the density of predators or intensity of threat. They theoretically considered thresholds of predator-threat below which prey might respond little (and suffer increased mortality, etc.), but above which prey might alter behaviors and therefore enjoy at least short-term gains in survival (Fig. 10.5). Gilliam and Fraser (1988) noted that a high risk of death does not preclude all use of an area—i.e., if energetic returns are sufficiently high, the potential prey will forage in a site (although not specifying how individual behavior interacts with this problem). They concluded that in size-structured populations, the interplay of differential predation threat (small individuals more vulnerable than large) with food availability or depletion could result either in the extremes of complete overlap of size-classes of the prey species or in their spatial segregation, as less vulnerable (larger) individuals could gain access to resources with less cost of predator threat.

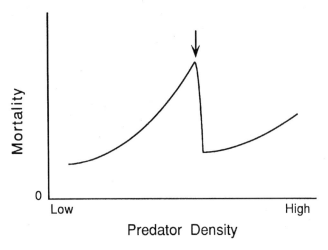

Figure 10.5. Hypothetical mortality rates for prey fish species (*y* axis) as a function of predator density (*x* axis) showing threshold (arrow) above which prey increase predator-avoidance behaviors. [Modified from Gilliam and Fraser (1988).]

Gilliam (1990) presented a summary of mathematical models of foraging choice, and extended classic energy-maximizing models to include mortality hazard to small fishes that are themselves vulnerable to predators. Gilliam argued that foraging animals can be affected just as much by upper trophic levels (their predators) as by availability of food, and cited 19 articles, most on aquatic systems, showing examples of species being restricted in their use of foraging habitats by the presence of a predator. Gilliam extended the question of predator influence on foraging behavior of a prey species to ask how a predator influences foraging choices by their prey species, even when the latter are already "confined" to certain habitats by the predator's presence. When handling prey is more hazardous than the searching for prey (his Case II), animals decrease diet breadth, but when searching is more hazardous (due to predator threat) than the handling of prey (his Case III), animals should broaden diets (i.e., not pass up prey once it is detected, because of the increased mortality risk of continuing to search).

In streams in Trinidad, Fraser and Gilliam (1987) examined foraging behaviors of upstream and downstream populations of guppies (*Poecilia reticulata*) and of giant rivulus (*Rivulus harti*) (called "Hart's rivulus" by Fraser and Gilliam) that lived under differing levels of predator threat. Fish in upstream sites lived with few predators and thus had no predator experience; fish from downstream sites were exposed to predation pressure. Fraser and Gilliam (1987) tested the "tenacity" of individuals from each location to continue feeding in the presence of a predator threat. They predicted from earlier work by Seghers (1973; original not seen), Giles and Huntingford (1984), and Magurran (1986) that individuals from predator-rich environments would be more wary of predators and show more antipredator behaviors than would individuals from low-predator-threat stream reaches.

Essentially, they predicted that individuals from low-predator environments would fail to recognize predators or take evasive actions, whereas the fish from high-predation environments should recognize predators, be predator-wary, and the like. However, they found exactly the opposite: Both guppies and rivulines from low-predator habitats (upstream) were extremely predator-wary (ceasing feeding, hiding, etc.), whereas individuals from predator-rich downstream habitats showed less response to the predator (continuing to feed, etc.) (i.e., showed much higher "tenacity").

The presence of a predator would interrupt feeding by guppies or rivulines from upstream (low-predator) sites. Individuals from downstream sites (where they had had experience behaviorally and presumably over evolutionary time with predators) would continue to feed. This seemed surprising, but Fraser and Gilliam (1987) offered the explanation that if animals live fulltime with high levels of predation pressure, it becomes counterproductive to overly respond to predators, ceasing activity because of their presence. This could, in effect, leave little time in which to feed or carry out other activities. Evolutionary mechanisms may lead to "boldness" or tenacity in these populations of small-bodied fishes in which predators are an ever-present threat. This could relate to predator-swamping: these are highly productive populations that produce many individuals in a short period of time (short intergenerational time). In the presence of predators, they may simply evolve responses that allow them to continue feeding, even though some individuals are apparently lost to predators in the process. The conclusion of Fraser and Gilliam (1987) was that "upstream fish (low-predator experience) readily trade off feeding for hiding and avoiding predation hazard, a likely response when predators appear infrequently, while downstream fish appear to be selected for boldness and tenacity while foraging under chronically high hazard."

Fraser and Gilliam (1987) explicitly used the term "selected for" (i.e., implying an evolutionary response of populations in high-predation-risk environments (although multigenerational tests of persistence of tenacity in a "common-garden" environment remain to be done)]. It would also be interesting to know the degree to which individuals from upstream versus downstream environments would actually be killed by predators in experiments like that of Fraser and Gilliam (1987). Are the ones with a predator-rich history (downstream) that continue foraging actually more likely to be killed, or is their increased tenacity in feeding matched by better capacity for predator avoidance in the event of an attack (evasive maneuvers, etc.)? Experiments in sufficient test arenas to ensure normal predator behavior would be interesting (either with the fishes from Trinidad streams, or perhaps by North American minnows, like *Campostoma*, that can be identified as being from predator-rich or predator-poor streams).

I suspect that a similar pattern of continuing to feed even in presence of a predator may characterize the behavior of the dense *Campostoma* populations that I have described from Baron Fork of Illinois River, Oklahoma (USA). Here, although the primary predator (smallmouth bass, *Micropterus dolomieu*) is abundant, the shoals of *Campostoma* (often with other minnow species intermingled) continue to feed actively when predators are in the same pool, only ceasing feeding to move away if the bass approaches within about a meter.

Nonlethal Effects of Predators on Growth, Reproduction, or Interactions of Prey Species

In streams of Trinidad, Fraser and Gilliam (1992) experimentally showed that presence of a piscivore (*Hoplias malabaricus*) could have important sublethal effects on prey fish species, altering net emigration, reproduction, and growth of individual guppies (*Poecilia reticulata*) and giant rivulus (*Rivulus harti*). Fraser and Gilliam (1992) found that in the presence of the predator, the *Rivulus* produced fewer eggs, and both the spatial and temporal patchiness of egg deposition was increased. When the predator was present, *Rivulus* usually laid eggs in pulses and placed them in pools lacking predators. Overall, the presence of the predator also reduced the growth rate of adult (but not of juvenile) *Rivulus*. Similarly, Werner et al. (1983b) showed depressed growth (27% less) for small bluegills (*Lepomis macrochirus*) that, in the presence of a predator (largemouth bass, *Micropterus salmoides*), shifted foraging to less profitable, vegetated microhabitats.

McDonald et al. (1992) showed that young lake trout (*Salvelinus namaycush*) in Toolik Lake, Alaska, avoided predators by occupying suboptimal inshore habitat, with feeding returns so low in this oligotrophic lake that recruitment could be affected. For adequate growth and recruitment, McDonald et al. suggested that young lake trout also used deeper-water foraging areas where rock structure provided some refuge from predators. In such an oligotrophic system, prey could apparently be particularly vulnerable to any trade-offs between foraging opportunities and predator protection. Would the mortality-minimization/energetics-maximization models of Gilliam and others work equally well in these very-low-productive arctic environments? Will fish take greater predation risks in low-productivity environments?

McDonald and Hershey (1992) also showed in Toolik Lake that as lake trout became smaller in body size over several years, they also were less effective predators on slimy sculpin (*Cottus cognatus*). As a result of the apparent release from predation pressure, the *Cottus* were able to forage with greater safety in soft sediments in which food was more abundant. As a result (McDonald and Hershey, 1992), the growth rate of sculpins increased. Thus, in this case history, a decrease in piscivory opened new foraging opportunities to the prey fish species, which responded by enhanced growth.

One example exists of predator-mediated shifts in thermal habitat relationships.

Bluegill in the presence of predatory largemouth bass showed a 3–4°C increase in upper avoidance temperatures in "shuttlebox" apparatus (Fischer et al., 1987). They predicted that ultimate consequences of modified thermoregulation by the bluegill might include lower conversion efficiency and reduced growth rate.

Mittlebach (1986) reviewed previous studies on foraging profitability versus habitat shifts by small fishes in the presence of predators and the effects of predators on intraspecific and interspecific competition among fishes. Responses of fish (and other prey taxa) to the presence of predators included behavioral movements to protected habitats (11 citations), increased vigilance or reducing foraging distances (3 citations), or limiting feeding time and intake (6 citations) (Mittlebach, 1986). Mittlebach pointed out that although it seemed logical that predator-mediated mutual use of sheltered habitats (e.g., vegetation) by small fishes could increase intraspecific competition, "no direct tests of this idea exist." He noted his earlier postulate (Mittlebach, 1981b, 1983) that predators like bass could "enforce habitat segregation between size classes and make open-water resources exclusively available to large fish." Mittlebach concluded with the caveat that predators, by causing changes in behavior of their prey species, have the potential to change many competitive interactions between and among fish species.

Gorman (1988b) suggested that in creeks, predation can be a major factor mediating interspecific interactions [(e.g., by causing more aggregation and schooling (including mixed-species schools) as an antipredator tactic]. There seems to be an important difference in the views of Mittlebach and of Gorman (for lakes and streams, respectively) on prey response to predators. Mittlebach implies prey "passivity"; that is, that small individuals (e.g., of bluegills) will be restricted to "safe" microhabitats (e.g., vegetation) by bass. Gorman implies, in contrast, that minnows in pools (where there may be no "safe" habitats, except at stream edge—where avian predators are a concern) will be "active" in their antipredator tactics (i.e., by aggregating to "challenge" the predation threat). However, Johannes (1993) offers a contrasting exception to the generalization I have suggested: He found that golden shiner minnows (*Notemigonus crysoleucas*) in lakes aggregated more at higher predator densities over a 7-year period of manipulation of predator densities in lakes. Johannes also showed an age-related aspect to the aggregations, with older golden shiners aggregating more in response to a predator than did young. Johannes concluded that the increased aggregation was likely an antipredator mechanism, increasing survival of prey (much like the "active" antipredator tactics suggested above by Gorman for stream minnows). Gelwick (1995) found in experimental streams that small green sunfish (*Lepomis cyanellus*) refuged passively within algae or under rocks in the presence of bass. Perhaps the appropriate generalization is that different taxa of prey (e.g., minnows versus sunfish) cope with predator threat differently (passively versus actively), rather than seeking difference between lentic and lotic systems. [Also, in many streams of the midwestern United States, individual pools may have many more of the physical characteristics of small ponds—slow currents, high structure, deep pool middle—and they are inhabited as much or more by centrarchids than by minnows (Matthews et al., 1994).]

Christensen and Persson (1993) showed that potential prey species used different antipredator behavior in different habitats. Persson (1993) also considered competition among prey species in refuges, from the perspective of resources available within the refuges. Persson showed that adult perch (*Perca fluviatilis*) restricted feeding by juvenile roach (*Rutilus rutilus*) to consumption of detritus and algae, whereas in the absence of the predator, they fed on zooplankton. As a result, juvenile roach in the presence of a predator grew less rapidly. Juvenile perch were less limited by predators, and their growth and habitat use was linked more to availability of their own invertebrate prey within the vegetation. Persson (1993) showed a complex interaction between juveniles of roach and perch, in that habitats with complex structure and invertebrates available on the structure provided a competitive refuge (from roach) for small perch, but that roach were better than perch at using the vegetation to evade predators. Persson concluded that "habitat dependent competitive abilities of juvenile perch and roach thus cause them to be affected differently when driven into prey refuges by predators." Thus, Persson showed that interpretation of predator effects on prey species can be very complex, and that the outcome can depend not only on the ability of the prey species to forage in the refuge but also on their ability to use a "leaky" refuge to evade the predator. Aquatic plants are a "leaky" refuge, not absolute, in that they do not completely preclude entry of a piscivore—thus, escape or evasion abilities within the refuge may also mediate outcomes of the piscivore × structural complexity effects on prey species.

Harvey (1991b) provided an additional example of the likelihood that cascading interactions involving piscivores could result in increased survival of fish larvae (and presumably in responses in population dynamics). In pools of a large stream (Baron Fork of Illinois River, Oklahoma, USA), Harvey showed, both by survey and by manipulation of piscivore densities, that predatory smallmouth bass (*Micropterus dolomieu*) created safe sites for fish larvae by discouraging the use of those pool habitats by intermediate-sized fish that eat larvae. Drift of fish larvae was also involved: more larvae accumulated in short-term experiments in bass-occupied pools within the main channel than in side pools. However, Harvey argued that this effect might be transitory, and that side pools, with minimal risk of "washout" and protection of larvae by resident bass, might actually be optimal nurseries.

Fragmentation of Populations by Predators?

In their Trinidad stream systems, Fraser et al. (1995) provided an interesting example of the potential of a piscivore to fragment local populations of small prey fish species. In two different streams, Fraser et al. have found sections in which natural barriers on the main channel result in a portion of the watershed with piscivores (*Hoplias*) present and others with them absent. In parts of watersheds lacking piscivores, small potential prey (*Rivulus harti*) expanded habitat use to include the main channel (e.g., fourth- and fifth-order streams). Where piscivores were present, the rivulines were confined more to tributaries of the main channel. However, in a relatively small-scale experiment on movement, Fraser et al. (1995) found an unexpected *increase* in movement of rivulines among experimental "tributaries" when the predator was present. They reasoned that in the experiment, the close proximity of available tributaries (2 m), coupled with predator stimulus to find a tributary once in the "main channel," may have resulted in more intertributary movement than likely in nature. They noted that in natural systems, many individual rivulines might move downstream into main channels, but if there was no predator, there was less impetus to again ascend into steep tributaries. Where a predator was present, they reasoned that many individuals descending from tributaries to a main channel could have been eaten, particularly if the actual distance between tributaries was substantial. Overall, they concluded that the presence of the predator in the mainstreams of the systems increased fragmentation of the rivuline population. This seems to be an ideal situation for the application of molecular techniques to evaluate actual levels of differentiation among rivulines of different tributaries.

In New Zealand, the introduction of piscivorous brown trout in the last century may have been responsible for the restriction of local species of galaxiids to headwaters stream, with apparent fragmentation of what was formerly continuous populations (although there are also many high-gradient barriers like waterfalls that may have naturally caused population fragmentation) (Townsend and Crowl, 1991; McIntosh et al., 1994). In this situation, there was little or no documentation of the status of the native species taxonomy or distribution before trout were introduced throughout the country, so the effects of the piscivore may never be known with certainty, but genetic (electrophoretic) assays of levels of population differences among galaxiids support the notion of population fragmentation by the piscivorous trout (Allibone and Wallis, 1993).

Quantifying Effects of Piscivory in Streams?

In addition to behavioral aspects of bass–prey interactions, it should be possible to make some reasonable calculations of the potential for predation by bass in a stream like the Baron Fork to alter the compostion or standing crop of minnow species. Table 16.1 of Matthews et al. (1987) provides a large data set on the numbers of stonerollers versus the numbers of smallmouth bass considered sufficiently large to eat adult stonerollers. Baron Fork was the most predator-saturated stream of the several that I surveyed in the Arkansas–Oklahoma uplands during 2 years. From my field notes, the approximate size of each pool is also available. How many individual prey (e.g., minnows) might be removed by smallmouth bass from the reaches of upland streams I studied by snorkeling for the Matthews et al. (1987) article? Limited experimental evidence that exists (Lewis and Helms, 1964) suggests that smallmouth bass are likely to eat no more than one minnow per day (particularly if alternative prey such as crayfish are available; crayfish are abundant in the upland streams we studied). Note, however, that in the confines of experimental stream channels, smallmouth bass ate an average of 3.8 minnows per day (Schlosser, 1988a). My snorkeling was in July 1984 and in September 1983, covering a total of about 1 km of river (both samples combined, for purposes of calculation).

In approximate numbers, there were present in 23 pools (some surveyed twice, but all surveys included herein) a total of about 160 bass large enough to eat minnows, and 25,000 large juvenile or adult *Campostoma*. Although not all other minnows were counted by me, a conservative estimate makes the *Campostoma* not much more abundant than all other minnows combined, so let us include (conservatively) 15,000 minnows of other species (e.g., *Luxilus cardinalis, Notropis boops,* and *Notropis nubilus*, all of which are common and abundant in that stream reach), for an estimated total of 40,000 minnows in that reach of Baron Fork. Assume also that water exceeds 15°C in all but three winter months (USGS records), and that smallmouth bass do not feed below 15°C (Carlander, 1977). Finally, assume no immigration and that the numbers of minnows are all that will be available to survive to reproduce next spring. Take an average snorkel survey date of August and an average reproduction date of May for all minnows included [actually, *Campostoma* spawn early (e.g., April); others spawn in late May or June, but an average of May is probably near correct].

Based on these assumptions, there should be about 150 days from August to the next reproduction with water sufficiently warm for smallmouth bass to eat. If the 160 smallmouth bass eat 1 minnow each per day, this results in removal of 160 × 150 = 24,000 minnows in the stream reach via bass predation. From our late summer average date total of 40,000 minnows, remove 24,000 killed by predation. Even though it may take only a relatively small number of breeding adults to repopulate in the next spring, these calculations suggest that piscivory by bass could exert a substantial control on numbers of minnows in this reach of Baron Fork, which, coupled with other sources of mortality (disease, winter stress, avian predation—which we have observed) could have a strong role in dynamics of minnow populations in the reach. Finally, if one uses Schlosser's (1988a) determination of smallmouth bass eating an average of 3.8 minnows per day, the total potential removal of fish by bass predation would equal more than 91,000 individuals, a number not possible from the initial standing crop described above. However, even if the truth lies between a low estimate of 1 and a high estimate of 3.8 minnows per day, one would conclude that smallmouth bass can eat a lot of small fishes from the end of one reproductive period to the beginning of the next in Baron Fork.

Similar calculations can be made for minnows in Brier Creek, Oklahoma, in the 1 km reach of 14 pools that I surveyed regularly by snorkeling (Matthews et al., 1994). There was an average of 34.1 bass and 1529 "minnows" (including some noncyprinids like *Fundulus notatus* that are "minnow-sized" prey) per pool. Perhaps more of interest, in September 1993 (i.e., after recruitment of most young-of-year to juvenile stage or small adulthood), there were 42 bass > 100 mm (i.e., with potential to eat fish) and 943 "minnows." Calculations similar to those for Baron Fork, assuming about 120 days of temperatures suitable for fish activity from September to April–May (spawning for these fishes), and again assuming eating of 1 minnow per day per bass, results in hypothetical removal of $120 \times 34 = 4080$ minnows to be eaten in this reach of Brier Creek from the time of observation (5 September) to spawning in the next spring, which is obviously impossible (with fewer than 1000 minnows available). Even if all other small fishes are included as potential prey for bass (i.e., 241 sunfish, *Lepomis*, < 75 mm in total length), it is obvious that the predation pressure, at least with the standing crops in Brier Creek in 1983, could be relatively more intense in Brier Creek than in Baron Fork in terms of the numbers of minnows "to be eaten" relative to the number actually available.

In a previous year in Brier Creek, there were more available "minnows" per predator than in 1983. In mid-November 1982 there were an average of 1938 minnows in the 14 pools and 30 bass. Assuming only 60 days of temperatures warm enough for active bass predation until spawning in the next spring results in $30 \times 60 = 1800$ minnows to be eaten, relative to approximately 1900 available in the stream reach. Hypothetically, at least, some minnows could survive this rate of predation, but not many. Regardless, it seems likely that Brier Creek minnows might be more wary, take more rapid evasive action, or avoid bass more (e.g., by interpool movement) than do minnows in the Baron Fork, where numbers of individual minnows could be large enough to sustain the effects of bass as predators. Similarly, it would be interesting to know what bass do eat in Brier Creek, in that we obviously have not collected these important predators from our study stream in order to do gut analyses. However, I suspect that a large portion of the food of the bass (and that of other centrarchids) in these pools comes from input of terrestrial insects and so forth, and that they do not subsist primarily on fish. In fact, most minnows present in November 1982 apparently survived the winter. In November 1982 (2 different surveys) there were 2054 and 1822 total minnows in the 14 pools. In the first survey next spring (March 1993), I recorded 1809 minnows—nearly as many as the previous autumn. Apparently from November to the next May, there was little attrition of the total minnow standing crop, and predation seems not to have removed nearly the numbers of fish per day that the logic above suggested that they might have. Clearly, the above calculations are fraught with possibilities for errors (e.g., overestimate or underestimate of predation rates, lack of knowledge about mortality from other sources, immigration of "new" minnows into pools, replacing any lost to predation), but they may be sufficient to serve as a basis for the prediction that predation in streams may be as substantial a controlling factor as it is in closed systems like lakes.

Growth as a Survival Strategy?

A different strategy of small-bodied fish species for survival of predation is suggested by another life-history trait of *Campostoma* from Brier Creek versus Baron Fork. As indicated above, the primary predators in Brier Creek are largemouth and spotted bass (*Micropterus punctulatus*), whereas the primary piscivore in Baron Fork is smallmouth bass. As the name implies, smallmouth bass have a relatively smaller mouth (gape width; jaw length) than either largemouth of spotted bass of comparable body size. It would appear impossible for minnows in Brier Creek to achieve a body size sufficiently large to provide a "size refuge" from predation by largemouth or spotted bass; that is, the larger individual bass in Brier Creek can readily eat even the largest *Campostoma* that are known. In contrast, *Campostoma* achieving large body size in Baron Fork might have a better chance of finding a size refuge from smallmouth bass predation; that is, they could become too large for a smallmouth bass to readily engulf (cf. Werner et al., 1983a). The *Campostoma* in Baron Fork are, on the whole, much larger in body size (total length and circumferential girth) than conspecifics in Brier Creek.

It might be a logical postulate that in Brier Creek, where largemouth or spotted bass are the primary predators, *Campostoma* have evolutionarily abandoned any strategy for fast growth to a size refuge, in that achieving such a size is apparently not possible. In contrast, rapid growth to large body size could make good evolutionary sense for a *Campostoma* in Baron Fork, in that it should be possible to reach a size at which predation from smallmouth bass is unlikely. However, by becoming large enough to lessen threats from piscivorous fishes, a fish like *Campostoma* might become a more attractive target for avian predators (Power, 1987), such as several species of herons, which are common throughout our study reach of the Baron Fork. Predation should have strong potential to be a selective force on body size, or reproductive tactics of these fishes.

10.5 Conclusions: Effects of Piscivory in Fish Assemblages

A large number of studies have evaluated the effects of piscivores on survival, habitat use, behaviors, or success of prey species. Many of these studies have been based on field sampling or observation of population responses. Many others have included experiments in seminatural or artificial stream systems, or in small lakes or ponds. The results of experiments and a considerable body of theory all indicate that predators, especially piscivorous fishes, can have a strong impact on individual prey species or their behavior, in both ecological and evolutionary time. Thus, the effects of particular predators on particular prey are well documented for specific species or systems.

What seems to be lacking is a larger view of the effects of piscivores on whole assemblages of local fishes. There are some field correlations between the presence of piscivores and the composition of local assemblages or the number of species present [e.g., Matthews et al. (1994), and Chapter 2]. However, there are few studies in which predator effects have been carefully documented for an array of predators (often three to five species present in natural systems; Chapter 2) on a larger array of prey species, from the perspective of strengths of predation in an entire food web (e.g., Chapter 11). Studies summarized in Chapter 11 do show evidence of trophic cascades, in which a top predator decreases the abundance of the next trophic level, but in those studies, it is also rare that *all* predators or prey species that are present are included in an analysis [but see Winemiller (1995)]. There is also limited focus, in studies of predator–prey systems, on the interactions of rare predator or rare prey species. There is theoretical prediction, and some empirical evidence, for predator "switching" (Murdoch et al., 1975) as prey items become too rare. However, in most studies, logistical and conceptual constraints have resulted in a focus on common or abundant predators or prey.

It would be nice to see a body of work develop in which a marriage of empirical and theoretical work incorporated predation strength functions among all members and size classes (including cannibalism and reciprocal predation between some species) of a local assemblage. This should ideally be based on field work in real systems, but might be augmented by trials in experimental systems in which presence–absence or density of a variety of predator and prey species is manipulated to determine which interactions are "strong effects" or "keystone" effects, *sensu* Paine (1969, 1980). What will happen to the dynamics of an entire fish assemblage when one (out of several) piscivore or prey species is deleted? This approach should, ideally, not be a "black box," but designed to include factors like ecomorphological traits or "functional morphology" of predators and prey, to ask how much of the strength of predation might be predicted from comparing theoretical abilities of predator and prey. Empirical and theoretical assessments of predation strengths and linkages in natural fish assemblages would seem to provide the kinds or rate predictions needed in theoretical food web assessments (e.g., Polis and Winemiller, 1995), and also yield practical information needed to assess outcomes of management decisions on the well-being of whole systems. Finally, it would seem ideal to develop a body of research on the effects of predation in combination with work on competition effects (Chapter 9) to provide a comprehensive view of the effects of biotic interactions in local fish assemblages. At present, the available knowledge on effects of predators remains based mostly on experiments with one or a few piscivore and one or a few prey species, and rather little on whole-assemblage patterns.

11

Effects of Fish in Ecosystems

> *Also the possibility should be mentioned that certain fish species may alter directly the environment.*
> —Nilsson (1967)

> *Following a long history of neglect, fishes are now commonly studied by limnologists.*
> —Winfield (1991)

11.1 Introduction

As early as the 1800s, Forbes recognized that some minnows were "mud eaters." Mud eaters ate material from stream or lake bottoms and processed it through their gut. There is also evidence by the late 1800s of concern about the role that imported carp (*Cyprinus carpio*) might play in disturbing sediments or increasing turbidity in streams and lakes of North America. However, through much of the mid-20th-century, the role of fish in general was neglected in general publications and in many professional journals in limnology and stream ecology [see review of the history of "fish in limnology" in Northcote (1988)]. In fact, one could read most of the literature in limnology in the middle decades of this century without being aware that limnologists thought it would matter in the ecology of a lake whether or not fish were present. Fishes simply were not a major part of the thought process during this period in limnology or stream ecology.

All the above is in sharp contrast to views now prevalent in basic (Brooks and Dodson, 1965; Wetzel, 1995) and applied (Kitchell, 1992) limnology, and in stream ecology (Power, 1990b, 1992a), in which fish are regarded as "strong interactors," "keystone species," or essential "top-down" or intermediate players of importance in food webs. Northcote (1988), in a comprehensive review of the effects of fish in freshwater ecosystems, listed 197 references which explicitly suggest in their titles the role of fishes or "vertebrate predators" (mostly fish) in streams, ponds, or lakes. Dr. Ray Drenner (Texas Christian University, personal communication) has compiled a bibliography of more than a 1500 references in which the effects of fish in freshwater ecosystems are considered.

The impetus to renew thinking about fish in aquatic ecosystems came with publications by Hrbacek et al. (1961) and Brooks and Dodson (1965) which showed that the presence or absence of fish in lentic systems could have a major influence on the size structure of the zooplankton assemblages, and the posing of the size-selective hypothesis by Brooks and Dodson (1965) explaining why large zooplankters dominated in fishless or low-predation systems. Large- or medium-sized fish tended, both by visual predation and particulate filter feeding, to remove the larger zooplankton, setting the stage for cascading effects on phytoplankton or water quality. Brooks and Dodson examined a suite of lakes in New England (USA), finding a distinct difference in size–frequency distribution of zooplankton in lakes with or without planktivorous fishes.

In a parallel history, stream ecologists gave short shrift to the role of fishes as well. There are again some exceptions to this. Juday et al. (1932) documented the importance of the death and decay of salmonids after spawning runs in streams in Alaska and measured the increases of the nutrients that were provided by input from the carcasses. In 1972, C.A.S. Hall (1972) published an often overlooked article (because of a cryptic title), in which he examined the role of suckers (catostomids) in transporting nutrients up and down large streams in the eastern seaboard of North America. However, it is relatively apparent that even as late as 1980, in the River Continuum Concept (RCC; Vannote et al., 1980), there was relatively little attention paid to the functional roles of fishes in stream ecosystems. Although the presence of different kinds of fish in different stream reaches was noted in the RCC, the influence of fish on the way a stream ecosystem functioned was not chronicled in any detail as recently as 1980. In reviewing the literature, it is probably safe to say that the major thought of fish as playing roles in lakes until 1965 (Brooks and Dodson) or in streams as recently as 1980 revolved around their potential as contributors of nutrients, such as Juday's study of nutrient inputs from the death and decay of fish carcasses. The first indications that fish could be strong functional interactors in stream systems by virtue of their trophic activity came for tropical streams in Power (1983a, 1983b, 1984b), and for temperate streams in Power and Matthews (1983) and Power et al. (1985).

In this chapter I first review the direct roles of fish in lakes and streams with respect to (1) direct predation on macroinvertebrates and zooplankton, (2) the role of herbivorous fishes (both those feeding on attached algae and on macrophytes), (3) the role of detritivorous fishes, (4) the role of fishes as bioengineers or bioturbators (i.e., in modifying the physical structure of ecosystems), and (5) their roles in inputs of nutrients or changing movements of particulate carbon or nutrients within stream or lake systems.

I also consider in Chapter 11 the indirect effects of fishes, as in predations of one fish—a large fish on a smaller fish—and the effect of that predation on stream invertebrates. I address second-order indirect interactions and trophic cascades set in motion by direct predation effect by fishes or activities that change their environment. Finally, all of the effects of fish in ecosystems, both direct and indirect, are combined in the application of fish manipulations to modification of whole lake ecosystems or food webs. Overall, this chapter will point out ways in which or the degree to which the underwater landscape, in both lakes and streams, would be a very different if fish were not present in the ecosystem.

11.2 Direct Effects of Fish in Ecosystems

There are many pathways by which fish have direct effects on structure or function of stream and lake ecosystems. Figure 11.1 depicts in bold outline the components from Figure 1.4 that have been shown to be directly affected by fishes. Note that in this case, the arrows flow *outward* from the "local assemblage," and in some cases, they do continue outward to a second trophic level (suggesting "indirect effects" discussed in Sections 11.6 and 11.7).

Effects of Fish on Abundance of Macroinvertebrates

Fish eat a lot of macroinvertebrates. A negative correlation has been shown between the densities of fish in streams and the size of stream insects (Hildrew et al., 1984). Some kinds of stream fish specialize on chironomids or other insect larvae (Matthews et al., 1992b; Fuller and Hynes, 1987; Todd and Stewart, 1985). Others include a wide range of macroinvertebrates in their diet and can be opportunistic (Matthews et al., 1978); others consume or specialize on mollusks or other hard-bodied prey (Robinson and Wellborn, 1988; Wainwright and Lauder, 1992). The feeding repertoires of freshwater fish, worldwide, include many macroinvertebrates (Gerking, 1994). Fish can be selective visual predators of macroinvertebrates (Hershey and Dodson, 1985), with an array of different search patterns that include complicated components of prey location and attack (e.g., Vogt and Coon, 1990), and active feeding on drifting invertebrates (Allan, 1981; Rincón and Lobón-Cerviá, 1995). Fish as predators of macroinvertebrates may function synergistically with predatory invertebrates, changing the prey consumption rate of both (Soluk and Collins, 1988a). Allan (1983) showed that trout consumed about twice as much benthic invertebrate prey as did carnivorous stoneflies. Strictly speaking, the eating of even a single aquatic insect larva changes the ecosystem—by killing that individual and modifying, ever so slightly, the transfer of energy in the food web. However, the first question addressed here is whether or not fish in streams or lakes change the density or structure of their macroinvertebrate prey assemblages. The effects of fish on invertebrates in streams is one of the oldest controversies in stream ecology. The first major estimate of amounts of macroinvertebrates consumed relative to amounts produced resulted in the now-famous "Allen's Paradox" (Allen, 1951), in which K. R. Allen calculated for a stream in New Zealand that trout ate, annually, from 40 to as much as 150 times as many invertebrates as estimates of macroinvertebrate production for the stream could provide, yet the stream transported excess macroinvertebrates as drift. Waters (1988) summarized that for studies of trout streams, worldwide,

568 / Patterns in Freshwater Fish Ecology

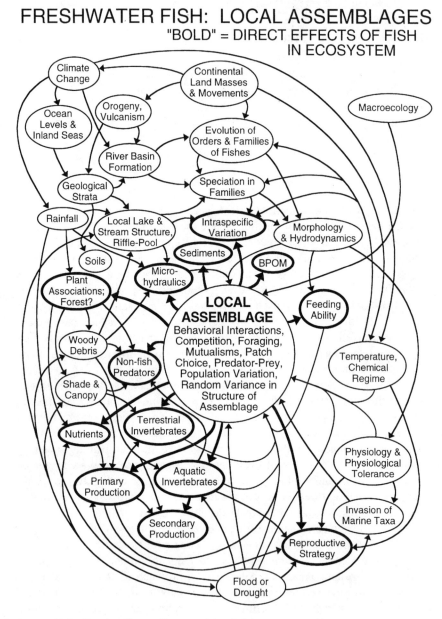

Figure 11.1 Known direct effects (bold) of fish on their ecosystems. Factors as in Figure 1.2.

only two showed sufficient production of invertebrates to support the apparent consumption by trout. Huryn (1996) reviewed all of the historical evidence for this problem and was able to "balance the budget" for a New Zealand trout stream only by including all sources of prey for trout (including hyporheic and surface invertebrates, terrestrial invertebrates, and cannibalism by trout). Huryn concluded that trout may consume, annually, 80% or more of the benthic macroinvertebrate production.

The problem of Allen's Paradox was made even more tantalizing for stream ecologists when J. D. Allan (1982) showed that in long sections of a Colorado (USA) trout stream, removal of most trout for up to 4 years resulted in no consistent pattern of change in abundance of benthic or drifting invertebrates. Allan (1982) summarized numerous earlier studies of the effect of trout or other fish on invertebrate populations, concluding that in several such studies, there was little effect, but suggested that the putative predator in those studies may have been a weak interactor [*sensu* Paine (1980)]. Allan (1982) suggested that either (1) trout might eat only a small fraction of the potential prey [unlikely, now given the findings of Huryn (1996)] or (2) that prey taxa might be highly adapted to coexist with trout predation and thus not be sensitive to the particular manipulation. Allan (1982) and others have also suggested that drift of invertebrates could ameliorate the effects of fishes on the density of their prey in any given reach of a stream, replacing those consumed or removing any surplus produced in manipulated absence of trout, and Hildrew (1990) argued that it will be difficult to detect reduction in prey by fish if macroinvertebrates are free (by experimental design) to move in and out of cages holding fish; that is, immigration from elsewhere in the streambed replacing those lost to predation.

Since Allan's (1982) now famous experiment, many studies have tested the effects of stream fishes or littoral fishes in lakes on standing crops of macroinvertebrates, most showing some effect. Gerking (1994) devoted an entire chapter to this issue and concluded that in some situations (or under certain experimental conditions), fish lower the densities of benthic invertebrates, but in others they do not. Gerking (1994) emphasized that the experimental conditions selected by the investigator can influence results, and the need for standardized protocols for comparative studies. Gerking noted consensus by many authors (e.g., Reice and Edwards, 1986) that trout may be poor predators on benthic invertebrates (unable to forage deep into the substrate), hence not "typical" of stream fishes in general. This is supported by the finding (Flecker, 1984) that sculpins (*Cottus bairdi* and *Cottus girardi*), strongly benthic species, lowered the densities of benthic invertebrates in an upland stream and the finding by Bechara et al. (1993) that trout are effective predators of epibenthic and drifting invertebrates, but have little effect on densities of infauna (invertebrates living within stream substrates). Gerking (1994) accepted that fish can be size-selective predators and can alter the size structure of prey populations, and that fish can have differential effects on different taxa of benthic invertebrates, depending on the life history or behavior of the prey. He summarized that invertivorous fishes could affect benthic invertebrates by (1) eliminating prey species, (2) changing prey-size distribution, (3) changing prey abundance or distributions, or (4) all of the above (Gerking, 1994, pp. 255–256).

Recent studies in lotic and lentic environments tend to support the idea that fish can have substantial effects on the total density of macroinvertebrates or at least on certain taxa (Table 11.1). Allan (1982), Flecker and Allan (1984), Reice (1983), and Reice and Edwards (1986) showed little response of stream invertebrates to fish predation. Allan (1982) involved large reaches of a stream, the other three were in small to large cages. In all other studies I have surveyed, the investigators reported at least moderate, and sometimes very strong, effects of fishes on stream invertebrates (Table 11.1), in designs ranging from small to large inclusion or exclusion units. On balance, of 20 cases taken from the literature for streams (Table 11.1), 16 show some reduction in macroinvertebrates. In some cases, total invertebrate abundance or biomass was depressed in the presence of fishes, and in other instances, only certain invertebrate taxa decreased, but at least modest effects have been widely detected. In general, it also seems that a large number of small fish might have more impact on benthic or drift invertebrates than would a lesser number of larger fish.

Therefore, it is desirable to consider the presence and abundance of all the fishes in a stream, beginning with the headwaters if one is to truly access the withdrawal of materials from a stream by the fish and the degree to which they might actually depress numbers of aquatic invertebrates in the stream relative to that if the stream were truly fishless. As far as I know, this has not been calculated across an entire local fish assemblage.

Do stream fish change the size structure to benthic aquatic invertebrates in streams, as planktivorous fish change the size of plankton in lakes? Most research on fish effects in streams has merely asked if fish change the number or assemblage structure of benthic macroinvertebrates. It would be desirable to know if fish directly influence the size of aquatic invertebrates that prevail in streams. If one assumes that most fishes like *Notropis boops, Luxilus pilsbryi, Notropis rubellus,* and other invertivores, are sight-feeding fishes, they should attack the items they can see best, particularly if a stream is turbulent or mildly turbid.

Finally, some fish are primarily molluscivores, particularly in lakes and, to a lesser degree, in streams. The black carp (*Mylopharyngodon piceus*), native to the Amur River basin of Asia, eats large numbers of mollusks up to the size at which gape limitation is reached, crushing the shells with powerful phayrngeal teeth (Shelton et al., 1995). In North America, the freshwater drum (*Aplodinotus grunniens*), redear sunfish (*Lepomis microlophus*), pumpkinseed (*Lepomis gibbosus*), copper redhorse (*Moxostoma hubbsi*), and river redhorse (*Moxostoma carinatum*) eat mollusks, crushing their shells with molariform pharyngeal teeth or pharyngeal teeth and dorsal chewing pads, and roach (*Rutilus rutilus*) of Europe have similar capabilities (French, 1993). Thus, this is another entire area in which

Table 11.1. *The Effect of Fishes on Macroinvertebrate Prey in Lotic (Streams) and Lentic (Lakes, Ponds, Wetlands) Systems.*

Author (Year); Predator	Change in Prey Density?			Effect of Fish & Comments
	Down	Up	None	
Streams				
Allan (1982); trout			X	Some groups decrease; but no pattern.
Flecker (1984); sculpins[a]	X			Chironomids decreased; no significant trend in total biomass.
Flecker & Allan (1984); all fish			X	Cages open to all native upland stream fishes.
Hemphill & Cooper (1984); trout	X			Decreased conspicuous prey.
Reice (1983); "darters & dace"			X	No effect in small exclusion cages.
Reice & Edwards (1986); trout			X	Only very minor effects of trout fry in cages.
Lancaster et al. (1988); trout	X			Decreased caddisfly larvae in summer.
Schofield et al. (1988); trout	X			Reduced caddisfly larvae.
Koetsier (1989); sculpin	X			Total density and some indiv. taxa; seasonal effect.
Schlosser & Ebel (1989); minnows	X			More effect in pools, less in riffles.
Gilliam et al. (1989); creek chub	X			Reduced total volume & numbers, and two taxa.
Harvey & Hill (1991); creek chub	X			Reduced biomass, not total numbers.
Power (1990); small fish fry	X			With damselfly nymphs, lowered midge colonization.
Holomuzki & Stevenson (1992); sunfish	X			"Modest" effects on two taxa; change in functional groups.
Flecker (1992a); characin species	X			Weak reduction overall; change in mayflies & caddis; grazing fish had more effect on inverts!
Andersen et al. (1993); trout	X			Decreased *Gammarus*.
Wiseman et al. (1993); trout	X			Decreased damselfly.
Dudgeon et al. (1993); loach	X			Lowered total abundance, esp. chironomid & mayfly.
Bechara et al. (1993); trout	X			Lowered density of epibenthic and drift inverts, but not infaunal taxa.
Flecker (1994); trout, galaxiid	X			Introduced trout lowered invertebrates more than did native galaxiids.

continued

Table 11.1. Continued.

Author (Year); Predator	Change in Prey Density?			Effect of Fish & Comments
	Down	Up	None	
Lentic systems: lakes, ponds and wetlands				
Thorp & Bergey (1981); "fish & turtles"			X	Littoral zone of cooling reservoir.
Crowder & Cooper (1982); sunfish	X			Reduced total benthic biomass.
Bohanan & Johnson (1983); sunfish	X			Midge larvae decreased; other taxa no response.
Morin (1984a); sunfish and bass	X			Total abundance, larval odonates.
Morin (1984b); sunfish and bass	X	X		Decreased total odonates; increased one species.
Post & Cucin (1984); yellow perch	X			Decrease total benthic biomass.
Gilinsky (1984); sunfish	X	X		Decrease in chironomids; others = no response or seasonal increase.
Mittelbach (1988); sunfish	X			Decrease invertebrate size.
Luecke (1990); trout	X			Decreased inverts offshore; not in littoral.
McPeek (1990); lakes w&wo/fish	X	X		Fish increase one damselfly group; decreased another.
Wellborn & Robinson (1991); sunfish	X			Decreased Trichoptera & amphipods, not chironomids.
Goyke & Hershey (1992); sculpin, burbot, trout	X			Decreased chironomids.
Hanson & Riggs (1995); fathead minnows	X			Decreased insects & crustaceans (wetland).

[a] And other native fish, mostly *Rhinichthys* in an "open-cage" control.

freshwater fish, both in streams and lakes, may have substantial direct effects on the biota, by killing, crushing, and eating of mussels. Because dense beds of mussels process huge volumes of water in their siphon/filtering feeding, changes in mussels due to fish predation could have strong indirect effects in stream or lake ecosystems.

Conversely, freshwater mussels depend on fishes as hosts for their glochidia, and many mussels have elaborate lures (C. Barnhardt, personal communication) to assure that a proper fish host will take glochida into the mouth or onto the gills and fins. Because many freshwater mussels are host-specific or use only a limited range of fish hosts, the distributions of mussels may be intimately linked to and influenced by the distribution of fishes. Information on the entire evolution of fish and mussels is scanty, but there appears to be strong potential for coevolutionary factors to have a strong effect on either or both participants in the host–parasite relationships (C. Vaughn, personal communication).

In lentic waters (mostly the littoral zone of ponds or lakes, and in wetlands), most studies have shown general decreases of taxa or in abundance of some macroinvertebrate taxa in the presence of fish. Thorp and Bergey (1981) appear to be unique in this respect [per Gerking (1994)], as they found "no effect." In the other 12 studies (Table 11.1) the presence of fish was associated with a reduction in at least one group of macroinvertebrates. In three of the lentic studies, there was a significant reduction in the total abundance or biomass of invertebrates. There also was an increase in at least one invertebrate taxon in the presence of fish in three studies, and complicated fish–fish interactions that suppressed effectiveness of one as a predator had an effect on the kinds of invertebrates that decreased, increased, or remained unchanged (e.g., Goyke and Hershey, 1992). Thus, in spite of the early negative finding of Thorp and Bergey (1981), it appears that, overall, fish as predators of macroinvertebrates have the potential to control all or part of the assemblage in littoral zones or other lentic habitats. Fish may differentially change the abundance or size of some, but not all invertebrate taxa (Crowder and Cooper, 1982; Mittlebach, 1988) alter species composition in invertebrate assemblages, or change size structure (Post and Cucin, 1984), and exacerbate patchiness in distribution of benthic invertebrates (Collins, 1989).

Effects of Fish on Life History or Behavior of Macroinvertebrates

Much like piscivores can alter the life history of smaller fish species (Section 8.3; Reznick and Endler, 1982; Reznick et al., 1990), fishes in streams or lakes may alter the life history or behavior of their invertebrate prey (Allan, 1982; Hildrew, 1990; Wiseman et al., 1993). Wellborn (1994, 1995a, 1995b) found that amphipods exposed to contrasting forms of size-selective predation showed life-history responses similar to those predicted by general life-history theory and paralleling the changes shown by Reznick and Endler (1982) for fishes. For amphipods in lakes with centrarchids, size-selective predation by the centrarchids increased the mortality of adults at larger body size; in fishless habitats, the mortality of juveniles exceeded that of adults. Amphipods in environments where larger individuals are most likely to be eaten by fish mature at smaller body sizes and have smaller adult body size, with small eggs and higher size-specific fecundity, with the traits genetically based. Amphipods in a nearby fishless environment had larger adult body size, later reproduction, and fewer, larger offspring. A broader survey of various habitats in Michigan (USA) confirmed this basic pattern (Wellborn, 1995b), with amphipods in fishless habitats or habitats with only sticklebacks as predators differing in fundamental life-history traits from those in habitats with centrarchids (which are more likely to consume adult amphipods). Thus, it appears that size-selective fish predation may be causing or maintaining life-history differences between the amphipod morphotypes. Scrimgeour and Culp (1994b) also showed life-history trade-offs by mayflies when they were reared with and without fish predators, changing larval growth, time to emergence, adult size, fecundity, and egg size.

Numerous articles have shown that macroinvertebrates change behavior in the presence of predation threat, including threats from fishes (Peckarsky, 1984; Feltmate et al., 1986; Culp et al., 1991), and that detection of threat by fishes can involve at least olfactory and hydrodynamic stimuli (Scrimgeour et al., 1994a). Size-dependent risk from fish can constrain macroinvertebrates to drift at night (Allan, 1978), with the basis of the behavior apparently fixed by selection (Flecker, 1992). However, responses of invertebrates to fishes can be complicated, and at least one study has suggested that large macroinvertebrates may *not* respond to the presence of fish (sculpins) because of a potential inability to detect the predators (Soluk and Collins, 1986b). Furthermore, the presence of more than one kind of predator can cause changes in the functional response of a fish predator to prey (Soluk, 1993). An isopod changed activity in the presence of green sunfish (*Lepomis cyanellus*), decreasing overall activity and spending more time in shelter (Huang and Sih, 1990). Culp and Scrimgeour (1993) showed a size-dependent impact of native stream fish on baetid mayfly larvae, causing large but not small baetids to forage more at night. Additionally, baetids traded risk for food rewards in the presence of fish predators (Scrimgeour and Culp, 1994a), and large baetids took more risk to feed than did small individuals, linked to a lesser fish threat to large baetids (Scrimgeour et al., 1994b). Finally, in assessing the response of macroinvertebrates to the threat of fish predation, it is important to know the life history or behavior of the potential prey: Macroinvertebrates can only exercise predator-avoidance options within the range of possibilities dictated by their life-style. For example, Baker and Ball (1995) showed that tube-building chironomids could not escape areas with fish because the necessity of building tubes limited the range of microhabitats they could use.

Fish Effects on Large Macroinvertebrates: Crayfish

Since Stein and Magnuson (1976), Stein (1977), it has been widely recognized that predatory fish can change behavior of crayfish, restricting their foraging and other activities. Fish predation can influence outcome of interactions between crayfish species (Mather and Stein, 1993). Resetarits (1991) showed experimentally that fish could decrease the growth of crayfish by causing them to spend more time in shelters and not foraging. DiDonato and Lodge (1993) and Garvey et al. (1994) showed that predatory fish have an important influence on the replacement of vulnerable native crayfish by non-native, less vulnerable species.

Effects of Fish on Vertebrates: Frogs and Salamanders

The effect of predation by fish on vertebrates is not nearly so pervasive in aquatic systems as are fish effects on invertebrates. However, fish can influence distribution of amphibians (Brönmaak and Edenhamn, 1994). Predaceous fish can lower abundances or outright prevent use of some habitats (e.g., small headwater stream pools or ponds) by amphibian larvae (e.g., Kats et al., 1988). Aquatic salamanders or anuran larvae can be important grazers or detritivores, responsible for substantial transfer of energy in a food web. Alternatively, some aquatic salamanders are themselves quite predaceous on invertebrates. Thus, fish lowering the abundances of these vertebrates have the potential to influence the ecosystems where they might otherwise be abundant.

For example, Sexton and Phillips (1986) found that invasion of 6 fish species lowered the number of amphibian species using 1 pond from 11 to 2 species, but that in another pond, invasions of minnows had little effect on amphibians. Sexton and Phillips (1986) considered that green sunfish (*Lepomis cyanellus*) were most responsible for demise of the amphibian assemblage. Fish can have negative effects on tadpole populations by direct predation (Woodward, 1983; Semlitsch and Gibbons, 1988), or through indirect effects (Hopey and Petranka, 1994). Fish in ponds in Sweden markedly reduced or eliminated the tree frog *Hyla arborea*. Bradford et al. (1993) showed that salmonids stocked into streams and lakes in California not only eliminated *Rana mucosa* (mountain yellow-legged frog) from streams and lakes but is now responsible for maintaining more fragmentation of the remaining populations than existed before fish additions.

Fish can also be detrimental to aquatic salamanders, via direct predation on eggs or larvae (Semlitsch, 1988), by interfering with their egg-laying (Sexton et al., 1994), or by increasing their sheltering at the expense of growth (Resetarits, 1991). Petranka (1983) showed that larval *Ambystoma texanum* in upland streams are restricted to fishless upper stream segments, and that the larvae are highly susceptible to predation by many of the native fish species. Stangel and Semlitsch (1986) and Semlitsch (1987) found that bluegill sunfish (*Lepomis macrochirus*) reduced body sizes, changed behavior, altered vertical migrations, and decreased survival of *Ambystoma* larvae by up to 97% in ponds, and Sih et al. (1992) showed that green sunfish (*Lepomis cyanellus*) increased the reduction in numbers of larval salamanders in stream pools in spite of substantial drift of larvae into and through pools. Resetarits (1995) showed that fingerling brook trout (*Salvelinus fontinalis*) reduced the survival of larval salamanders in experimental streams, and sharply decreased the growth of the salamanders.

11.3 Planktivory

Effects of Fish on Zooplankton

In lakes, many fishes feed primarily on zooplankton and to some degree on the larvae and nymphs of aquatic invertebrate such as mayflies or the pupae of invertebrates as they rise to the surface (Keast, 1985c, 1985d, 1988). Most of the effect of planktivorous fish on invertebrates in lakes probably comes directly from the feeding of both large and small fishes on zooplankton (although fish can also release or transport nutrients, potentially influencing primary and secondary production). One of the most abundant planktivorous fish in large southern reservoirs is the gizzard shad (*Dorosoma cepedianum*). Gizzard shad are pump filter feeders and do not visually feed as large adults (R. Drenner, personal communication). In northern lakes, yellow perch (*Perca flavescens*), cisco (*Coregonus artedii*) (Luecke et al., 1992), various centrarchids, and alewife (*Alosa pseudoharengus*) can be important, as visual, particle-feeding planktivores or (the latter two) as filtering planktivores. Many small-bodied fishes (and juveniles of other large-bodied species) also are visual particulate feeders on plankton, such as various minnows, atherinids, or *Gambusia*. Very small fish (e.g., young-of-year or larvae) may be gape-limited on large zooplankton and thus take smaller particles, but most fish selectively remove larger zooplankton from the water column.

There are many studies dealing with how and why fishes detect their prey. John O'Brien and collaborators have studied the ways in which fish detect plankton prey and how they choose to attack or pass up a particular particle during foraging (Luecke and O'Brien, 1981; Wright and O'Brien, 1982; O'Brien et al., 1985; Evans and O'Brien, 1986; O'Brien et al., 1986). Most prey items taken by fish in standing water are those that are most visible (Hessen, 1985), either by their size, coloration, or contrast with their surroundings. As a result of their visibility, or retention in filtering structures, larger items tend to be eaten. Thus, zooplankton size structure differs in lakes with and without fish, with smaller zooplankton dominating where fish predation is intense. Overall, fish have pervasive effects on zooplankton in many systems (Northcote, 1988). (However, this may be less so in large southern reservoirs; see below.)

Many fish species are planktivorous as adults, and virtually all freshwater fish pass through a plankton-feeding stage during ontogeny (Winfield, 1991). Fish can consume a large amount of zooplankton production, and Rand et al. (1995) calculated removal of zooplankton in Lake Ontario to exceed the annual production by a factor of about 1.7, in a fashion analogous to Allen's Paradox for fish in streams. Hewett and Stewart (1989) estimated that alewife (*Alosa pseudoharengus*) in Lake Michigan could consume as much as 20% of the zooplankton biomass in the lake per day, and Evans (1986) considered that *Alosa* predation probably was responsible for severe reduction in zooplankton densities in the lake. Allen et al. (1995) showed the likelihood that dense populations of *Menidia menidia* could lower zooplankton even in some open systems like salt-marsh creeks. Gophen (1988) suggested that long-term changes in large versus small zooplankters in Lake Kinneret (Israel) were mediated by particulate-feeding and filter-feeding fishes, respectively. Hairson (1988) showed that the effects of fish on zooplankton was related to springtime warming. Threlkeld and Søballe (1988) showed that the effects of fish on zooplankton in experimental tanks were not overridden by changes in turbidity, with initial conditions generally prevailing throughout a mineral-addition experiment.

Planktivorous fish can have strong effects on the size structure of zooplankton populations (Brooks and Dodson, 1965), with cascading effects to standing crops of phytoplankton. In essence, when size-selective fishes eat mostly large zooplankton, lowering their density, smaller zooplankters are released from competition (or potential predation by larger carnivorous taxa), and taxa like *Bosmina* move from the littoral to the limnetic zone (Brooks and Dodson, 1965) and become more abundant. Brooks and Dodson (1965) based their conclusions on surveys of numerous lakes with and without planktivorous fishes, and on one lake in which a known introduction of *Alosa* had occurred. The fishless lakes and lakes with fish differed markedly in zooplankton size, with few plankters more than 1 mm long occurring in the water column of the latter, and the one lake changing in zooplankton size structure after the introduction of fishes. Because smaller zooplankton, like *Bosmina* are less efficient filterers of phytoplankton than are large herbivores like *Daphnia* [size-efficiency hypothesis of Brooks and Dodson (1965)], the standing crop of phytoplankton and primary production in the water column will be higher when small zooplankton dominate (Hrbácek et al., 1961; Brooks and Dodson, 1965; Carpenter et al., 1987). Thus, to summarize, if size-selective fish predators lower standing crop of large zooplankton (which are more efficient grazers of algae), the zooplankton assemblage will shift to small-bodied taxa (which are less efficient grazers of algae), and phytoplankton standing crops should increase. This cascade of effects is discussed more fully in Section 11.6.

Hall et al. (1976) reviewed the assumptions of Brooks and Dodson's (1965) size–efficiency hypothesis, and the generality of size-selective predation by fishes, noting that "an overwhelming body of empirical evidence has documented size-selection of zooplankton prey by vertebrate predators of sufficient intensity to drastically reduce or eliminate large-bodied prey," citing at least a dozen studies up to that time supporting these findings. Hall et al. (1976) concluded that predation by both fishes and invertebrates, combined with food limitation, could control zooplankton assemblages, but that the interactions among factors likely were complex. However, they concluded that vertebrate predation (fish) sets an upper limit on the body size of zooplankton that can be successful in a lake or pond. Taylor (1980) showed that there may be complex responses of zooplankton populations to size-selective predation by fishes, because life-history phenomena of the zooplankton may vary in an age-specific fashion.

The comprehensive review of planktivorous fishes by Lazzaro (1987) noted the existence of four different kinds of planktivory (including gape-limited small fishes; particulate feeders; pump filter feeders; and "tow-net" filter feeders), each with its own set of feeding styles and constraints and consequent effects on zooplankton. Drenner et al. (1982) demonstrated the pump filtering rate capacity of gizzard shad (*Dorosoma cepedianum*) and estimated the time in hours that populations of this omnivore would require to hypothetically filter a volume of water equivalent to that of a lake (ranging from about 56 to 130 h of feeding. Lazzaro (1987) and McQueen et al. (1986) noted the decreasing strength of top-down effects of fish at trophic levels below zooplankton (i.e., phytoplankton), likely because direct effects of nutrients or other variables can have greater impact in food webs than indirect, cascading effects of planktivorous fish mediated through zooplankton.

Since Hrbácek et al. (1961), Brooks and Dodson (1965), and Hall et al. (1976), many other studies have shown the effects of fish predation on the size structure or composition of the plankton. However, predation by invertebrate planktivores may also be important in regulating zooplankton size structure (Dodson, 1974; Morin, 1988). O'Brien et al. (1992) reviewed the evidence for factors controlling lake productivity and zooplankton structure, [i.e., top-down (predation) versus bottom-up (nutrients)]. Fernando (1994) emphasized the role of fishes in the evolution and global distribution of zooplankton and related Brachiopod groups, and the generalization that cladocera existing with fish have smaller sizes. Gliwicz (1994) emphasized the potential of indirect effects of zooplanktivorous fish on distribution of zooplankton because of influences on behavior and physiology, in contrast to the removal of zooplankton by direct predation.

Selected examples from the many articles in the literature addressing fish effects on zooplankton assemblage structure show a range of responses including (1) reduction of the mean body size of zooplankton in the system, (2) lowering the abundance overall, and (3) changing the overall zooplankton community composition (Table 11.2). However, not all fish consume large-bodied zooplankters. Keast (1985d) showed that fish species differed with respect to size selectivity, with larger bluegills eating large-bodied zooplankton and ignoring small taxa like *Bosmina,* whereas small bluegills and cyprinids ate smaller-bodied zooplankton, presumably due to gape-limitation. Azoulay and Gophen (1992) showed that larval Lake Kinneret (Israel) sardine (or bleak) (*Mirogrex terraesanctae*) selectively ate small zooplankton, and Landau et al. (1988) suggested that larval *Mirogrex* are sufficiently dense to have negative impacts on rotifer biomass. Thus, conclusions about size selectivity or zooplankton reduction by fishes should ideally consider trophic differences among species or age-classes.

Additionally, fish do not act alone as predators of zooplankton in most natural waters, and physical structure can affect the outcomes of fish predation on zooplankton. Hanazato and Yasuno (1989) showed that fish (gudgeon) alone in enclosures reduced large zooplankton and resulted in a large increase in the density of *Bosmina.* However, when gudgeon and the predaceous midge *Chaborous* were both present in enclosures, *Bosmina* also remained in low density. Black and Hairston (1988) also have shown that *Chaoborus* and fish affect plankton differently, and a change in zooplankton assemblages when one replaced the other. Beklioglu and Moss (1996) showed that fish predation markedly changed open-

Table 11.2. Direct Effects of Fish Predation on Zooplankton Assemblages

Author (Year); Predator	Changed Zooplankton			Kind of Study?
	Mean Body Size	Total Abundance	Taxonomic Composition	
Hurlbert & Mulla (1981); mosquitofish	X	X	X	Expt. ponds
Langeland (1982); arctic char	X		X	Lake expt.
Morin (1984a); centrarchids	X	X	X	Pond exclosure
Tátrai et al. (1985); bream	X	X	X	Lake enclosure
Drenner et al. (1986); *Dorosoma* and *Menidia*	X	X	X	Tank expts.
Vanni (1986); *Lepomis* spp.	X	X		Lake enclosure
Gophen et al. (1988); Fingerling *Clarias*	X			70-L tanks
Drenner et al. (1988); mosquitofish	X			Expt. tanks
Hanazato & Yasuno (1989); gudgeon	X		X	Expt. ponds
Gophen and Threlkeld (1989); bleak	X	X	X	Tank expts.
Campbell & Knoechel (1990); sticklebacks		X		Distribution patterns
Gophen et al. (1990); bleak + other spp.	X	X		Long-term lake records
Guest et al. (1990); *Dorosoma petenense*	X			Pond expts.
DeVries & Stein (1992); *Dorosoma cepedianum*		X		Enclosure expt.
O'Brien et al. (1992); arctic grayling	X		X	Limnocorrals
Alimov et al. (1992); *Coregonus*		X	X	Comparative—2 lakes
Richardson & Threkeld (1993); bass and *Menidia*	X			Experimental tanks
Visman et al. (1994); northern lake species	X	X	X	Comparative—2 lakes
Seda and Duncan (1994); "anti-fish" reservoirs	X		X	Lack of fish in reservoirs
Hessen et al. (1995); lake fish community	X		X	Comparative in 342 lakes
Daldorph & Thomas (1995); sticklebacks		X		Enclosure expt.
Schriver et al. (1995); stickleback & roach	X	X		Enclosure expt.
Beklioglu and Moss (1996); roach & perch	X		X	Mesocosms

water zooplankton but had little effect on cladocerans in littoral plant beds. Drenner et al. (1978) and Drenner and McComas (1980) showed that the capture of zooplankton depended not only on the kind of predator but also on the abilities of individual zooplankton taxa to evade or escape predation. Dodson (1988) showed that at least some zooplankton could respond with predator-avoidance behaviors to chemical stimuli from sunfish (*Lepomis*), and Stenson (1980) showed that pigmentation of zooplankton can make them differentially susceptible to predators by increasing their visibility.

To summarize Table 11.2, of 23 studies ranging from manipulations in enclosures or tanks to broad comparative surveys of whole lakes, 19 showed or implied smaller mean body size of zooplankton when zooplanktivorous fish were present, 13 showed lowering of total abundance of zooplankton in the presence of fish (although some showed increases, because of increases in small-bodied taxa like *Bosmina*) and 13 reported or implied changes in overall zooplankton community structure if fish were present.

In lakes, fish have a second major effect on the zooplankton, particularly mobile forms like *Daphnia*. There is a very well known and lengthy history on the vertical migration of zooplankton, with upward movement at night and, by and large, downward to deeper depths in the daytime (Hutchinson, 1967). By being in the surface waters only at night, they escape some of the predation pressure from largely visual predators. Diurnal occupancy of cooler, darker, deep waters, which are often less productive, reduces predation but slows growth rates (Wright et al., 1980). Many limnologists, worldwide, have shown vertical migration of zooplankton, which can be explained by various alternative hypotheses. However, fish predation probably remains the most likely explanation for this pervasive pattern of zooplankton migration. Simulation models combined with data from real lakes showed that for large-bodied, predator-vulnerable zooplankton taxa, the strategy of migrating to avoid predation typically "wins" (Wright et al., 1980). However, Wright et al. showed that for *Bosmina,* which is too small for predation by most fish, vertical migration offered no benefit. It should also be noted that some fish (e.g., *Menidia beryllina*) do appear to migrate in order to feed in zones with high zooplankton abundance (Wurtsbaugh and Li, 1985). It would seem that a complete understanding of the vertical migration of zooplankton to escape predation would require two things. First, it would be interesting to know the degree to which coevolution of zooplankters and fish influence migration, thus studies matching fish and zooplankton with histories of coexistence or not would be instructive. Second, studies focusing on the combined searching and feeding behaviors of all members of a native lacustrine fish assemblage with respect to zooplankton prey should be useful. Zooplankters may escape visual, particulate predators by avoiding lighted surface waters, but to what degree do they thereby encounter predation pressure from deeper-swimming filter-feeding predators? For example, planktivorous alewives (*Alosa pseudoharengus*) and ciscoes (*Coregonus artedii*) can feed by filtering in the dark (Janssen, 1980).

Finally, in lakes, there is good evidence that the tremendous influx of larval fishes, typically in late spring to early summer, may have a major influence on the late spring or early summer population declines of zooplankton in lakes (e.g., Work and Gophen, 1995). Kairesalo and Seppälä (1987) and references therein showed evidence that the annual influx of young fishes in a lake in southern Finland might directly lower densities of *Bosmina*. McQueen and Post (1988) showed in limnocorrals that when young yellow perch (*Perca flavescens*) densities reached 30–40 kg per hectare, the fish were responsible for the collapse of *Daphnia* populations. Evidence is largely circumstantial, but it is apparent from May and into June that, in many lakes, particularly Lake Texoma, there is a major decline in the numbers of zooplankton per cubic meter or per liter of water. This decline, coincidence or not, comes at about the same time as peaks in density of larval fishes (Matthews, 1984), which feed on zooplankton. However, it should be noted that the annual "crash" of zooplankton in Lake Texoma (Work and Gophen, 1995) also is typically at a time of rapid warming of the water, and experiments by Kirsten Work (1997) demonstrate a limited tolerance of native zooplankton with respect to reproduction and survival at those temperatures. Whether or not it is causal, the increase in larval fish and the decline or crash of zooplankton comes at such a time that the direct influence of predation by larval and small juvenile fish cannot be ruled out as at least helping to drive zooplankton populations to their mid-summer low in lakes and in reservoirs in much of the temperate zone.

Effects of Fish on Phytoplankton

There has been substantial evidence in many of the studies cited above that fish predation, by eliminating large zooplankters (which tend to be efficient herbivores), can indirectly enhance phytoplankton density or primary productivity. It is also possible that fish can directly stimulate or enhance phytoplankton densities via nutrient supplementation, instead of by suppression of zooplankton (Drenner et al., 1986). These two contrasting models for the indirect effect of fish on phytoplankton remain in competition, and evidence from numerous experiments suggest important interactions among zooplankton suppression, nutrient enhancement, and experimental conditions (Drenner et al., 1986; Richardson and Threlkeld, 1993). Additionally, there can be persistent effects of fish on phytoplankton assemblages, presumably mediated through nutrient relationships, even after fish are removed (Threlkeld and Drenner, 1987).

Drenner et al. (1996) experimentally tested the effects of omnivorous gizzard shad (*Dorosoma cepedianum*) on a variety of limnological parameters in mesocosm experiments with water from two lakes—one eutrophic and one oligotrophic—and found that shad caused increases in total phosphorus, small-particulate phosphorus, primary productivity, and chlorophyll. There were concomitant decreases in water clarity, cladocerans, copepods, and large-particulate phosphorus. However, effects on six of the measured parameters were more intense in the eutrophic than in the oligotrophic system. Drenner et al. (1996) also reviewed 48 studies of the effects of omnivorous fishes on phytoplankton density and found that in all but five (with fish at low densities), there were significant effects of omnivorous fish on phytoplankton biomass. Most of the studies detecting a significant effect showed that omnivorous fish enhanced the total phytoplankton abundance, biomass, or primary productivity, particularly by enhancing nannoplankton, which is too small to be grazed efficiently by these fishes.

It is also necessary to consider the potential effects of various fishes via direct consumption of phytoplankton. Some freshwater fish are specialized for feeding on phytoplankton [e.g., the pharyngeal teeth of *Tilapia esculenta* modified to rake filamentous algae into the esophagus while breaking up long strands (Greenwood, 1953, as summarized by Gerking (1994)]. Omnivorous filter-feeding fishes, such as some *Tilapia* (Drenner et al. 1987a, 1987b; Gophen and Spataru, 1989; Vinyard et al., 1988) and *Dorosoma cepedianum* (gizzard shad) (Drenner et al., 1982, 1986; DeVries and Stein, 1992), not only eat zooplankton but also consume phytoplankton by straining it from the water column, with trade-off between phytoplankton and zooplankton consumption depending on gill-raker spacing, which changes as the fish grow.

Drenner (1977) estimated gill-raker retention from interraker distances, and Drenner et al. (1984a) showed that filter-feeding gizzard shad were increasingly efficient at filtering particles up to about 60 µm. Drenner et al. (1986) showed that small *Dorosoma* fed more efficiently on smaller particle sizes than did large *Dorosoma*, with the shift in feeding efficiency related to increases in gill-raker spaces with fish size (Mummert and Drenner, 1986). In ponds, Drenner et al. (1984a) showed that gizzard shad suppressed *Ceratium*, which was the only phytoplankter sufficiently large to be ingested efficiently (while not having significant effect on some smaller phytoplankton taxa, and enhancing other small taxa—through nutrient input?). Drenner et al. (1986) also showed in a laboratory experiment that gizzard shad did not reduce the ambient abundance or size–frequency distribution of algal filaments that were too small (< 40 µm) to be filtered efficiently by gill rakers. DeVries and Stein (1992) showed that gizzard shad did influence zooplankton assemblages by predation, but also reduced edible-sized phytoplankton by direct consumption (and apparently by also changing availability of nutrients to the phytoplankton).

Drenner et al. (1984b) showed that small blue tilapia (*Tilapia aurea*) most effectively fed on particles > 25 µm in diameter, and that this species in ponds suppressed larger-sized algae, but enhanced populations of smaller algae. Drenner et al. (1984b) concluded that both gizzard shad and blue tilapia functioned as a size-selective grazers of phytoplankton (and escape-selective zooplankton filterers, capturing zooplankters lacking escape abilities). Drenner et al. (1987b) showed that *Tilapia galilaea*) suppressed *Peridinium* in tank experiments, thereby lowering primary productivity and total chlorophyll. In tank experiments comparing the effects of fish with regard to eating zooplankton versus directly eating phytoplankton, Vinyard et al. (1988) showed suppression of the dominant alga (*Peridinium*) in Lake Kinneret by grazing by *Tilapia* species, with implications for changes in water quality of the lake.

In an important sequel to these studies, Mummert and Drenner (1986) modeled filtering efficiency of different sizes of gizzard shad, showing that it was possible to use a cumulative frequency distribution of interraker distances to predict sizes of particles that would be used by the fish in nature. The predictions were corroborated in feeding trials, which showed ontogenetic changes in food use that matched changes in raker measurements during growth of individual fish. As a result, small gizzard shad are able to effectively filter small phytoplankton cells 20–30 μm in diameter, whereas larger individuals shift to a diet with more zooplankton (Mummert and Drenner, 1986). However, there have been several recent studies [including Drenner et al. (1987a)] that question the generality of gill raker spacing as the primary determinant of food size, or whether other ancillary structures or mucus-dependent traps for food items in the buccal cavity of filter-feeding fish are most important, as reviewed by Gerking (1994, pp. 189–192).

11.4 Herbivory

Algivory

The fact that many freshwater fishes eat attached algae (periphyton) or vascular plant material has been known for a long time in ichthyology (Jordan, 1905) or in aquarium studies (Kraatz, 1923). Fryer and Iles (1972, pp. 66–74), for example, showed detailed morphological adaptations of cichlid fishes of the African rift lakes for eating attached algae from stony substrates or from stems or leaves of vascular plants. Although Fryer and Iles (1972) described in detail various morphological adaptations of the cichlids for scraping algae from rocks, they did not provide any indication that this algivory had a measurable effect on standing crops of algae. Horn (1992) provides a valuable summary of morphology and mechanisms associated with herbivory (although focused on marine species). Reinthal (1990b) has shown that herbivory by rock-dwelling cichlids in Lake Malawi is related to their morphology. About 45 freshwater fish species in North America include attached algae as an important component of their diet, with many species having mouths specialized for scraping or dislodging attached algae from hard substrates (Lee et al., 1980; Matthews et al., 1986, 1987).

Although it was widely known in ichthyology and natural history that many fishes ate attached algae or vascular plants, what was not known was the extent to which freshwater fishes can, by algivory, change the structure or composition of the attached algae in streams or lakes. A large number of marine fish are known to maintain algal "turfs" by their grazing, changing taxonomic and structural composition of the algae [many references, reviewed by Gerking (1994, pp. 70–76)], with "profound effects on primary productivity, standing crops, taxa or growth forms" [references in Power et al. (1985) and Matthews et al. (1986, 1987)]. The fact that freshwater fish could directly control standing crops, distribution or taxonomic composition of attached algae was not described until the ability of tropical (Power, 1983, 1984b), and temperate (Power and Matthews, 1983; Power et al., 1985) stream fishes was quantified in the early 1980s (Winfield and Townsend, 1991).

M. E. Power's doctoral dissertation focused on the interactions of algivorous loricariid catfishes in Panamanian streams. Power (1983) described the mouths and algae-removing capability of four loricariid species from the Rio Frijoles in Panama and showed that the loricariids were heterogeneously distributed on depth gradients in stream pools, avoiding depths < 20 cm, presumably to avoid avian predators (Power, 1984a). More recently Power et al. (1989) confirmed experimentally that the loricariids were indeed responsible for the differences in algal standing crops initially observed along depth gradients. The loricariid species differed in substrates they grazed, with habitat use and grazing activitites corresponding to mouth and body morphology. The most common loricariid in the Panamanian streams, *Ancistrus,* grazed periphyton from flat surfaces on wood, bedrock, and clay substrates in pools, whereas others specialized in grazing other substrates (e.g., pebbles in riffles). Grazing by these catfishes continues in both rainy and dry seasons, and Power (1983) considered them food-limited in the dry season.

Power (1983, 1984c) showed that the grazing activities of loricariids created patchiness in distribution of algae on stream substrates at a within-pool scale, and that these fishes could either keep patches open (sediment-free) by continued grazing, or open new patches by vigorously removing sediments. All of these activities create variation in distribution of algae and sediments that influenced the fishes themselves and could have consequences for habitat use by other small fishes or invertebrates in these streams. On a larger scale, the loricariids sampled and tracked by their movements the availability of periphyton from pool to pool (Power, 1984b), and by their grazing activities, they modified standing crops of algae in a given pool (Power, 1983). Power (1984b) used behavioral observations combined with estimates of mouth size to estimate rates of algae removal by grazing of loricariids. Power (1984b) demonstrated a balance between primary productivity and grazing of periphyton by the loricariids, resulting in similar standing crops of periphyton among pools. Power (1983) interpreted this to indicate that the loricarriids thereby "damped incipient pool-to-pool variation in periphyton standing crop as it arose." This was the first documented quantification of the ability of algae-grazing fishes to control standing crops of algae in their ecosystems. In subsequent experiments, Power (1990a) was able to refine the role of algivorous loricariids, showing that at high densities, the fish depleted algae, but enhanced it at low fish densities by removal of sediments that, in the absence of fishes, became limiting to algae.

In a stony-bottomed tropical stream in Venezuela, Flecker (1992a) compared the effects of insectivorous characins, algivorous armored catfish (*Chaetostoma*), and detritivorous *Prochilodus mariae* on the abundance of benthic invertebrates and amounts of sediment and detritus on stream substrates. Flecker (1992a) found significant increases in algal standing crops and altered algal community composition in fish exclusion cages. He concluded overall that direct effects of the insectivores were relatively weak, but that the indirect effects of the algivores and detritivores were strong, with sediment removal or algal depletion altering invertebrate standing crops, with strong impacts on 11 benthic taxa and a reduction in total invertebrates in grazed cages.

In a second experiment, Flecker (1992b) found that removal of sediments by fishes indirectly reduced the numbers of invertebrates, but that direct intimidation of invertebrates by some activities of the grazing fishes also contributed to lowered invertebrate densities. Flecker (1992b) speculated that algivory or detrivivory, by reducing the quality of habitat patches for invertebrates, might have more influence on densities than does direct predation, in that invertebrates will not immigrate to or remain for long in patches of poor habitat quality, whereas immigration (e.g., drift) could swamp apparent effects of predaceous fishes if habitat quality is high.

Subsequent experiments (Flecker, 1996) with the exclusion of *Prochilodus* showed that the fish caused major changes in invertebrate assemblages, in patterns of sediment accrual as ash-free dry mass, and in composition of algal assemblages. assemblages. *Prochilodus* decreased diatoms and facilitated cyanobacteria (bluegreens). Flecker (1996) concluded that *Prochilodus,* by virtue of its sediment processing, is a "functionally dominant species" in these tropical streams. Color photographs in Flecker (1996) of the streambed where *Prochilodus* was versus was not excluded show an overall striking difference in standing crops of sediment and in apparent quality (green color, etc.) of attached algae where the fish are present. Gelwick et al. (1997), as described below, found almost identical, visually striking differences in appearance of gravel streambed areas exposed or not exposed to algae-grazing minnows in an Oklahoma river.

Many other tropical fishes consume algae (Roberts, 1972), at least seasonally (Prejs and Prejs, 1987). Prejs and Prejs (1987) showed that numerous fishes in natural or artificial ponds in the Venezuelan savannah switched from diets of invertebrates to algae and detritus during the dry season. Wootton and Oemke (1992) showed in a Costa Rican stream that fish consumption of macrophytes and attached algae was common, and that grazing sharply reduced standing crops of periphyton. They concluded that a higher proportion of fish in the tropics are herbivorous and that as a consequence fish grazing is more likely to have important influences on biotic communities in tropical than in temperate streams. However, our work (below) with one widespread and highly abundant temperate-stream algivorous species (*Campostoma anomalum*) suggests that even one species of active grazer can have major impacts on the structure of a stream ecosystem. Additionally, Grimm (1988) has shown that a single species of algivore–omnivore (*Agosia chrysogaster*) consume much algae and have strong effects on nitrogen budgets and cycling in a temperate-zone desert stream.

Power's work with tropical algae-grazing fishes in about 1980 suggested that grazing fishes in temperate streams might similarly regulate standing crops or structure of algae in streams. In autumn 1982, we (Power and Matthews, 1983) found that *Campostoma anomalum,* which is a highly abundant cyprinid that is widespread in eastern North America, could control standing crops of algae at the scale of whole pools in Brier Creek, Oklahoma. Our initial surveys revealed that pools that contained schools (hundreds) of *Campostoma* had very low standing crops of attached algae, with the exception of shallow-margin fringes (as Power had observed in Panama), whereas pools lacking *Campostoma* (due to a predator threat from *Micropterus* basses) had dense accumulations of attached filamentous algae. In an initial transfer of algae-coated substrates into pools with *Campostoma,* the schools ate the attached algae in a matter of minutes. On average, *Campostoma* reduced the ash-free dry mass of algae on introduced cobbles to less than one-third within 24 h. Experimental regulation of *Campostoma* densities showed that pools or areas in pools with these grazers had low standing crops of attached algae, whereas algae markedly increased over time in areas without *Campostoma* (Power et al., 1985). Where *Campostoma* were absent, dense standing crops of *Spirogyra* accumulated over time, whereas areas lacking *Campostoma* had scant standing crops of *Spirogyra* or *Rhizoclonium* and much feces and detritus (Power et al., 1985).

Subsequent observations and trials in the larger Baron Fork drainage of the Illinois River, Oklahoma, showed that *Campostoma* produced large numbers of grazing scars on stone substrates or wood (Matthews et al., 1986), similar to those found by D. L. Kramer [in Power (1983)] to be produced by *Labeo* in Ghana, to the grazing scars of chiselmouth (*Acrocheilus alutaceus*) in western North America (Moodie and Lindsey, 1972), or to grazing scars of detritivorous *Prochilodus mariae* in Venezuelan streams (Flecker, 1992a, 1992b). All such grazing scars represent modification of structure of algae at small spatial scales, with potential influence on microhabitat use by invertebrates or the fishes themselves. Matthews et al., (1987) summarized the known consequences of *Campostoma* for stream ecosystems to that date, including information on standing crop of the fishes (> 5000 individuals per pool), rates of periphyton removal, modification of taxonomic composition of algal communities subjected to grazing, grazing patterns and feeding behaviors, and speculated on stimulation of primary productivity by grazing, potential regeneration of algae by gut passage in *Campostoma,* maintenance of grazing "lawns" by *Campostoma,* and consequences of *Campostoma* to nutrient relationships in stream ecosystems.

Subsequently, Stewart (1987) showed that *Campostoma* in Brier Creek could control standing crops of attached algae even when algal growth was stimulated by the addition of limiting nutrients, and that *Campostoma* selectively grazed certain kinds of attached algae. Furthermore, in the Baron Fork of the Illinois River, where huge numbers of *Campostoma* actively graze throughout warm months, Power et al. (1988) showed evidence that the "normal" condition of algal "felts" most commonly observed on the streambed was maintained by grazing of *Campostoma*, and that substrates protected from these fishes were rapidly overgrown by long chains of benthic diatoms. The experiments by Power et al. (1988) suggested that in the presence of dense *Campostoma*, the algae was dominated by forms having capability of basal regeneration.

Gelwick and Matthews (1992) and Gelwick et al. (1997) have shown a large number of other direct consequences of grazing by *Campostoma* in small and large streams of the midwestern United States. In Brier Creek, Oklahoma, a major spate reset algal standing crops and scoured the streambed. We maintained for the next 61 days (April–June), four pools with and four without schools of *Campostoma*, and then conducted a partial changeover of nongrazed to grazed pools. Direct effects of *Campostoma* that differed between grazed and nongrazed pools or that showed a grazer × time interaction included the following: decreased standing crop of algae (as ash-free dry mass and algal height) in grazed pools; percentage composition of algae visually detectable as dominants in cross-stream transects (dominance of *Spirogyra* and *Rhizoclonium* in nongrazed pools; dominance of diatom/bluegreens and feces/detritus in grazed pools); changes in proportion of green, bluegreen, and diatom cells at the microscopic level; decrease in net primary productivity per unit area in grazed pools, but an increase in net primary productivity per unit algal biomass; conversion of benthic particulate organic matter (BPOM) to smaller size fractions; and changes in C:N ratios and nontanypodine chironomids. The latter effects (C:N ratios and chironomid standing crops) are likely mediated by the indirect as well as direct effects of *Campostoma*, and the mechanisms are unknown; the other effects can be accounted for by direct grazing effects of these fishes.

Gelwick et al. (in press, 1997) carried out experiments in 4.6-m^2 pens in the Baron Fork of the Illinois River, Oklahoma, with closed pens excluding *Campostoma* and other common fishes, and open pens simulating "pen effects" but allowing access to the fishes. We interpret the results as primarily due to *Campostoma*, which are highly abundant in the study pools, but note that insectivorous fishes (e.g., the abundant *Luxilus cardinalis*) likely have effects on some of the stream invertebrates, and that other fishes that are present in the stream (e.g., *Moxostoma* spp. can also alter benthic conditions). In closed stream pools, rapid growth of an algal overstory [e.g., suggested in Power et al. (1988)] resulted in very dense standing crops of attached algae and of organic and inorganic silt-detritus within 15 days in late summer. Algae subsequently senesced, and sloughed from many of the substrates that were protected from grazing. In a pattern that we also have observed in artificial streams over a period of months (Gelwick and Matthews, in press), attached algae on grazed substrates in Baron Fork remained consistently of low growth form, and throughout the summer months, appeared bright green and actively growing. Exclusion of fishes (Baron Fork) or lack of *Campostoma* (unpublished artificial stream data) resulted in a rapid growth–senescence cycle for attached algae, very different from that observed in the presence of grazing fishes. We conclude that *Campostoma* have strong effects on long-term as well as short-term dynamics of algae in a wide variety of streams (Gelwick and Matthews, 1992; Gelwick et al., 1997). The generality of the importance of *Campostoma* as an algivore is supported also by the findings by E. D. Ballard and T. E. Wissing (Miami University of Ohio, unpublished data) that the species altered standing crops and kinds of algae and changed macroinvertebrate communities in experimental enclosures in an Ohio (USA) stream.

In addition to direct effects on algae growth form, Gelwick et al. (1997) also showed that access to stream substrates by the natural fish assemblage (dominated by *Campostoma*) resulted in the following: lowered standing crop of attached algae; increased primary productivity per unit algal biomass; decreased size fractions of BPOM; lowered standing crops of some benthic invertebrate taxa; changes in functional groups of benthic invertebrates dominating patches; and changes in C:N ratios and bacterial standing crops. Additionally, Gelwick et al. (1997) showed that schools of *Campostoma* had the capability to remove in an hour an amount of algae greater than the total production per unit area of attached algae in a day. In addition, there were spatial differences (pool to pool) in some of the effects shown in open versus closed pens, due to stream pools and creation of shallow-water habitats, and change in habitat use by fish, during a period of decreased water levels.

Additional work with *Campostoma* has suggested their potential importance in processing or movement of materials in stream ecosystems. Partridge (1991) showed that grazing by *Campostoma* in artificial stream troughs resulted in increased downstream export of coarse particulate organic matter (CPOM, > 925 μm), suggesting that the fish were "sloppy eaters." This exported material consisted mostly of fragments apparently broken loose from attached algae filaments by feeding activities and did not represent feces (which settled to the bottom in these streams, which flowed at ~ 7 cm/s near the substrate). Gardner (1993) showed that *Campostoma*-grazed stream mesocosms accumulated more fine particulate organic matter (FPOM; 41–156 μm) than did ungrazed units. [Taylor and Hendricks (1987) also suggested that fish could influence fragmentation of leaf litter directly or indirectly.] Vaughn et al. (1993) demonstrated that grazing by *Campostoma* in artificial mesocosms negatively impacted crayfish (by resource monopolization by the fish) and enhanced snail production (indirectly) by apparently making nonfilamentous algae more available to the snails. Unpublished data by Matthews, F. P. Gelwick, and M. S. Stock from other mesocosm experiments suggest that crayfish and *Campostoma* are both effective grazers of stream algae, but when they were housed together, *Campostoma* apparently dominated the acquisition of algae (Vaughn et al., 1993). Gelwick (1995) has shown in large, outdoor artificial streams that grazing by *Campostoma* and by crayfish both have strong consequences for spatial dependence of standing crops of algae. However, the two contrasting kinds of grazers establish patchiness in standing crops of attached algae at different spatial scales: *Campostoma* effects tend to be poolwide or to differ between pools with or without piscivorous bass, whereas algivory by crayfish is localized in deeper habitats, often in close proximity to their burrows, and does not differ in pools with and without bass.

Power (1990b) has also showed in rivers of the western United States, with generally depauperate fish faunas, the potential importance of algivory of fishes in the entire food web. Power (1990b) showed that adult California roach (*Hesperoleucas symmetricus*) could significantly reduce standing crops of attached algae by direct herbivory, and that fish are critical as upper-level consumers regulating food webs (Section 11.6).

At about the same time that the effects of algivorous fishes on periphyton was being quantified in temperate and tropical streams (Power, 1983, 1984b; Power and Matthews, 1983), Goulding (1980) showed that consumption of vascular plant material (seeds, nuts) was intimately related to propagation of many floodplain trees in the tropics. For floodplains of the Amazon basin, Goulding (1980) documented in detail the links between the morphology of large plant-eating fishes and the kinds of plants they consume. Many fish of large body size in South America are capable of eating and metabolizing (Roubach and Saint-Paul, 1994) materials like fruits and seeds that fall into the water from terrestrial plants. Some terrestrial plants and trees in South America have seeds covered with a fleshy external covering, that are eaten by fishes, but because the actual seed passes through the gut undigested, the fishes help to redistribute the viable seeds in the riverine system (Goulding, 1980). However, the precise consequence of fishes to the flooded forest trees is unknown, although Goulding (1980) suggested the likelihood that seed-eating/dispersing fishes and the forest tree might be an example of coevolution.

Effect of Fish on Vascular Plants

There is little consumption of macrophytes or macrophyte materials by fish in temperate North America. Most macrophyte-eating fish are native to tropical waters, or to rivers in Asia or Europe (Prejs, 1978), although some, like carp (*Cyprinus carpio*) that may reduce soft-leaved aquatic vegetation [Fletcher et al., 1985] have been stocked worldwide. For European waters, Prejs (1984) considered ide (*Leuciscus idus*), roach (*Rutilus rutilus*), and rudd (*Scardinus erythrophthalmus*) to be highly herbivorous, with at least occasional consumption of plants by 12 other species. Because of the lack of native herbivores in North America that consume macrophytes, grass carp (*Ctenopharyngodon idella* have been imported for aquatic weed control in ponds or lakes. Grass carp are true to their name, eating large amounts of rooted aquatic vegetation. There is huge literature on herbivory by grass carp, addressing the direct and indirect effects of this species on ecosystems. Removal of vegetation by grass carp can have a negative pronounced effect on other fishes in the system, like centrarchids that benefit from weed beds. Prejs (1984) noted that whereas roach and rudd eat individual leaves or fragments (possibly stimulating growth of submerged macrophytes), grass carp actually pull whole plants out of substrates and destroy fringing vegetation of a lake.

11.5 Nutrient Effects and Ecosystem "Engineering"

Nutrient Changes by Fish in Ecosystems

Fish seem to have general kinds of potential impact on nutrient relationships in aquatic ecosystems, including (1) transport of nutrients within their body and subsequent release at a place different than the place where materials were obtained and (2) altering the cycling of nutrients within a system or at a place in a system. Within the general category of "transporting" could be considered release by death and decay versus release by metabolic waste releases. Drenner et al. (1986) have suggested that dead fish in lentic systems can have a strong influence on production and standing crops of phytoplankton.

There has been a long history of interest in the effects of migrating fishes on nutrient transport or influx into systems. Particularly for migratory species that are abundant or that mostly die after spawning, there is substantial potential for nutrient additions by their decay. Juday et al. (1932) evaluated the influence of salmonid carcasses in release of phosphorus in an Alaskan lake. Krokhin (1975) showed from long-term data that carcasses of spawned-out sockeye salmon (*Oncorhynchus nerka*) provided about one-fourth to one-fifth of the phosphorus budget of two lakes in Kamchatka (formerly the Soviet Union).

Cederholm and Peterson (1985) showed that debris within stream channels played a significant role in the retention of decaying salmon carcasses in streams of the northwestern United States. Richey et al. (1975) showed that nutrients (nitrates and phosphorus) from decomposition of salmonids represented a substantial subsidy in a tributary creek of Lake Tahoe, Nevada (USA). Durbin et al. (1979) showed that spawning migrations of alewife (*Alosa pseudoharengus*) into freshwater systems could provide a substantial and positive (i.e., exceeded export of nutrients by juveniles) nutrient input through mortality and, secondarily, via excretion. They also showed that the influx of alewife resulted indirectly in a higher rate of leaf-litter respiration, as a result of nutrient supplements from the fish. Kline et al. (1990) found that pink salmon (*Onchrhynchus gorbuscha*) migrating into an Alaskan creek provided most of the nitrogen in the food web and was a substantial supplement to carbon in the lower portion of the stream (below a waterfall that blocked further inland movement of the salmon). They concluded that resident fishes are in part dependent on nutrients transported to the lower part of the stream by the annual salmon spawning run, and noted several other references documenting increases in nutrients in streams following spawning die-offs of salmon.

An important article on movements of fish in a freshwater stream and their potential effects on nutrient balance in the stream was by Hall (1972). In a North Carolina (USA) creek, there was a net upstream movement of fish biomass (but not individuals) in the spring, with fewer, but larger fish moving upstream and many smaller fish moving downstream (Hall, 1972). The result was upstream movement of fish biomass 3.58 times greater than downstream movement of fish biomass. Hall (1972) concluded that the differential migration of large fish upstream and small fish downstream was adaptive in terms of reproductive and of hydrodynamics of different-sized fishes. He also suggested that the net upstream movement of fish biomass compensated for downstream transport of nutrients in the system. Although Hall concluded that the contribution of fish to headwaters phosphorus concentrations was small, the fish should comprise an important reservoir of phosphorus, and that their migration upstream maintained this reservoir in the system.

It is clear that much remains to be learned about the role of fish movement in transport of nutrients in lakes and streams. A potentially important factor that has largely been overlooked in freshwater system is the fate of feces, or the potential that fish might forage in one place and defecate elsewhere, resulting in a net movement of nutrients in some particular vector (e.g., up or downstream) or locally (e.g., from riffles to pools, etc.). Ecologists in fresh water might find relationships like those shown by Meyer et al. (1983) and Meyer and Schultz (1985) who showed that the daily cycle of feeding–movement–defecation for marine fishes had important implications for their ecosystem.

Algae-grazing *Campostoma* produce copious quantities of feces with a characteristic shape that makes it identifiable in stream bottoms where the fish is present. In experimental streams, *Campostoma* feces (W. J. Matthews and F. P. Gelwick, unpublished data) tended to accumulate in the more downstream of paired pools (connected by a riffle ~ 25 cm/s), but within pools, feces were commonly found at points with measured near-bottom current speeds up to about 7 cm/s. Partridge (1991) found that current speeds of 7 cm/s were not sufficient to sweep feces downstream, and our observations in Brier Creek suggest that elevated flows are required to transport the feces downstream—otherwise it tends to accumulate on pool bottoms within the pool where fish are feeding at base flow.

Fresh *Campostoma* feces placed in distilled water rapidly leached phorophorus into the water (A. J. Stewart, personal communication). Prejs (1984) noted the rapid release of organic matter from cyprinid feces and suggested that fish feces might contribute significantly to the metabolism of lake systems. Also, it appears on visual inspection that a fairly significant portion of the algae that passes through the gut of *Campostoma* remains viable. This varies, depending on the kind of algae, but A. J. Stewart and I carried out one experiment in which we fed *Campostoma* cultures of various kinds of alga like *Rhizoclonium*, *Cladophora*, and *Spirogira*. We then placed the fish in plastic bags, collected the feces, and examined feces for live cells. Depending on the kind of algae, about 20–60%

of the cells appeared to be live, still with chloroplast and outer walls intact after they had passed through the gut of the *Campostoma*. Thus, although *Campostoma* may feed on various kinds of alga it may be digest some algae relatively inefficiently; thus, it may be possible that feces of the *Campostoma* comprise packets of nutrients and live cells that would then be subject to transport if current speeds increase above base flow. If so, then this material would be transported downstream out of the immediate feeding area by water currents and might be a propagule for beginning growth of new algae. Finally, if *Campostoma* selectively digest algae, large numbers of individual feeding in a stream might directly influence the types of algae that dominate a stream by their variable killing of different kinds of cells in their gut.

Overall, with regard to nutrients and feces, it would be desirable for a whole assemblage of stream fishes to know what the fate of their feces is, how much is transported, the degree to which this actually does provide nutrients for the stream, and the degree to which the feces become substrate for the microorganisms. Feces of *Campostoma* were rapidly colonized by rotifers (A. J. Stewart, personal communication). The feces may not just be a nutrient source; they may provide a point of attachment for biofilm flora and fauna.

In addition to affecting nutrient movements in lakes or streams by migration or by release of feces downstream, fish may play a substantial role in nutrient cycling. For example, Hansson et al. (1987) showed that omnivory by rudd (*Scardinius erythrophthalmus*) selectively reduced macrophytes, released phosphorus that the macrophytes had taken up from the sediments, and thereby promoted increased growth of eiphytic algae. Nriagu (1983) quantified decomposition of bones of fishes in a large temperate lake and suggested that fish carcasses accounted for as much as 10–20% of the phosphorus flux to sediments in littoral or near-shore waters. Penczak (1985) calculated total carbon, phosphorus, and nitrogen consumed by fishes in rivers in Poland and their assimilation, finding that young fish were highly effective in nutrient uptake. Penczak and Tátrai (1985) estimated the standing crop of carbon, nitrogen, and phosphorus in the bream (*Abramis brama*) population in a European lake, suggesting that up to about 3% of the phosphorus entering the lake could be removed by harvest of the species. Tátrai and Istvánovics (1986) showed in enclosure experiments that bream increased nitrogen concentration and decreased phosphorus concentrations, presumably through promotion of recycling of the latter. Schindler (1992) demonstrated that fish-regenerated nutrients caused a major increase in some taxa of phytoplankton in lake mesocosms, and that enhancement by nutrients from fish was much more important than grazer control in regulating algal production. Schindler et al. (1993) showed that piscivore-dominated lakes had little recycling of phosphorus through "fish," but that planktivore-dominated systems had a greatly enhanced cycling of phosphorus through the fish compartment. They showed more than a 40% increase in phosphorus recycling attributable to planktivore dominance in the systems. Schindler et al. (1993) argued that it is not just "fish" or "no fish", but the structure of the food web that determines dynamics of phosphorus cycling.

Ecosystem "Engineering"

Ecosystem engineering is a general term suggesting that animals by their activities cause changes in physical conditions of the environment, thus modifying, creating, or maintaining habitats (Jones et al., 1994). Levinton (1995) referred to similar roles of organisms as their being "habitat fabric interactors." In the scheme of Jones et al., fish, by their feeding or other activities, mechanically disturb things in or on the substrate, hence would be considered "allogenic" engineers. For example, if a detritivorous fish consumes material from the bottom, the mechanical activity of feeding may also resuspend this material, allowing it to be entrained in the current and potentially carried out of the immediate area to be deposited elsewhere. This activity thus changes the nature of the habitat for other fishes or for invertebrates both in the "donating" and the "receiving" pool. Both *Prochilodus* (Flecker, 1996, 1997) and *Campostoma* (Gelwick et al., 1997) may play similar roles in streams by keeping the areas they graze relatively silt-free and by modifying the standing crop of algae, which changes the habitat for all fishes and invertebrates in a stream. Sediment removal may be by action of mouth or pectoral or pelvic fins. Flecker (1996) also pointed out that *Prochilodus* was apparently unique as an ecosystem engineer for the streams in which it is abundant; that is, its role is not readily replaced by other fishes in the system. The same is probably true for *Campostoma,* in that trails in experimental streams (Matthews, F. Gelwick, and T. Gardner, unpublished data) showed that the next most abundant algivore in upland streams of the Midwest (*Notropis nubilus*) was not nearly as effective as *Campostoma* at regulating growth of algae.

Power (1984c) also showed that loricariid catfish would use the whole body, wiggling onto sediment to remove it from patches of streambed. All of these actions may farm the attached algae by protecting it from an overstory of filaments and silt that might block sunlight and nutrients, allowing grazed algae to grow better or be more productive (Gelwick and Matthews, 1992).

There can also be deep bioturbation by fishes. Carp (*Cyprinus carpio*) root in soft streambeds. I have observed individual carp with the snout buried up to the eyes, pumping a steady stream of silt and mud through the gill openings as they apparently foraged for food items in soft substrates. Thus, there is apparently a large amount of material being put back into the water column by these fishes.

The role of bioturbation in transport of silt is related to current speed. When fish disturb stream bottoms by their feeding activity, the resuspended material can either settle to the bottom of the pool where they are located or be carried downstream to adjacent pools. Although I am aware of no studies testing this model, it would be important to know how much fish activity intereacts with current speeds to produce within-pool or among-pool effects on transport of materials in stream ecosystems.

Fish also disturb substrates by building or maintaining nests. Some fish build extensive nests as depressions on the stream bottom [e.g., *Lepomis* spp. (Johnston and Page, 1992) and salmonids (Field-Dodgson, 1987)], and others build substantial gravel mounds (e.g., nests of *Nocomis* in upland gravel-bottomed streams). Field-Dodgson (1987) showed that excavation of redds by salmon in a New Zealand stream had the effect of decreasing macrophytes, algae, and mosses within the stream, reducing fine particulate matter and detritus and modifying the pool–riffle structure of the stream. In addition, the activity of excavating redds in the streambed left the areas with markedly reduced benthic macroinvertebrate standing crops (Field-Dodgson, 1987). Many fishes will recognize and use gravel-mound nests of *Nocomis*. Other minnows, and sometimes darters, will place their eggs in these nests where they will also be protected by the nest owner. Additional nesting activities of fishes that alter the structure of underwater landscapes include digging of deep nests (i.e., to almost 1 m deep in the substrate of littoral zones by some Tilapia in African lakes) (M. Gophen, personal communication).

11.6 Indirect Effects of Fish in Ecosystems

This section considers the various ways that fish affect ecosystems through indirect effects. Proper identification of direct versus indirect effects can be problematic and I make no attempt to partition overall fish effects into the two categories. See Pilette (1989) for a detailed analysis of direct versus indirect effects for a marine food web. I will merely use "indirect" to include any pathway of fish effects that appears more complicated than the direct killing, eating, or displacing of organisms, or direct movement of materials by fishes. The literature suggests that fish do have the capacity to alter indirectly many aspects of the structure or function of ecosystems, and/or influence or change interactions that take place between particular organisms, or the organisms and the habitat. To organize this body of information on indirect fish effects in ecosystems, I consider three levels of influence of the fishes (Fig. 11.2):

1. "Second-order" effects
2. Cascading trophic effects
3. Food web effects

I consider second-order effects (Fig. 11.2A) to be cases in which the activity of a given fish species directly changes some component of the ecosystem such that some other fish species or organism is changed in distribution or abundance. (I consider this "second order" in contrast to the direct effects of Sections 11.2–11.5, which would be "first-order" effects). Second-order effects are those that can clearly be defined to extend in a measurable way to some other organism in the ecosystem because of the indirect influence of the fish in question, but I do

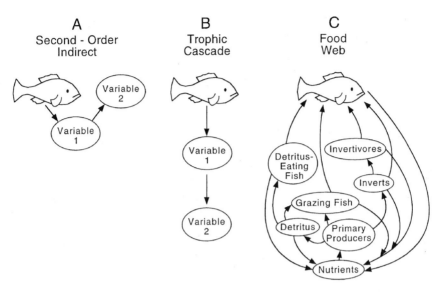

Figure 11.2. Hypothetical interactions of fish in ecosystems at the level of second-order indirect effects, a trophic cascade, and complex interactions "up" and "down" in food webs.

not include as second-order effects the more complicated "cascading trophic effects," which involve at least three trophic levels (Fig. 11.2B) connected by alternating increases or decreases in compartments linked by either threat or outright predation. Cascading trophic effects are a special, linear case that obviously occur in the context of food webs, but in the section on cascading effects, I include only studies that have shown by direct measurement, a linear set of *downward* effects of a member of the upper trophic level. At a third level of complexity (Fig. 11.2C), I consider the influence of fish either *upward* or *downward* in "whole" food webs [acknowledging that no "whole" food web has ever really been elucidated (Polis and Winemiller, 1995)] with respect to effects in numbers of links, and with effects between either adjacent trophic level (e.g., herbivores eating plants) or effects "jumping" across trophic levels (e.g., the release of nutrients by planktovorous or piscivorous fishes). Finally, some fish, by virtue of their activities, may have "strong effects" in ecosystems [sensu Paine (1980)] or be "keystone" species, whose presence (or absence) has a particularly strong influence (relative to other species) on a numerous processes or structural compartments of an ecosystem. These will be considered in a final section of this chapter.

Second-Order Effects

Second-order effects include cases in which fish, by virtue of actions such as bioengineering (Section 11.5) change the habitat, thus modifying its availability to and standing crop of other organisms, such as other fishes or invertebrates. For example, where *Campostoma* graze algae, maintaining a lower growth form of algae that has high primary productivity, grazing opportunities for snails may be improved (Vaughn et al., 1993), enhancing production of the latter. Additionally, the "scars" in algae left from grazing by *Campostoma* (Matthews et al., 1986), similar scars by *Prochilodus* (Flecker, 1997; his Fig. 1), or depressions in sand produced by herbivorous cichlids (Bowen, 1979) might serve as spatial microhabitats for various invertebrates. Flecker (1992, 1996, 1997) has documented that by removal of sediment or detritus by feeding activities, *Prochilodus* or loricariid catfish in South American streams change the habitat and cause a lowered total standing crop of macroinvertebrates (although enhancing some taxa while lowering others). Gelwick et al. (1997) have shown an identical function of *Campostoma* in an Ozark Mountain river—that is, with removal of overstory of algal growth apparently changing total densities of macroinvertebrates, yet enhancing some while decrementing others. In the examples above, the fish have changed the habitat or feeding environment for some other organism, which was then changed in abundance. The effect may also be by causing emigration of existing invertebrates or by lowering the tendency of potential immigrants to settle in a particular habitat (Flecker, 1992). This latter effect (grazers make habitat unsuitable for settlement) appeared to have a stronger indirect effect on macroinvertebrates than did the direct effect of insectivory by a variety of small fishes in a stream in Venezuela (Flecker, 1992).

Other cases of second-order effects include those in which actions of a fish cause the interaction between other organisms to be changed. For example, if a fish like a carp (*Cyprinus*) or a hogsucker (*Hypentelium*) feeds by disturbing gravel to a depth of several centimeters, the mechanical activity of the fish digging into the gravel may release invertebrates into the water column, making them available as prey for smaller, insectivorous fishes. Reighard (1920) noted the frequency with which catostomids feeding by disturbing gravel were accompanied by an entourage of insectivorous minnows, apparently picking from the water column the invertebrates disturbed by the larger fish. In Piney Creek, Arkansas, I have frequently observed that a northern hogsucker (*Hypentelium nigricans*) feeding in clear, shallow water by rooting in gravel substrates will be accompanied by one (rarely more than one) juvenile smallmouth bass (*Micropterus dolomieui*), and numerous insectivorous minnows like *Luxilus pilsbryi* or *Notropis boops*, apparently feeding actively on upturned invertebrates that drift downstream from the hogsucker.

Another category of second-order effects involves the conversion of an otherwise unavailable food resource to a form making it available to other organisms. As an example, it appeared that the feeding and the formation of feces by *Campostoma* in Brier Creek, Oklahoma, converted coarse particulate organic matter CPOM (> 1 mm in sieve size) to medium-sized MPOM of a size appropriate for collection by herbivorous chironomids, whose density increased in the presence of *Campostoma* (Gelwick and Matthews, 1992). Similarly, the production of feces by *Campostoma* or other fishes may provide a highly significant substrate for colonization by stream microorganisms [e.g., rotifers (A. J. Stewart, personal communication)].

Trophic-Level Cascade Effects

Both three- and four-level cascading food chains occur in aquatic ecosystems, with a calculated "mean chain length" (roughly, average trophic level number) of 3.6 in pelagic and 2.6 in riverine ecosystems [(Hairston and Hairston (1993), calculated from Briand and Cohen (1987)]. To examine trophic cascade effects for freshwater systems, I reviewed articles that I considered to show a direct downward effects of some fish, through an intermediate trophic level to a lower trophic level. At least three trophic levels, with direct predator–prey links, are involved or examined, and in some cases, the effect is four trophic levels in length (Table 11.3). Cascading trophic effects are anticipated in the well-known Hairston–Smith–Slobodkin–(1960) hypothesis (HSS), in which "the world is green" because predators keep herbivore numbers low, releasing plants to dominate the landscape (in a three-level interaction). When a fourth level is added (i.e., a top carnivore), the original top predator is suppressed (e.g., Power, 1990b), releasing the herbivores, which then by their feeding maintain a low standing crop of plants (turning that kind of a system "brown" for lack of vegetation). This postulate has had wide exposure and test in ecology, but the fundamental strength of the "alternating level effects" model remains largely intact. This is not to imply that upper-level players cannot "jump" a level, and for many fish species, proceeding ontogenetically from feeding on plankton to larger invertebrates to (possibly) other fish, a single individual may fit at least two and perhaps three trophic levels within its lifetime. Thus, the role of fish in trophic cascades is strongly dependent on body size or life stage, not just on taxonomic identity.

I included in Table 11.3 a total of 28 studies showing or testing for multitrophic levels, downward cascades of effects. These could be mediated, especially at the uppermost level, by either outright predation (as in the case of planktivorous fishes consuming large numbers of herbivorous cladocerans), or by the threat of a large-bodied predator causing smaller organisms (like fishes) to leave or avoid some part of the habitat, conveying protection on some even smaller organism (Fig. 11.3). For example, He and Kitchell (1990) showed that the effects of introduced northern pike (*Esox lucius*) in lowering densities of prey fish species (native minnows and other small fish species) was as much by causing emigration of the small fishes as by outright predation. This distinction may, however, be blurred over evolutionary time; that is, if cladocerans only migrate to upper levels of the water column during hours of darkness, in order to avoid surface-feeding visual predators (fish), the fish, in effect, create a "safe zone" for the planktivorous prey of large cladocerans during daylight. In this case, the avoidance of a potential

Table 11.3. Effects of Fishes Through Various Trophic Levels in Aquatic Ecosystems

Study No.[a]	Level of Effect[b]						
	Terr	BigFish	Smallfish	CarnInvt	YOYfish	Herbiv	pp
1			XX			X	X
2			XX			X	X
3			XX	X		X	
4			XX			X	X
5		XX	X			X	X
6		XX				X	X
7		XX				X	X
8		XX	X			X	
9			X			X	X
10			XX			X	X
11	XX					X	X
12			XX			X	X
13			XX			X	X
14			XX			X	X
15			XX			X	X
16			XX			X	X
17		XXXXXXXXXXXXXX				X	X
18			XX			X	—
19		XX	X			X	
20		XX	X		X	X	
21		XX	X		X		
22			XX	XXXXXXXX		X	X
23		XX	X			X	
24		XX	X			X	
25		XX				X	X
26		XX	X			X	X
27		XXXXXXXXXXX		X	X		
28			XX			X	X

[a] Study numbers: (1) Hurlbert et al., 1972; (2) Hurlbert and Mulla, 1981; (3) Crowder and Cooper, 1982; (4), (5) Spencer and King, 1984; (6) Power and Matthews, 1983; (7) Power et al., 1985; (8) Mittlebach, 1988; (9) Drenner et al., 1988; (10) McQueen and Post, 1988; (11) Power et al., 1989; (12) Vanni and Findlay, 1990; (13) Dewey, 1990; (14) Jeppssen et al., 1990; (15) McCormick, 1990; (16) Northcote et al., 1990; (17) Vanni et al., 1990; (18) Hill and Harvey, 1990; (19) Huang and Sih, 1991; (20) Harvey, 1991a; (21) Harvey, 1991b; (22) Power, 1990; Power et al., 1992; (23) Goyke and Hershey, 1992; (24) Trippel and Beamish, 1993; (25) Goldschmidt et al., 1993; (26) Wootton and Power, 1993; (27) Jackson and Harvey, 1993; (28) Flecker and Townsend, 1994).

[b] XX = level added or variably present; X = effect confirmed or measured. Terr = terrestrial or avian predators; BigFish = large-bodied, generally piscivorous fish; SmallFish = small-bodied fishes, generally insectivorous but also including omnivorous/planktivorous fish like shad (*Dorosoma*); CarnIvnt = carnivorous invertebrates; YOYfish = larval or small young-of-year fish; Herbiv = any herbivore, ranging from fish (e.g., *Campostoma*) to herbivorous cladocerans; PP = primary producer.

"Trophic" Cascade

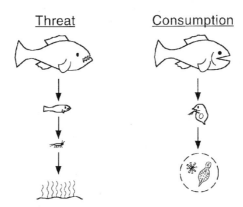

Figure 11.3. Two types of effects of fish in trophic cascades, including "threat" and actual "consumption of prey."

predaceous fish by the cladocerans is not likely as much in response to a perceived extant threat as it is to an innate propensity of the cladoceran evolved through millenia of enhanced fitness for those forms that do vertically migrate.

Additional examples of threats rather than frequent predation mediating a "trophic" cascade are by Power and Matthews (1983) and Power et al. (1985), in which threat (and occasional predation) by piscivorous bass (*Micropterus* spp.) in Brier Creek, Oklahoma, pools kept numbers of algivorous *Campostoma* low, releasing attached algae, which proliferated. Similarly, Harvey (1991a, 1991b) showed in two Oklahoma streams that adult bass by threat or predation kept areas of stream pools free of smaller fish, which, in turn, ate larval fish. Thus, the adult bass created "safe zones" for larval fish by their indirect effects mediated through larval predators like small sunfish (*Lepomis*) and minnows. McIntosh and Townsend (1996) showed cascading effects of threat from fishes restricting habitat use by a grazing mayfly, and resulting in increased standing crop of periphyton.

A summary of the 28 studies included in Table 11.3 suggests the generality of multitrophic-level cascades initiated by piscivorous, insectivorous, or planktivorous fishes. In most of these studies there was a measured pattern of increased fish = decrease standing crop next lower level = increased standing crop at the lowest level (often algae or phytoplankton). In 24 of the 28 studies, there was a demonstrated three-level effect of fish. Cascading fish effects extended through four trophic levels in only two examples. In only one of the experimental studies was there no effect of fish at the algae trophic level (Hill and Harvey, 1990), and in this case the authors noted a potential lack of enough grazing invertebrates to have mediated a cascading effect from fish downward to the attached algae.

Hill and Harvey (1990) also showed that light levels had more influence on attached algae growth than did fish or snails. Four-level effects were shown by Power (1990b) and Power et al. (1992a) in a California (USA) stream, where intermediate-sized roach and juvenile steelhead trout lowered the numbers of carnivorous invertebrates and fry of fishes (= one level, combined), which allowed an increase in algae-grazing midges, and ultimately a decrease in attached algae; Wootton and Power (1993) showed four-level effects using artificial channels in a California stream.

One of the largest examples of three-level effects may be taking place in Lake Victoria. The entire world is aware that the introduction of nile perch (*Lates niloticus*) has caused the extinction of a huge number of endemic cichlid species (Stiassny, 1996) in what has been called the worst case of human-mediated mass extinctions in history. Equally devastating to the lake, although less publicized, may be the cascading effect of increased piscivory (nile perch) reducing haplochromine cichlids (phytoplanktivores and detritivores), thereby contributing to major algae blooms (Goldschmidt et al., 1993). However, Reinthal and Kling (1994) suggest that it is not only predator–prey effects but also its effects of increased nutrient inputs and climate change that collectively are leading to eutrophication of Lake Victoria.

Not all of the other studies tallied in Table 11.3 showed the effects of fish at lower levels, and the strength of the connection can be tenuous at levels further from the fish [as noted by McQueen et al. (1986) and Ramcharan et al. (1995)]. For example, in the limnocorral study by McQueen and Post (1988), there was a collapse of *Daphnia* populations when growing young yellow perch reached densities of 30–50 kg per hectare. However, in only two of four limnocorrals was the decrease in *Daphnia* coincident with increases in phytoplankton, and there was evidence that some phytoplankton "bloomed" weeks before the collapse of *Daphnia* populations. Thus, in this study, the first-level effect was strong, and the second-level cascade was tenuous. In the lakes studied by Jackson and Harvey (1993), there were correlations among fish presence, reductions in predaceous *Chaoborus,* and size structure of cladocerans, but these relationships were confounded with lake morphometry and water chemistry. An additional confounding problem for clear-cut assessment of consumptive trophic cascades comes when the "top carnivore" feeds directly on the next two trophic levels, including the intermediate predator and its prey. Diehl (1995) showed that perch (*Perca fluviatilus*) can influence standing crops of chironomids in complex interactions that involve direct predation on the chironomids or indirect effects mediated through their predators (odonates and megalopterans).

Additionally, some workers who have found a trophic cascade effect to exist among planktivorous fish, cladocerans, and phytoplankton (top-down cascade) have also shown the potential importance of the "bottom-up" effects of nutrients (increased by fish presence) on the growth of phytoplankton (e.g., Vanni and Findlay, 1990). McCormick (1990) argued that fish could both suppress algivorous snails directly (enhancing algae) and cause release of nutrients from snails to algae (also enhancing algae). Thus, although a top-down cascade can be detected in virtually all the studies, the question of cascading effects versus nutrient enhancements by fish as major factors in plankton or attached algae production remains in question and may well depend on the specifics of particular systems. One apparently clear-cut demonstration of top-down cascading effects dominating came when a massive fish kill in Lake Mendota, Wisconsin (USA), releasing grazing pressure on zooplankton, resulted in increases in large-bodied cladocerans and "dramatic" decreases in phytoplankton, in spite of the fact that nutrients were not demonstrably changed before and after the fish mortality. Thus, this "natural experiment" controlled for nutrients, removed zooplanktivorous fish lakewide, and appeared to produce a pervasive effect of a three-level trophic cascade.

There can be further complicating variables in the field that regulate changes in fish densities, setting cascades in motion. One very interesting case involved Lake Sobygard, Denmark (Jeppesen et al., 1990), in which unusually high pH for several summers apparently resulted in a lack or loss of year-classes of roach (*Scardinius erythropthalmus*), which normally dominate the fish assemblage. As the roach population came to consist more and more of older and larger individuals (which have relatively less effect as grazers on large zooplankton), large cladocerans increased, putting more grazing pressure on phytoplankton, and resulting in several episodes of whole-lake phytoplankton collapses. Thus, in this case, the cascade was at three trophic levels, but putatively set in progress by a physicochemical change in the system.

In spite of the complexity in many studies, disagreement or uncertainty remains as to what the most common mechanism is by which the presence of omnivorous or insectivorous fish commonly reduce standing crops of attached algae or phytoplankton at two trophic levels lower in the ecosystem. One school argues that the mechanism is, indeed, the trophic cascade: Fish suppress grazing invertebrates, which release algae to proliferate. The opposing view holds that the most important effect of fish is to directly return or make nutrients available to the algae, which proliferate as a result. In numerous experiments, both pathways may have had an effect, and careful experiments partitioning relative effects of the two pathways, particularly as those mechanisms apply in real-world situations, remain needed. However, from an empirical perspective of ecosystems, one thing seems apparent: In many cases, the presence of fish *does* make a difference at two or more trophic levels lower in the ecosystem. In that respect, regardless of the mechanisms, fish "do matter" and need to be considered in any paradigm purporting to give general insight into trophic dynamics of aquatic systems.

Fish Effects in Food Webs

Two perspectives on fish in food webs have been emphasized recently, including (1) basic ecological roles of fish in food webs (e.g., Winemiller, 1995; Power et al., 1995) and (2) applied ecology of fish in food webs for purposes of biomanipulation of lake conditions (i.e., Shapiro and Wright, 1984; McQueen et al., 1986; Carpenter et al., 1987; Gophen, 1990; Schindler et al., 1995; Romo et al., 1996). Basic questions that are being asked include the pervasiveness of fish effects in food webs, the number or percentage of links affected by fish, redundancy or "substitutability" of species of fish in food webs, variation in fish effects in space or time or as fish grow, and other fundamental questions about the ways fish make a difference in the ways ecosystems function. Applied questions focused on biomanipulation typically ask if three- or four-level interactions that ultimately drive water quality can be altered by the addition (or removal) of fishes. For example, from classic top-down concepts, can the addition of a piscivore cause a reduction of planktivorous fishes, subsequent increase in large-bodied cladocerans that are efficient grazers of phytoplankton, and a decrease in phytoplankton (resulting in lowered propensity for biogenic turbidity or nuisance algal blooms)?

Food webs are not easy to understand or to study (Power, 1992). They can be incredibly complex (Winemiller, 1990), comparisons are fraught with methodological problems (Closs and Lake, 1994), and many disturbance phenomena or compensatory mechanisms confound sharp definitions of compartments or links in food webs (Persson et al., 1995). Realistic assessment of food-web dynamics is challenging (Hairston and Hairston, 1993; Winemiller and Polis, 1995) and quantification of the totality of effects of fish in whole aquatic ecosystems appears to be a near-Herculean task (Winemiller, 1995). Hairston and Hairston (1993) considered three general difficulties with assessment of food web dynamics, including (1) assessment of relative strengths of links, (2) competition is either ignored or assumed to be of equal strength in any links, and (3) pooling of biological species into "trophic species." Winemiller and Polis (1995) distinguished three fundamental kinds of food web studies, including (1) topological, descriptive approaches (which tend to be static and qualitative), (2) bioenergetic (which quantify transport of matter or energy via predation), and (3) functional or interaction approaches, that attempt to identify species or feeding links that most influence community composition and structure, including variation in time. Winemiller and Polis (1995) argue that most food-web theory has been based on topological webs (e.g., linkage density approaches), but that this approach often suffers from [in agreement with Hairston and Hairston (1993)] the following: inclusion of too few (1) species, (2) feeding links, (3) omnivores, (4) loops, and (5) cannibalism; too much lumping of species or taxa at lower trophic levels; and ignoring of temporal or spatial variation. In addition, the relative strength of top-down food-web regulation by predators versus bottom-up regulation by nutrients remains undecided. O'Brien et al. (1992) review in detail the varying perspectives on upward versus downward control of freshwater ecosystems.

In food webs including fish, there are particular difficulties in dealing adequately with ontogenetic changes in food habits of fishes (Gerking, 1994), detritivory (Bowen, 1983), omnivory (Vadas, 1990), cannibalism loops between life stages of a single species (Helfman, 1985), or links between individual species that can "flow" in both directions as a result of size- or age-class-dependent predation (Helfman, 1985). A review by Vadas (1990) suggested that omnivory (typically age-class dependent) is pervasive and often important in freshwater ecosystems, such that fish do not always control the abundance of their prey. If fish predators typically switch to alternate prey (including changes to different trophic levels) as a primary prey becomes scarce, or are highly opportunistic (Matthews et al., 1978), there should be increased blurring of links in food webs. There are well-known exceptions to the generality that top predators will switch [e.g., failure of striped bass, *Morone saxatilis* to use available alternative fish as prey, even in the face of starvation (Matthews et al. 1988b)], but it has long been known that a preponderance of fish in many assemblages have a propensity to be omnivorous or to feed widely from the prey base (e.g., Hartley, 1948). Finally, Power (1992a) showed that the effects of fish in a riverine food web differed, depending on the kind of substrate in a stream reach (e.g., gravel versus boulder–bedrock). Persson et al. (1992) also showed that the degree of food-web regulation by piscivores depended on levels of system productivity (less important in highly productive systems) and with habitat structure or heterogeneity. Closs and Lake (1994) showed strong temporal (but not spatial) variation in structure of a river food web.

In spite of the problems with a robust evaluation of fish in food webs, there are noteworthy recent studies that consider variation in the roles of fish in food webs across space or time. Power (1990b) quantified the various links provided by fish in one relatively complicated riverine food web, and Power (1995) and Power et al. (1995a, 1995b) reviewed various ways that hydrologic variation (flood/drought) could modify food-web relationships or food chain lengths involving stream fishes. "Top-down" versus "bottom-up" effects are both important in food webs (Power, 1992b). Power's studies show a strong tendency for top-down control of the food web by fishes at the upper-predator and intermediate-predator levels, although these effects differ among systems and with hydraulic changes over time. Winemiller (1990, 1995) demonstrated similarities and differences in food-web dynamics that include fish for comparable systems across considerably varying geographic and local landscapes. In Costa Rica and Venezuela, four study systems (Winemiller, 1990) differed markedly in number of taxonomic or trophic links but were relatively similar in overall structure. For widely differing systems ranging from east Texas (USA) rivers to African rivers to South American floodplains, Winemiller (1995) demonstrated the importance of disturbance and of life history. Winemiller (1990, 1995) emphasized the potential importance of detritus in tropical food webs and the propensity of fishes to switch rapidly from consuming algae to detritus as environmental conditions changed. Winemiller (1995) concluded that freshwater food webs are driven by a "combination of predictable abiotic and biotic processes, modified by change events" and urged continued efforts to incorporate temporate variation, environmental heterogeneity, and life-history responses into food-web dynamics.

Empirical studies in lakes have had markedly differing outcomes with respect to influence of fish throughout food webs. McQueen et al. (1986) showed that top-down effects of fish did not reach effectively to the level of phytoplankton in food webs of eutrophic lakes, whereas the influence of fish through zooplankton to phytoplankton could be significant in oligotrophic lakes. This study may also have relevance to food webs in warm monomictic reservoirs, most of which are moderately to strongly euthropic. Ramcharan et al. (1995) demonstrated that two northern lakes (Ontario, Canada) with long-term differences at the top predator level (presence or absence of piscivores) differed in intermediate food-web levels (e.g., body size and composition of herbivorous zooplankton), but differed little at the level of phytoplankton. Ramcharan et al. posited that compensatory mechanisms lower in the food web blurred the effects of top predators over the long term. Johnson et al. (1996a) also found strong effects of fish at the first downward trophic level (small dragonfly larvae) but decreasing effects of fish at the indirect level of smaller benthic invertebrates (that are prey of the odonates). Finally, studies by Sprules et al. (1991), Plante and Downing (1993), and Roell and Orth (1994) suggest stronger upward than downward effects in stream or lake food webs. Sprules et al. (1991) suggested the interesting possibility that increased upward flow of energy to top predators (salmonids) could be achieved if the forage fish assemblage were modified to include species more available to the piscivores.

Biomanipulation

Hurlbert et al. (1972) were apparently the first modern authors to suggest manipulation of water quality or phytoplankton by "top-down" manipulation lake food webs (instead of by "bottom-up" control of nutrient inputs. In one of the first long-term (multiyear) measurements of food-web response to fish manipulation, Shapiro and Wright (1984) showed that a manipulated increase of piscivores and decrease in planktivores led to increased large grazing cladocerans, and a reduction in phytoplankton, concordant with food-web theory. They also concluded that changes in nutrient levels after biomanipulation were not sufficient to explain the lowered phytoplankton and that it was the downward cascade of fish effects that had caused the changes in the system.

Carpenter et al. (1987) found in experimental manipulations of whole small northern lakes that introductions of piscivores had the effects predicted by food-web theory (i.e., that piscivores reduced planktivores, caused increases in herbivorous large cladocerans and a subsequent decrease in algal biomass, and reduced primary production. They also found that the rate of sedimentation of phytoplankton pigments, which serve as markers of long-term change in lakes, increased following biomanipulation. Carpenter et al. (1987) concluded that hierarchical control by fish and abiotic factors were "equally potent regulators of primary production" in these lakes. Alimov et al. (1992) demonstrated that pervasive downward effects of planktivorous or benthivorous fishes altered phytoplankton components in Russia. O'Brien et al. (1992) carried out limnocorral experiments that included additions of nutrients and of planktivorous Arctic grayling (*Thymallus namaycush*). Nutrient additions dramatically stimulated primary production, which passed upward through the food web. However, fish predation strongly reduced large-bodied cladocerans and modified the response of zooplankton to the nutrient additions. Hence, in this experiment, both top-down and bottom-up phenomena influenced the food web. Schindler et al. (1995) model and emphasize the importance of fish predation in pelagic food webs but also note the interactions of nutrients with fish effects. The message from all of these basic studies, from focused experiments to the broad review by Winemiller (1995), seems to be that fish clearly are of importance in freshwater food webs, linking energy upward and predatory effects downward, but that food webs remain minimally understood from a predictive perspective because of their incredible complexity and plasticity in time. This high degree of variability, often from uncontrollable phenomena, must be considered in attempts to apply food-web theory to biomanipulation.

Gophen (1990) summarized results of a biomanipulation workshop to include agreement on the definition of "biomanipulation" to be, "manipulation of food-web organisms and nutrient control to improve water quality." Gophen (1990) noted the history of biomanipulation, with increased measurement of complicated relationships and more food-web compartments, and reliance on stocking of piscivores during the 1980s. He summarized the cautionary note of participants that biomanipulation of large and small water bodies could be very different, and suggested the potential environmental problems of incautious introductions of fishes into large systems where their control becomes impossible once they are introduced. Any fish additions to natural systems for biomanipulation should be considered very carefully before an introduction, in light of the incredible problems that introduced fish have caused in systems worldwide. In contrast, biomanipulation by fish removal or control, although more challenging than merely stocking of a new species, is likely to be much more environmentally safe. From an ethical and practical perspective, all lake managers should be wary of introductions of exotic species, which have a long history of escape even from "escape-proof" systems. However, the concept of biomanipulation, particularly if species native to a local area can be used, has a strong potential for modifications of food webs under certain conditions.

Since the mid-1980s there have been many trials of biomanipulation as an approach for regulating water quality, nuisance algae, or other limnological aspects of aquatic ecosystems. The biomanipulation approach contrasts with the older, and often successful, approach of nutrient regulation for ecosystem control. There obviously are many examples of nutrient control (e.g., phosphorus or sewage abatement) that have had positive effects on maintaining or restoring aquatic ecosystems—nobody would argue for abandoning the regulation of nutrient inputs into lake or river systems. However, proponents of biomanipulation would argue that management of fish at the top predator or planktivore level has strong theoretical support and also can be more logistically practical than some other solutions to water-quality problems. In at least some systems, fish populations can be manipulated (added/deleted) more readily than nutrient inputs can be controlled [particularly when nonpoint source pollutants are widespread in whole basins (e.g., in agricultural regions)].

There have also been controversies over comparisons between enclosure or tank experiments and results that take place in natural lakes or whole systems (Mazumder et al., 1990). However, Mazamuder et al. (1990) demonstrated concordance of results in large enclosures and whole lakes with respect to fish-mediated food-web differences. Moreover, it seems obvious that whole-lake versus "container" or enclosure experiments should be mutually useful, and not exclusive. "Tank" experiments like those of Drenner et al. (1986) or Threlkeld (1987, 1988) seem likely to allow rather precise partitioning of mechanisms of control in food webs, whereas the whole-lake manipulations of Carpenter, Schindler, and others seem most appropriate for ensuring that linkages found in small, replicated experiments actually apply in the real (and usually nonreplicated) world of whole systems.

What are the empirical findings in trials of biomanipulation, and under what conditions might biomanipulation by fish additions (or deletions) be difficult or impractical logistically? It seems intuitive that small, relatively closed systems (e.g., small lakes) might be the best candidates for biomanipulation of water quality via fish management. Here, species can be stocked with reasonable expectation that the introduction will succeed, and in small lakes, removal of a species is relatively possible if things go awry. Many of the experiments on biomanipulation are for small lentic systems. At the other extreme of the spectrum for biomanipulation as a potential tool would seem to be large rivers and large man-made reservoirs. The open nature of rivers seems almost to defy effective biomanipulation, because the same fish species that are often top carnivores or planktivores also can be highly mobile; thus, any biomanipulation of a large river by fish additions or removals might have severe logistical limitations. Rivers with the best potential for biomanipulation might be cold-water "trout" streams of modest size, where food-web effects of fish have been tested in basic studies (e.g., Power, 1990a, 1990b), or where thermal gradients might cause salmonids to tend to remain in the desired portions of a target system.

Large reservoirs may be slightly better candidates for biomanipulation, but most also offer potential for fish to leave the system, and some may have operating paradigms that occasionally cause catastrophic loss of fish from the system. For example, in Lake Texoma (Oklahoma–Texas), release of water through floodgates of Denison Dam for approximately a month in May 1982 resulted in irreplaceable loss of many of the largest broodstock of striped bass (*Morone saxatilis*) which were congregated near the dam at that time, and my multiyear gill net monitoring showed a trenchant decrease in catch of large striped bass even in the uplake region, that never recovered during the next 4 years of sampling. Thus, any stocking of piscivores (or planktivores) for the purpose of biomanipulation in large reservoirs is fraught with potentially insurmountable, catastrophic events. In addition, many reservoirs have a combination of planktivorous fishes that are small-bodied, and potentially vulnerable to most piscivores, and deep-bodied planktivores (like shad, *Dorosoma*) that can grow sufficiently large to achieve a size refuge from gape-limited piscivores such as largemouth bass (Hambright et al., 1991). Finally, most large reservoirs contain fish assemblages, and often zooplankton, with a high percentage of exotic species or species that could not have coevolved. Such a lack of history of coexistence dictates that tight or predictable links in a reservoir food web will be more by coincidence of evolutionary history of individual species than by any coevolution of traits linking species within the system. The lack of a mutual evolutionary history makes prediction of the ways one species will affect another problematic, and the reality is that most of what has been learned has been by trial and error.

Additionally, many river mainstem reservoirs are extremely dynamic because of variability in inflow conditions. For example, in a large run-of-the-river like Lake Texoma (Oklahoma–Texas), even if planktivores (e.g., shad, *Dorosoma*) could be regulated by biomanipulation and even if their grazing did influence zooplankton, phytoplankton, and water clarity or quality, any such effects would be rapidly lost during even a modest river rainfall–inflow event, when secchi depth can go, literally in hours, from more than 1 m to less than 15 cm (W. Matthews, personal observations; Matthews, 1984). Alteration of water clarity by such events, combined with summer winds, can keep silt in suspension for weeks or even months in wide, shallow reservoirs, with major consequences to primary productivity. Furthermore, such inflow events introduce large quantities of nutrients from the river basin, and the biomanipulator has no control over such events. Thus, it is probably best to focus on the potential of biomanipulation in more manageable aquatic systems.

Turning to more tractable systems, how well has biomanipulation of food webs for regulation of water quality worked in empirical tests? Findlay et al. (1994) noted the large number of lake manipulation experiments during the last two decades and reviewed the competing theories of direct and indirect predator control versus upward control by nutrients. Full review of all the many biomanipulation studies is beyond the scope of this chapter, but some examples and generalizations are possible. In spite of early competition of top-down cascade versus upward nutrient control as exclusive theories for lake control, many studies now acknowledge that both predation and nutrient enhancement pathways can result from the addition of fishes and be important in food webs. In natural systems, there can be unpredictable effects of biomanipulations. For example, removal of one dominant planktivorous fish species resulted in compensatory changes in abundance of nonmanipulated fish species (i.e., redundancy in the system), leading to food-web changes that would have been difficult to predict (Elser et al., 1995).

Mazumder et al. (1990) showed changes from planktivorous fish manipulation that cascaded through three trophic levels, commensurate with food-web theory. However, they also showed that in the course of these changes there was a change in nutrient limitation, which could also be of importance in the food web.

Reinertsen et al. (1990) reported results of a 5-year study following a rather harsh fish removal experiment (i.e., elimination of all fish in a eutrophic lake by rotenone). After fish removal, the zooplankton shifted per food-web theory to larger-bodied and more abundant herbivorous cladocerans, whose grazing then resulted in a fourfold decrease in algal biomass, and lowered phosphorus levels in the lake as a result of sedimentation. Although this "biomanipulation" seems to have effectively lowered the state of eutrophication in this lake, it would seem that such harsh approaches should only be taken under extreme conditions of necessity, and any extrapolation to suggest "remove all fish = improve water quality" would seem badly misplaced. Biomanipulation that removes "all" fish from a system would seem in most cases to be a poor lake management strategy!

By 1990, there was a decade of experiments in biomanipulation, summarized by McQueen (1990) by several generalities, including the following: (1) the downward trophic cascade is strongly damped, thus strong manipulation of fish is necessary to ensure that effects cascade to the desired level; (2) fish can influence food webs more in shallow than in deep lakes; (3) fish can have direct and indirect effects through nutrient and bioturbation pathways, in addition to cascading effects; and (4) fish manipulations are more likely to have desired effects on phytoplankton in shallow than in deep lakes, and success in deep lakes is best achieved through concomittant efforts to also control physical or chemical influences on the systems.

Persson et al. (1993) showed in controlled experiments and in a winterkill situation that trophic effects from piscivores cascading from *Stizostedion* to plankton produced effects at intermediate levels that could be predicted in part from classic food-web theory, but that reduction in planktivore pressure (roach, *Rutilus*) resulted in increased numbers of but not increased body size of cladocerans, and that increased numbers of *Daphnia* resulted in increased variation in phytoplankton abundance instead of its outright suppression. Persson et al. posited that disturbance in a system might influence the outcome of trophic cascades, making prediction from classic theory less precise, and that over the long-term changes from disturbance might lead to establishment of alternative steady states in producer structure (e.g., from phytoplankton- to macrophyte-dominated lakes).

A long-term study (Findlay et al., 1994) of effects of piscivore introduction to a 9-hectare lake in the Experimental Lakes Area (Ontario, Canada) showed that addition of northern pike (*Esox lucius*) resulted in decreased predation pressure on and changes in the zooplankton, with increases of large cladocerans, leading to initial reduction in biomass and primary productivity of phytoplankton. However, after several years, phytoplankton productivity returned to premanipulation levels. Findlay et al. (1994) found evidence of both direct and indirect effects of fish through cascading and nutrient pathways. In the postmanipulation lake environment, chrysophytes dominated, and a principal components analysis of the phytoplankton community structure showed a distinct difference between the taxonomy of the algae before and after additions of piscivores. However, changes developed over multiyear scales, again suggesting the need to extrapolate with caution any foodweb changes from short-term experiments to long-term patterns in whole lakes.

Baca and Drenner (1995) found that largemouth bass (*Micropterus salmoides*) addition (piscivore) to a 15-hectare reservoir had strong effects on planktivorous fishes, but that effects did not cascade to the phytoplankton or to nutrient levels. Their results supported the contention of McQueen et al. (1986) that top-down effects are substantially dampened as the cascade proceeds through the food chain to lower tropic levels. Baca and Drenner (1995) concluded that "top-down effects of piscivorous fish are strong near the top of the food chain but weaken with every step down the food web. In such communities, piscivorous fish have weak effects on phytoplankton and play a small role in the regulation of water quality of lakes."

Mittlebach et al. (1995) showed that largemouth bass (*Micropterus salmoides*) could be "keystone" piscivores in small lakes. After largemouth bass were eliminated in Wintergreen Lake, Michigan, populations of small-bodied planktivorous fishes decreased and large *Daphnia* species disappeared from the lake. This condition remained stable until bass were reintroduced 8 years later, after which the structure of the planktivore and zooplankton populations reverted to original conditions. Thus, at least in some situations (particularly in environmentally stable systems), manipulations of piscivores can have an effect at least to the second lower trophic level.

Romo et al. (1996) consider that fish biomanipulations have been documented for approximately 16 freshwater systems, the majority of which have been in shallow lakes. For a lake in The Netherlands, Romo et al. showed results of 7 years of following a biomanipulation that consisted of removing omnivorous/planktivorous bream (*Abramis*), stocking piscivorous northern pike (*Esox lucius*), and adding macrophytes. A multivariate analysis of the phytoplankton assemblages over 8 years (including premanipulation) showed that there were initial major changes in the phytoplankton (a "clear-water" phase) after manipulation, but that over time, the phytoplankton assemblage has begun to revert to its premanipulation condition. This recent study seems to agree with several earlier ones to suggest that biomanipulation may have immediate effects that last for several years, but that careful management (or perhaps continued manipulation) may be required to maintain the desired effects in perpetuity.

On balance, as of 1997 it appears from the literature that the careful manipulation of fish as components of the food web has the potential for application as one tool in the efforts of lake managers to regulate water quality. However, it is my impression from the literature that results may be difficult to predict, particularly with respect to quantitative outcomes, and that there can be great variation from system to system with respect to the actual results that are achieved. Long-term, sustained results seem more difficult to achieve than short-term changes, and the simple fact that food webs in many lakes and streams ultimately reflect the environmental (meterological and geological) template as well as the zoogeographic history of the fish and aquatic assemblages seems to provide a formidable challenge to humans who wish to override in perpetuity such influences. There is no question that fish can have powerful influences in food webs and that biomanipulation can achieve predicted changes with some level of success in some systems. The greater challenge seems to be maintaining such changes in keeping with societal goals. Clean water at the expense of a healthy fishery seems as unwise as a healthy fishery at the expense of clean water. Biomanipulation seems likely to succeed only when there can be agreement among diverse human entities as to what influence they would desire from adding or removing fish from food webs.

11.7 Keystone Species and Strong Interactors

Power et al. (1996) point out that there are generally two ways in which a species can play an important role in an ecosystem. It may be one of the community dominants (in numbers or biomass), and by virtue of processing large amounts of materials and so forth, it may have an important effect on the way the ecosystem functions. Alternatively, a species may not be necessarily abundant in an ecosystem, yet have a large effect on the way the biotia is structured because of its "keystone" role in the system (e.g., Paine, 1969, 1974). Power et al. (1996) point out that keystone species "differ from dominant species in that their effects are much larger than would be predicted from their abundance." Menge (1995) has also reviewed evidence for keystone species in the rocky marine intertidal, and suggested keystone predation in 35% of the 83 food-web subtypes examined.

Power et al. (1996) suggested expanding the original meaning of "keystone" species, which was restricted to predators, to include what they now call "community importance" (CI), which seeks to measure the change in a system trait per change in the abundance of the species of interest. Thus, changes can be mediated through many pathways. Power et al. (1996, their Table 1) listed six cases of "likely" keystone species that included fish, of which five involved piscivores, and two (there were two species in one study) that involved planktivores and omnivores. They also considered that "keystone species" might or might not play a keystone role, depending on the "context" of the conditions. Differences in predators at the species level can also be important. For example, Power et al.

(1996) considered largemouth bass and spotted bass (*Micropterus salmoides* and *M. punctulatus*) to be keystone species in small streams in south Oklahoma, strongly affecting the composition of other fish species in stream pools, but the congeneric smallmouth bass (*Micropterus dolumieu*) has much less effect on the same system (Harvey et al., 1988). He and Kitchell (1990) considered that piscivores in numerous studies in northern lakes functioned as keystone predators, having strong effects on ecosystems and sometimes driving components to local extinction.

Strong Interactors

Paine (1980) suggested a related concept of "strong interactions" for species that were generally important in strengths of linkages in food webs—whether through predation or other pathways. Paine (1980) suggested, in particular, that "strong links tend to characterize some of the plant–herbivore relationships." Power et al. (1985) concluded that the algivorous minnow (*Campostoma anomalum*) was a strong interactor in the sense of Paine (1980) in small stream of south Oklahoma, as described in Section 11.4. Flecker (1996, 1997) has shown the South American *Prochilodus* to be a strong interactor in river food webs, and that, as predicted by Power et al. (1996) for keystone species, the effect of *Prochilodus* is strong in some habitats (pools) and less so in others (riffles). Thus, the effect of this species is "context dependent" (Power et al., 1996). The fact that algivorous fish, in particular, might be strong interactors in food webs has already been suggested in the review of their activities and effects in Sections 11.4 and 11.5. Below, I expand this postulate for one species (*Campostoma anomalum*) for which our research group has amassed considerable information and speculate about other mechanisms, most as yet untested, whereby the species might have strong effects in stream systems in temperate North America. Other algivorous species in temperate and tropical streams might well yield similarly strong effects.

An example of a Strongly Interacting Fish Species

Although there are 50 known taxa in North America that primarily make a living trophically by eating attached algae—periphyton—only two species have been studied intensively. Nancy Grimm, Arizona State University, has documented the strong effects of *Agosia chrysogaster* on nutrient dynamics in Sycamore Creek, Arizona (Grimm, 1988). Since 1982, our laboratory at the University of Oklahoma Biological Station has studied *Campostoma anomalum* extensively and we have considerable information on the direct effects of *Campostoma* on algae. *Campostoma* are substantially large minnows, up to about 150 mm in standard length; they are extremely abundant, often in schools numbering in the thousands. They are very widespread in the streams of North America east of the Continental Divide (Lee et al., 1980). There are three different species, the most common being the central stoneroller (*Campostoma anomalum*) to which the rest of this discussion refers. *Campostoma* are completely algivorous as adults, with a very long intestine. The intestine is actually many times longer than the body of the fish itself.

Campostoma graze very actively and vigorously throughout the day in warm weather, in streams ranging from fairly small to substantially large rivers throughout eastern North America. It is well documented that in some streams they can be not only the most abundant species, but, can actually make up the majority of the biomass of fishes in streams. In the Baron Fork of the Illinois River, I have counted more than 6000 individuals in a relatively small pool.

There is abundant evidence that grazing by *Campostoma* is involved in multitrophic-level cascading interactions (Power and Matthews, 1983; Power et al., 1985) and substantial direct and indirect effects in ecosystems (Table 11.4, Fig. 11.4). Matthews et al. (1987b) summarized and speculated about some of these effects. *Campostoma* have strong direct effects on algae. Numerous studies (Power et al., 1985; Power and Matthews, 1983; Gelwick and Matthews 1992; Stewart 1987; Power et al., 1988) show that *Campostoma*, by grazing, decrease the standing crop of attached algae on surfaces in streams. They can have a direct effect in changing the kinds of algae that dominate, not only taxonomically but also in terms of the growth form of the algae. *Campostoma* can alter the schedule of growth senescence and the sloughing of algae on stony surfaces (Gelwick et al., 1997). Thus, the grazing of *Compostoma* will dampen the temporal variation in algae. It will also dampen spatial variation in algae and make it more uniform across surfaces that are being grazed.

Campostoma decrease the amount of algae; they change the kinds of algae that persist and are present. They result in an algal community on substrates that is more tightly adherent and more resistant to scour. This algae will also have greater primary productivity per gram relative to ungrazed algae. Because large schools of *Campostoma* remove so much total algae they can actually lower the primary productivity in a stream per unit area, because they maintain low standing crops (although the algae may be more productive per gram in grazed conditions).

Campostoma also may have negative or positive effects on the invertebrates in streams. This is more speculative and also includes indirect effects (Fig. 11.4). If *Campostoma* are removing algae which invertebrates use for attachment or shelter, this can lower standing crop of invertebrates. Additionally, mechanical action of *Campostoma* in grazing actively taking in algae probably dislodges microinvertebrates, causing them to enter the drift by virtue of the mechanical disturbance.

Groups of minnows often seen following *Campostoma* may do so because of improved feeding opportunities, if grazing by *Campostoma* mechanically removes invertebrates from algae and puts them in the water column as part of the drift,

Table 11.4. Known Effects of Campostoma in Ecosystems

Known Campostoma Effects	1	2	3	4	5	6	7	8	9	10	11	12	13	14
Decrease algae standing crop	x	x								x	x	x	x	x
Change spatial patterns of algae	x	x	x								x	x	x	x
Change algal community composition or growth form		x									x	x	x	x
Decrease temporal variability in algae				x										
Differentially ingest algae											x			
Graze by marginal value theorem						x								
Produce feces with live algae = potential downstream transport of algae and nutrients												x		
Have strong effects in predator–grazer–algae cascade	x	x											x	x
Decrease size of particulate organic matter								x					x	x
Increase uptake of dissolved organic matter									x					
Lower bacterial standing crops													x	x
Lower percent organic matter					x								x	x
Lower PPR/area; increase PPR/unit biomass													x	x
Change carbon : nitrogen ratios													x	x
Change benthic invertebrate community										x				
Increase snail productivity; lower crayfish productivity										x				
Alter snail life history											x			
Impact on algae ~ crayfish > snails											x			
Move CPOM downstream							x					x		x
Alter heterogeneity of algae structure in pools														x

[a] (1) Power and Matthews (1983); (2) Power et al. (1985); (3) Power et al. (1988); (4) Power and Stewart (1987); (5) Stewart (1987); (6) Stewart and Matthews (unpublished data); (7) Partridge (1991); (8) Gardner (1993); (9) Gardner et al. (unpublished data); (10) Vaughn et al. (1993); (11) Gelwick and Matthews (unpublished data); (12) Gelwick (1995); (13) Gelwick and Matthews (1992); (14) Gelwick et al. (in press).

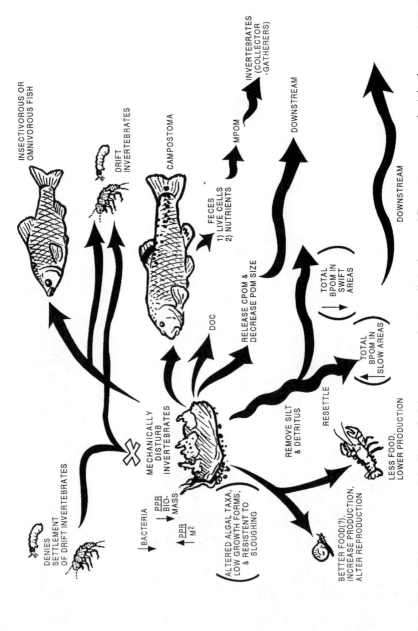

Figure 11.4. Known or potential effects of algivorous central stoneroller minnows (*Campostoma anomalum*) in the ecosystem of a small Midwest (USA) stream.

giving the insectivorous fish a better opportunity to feed. Thus, feeding action of *Campostoma* might directly increase the entry of invertebrates into the drift and their downstream transport, although they obviously may be eaten by accompanying fishes.

Campostoma increases the downstream transport of materials (Fig. 11.4). Partridge (1991) showed that grazing by *Campostoma* in artificial streams significantly increased the downstream transport of coarse particulate organic matter (CPOM). However, *Campostoma* had no significant effect on smaller-sized fractions of particulate matter, but the effect might be context dependent (depending on the kinds of periphyton available). In a natural stream, this could be an mechanism for moving materials, because attached algae that is growing in place in long strands is not available as food to microinvertebrates that require smaller particles of algae. These invertebrates might more readily eat algal particulates broken loose and dropped to the stream bottom or became buried there.

A related effect of *Campostoma* that is probably, at least in part, a direct effect of the fish is that they reduce, overall, the size of particulate organic matter (POM) in streams. Gelwick and Matthews (1992) and Gelwick et al. (1997) showed that in areas protected from grazing by *Campostoma,* the relative size fraction of POM is larger than where these fish graze. There is apparently direct ingestion and breaking down of the size of these materials, and the feces produced is in a size range that is approximate to medium POM (which is an appropriately sized food for collector–gatherer invertebrates).

Finally, *Campostoma* appear to have numerous demonstrated or speculative indirect effects on other trophic levels (Fig. 11.4). Vaughn et al. (1993) showed that snails housed with *Campostoma* exhibited shifts in life-history traits and in production, apparently benefitting from the grazing of *Campostoma* that made algae more available. In contrast, crayfish experienced direct competition for food from *Campostoma* (Vaughn et al., 1993). More speculative aspects of Figure 11.4 include the postulates that *Campostoma* might deny access of drifting invertebrates to settling in grazed patches, or rapidly resuspend them to the water column, that they will alter the distribution of silt and detritus, or change the overall distribution of particulate organic matter sizes and location in the stream. Finally, feces of *Campostoma* may contain many live cells and concentrated nutrients, and the species may selectively kill different kinds of algae by its grazing. If this is broadly true, *Campostoma* could influence distribution and kinds of algae in more ways than we now know.

To summarize, *Campostoma* is a well-studied strong interactor in many streams in eastern North America. Our group has documented at least 21 kinds of effects of this species in stream ecosystems, and numerous other direct and indirect pathways of effects remain speculative. We also know little to date about the ways ontogeny, population density, or local variation influence the effect of this species in ecosystems. Additionally, as indicated above, *Campostoma* is only one of several dozen strongly algivorous species in temperate North America, and the numbers of potentially strong interactors in the tropics or elsewhere worldwide is probably unknown. I offer this species not as an indication that it is an unusually strong interactor, but merely to suggest that as more and more studies address different species of algivores, detritivores, and all other fish functional groups, it seems likely that the role of fish in stream and lake ecosystems will continue to be recognized as broadly important.

Literature Cited

Aadland, L. P. (1993) Stream habitat types: their fish assemblages and relationship to flow. North Am. J. Fish. Mangmt., **3**, 790–806.

Abbott, I. (1983) The meaning of z in species-area regressions and the study of species turnover in island biogeography. Oikos **41**, 385–90.

Abrahams, M. V. and P. W. Colgan. (1985) Risk of predation, hydrodynamic efficiency and their influence on school structure. Environ. Biol. Fish., **13**, 195–202.

Adams, S. M., B. L. Kimmel, and G. R. Ploskey. (1983) Sources of organic matter for reservoir fish production: a trophic–dynamics analysis. Can. J. Fish. Aquat. Sci., **40**, 1480–95.

Adamson, S. W. and T. E. Wissing. (1977) Food habits and feeding periodicity of the rainbow, fantail, and banded darters in Four Mile Creek, Ohio J. Sci., **77**, 164–69.

Aerts, P. (1992) Fish biomechanics: purpose or means? Netherlands J. Zool., **42**, 430–44.

Aleev, Y. G. (1969) *Function and Gross Morphology in Fish* (translated from Russian, 1969). Israel Prog. Sci. Translation, Jerusalem, 1969.

Alevizon, W. S. (1976) Mixed schooling and its possible significance in a tropical western Atlantic parrotfish and surgeonfish. Copeia, **1976**, 796–98.

Alexander, G. R. (1979) Predators of fish in coldwater streams, in *Predator–Prey Systems in Fisheries Management* (ed. H. Clepper). Sport Fishing Institute, Washington, DC. pp. 153–170.

Alexander, R. M. N. (1966) The functions and mechanisms of the protrusible upper jaws of two species of cyprinid fish. J. Zool. Lond., **149**, 288–96.

Alexander, R. M. N. (1967) *Functional Design of Fishes*. Hutchinson and Co., London.

Alimov, A. F., L. E. Anokhina, E. V. Balushkina, S. M. Golubkov, T. V. Khlebovich, P. I. Krylov, E. B. Paveljeva, I. V. Telesh, and L. P. Umnova. (1992) Fish impact on the structure, abundance, and biomass of plankton and benthos of two small shallow lakes: a three-year study. Russ. J. Aquat. Ecol., **1**, 9–37.

Allan, J. D. (1978) Trout predation and the size composition of stream drift. Limnol. Oceanogr., **23,** 1231–37.

Allan, J. D. (1981) Determinants of diet of brook trout (*Salvelinus fontinalis*) in a mountain stream. Can. J. Fish. Aquat. Sci., **38,** 184–92.

Allan, J. D. (1982) The effects of reduction of trout density on the invertebrate community of a mountain stream. Ecology, **63,** 1444–55.

Allan, J. D. (1983) Food consumption by trout and stoneflies in a rocky mountain stream, with comparison to prey standing crop, in *Dynamics of Lotic Systems* (eds. T. D. Fontaine III and S. M. Bartell), Ann Arbor Science, Ann Arbor, MI.

Allan, J. D. (1995) *Stream Ecology—Structure and Function of Running Water.* Chapman & Hall, New York.

Allan, J. R. (1986) The influence of species composition on behavior in mixed-species cyprinid shoals. J. Fish. Biol., **29,** (Supplement A), 97–106.

Allan, J. R. and T. J. Pitcher. (1986) Species segregation during predator evasion in cyprinid fish shoals. Freshwater Biol., **16,** 653–59.

Allee, W. C., O. Park, A. E. Emerson, and T. Park. (eds.) (1949) *Principles of Animal Ecology.* W. B. Saunders Company, Philadelphia.

Allen, D. M., W. S. Johnson, and V. Ogburn-Matthews. (1995) Trophic relationships and seasonal utilization of salt-marsh creeks by zooplanktivorous fishes. Environ. Biol. Fish., **42,** 37–50.

Allen, K. R. (1951) The Horokiwi Stream: A study of a trout population. New Zealand Marine Department, Fisheries Bulletin No. 10, Wellington, New Zealand.

Allibone, R. M. and G. P. Wallis. (1993) Genetic variations and diadromy in some native New Zealand galaxiids (Teleostei: Galaxiidae). Biol. J. Linn. Soc., **50,** 19–33.

Amemiya, C. T., P. K. Powers and J. R. Gold. (1992) Chromosomal evolution in North American cyprinids, in *Systematics, Historical Ecology, & North American Freshwater Fishes,* (ed. R. L. Mayden). Stanford University Press, pp. 515–33.

Andersen, T. H., N. Friberg, H. O. Hansen, T. M. Iversen, D. Jacobsen, and L. Krøjgaard. (1993) The effects of introduction of brown trout (*Salmo trutta* L.) on *Gammarus pulex* L. drift density in two fishless Danish streams. Arch. Hydrobiol., **126,** 361–71.

Anderson, A. A., C. Hubbs, K. O. Winemiller, and R. J. Edwards. (1995) Texas freshwater fish assemblages following three decades of environmental change. Southwest. Nat., **40,** 314–21.

Anderson, C. S. (1985) The structure of sculpin populations along a stream size gradient. Environ. Biol. Fish., **13,** 93–102.

Andrewartha, H. G. and L. C. Birch. (eds.) (1954) *The Distribution and Abundance of Animals.* Chicago University Press, Chicago.

Andrews, E. D. (1994) Marginal bed load transport in a gravel bed stream, Sagehen Creek, California. Water Resources Research, **30,** 2241–50.

Andrus, C. W., B. A. Long, and H. A. Froehlich. (1988) Woody debris and its contribution to pool formation in a coastal stream 50 years after logging. Can. J. Fish. Aquat. Sci., **45,** 2080–86.

Angermeier, P. L. (1982) Resource seasonality and fish diets in an Illinois stream. Environ. Biol. Fish., **74,** 251–64.

Angermeier, P. L. (1985) Spatio-temporal patterns of foraging success for fishes in an Illinois stream. Am. Midl. Natl., **114,** 343–59.

Angermeier, P. L. (1987) Spatio-temporal variation in habitat selection by fishes in small Illinois streams, in *Community and Evolutionary Ecology of North American Stream Fishes* (eds. W. J. Matthews and D. C. Heins). University of Oklahoma Press, Norman, pp. 52–60.

Angermeier, P. L. and J. R. Karr. (1983) Fish communities along environmental gradients in a system of tropical streams. Environ. Biol. Fish., **9,** 117–35.

Angermeier, P. L. and J. R. Karr. (1984) Relationships between woody debris and fish habitat in a small warmwater stream. Trans. Am. Fish. Soc., **113,** 716–26.

Angermeier, P. L. and I. J. Schlosser. (1989) Species-area relationships for stream fishes. Ecology, **70,** 1450–62.

Angermeier, P. L. and R. A. Smogor. (1995) Estimating number of species and relative abundances in stream-fish communities: effects of sampling effort and discontinuous spatial distributions. Can. J. Fish. Aquat. Sci., **52,** 936–49.

Appenzeller, A. R. and W. C. Leggett. (1995) An evaluation of light-mediated vertical migration of fish based on hydroacoustic analysis of the diel vertical movements of rainbow smelt (*Osmerus mordax*). Can. J. Fish. Aquat. Sci., **52,** 504–11.

Ashley, E. J., L. B. Kats, and J. W. Wolfe. (1993) Balancing trade-offs between risk and changing shoal size in northern redbelly dace *Phoxinus eos*. Copeia, **1993,** 540–42.

Åström, M. (1994) Travel cost and the ideal free distribution. Oikos, **69,** 516–19.

Auerbach, M. and A. Shmida. (1993) Vegetation change along an altitudinal gradient on Mt. Hermon, Israel—no evidence for discrete communities. J. Ecol., **81,** 25–33.

Auerbach, M. J. (1984) Stability, probability, and the topology of food webs, in *Ecological Communities—Conceptual Issues and the Evidence* (eds. D. R. Strong, D. Simberloff, L. G. Abele, and A. B. Thistle. Princeton University Press, Princeton, NJ, pp. 413–36.

Auster, P. J., C. A. Griswold, M. J. Youngbluth, and T. G. Bailey. (1992) Aggregations of myctophid fishes with other pelagic fauna. Environ. Biol. Fish., **35,** 133–39.

Azoulay, B. and M. Gophen. (1992) Feeding habits of larval *Mirogrex terraesanctae* (Steinitz, 1952) in Lake Kinneret (Israel) II. experimental study. Hydrobiologia, **246,** 251–58.

Baca, R. M. and R. W. Drenner. (1995) Do the effects of piscivorous largemouth bass cascade to the plankton? Hydrobiologia, **316,** 139–151.

Bailey, R. G. (1989) Explanatory supplement to ecoregions map of the continents. Environ. Conserv., **16,** 307–10 (with map).

Bailey, R. G. (1995) *Descriptions of the Ecoregions of the United States,* 2nd ed. USDA Forest Services Miscellaneous Publication, No. 1391, (with included map).

Bailey, R. M. (1955) Differential mortality from high temperatures in a mixed-population of fishes in southern Michigan. Ecology, **36,** 526–28.

Bagenal, T. B. The interrelation of the size of fish eggs, the date of spawning and the production cycle. J. Fish. Biol., **3**, 207–19.

Baker, J. A. and S. L. Ball. (1995) Microhabitat selection by larval *Chironomus tentans* (Diptera: Chironomidae): effects of predators, food, cover and light. Freshwater Biol., **34**, 101–06.

Baker, J. A. and S. T. Ross. (1981) Spatial and temporal resource utilization by southeastern Cyprinids. Copeia, **1981**, 178–79.

Baker, J. A., S. A. Foster, and M. A. Bell. (1995) Armor morphology and reproductive output in threespine stickleback, *Gasterosteus aculeatus.* Environ. Biol. Fish., **44**, 225–33.

Baker-Dittus, A. M. (1978) Foraging patterns of three sympatric killifish. Copeia, **1978**, 383–89.

Bakker, T. C. M. (1993) Positive genetic correlation between female preference and preferred male ornament in sticklebacks. Nature, **363**, 255–57.

Balon, E. K. (1975) Reproductive guilds of fishes: a proposal and definition. J. Fish. Res. Bd. Canada, **32**, 821–64.

Balon, E. K. (1984) Patterns in the evolution of reproductive styles in fishes, in *Fish Reproduction: Strategies and Tactics,* (eds. G. W. Potts and R. J. Wootton). Academic Press, London. pp. 35–53.

Balon, E. K. and D. J. Stewart. (1983) Fish assemblages in a river with unusual gradient (Luongo, Africa–Zaire system), reflections on river zonation, and description of another new species. Environ. Biol. Fish., **9**, 225–52.

Baltz, D. M. and P. B. Moyle. (1982) Life history characteristics of tule perch (*Hysterocarpus traski*) populations in contrasting environments. Environ. Biol. Fish., **7**, 229–42.

Baltz, D. M. and P. B. Moyle. (1984) Segregation by species and size classes of rainbow trout, *Salmo gairdneri,* and Sacramento sucker, *Catostomus occidentalis,* in three California streams. Environ. Biol. Fish., **10**, 101–10.

Baltz, D. M., P. B. Moyle, and N. J. Knight. (1982) Competitive interactions between benthic stream fishes, riffle sculpin *Cottus gulosus,* and speckled dace, *Rhinichthys osculus.* Can. J. Fish. Aquat. Sci., **39**, 1502–11.

Baltz, D. M., B. Vondracek, L. R. Brown, and P. B. Moyle. (1991) Seasonal changes in microhabitat selection by rainbow trout in a small stream. Trans. Am. Fish. Soc., **120**, 166–76.

Banarescu, P. and B. W. Coad. (1991) Cyprinids of Eurasia, in *Cyprinid Fishes: Systematics, Biology and Exploitation* (eds. I. J. Winfield and J. S. Nelson). Chapman & Hall, London, pp. 127–55.

Bannister, R. C. A. (1977) North Sea plaice, in *Fish Population Dynamics,* (ed. J. A. Guilland). John Wiley and Sons, London, pp. 243–82.

Barber, F. G. and T. S. Murty. (1977) Perennial sea ice: speculation concerning physical and biological consequences, in *Polar Oceans,* (ed. M. J. Dunbar). Arctic Institute of North America, Montreal, pp. 257–68. (Original not seen, cited from Crossman and McAllister, 1986).

Barbour, C. D. and B. Chernoff. (1984) Comparative morphology and morphometrics of the pescados blancos (genus *Chirostoma*) from Lake Chapala, Mexico, in *Evolution of Fish Species Flocks* (eds. A. A. Echelle and I. Kornfield). University of Maine Press, Orono, pp. 111–27.

Barbour, C. D. and J. H. Brown. (1974) Fish species diversity in lakes. Am. Nat., **108,** 473–88.

Barica, J. and J. A. Mathias. (1979) Oxygen depletion and winterkill risk in small prairie lakes under extended ice cover. J. Fish. Res. Bd. Can., **36,** 980–86.

Barila, T. Y., J. R. Stauffer, Jr., and C. H. Hocutt. (1982) Temperature preference and avoidance response of the common shiner, *Notropis cornutus* (Mitchill). Arch. Hydrobiol., **96,** 112–19.

Barlow, G. W. (1974) Extraspecific imposition of social grouping among surgeonfishes (Pisces: Acanthuridae). J. Zool. Lond., **174,** 333–40.

Barrett, J. C., G. D. Grossman, and J. Rosenfeld. (1992). Turbidity-induced changes in reactive distance of rainbow trout. Trans. Am. Fish. Soc., **121,** 434–43.

Bart, H. A. Jr. (1989) Fish habitat association in an Ozark stream. Environ. Biol. Fish., **24,** 173–86.

Bart, H. A., Jr. and L. M. Page (1992) The influence of size and phylogeny on life history variation in North American percids, in *Systematics, Historical Ecology and North American Freshwater Fishes* (ed. R. L. Mayden). Stanford University Press, Stanford, CA, pp. 553–72.

Barton, M. (1984) The fishes of Jessamine Creek, Jessamine County, Kentucky. Trans. KY Acad. Sci., **45,** 30–2.

Bass, D. (1982) The importance of a stratified design in sampling stream benthos (ed. J. Davis). Proc. Symp. on Recent Benthol. Invest. in Texas and Adjacent States, pp. 3–9.

Bauchot, R., M. L. Bauchot, R. Platel, and J. M. Ridet. (1977) Brains of Hawaiian tropical fishes: brain size and evolution. Copeia, **1977,** 42–6.

Bauer, B. H., B. A. Branson, and S. T. Colwell. (1978) Fishes of Paddy's Run Creek and the Dry Fork of the Whitewater River, Southwestern Ohio, Ohio J. Sci., **78,** 144–48.

Bayley, P. B. and J. W. Li. (1992) Riverine fishes, in *The Rivers Handbook—Hydrological and Ecological Principles* (eds. P. Calow and G. E. Petts). Blackwell Scientific Publications, Oxford, Vol. 1, pp. 251–81.

Bayley, P. B. and L. L. Osborne. (1993) Natural rehabilitation of stream fish populations in an Illinois catchment. Freshwater Biol., **29,** 295–300.

Baylis, J. R., D. R. Wiegmann, and M. H. Hoff. (1993) Alternating life histories of smallmouth bass. Trans. Am. Fish. Soc., **122,** 500–10.

Beall, E., M. Heland, and C. Marty. (1989) Interspecific relationships between emerging Atlantic salmon, *Salmo salar* and coho salmon, *Oncorhynchus kisutch,* juveniles. J. Fish Biol., **35a,** 285–93.

Beaumont, P. (1975) Hydrology, in *River Ecology,* (ed. B. A. Whitton). University of California Press, Berkeley, pp. 1–38.

Bechara, J. A., G. Moreau, and L. Hare. (1993) The impact of brook trout (*Salvelinus fontinalis*) on an experimental stream benthic community: the role of spatial and size refugia. J. Animal Ecol., **62**, 451–64.

Becker, C. D., D. A. Neitzel, and D. H. Fickeisen. (1982) Effects of dewatering on chinook salmon redds: tolerance of four developmental phases to daily dewaterings. Trans. Am. Fish. Soc., **111**, 624–37.

Beebe, W. (1945) Vertebrate fauna of a tropical dry season mudhole. Zoologia (New York), **30**, 81–88.

Beecher, H. A. and R. F. Fernau. (1982) Fishes of Oxbow Lakes of Washington. Northwest Sci. **57**, 125–131.

Beecher, H. A., E. R. Dott, and R. F. Fernau. (1988) Fish species richness and stream order in Washington State streams. Environ. Biol. Fish., **22**, 193–209.

Beitinger, T. L. and L. C. Fitzpatrick. (1979) Physiological and ecological correlates of preferred temperature in fish. Am. Zool., **19**, 319–30.

Beitinger, T. L. and M. J. Pettit. (1984) Comparison of low oxygen avoidance in a bimodal breather, *Erpetoichthys calabaricus* and an obligate water breather, *Percina* caprodes. Environ. Biol. Fish., **11**, 235–40.

Beitinger, T. L., J. J. Magnuson, W. H. Neill, and W. R. Shaffer. (1975) Behavioral thermoregulation and activity patterns in the green sunfish, *Lepomis cyanellus*. Animal Behav., **23**, 222–29.

Beklioglu, M. and B. Moss. (1996) Mesocosm experiments on the interaction of sediment influence, fish predation and aquatic plants with the structure of phytoplankton and zooplankton communities. Freshwater Biol., **36**, 315–25.

Belk, M. C. (1995) Variation in growth and age at maturity in bluegill sunfish: genetic or environmental effects? J. Fish. Biol., **47**, 237–47.

Belk, M. C. and C. Lydeard. (1994) Effect of *Gambusia holbrooki* on a similar-sizes, syntopic poeciliid, *Heterandria formosa:* competitor or predator? Copeia, **1994**, 296–302.

Bell, D. E. and R. D. Hoyt. (1980) Temporal and spatial abundance and diversity of fishes in a Kentucky Stream. Trans. KY Acad. Sci., **41**, 35–44.

Bell-Cross, G. and J. L. Minshull. (1988) *The Fishes of Zimbabwe*. Trustees of the National Museums and Monuments of Zimbabwe, Harare.

Bence, J. R. and W. W. Murdock. (1986) Prey size selection by the mosquito fish: relation to optimal diet theory. Ecology, **67**, 324–36.

Bendell, B. E. and D. K. McNicol. (1987) Cyprinid assemblages, and the physical and chemical characteristics of small northern Ontario lakes. Environ. Biol. Fish., **19**, 229–34.

Bengtson, S. A. and P. H. Enckell. (1983) Preface to symposium on "Island Ecology." Oikos, **41**, 296–98.

Benke, A. C., R. L. Henry, III, D. M. Gillespie, and R. J. Hunter. (1985) Importance of snag habitat for animal production in southeastern streams. Fisheries, **10**, 8–13.

Benke, A. C., T. C. Van Arsdall, D. M. Gillespie, and F. K. Parrish. (1984) Invertebrate productivity in a subtropical backwater river: the importance of habitat and life history. Ecol. Monogr., **54,** 25–63.

Bennett, G. W. (1958) Aquatic biology, in *A Century in Biological Research* (eds. H. B. Mills et al.). IL Nat. Hist. Surv. Bull., **27,** 163–78.

Bennett, G. W. (1970) *Management of Lakes and Ponds* (ed. G. W. Bennett). Van Nostrand Reinhold Company, New York.

Bennet, W. A. and T. L. Beitinger. (1996) Extreme thermal tolerance of the sheepshead minnow, *Cyprinodon variegatus.* Abstract ASIH Annual Meeting, New Orleans, LA.

Berg, L. S. (1940) (English translation, 1947). *Classification of Fishes, Both Recent and Fossil* J. W. Edwards Co., Ann Arbor, MI.

Berg, L. S. (1949) *Freshwater Fishes of the U.S.S.R. and Adjacent Countries,* Academy of Sciences, Moscow, Vol. III. [Translated from Russian (1965) by Israel Program for Scientific Translations, Jerusalem.]

Bergman, E. (1990) Effects of roach *Rutilus rutilus* on two percids, *Perca fluviatilis* and *Gymnocephalus cernua:* importance of species interaction for diet shifts. Oikos, **57,** 241–49.

Berkman, H. E. and C. F. Rabeni. (1987) Effect of siltation on stream fish communities. Environ. Biol. Fish., **18,** 285–94.

Berra, T. M. (1981) *An Atlas of Distribution of the Freshwater Fish Families of the World.* University of Nebraska Press, Lincoln.

Berra, T. M. (1987) Speculations on the evolution of life history tactics of the Australian grayling. Am. Fish. Soc. Symp., **1,** 519–30.

Berra, T. M. and G. E. Gunning. (1970) Repopulation of experimentally decimated sections of streams by longear sunfish *Lepomis megalotis megalotis* (Rafinesque). Trans. Amer. Fish. Soc., **99,** 776–81.

Berra, T. M. and G. E. Gunning. (1972) Seasonal movement and home range of the longear sunfish, *Lepomis megalotis* (Rafinesque) in Louisiana. Am. Midl. Nat., **88,** 368–75.

Berven, K. A. and D. E. Gill. (1983) Interpreting geographic variation in life-history traits. Am. Zool., **23,** 85–97.

Biggs, B. J. F. (1995) The contribution of flood disturbance, catchment geology and land use to the habitat template of periphyton in stream ecosystems. Freshwater Biol., **33,** 419–38.

Bilby, R. E. and J. W. Ward. (1989) Changes in characteristics and function of woody debris with increasing size of streams in western Washington. Trans. Am. Fish. Soc., **118,** 368–78.

Binderim, G. E. (1977) Fishes of Mill Creek, a tributary of the Washita River, Johnston and Murray counties, Oklahoma. Proc. OK Acad. Sci., **57,** 1–11.

Bisson, P. A. and J. R. Sedell. (1984) Salmonid populations in streams in clearcut vs. old-growth forests of western Washington, in *Fish and Wildlife Relationships in Old-Growth Forest* (eds. W. R. Meehan, T. R. Merrell, Jr., and T. A. Hanley). American Institute of Fisheries Research Biologists, Juneau, AK, pp. 121–29.

Blaber, S. J. M., D. T. Brewer, and J. P. Salini. (1994) Diet and dentition in tropical ariid catfishes from Australia. Environ. Biol. Fish., **40**, 159–74.

Blachuta, J. and A. Witkowski. (1990) The longitudinal changes of fish community, in the Nysa Klodzka River (Sudety Mountains) in relation to stream order. Pols. Arch. Hydrobiol., **37**, 325–42.

Black, E. C. (1953) Upper lethal temperatures of some British Columbia freshwater fishes. J. Fish. Res. Bd. Can., **10**, 196–210.

Black, J. D. (1940) The distribution of the fishes of Arkansas. Unpublished Ph.D. dissertation, University of Michigan, Ann Arbor.

Black, J. D. (1954) *Biological Conservation—With Particular Emphasis on Wildlife.* Mc-Graw-Hill, New York.

Black, R. W., II and N. G. Hairston, Jr. (1988) Predator driven changes in community structure. Oecologia, **77**, 468–79.

Black, T. and R. V. Bulkley. (1985) Preferred temperature of yearling Colorado squawfish. Southwest. Nat. **30**, 95–100.

Blackburn, T. M., V. K. Brown, B. M. Doube, J. J. D. Greenwood, J. H. Lawton, and N. E. Stork. (1993) The relationship between abundance and body size in natural animal assemblages. J. Anim. Ecol., **62**, 519–28.

Blair, A. P. (1959) Distribution of the darters (Percidae, Etheostomatinae) of northeastern Oklahoma. Southwest. Nat., **4**, 1–13.

Blinn, D. W., C. Runck, D. A. Clark, and J. N. Rinne. (1993) Effects of rainbow trout predation on Little Colorado spinedace. Trans. Am. Fish. Soc., **122**, 139–43.

Block, B. A. and J. R. Finnerty. (1994) Endothermy in fishes: a phylogenetic analysis of constraints, predispositions, and selection pressures. Environ. Biol. Fish., **40**, 283–302.

Blumer, L. S. (1979) Male parental care in the bony fishes. Quart. Rev. Biol., **54**, 149–61.

Blumer, L. S. (1985a) Reproductive natural history of the brown bullhead *Ictalurus nebulosus* in Michigan. Am. Midl. Natl., **114**, 318–30.

Blumer, L. S. (1985b) The significance of biparental care in the bullhead, *Ictalurus nebulosus.* Environ. Biol. Fish., **12**, 231–36.

Bohanan, R. E. and D. M. Johnson. (1983) Response of littoral invertebrate populations to a spring fish exclusion experiment. Freshwat. Invertebr. Biol., **2**, 28–40.

Bonacci, O. (1993) Hydrological identification of drought. Hydrol. Process., **7**, 249–62.

Bond, C. E. (1996) *Biology of Fishes,* 2nd ed. Saunders College Publ., Philadelphia.

Bonetto, A. A. (1975) Hydrologic regime of the Paraná River and its influence on ecosystems, in *Coupling of Land and Water Systems* (ed. A. D. Hasler). Springer-Verlag, New York, pp. 175–97.

Borges, M. H. (1950) Fish distribution studies, Niangua Arm of the Lake of the Ozarks, Missouri. J. Wildl. Mangmt., **14**, 16–33.

Borjeson, H. and L. B. Hoglund. (1976) Swimbladder gas and Root effect in young salmon during hypercapnia. Comparative Biochemistry and Physiology, **54A**, 335–39.

Bornbusch, A. H. and J. G. Lundberg. (1989) A new species of *Hemisilurus* (Siluriformes,

Siluridae) from the Mekong River, with comments on its relationships and historical biogeography. Copeia, **1989**, 434–44.

Boschung, H. T. and P. O'Neil. (1981) The effects of forest clearcutting on fishes and macroinvertebrates in an Alabama stream, in *The Warmwaters Stream Symposium* (eds. C. F. Bryan, G. E. Hall, and G. B. Pardue). Southern Div., American Fisheries Society, Lawrence, KS, pp. 200–17.

Boucher, D. H., S. James, and K. H. Keeler. (1982) The ecology of mutualism. Annu. Rev. Ecol. Syst., **13**, 315–47.

Boulton, A. J. and E. H. Stanley. (1995) I. Hyporheic processes during flooding and drying in a Sonoran Desert stream. II. Faunal dynamics. Arch. Hydrobiol., **134**, 27–52.

Boulton, A. J., C. G. Peterson, N. B. Grimm, and S. G. Fisher. (1992) Stability of an aquatic macroinvertebrate community in a multiyear hydrologic disturbance regime. Ecology, **73**, 2192–2207.

Bowen, S. H. (1979) A nutritional constraint in detritivory by fishes: the stunted population of *Sarotherodon mossambicus* in Lake Sibaya, South Africa. Ecol. Monogr., **49**, 17–31.

Bowen, S. H. (1983) Detritivory in neotropical fish communities. Environ. Biol. Fish., **9**, 137–44.

Bowers, M. A. and J. A. Brown. (1982) Body size and coexistence in desert rodents: chance or community structure? Ecology, **63**, 391–400.

Brabrand, A. (1985) Food of roach *Rutilus rutilus* and ide *Leusiscus idus:* significance of diet shift for interspecific competition in omnivorous fishes. Oecologia, **66**, 461–67.

Bradford, D. F., F. Tabatabai, and D. M. Graber. (1993) Isolation of remaining populations of the native frog, *Rana muscosa,* by introduced fishes in Sequoia and Kings Canyon National Parks, California. Conserv. Biol., **7**, 882–88.

Bramblett, R. G. and K. D. Fausch. (1991a) Fishes, macroinvertebrates, and aquatic habitats of the Purgatorie River in Pinion Canyon, Colorado. Southwest. Nat., **36**, 281–94.

Bramblett, R. G. and K. D. Fausch. (1991b) Variable fish communities and the index of biotic integrity in a western Great Plains river. Trans. Am. Fish. Soc., **120**, 752–69.

Brandt, S. B. and V. A. Wadley. (1981) Thermal fronts as ecotones and zoogeographic barriers in marine and freshwater systems. Proc. Ecol. Soc. Aust., **11**, 13–26.

Brandt, S. B., J. J. Magnuson, and L. B. Crowder. (1980) Thermal habitat partitioning by fishes in Lake Michigan. Can. J. Fish. Aquat. Sci., **37**, 1557–64.

Breden, F. and G. Stoner. (1987) Male predation risk determines female preference in the Trinidad guppy. Nature, **329**, 831–33.

Breder, C. M., Jr. (1926) The locomotion of fishes. Zoologica, **4**, 159–297.

Breder, C. M., Jr. and D. E. Rosen (eds.) (1966) *Modes of Reproduction in Fishes.* Natural History Press, Garden City, NY.

Bretschko, G. and W. E. Klemens. (1986) Quantitative methods and aspects in the study of the interstitial fauna of running waters. Stygologia, **2**, 297–316.

Brett, J. R. (1956) Some principles in the thermal requirements of fish. Quart. Rev. Biol., **31**, 75–87.

Brett, J. R. (1971) Energetic responses of salmon to temperature: a study of some thermal relations in the physiology and freshwater ecology of sockeye salmon (*Onchorhychus nerka*). Am. Zool., **11**, 91–113.

Brian, M. V. (1956) Segregation of species of the ant genus Myrmica. J. Animal Ecol., **25**, 319–37.

Briand, F. and J. E. Cohen. (1987) Environmental correlates of food chain length. Science, **238**, 956–60.

Bridcut, E. E. and P. S. Giller. (1993) Diet variability in relation to season and habitat utilization in brown trout, *Salmo trutta* L., in a southern Irish stream. Spec. Publ. Can. Fish. Aquat. Sci., **118**, 17–24.

Brietburg, D. L. and T. Loher. (1994) Effects of physical disturbance on fish trophic interactions: The importance of consumer mobility, in *Theory and Application in Fish Feeding Ecology*, (eds. D. J. Stouder, K. L. Fresh, and R. J. Feller). University of South Carolina Press, Columbia, SC, pp. 242–53.

Bretschko, G. (1995) River/land ecotones: scales and patterns. Hydrobiologia, **303**, 83–91.

Briggs, J. C. (1979) Ostariophysian zoogeography: an alternative hypothesis. Copeia, **1979**, 111–18.

Briggs, J. C. (1984) Freshwater fishes and biogeography of Central America and the Antilles. Syst. Zool., **33**, 428–35.

Briggs, J. C. (1986) Introduction to the zoogeography of North American fishes, in *The Zoogeography of North American Freshwater Fishes* (eds. C. H. Hocutt and E. O. Wiley). John Wiley and Sons, New York, pp. 1–16.

Briggs, J. C. (1987) Antitropical distribution and evolution in the Indo-West Pacific Ocean. Syst. Zool., **36**, 237–47.

Brittain, J. E. and T. J. Eikeland. (1988) Invertebrate drift—a review. Hydrobiologia, **166**, 77–93.

Brönmark, C. and P. Edenhamn. (1994) Does the presence of fish affect the distribution of tree frogs (*Hyla arborea*)? Conserv. Biol., **8**, 841–45.

Brooks, D. R. and D. A. McLennan. (1992) Historical ecology as a research program, in *Systematics, Historical Ecology and North American Freshwater Fishes* (ed. R. L. Mayden). Stanford University Press, Stanford, CA., pp. 76–113.

Brooks, D. R. and D. A. McLennan. (1993) Historical ecology examining phylogenetic components of community evolution, in *Species Diversity in Ecological Communities— Historical and Geographical Perspectives* (eds. R. E. Ricklefs and D. Schluter). Chicago University Press, Chicago, pp. 267–80.

Brooks, J. L. and S. I. Dodson. (1965) Predation, body size and composition of plankton. Science, **150**, 28–35.

Brown, A. V. and W. J. Matthews. (1995) Stream ecosystems of the central United States, in *Ecosystems of the World*, Vol. 22, *River and Stream Ecosystems* (eds. C. E. Cushing, K. W. Cummins, and G. W. Minshall). Elsevier, Amsterdam, pp. 89–116.

Brown, G. E. and J. A. Brown. (1993) Social dynamics in salmonid fishes: do kin make better neighbors? Anim. Behav., **45**, 863–71.

Brown, J. H. (1971) The desert pupfish. Sci. Am., **225**, 104–10.

Brown, J. H. (1995a) *Macroecology.* University of Chicago Press, Chicago, IL.

Brown, J. H. (1995b) Organisms as engineers: a useful framework for studying effects on ecosystems? Trends Ecol. Evol., **10,** 51–2.

Brown, J. H. and B. A. Maurer. (1989) Macroecology: the division of food and space among species on continents. Science, **143:** 1145–50.

Brown, J. H. and C. R. Feldmeth. (1971) Evolution in constant and fluctuating environments: thermal tolerances of desert pupfish (*Cyprinodon*). Evolution, **25,** 390–98.

Brown, J. H. and M. A. Kurzius. (1987) Composition of desert rodent faunas: combinations of coexisting species. Annals Zool. Fennici, **24,** 227–37.

Brown, J. H., P. A. Marquet, and M. L. Taper. (1993) Evolution of body size: consequences of an energetic definition of fitness. Am. Nat., **142,** 573–84.

Brown, J. H. and P. F. Nicoletto. (1991) Spatial scaling of species composition: body masses of North American land mammals. Am. Nat., **138,** 1478–512.

Brown, L. R. and P. B. Moyle. (1991) Changes in habitat and microhabitat partitioning within an assemblage of stream fishes in response to predation by Sacramento squawfish *Ptychocheilus grandis.* Can. J. Fish. Aquat. Sci., **48,** 840–56.

Brown, R. S. and W. C. Mackay. (1995) Fall and winter movements of and habitat use by cutthroat trout in the Ram River, Alberta. Trans. Am. Fish. Soc., **124,** 873–85.

Browne, R. A. (1981) Lakes as islands: biogeographic distribution, turnover rates, and species composition in the lakes of central New York. J. Biogeog., **8,** 75–83.

Brunet, L. A. and M. J. Sabo. (1996) Effects of hypoxia on ovary development of three *Lepomis* species in the Atchafalaya Louisiana. Abstracts ASIH An. Meeting, New Orleans, LA, pp. 95–6.

Bryan, J. E. and P. A. Larkin. (1972) Food specialization by individual trout. J. Fish. Res. Bd. Can., **29,** 1615–24.

Buckley, R. V. and R. Pimentel. (1983) Temperature preference and avoidance by adult razorback suckers. Trans. Am. Fish. Soc., **112,** 601–07.

Bulkley, R. V., C. R. Berry, R. Pimental, and T. Black. (1981) Tolerances and preferences of Colorado River endangered fishes to select habitat parameters. UT Coop. Fish. Res. Unit, Utah State University, Logan, UT.

Burr, B. M. (1991) The fishes of Illinois: an overview of a dynamic fauna. IL Nat. Hist. Survey Bull. **34,** 417–427.

Burr, B. M. and R. L. Mayden. (1992) Phylogenetics and North American freshwater fishes, in *Systematics, Historical Ecology, and North American Freshwater Fishes* (ed. R. L. Mayden). Stanford University Press, Stanford, CA, pp. 287–324.

Burr, B. M. and M. A. Morris. (1977) Spawning behaviour of the shorthead redhorse, *Moxostoma macrolepidotum,* in Big Rock Creek, Illinois. Trans. Am. Fish. Soc., **106,** 80–2.

Burr, B. M. and L. M. Page. (1986) Zoogeography of fishes of the lower Ohio–upper Mississippi basin, in *The Zoogeography of North American Freshwater Fishes* (eds. C. H. Hocutt and E. O. Wiley). John Wiley and Sons, New York, pp. 287–324.

Burton, D. T. and A. G. Heath. (1980) Ambient oxygen tension (PO_2) and transition to anaerobic metabolism in three species of freshwater fish. Can. J. Fish. Aquat. Sci., **37**, 1216–24.

Burton, G. W. and E. P. Odum. (1945) The distribution of stream fish in the vicinity of Mountain Lake, Virginia. Ecology, **26**, 182–94.

Bye, V. J. (1984) The role of environmental factors in the timing of reproductive cycles, in *Fish Reproduction: Strategies and Tactics,* (eds. G. W. Potts and R. J. Wootton). Academic Press, London, pp. 187–205.

Cadwallader, P. L. (1975a) Feeding relationships of galaxiids bullies, eels, and trout in a New Zealand river. Aust. J. Mar. Freshwater Res., **26**, 299–316.

Cadwallader, P. L. (1975b) Feeding habits of two fish species in relation to invertebrate drift in a New Zealand river. N.Z. J. Mar. Freshwater Res., **9**, 11–26.

Cady, E. R. (1945) Depth distribution of fish in Norris reservoir. *J. Tenn. Acad. Sci.* **20**, 103–114.

Cady, E. R. (1945) Fish distribution, Norris Reservoir, Tennessee, 1943. J. Tenn. Acad. Sci., **20**, 114–135.

Caldwell, J. P. (1966) Diversity of Amazonian Anurans: the role of systematics and phylogeny in identifying macroecological and evolutionary patterns, in *Neotropical Biodiversity and Conservation,* (ed. A. C. Gibson). University of California, Los Angeles, CA., pp. 73–88.

Cambray, J. A. (1994) The comparative reproductive styles of two closely related African minnows (*Pseudobarbus afer* and *P. asper*) inhabiting two different sections of the Gamtoos River system. Environ. Biol. Fish., **41**, 247–68.

Cambray, J. A. and M. N. Bruton. (1984) The reproductive strategy of a barb, *Barbus anoplus* (*Pisces: Cyprinidae*), colonizing a man-made lake in South Africa. J. Zool. Lond. **204**, 143–68.

Campagna, C. G. and J. J. Cech, Jr. (1981) Gill ventilation and respiratory efficiency of Sacramento blackfish, *Orthodon microlepidotus* Ayes, in hypoxic environments. J. Fish. Biol., **19**, 581–91.

Campbell, C. E. and R. Knoechel. (1990) Distribution patterns of vertebrate and invertebrate planktivores in Newfoundland lakes with evidence of predator-prey and competitive interactions. Can. J. Zool., **68**, 1559–67.

Campbell, K. P. (1979) Predation principles in large rivers: a review, in *Predator–Prey Systems in Fisheries Management* (ed. H. Clepper). Sport Fishing Institute, Washington, DC, pp. 181–91.

Canton, S. P., L. D. Cline, R. A. Short, and J. V. Ward. (1984) The macroinvertebrates and fish of a Colorado stream during a period of fluctuating discharge. Freshwater Biol., **14**, 311–16.

Capone, T.A. and J. A. Kushlan. (1991) Fish community structure in dry-season stream pools. Ecology, **72**, 983–92.

Carghelli, L. M. and M. R. Gross. (1997) Fish energetics: larger individuals emerge from winter in better condition. Trans. Am. Fish. Soc., **126**, 153–56.

Carl, L. M. (1983) Density, growth, and change in density of coho salmon and rainbow trout in three Lake Michigan tributaries. Can. J. Zool., **61**, 1120–27.

Carlander, K. D. (1969) *Handbook of Freshwater Fishery Biology,* Iowa State University Press, Ames, Vol. 1.

Carlander, K. D. (1977) *Handbook of Freshwater Fishery Biology,* Iowa State University Press, Ames, Vol. 2.

Carline, R. F. (1986) Indices as predictors of fish community traits, in *Reservoir Fisheries Management: Strategies for the 80's* (eds. G.E. Hall and M. J. Van Den Avyle). American Fisheries Society, Bethesda, MD, pp. 45–56.

Carlson, A. R. and R. E. Siefert. (1974) Effects of reduced oxygen on the embryos and larvae of lake trout (*Salvelinus namaycush*) and largemouth bass (*Micropterus salmoides*). J. Fish. Res. Bd. Can., **31**, 1393–96.

Carpenter, S. R., P. R. Leavitt, J. J. Esler, and M. M. Esler. (1988) Chlorophyll budgets: response to food web manipulation. Biogeochemistry, **6**, 79–90.

Carpenter, S. R., J. F. Kitchell, J. R. Hodgson, P. A. Cochran, J. J. Esler, M. M. Esler, D. M. Lodge, D. Kretchmer, X. He, and C. N. von Ende. (1987) Regulation of lake primary productivity by food web structure. Ecology, **68**, 1863–76.

Carr, A. F., Jr. (1941) The Carr Key Number. Dopeia, Ser. B., **3**, Part Q, No. X., The American Society of Fish Prevaricators and Reptile Fabricators (ASIH Annual Meeting).

Case, T. J., J. Faaborg, and R. Sidell. (1983) The role of body size in the assembly of West Indian bird communities. Evolution, **37**, 1062–74.

Cashner, R. C., F. P. Gelwick, and W. J. Matthews. (1994) Spatial and temporal variation in the distribution of lake fishes of the LaBranche wetlands area of the Lake Ponchartrain Estuary, Louisiana. Northeast Gulf Sci., **13**, 107–20.

Cashner, R. C. and W. J. Matthews. (1988) Changes in the known Oklahoma fish fauna from 1973 to 1988. Proceedings of the Oklahoma Academy of Science, **68**, 1–7.

Cashner, R. C., J. S. Rogers and J. M. Grady. (1992) Phylogenetic studies of the genus *Fundulus,* in *Systematics, Historical Ecology, & North American Freshwater Fishes,* (ed. R. L. Mayden). Stanford University Press, pp. 421–37.

Casterlin, M. E. and W. W. Reynolds. (1977) Aspects of habitat selection in the mosquitofish *Gambusia affinis.* Hydrobiologia, **55**, 125–27.

Castleberry, D. T. and J. J. Cech, Jr. (1986) Physiological responses of a native and an introduced desert fish to environmental stressors. Ecology, **67**, 912–18.

Castleberry, D. T. and J. J. Cech, Jr. (1992) Critical thermal maxima and oxygen minima of five fishes from the upper Klamath basin. CA Fish Game, **78**, 145–52.

Castonguay, M., G. J. FitzGerald and Y. Côté. (1982) Life history and movements of anadromous brook charr, *Salvelinus fontinalis,* in the St-Jean River, Gaspé, Quebec. Can. J. Zool., **60**, 3084–91.

Cavender, T. M. (1986) Review of the fossil history of North American freshwater fishes, in *The Zoogeography of North American Freshwater Fishes* (eds. C. H. Hocutt and E. O. Wiley), John Wiley and Sons, New York, pp. 701–24.

Cavender, T. M. (1991) The fossil record of the Cyprinidae, in *Cyprinid Fishes: Systematics, Biology and Exploitation* (eds. I. J. Winfield and J. S. Nelson). Chapman & Hall, London, pp. 34–54.

Cavender, T. M. and M. M. Coburn. (1992) Phylogenetic relationships of North American Cyprinidae, in *Systematics, Historical Ecology, & North American Freshwater Fishes*, (ed. R. L. Mayden). Stanford University Press, pp. 293–327.

Cech, J. J., Jr., S. J. Mitchell, and M. J. Massingill. (1979) Respiratory adaptations of Sacramento blackfish, *Orthodon microlepidotus* (Ayres), for hypoxia. Comp. Biochem. Physiol., **63A**, 411–15.

Cederholm, C. J. and N. P. Peterson. (1985) The retention of coho salmon (*Oncorchynchus kisutch*) carcasses by organic debris in small streams. Can. J. Fish. Aquat. Sci., **42**, 1222–25.

Cerri, R. D. (1983) The effect of light intensity on predator and prey behavior in cyprinid fish: factors that influence prey risk. Animal Behav., **31**, 736–42.

Cerri, R. D. and D. F. Fraser. (1983) Predation and risk in foraging minnows: balancing conflicting demands. Am. Nat., **121**, 552–61.

Chadwick, E. M. P. (1976) Ecological fish production in a small Precambrian shield lake. Environ. Biol. Fish., **1**, 13–60.

Chapman, C. A. and W. C. Mackay. (1984) Versatility in habitat use by a top aquatic predator, *Esox lucius* L. J. Fish. Biol., **25**, 109–15.

Chapman, L. J. and D. L. Kramer. (1991) The consequences of flooding for the dispersal and fate of poeciliid fishes in an intermittant tropical stream. Oecologia, **87**, 299–306.

Charnov, E. L. (1976) Optimal foraging: the marginal value theorem. Theor. Pop. Biol., **9**, 129–36.

Chereshnev, I. A. (1990) Ichthyofauna composition and features of freshwater fish distribution in the Northeastern USSR. J. Ichthyol., **30**, 110–21.

Chernov, B. (1982) Character variation among populations and the analysis of biogeography. Am. Zool., **22**, 425–39.

Cherry, D. S., K. L. Dickson, J. Cairns, Jr., and J. R. Stauffer. (1977) Preferred, avoided, and lethal temperatures of fish during rising temperature conditions. J. Fish. Res. Bd. Can., **34**, 239–46.

Chipps, S. R., W. B. Perry, and S. A. Perry. (1994) Fish assemblages of the central Appalachian Mountains: an examination of trophic group abundance in nine West Virginia streams. Envir. Biol. Fish., **40**, 91–98.

Clemens, W. A., J. R. Dymond, and N. K. Bigelow. (1924) Food studies of Lake Nipigon fishes. Ontario Fish. Res. Lab., **25**, 103–65.

Clements, F. E. (1916) *Plant Succession: An Analysis of the Development of Vegetation.* Carnegie Institute Publ. No. **242**, Washington, DC., xiii + 512 pp.

Closs, G. P. (1994) Feeding of *Galaxias olidus* (Günther) (*Pisces: Galaxiidae*) in an intermittant stream. Aust. J. Mar. Freshwater Res., **45**, 227–32.

Closs, G. P. and P. S. Lake. (1994) Spatial and temporal variation in the structure of an intermittent-stream food web. Ecol. Monogr., **64**, 1–21.

Closs, G. P. and P. S. Lake. (1996) Drought, differential mortality and the coexistance of a native and an introduced fish species in a south east Australian intermittent stream. Environ. Biol. Fish., **47**, 17–26.

Coates, D. (1993) Fish ecology and management of the Sepik-Ramu, New Guinea, a large contemporary tropical river basin. Environ. Biol. Fish., **38,** 345–68.

Coble, D. W. (1982) Fish populations in relation to dissolved oxygen in the Wisconsin River. Trans. Am. Fish. Soc., **111,** 612–23.

Coburn, M. M. and T. M. Cavender. (1992) Interrelationships of North American cyprinid minnows, in *Systematics, Historical Ecology, and North American Freshwater Fishes* (ed. R. L. Mayden). Stanford University Press, Stanford, CA, pp. 328–73.

Cohen, J. E. (1977) Ratio of prey to predators in community food webs. Amer. Naturalist, **270,** 165–67.

Coke, M. (1968) Depth distribution of fish on a bush-cleared area of Lake Kariba, Central Africa. Trans. Am. Fish. Soc., **97,** 460–65.

Coker, R. E. (1925) Observations of hydrogen-ion concentration and of fishes in waters tributary to the Catawba River, North Carolina (with supplemental observations in some waters of Cape Cod, Massachusetts). Ecology, **6,** 52–65.

Cole, G. A. (1966) The American Southwest and Middle America, in *Limnology in North America,* (ed. D. G. Frey). University of Wisconsin Press, Madison, pp. 393–434.

Cole, G. A. (1994) *Textbook on Limnology,* 4th ed. Waveland Press, Prospect Heights, IL.

Cole, L. C. (1954) The population consequences of life history phenomena. Quart. Rev. Biol. **29,** 103–37.

Colebrook, J. M. (1979) Continuous plankton records: seasonal cycles of phytoplankton and copepods in the North Atlantic Ocean and the North Sea. Marine Biology, **51,** 23–32.

Coleman, J. S., S. A. Heckathorn, and R. L. Hallberg. (1995) Heat-shock proteins and thermo-tolerance: linking molecular and ecological perspectives. Trends Ecol. Evol., **10,** 305–06.

Colinvaux, C. A. (1973) *Introduction to Ecology.* John Wiley and Sons, New York.

Collins, J. P., C. Young, J. Howell, and W. L. Minckley. (1981) Impact of flooding in a Sonoran desert stream, including elimination of an endangered fish population. Southwest. Nat., **26,** 415–23.

Collins, N. C. (1989) Daytime exposure to fish predation for littoral benthic organisms in unproductive lakes. Can. J. Fish. Aquat. Sci., **46,** 11–5.

Connell, J. H. (1975) Some mechanisms producing structure in natural communities: a model and evidence from field experiments, in *Ecology and Evolution of Communities* (eds. M. L. Cody and J. M. Diamond). Harvard University Press, Cambridge, MA, pp. 460–90.

Connell, J. H. (1978) Diversity of tropical rain forests and coral reefs. Science, **199,** 1302–09.

Connell, J. H. (1980) Diversity and the coevolution of competitors, or the ghost of competition past. Oikos, **35,** 131–38.

Connell, J. H. (1983) On the prevalence and relative importance of interspecific competition: evidence from field experiments. Am. Nat. **122,** 661–96.

Connell, J. H. and W. P. Sousa. (1983) On the evidence needed to judge ecological stability or persistance. Amer. Nat., **121,** 789–824.

Conner, J. V. and R. D. Suttkus. (1986) Zoogeography of freshwater fishes of the Western Gulf Slope, in *The Zoogeography of North American Freshwater Fishes* (eds. C. H. Hocutt and E. O. Wiley). John Wiley and Sons, New York, pp. 413–56.

Connor, E. F. and D. Simberloff. (1979) The assembly of species communities: chance or competition? Ecology, **60,** 1132–40.

Connor, E. F. and D. Simberloff. (1984) Neutral models of species' co-occurrence patterns, in *Ecological Communities—Conceptual Issues and the Evidence* (eds. D. R. Strong, Jr., D. Simberloff, L. G. Abele, and A. B. Thistle). Princeton University Press, Princeton, NJ, pp. 316–31.

Conover, D. O. (1992) Seasonality and the scheduling of life history at different latitudes. J. Fish. Biol., **41**(Suppl. B), 161–78.

Constantz, G. (1981) Life history patterns of desert fishes, in *Fishes in North American Deserts* (eds. R. Naiman and D. Soltz). John Wiley and Sons New York, pp. 237–90.

Constanz, G. D. (1985) Allopaternal care in the tessellated darter, *Etheostoma olmstedi* (Pisces: Percidae). Environ. Biol. Fish., **14,** 175–83.

Coon, T. G. (1987) Responses of benthic riffle fishes to variation in stream discharge and temperature, in *Community and Evolutionary Ecology of North American Stream Fishes* (eds. W. J. Matthews and D. C. Heins). University of Oklahoma Press, Norman, pp. 77–85.

Cooper, G. P. and G. N. Washburn. (1946) Relation of dissolved oxygen to winter mortality of fish in Michigan lakes. Trans. Am. Fish. Soc., **76,** 23–33.

Cooper, S. D. (1984) The effects of trout on water striders in stream pools. Oecologia, **61,** 376–79.

Copes, F. A. (1983) The longnose dace *Rhinichthys cataractae* Valenciennes in Wisconsin and Wyoming waters. Univ. Wisconsin, Stevens Point Museum Nat. Hist., **19,** 1–11.

Copp, G. H. and P. Jurajda. (1993) Do small riverine fish move inshore at night? J. Fish. Biol., **43**(Suppl. A), 229–41.

Cornell, H. V. (1985a) Local and regional richness of cynipine gall wasps on California oaks. Ecology, **66,** 1247–60.

Cornell, H. V. (1985b) Species assemblages of cynipid gall wasps are not saturated. Am. Nat., **126,** 565–69.

Cornell, H. V. and J. H. Lawton. (1992) Species interactions, local and regional processes, and limits to the richness of ecological communities: a theoretical perspective. J. Animal Ecol., **61,** 1–12.

Cotgreave, P. (1993) The relationship between body size and population abundance in animals. Trends Ecol. Evol., **8,** 456–65.

Coulter, G. W. (1981) Biomass, production, and potential yield of the Lake Tanganyika pelagic fish community. Trans. Am. Fish. Soc., **110,** 325–35.

Coutant, C. C. (1985) Striped bass, temperature, and dissolved oxygen: a speculative hypothesis for environmental risks. Trans. Am. Fish. Soc., **114,** 31–61.

Coutant, C. C. and D. L. Benson. (1990) Summer habitat suitability for striped bass in Chesapeake Bay: reflections on a population decline. Trans. Am. Fish. Soc., **119,** 757–78.

Coutant, C. C. and D. S. Carroll. (1980) Temperatures occupied by ten ultrasonic-tagged striped bass in freshwater lakes. Trans. Am. Fish. Soc., **109**, 195–202.

Coutant, C. C. and S. S. Talmadge. (1977) Thermal effects. J. Water Pollut. Contr. Fed., **49**, 1369–425.

Cowell, B. C. and W. C. Carew. (1976) Seasonal and diel periodicity in the drift of aquatic insects in a subtropical Florida stream. Freshwater Biol., **6**, 587–94.

Cowley, D. E. and J. E. Sublette. (1987) Distribution of fishes in the Black River drainage, Eddy County, New Mexico. Southwest. Nat., **32**, 213–21.

Cowx, I. G. (1989) Interaction between the roach, *Rutilus rutilus*, and dace, *Leuciscus leuciscus*, populations in a river catchment in south-west England. J. Fish. Biol., **35**, 279–84.

Cowx, I. G. (1990) The reproductive tactics of roach, *Rutilus rutilus* (L.) and dace, *Leuciscus leuciscus* (L.) populations in the Rivers Exe and Culm, England. Pol. Arch. Hydrobiol., **37**, 193–208.

Cowx, I. G., W. O. Young, and J. M. Hellawell. (1984) The influence of drought on the fish and invertebrate populations of an upland stream in Wales. Freshwater Biol., **14**, 165–77.

Cracraft, J. (1974) Continental drift and vertebrate distribution. Annu. Rev. Syst. Ecol., **5**, 215–61.

Craig, G. R. and W. F. Baksi. (1977) The effects of depressed pH on flagfish reproduction, growth and survival. Water Res., **11**, 621–26.

Craig, N. E. (1996) Effects of vegetation on population structure of the largespring gambusia, *Gambusia geiseri*. M.S. thesis, Angelo State University, San Angelo, TX.

Crawshaw, L. I. (1975) Attainment of the final thermal preferendum in brown bullheads acclimated to different temperatures. Comp. Biochem. Physiol., **52A**, 171–73.

Crawshaw, L. I. and H. T. Hammel. (1974) Behavioral regulation of internal temperature in the brown bullhead, *Ictalurus nebulosus*. Comp. Biochem. Physiol., **47A**, 51–60.

Crawshaw, L. T. (1979) Responses to rapid temperature change in vertebrate ecotherms. Am. Zool., **19**, 225–37.

Crisp, D. T. (1993) Population densities of juvenile trout (*Salmo trutta*) in five upland streams and their effects upon growth, survival and dispersal. J. Applied Ecology, **30**, 759–71.

Cristensen, B. and L. Persson. (1993) Species-specific antipredatory behaviors: effects on prey choice in different habitats. Behav. Ecol. Sociobiol., **32**, 1–9.

Croizat, L., G. Nelson, and D. E. Rosen. (1974) Centers of origin and related concepts. Syst. Zool., **23**, 265–87.

Crombie, A. C. (1947) Interspecific competition. J. Animal Ecol., **16**, 44–73.

Cross, F. B. (1967) *Handbook of Fishes of Kansas*. Museum of Natural History, Univ. of Kansas, Miscellaneous Publ., **45**, 1–357.

Cross, F. B. and M. Brasch. (1968) Qualitative changes in the fish–fauna of the Upper Neosho River system, 1952–1967. Trans. KS Acad. Sci., **71**, 350–360.

Cross, F. B. and L. M. Cavin. (1971) Effects of pollution, especially from feedlots, on fishes of the upper Neosho River basin. KS Water Resources Research Institute, Contribution No. 79.

Cross, F. B. and R. E. Moss. (1987) Historic changes in fish communities and aquatic habitats in plains streams of Kansas, in *Community and Evolutionary Ecology of North American Stream Fishes* (eds. W. J. Matthews and D. C. Heins). University of Oklahoma Press, Norman) pp. 155–65.

Cross, F. B., R. L. Mayden, and J. D. Stewart. (1986) Fishes in the western Mississippi drainage, in *The Zoogeography of Northern American Freshwater Fishes* (eds. C. H. Hocutt and E. O. Wiley). John Wiley and Sons, New York, pp. 363–412.

Cross, F. B., R. E. Moss, and J. T. Collins. (1985) Assessment of dewatering impacts on stream fisheries in the Arkansas and Cimmaron Rivers. KS. Fish and Game Commission Nongame Contract, **46,** 1–161.

Cross, J. N. (1985) Distribution of fish in the Virgin River, a tributary of the lower Colorado River. Environ Biol. Fish., **12,** 13–21.

Crossman, E. J. and D. E. McAllister. (1986) Zoogeography of freshwater fishes of the Hudson Bay drainage, Ungava Bay and the Arctic Archipelago, in *The Zoogeography of North American Freshwater Fishes* (eds. C. H. Hocutt and E. O. Wiley). John Wiley and Sons, New York, pp. 53–104.

Crowder, L. B. (1980) Alewife, rainbow smelt and native fishes in Lake Michigan: competition or predation. Environ. Biol. Fish., **5,** 225–33.

Crowder, L. B. and W. E. Cooper. (1979) Structural complexity and fish–prey interactions in ponds: a point of view, in *Response of Fish to Habitat Structure in Standing Water* (eds. D. L. Johnson and R. Stein). North Central Division., Spec. Publ. **6,** American Fisheries Society, **6,** 2–10.

Crowder, L. B. and W. E. Cooper. (1982) Habitat structural complexity and the interaction between bluegills and their prey. Ecology, **63,** 1802–13.

Crowl, T. A. (1989) Effects of crayfish size, orientation, and movement on the reactive distance of largemouth bass foraging in clear and turbid water. Hydrobiologia, **183,** 133–40.

Crowl, T. A. and J. Boxrucker. (1988) Possible competitive effects of two introduced planktivores on white crappie. Proc. Annu. Conf. S.E. Assoc. Fish Wildl. Agency, **42,** 185–92.

Crowl, T. A., C. R. Townsend, and A. R. McIntosh. (1992) The impact of introduced brown and rainbow trout on native fish: the case of Australasia. Rev. Fish Biol. Fisher., **2,** 217–41.

Cuffney, T. F. and J. B. Wallace. (1989) Discharge–export relationships in headwater streams: the influence of invertebrate manipulations and drought. J. North Am. Benthol. Soc., **8,** 331–41.

Culp, J. M. and G. J. Scrimgeour. (1993) Size-dependent diel foraging periodicity of a mayfly grazer in streams with and without fish. Oikos, **68,** 242–50.

Culp, J. M., N. E. Glozier, and G. J. Scrimgeour. (1991) Reduction of predation risk under the cover of darkness: avoidance responses of mayfly larvae to a benthic fish. Oecologia, **86,** 163–69.

Cummins, K. W. (1978) Ecology and distribution of aquatic insects, in *An Introduction to the Aquatic Insects of North America* (eds. R. W. Merritt and K. W. Cummins). Kendall-Hunt, Dubuque, IA, pp. 29–31.

Cummins, K. W. and M. J. Klug. (1979) Feeding ecology of stream invertebrates. Ann. Rev. Ecol. Syst., **10**, 127–72.

Cunningham, J. E. R. and E. K. Balon. (1986) Early ontogeny of *Adinia xenica* (Pisces: Cyprinodontiformes): 3. Comparison and evolutionary significance of some patterns in epigenesis of egg-scattering, hiding and bearing cyprinodontiforms. Environ. Biol. Fish., **15**, 91–105.

Curtis, J. T. and R. P. McIntosh. (1951) An upland forest continuum in the prairie–forest border region of Wisconsin. Ecology, **32**, 476–96.

Daget, J. (1968) Diversite des faunes de poissons dans les cours d'eau du Portugal. Arq. Mus. Bocage **2**, 21–6. [In French, original not seen, cited in Hugueny (1989).]

Daget, J. and P. S. Economidis. (1975) Richesse specifique de l'ichthyofaune de Macedoine orientale et de Thrace occidentale (Grece). Bull. Mus. Nat. Hist., Paris Ecol. Gen. **346**, 81–4. [In French, original not seen, cited in Hugueny (1989).]

Daldorph, P. W. G. and J. D. Thomas. (1995) Factors influencing the stability of nutrient-enriched freshwater macrophyte communities: the role of sticklebacks *Pungitius pungitius* and freshwater snails. Freshwater Biol., **33**, 271–89.

Dalquest, W. W. (1957) Flood dispersal of brook silverside, *Labidesthes sicculus*. Southwest. Nat., **2**, 173–74.

Darlington, P. J., Jr. (1957) *Zoogeography: The Geographical Distribution of Animals*. John Wiley and Sons, New York.

Davis, J. C. (1975) Minimal dissolved oxygen requirements of aquatic life with emphasis on Canadian species: a review. J. Fish. Res. Bd. Can., **32**, 2295–332.

Davis, J. R. and D. E. Louder. (1971) Life history and ecology of the cyprinid fish *Notropis petersoni* in North Carolina waters. Trans. Am. Fish. Soc., **100**, 726–33.

Dawson, W. R. (1967) Interspecific variation in physiological responses of lizards to temperature, in *Lizard Ecology: A Symposium* (ed. W. W. Milstead). University of Missouri Press, Columbia, pp. 230–57.

Daxboeck, D. and G. F. Holeton. (1978) Oxygen receptors in the rainbow trout, *Salmo gairdneri*. Can. J. Zool., **56**, 1254–59.

Day, J. A., B. R. Davies, and J. M. King. (1986) Riverine ecosystems, in *The Conservation of South African Rivers* (ed. J. H. O'Keeffe). South Afr. Nat. Sci. Prog. Rep., **131**, pp. 1–18.

Deacon, J. E. (1961) Fish populations, following a drought, in the Neosho and Marais des Cynges River. University of Kansas Publ., Museum Nat. Hist., **13**, 359–427.

Deacon, J. E., F. R. Taylor and J. W. Pedretti. (1995) Egg viability and ecology of Devils Hole pupfish: insights from captive propagation. Southwest. Nat., **40**, 216–23.

DeAngelis, D. L., K. A. Rose, L. B. Crowder, E. A. Marschall, and D. Lika. (1993) Fish cohort dynamics: application of complementary modeling approaches. Am. Nat., **142**, 604–22.

Death, R. G. (1995) Spatial patterns in benthic invertebrate community structure: products of habitat stability or are they habitat specific? Freshwater Biol., **33,** 455–67.

Death, R. G. and M. J. Winterbourn. (1995) Diversity patterns in stream benthic invertebrate communities: the influence of habitat stability. Ecology, **76,** 1446–60.

Debrot, A. O. and A. A. Myrberg, Jr. (1988) Intraspecific avoidance as a proximate cause for mixed-species shoaling by juveniles of a western Atlantic sturgeonfish, *Acanthurus bahianus.* Bull. Mar. Sci., **43,** 104–06.

DeCamps, H. and R. J. Naimen. (1989) L'ecologie des fleuves. La Recherche, **20,** 310–39.

DeHaven, J. E., D. J. Stouder, R. Ratajczak, T. J. Welch, and G. D. Grossman. (1992) Reproductive timing in three southern Appalachian stream fishes. Ecol. Freshwater Fish, **1,** 104–11.

Delbeek, J. C. and D. D. Williams. (1987) Food resource partitioning between sympatric populations of brackish-water sticklebacks. J. Animal Ecol., **56,** 949–97.

Delbeek, J. C. and D. D. Williams. (1988) Feeding selectivity of four species of sympatric stickleback in brackish-water habitats in eastern Canada. J. Fish. Biol., **32,** 41–62.

Derksen, A. J. (1989) Autumn movements of underyearling northern pike, *Esox lucius,* from a large Manitoba marsh. Can. Field-Nat., **103,** 429–31.

Desselle, W. J., M. A. Poirrier, J. S. Rogers, and R. C. Cashner. (1978) A discriminant functions analysis of sunfish *Lepomis* food habits and feeding niche segregation in the Lake Ponchartrain, Louisiana, Estuary. Trans. Am. Fish. Soc., **107,** 713–19.

De Staso, J., III and F. J. Rahel. (1994) Influence of water temperature on interactions between juvenile Colorado River cutthroat trout and brook trout in a laboratory stream. Trans. Am. Fish. Soc., **123,** 289–97.

Detenbeck, N. E., P. W. DeVore, G. J. Niemi, and A. Lima. (1992) Recovery of temperate-stream fish communities from disturbance: a review of case studies and synthesis of theory. Environ. Mangmt., **16,** 33–53.

DeVries, D. R. and R. A. Stein. (1992) Complex interactions between fish and zooplankton: quantifying the role of an open-water planktivore. Can. J. Fish. Aquat. Sci., **49,** 1216–27.

Dewey, M. R. and T. E. Moen. (1978) Fishes of the Caddo River, Arkansas after impoundment of DeGray Lake, Arkansas. Proc. AR Acad. Sci., **32,** 39–42.

Dewey, S. (1990) Cascading trophic interactions in experimental stream benthic communities. Ph.D. dissertation, University of Kansas, Lawrence.

Dial, K. P. and J. M. Marzluff. (1989) Nonrandom diversification within taxonomic assemblages. Systematic Zoology, **38,** 26–37.

Diamond, J. M. (1969) Avifaunal equilibria and species turnover rates on the Channel Islands of California. Proc. Nat. Acad. Sci. USA, **64,** 57–63.

Diamond, J. M. (1975) Assembly of species communities, in *Ecology and Evolution of Communities* (eds. M. L. Cody and J. M. Diamond). Belknap Press of Harvard University, Cambridge, MA, pp. 342–44.

Diamond, J. M. (1978) Niche shifts and the rediscovery of interspecific competition. Am. Sci., **66,** 322–31.

Diamond, J. M. and R. M. May. (1976) Island biogeography and the design of natural reserves, in *Theoretical Ecology,* 2nd ed. (ed. R. M. May). Sinauer Publ., Sunderland, MA, pp. 162–86.

Dickie, L. M., S. R. Kerr, and P. Schwinghamer. (1987) An ecological approach to fisheries assessment. Can. J. Fish. Aquat. Sci., **44,** 68–74.

Dickins, J. M. (1994) What is Pangaea? In *Pangea: Global Environments and Resources* (eds. A. F. Embry, B. Beauchamp, and D. J. Glass). Memoir **17,** Canada Society of Petroleum Geology, Calgary, Alberta, pp. 67–80.

DiDonato, G. T. and D. M. Lodge. (1993) Species replacements among *Orconectes* crayfishes in Wisconsin lakes: the role of predation by fish. Can. J. Fish. Aquat. Sci., **50,** 1484–88.

Diehl, S. (1995) Direct and indirect effects of omnivory in a littoral lake community. Ecology, **76,** 1727–40.

Dietrich, W. E. and T. Dunne. (1993) The channel head, in *Channel Network Hydrology* (eds. K. Beven and M. J. Kirkby). John Wiley and Sons, New York, pp. 175–219.

Dirnberger, J. M. (1983) The influence of wind mixing and river inflow on zooplankton vertical distribution and abundance in a large impoundment. Master's thesis. University of Oklahoma, Norman.

Dirnberger, J. M. and S. T. Threlkeld. (1986) Advective effects of a reservoir flood on zooplankton abundance and dispersion. Freshwater Biol., **16,** 387–96.

Dodds, W. K. (1988) Community structure and selection for positive or negative species interactions. Oikos, **53,** 387–90.

Dodson, J. J. and J. C. Young. (1977) Temperature and photoperiod regulation of rheotrophic behavior in prespawning common shiners, *Notropis cornutus*. J. Fish. Res. Bd. Can., **34,** 341–46.

Dodson, S. (1988) The ecological role of chemical stimuli for the zooplankton: predator-avoidance behavior in *Daphnia*. Limnol. Oceanogr., **33,** 1431–39.

Dodson, S. I. (1974) Zooplankton competition and predation: an experimental test of the size–efficiency hypothesis. Ecology, **55,** 605–13.

Dolloff, C. A. and G. H. Reeves. (1990) Microhabitat partitioning among stream-dwelling juvenile coho salmon, *Oncorhynchus kisutch* and Dolly Varden, *Salvelinus malma*. Can. J. Fish. Aquat. Sci., **47,** 2297–306.

Donald, D. B. and D. J. Alger. (1993) Geographic distribution, species displacement, and niche overlap for lake trout and bull trout in mountain lakes. Can. J. Zool., **71,** 238–47.

Dorsey, L. (1990) Variation in size and number of embryos in the largespring gambusia, *Gambusia geiseri*. MS thesis, Angelo State University, San Angelo, TX.

Doudoroff, P. (1938) Reactions of marine fishes to temperature gradients. Biol. Bull., **75,** 494–505.

Doudoroff, P. (1957) Water quality requirements of fishes and effects of toxic substances, in *The Physiology of Fishes* (ed. M. E. Brown). Academic Press, New York, pp. 403–30.

Doudoroff, P. and M. Katz. (1950) Critical review of literature on the toxicity of industrial wastes and their components to fish. I. Alkalies, acids, and inorganic gases. Sewage Ind. Wastes, **22,** 1432–56.

Douglas, M. E. (1987) An ecomorphological analysis of niche packing and niche dispersion in stream-fish clades, in *Community and Evolutionary Ecology of North American Stream Fishes* (eds. W. J. Matthews and D. C. Heins). University of Oklahoma Press, Norman, pp. 144–49.

Douglas, M. E. and W. J. Matthews. (1992) Does morphology predict ecology? Hypothesis testing within a freshwater fish assemblage. Oikos, **65,** 213–24.

Douglas, M. E., P. C. Marsh, and W. L. Minckley. (1994) Indigenous fishes of western North America and the hypothesis of competitive displacement: *Meda fulgida* (Cyprinidae) as a case study. Copeia, **1994,** 9–19.

Downhower, J. F., L. S. Blumer, and L. Brown. (1987) Seasonal variation in sexual selection in the mottled sculpin. Evolution, **41,** 1386–94.

Downing, J. A. and C. Plante. (1993) Production of fish populations in lakes. Can. J. Fish. Aquat. Sci., **50,** 110–20.

Drenner, R. W. (1977) The feeding mechanics of the gizzard shad (*Dorosoma cepedianum*). Unpublished Ph.D. dissertation, University of Kansas, Lawrence, KS.

Drenner, R. W. and S. R. McComas. (1980) The roles of zooplankter escape ability and fish size selectivity in the selective feeding and impact of planktivorous fish, in *Evolution and Ecology of Zooplankton Communities* (ed. W. C. Kerfoot). University Press of New England, Hanover, NH, pp. 587–93.

Drenner, R. W., J. R. Mummert, F. deNoylles, Jr., and D. Kettle. (1984a) Selective particle ingestion by a filter-feeding fish and its impact on phytoplankton community structure. Limnol. Oceanogr., **29,** 941–48.

Drenner, R. W., S. B. Taylor, X. Lazzaro, and D. Kettle. (1984b) Particle-grazing and plankton community impact of an omnivorous cichlid. Trans. Am. Fish. Soc., **113,** 397–402.

Drenner, R. W., W. J. O'Brien, and J. R. Mummert. (1982) Filter-feeding rates of gizzard shad. Trans. Am. Fish. Soc., **111,** 210–15.

Drenner, R. W., J. D. Smith, and S. T. Threlkeld. (1996) Lake trophic state and the limnological effects of omnivorous fish. Hydrobiologia, **319,** 213–23.

Drenner, R. W., J. R. Strickler, and W. J. O'Brien. (1978) Capture probability: the role of zooplankter escape in the selective feeding of planktivorous fish. J. Fish. Res. Bd., Can., **35,** 1370–73.

Drenner, R. W., S. T. Threlkeld, and M. D. McCracken. (1986) Experimental analysis of the direct and indirect effects of an omnivorous filter-feeding clupeid on plankton community structure. Can. J. Fish. Aquat. Sci., **43,** 1935–45.

Drenner, R. W., G. L. Vinyard, K. D. Hambright, and M. Gophen. (1987a) Particle ingestion by *Tilapia galilaea* is not affected by removal of gill rakers and microbranchiospines. Trans. Am. Fish. Soc., **116,** 272–76.

Drenner, R. W., K. D. Hambright, G. L. Vinyard, M. Gophen, and U. Pollinghen. (1987b) Experimental study of size-selective phytoplankton grazing by a filter-feeding cichlid and the cichlid's effects on plankton community structure. Limnol. Oceanogr., **32,** 1138–44.

Drenner, R. W., S. T. Threlkeld, J. D. Smith, J. R. Mummert and P. A. Cantrell. (1989)

Interdependence of phosphorus, fish and site effects on phytoplankton biomass and zooplankton. Limnol. Oceanogr., **34**, 1315–21.

Dudgeon, D. (1987) Niche specifications of four species *Homalopteridae, Cobitidae* and *Gobiidae* in a Hong Kong forest stream. Arch. Hydrobiol., **108**, 349–64.

Dudgeon, D. (1993) The effects of spate-induced disturbance, predation and environmental complexity on macroinvertebrates in a tropical stream. Freshwater Biol., **30**, 189–97.

Duellman, W. E. (1978) The biology of an equatorial herpetofauna in Amazonian Ecuador. University of Kansas, Museum Nat. Hist. Misc. Publicat., **65**, 1–352.

Dugan, P. J. and R. J. Livingston. (1982) Long-term variation of macroinvertebrate assemblages in Apalachee Bay, Florida. Estuarine, Coastal and Shelf Science, **14**, 391–403.

Dugatkin, L. A. and M. Alfieri. (1992) Inter-populational differences in the use of the tit-for-tat strategy during predator inspection in the guppy, *Poecilia reticulata*. Evol. Ecol., **6**, 519–26.

Dunbar, N. J. (1980) The blunting of Occam's Razor, or to hell with parsimony. Can. J. Zool., **58**, 123–28.

Dupuis, H. M. and M. H. A. Keenleyside. (1982) Egg-care behavior of *Aequidens paraguayensis* (Pisces, Cichlidae) in relation to predation pressure and spawning substrate. Can. J. Zool., **60**, 1794–99.

Durbin, A. G., S. W. Noxon, and C. A. Oviatt. (1979) Effects of the spawning migration of the alewife, *Alosa pseudoharengus,* on freshwater ecosystems. Ecology, **60**, 8–17.

Dutta, H. M. (1979) Form and function of jaws during feeding: *Ctenopoma acutirostre, Anabas testudineus* and *Macropodus opercularis*. Acta Morphol. Neerl.-Scand., **17**, 119–32.

Dutta, H. M. and E-K. Chen. (1983) Structural basis of jaw protrusion in the largemouth bass, *Micropterus salmoides:* a microscopic analysis. Can. J. Zool., **61**, 1251–64.

Eastman, J. T. (1977) The pharyngeal bones and teeth of catostomid fishes. Am. Midl. Nat., **97**, 68–88.

Eaton, J. G., J. H. McCormick, H. G. Stefan, and M. Hondzo. (1995) Extreme value analysis of a fish/temperature field database. Ecol. Eng., **4**, 289–305.

Ebeling, A. W., S. J. Holbrook, and R. J. Schmitt. (1990) Temporally concordant structure of a fish assemblage: bound or determined. Am. Nat., **135**, 63–7.

Echelle, A. A. (1973) Behavior of the pupfish, *Cyprinodon rubrofluviatilis*. Copeia, **1973**, 68–76.

Echelle, A. A. and A. F. Echelle. (1984) Evolutionary genetics of a "species flock": atherinid fishes on the Mesa Central of Mexico, in *Evolution of Fish Species Flocks* (eds. A. A. Echelle and I. Kornfield). University of Maine Press, Orono, pp. 93–110.

Echelle, A. A. and A. F. Echelle. (1992) Mode and pattern of speciation in the evolution of inland pupfishes in the *Cyprinodon variegatus* complex (Teleostei: Cyprinodontidae): an ancestor–descendant hypothesis, in *Systematics, Historical Ecology & North American Freshwater Fishes* (ed. R. L. Mayden). Stanford University Press, Stanford, CA, pp. 691–709.

Echelle, A. A. and I. Kornfeld (eds.) (1984) *Evolution of Fish Species Flocks.* University of Maine Press, Orono.

Echelle, A. A. and G. D. Schnell. (1976) Factor analysis of species associations among fishes of the Kiamichi River, Oklahoma. Trans. Am. Fish. Soc., **105,** 17–31.

Echelle, A. A., A. F. Echelle, and L. G. Hill. (1972a) Interspecific interactions and limiting factors of abundance and distribution in the Red River pupfish, *Cyprinodon rubrofluviatilis.* Am. Midl. Nat., **88,** 109–30.

Echelle, A. A., C. Hubbs, and A. F. Echelle. (1972b) Developmental rates and tolerances of the Red River pupfish, *Cyprinodon rubrofluviatilis.* Southwest. Nat., **17,** 55–60.

Edds, D. R. (1989) Multivariate analysis of fish assemblage composition and environmental correlates in a Himalayan river—Nepal's Kali Gandaki/Narayani. Ph.D. dissertation, Oklahoma State University, Stillwater.

Edds, D. R. (1993) Fish assemblage structure and environmental correlates in Nepal's Gandaki River. Copeia, **1993,** 48–60.

Eddy, F. B. (1976) Acid–base balance in rainbow trout (*Salmo gairdneri*) subjected to acid stresses. J. Expt. Biol., **64,** 159–71.

Edlund, A.-M. and C. Magnhagen. (1981) Food segregation and consumption suppression in two coexisting fishes, *Pomatoschistus minutus* and *P. microps:* an experimental demonstration of competition. Oikos, **36,** 23–27.

Edmondson, W. T. (1966) Pacific Coast and Great Basin, in *Limnology in North America* (ed. D. G. Frey). University of Wisconsin Press, Madison, pp. 371–92.

Edwards, L. F. (1926) The protractile apparatus of the mouth of the catostomid fishes. Anat. Rec., **33,** 257–70.

Ehrlich, P. R. and P. H. Raven. (1964) Butterflies and plants: a study in coevolution. Evolution, **18,** 586–608.

Eigenmann, C. H. (1920) The Magdalena Basin and the horizontal and vertical distribution of its fishes. Indiana Univ. Studies, **7,** 21–34.

Eigenmann, C. H. and W. R. Allen. (1942) *Fishes of Western South America.* University of Kentucky, Lexington.

Eklöv, P. and L. Perrson. (1995) Species-specific antipredator capacities and prey refuges: interactions between piscivorous perch (*Perca fluviatilis*) and juvenile perch and roach (*Rutilus rutilus*). Behav. Ecol. Sociobiol., **37,** 169–78.

Elliott, J. M. (1970) Diel changes in invertebrate drift and the food of trout *Salmo trutta* L., J. Fish. Biol., **2,** 161–65.

Elliott, J. M. and J. A. Elliott. (1995) The critical thermal limits for the bullhead, *Cottus gobio,* from three populations in north-west England. Freshwater Biol., **33,** 411–18.

Elliott, J. M., J. A. Elliott, and J. D. Allonby. (1994) The critical thermal limits for the stone loach, *Noemacheilus barbatulus,* from three populations in northwest England. Freshwater Biol., **32,** 593–601.

Elrod, J. H. and C. P. Schneider. (1987) Seasonal bathythermal distribution of juvenile lake trout in Lake Ontario. J. Great Lakes Res., **13,** 121–34.

Elrod, J. H., W.-D. N. Busch, B. L. Griswold, C. P. Schneider, and D. R. Wolfert. (1981)

Food of white perch, rock bass and yellow perch in eastern Lake Ontario. J. NY Fish Game, **28**, 191–201.

Elser, J. J., C. Luecke, M. T. Brett, and C. R. Goldman. (1995) Effects of food web compensation after manipulation of rainbow trout in an oligotrophic lake. Ecology, **76**, 52–69.

Elton, C. S. (1927) *Animal Ecology*. Sidgwick and Jackson, London.

Elwood, J. W. and T. F. Waters. (1969) Effects of floods on food consumption and production rates in a stream brook trout population. Trans. Am. Fish. Soc., **98**, 253–62.

Endler, J. A. (1983) Natural and sexual selection on color patterns in poeciliid fishes. Environ. Biol. Fish., **9**, 173–90.

Endler, J. A. and A. E. Houde. (1995) Geographic variation in female preferences for male traits in *Poecilia reticulata*. Evolution, **49**, 456–68.

Engle, S. and J. J. Magnuson. (1976) Vertical and horizontal distributions of coho salmon *Oncorhynchus kisutch,* yellow perch *Perca flavescens,* and cisco *Coregonus artedii* in Pallette Lake, Wisconsin. J. Fish. Res. Bd. Can., **33**, 2710–15.

Englert, J. and B. H. Seghers. (1983) Habitat segregation by stream darters (Pisces: Percidae) in the Thames River watershed southwestern Ontario. Can. Field-Natl., **97**, 177–80.

Ensign, W. E., J. W. Habera, and R. J. Strange. (1989) Food resource competition in southern Appalachian brook and rainbow trout. Proc. Annu. Conf. SE Assoc. Fish Wildl. Agency, **43**, 239–47.

Erman, D. C. (1986) Long-term structure of fish populations in Sagehen Creek, California. Trans. Am. Fish. Soc., **115**, 682–92.

Erman, D. C., E. D. Andrews, and M. Yoder-Williams. (1988) Effects of winter floods on fishes in the Sierra Nevada. Can. J. Fish. Aquat. Sci., **45**, 2195–200.

Erskine, D. J. and J. R. Spotila. (1977) Heat-energy-budget analysis and heat transfer in the largemouth blackbass (*Micropterus salmoides*). Physiol. Zool., **50**, 157–69.

Eschmeyer, W. N. (1990) *Catalog of the Genera of Recent Fishes*. California Academy of Sciences, San Francisco.

Etnier, D. A. and W. C. Starnes. (eds.) (1993) *The Fishes of Tennessee*. University of Tennessee Press, Knoxville.

Evans, B. I. (1986) An analysis of the feeding rate of white crappie, in *Contemporary Studies on Fish Feeding* (eds. C. A. Simenstad and G. M. Cailliet). Dr. W. Junk Publishers, Dordrecht, pp. 299–306.

Evans, J. W. and R. L. Noble. (1979) The longitudinal distribution of fishes in an east Texas stream. Am. Midl. Natl., **101**, 333–34.

Evans, M. S. and W. J. O'Brien (1986) Recent major declines in zooplankton populations in the inshore region of Lake Michigan: probable causes and implications. Can. J. Fish. Aquat. Sci., **43**, 154–59.

Everett, R. A. and G. M. Ruiz. (1993) Coarse woody debris as a refuge from predation in aquatic communities—an experimental test. Oecologia, **93**, 475–86.

Facey, D. E. and G. D. Grossman (1990) The metabolic cost of maintaining position for four North American stream fishes: effects of season and velocity. Physiol. Zool., **63**, 757–76.

Facey, D. E. and G. D. Grossman. (1992) The relationship between water velocity, energy costs, and microhabitat use in four North American stream fishes. Hydrobiologia, **239**, 1–6.

Fader, S. C., Z. Yu, and J. R. Spotila. (1994) Seasonal variation in heat shock proteins (hsp70) in stream fish under natural conditions. J. Therm. Biol., **5**, 335–41.

Fahy, W. E. (1954) The life history of the northern greenside darter, *Etheostoma blennioides blennioides* Rafinesque. Proc. Elisha Mitchell Sci., **70**, 139–205.

Fajen, O. F. 1962. The influence of stream stability on homing behavior of smallmouth bass populations. Trans. Amer. Fish. Soc. **91**, 346–49.

Farrell, B. D. and C. Mitter. (1993) Phylogenetic determinants of insect/plant community diversity, in *Species Diversity in Ecological Communities* (eds. R. E. Ricklefs and D. Schulter). University of Chicago Press, Chicago, pp. 253–66.

Fausch, K. D. (1984) Profitable stream positions for salmonids: relating specific growth rate to net energy gain. Can. J. Zool., **62**, 441–51.

Fausch, K. D. (1988) Tests of competition between native and introduced salmonids in streams: what have we learned? Canadian J. Fish. Aquat. Sci., **45**, 2238–46.

Fausch, K. D. (1989) Do gradient and temperature affect distributions of, and interactions between, brook charr *Salvelinus fontinalis* and other resident salmonids in streams? Physiology and Ecology Japan Special Volume, **1**, 303–22.

Fausch, K. D. and T. G. Northcote. (1992) Large woody debris and salmonid habitat in a small coastal British Columbia stream. Can. J. Fish. Aquat. Sci., **49**, 682–93.

Fausch, K. D. and K. R. Bestgen. (1996) Ecology of fishes indigenous to the central and southwestern Great Plains, in *Ecology of Great Plains Vertebrates and Their Habitats* (eds. F. L. Knopf and F. B. Samson). Springer-Verlag, New York.

Fausch, K. D. and R. G. Bramblett. (1991) Disturbance and fish communities in intermittent tributaries of a western Great Plains river. Copeia, **1991**, 208–18.

Fausch, K. D. and R. J. White. (1981) Competition between brook trout (*Salvelinus fontinalis*) and brown trout (*Salmo trutta*) in a Michigan stream. Can. J. Fish. Aquat. Sci., **38**, 1220–27.

Fausch, K. D. and R. J. White. (1986) Competition among juveniles of coho salmon, brook trout, and brown trout in a laboratory stream, and implications for Great Lakes tributaries. Trans. Am. Fish. Soc., **115**, 363–81.

Fausch, K. D. and M. K. Young. (1995) Evolutionary significant units and movement of resident stream fishes: a cautionary tale. Am. Fish. Soc. Symp., **17**, 360–70.

Fausch, K. D., J. R. Karr, and P. R. Yant. (1984) Regional application of an index of biotic integrity based on stream fish communities. Trans. Am. Fish. Soc., **113**, 39–55.

Felley, J. D. (1984) Multivariate indentification of morphological–environmental relationships within the Cyprinidae (Pisces). Copeia, **1984**, 442–55.

Felley, J. D. and S. M. Felley. (1986) Habitat partitioning of fishes in an urban, estuarine bayou. Estuaries, **9**, 208–18.

Felley, J. D. and S. M. Felley. (1987) Relationships between habitat selection by individuals of a species and patterns of habitat segregation among species: fishes of the Calcasieu Drainage, in *Community and Evolutionary Ecology of North American Stream Fishes* (eds. W. J. Matthews and D. C. Heins). University of Oklahoma Press, Norman, pp. 61–68.

Feltmate, B. W., R. L. Baker, and P. J. Pointing. (1986) Distribution of the stonefly nymph *Paragnetina media* (Plecoptera: Perlidae): influence of prey, predators, current speed, and substrate composition. Can. J. Fish. Aquat. Sci., **43**, 1582–87.

Feminella, J. W. and W. J. Matthews. (1984) Intraspecific differences in thermal tolerance of *Etheostoma spectabile* (Agassiz) in constant versus fluctuating environments. J. Fish. Biol., **25**, 455–61.

Ferguson, R. G. (1958) The preferred temperature of fish and their midsummer distribution in temperate lakes and streams. J. Fish. Res. Bd. Can., **15**, 607–24.

Fernando, C. H. (1994) Zooplankton, fish and fisheries in tropical freshwaters. Hydrobiologia, **272**, 105–23.

Feunteun, E. and L. Marion. (1994) Assessment of grey heron predation on fish communities: the case of the largest European colony. Hydrobiologia, **279/280**, 327–44.

Field-Dodgson, M. S. (1987) The effect of salmon redd excavation on stream substrate and benthic community of two salmon spawning streams in Canterbury, New Zealand. Hydrobiologia, **154**, 3–11.

Findlay, D. L., S. E. M. Kasian, L. L. Hendzel, G. W. Regehr, E. U. Schindler, and J. A. Shearer. (1994) Biomanipulation of lake 221 in the experimental lakes area (ELA): effects on phytoplankton and nutrients. Can. J. Fish Aquat. Sci., **51**, 2794–807.

Findley, J. S. (1973) Phenetic packing as a measure of faunal diversity. Am. Nat., **107**, 580–84.

Findley, J. S. (1976) The structure of bat communities. Am. Nat., **110**, 129–39.

Finger, T. R. (1982) Interactive segregation among three species of sculpins *Cottus*. Copeia, **1982**, 680–94.

Finger, T. R. and E. M. Stewart. (1987) Responses of fishes to flooding regime in lowland hardwood wetlands, in *Community and Evolutionary Ecology of North American Stream Fishes* (eds. W. J. Matthews and D. C. Heins). University of Oklahoma Press, Norman, pp. 86–92.

Fink, S. V. and W. L. Fink. (1981) Interrelationships of the ostariophysian fishes (Teleostei). Zool. J. Linn. Soc., **72**, 297–353.

Fink, S. V. and W. L. Fink. (1996) Interrelationships of ostariophysan fishes (Teleostei), in *Interrelationships of Fishes* (eds. M. L. J. Stiassny, L. R. Parenti, and G. D. Johnson). Academic Press, San Diego, CA, pp. 209–49.

Fischer, R. U., Jr., E. A. Standora, and J. R. Spotila. (1987) Predator-induced changes in thermo-regulation of bluegill, *Lepomis macrochirus*, from a thermally altered reservoir. Can. J. Fish. Aquat. Sci., **44**, 1629–34.

Fishelson, L. (1980) Partitioning and sharing of space and food resources by fishes, in *Fish Behavior and Its Use in the Capture and Culture of Fishes*. ICLARM, Manila, Vol. 5, 415–45.

Fisher, R. A., A. S. Corbett, and C. B. Williams. (1943) The relation between the number of species and the number of individuals in a random sample of an animal population. J. Animal Ecol., **12,** 42–58.

Fisher, S. G. and G. E. Likens. (1973) Energy flow in Bear Brook, New Hampshire: an integrative approach to stream ecosystem metabolism. Ecol. Monogr., **43,** 421–39.

Fisher, S. G., L. J. Gray, N. B. Grimm, and D. E. Busch. (1982) Temporal succession in a desert stream ecosystem following flash flooding. Ecol. Monogr., **52,** 93–110.

Fisher, W. L. and W. D. Pearson. (1987) Patterns of resource utilization among four species of darters in three central Kentucky streams, in *Community and Evolutionary Ecology of North American Stream Fishes* (eds. W. J. Matthews and D. C. Heins). University of Oklahoma Press, Norman, pp. 69–76.

FitzGerald, G. J. and N. Caza. (1993) Parental investment in an anadromous population of three sticklebacks: an experimental study. Evol. Ecol., **7,** 279–86.

FitzGerald, G. J. and S. Lachance. (1993) Paternal investment in the blackspotted stickleback *Gasterosteus wheatlandi*. Acta Ecol., **14,** 17–22.

Fitzsimmons, J. M. and R. T. Nishimoto. (1995) Use of fish behavior in assessing the effects of Hurricane Iniki on the Hawaiian island of Kaua'i. Environ. Biol. Fish., **43,** 39–50.

Flebbe, P. A. and C. A. Dolloff. (1995) Trout use of woody debris and habitat in Appalachian wilderness streams of North Carolina. North Am. J. Fish. Mangmt., **15,** 570–90.

Flecker, A. S. (1984) The effects of predation and detritus on the structure of a stream insect community: a field test. Oecologia, **64,** 300–05.

Flecker, A. S. (1992a) Fish trophic guilds and the structure of a tropical stream: weak direct vs. strong indirect effects. Ecology, **73,** 927–40.

Flecker, A. S. (1992b) Fish predation and the evolution of invertebrate drift periodicity: evidence from neotropical streams. Ecology, **73,** 438–48.

Flecker, A. S. (1994) Community-level consequences of fish species replacement in New Zealand streams. Abstracts Annual Meet. ASIH, University of Southern California.

Flecker, A. S. (1996) Ecosystem engineering by a dominant detritivore in a diverse tropical stream. Ecology, **77,** 1845–54.

Flecker, A. S. (1997) Habitat modification by tropical fishes: environmental heterogeneity and the variability of interaction strength. J. North Am. Benthol. Soc., **16,** 286–95.

Flecker, A. S. and J. D. Allen. (1984) The importance of predation, substrate and spatial refugia in determining lotic insect distributions. Oecologia, **64,** 306–13.

Flecker, A. S. and B. Feifarek, (1994) Disturbance and the temporal variability of invertebrate assemblages in the Andean streams. Freshwater Biol., **31,** 131–42.

Flecker, A. S. and C. R. Townsend. (1994) Community-wide consequences of trout introduction in New Zealand streams. Ecol. Appl., **4,** 798–807.

Fletcher, A. R., A. K. Morrison, and D. J. Hume. (1985) Effects of carp, *Cyprinus carpio* L., on communities of aquatic vegetation and turibity of waterbodies in the Lower Goulburn River Basin. Aust. J. Mar. Freshwater Res., **36,** 311–27.

Fletcher, D. E. and B. M. Burr. (1992) Reproductive biology, larval description, and diet of the North American bluehead shiner, *Pteronotropis hubbsi* (Cypriniformes: Cyprinidae), with comments on conservation status. Ichthyol. Explor. Freshwater, **3,** 193–218.

Floyd, K. B., R. D. Hoyt, and S. Timbrook. (1984) Chronology of appearance and habitat partitioning by stream larval fishes. Trans. Am. Fish. Soc., **113,** 217–23.

Foerster, R. E. and W. E. Ricker. (1941) The effect of reduction of predaceous fish on survival of young sockeye salmon at Cultus Lake. J. Fish. Res. Bd. Can., **5,** 315–36.

Foote, C. J., J. W. Clayton, C. C. Lindsey, and R. A. Bodaly. (1992) Evolution of lake whitefish (*Coregonus clupeaformis*) in North America during the Pleistocene: evidence for a Nahanni glacial refuge race in the northern Cordillera region. Can. J. Fish. Aquat. Sci., **49,** 760–68.

Forbes, S. A. (1878) The food of Illinois fishes. Bull. Ill. State Lab. Nat. Hist., **I,** 71–89.

Forbes, S. A. (1880a) The food of fishes. Bull. Ill. State Lab. Nat. Hist., **3,** 19–70.

Forbes, S. A. (1880b) The food at the darters. Am. Nat., **14,** 697–703.

Forbes, S. A. (1883) The food of smaller fresh-water fishes. Bull. Ill. State Lab. Nat. Hist., **VI,** pp. 65–94.

Forbes, S. A. (1888a) Studies of the food of the freshwater fishes. Bull. Ill. State Lab. Nat. Hist., **II,** 433–473.

Forbes, S. A. (1888b) On the food relations of fresh-water fishes: a summary and discussion. Bull. Ill. State Lab. Nat. Hist., **II,** 475–555.

Forbes, S. A. (1907) On the local distribution of certain Illinois fishes: an essay in statistical ecology. Bull. Ill. State Nat. Hist., **VII,** 273–303.

Ford, D. E. (1990) Reservoir transport processes, in *Reservoir Limnology: Ecological Perspectives* (eds. K. W. Thornton, B. L. Kimmel, and F. E. Payne). John Wiley and Sons, New York, pp. 15–41.

Foster, S. A. and J. A. Baker. (1995) Evolutionary interplay between ecology, morphology and reproductive behavior in threespine stickleback, *Gasterosteus aculeatus*. Environ. Biol. Fish., **44,** 213–23.

Foster, S. A., J. A. Baker, and M. A. Bell. (1992) Phenotypic integration of life history and morphology: an example from three-spined stickleback, *Gasterosteus aculeatus* L. J. Fish. Biol., **41**(Suppl. B), 21–35.

Fowler, C. L. and G. L. Harp. (1974) Ichthyofaunal diversification and distribution in Jane's Creek watershed, Randolph County, Arkansas. Proc. AR Acad. Sci., **28,** 13–8.

Fox, L. (1988) Diffuse coevolution within complex communities. Ecology, **69,** 906–07.

Fraser, D. F. (1983) An experimental investigation of refuging behavior in a minnow. Can. J. Zool., **61,** 666–72.

Fraser, D. F. and R. D. Cerri. (1982) Experimental evaluation of predator–prey relationships in a patchy environment: consequences for habitat use patterns in minnows. Ecology, **63,** 307–13.

Fraser, D. F. and E. E. Emmons. (1984) Behavioral responses of juvenile blacknose dace *Rhinichthys atralulus* to varying densities of predatory creek chub *Semotilus atromaculatus*. Can. J. Fish. Aquat. Sci., **41,** 364–70.

Fraser, D. F. and J. F. Gilliam. (1987) Feeding under predation hazard: response of the guppy and Hart's rivulus from site with contrasting predation hazard. Behav. Ecol. Sociobiol., **21**, 203–09.

Fraser, D. F. and J. F. Gilliam. (1992) Nonlethal impacts of predator invasion: facultative suppression of growth and reproduction. Ecology, **73**, 959–70.

Fraser, D. F. and F. A. Huntington. (1986) Feeding and avoiding predation hazard: behavioral response of the prey. Ethology, **73**, 57–68.

Fraser, D. F. and T. E. Sise. (1980) Observations on stream minnows in a patchy environment: a test of a theory of habitat distribution. Ecology, **61**, 790–97.

Fraser, D. F., D. A. DiMattia, and J. D. Duncan. (1987) Living among predators: the response of a stream minnow to the hazard of predation, in *Community and Evolutionary Ecology of North American Stream Fishes* (eds. W. J. Matthews and D. C. Heins). University of Oklahoma Press, Norman, pp. 121–27.

Fraser, D. F., J. F. Gilliam, and T. Yip-Hoi. (1995) Predation as an agent of population fragmentation in a tropical watershed. Ecology, **76**, 1461–72.

Freeman, M. C. and G. D. Grossman. (1992) A field test for competitive interactions among foraging stream fishes. Copeia, **1992**, 898–902.

Freeman, M. C. and G. D. Grossman. (1993) Effects of habitat availability on dispersion of a stream cyprinid. Environ. Biol. Fish., **37**, 121–30.

Freeman, M. C. and D. J. Stouder. (1989) Intraspecific interactions influence size specific depth distribution in *Cottus bairdi*. Environ. Biol. Fish., **24**, 231–36.

Freeman, M. C., M. K. Crawford, J. C. Barrett, D. E. Facey, M. G. Flood, J. Hill, D. J. Stouder, and G. D. Grossman. (1988) Fish assemblage stability in a southern Appalachian stream. Can. J. Fish. Aquat. Sci., **45**, 1949–58.

French, J. R. P., III. (1993) How well can fishes prey on zebra mussels in eastern North America? Fisheries, **18**, 13–9.

Fretwell, S. D. (1972) *Populations in a Seasonal Environment*. Princeton University Press, Princeton, NJ.

Fretwell, S. D. and H. L. Lucas, Jr. (1970) On territorial behavior and other factors influencing habitat distribution in birds. I. Theoretical development. Acta Biotheoret., **19**, 16–36.

Friedman, J. M., W. R. Osterkamp, and W. M. Lewis, Jr. (1996) Channel narrowing and vegetation development following a Great Plains Flood. Ecology, **77**, 2167–81.

Frissel, C. A., W. L. Liss, C. E. Warren, and M. D. Hurley. (1986) A hierarchical framework for stream habitat classification: viewing streams in a watershed context. Environ. Mangmt., **10**, 199–214.

Fritz, S. C., D. R. Engstrom, and B. J. Haskell. (1994) "Little Ice Age" aridity in the North American Great Plains: a high resolution reconstruction of salinity fluctuations from Devils Lake, North Dakota, (USA). Holocene, **4**, 69–73.

Fromm, P. O. (1980) A review of some physiological and toxicological responses of freshwater fish to acid stress. Environ. Biol. Fish., **5**, 79–93.

Fry, F. E. J. (1947) Effects of the environment on animal activity. University Toronto Studies, Ontario Fish. Res. Lab. Publ., **68**, 1–62.

Fry, F. E. J., J. R. Brett, and G. H. Clawson. (1942) Lethal limits of temperature for young goldfish. Rev. Can. Biol., **1**, 50–6.

Fryer, G. and T. D. Iles. (1972) The chiclid fishes of the great lakes of Africa—their biology and evolution. T.F.H. Publications, Neptune City, NJ.

Fuller, R. L. and H. B. N. Hynes. (1987) Feeding ecology of three predacious aquatic insects and two fish in a riffle of the Speed River, Ohio. Hydrobiologia, **150**, 243–55.

Fulling, G. (1993). Variation in population size and structure of the largespring gambusia, *Gambusia geiseri*, in the headwaters of the South Concho River, Tom Green County, Texas. M.S. thesis, Angelo State University, San Angelo, Texas.

Funk, J. L. (1955) Movement of stream fishes in Missouri. Trans. Amer. Fish. Soc., **85**, 39–57.

Fuselier, L. and D. Edds. (1994) Seasonal variation in habitat use by the Neosho madtom (*Teleostei: Ictaluridae: Noturus placidus*). Southwest. Nat. **39**, 217–23.

Fuselier, L. and D. Edds. (1995) An artificial riffle as restored habitat for the threatened Noesho madtom. North Am. J. Fish. Mangmt., **15**, 499–503.

Futuyma, D. J. (1986) *Evolutionary Biology*, 2nd ed., Sinauer Associates, Sunderland, MA.

Galacatos, K., D. J. Stewart, and M. Ibarra. (1996) Fish community patterns of lagoons and associated tributaries in the Ecuadorian Amazon. Copeia, **1996**, 875–94.

Galat, D. L. and N. Vucinich. (1983) Food partitioning between young of the year of two sympatric tui chub morphs. Trans. Am. Fish. Soc., **112**, 486–97.

Gale, W. F. and G. L. Buynak. (1978) Spawning frequency and fecundity of satinfin shiner (*Notropis analostanus*)—a fractional, crevice spawner. Trans. Am. Fish. Soc., **107**, 460–63.

Gale, W. F. and C. Gale. (1977) Spawning habits of spotfin shiner (*Notropis spilopterus*)—a fractional, crevice spawner. Trans. Am. Fish. Soc., **106**, 170–7.

Gammon, J. R. and J. M. Reidy. (1981) The role of tributaries during an episode of low dissolved oxygen in the Wabash River, Indiana. Am. Fish. Soc. Warmwater Streams Symp., pp. 396–407.

Gard, R. and G. A. Flitner. (1974) Distribution and abundance of fishes in Sagehen Creek, California. J. Wildl. Mangmt., **38**, 347–58.

Gardner, T. J. (1993) Grazing and the distribution of sediment particle sizes in artificial stream systems. Hydrobiologia, **252**, 127–32.

Garman, G. C. and L. A. Neilsen. (1982) Piscivority by stocked brown trout, *Salmo trutta* and its impact on the nongame fish community of Bottom Creek, Virginia. Can. J. Fish. Aquat. Sci., **39**, 862–69.

Garman, G. C., T. L. Thorn, and L. A. Nielsen. (1982) Longitudinal variation in the fish community of Brumley Creek, Virginia, and implications for sampling. Proc. Annual Conf. SE Assoc. Fish Wildl. **36**, 386–393.

Garrett, G. P. (1981) Variation in reproductive strategy in the Pecos pupfish, *Cyprinodon pecosensis*. Ph.D. dissertation, University of Texas, Austin.

Garrett, G. P. (1982) Variation in the reproductive traits of the Pecos pupfish, *Cyprinidon pecosensis*. Am. Midl. Nat., **108**, 355–63.

Garvey, J. E., R. A. Stein, and H. M. Thomas. (1994) Assessing how fish predation and interspecific prey competition influence a crayfish assemblage. Ecology, **75**, 532–47.

Gascon, D. and W. C. Leggett. (1977) Distribution, abundance, and resource utilization of littoral zone fishes in response to a nutrient-production gradient in Lake Memphremagog. J. Fish. Res. Bd. Can., **34**, 1105–17.

Gatz, A. J., Jr. (1979a) Community organization in fishes as indicated by morphological features. Ecology, **60**, 711–18.

Gatz, A. J., Jr. (1979b) Ecological morphology of freshwater stream fishes. Tulane Studies Zool. Bot., **21**, 91–124.

Gatz, A. J., Jr. (1981) Morphology inferred niche differentiation in stream fishes. Am. Midl. Nat., **106**, 10–21.

Gatz, A. J., Jr., M. J. Sale, and J. M. Loar. (1987) Habitat shifts in rainbow trout: competitive influence of brown trout. Oecologia, **74**, 7–19.

Gauch, H. G., Jr. (1982) *Multivariate Analysis in Community Ecology*. Cambridge University Press, Cambridge, MA.

Gaudreault, A., T. Miller, W. L. Montgomery, and G. J. FitzGerald. (1986) Interspecific interactions and diet of sympatric juvenile brook charr, *Salvelinus fontinalis* and adult ninespine sticklebacks, *Pungitius pungitius*. J. Fish. Biol., **28**, 133–40.

Gause, G. F. (1934) *The Struggle for Existence*. Macmillan (Hafner Press), New York. (Reprinted 1964.)

Gee, J. H. and T. G. Northcote. (1963) Comparative ecology of two sympatric species of dace *Rhinichthys* in the Fraser River System, British Colombia. J. Fish. Res. Bd. Can., **20**, 117–19.

Gee, J. H., R. F. Tallman, and H. J. Stone. (1978) Reactions of some Great Plains fishes to progressive hypoxia. Can. J. Zool., **56**, 1962–66.

Gehring, W. J. and R. Wehner. (1995) Heat shock protein synthesis and thermotolerance in Cataglyphis, and ant from the Sahara Desert. Proc. Nat. Acad. Sci. USA, **92**, 2994–98.

Gelwick, F. P. (1990) Longitudinal and temporal comparisons of riffle and pool fish assemblages in a northeastern Oklahoma Ozark stream. Copeia, **1990**, 1072–82.

Gelwick, F. P. (1995) Effects of grazers and their shared predator on algal heterogeneity: consequences for streamfish interactions. Ph.D. dissertation, University of Oklahoma, Norman.

Gelwick, F. P. and J. A. Gore. (1990) Fishes of Battle Branch, Delaware County, in northeastern Oklahoma. Proceedings of the Oklahoma Academy of Science, **70**, 13–18.

Gelwick, F. P. and W. J. Matthews. (1990) Temporal and spatial patterns in littoral-zone fish assemblages of a reservoir, Lake Texoma, Oklahoma–Texas, Environ. Biol. Fish., **27**, 107–20.

Gelwick, F. P. and W. J. Matthews. (1992) Effects of an algivorous minnow on temperate stream ecosystem properties. Ecology, **73**, 1630–45.

Gelwick, F. P. and W. J. Matthews. (1993) Artificial streams for studies of fish ecology. J. North Am. Benthol. Soc., **12**, 343–47.

Gelwick, F. P. and W. J. Matthews. Effects of algivorous minnows *Campostoma* on spatial and temporal heterogeneity of stream periphyton. Oecologia, (in press).

Gelwick, F. P., M. S. Stock, and W. J. Matthews. (1997) Effects of fish, water depth, and predation risk on patch dynamics in a north-temperate river system. Oikos, **80**, 382–389.

George, E. L. and W. F. Hadley. (1979) Food and habitat partitioning between rock bass *Ambloplites rupestris* and smallmouth bass *Micropterus dolomieui* young of the year. Trans. Am. Fish. Soc., **108**, 253–61.

Gerking, S. D. (1949) Characteristics of stream fish populations. Investigations of Indiana Lakes and Streams, **3**, 283–309.

Gerking, S. D. (1950) Stability of a stream fish population. J. Wildl. Mangmt. **14**, 193–202.

Gerking, S. D. (1953) Evidence for the concepts of home range and territory in stream fishes. Ecology, **34**, 347–65.

Gerking, S. D. (1959) The restricted movement of fish populations. Biol. Rev., **34**, 221–42.

Gerking, S. D. (1994) *Feeding Ecology of Fish.* Academic Press, San Diego, CA.

Gery, J. (1969) The freshwater fishes of South America, in *Biogeography and Ecology in South America* (eds. E. J. Fittkau, et al.). Dr. W. Junk Publishers, The Hague, pp. 828–48. [Original not seen, cited by Lowe-McConnell (1987)].

Gido, K., D. L. Propst, and M. C. Molles, Jr. (1997) Spatial and temporal variation of fish communities in secondary channels of the San Juan River, New Mexico and Utah, USA. Envir. Biol. Fish., **49**, 417–434.

Gilbert, C. R. (1976) Composition and derivation of the North American freshwater fish fauna. FL Sci., **39**, 104–11.

Gilbert, C. R. (1980) Zoogeographic factors in relation to biological monitoring of fish, in *Biological Monitoring of Fish* (eds. C. H. Hocutt and J. R. Stauffer, Jr.). D. C. Heath and Co., Lexington, MA, pp. 309–55.

Giles, N. and F. A. Huntingford. (1984) Predation risk and inter-population variation in anti-predator behavior in the three spines stickleback, *Gasterosteus aculeatus* L. Animal Behav., **32**, 264–75.

Gillen, A. L. and T. Hart. (1980) Feeding interrelationships between the sand shiner and the striped shiner. Ohio J. Sci., **80**, 71–6.

Gilliam, J. F. (1990) Hunting by the hunted: optimal prey selection by foragers under predation hazard, in *Behavioral Mechanisms of Food Selection* (ed. R. N. Hughes). University College of North Wales, Bangor, pp. 797–19.

Gilliam, J. F. and D. F. Fraser. (1987) Habitat selection under predation hazard: test of a model with foraging minnows. Ecology, **68**, 1856–62.

Gilliam, J. F. and D. F. Fraser. (1988) Resource depletion and habitat segregation by competitors under predation hazard, in *Size-Structured Populations* (eds. B. Ebenman and L. Persson). Springer-Verlag, Berlin, pp. 173–84.

Gilliam, J. F., D. F. Fraser, and A. M. Sabat. (1989) Strong effects of foraging minnows on a stream benthic invertebrate community. Ecology, **70**, 445–52.

Gilpin, M. E. and J. M. Diamond. (1984) Are species co-occurrences on islands nonrandom, and are null hypotheses useful in community ecology, in *Ecological Communities—Conceptual Issues and the Evidence* (eds. D. R. Strong, Jr., D. Simberloff, L. G. Abele, and A. B. Thistle). Princeton University Press, Princeton, NJ., pp. 297–315.

Gilinsky, E. (1984) The role of fish predation and spatial heterogeneity in determining benthic community structure. Ecology, **65,** 455–68.

Giussani, G. (1989) Lago Maggiore fish community evolution. Memor. Inst. Ital. Idrobiol., **46,** 125–35.

Glasscock, S. N. (1989). Ecology of the fish community in the headwaters of the South Concho River, Tom Green County, Texas. M.S. thesis, Angelo State University, San Angelo, TX.

Gleason, C. A. and T. M. Berra. (1993) Demonstration of reproductive isolation and observation of mismating in *Luxilus cornutus* and *L. chrysocephalus* in sympatry. Copeia, **1993,** 614–28.

Gleason, H. A. (1917) The structure and development of the plant association. Torrey Botan. Club, **44,** 463–81.

Gleason, H. A. (1926) The individualistic concept of plant association. Torrey Botan. Club, **53,** 7–26.

Gliwicz, Z. M. (1994) Relative significance of direct and indirect effects of predation by planktivorous fish on zooplankton. Hydrobiologia, **272,** 201–10.

Glover, R. S. (1967) The continuous plankton recorder survey of the North Atlantic. Symp. Zool. Soc. London, **19,** 189–210.

Godin, J.-G. J., I. J. Classon, and M. V. Abrahams. (1988) Group vigilance and shoal size in a small characin fish. Behaviour, **104½,** 29–40.

Golding, M. (1980) *The Fishes and the Forest.* University of California Press, Stanford.

Goldschmidt, T., F. Witte, and J. de Visser. (1990) Ecological segregation in zooplanktivorous haplochromine species (Pisces: Cichlidae) from Lake Victoria. Oikos, **58,** 343–55.

Goldschmidt, T., F. Witte, and J. Wanink. (1993) Cascading effects of the introduced Nile perch on the detritivorous/phytoplanktivorous species in the sublittoral areas of Lake Victoria. Cons. Biol., **7,** 686–700.

Goldstein, R. M. (1978) Quantitative comparison of seining and underwater observation for stream fishery surveys. Prog. Fish. Cult., **40,** 108–11.

Golladay, S. W. and C. L. Hax. (1995) Effects of an engineered flow on distribution of meiofauna in a north Texas prairie stream. J. North Am. Benthol. Soc., **14,** 404–13.

Golonka, J., M. I. Ross, and C. R. Scotese. (1994) Phanerozoic paleogeographic and paleoclimatic modeling maps in *Pangea: Global Environments and Resources* (eds. A. F. Embry, B. Beauchamp, and D. J. Glass). Canadian Society of Petroleum Geology, Calgary, Alberta, pp. 1–47.

Gonzalez, R. J. and W. A. Dunson. (1989) Differences in low pH tolerance among closely related sunfish of the genus *Enneacanthus*. Environ. Biol. Fish., **26,** 303–10.

Goodman, D. (1975) The theory of diversity–stability relationships in ecology. Quart. Rev. Biol., **50,** 237–66.

Gophen, M. (1988) Changes of Copepoda populations in Lake Kinneret during 1969–1985. Hydrobiologia, **167/168**, 375–79.

Gophen, M. (1990) Summary of the workshop on perspectives of biomanipulation in inland waters. Hydrobiologia, **191**, 315–18.

Gophen, M. and P. Spataru. (1989) Feeding habits of cichlids in Lake Kenneret (Israel) and their impact on ecosystem structure. Annals Museum Roy. Afr. Centr., Sci. Zool., **257**, 133–38.

Gophen, M. and S. Threlkeld. (1989) An experimental study of zooplankton consumption by the Lake Kinneret sardine. Arch. Hydrobiol., **115**, 91–5.

Gophen, M., B. Azoulay, and M. N. Bruton. (1988) Selective predation of Lake Kinneret zooplankton by fingerlings of *Clarias gariepinus*. Verh. Int. Verein. Limnol., **23**, 1763–65.

Gophen, M., P. B. O. Ochumba, and L. Kaufman. (1995) Some aspects of perturbation in the aquatic structure and biodiversity of the ecosystem of Lake Victoria. Aquat. Living Res., **8**, 000–000.

Gophen, M., S. Serruya, and P. Spataru. (1990) Zooplankton community changes in Lake Kinneret (Israel) during 1969–1985. Hydrobiologia, **191**, 39–46.

Gophen, M., P. B. O. Ochumba, U. Polllinger, and L. S. Kaufman. (1993) Nile perch, *Lates niloticus* invasion in Lake Victoria (East Africa). Verh. Int. Verin. Limnol., **25**, 856–59.

Gorman, O. T. (1986) Assemblage organization of stream fishes: the effects of rivers on adventitious streams. Am. Nat., **128**, 611–16.

Gorman, O. T. (1987) Habitat segregation in an assemblage of minnows in an Ozark stream, in *Community and Evolutionary Ecology of North American Stream Fishes* (eds. W. J. Matthews and D. C. Heins). University of Oklahoma Press, Norman, pp. 33–41.

Gorman, O. T. (1988a) An experimental study of habitat use in an assemblage of Ozark minnows. Ecology, **69**, 1239–50.

Gorman, O. T. (1988b) The dynamics of habitat use in a guild of Ozark minnows. Ecol. Monogr., **58**, 1–18.

Gorman, O. T. (1992) Evolutionary ecology and historical ecology: assembly, structure, and organization of stream fish communities, in *Systematics, Historical Ecology and North American Freshwater Fishes,* (ed. R. L. Mayden). Stanford University Press, Stanford, CA, pp. 659–88.

Gorman, O. T. and J. R. Karr. (1978) Habitat structure and stream fish communities. Ecology, **59**, 507–15.

Gosline, W. A. (1971) *Functional Morphology and Classification of Teleostean Fishes.* University of Hawaii Press, Honolulu.

Gosline, W. A. (1977) The structure and function of the dermal pectoral girdle in bony fishes with particular references to ostariophysines. J. Zool. Lond., **183**, 329–38.

Gosline, W. A. (1985) A possible relationship between aspects of dentition and feeding in the centrarchid and anabantoid fishes. Environ. Biol. Fish., **12**, 161–68.

Gosline, W. A. (1994) Function and structure in the paired fins of scorpaeniform fishes. Environ. Biol. Fish., **40**, 219–26.

Gotceitas, V. and J. G. J. Godin. (1992) Effects of location of food delivery and social status on foraging-site selection by juvenile Atlantic salmon. Envir. Biol. Fish., **35**, 291–300.

Gotceitas, V. and P. Colgan. (1988) Individual variation in learning by foraging juvenile bluegill sunfish Lepomis *macrochirus*. J. Compar. Psychol., **102**, 294–99.

Gotceitas, V. and P. Colgan. (1989) Predator foraging success and habitat complexity: quantitative test of the threshold hypothesis. Oecologia, **80**, 158–66.

Goulding, M. (1980) *The Fishes and the Forest—Explorations in Amazonian Natural History*. University of California Press, Berkeley.

Goulding, M., M. L. Carvalho, and E. G. Ferreira. (1988) Rio Negro: rich life in poor water: Amazonian diversity and food chain ecology as seen through fish communities. SPB Academic Publ., The Hague.

Gowan, C. and K. D. Fausch. (1996) Mobile brook trout in two high-elevation Colorado streams: re-evaluating the concept of restricted movement. Can. J. Fish. Aquat. Sci., **53**, 1370–81.

Gowan, C., M. K. Young, K. D. Fausch, and S. C. Riley. (1994) Restricted movement in resident stream salmonids: a paradigm lost? Can. J. Fish. Aquat. Sci., **51**, 2626–37.

Goyke, A. P. and A. E. Hershey. (1992) Effects of fish predation on larval chironomid (Diptera: Chironomidae) communities in an arctic ecosystem. Hydrobiologia, **240**, 203–11.

Grady, J. M., R. C. Cashner, and J. S. Rogers. (1983) Fishes of the Bayou Sara drainage, Louisiana and Mississippi, with a discriminant functions analysis of factors influencing species distribution. Tulane Studies Zool. Bot., **24**, 83–100.

Graham, R. W., et al. (1996) Spatial response of mammals to late quarternary environment fluctuations. Science, **272**, 1601–06.

Grande, L. (1984) Paleontology of the Green River Formation, with a review of the fish fauna (2nd ed.). Geol. Surveys WY Bull., **63**, pps.

Grande, L. (1990) Vicariance biogeography, in *Palaeobiology: A Synthesis* (eds. D. E. G. Briggs and P. Crowther). Blackwell Scientific Publ., Oxford, pp. 448–51.

Grande, L. and W. E. Beamis. (1991) Osteology and phylogenetic relationships of fossil and recent paddlefishes (Polyodontidae) with comments on the interrelationships of Acipenseriformes. J. Vertebr. Paleontol. Memoir I, **11**, (Suppl. 1), 1–121.

Grant, J. W. A. and D. L. Noakes. (1987) A simple model of optimal territory size for drift-feeding fish. Can. J. Zool., **65**, 270–76.

Grant, P. R. (1994) Ecological character displacement. Science, **266**, 746–47.

Gray, L. J. (1981) Species composition and life histories of aquatic insects in a lowland Sonoran Desert stream. Am. Midl. Nat., **102**, 229–42.

Gray, J. S. (1977) The stability of benthic ecosystems. Helogol. Wiss. Meeresunters, **30**, 427–44.

Gray, J. S. (1989) Effects of environmental stress on species rich assemblages. Biol. J. Linn. Soc., **37**, 19–32.

Greenberg, L. A. (1988) Interactive segregation between stream fishes *Etheostoma simoterum* and *E. rufilineatum.* Oikos, **51,** 193–202.

Greenberg, L. A. (1991) Habitat use and feeding behavior of thirteen species of benthic stream fishes. Environ. Biol. Fish., **31,** 389–401.

Greenberg, L. A. and R. A. Stiles. (1993) A descriptive and experimental study of microhabitat use by young-of-the-year benthic stream fishes. Ecol. Freshwater Fish., **2,** 40–9.

Greenfield, D. W., T. A. Greenfield, and S. L. Brinton. (1983a). Spatial and trophic interactions between *Gambusia sexradiata* and *Gambusia puncticulata yucatana* (Pisces: Poeciliidae) in Belize, Central America. Copeia, **1983,** 598–607.

Greenfield, D. W., C. F. Rakocinski, and T. A. Greenfield. (1983b) Spatial and trophic interactions in wet and dry seasons between *Gambusia luma* and *Gambusia sexradiata* (Pisces: Poeciliidae) in Belize, Central America. Fieldiana Zool., **14,** 1–16.

Greenwood, P. H. (1953) Feeding mechanism of the cichlid fish, *Tilapia esculenta* Graham. Nature (Lond.), **172,** 207–08.

Greenwood, P. H. (1974) The cichlid fishes of Lake Victoria, East Africa: the biology and evolution of a species flock. Bull. Br. Museum Nat. Hist., Zool. Suppl. **6,** 1–134.

Greenwood, P. H. (1981) *The Haplochromine Fishes of the East African Lakes,* Kraus International, Muenchen. [Original not seen, cited by Witte (1984)].

Greenwood, P. H. (1983) The zoogeography of African freshwater fishes: bioaccountancy or biogeography? in *Evolution, Time and Space: The Emergence of the Biosphere* (eds. R. W. Sims, J. H. Price, and P. E. S. Whalley). Academic Press, London (pp. 179–99. [Original not seen; cited by Lowe-McConnell (1987)].

Greenwood, P. H. (1984) African cichlids and evolutionary theories, in *Evolution of Fish Species Flocks* (eds. A. A. Echelle and I. Kornfeld). University of Maine Press, Orono, pp. 141–54.

Greenwood, P. H., D. E. Rosen, S. H. Weitzman, and G. S. Myers. (1966) Phyletic studies of teleostean fishes, with a provisional classification of living forms. Bull. Am. Museum Nat. Hist., **131,** 339–456.

Greger, P. D. and J. E. Deacon. (1988) Food partitioning among fishes of the Virgin River. Copeia, **1988,** 314–23.

Gregory, R. S. (1994) The influence of ontogeny, perceived risk of predation, and visual ability on the foraging behavior of juvenile chinook salmon, in *Theory and Application in Fish Feeding Ecology* (eds. D. L. Stouder, K. L. Fresh, and R. L. Feller). Belle W. Baruch Institute Marine Biology Coastal Research No. 18, University of South Carolina Press, Columbia, SC.

Gregory, S. V., F. J. Swanson, W. A. McKee, and K. W. Cummins. (1991) An ecosystem perspective of riparin zones. BioScience, **41,** 540–51.

Gregory, W. K. (1933) Fish skulls: A study of the evolution of natural mechanisms. Trans. Amer. Phil. Soc. (Philadelphia), **23,** 75–481.

Griffiths, D. (1986) Size-abundance relations in communities. Am. Nat., **127,** 140–66.

Grimm, N. B. (1988) Feeding dynamics, nitrogen budgets, and ecosystem role of a desert stream omnivore, *Agosia chrysogaster* (Pisces: Cyprinidae). Environ. Biol. Fish., **21,** 143–52.

Grimm, N. B. and S. G. Fisher. (1986) Nitrogen limitation in a Sonoran Desert stream. J. North Am. Benthol. Soc., **5,** 2–15.

Grimm, N. B. and S. G. Fisher. (1989) Stability of periphyton and macro-invertebrated to disturbance by flash floods in a desert stream. J. North Am. Benthol. Soc., **8,** 293–307.

Grinnel, J. (1917) The niche relationships of the California thrasher. Auk, **21,** 364–82.

Grinnel, J. (1924) Geography and evolution. Ecology, **5,** 225–29.

Griswold, B. L., C. J. Edwards, and L. C. Woods. III. (1982) Recolonization of macroinvertebrates and fish in an channelized stream after a drought. Ohio J. Sci., **82,** 96–102.

Gross, M. R. (1979) Cuckoldry in sunfishes (*Lepomis:* Centrarchidae). Can. J. Zool., **57,** 1507–09.

Gross, M. R. (1984) Sunfish, salmon, and the evolution of alternative reproductive strategies and tactics in fishes, in *Fish Reproduction: Strategies and Tactics,* (eds. G. W. Potts and R. J. Wootton). Academic Press, London, pp. 55–75.

Gross, M. R. (1985) Disruptive selection for alternative life histories in salmon. Nature, **313,** 47–8.

Grossman, G. D. and V. Boulé. (1991) Effects of rosyside dace *Clinostomus funduloides* on microhabitat use of rainbow trout *Oncorhynchus mykiss*. Can. J. Fish. Aquat. Sci., **48,** 1235–43.

Grossman, G. D. and M. C. Freeman. (1987) Microhabitat use in a stream fish assemblage. J. Zool. Lond., **212,** 151–76.

Grossman, G. D., J. F. Dowd, and M. Crawford. (1990) Assemblage stability in stream fishes: a review. Environ. Mangmt. **14,** 661–71.

Grossman, G. D., P. B. Moyle, and J. O. Whitaker, Jr. (1982) Stochasticity in structural and functional characteristics of an Indiana stream fish assemblage: a test of community theory. Am. Nat., **120,** 423–54.

Grossman, G. D., P. B. Moyle, and J. O. Whitaker. Jr. (1985) Stochasticity and assemblage organization in an Indiana stream fish assemblage. Am. Nat., **126,** 275–85.

Grossman, G. D., A. de Sostoa, M. C. Freeman, and J. Lobon-Cervia. (1987) Microhabitat use in a Mediterranean riverine fish assemblage. Oceologia, **73,** 490–500.

Guest, W. C. (1985) Temperature tolerance of Florida and northern largemouth bass: effects of subspecies, fish size, and season. TX J. Sci., **XXXVII,** 75–84.

Guest, W. C., R. W. Drenner, S. T. Threlkeld, F. D. Martin, and J. D. Smith. (1990) Effects of gizzard shad and threadfin shad on zooplankton and young-of-year white crappie production. Trans. Am. Fish. Soc., **119,** 529–36.

Guill, J. M. and D. C. Heins. (1996) Clutch and egg size variation in the banded darter, *Etheostoma zonale,* from three sites in Arkansas. Environ. Biol. Fish., **46,** 409–13.

Gunning, G. E. and R. D. Suttkus. (1991) Species dominance in the fish populations of the Pearl River at two study areas in Mississippi and Louisiana: 1966–1988. SFC Proc., **23,** 7–15.

Gunter, G. (1942) A list of the fishes of the mainland of North and Middle America recorded from both freshwater and seawater. Am. Midl. Nat., **28,** 305–26.

Günther, A. C. L. G. (1880) *An Introduction to the Study of Fishes.* Adams and Charles Black, Edinburgh. (Reprinted 1963.)

Guo-Qing, L. and M. V. H. Wilson. (1994) An Eocene species of *Hiodon* from Montana, its phylogenetic relationships, and the evolution of the postcranial skeleton in the Hiodontidae (Teleostei). J. Vertebr. Paleontol., **14,** 153–67.

Hackney, P. A. (1979) Influence of piscivorous fish on fish community structure of ponds, in *Predator–Prey Systems in Fisheries Management* (ed. H. Clepper). Sport Fishing Institute, Washington, DC., pp. 111–22.

Hagen, D. W. (1964) Evidence of adaptation to environmental temperatures in three species of *Gambusia* (Poeciliidae). Southwest. Nat., **9,** 6–19.

Haines, T. A. (1981) Acidic precipitation and its consequences for aquatic ecosystems: a review. Trans. Am. Fish. Soc., **110,** 669–707.

Hairston, N. G., F. E. Smith, and L. B. Slobodkin. (1960) Community structure, population control, and competition. Amer. Naturalist, **94,** 421–25.

Hairston, N. G., Jr. (1988) Interannual variation in seasonal predation: its origin and ecological importance. Limnol. Oceanogr., **33,** 1245–53.

Hairston, N. G., Jr. and N. G. Hairston, Sr. (1993) Cause–effect relationships in energy flow, trophic structure, and interspecific interactions. Am. Nat., **142,** 379–411.

Hall, C. A. S. (1972) Migration and metabolism in a temperate stream ecosystem. Ecology, **53,** 585–604.

Hall, D. J. and E. E. Werner. (1977) Seasonal distribution and abundance of fishes in the littoral zone of a Michigan Lake. Trans. Am. Fish. Soc., **106,** 445–55.

Hall, D. J., S. T. Threlkeld, C. W. Burns, and P. H. Crowley. (1976) The size-efficiency hypothesis and the size structure of zooplankton communities. Ann. Rev. Ecol. Syst., **7,** 177–208.

Hall, D. J., E. E. Werner, J. F. Gilliam, G. G. Mittlebach, D. Howard, and C. G. Doner. (1979) Diel foraging behavior and prey selection in the golden shiner (*Notemigonus crysoleucas*). J. Fish. Res. Bd. Can., **36,** 1029–39.

Hall, L. W., Jr., C. J. Hocutt, and J. R. Stauffer, Jr. (1978) Implication of geographic location on temperature preference of white perch, *Morone americana.* J. Fish. Res. Bd. Can., **35,** 1464–68.

Hambright, K. D., R. W. Drenner, S. R. McComas, and N. G. Hairston, Jr. (1991) Gape-limited piscivores, planktivore size refuges, and the trophic cascade hypothesis. Arch. Hydrobiol., **4,** 389–404.

Hamrin, S. F. (1986) Vertical distribution and habitat partitioning between different size classes of vendace, *Coregonus albula,* in thermally stratified lakes. Can. J. Fish. Aquat. Sci., **43,** 1617–25.

Hamrin, S. F. and L. Persson. (1986) Asymmetrical competition between age classes as a factor causing population oscillations in an obligate planktivorous fish species. Oikos, **47,** 223–32.

Hanazato, T. and M. Yasuno. (1989) Zooplankton community structure driven by vertebrate and invertebrate predators. Oecologia, **81,** 450–58.

Hanchet, S. M. (1990) Effect of land use on the distribution and abundance of native fish in tributaries of the Waikato River in the Hakarimata Range, North Island, New Zealand, NZ J. Mar. Freshwater Res., **24**, 159–91.

Hansen, M. J. and C. W. Ramm. (1994) Persistence and stability of fish community structure in a southwest New York stream. Am. Midl. Nat., **132**, 52–67.

Hanski, I. (1982) Dynamics of regional distribution: the core and satellite species hypothesis. Oikos, **38**, 210–21.

Hansen, M. J., D. Boisclair, S. B. Brandt, S. W. Hewett, J. F. Kitchell, M. C. Lucas, and J. J. Ney. (1983) Applications of bioenergetics models to fish ecology and management: Where do we go from here? Trans. Amer. Fish. Soc. **122**, 1019–30.

Hanson, M. A. and M. R. Riggs. (1995) Potential effects of fish predation on wetland invertebrates: a comparison of wetlands with and without fathead minnows. Wetlands, **15**, 167–75.

Hansson, S. (1985) Local growth differences in perch (*Perca fluviatilis* (L.) in a Baltic archipelago. Hydrobiologia, **121**, 3–10.

Hansson, S., et al. (1987) Effects of fish grazing on nutrient release and succession of primary producers. Limnol. Oceanogr., **32**, 723–29.

Haraldstad, O. and B. Jonsson. (1983) Age and sex segregation in habitat utilization by brown trout in a Norwegian Lake. Trans. Am. Fish. Soc., **112**, 27–37.

Harcup, M. F., R. Williams, and D. M. Ellis. (1984) Movements of brown trout, *Salmo trutta* L., in the River Gwyddon, South Wales. J. Fish. Biol., **24**, 415–26.

Hardin, G. (1960) The competitive exclusion principle. Science, **131**, 1292–97.

Harima, H. and P. R. Mundy. (1974) Diversity indices applied to the fish biofacies of a small stream. Trans. Am. Fish. Soc., **103**, 457–61.

Harper, J. L. and H. E. Namminga. (1986) Fish population trends in Texoma Reservoir following establishment of striped bass, in *Reservoir Fisheries Management: Strategies for the 80's* (eds. G. E. Hall and M. J. Van Doren). American Fisheries Society, Bethesda, MD, pp. 156–65.

Harrel, R. C., N. J. Davis, and T. C. Dorris. (1967) Stream order and species diversity of fishes in an intermittent Oklahoma stream. Am. Midl. Nat., **78**, 428–35.

Harrell, H. L. (1978) Responses of the Devils River (Texas) fish community to flooding. Copeia, **1978**, 60–8.

Harrington, R. W. Jr. (1955) The osteocranium of the American cyprinid fish, *Notropis bifrenatus*, with an annotated synonymy of teleost skull bones. Copeia, **1955**, 267–90.

Harris, J. L. and N. H. Douglas. (1978) Fishes of the Mountain Province section of the Ouachita River. Proc. AR Acad. Sci., **32**, 55–9.

Hart, J. S. (1947) Lethal temperature relations of certain fish of the Toronto region. Trans. Roy. Soc. Can. Ser. 3, **41**, (5), 57–71.

Hart, J. S. (1952) Geographic variations of some physiological and morphological characters in certain freshwater fish. University of Toronto Biol. Ser., **LXXII**, 1–79.

Hartley, P. H. T. (1948) Food and feeding relationships in a community of fresh-water fishes. J. Animal Ecol., **17**, 1–14.

Harvey, B. C. (1987) Susceptibility of young-of-the-year fishes to downstream displacement by flooding. Trans. Am. Fish. Soc., **116,** 851–55.

Harvey, B. C. (1991a) Interactions among stream fishes: predator-induced habitat shift and larval survival. Oecologia, **87,** 29–36.

Harvey, B. C. (1991b) Interaction of abiotic and biotic factors influences larval fish survival in an Oklahoma stream. Can. J. Fish. Aquat. Sci., **48,** 1476–80.

Harvey, B. C. and W. R. Hill. (1991) Effects of snails and fish on benthic invertebrate assemblages in a headwater stream. J. North Am. Benthol. Soc., **10,** 263–70.

Harvey, B. C. and A. J. Stewart. (1991) Fish size and habitat depth relationships in headwater streams. Oceologia, **87,** 336–42.

Harvey, B. C., R. C. Cashner, and W. J. Matthews. (1988) Differential effects of largemouth and smallmouth bass on habitat use by stoneroller minnows in stream pools. J. Fish. Biol., **33,** 481–87.

Hasler, A. D. (1966) *Underwater Guideposts.* University of Wisconsin Press, Madison, p. 155.

Hasler, A. D. and W. J. Wisby. (1958) The return of displaced largemouth bass and green sunfish to a "home" area. Ecology, **39,** 289–93.

Hasler, A. D., A. T. Scholz, and R. W. Goy (eds.). (1983) *Olfactory Imprinting and Homing in Salmon.* Springer-Verlag, Berlin, p. 134.

Hastings, R. W. and R. E. Good. (1977) Population analysis of the fishes of a freshwater tidal tributary of the lower Delaware River. NJ Acad. Sci. Bull., **22,** 13–20.

Hatch, D. H., W. H. Baltosser, and C. G. Schmitt. (1985) Life history and ecology of the bluntnose shiner *Notropis simus pecosensis* in the Pecos River in New Mexico. Southwest. Nat., **30,** 555–62.

Hawkes, H. A. (1975) River zonation and classification, in *River Ecology* (ed. B. A. Whitton). University of California Press, Berkeley, pp. 312–74.

Hawkins, B. A. and S. G. Compton. (1992) African fig wasp communities: undersaturation and latitudinal gradients in species richness. J. Animal Ecol., **61,** 361–72.

Hawkins, C. P., J. L. Kershner, P. A. Bisson, M. D. Bryant, L. M. Decker, S. V. Gregory, D. A. McCullough, C. K. Overton, G. H. Reeves, R. J. Steedman, and M. K. Young. (1993) A hierarchical approach to classifying stream habitat features. Fisheries, **18,** 3–10.

Hawkins, D. K. and T. P. Quinn. (1996) Critical swimming velocity and associated morphology of juvenile coastal cutthroat trout (*Oncorhynchus clarki clarki*), steelhead trout (*Oncorhynchus mykiss*), and their hybrids. Can. J. Fish. Aquat. Sci., **53,** 1487–96.

He, X. and J. F. Kitchell. (1990) Direct and indirect effects of predation on a fish community: a whole-lake experiment. Trans. Am. Fish. Soc., **119,** 825–35.

Healey, M. C. and W. R. Heard. (1984) Inter- and intra-population variation in the fecundity of chinook salmon (*Oncorhynchus tshawytscha*) and its relevance to life history theory. Can. J. Fish. Aquat. Sci., **41,** 476–83.

Heard, W. R. (1962) The use and selectivity of small-meshed gill nets at Brooks Lake, Alaska. Trans. Am. Fish. Soc., **91,** 263–68.

Hearn, W. E. and B. E. Kynard. (1986) Habitat utilization and behavioral interaction of juvenile Atlantic salmon *Salmo salar* and rainbow trout *S. gairdreni* in tributaries of the White River of Vermont. Can. J. Fish. Aquat. Sci., **43**, 1988–98.

Hecky, R. E. (1993) Peter Kilham Memorial Lecture: The eutrophication of Lake Victoria. Verh. Int. Verin. Limnol., **25**, 39–48.

Hefley, H. M. (1937) Ecological studies on the Canadian River floodplain in Cleveland County, Oklahoma. Ecol. Monogr., **7**, 345–402.

Heggenes, J. (1988a) Effects of experimentally increased intraspecific competition on sedentary adult brown trout *Salmo trutta* movement and stream habitat choice. Can. J. Fish. Aquat. Sci., **45**, 1163–72.

Heggenes, J. (1988b) Effects of short-term flow fluctuations on small displacement of, and habitat use by, brown trout in a small stream. Trans. Am. Fish. Soc., **117**, 336–44.

Heggenes, J. and T. Traaen. (1988) Daylight responses to overhead cover in stream channels for fry of four salmonid species. Holarctic Ecol., **11**, 194–201.

Heins, D. C. (1991) Variation in reproductive investment among populations of the longnose shiner, *Notropis longirostris*, from contrasting environments. Copeia, **1991**, 736–44.

Heins, D. C. and J. A. Baker. (1987) Analysis of factors associated with intraspecific variation in propagule size of a stream-dwelling fish, in *Community and Evolutionary Ecology of North American Stream Fishes* (eds. W. J. Matthews and D. C. Heins). University of Oklahoma Press, Norman, pp. 223–31.

Heins, D. C. and M. D. Machado. (1993) Spawning season, clutch characteristics, sexual dimorphism and sex ratio in the redfin darter *Etheostoma whipplei*. Am. Midl. Nat., **129**, 161–71.

Heins, D. C. and W. J. Matthews. (1987) Historical perspectives on the study of community and evolutionary ecology of North American stream fishes, in *Community and Evolutionary Ecology of North American Stream Fishes* (eds. W. J. Matthews and D. C. Heins). University of Oklahoma Press, Norman, pp. 3–7.

Helfman, G. S. (1979) Fish attraction to floating objects in lakes, in *Response of Fish to Habitat Structure in Standing Water* (eds. D. L. Johnson and R. A. Stein). Special Publication 6. American Fisheries Society, Bethesda, MD.

Helfman, G. S. (1981a) Twilight activities and temporal structure in a freshwater fish community. Can. J. Fish. Aquat. Sci., **38**, 1405–20.

Helfman, G. S. (1981b) The advantage to fishes of hovering in shade. Copeia, **1981**, 392–400.

Helfman, G. S. (1985) Fishes, in *An Ecosystem Approach to Aquatic Ecology—Mirror Lake and Its Environment* (ed. G. E. Likens). Springer-Verlag, New York, pp. 236–45.

Helfman, G. S. (1994) Adaptive variability and mode choice in foraging fishes, in *Theory and Application in Fish Feeding Ecology*, (eds. D. J. Stouder, K. L. Fresh, and R. J. Feller). University of South Carolina Press, pp. 3–17.

Hemphill, N. and D. S. Cooper. (1984) Differences in the community structure of stream pools containing or lacking trout. Verh. Int. Verein. Limnol., **22**, 1858–61.

Henderson, B. A. and F. E. J. Fry. (1987) Interspecific relations among fish species in South Bay, Lake Huron, 1949–84. Can. J. Fish. Aquat. Sci., **44**, 10–4.

Henry, C. P., G. Bornette, and C. Amoros. (1994) Differential effects of flood on the aquatic vegetation of braided channels of the Rhône River. J. North Am. Benthol. Soc., **13**, 439–67.

Herbold, B. (1984) Structure of an Indiana stream fish association: choosing an appropriate model. Am. Nat., **124**, 561–72.

Herbold, B. and P. B. Moyle. (1986) Notes and Comments. Am. Nat., **128**, 751–60.

Hershey, A. E. and S. I. Dodson. (1985) Selective predation by a sculpin and a stonefly on two chironomids in laboratory feeding trials. Hydrobiologia, **124**, 269–73.

Hessen, D. O. (1985) Selective zooplankton predation by pre-adult roach (*Rutilus rutilus*): the size-selective hypothesis versus the visibility-selective hypothesis. Hydrobiologia, **124**, 73–9.

Hessen, D. O., B. A. Faafeng, and T. Anderson. (1995) Replacement of herbivore zooplankton species along gradients of ecosystem productivity and fish predation pressure. Can. J. Fish. Aquat. Sci., **52**, 733–42.

Hesthagen, T. (1988) Movements of brown trout, *Salmo trutta,* and juvenile Atlantic salmon, *Salmo salar,* in a coastal stream in northern Norway. J. Fish. Biol., **32**, 639–53.

Hesthagen, T. (1990) Home range of juvenile Atlantic salmon, *Salmo salar,* and brown trout, *Salmo trutta,* in a Norwegian stream. Freshwater Biol., **24**, 63–67.

Hewett, S. W. and D. J. Stewart. (1989) Zooplanktivory by alewives in Lake Michigan: ontogenetic, seasonal, and historical patterns. Trans. Am. Fish. Soc., **118**, 581–96.

Hildrew, A. G. (1990) Fish predation and the organisation of invertebrate communities in streams. Pol. Arch. Hydrobiol., **37**, 95–107.

Hildrew, A. G., C. R. Townsend, and J. Francis. (1984) Community structure in some southern English streams: the influence of species interactions. Freshwater Biol., **14**, 297–310.

Hill, J. and G. D. Grossman. (1987) Home range estimates for three North American stream fishes. Copeia, **2**, 376–80.

Hill, J. and G. D. Grossman. (1993) An energetic model of microhabitat use for rainbow trout and rosyside dace. Ecology, **74**, 685–98.

Hill, L. G. (1968) Oxygen preference in the spring cavefish, *Chologaster agassizi.* Trans. Am. Fish. Soc., **97**, 448–54.

Hill, W. R. and B. C. Harvey. (1990) Periphyton responses to higher trophic levels and light in a shaded stream. Can. J. Fish. Aquat. Sci., **47**, 2307–14.

Hinch, S. G. and N. C. Collins. (1991) Relative abundance of littoral zone fishes: biotic interactions, abiotic factors, and postglacial colonization. Ecology, **72**, 1314–24.

Hinch, S. G., K. M. Somers, and N. C. Collins. (1994) Spatial autocorrelation and assessment of habitat–abundance relationships in littoral zone fish. Can. J. Fish. Aquat. Sci., **51**, 701–12.

Hindar, K. and B. Jonsson. (1982) Habitat and food segregation of dwarf and normal Arctic charr *Salvelinus alpinus* from Vangsvatnet Lake, Western Norway. Can. J. Fish. Aquat. Sci., **39**, 1030–45.

Hindar, K., B. Jonsson, J. H. Andrew, and T. G. Northcote. (1988) Resource utilization of sympatric and experimentally allopatric cutthroat trout and Dolly Varden charr. Oecologia (Berlin), **74**, 481–91.

Hirshfield, M. F., C. R. Feldmeth and D. L. Soltz. (1980) Genetic differences in physiological tolerances of Amargosa pupfish (*Cyprinodon nevadensis*) populations. Science, **207**, 999–1001.

Hlohowskyj, I. and A. M. White. (1983) Food resource partitioning and selectivity by the greenside, rainbow, and fantail darters (Pisces: Percidae). Ohio J. Sci., **83**, 201–08.

Hlohowskyj, I. and T. E. Wissing. (1987) Seasonal changes in the thermal preferences of fantail (*Etheostoma flabellare*), rainbow (*E. caeruleum*), and greenside (*E. blennioides*) darters, in *Community and Evolutionary Ecology of North American Stream Fishes* (eds. W. J. Matthews and D. C. Heins). University of Oklahoma Press, Norman, pp. 105–10.

Hoagland, B. W. (1995) Gradient structure and heterogeneity in wetland plant communities. Ph.D. dissertation University of Oklahoma, Norman.

Hobbs, H. H., III. (1992) Caves and springs, in *Biodiversity of the Southeastern United States: Aquatic Communities* (eds. C. T. Hackney, S. M. Adams, and W. H. Martin). John Wiley and Sons, New York, pp. 59–131.

Hocutt, C. H. and J. R. Stauffer. (1975) Influence of gradient on the distribution of fishes in Conowingo Creek, Maryland and Pennsylvania. Chesapeake Sci., **16**, 143–47.

Hocutt, C. H. and E. O. Wiley (eds.). (1986) *The Zoogeography of North American Freshwater Fishes*. John Wiley and Sons, New York.

Hocutt, C. H., R. E. Jenkins, and J. R. Stauffer, Jr. (1986) Zoogeography of the fishes of the Central Appalachians and Central Atlantic Coastal Plain, in *The Zoogeography of North American Freshwater Fishes* (eds. C. H. Hocutt and E. O. Wiley). John Wiley and Sons, New York, pp. 161–211.

Höglund, L. B. and J. Härdig. (1969) Reactions of young salmonids to sudden changes of pH, carbon-dioxide tension and oxygen content. Inst. Freshwat. Res., Drottningholm, **49**, 77–115.

Holomuzki, J. R. and R. J. Stevenson. (1992) Role of predatory fish in community dynamics of an ephemeral stream. Can. J. Fish. Aquat. Sci., **49**, 2322–30.

Hoopes, R. (1975) Flooding as a result of Hurricane Agnes, and its effects on a native brook trout population in an infertile headwaters stream in central Pennsylvania. Trans. Am. Fish. Soc., **104**, 96–9.

Hopey, M. E. and J. W. Petranka. (1994) Restriction of wood frogs to fish-free habitats: how important is adult choice? Copeia, **1994**, 1023–25.

Hora, S. L. (1952) The Himalayan fishes. Himalaya, **1**, 66–74.

Hori, M., K. Yamaoka, and K. Takamura. (1983) Abundance and micro-distribution of cichlid fishes on a rocky shore of Lake Tanganyika. Afr. Study Monogr., **3**, 25–38.

Hori, M., M. M. Gashagaza, M. Nshombo, and H. Kawanabe. (1993) Littoral fish communities in Lake Tanganyika: irreplacable diversity supported by intricate interactions among species. Conserv. Biol., **7**, 657–66.

Horn, M. H. (1992) Herbivorous fishes: feeding and digestive mechanisms, in *Plant-*

Animal Interactions in the Marine Benthos, (eds. D. M. John, S. J. Hawkins, and J. H. Price). Systematics Association Special Volume No. 46, Clarendon Press, Oxford. pp. 339–62.

Horton, R. E. (1945) Erosional development of streams and their drainage basins. Bull. Geol. Soc. Am., **56,** 275–370.

Horwitz, R. J. (1978) Temporal variability patterns and the distributional patterns of stream fishes. Ecol. Monogr., **48,** 307–21.

Houde, A. E. and A. J. Torio. (1992) Effect of parasitic infection on male color pattern and female choice in guppies. Behav. Ecol., **3,** 346–51.

Houde, A. E. and M. A. Hankes. (1997) Evolutionary mismatch of mating preferences and male colour patterns in guppies. Anim. Behav. **53,** 343–51.

Howes, G. J. (1991) Systematics and biogeography: an overview, in *Cyprinid Fishes: Systematics, Biology and Exploitation* (eds. I. J. Winfield and J. S. Nelson). Chapman & Hall, London, pp. 1–33.

Hoyt, R. D., S. E. Neff, and V. H. Resh. (1979) Distribution, abundance, and species diversity of fishes of the upper Salt River drainage, Kentucky. Trans. KY Acad. Sci., **40,** 1–20.

Hrbácek, J., M. Dvořakova, V. Kořinek, and L. Procházkóva. (1961) Demonstration of the effect of the fish stock on the species composition of zooplankton and the intensity of metabolism of the whole plankton association. Verh. Int. Verein. Limnol., **14,** 192–95.

Huang, C. and A. Sih. (1990) Experimental studies on behaviorally mediated, indirect interactions through a shared predator. Ecology, **71,** 1515–22.

Huang, C. and A. Sih. (1991) Experimental studies on direct and indirect interactions in a three trophic-level stream system. Oecologia, **85,** 530–36.

Hubbs, C. (1958) Geographic variations in egg complement of *Percina caprodes* and *Etheostoma spectabile.* Copeia, **1958,** 102–05.

Hubbs, C. (1961a). Developmental temperature tolerances of four Etheostomatine fishes occurring in Texas. Copeia, **1961,** 195–98.

Hubbs, C. (1961b) Differences in the incubation period of two populations of *Etheostoma lepidum.* Copeia, **1961,** 198–200.

Hubbs, C. (1964) Effects of thermal fluctuations on the relative survival of greenthroat darter young from stenothermal and eurythermal waters. Ecology, **45,** 376–79.

Hubbs, C. (1965) Developmental temperature tolerance and rates of four southern California fishes, *Fundulus parvipinnis, Atherinops affinis, Leuresthes tenuis,* and *Hypsoblennius* sp. CA Fish Game, **51,** 113–22.

Hubbs, C. (1984) Changes in fish abundance with time of day and among years at a station in Lake Texoma. Proc. TX Chapter, Am. Fish. Soc. (TX/OK Ch. Meet. 1983), pp. 42–57.

Hubbs, C. (1985) Darter reproductive seasons. Copeia, **1985,** 56–68.

Hubbs, C. (1990) Declining fishes of the Chihuahuan Desert, in (eds. A. M. Powell et al.) *Third Symposium on Resources of the Chihuahuan Desert Region, United States and Mexico,* Chihuahuan Desert Research Institute, Alpine, TX, pp. 89–96.

Hubbs, C. (1995) Springs and spring runs as unique aquatic systems. Copeia, **1995,** 989–91.

Hubbs, C. (1996) Geographic variation in life history traits of *Gambusia* species. Proc. Desert Fish. Council, 1995 Symp., **XXVII,** pp. 1–12 and 21.

Hubbs, C. and N. E. Armstrong. (1962) Developmental temperature tolerance of Texas and Arkansas–Missouri *Etheostoma spectabile* (Percidae, osteichthyes). Ecology, **43,** 742–44.

Hubbs, C. and H. H. Bailey. (1977) Effects of temperature on the termination of breeding season of *Menidia audens.* Southwest. Nat., **22,** 544–47.

Hubbs, C. and C. Bryan. (1973) Effect of parental temperature experience on thermal tolerance of eggs of *Menidia audens,* in *The Early Life History of Fish* (ed. J. H. S. Blaxter). Springer-Verlag, Verlin, pp. 431–35.

Hubbs, C. and D. F. Burnside. (1972) Developmental sequences of *Zygonectes notatus* at several temperatures. Copeia. **1972,** 862–65.

Hubbs, C. and E. A. Delco, Jr. (1960) Geographic variations in egg complement of *Etheostoma lepidum.* TX J. Sci., **XII,** 3–7.

Hubbs, C., R. J. Edwards, and G. P. Garrett. (1991) An annotated checklist of the freshwater fishes of Texas, with keys to identification of species. Texas Journal of Science **43**(Supplement), 1–56.

Hubbs, C. and W. F. Hettler. (1958) Fluctuations of some central Texas fish populations. Southwest. Nat., **3,** 13–16.

Hubbs, C. and W. F. Hettler. (1964) Observations on the toleration of high temperatures and low dissolved oxygen in natural waters by *Crenichthys baileyi.* Southwest. Nat., **9,** 245–48.

Hubbs, C. and D. T. Mosier. (1985) Fecundity of *Gambusia gaigei.* Copeia, **1985,** 1063–64.

Hubbs, C. and J. Pigg. (1976) The effects of impoundments on threatened fishes of Oklahoma. Annals. OK Acad. Sci., **5,** 113–17.

Hubbs, C. and K. Strawn. (1957) The effects of light and temperature on the fecundity of the greenthroat darter, *Etheostoma lepidum* (Girard). Ecology, **38,** 596–602.

Hubbs, C. and K. Strawn. (1963) Differences in the developmental temperature tolerance of central Texas and more northern stocks of *Percina caprodes* (Percidae: Osteichthyes). Southwest. Nat., **8,** 43–5.

Hubbs, C., A. E. Peden, and M. M. Stevenson (1969) The development rate of the greenthroat darter, *Etheostoma lepidum.* Am. Midl. Nat., **81,** 182–88.

Hubbs, C., M. M. Stevenson, and A. E. Peden. (1968) Fecundity and egg size in two central Texas darter populations. Southwest. Nat., **13,** 301–24.

Hubbs, C., E. Marsh-Matthews, W. J. Matthews, and A. A. Anderson. (1997) Changes in fish assemblages in East Texas streams from 1953 to 1986. TX J. Sci. (in press).

Hubbs, C. L. (1940) Fishes of the desert. Biologist, **XXII,** 61–9.

Hubbs, C. L. (1941) The relation of hydrological conditions to speciation in fishes, in *A Symposium on Hydrobiology.* University of Wisconsin Press, Madison, pp. 182–95.

Hubbs, C. L. and R. W. Eschmeyer. (1938) The improvement of lakes for fishing. A method of fish management. Fish. Res. Bull., of the Michigan Dept. Conserv., **2,** 1–233.

Hubbs, C. L. and A. I. Ortenburger. (1929) Fishes collected in Oklahoma and Arkansas in 1927. Univ. OK Biol. Surv., **1,** 45–112.

Hubbs, C. L., R. R. Miller, and L. C. Hubbs. (1974) Hydrographic history and relict fishes of the north-central Great Basin. Mem. CA Acad. Sci., **7,** 1–259.

Huber, R. and M. K. Rylander. (1992) Quantitative histological study of the optic nerve in species of minnows (Cyprinidae, Teleostei) inhabiting clear and turbid water. Brain Behav. Evol., **40,** 250–55.

Huber, R. and M. K. Rylander. (1992) Brain morphology and turbidity preference in *Notropis* and related genera (Cyprinidae, Teleostei). Environ. Biol. Fish., **33,** 153–65.

Huet, M. (1949) Apercu des relations entre la pente et les populations piscicoles des eaus courantes. Schweiz. Z. Hydrol., **11,** 332–51. [In French, original not seen, cited from Balon and Stewart, (1983)].

Huet, M. (1954) Biologie, profils en long et en travers des eaux courantes. Bull. franc. Piscic., **175,** 41–53. [In French, original not seen, cited from Balon and Stewart, (1983)].

Huet, M. (1959) Profiles and biology of Western European streams as related to fish management. Trans. Am. Fish. Soc., **88,** 155–63.

Huets, M. J. (1947) Experimental studies on adaptive evolution in *Gasterosteus aculeatus* L., Evolution, **1,** 89–102.

Huey, R. B. and M. Slatkin. (1976) Costs and benefits of lizard thermoregulation. Quart. Rev. Biol., **51,** 363–84.

Hughes, A. L. (1985) Male size, mating success, and mating strategy in the mosquitofish *Gambusia affinis* (Poeciliidae). Behav. Ecol. Sociobiol., **17,** 271–78.

Hughes, G. M. (1972) Morphometrics of fish gills. Respir. Physiol., **14,** 1–25.

Hughes, G. M. (1995) Preliminary morphometric study of the gills of *Oreochromis alcalicus grahami* from Lake Magadi and a comparison with *O. niloticus*. J. Fish. Biol., **47,** 1102–05.

Hughes, M. K. and H. F. Diaz. (1994) Was there a "Medieval Warm Period," and if so, where and when? Climatic Change, **26,** 109–42.

Hughes, N. F. (1992) Ranking and feeding positions by drift-feeding arctic grayling *Thymallus arcticus* in dominance hierarchies. Can. J. Fish. Aquat. Sci., **49,** 1994–98.

Hughes, N. F. and L. M. Dill. (1990) Position choice by drift-feeding salmonids: model and test for Arctic grayling *Thymallus arcticus* in subarctic mountain streams, interior Alaska. Can. J. Fish. Aquat. Sci., **47,** 2039–48.

Hughes, R. M. and J. R. Gammon. (1987) Longitudinal changes in fish assemblages and water quality in the Williamette River, Oregon. Trans. Am. Fish. Soc., **116,** 196–209.

Hughes, R. M. and D. P. Larsen. (1988) Ecoregions: an approach to surface water protection. J. Water Pollut. Control. Fed., **60,** 486–93.

Hughes, R. M. and J. M. Omernik. (1981) Use and misuse of the terms watershed and stream order. Am. Fish. Soc. Warmwater Streams Symp., pp. 320–26.

Hughes, R. M. and J. M. Omernik. (1983) An alternative for characterizing stream size, in *Dynamics of Lotic Ecosystems* (eds. T. D. Fontaine III and S. M. Bartel). Ann Arbor Science Publ., Ann Arbor, pp. 87–101.

Hughes, R. M., E. Rexstad, and C. E. Bond. (1987) The relationship of aquatic ecoregions, river basins and physiographic provinces to the ichthyogeographic regions of Oregon. Copeia, **1987,** 423–32.

Hughes, R. M., S. A. Heiskary, W. J. Matthews, and C. O. Yoder. (1994) Use of ecoregions in biological monitoring, in *Biological Monitoring of Aquatic Systems* (eds. S. L. Loeb and A. Spacie). Lewis Publ., Boca Raton, FL, pp. 125–51.

Hugueny, B. (1989) West African rivers as biogeographic islands: species richness of fish communities. Oecologia, **79,** 236–43.

Hugueny, B. (1990) Geographic range of west African freshwater fishes: role of biological characteristics and stochastic processes. Acta Œcolog., **11,** 351–75.

Hugueny, B. and C. Lévêque. (1994) Freshwater fish zoogeography in west Africa: faunal similarities between river basins. Environ. Biol. Fish., **39,** 365–80.

Hugueny, B. and D. Paugy. (1995) Unsaturated fish communities in African rivers. Am. Nat., **146,** 163–69.

Hulsman, P. F., P. M. Powles, and J. M. Gunn. (1983) Mortality of walleye eggs and rainbow trout yolk-sac larvae in low-pH waters of the LaCloche Mountain area, Ontario, Trans. Am. Fish. Soc., **112,** 680–88.

Hume, J. M. B. and T. G. Northcote. (1985) Initial changes in use of space and food by experimentally segregating populations of Dolly Varden *Salvelinus malma* and cutthroat trout *Salmo clarki.* Can. J. Fish Aquat. Sci., **42,** 101–09.

Humphries, J. M. (1984) Genetics of speciation in pupfishes from Laguna Chichancanab, Mexico, in *Evolution of Fish Species Flocks* (eds. A. A. Echelle and I. Kornfeld), University of Maine Press, Orono, pp. 129–39.

Humphries, J. M. and R. C. Cashner. (1994) *Notropis suttkusi,* a new cyprinid fish from the Ouachita uplands of Oklahoma and Arkansas, with comments on the status of Ozarkian populations. Copeia, **1994,** 82–90.

Humphries, P. (1993) A comparison of the mouth morphology of three co-occurring species of arthernid. J. Fish. Biol., **42,** 585–93.

Hunt, C. B. (1974) *Natural Regions of the United States and Canada,* W. C. Freeman, San Francisco.

Hupp, C. R. (1986) Upstream variation in bottomland vegetation patterns, northwestern Virginia. Torry Botan. Club Bull., **113,** 421–30.

Hurlbert, S., J. Zedler, and D. Fairbanks. (1972) Ecosystem alteration by mosquitofish (*Gambusia affinis*) predation. Science, **175,** 639–41.

Hurlbert, S. H. and M. S. Mulla. (1981) Impacts of mosquitofish (*Gambusia affinis*) predation on plankton communities. Hydrobiologia, **83,** 125–51.

Huryn, A. D. (1996) An appraisal of the Allen paradox in a New Zealand trout stream. Limnol. Oceanogr., **41,** 243–52.

Hutchinson, G. E. (1939) Ecological observations on the fishes of Kashmir and Indian Tibet. Ecol. Monogr., **9,** 146–82.

Hutchinson, G. E. (1957a) Concluding remarks. Cold Spring Harbor Symp. Quant. Biol., **22,** 415–27.

Hutchinson, G. E. (1957b) Geography, physics, and chemistry, in *A Treatise on Limnology,* John Wiley and Sons, New York, Vol. 1.

Hutchinson, G. E. (1959) Homage to Santa Rosalia, or "Why are there so many kinds of animals?" Am. Nat., **93,** 137–46.

Hutchinson, G. E. (1967) Introduction to lake biology and the limnoplankton, in *A Treatise on Limnology,* John Wiley and Sons, New York, Vol. 2.

Hutchinson, G. E. (1975) Limnological botany, in *A Treatise on Limnology,* John Wiley and Sons, New York, Vol. 3.

Hutchison, V. H. (1961) Critical thermal maxima in salamanders. Physiol. Zool., **34,** 92–125.

Hutchison, V. H. (1976) Factors influencing thermal tolerances of individual organisms, in *Thermal Ecology II* (eds. G. W. Esch and R. W. McFarlane), ERDA Symposium Series No. 40, pp. 10–26. Tech. Info. Ctr. ERDA, **40,** 10–2.

Hutchison, V. H. and J. D. Maness. (1979) The role of behavior in temperature acclimation and tolerance in ectotherms. Am. Zool., **19,** 367–84.

Hynes, H. B. N. (1970) *The Ecology of Running Waters.* University of Toronto Press, Toronto.

Hynes, H. B. N. (1975) The stream and its valley. Verh. Int. Vertibr. Theor. Ang. Limnol., **19,** 1–15.

Ibarra, M. and D. J. Stewart. (1989) Longitudinal zonation of sandy beach fishes in the Napo River Basin, eastern Ecuador. Copeia, **1989,** 364–81.

Imhof, J. G., J. Fitzgibbon, and W. K. Annable. (1996) A hierarchical evaluation system for characterizing watershed ecosystems for fish habitat. Can. J. Fish. Aquat. Sci., **53,** 312–26.

Inger, R. F. (1955) Ecological notes on the fish fauna of a coastal drainage of North Borneo. Fieldiana Zool., **37,** 47–90.

Ingersoll, C. G. and D. L. Claussen. (1984) Temperature selection and critical thermal maxima of the fantail darter, *Etheostoma flabellare,* and johnny darter, *E. nigrum,* related to habitat and season. Environ. Biol. Fish., **2,** 131–38.

Ingersoll, C. G., I. Hlohowskyj, and N. D. Mundahl. (1984) Movements and densities of the darters *Etheostoma flabellare, E. spectabile,* and *E. nigrum* during spring spawning. J. Freshwater Ecol., **2,** 345–51.

Isascs, E. H. and R. M. Srivastava. (eds.) (1989) *An Introduction to Applied Geostatistics.* Oxford University Press, New York.

Itzkowitz, M. (1977) Social dynamics of mixed-species groups of Jamaican reef fishes. Behav. Ecol. Sociobiol., **2,** 361–84.

Itzkowitz, M. and J. Nyby (1982) Field observations of parental behavior of the Texas cichlid *Cichlasoma cyanoguttatum.* Am Midl. Nat., **108,** 364–68.

Itzkowitz, M. (1984) Parental division of labor in a monogamous fish. Behavior, **89,** 251–60.

Ivlev, V. S. (1961) *Experimental Ecology of the Feeding of Fishes.* Yale University Press, New Haven, CT.

Jackson, D. A. and H. H. Harvey. (1989) Biogeographic associations in fish assemblages: local vs. regional processes. Ecology, **70**, 1472–84.

Jackson, D. A. and H. H. Harvey. (1993) Fish and benthic invertebrates: community concordance and community-environment relationships. Can. J. Fish. Aquat. Sci., **50**, 2641–51.

Jackson, D. A., H. H. Harvey, and K. M. Somers. (1990) Ratios in aquatic sciences: statistical shortcomings with mean depth and the morphoedaphic index. Can. J. Fish. Aquat. Sci., **47**, 1788–95.

Jackson, P. B. N. (1961) The impact of predation, especially by the tiger-fish (*Hydrocyon bittatus* Cast.) on African freshwater fishes. Proc. Zool. Soc. Lond., **136**, 603–22.

Jackson, W. D. and G. L. Harp. (1974) Ichthyofaunal diversification and distribution in an Ozark stream in northcentral Arkansas. Proc. AR Acad. Sci., **27**, 42–6.

Jaksic, F. M. (1981) Recognition of morphological adaptations in animals: the hypothetico-deductive method. BioScience, **31**, 667–70.

James, M. C. (1934) Effect of 1934 drought on fish life. Trans. Am. Fish. Soc., **64**, 57–62.

Janssen, J. (1980) Alewives (*Alosa pseudoharengus*) and ciscoes (*Coregonus artedii*) as selective and non-selective planktivores, in *Evolution and Ecology of Zooplankton Communities* (ed. W. C. Kerfoot). University Press of New England, Hanover, NH, pp. 580–86.

Javier, L.-C., A. de Sostoa, and C. Moñtanes. (1986) Fish production and its relation with the community structure in an aquifer-fed stream of Old Castile (Spain). Pol. Arch. Hydrobiol., **33**, 333–43.

Jayaram, K. C. (1977) Zoogeography of Indian freshwater fishes. Proc. Indian Acad. Sci., **86B**, 265–74.

Jefferies, M. J. and J. H. Lawton. (1985) Predator–prey ratios in communities of freshwater invertebrates: the role of enemy free space. Freshwater Biol., **15**, 105–12.

Jenkins, R. E. and N. M. Burkhead. (1994) *Freshwater Fishes of Virginia.* American Fisheries Society, Bethesda, MD.

Jenkins, R. E. and C. A. Freeman. (1972) Longitudinal distribution and habitat of the fishes of Mason Creek, an upper Roanoke River drainage tributary, Virginia. VA J. Sci., **23**, 194–202.

Jenkins, R. E., E. A. Lachner, and F. J. Schwartz. (1972) Fishes of the central Appalachian drainages: their distribution and dispersal, in *The Distributional History of the Biota of the Southern Appalachians, Part III: Vertebrates* (ed. P. C. Holt). Virginia Polytechnic Institute and State University, Blacksburg, VA., pp. 43–117.

Jenkins, R. M. (1982) The morphoedaphic index and reservoir fish production. Trans. Am. Fish. Soc., **111**, 133–40.

Jennings, M. J. and D. P. Philipp. (1992) Reproductive investment and somatic growth rates in longear fish. Environ. Biol. Fish., **35**, 257–71.

Jennings, M. J. and D. P. Philipp. (1994) Biotic and abiotic factors affecting survival of early life history invervals of a stream-dwelling sunfish. Environ. Biol. Fish., **39**, 153–59.

Jeppesen, E., M. Søndergaard, O. Sortkjær, E. Mortensen, and P. Kristensen. (1990)

Interactions between phytoplankton, zooplankton and fish in a shallow, hypertrophic lake: a study of phytoplankton collapses in Lake Søbygård, Denmark. Hydrobiologia, **191,** 149–64.

Jester, D. B. (1971) Effects of commercial fishing, species introductions, and drawdown control on fish populations in Elephant Butte Reservoir, New Mexico, in *Reservoir Fisheries and Limnology* (ed. G. E. Hall). American Fisheries Society, Bethesda, MD, pp. 265–85.

Jobling, M. (1985) Physiological and social constraints on growth of fish with special reference to arctic charr, *Salvelinus alpinus* L. Aquaculture, **44,** 83–90.

Jobling, M. (1994) *Fish Bioenegetics.* Chapman & Hall, London.

Johannes, M. R. S. (1993) Prey aggregation is correlated with increased predation pressure in lake fish communities. Can. J. Fish. Aquat. Sci., **50,** 66–73.

Johannes, R. E. and P. A. Larkin. (1961) Competition for food between redside shiners (*Richardsonius balteatus*) and rainbow trout (*Salmo gairdneri*) in two British Columbia lakes. J. Fish. Res. Bd. Can., **18,** 203–20.

Johansson, L. (1987) Experimental evidence for interactive habitat segregation between roach *Rutilus rutilus* and rudd *Scardinius erythrophthalmus* in a shallow eutrophic lake. Oecologia (Berlin), **73,** 21–7.

John, K. R. (1964) Survival of fish in intermittant streams of the Chiricahua Mountains, Arizona. Ecology, **45,** 112–19.

Johnsen, P. B. and A. D. Hasler. (1977) Winter aggregations of carp (*Cyprinus carpio*) as revealed by ultrasonic tracking. Trans. Am. Fish. Soc., **106,** 556–59.

Johnson, D. M., T. H. Martin, P. H. Crowley, and L. B. Crowder. (1996a) Link strength in lake littoral food webs: net effects of small sunfish and larval dragonflies. J. North Am. Benthol. Soc., **15,** 271–88.

Johnson, J. H. (1981) Comparative food selection by coexisting subyearling coho salmon, chinook salmon and rainbow trout in a tributary of Lake Ontario. NY Fish Game J., **28,** 150–61.

Johnson, S. L. and C. C. Vaughn. (1995) A hierarchical study of macro-invertebrate recolonization of disturbed patches along longitudinal gradient in a prairie river. Freshwater Biol., **34,** 531–40.

Johnson, S. W., J. Heifetz, and K. V. Koski. (1986) Effects of logging on the abundance and seasonal distribution of juvenile steelhead in some Alaska streams. N. Amer. J. Fish. Management. **6,** 532–537.

Johnson, T. C., C. A. Scholz, M. R. Talbot, K. Kelts, R. D. Ricketts, G. Ngobi, K. Beuning, I. Ssemmanda, and J. W. McGill. (1996b) Late Pleistocene desiccation of Lake Victoria and rapid evolution of cichlid fishes. Science, **273,** 1091–93.

Johnston, C. E. and L. M. Page. (1992) The evolution of complex reproductive strategies in North American minnows (Cyprinidae), in *Systematics, Historical Ecology, and North American Freshwater Fishes,* (ed. R. L. Mayden). Stanford University Press, Stanford, CA, pp. 600–21.

Jones, A. N. (1975) A preliminary study of fish segregation in salmon spawning streams. J. Fish. Biol., **7,** 95–104.

Jones, C. G., J. H. Lawton, and M. Shachak. (1994) Organisms as ecosystem engineers. Oikos, **69**, 373–86.

Jones, D. T. (1925) The protractile apparatus of the mouth of the pumpkin-seed sunfish, *Eupomotus gibbosus* L. Anat. Rec., **31**, 173–91.

Jones, J. R. and M. V. Hoyer. (1982) Sportfish harvest predicted by summer chlorophyll-a concentration in midwestern lakes and reservoirs. Trans. Am. Fish. Soc., **111**, 176–79.

Jones, J. R. E. (1952) The reactions of fish to water of low oxygen concentration. J. Expt. Biol., **XXIX**, 403–15.

Jones, K. A., S. B. Brown, and T. J. Hara. (1987) Behavioral and biochemical studies of onset and recovery from acid stress in arctic char (*Salvelinus alpinus*). Can. J. Fish. Aquat. Sci., **44**, 373–81.

Jones, K. A., T. J. Hara, and E. Scherer. (1985a) Behavioral modifications in arctic char (*Salvelinus alpinus*) chronically exposed to sublethal pH. Physiol. Zool., **58**, 400–12.

Jones, K. A., T. J. Hara, and E. Scherer. (1985b) Locomotor response by arctic char (*Salvelinus alpinus*) to gradients of H^+ and CO_2. Physiol. Zoo., **58**, 413–20.

Jonsson, B. and F. R. Gravem. (1985) Use of space and food by resident and migrant brown trout, *Salmo trutta*. Environ. Biol. Fish., **14**, 281–93.

Jonsson, B. and K. Hindar. (1982) Reproductive strategy of dwarf and normal arctic charr (*Salvelinus alpinus*) from Vangsvatnet Lake, western Norway. Can. J. Fish. Aquat. Sci., **39**, 1404–13.

Jonsson, B. and O. T. Sandlund. (1979) Environmental factors and life histories of isolated river stocks of brown trout (*Salmo trutta* m. *fario*) in Søre Osa river system, Norway. Environ. Biol. Fish., **4**, 43–54.

Jordan, D. S. (1989) Report of exploration made during the summer and autumn of 1888, in the Allegheny region of Virginia, North Carolina and Tennessee, and in western Indiana, with an account of the fishes found in each of the river basins of those regions. Bull. U.S. Fish Comm., **8**, 97–173.

Jordan, D. S. (1905) *Guide to the Study of Fishes*. Henry Holt, New York, Vols. 1 and 2.

Jowett, I. G. and J. Richardson. (1989) Effects of a severe flood on instream habitat and trout populations of seven New Zealand rivers. NZ J. Mar. Freshwater Res., **23**, 11–7.

Jubb, R. A. (1977) Comments on Victoria Falls as a physical barrier for downstream dispersal of fishes. Copeia, **1977**, 198–99.

Juday, C., W. H. Rich, G. I. Kemmerer, and A. Mann. (1932) Limnological studies of Karluk Lake, Alaska 1926–1930. US Bur. Fish., **47**, 407–36.

Jungwirth, S. M. and S. Schmutz. (1995) The effects of recreated instream and ecotone structures on the fish fauna of an epipotamal river. Hydrobiologia, **303**, 195–206.

Junk, W. J. (1970) Investigations on the ecology and production-biology of the "floating meadows" (*Paspalo-Echinochloetum*) on the Middle Amazon. Part 1: The floating vegetation and its ecology. Amazoniana, **2**, 449–95.

Junk, W. J. (1973) Investigations on the ecology and production-biology of the "floating meadows" (*Paspalo-Eschinochloetum*) on the Middle Amazon. Part II. The aquatic fauna in the root zone of the floating vegetation. Amazoniana, **4**, 9–102.

Junk, W. J., P. B. Bayley, and R. E. Sparks. (1989) The flood pulse concept in river–floodplain systems, in *Proceedings Of the International Large River Symposium* (ed. D. P. Dodge). Can. Spec. Publ. Fish Aquat. Sci., **106**, 110–27.

Kairesalo, T. and T. Seppälä. (1987) Phosphorus flux through a littoral ecosystem: the importance of cladoceran zooplankton and young fish. Int. Rev. Hydrobiol., **72**, 385–403.

Kalikhman, T., P. Walline, and M. Gophen. (1992) Simultaneous patterns of temperature, oxygen, zooplankton and fish distribution in Lake Kinneret, Israel. Freshwater Biol., **28**, 337–47.

Kamler, E. 1992. *Early Life History of Fish: An Energetics Approach.* Chapman and Hall, London.

Karlstrom, O. (1977) Habitat selection and population densities of salmon and trout parr in Swedish rivers. Inst. Freshwater Res., Drottingholm, **6**, 1–72.

Karr, J. R. (1994) Defining disturbance, in *Theory and Application in Fish Feeding Ecology,* (eds. D. J. Stouder, K. L. Fresh, and R. J. Feller). University of South Carolina Press, pp. 285–291.

Karr, J. R. and K. E. Freemark. (1985) Disturbance and vertebrates: an integrative perspective, in *The Ecology of Natural Disturbance and Patch Dynamics* (eds. S. T. A. Pickett and P. S. White). Academic Press, New York, pp. 154–68.

Karr, J. R., K. D. Fausch, P. L. Angermeier, P. R. Yant, and I. J. Schlosser. (1986) Assessing biological integrity in running waters—a method and its rationale. Ill. Nat. Hist. Surv. Spec. Pub., **5**.

Kashuba, S. A. and W. J. Matthews. (1984) Physical condition of larval shad during spring-summer in a southwestern reservoir. Trans. Amer. Fish. Soc., **113**, 199–204.

Kats, L. B., J. W. Petranka, and A. Sih. (1988) Antipredator defenses and the persistence of amphibian larvae with fishes. Ecology, **69**, 1865–70.

Kaufman, L. (1992) Catastrophic change in species-rich freshwater ecosystems: the lessons of Lake Victoria. BioScience, **42**, 846–58.

Kaya, C. M. (1991) Rheotactic differentiation between fluvial and lacustrine populations of Arctic grayling (*Thymallus arcticus*), and implications for the only remaining indigenous populations of fluvial "Montana grayling". Can. J. Fish. Aquat. Sci., **48**, 53–59.

Keast, A. (1965) Resource subdivision amongst cohabitating fish species in a bay, Lake Opincon, Ontario. University of Michigan Great Lakes Res. Div., **13**, 106–32.

Keast, A. (1966) Trophic interrelationships in the fish fauna of a small stream. Great Lakes Research Division, University of Michigan Publ. No., **15**, 51–79.

Keast, A. (1977) Mechanisms expanding niche width and minimizing intraspecific competition in two centrarchid fishes. Evol. Biol., **10**, 333–95.

Keast, A. (1978a) Trophic and spatial interrelationships in the fish species of an Ontario temperate lake. Environ. Biol. Fish., **3**, 7–31.

Keast, A. (1978b) Feeding interrelations between age-groups of pumpkinseed *Lepomis gibbosus* and comparisons with bluegill (*L. macrochirus*). J. Fish. Res. Bd. Can., **35**, 12–27.

Keast, A. (1985a) The piscivore feeding guild of fishes in small freshwater ecosystems. Environ. Biol. Fish., **12**, 119–29.

Keast, A. (1985b) Growth response of the brown bullhead *Ichtalurus nebulosus* to temperature. Can. J. Zool., **63**, 1510–15.

Keast, A. (1985c) Development of dietary specializations in a summer community of juvenile fishes. Environ. Biol. Fish., **13**, 211–24.

Keast, A. (1985d) Planktivory in a littoral-dwelling lake fish association: prey selection and seasonality. Can. J. Fish. Aquat. Sci., **42**, 1114–26.

Keast, A. (1988) Planktivory in larval, juvenile, and adult planktivores: resource division in a small lake. Verh. Int. Ver. Limnol., **23**, 1692–97.

Keast, A. (1991) Panbiogeography: then and now. Quart. Rev. Biol., **66**, 467–72.

Keast, A. and M. G. Fox. (1990) Fish community structure, spatial distribution and feeding ecology in a beaver pond. Environ. Sci., **27**, 201–14.

Keast, A. and J. Harker. (1977) Fish distribution and benthic invertebrate biomass relative to depth in an Ontario Lake. Environ. Biol. Fish., **2**, 235–40.

Keast, A., J. Harker, and D. Turnbull. (1978) Nearshore fish habitat utilization and species associations in Lake Opinicon (Ontario, Canada). Environ. Biol. Fish., **3**, 173–84.

Keast, A. and J. McA. Eadie. (1985) Growth depensation in year-0 largemouth bass: the influence of diet. Trans. Am. Fish. Soc. **114**, 204–13.

Keast, A. and D. Webb. (1966) Mouth and body form relative to feeding ecology in the fish fauna of a small lake, Lake Opinicon, Ontario J. Fish. Res. Bd. Can., **23**, 1845–74.

Keenleyside, M. H. A. (1972) Intraspecific intrusions into nests of spawning longear sunfish (Pisces: Centrarchidae). Copeia, **1972**, 272–78.

Keenleyside, M. H. A. (1979) *Diversity and Adaption in Fish Behavior.* Springer-Verlag, Berlin.

Kellogg, R. L. and J. J. Gift. (1983) Relationship between optimum temperatures for growth and preferred temperatures for the young of four fish species. Trans. Am. Fish. Soc., **112**, 424–30.

Kelso, J. R. M. and J. H. Lipsit. (1988) Young-of-the-year fish community in nine lakes, varying in pH, on the Canadian shield. Can. J. Fish. Aquat. Sci, **45**, 121–26.

Kelso, J. R. M. and C. K. Mimms. (1996) Is fish species richness at sites in the Canadian Great Lakes the result of local or regional factors? Can. J. Fish. Aquat. Sci., **53**(Suppl. 1), 175–93.

Kendall, A. W., Jr., E. H. Ahlstrom, and H. G. Moser. (1984) Early life history of fishes and their characters, in *Ontogeny and Systematics of Fishes,* Special Publication 1, American Society of Ichthyologists and Herpetologists, LaJolla, CA.

Kennedy, M. and R. D. Gray. (1993) Can ecological theory predict the distribution of foraging animals? A critical analysis of experiments on the ideal free distribution. Oikos, **68**, 158–66.

Kerr, S. R. and R. A. Ryder. (1977) Niche theory and percid community structure. J. Fish. Res. Bd. Can., **34**, 1952–58.

Kessler, R. K. and J. H. Thorp. (1993) Microhabitat segregation of the threatened spotted darter *Etheostoma maculatum* and closely related orangefin darter *E. bellum.* Can. J. Fish. Aquat. Sci., **50**, 1084–91.

Kessler, R. K., A. F. Casper, and G. K. Weddle. (1995) Temporal variation in microhabitat use and spatial relations in the benthic fish community of a stream. Am. Midl. Nat., **134**, 361–70.

Kindscher, K. and P. V. Wells. (1995) Prairie plant guides: a multivariate analysis of prairie species based on ecological and morphological traits. Vegetatio, **117**, 20–50.

King, C. E. (1964) Relative abundance of species and MacArthur's model. Ecology, **45**, 716–27.

King, L. R. (1973) Comparison of the distribution of minnows and darters collected in 1947 and 1972 in Boone County, Iowa. Proc. IA Acad. Sci., **80**, 133–35.

Kirchhofer, A. (1995) Morphological variability in the ecotone—an important factor for the conservation of fish species richness in Swiss rivers. Hydrobiologia, **303**, 103–10.

Kitchell, J. R. (1992) *Food Web Management—A Case Study of Lake Mendota.* Springer-Verlag, New York.

Kleckner, N. W. and B. D. Sidell. (1985) Comparison of maximal activities of enzymes from tissue of thermally acclimated and naturally acclimatized chain pickeral (*Esox niger*). Physiol. Zool., **58**, 18–28.

Kline, T. C., Jr., J. J. Goering, O. E. Mathisen, P. H. Poe, and P. L. Parker. (1990) Recycling of elements transported upstream by runs of Pacific Calmon: I. $\delta^{15}N$ and $\delta^{13}C$ evidence in Sashin Creek, southeastern Alaska. Can. J. Fish. Aquat. Sci., **47**, 136–44.

Klinger, S. A., J. J. Magnuson, and G. W. Gallepp. (1982) Survival mechanisms of the central mudminnow (*Umbra lima*), fathead minnow (*Pimephales promelas*) and brook stickleback (*Culaea inconstans*) for low oxygen in winter. Environ. Biol. Fish., **2**, 113–20.

Knapp, R. A. and R. C. Sargent. (1989) Egg-mimicry as a mating strategy in the fantail darter, *Etheostoma flabellare:* females prefer males with eggs. Behav. Ecol. Sociobiol., **25**, 321–26.

Kneib, R. T. (1994) Spatial pattern, spatial scale, and feeding in fishes in *Theory and Application in Fish Feeding Ecology* (eds. D. J. Stouder, K. L. Fresh, R. J. Feller, and M. Duke). University of South Carolina Press, Columbia, SC, pp. 171–85.

Knight, J. G. and S. T. Ross. (1992) Reproduction, age and growth of the bayou darter *Etheostoma rubrum* (Pisces, Percidae): an endemic of Bayou Pierre. Am. Midl. Nat., **127**, 91–105.

Kobayashi, D., K. Susuki, and M. Nomura. (1990) Diurnal fluctuation in stream flow and in specific electric conductance during drought periods. J. Hydrol., **115**, 105–14.

Kodric-Brown, A. (1993) Female choice of multiple male criteria in guppies: interacting effects of dominance, coloration and courtship. Behav. Ecol. Sociobiol., **32**, 415–20.

Kodric-Brown, A. and M. E. Hohmann. (1990) Sexual selection is stabilizing selection in pupfish (*Cyprinodon pecosensis*). Biol. J. Linn. Soc., **40**, 113–23.

Koebele, B. P. (1985) Growth and the size hierarchy effect: an experimental assessment of three proposed mechanisms; activity differences, disproportional food acquisition, physiological stress, Environ. Biol. Fish., **12**, 181–88.

Koetsier, P. (1989) The effects of fish predation and algal biomass on insect community structure in an Idaho stream. J. Freshwater Ecol., **5,** 187–96.

Kohler, S. L. (1985) Identification of stream drift mechanisms: an experimental and observational approach. Ecology, **66,** 1749–61.

Kolasa, J. (1989) Ecological system in hierarchical perspective: breaks in community structure and other consequences. Ecology, **70,** 36–47.

Kolok, A. S. (1991) Temperature compensation in two centrarchid fishes: do winter-quiescent fish undergo cellular temperature compensation? Trans. Am. Fish. Soc., **120,** 52–7.

Kondolf, G. M., G. F. Cada, M. J. Sale, and T. Felando. (1991) Distribution and stability of potential salmonid spawning gravels in steep boulder-bed streams of the Eastern Sierra Nevada. Trans. Am. Fish. Soc., **120,** 177–86.

Kornfield, I. and K. E. Carpenter. (1984) Cyprinids of Lake Lanao, Philippines: taxonomic validity, evolutionary rates and speciation scenarios, in *Evolution of Fish Species Flocks* (eds. A. A. Echelle and I. Kornfield). University of Maine at Orono, Orono. pp. 69–84.

Kowalski, K. T., J. P. Schubauer, C. L. Scott, and J. R. Spotila. (1978) Interspecific and seasonal differences in the temperature tolerance of stream fish. J. Therm. Biol., **3,** 105–08.

Kraatz, W. C. (1923) A study of the food of the minnow, *Campostoma anomalum.* Ohio J. Sci., **23,** 265–83.

Kraft, C. E. and J. F. Kitchell. (1986) Partitioning of food by sculpins in Lake Michigan. Environ. Biol. Fish., **16,** 309–16.

Kramer, B. (1990) Sexual signals in electric fishes. Trends Ecol. Evol., **5,** 247–50.

Kramer, D. L. (1978) Reproductive seasonality in the fishes of a tropical stream. Ecology, **59,** 976–85.

Kramer, D. L. (1983) The evolutionary ecology of respiratory mode in fishes: an analysis based on the costs of breathing. Environ. Biol. Fish., **9,** 145–58.

Kramer, D. L. (1987) Dissolved oxygen and fish behavior. Environ. Biol. Fish., **2,** 81–92.

Kramer, D. L. and E. A. Braun. (1983) Short-term effects of food availability on air-breathing frequency in the fish *Corydoras aeneus* (Callichthyidae). Can. J. Zool., **61,** 1964–67.

Krebs, C. J. (1994) The experimental analysis of distribution and abundance in *Ecology,* 4th ed. HarperCollins, College, New York.

Krokhin, E. M. (1975) Transport of nutrients by salmon migrating from the sea into lakes, in *Coupling of Land and Water Systems* (ed. A. D. Hasler). Springer-Verlag, New York.

Kubb, R. N., J. R. Spotila, and D. R. Pendergast. (1980) Mechanisms of heat transfer and time-dependent modeling of body temperature in the largemouth bass (*Micropterus salmoides*). Physiol. Zool., **53,** 222–39.

Kuehne, R. A. (1962) A classification of streams, illustrated by fish distribution in an eastern Kentucky creek. Ecology, **43,** 608–14.

Kushlan, J. A. (1976) Environmental stability and fish community diversity. Ecology, **57,** 821–25.

Kwain, W., R. W. McCauley, and J. A. MacLean. (1984) Susceptibility of starved, juvenile smallmouth bass, *Micropterus dolomieui* (Lacépède) to low pH. J. Fish. Biol., **25**, 501–04.

Kwak, T. J. (1988) Lateral movement and use of floodplain habitat by fishes of the Kankakee River, Illinois. Am. Midl. Nat., **120**, 241–49.

Lack, D. (1954) *The Natural Regulation of Animal Numbers.* Claredon Press, Oxford.

Ladle, M. and J. A. B. Bass (1981) The ecology of a small chalk stream and its responses to drying during drought conditions. Arch. Hydrobiol. **90**, 448–66.

Lagler, K. F. (1944) Problems of competition and predation. Trans. North Am. Wildl. Conf., **9**, 212–19.

Lagler, K. F., J. E. Bardach, and R. R. Miller. (eds.) (1962) *Ichthyology.* John Wiley and Sons, New York.

Lancaster, J. and A. G. Hildrew. (1993) Characterizing in-stream flow refugia. Can. J. Fish. Aquat. Sci., **50**, 1663–75.

Lancaster, J., A. G. Hildrew, and C. R. Townsend. (1988) Competition for space by predators in streams: field experiments on a net-spinning caddisfly. Freshwater Biol., **20**, 185–93.

Landau, R., M. Gophen, and P. Walline. (1988) Larval *Mirogrex terraesanctae* (Cyprinidae) of Lake Kinneret (Israel): growth rate, plankton selectivities, consumption rates and interaction with rotifers. Hydrobiologia, **169**, 91–106.

Langeland, A. (1982) Interactions between zooplankton and fish in a fertilized lake. Holarc. Ecol., **5**, 273–310.

Larimore, R. W. (1952) Home pools and homing behavior of smallmouth black bass in Jordan Creek. Biol. Notes, Urbana, **28**, 3–12.

Larimore, R. W. (1975) Visual and tactile orientation of smallmouth bass fry under floodwater conditions, in *Black Bass Biology Management,* (ed. H. Clepper). Sport Fishing Institute, Washington, DC, pp. 323–32.

Larimore, R. W. and M. J. Duever. (1968) Effects of temperature acclimation on the swimming ability of smallmouth bass fry. Trans. Am. Fish. Soc., **97**, 175–84.

Larimore, R. W., W. F. Childers and C. Heckrotte. (1959) Destruction and reestablishment of stream fish and invertebrates affected by drought. Trans. Am. Fish. Soc., **88**, 261–85.

Larimore, R. W. and P. W. Smith. (1963) Fishes of Champaign County. Ill. Nat. Hist. Surv. Bull., **28**, 299–382.

Larkin, P. A. (1956) Interspecific competition and population control in freshwater fish. J. Fish. Res. Bd. Can., **13**, 327–42.

Larsen, D. P., D. R. Dudley and R. M. Hughes. (1988) A regional approach for assessing attainable surface water quality: an Ohio case study. J. Soil & Wat. Consv. **43**, 171–76.

Lauder, G. V. (1982) Structure and function in the tail of the pumpkinseed sunfish (*Lepomis gibbosus*). J. Zool. Lond., **197**, 483–95.

Lauder, G. V. (1983) Neuromuscular patterns and the origin of trophic specialization in fishes. Science, **219**, 1235–37.

Lauder, G. V. (1985) Functional morphology of the feeding mechanism in lower vertebrates in *Vertebrate Morphology,* (eds., Duncker and Fleischer). Gustav Fischer Verlag, Stuttgart & New York, pp. 179–88.

Lauder, G. V. and P. C. Wainwright. (1992) Function and history: the pharyngeal jaw apparatus in primitive ray-finned fishes, in *Systematics, Historical Ecology and North American Freshwater Fishes* (ed. R. L. Mayden). Stanford University Press, Stanford CA., pp. 455–71.

Lauder, G. V., R. B. Huey, R. K. Monson, and R. J. Jensen. (1995) Systematics and the study of organismal form and function. BioScience, **45**, 696–704.

Laughlin, D. R. and E. E. Werner. (1980) Resource partitioning in two coexisting sunfish: pumpkinseed *Lepomis gibbosus* and northern longear sunfish *Lepomis megalotis peltastes*, Can. J. Fish. Aquat. Sci., **37**, 1411–20.

Laurent, P., J. N. Maina, H. L. Bergman, A. Narahara, P. J. Walsh, and C. M. Wood. (1995) Gill structure of a fish from an alkaline lake: effect of short-term exposure to neutral conditions. Can. J. Zool., **73**, 1170–81.

Lawton, J. H. (1990) Species richness and population dynamics of animal assemblages. Patterns in body size: abundance space. Phil. Trans. Roy. Soc. Lond. B, **330**, 283–91.

Lazzaro, X. (1987) A review of planktivorous fishes: their evolution, feeding behaviours, selectivities, and impacts. Hydrobiologia, **146**, 97–167.

Lee, D. S., C. R. Gilbert, C. H. Hocutt, R. E. Jenkins, D. E. McAllister, and J. R. Stauffer, Jr. (eds.). (1980) *Atlas of North American Freshwater Fishes*. North Carolina State Museum of Natural History, Raleigh, NC.

Leggett, W. C. and J. E. Carscadden. (1978) Latitudinal variation in reproductive characteristics of American shad (*Alosa sapidissima*): evidence for population specific life history strategies in fish. J. Fish. Res. Bd. Can., **35**, 1469–78.

Lehtinen, S. and A. A. Echelle. (1979) Reproductive cycle of *Notropis boops* (Pisces: Cyprinidae) in Brier Creek, Marshall County, Oklahoma. Am. Midl. Nat., **102**, 237–43.

Leopold, L. B. and W. B. Langbein. (1966) River meanders. Sci. Am., **214**, 60–70.

Letcher, B. H. and D. A. Bengston. (1993) Effects of food density and temperature on feeding and growth of young inland silversides (*Menidia beryllina*). J. Fish Biol., **43**, 671–86.

Levins, R. (1968) *Evolution in Changing Environments: Some Theoretical Explorations.* Monographs on Population Biology 2, Princeton University Press, Princeton, NJ.

Levins, R. (1969) Thermal acclimation and heat resistance in *Drosophila* species. Am. Nat., **103**, 483–99.

Levinton, J. (1995) Bioturbators as ecosystem engineers: control of the sediment fabric, inter-individual interactions and material fluxes, in *Linking Species and Ecosystems* (eds. C. G. Jones and J. H. Lawton). Chapman & Hall, New York, pp. 29–36.

Lewis, W. M., Jr. (1970) Morphological adaptations of cyprinodontids for inhabiting oxygen deficient waters. Copeia, **1970**, 319–26.

Lewis, W. M. and S. Flickinger. (1967) Home range tendency of the largemouth bass (*Micropterus salmoides*). Ecology, **48**, 1020–23.

Lewis, W. M. and D. R. Helms. 1963. Vulnerability of forage organisms to largemouth bass. Trans. Amer. Fish. Soc. 93:315–18.

Li, H. W. and R. W. Brocksen. (1977) Approaches to the analysis of energetic costs of intraspecific competition for space by rainbow trout *Salmo gairdneri*. J. Fish. Biol., **11**, 329–41.

Licht, P., W. R. Dawson, and V. H. Shoemaker. (1966a) Heat resistance of some Australian lizards. Copeia, **1966**, 162–69.

Licht, P., W. R. Dawson, V. H. Shoemaker, and A. R. Main. (1966b) Observations on the thermal relations of western Australian lizards. Copeia, **1966**, 97–110.

Liem, K. F. (1967) Functional morphology of the head of the Anabantoid teleost fish, *Helostoma temminchi*. J. Morphol., **121**, 135–58.

Liem, K. F. (1980) Adaptive significance of intra- and interspecific differences in the feeding repertoires of cichlid fishes. Am. Zool., **20**, 295–314.

Liem, K. F. (1986) The pharyngeal jaw apparatus of the Embiotocidae (Teleostei): a functional and evolutionary perspective. Copeia, **1986**, 311–23.

Lindsey, C. C. (1996) Body size of poikilotherm vertebrates at different latitudes. Evolution, **20**, 456–65.

Lindsey, C. C. and J. D. McPhail. (1986) Zoogeography of fishes of the Yukon and Mackenzie basins, in *The Zoogeography of North American Freshwater Fishes* (eds. C. H. Hocutt and E. O. Wiley). John Wiley and Sons, New York, pp. 639–74.

Linton, L. R., R. W. Davies, and F. J. Wrona. 1981. Resource utilization indices: an assessment. J. Animal Ecology 50:283–92.

Liu, K., J. D. Cruzan, and R. J. Saykally. (1996a) Water clusters. Science, **271**, 929–33.

Liu, K., M. G. Brown, J. D. Cruzan, and R. J. Saykally. (1996b) Vibration–rotation tunneling spectra of the water pentameter: structure and dynamics. Science, **271**, 62–4.

Lobón-Cerviá, J. (1996) Response of a stream fish assemblage to a severe spate in Northern Spain. Trans. Am. Fish. Soc., **125**, 913–19.

Loiselle, P. V. (1982) Male spawning-partner preference in an arena-breeding teleost *Cyprinodon macularius californiensis* Girard (Atherinomorpha: Cyprinodontidae). Am. Nat., **120**, 721–32.

Lomolino, M. V., J. H. Brown, and R. Davis. (1989) Island biogeography of montane forest mammals in the American Southwest. Ecology, **70**, 180–94.

Longwell, C. R., A. Knopf, and R. F. Flint. (eds.) (1948) *Physical Geology*, 3rd ed. John Wiley and Sons, London.

Lotrich, V. A. (1973) Growth, production, and community composition of fishes inhabiting a first-, second-, and third-order stream of eastern Kentucky. Ecol. Monogr., **43**, 377–97.

Lowe, C. H., D. S. Hinds, and E. A. Halpern. (1967) Experimental catastrophic selection and tolerances to low oxygen concentration in native Arizona freshwater fishes. Ecology, **48**, 1013–17.

Lowe, R. H. (McConnell). (1964) The fishes of the Rupununi savanna district of British Guiana. S. Am. J. Linn. Soc. (Zool.), **45**, 103–44.

Lowe-McConnell, R. H. (1987) *Ecological Studies in Tropical Fish Communities*. Cambridge University Press, Cambridge.

Lowe-McConnell, R. H. (1991) Natural history of fishes in Araguaia and Xingu Amazonian tributaries, Serra do Roncador, Mato Grosso, Brazil. Ichthyol. Explor. Freshwaters, **2**, 63–82.

Lowe-McConnell, R. H. (1975) *Fish Communities in Tropical Freshwaters, Their Distribution, Ecology and Evolution*. Longman, London.

Luczkovich, J. J., S. F. Norton, and R. G. Gilmore, Jr. (1995) The influence of oral anatomy on prey selection during the ontogeny of two percoid fishes, *Lagodon rhomboides* and *Centropomus undecimalis*. Environ. Biol. Fish., **44**, 79–95.

Luecke, C. (1990) Changes in abundance and distribution of benthic macroinvertebrates after introduction of cutthroat trout into a previously fishless lake. Trans. Am. Fish. Soc., **119**, 1010–21.

Luecke, C. and W. J. O'Brien. (1981) Prey location volume of a planktivorous fish: a new measure of prey vulnerability. Can. J. Fish. Aquat. Sci., **38**, 1264–70.

Luecke, C., I. G. Rudstam, and Y. Allen. (1992) Interannual patterns of planktivory 1987–89: an analysis of vertebrate and invertebrate planktivores, in *Food Webs Management* (ed. J. F. Kitchell). Springer-Verlag, New York, pp. 275–301.

Lundberg, J. G. (1992) The phylogeny of ictalurid catfishes: A synthesis of recent work, in *Systematics, Historical Ecology, and North American Freshwater Fishes* (ed. R. L. Mayden). Stanford University Press, Stanford, CA, pp. 392–20.

Lundberg, J. G. (1993) African–South American freshwater fish clades and continental drift: problems with a paradigm, in *The Biotic Relationships between Africa and South America* (ed. P. Goldblatt). Yale University Press, New Haven, CT, pp. 156–99.

Lundberg, J. G. (1996) A new and unusually large catfish from the Eocene of Arkansas, and its implications for siluriform phylogeny and biogeography. Abstracts Annual Conf. ASIH, New Orleans, LA.

Lundberg, J. G. and E. Marsh. (1976) Evolution and functional anatomy of the pectoral fin rays in cyprinoid fishes, with emphasis on the suckers (family Catostomidae). Am. Midl. Nat., **96**, 332–49.

Lundberg, J. G., A. Machado-Allison, and R. F. Kay. (1986) Miocene characid fishes from Colombia: evolutionary stasis and extirpation. Science, **234**, 208–09.

Lurie, E. (1960) *Louis Agassiz—A Life in Science*. University of Chicago Press, Chicago.

Lydeard, C. and M. C. Belk. (1993) Management of indigenous fish species impacted by introduced mosquitofish: an experimental approach. Southwest. Nat., **38**, 370–73.

Lyons, J. (1987) Distribution, abundance and mortality of small littoral-zone fishes in Sparkling Lake, Wisconsin. Environ. Biol. Fish., **18**, 93–107.

Lyons, J. (1989a) Changes in the abundance of small littoral-zone fishes in Lake Mendota, Wisconsin. Can. J. Zool., **67**, 2910–16.

Lyons, J. (1989b) Correspondence between the distribution of fish assemblages in Wisconsin streams and Omernik's ecoregions. Am. Midl. Nat., **122**, 163–72.

Lyons, J. (1992) The length of stream to sample with a towed electrofishing unit when fish species richness is estimated. North Am. J. Fish. Mangmt., **12**, 198–203.

Lyons, J. (1996) Patterns in the species composition of fish assemblages among Wisconsin streams. Envir. Biol. Fish. 45: 329–341.

Lyons, J. (1997) Influence of winter starvation on the distribution of smallmouth bass among Wisconsin streams: a bioenergetics modeling assessment. Trans. Am. Fish. Soc., **126**, 157–62.

Lyons, J. and S. Navarro-Perez. (1990) Fishes of the Sierra De Manantlan, west-central Mexico. Southwest. Nat., **35**, 32–46.

Macan, T. T. (1963) *Freshwater Ecology*. John Wiley and Sons, New York.

MacArthur, R. H. (1957) On the relative abundance of bird species. Proc. Nat. Acad. Sci. USA, **43**, 293–95.

MacArthur, R. H. (1958) Population ecology of some warblers of northeastern coniferous forests. Ecology, **39**, 599–619.

MacArthur, R. H. (1968) The theory of the niche, in *Population Biology and Evolution* (ed. R. C. Lewonton). Syracuse University Press, Syracuse, NY., pp. 159–76.

MacArthur, R. H. (1972) *Geographical Ecology*. Harper and Row, New York.

MacArthur, R. H. and E. O. Wilson. (1967) *The Theory of Island Biogeography*. Princeton Monograph in Population Biology **1**, Princeton University Press, Princeton, NJ.

MacCrimmon, H. R. and W. H. Robbins. (1981) Influence of temperature, water current, illumination, and time on activity and substrate selection in juvenile smallmouth bass (*Micropterus dolomieui*). Can. J. Zool., **59**, 2322–30.

MacLean, J. and J. J. Manguson. (1977) Species interactions in percid communities. J. Fish. Res. Bd. Can., **34**, 1941–51.

Magnan, P. (1988) Interactions between brook charr *Salvelinus fontinalis* and nonsalmonid species: ecological shift, morphological shift, and their impact on zooplankton communities. Can. J. Fish. Aquat. Sci., **45**, 999–1009.

Magnan, P. and G. J. FitzGerald. (1982) Resource partitioning between brook trout *Salvelinus fontinalis* Mitchill and creek chub *Semotilus atromaculatus* Mitchill in selected oligotrophic lakes of southern Quebec. Can. J. Zool., **60**, 1612–17.

Magnhagen, C. (1988) Predation risk and foraging in juvenile pink, *Oncorhynchus gorbuscha* and chum salmon, *O. keta*. Can. J. Fish. Aquat. Sci., **45**, 592–96.

Magnuson, J. J. (1976) Managing with exotics—a game of chance. Trans. Am. Fish. Soc., **105**, 1–9.

Magnuson, J. J. (1991) Fish and fisheries ecology. Ecol. Applicat., **1**, 13–26.

Magnuson, J. J., L. B. Crowder, and P. A. Medvick. (1979) Temperature as an ecological resource. Am. Zool., **19**, 331–43.

Magnuson, J. J. and R. C. Lathrop. (1992) Historical changes in the fish community, in *Food Web Management—a Case Study of Lake Mendota* (ed. J. E. Kitchell), pp. 193–231.

Magurran, A. E. (1986) Predator inspection behavior in minnow shoals: differences between populations and individuals. Behav. Ecol. Sociobiol., **19**, 267–73.

Magurran, A. E. and A. Higham. (1988) Information transfer across fish shoals under predator threat. Ethology, **78**, 153–58.

Mahon, R. (1984) Divergent structure in fish taxocenes of north temperate streams. Can. J. Fish. Aquat. Sci., **41**, 330–50.

Mahon, R. and E. K. Balon. (1976) Fish community structure in lakeshore lagoons on Long Point, Lake Erie, Canada. Environ. Biol. Fish., **2**, 71–82.

Mahon, R. and C. B. Portt. (1985) Local size related segregation of fishes in streams. Arch. Hydrobiol., **103**, 267–71.

Maiorana, V. C. (1977) Density and competition among sunfish: some aternatives. Science, **195**, 94–5.

Maitland, P. S. (1969) A preliminary account of the mapping of the distribution of freshwater fish in the British Isles. J. Fish. Biol., **1**, 45–58.

Makarewicz, J. C. and R. I. Baybutt. (1981) Long-term (1927–1978) changes in the phytoplankton community of Lake Michigan at Chicago. Bull. Torrey Botanical Club. 108:240–54.

Mandrak, N. E. (1995) Biogeographic patterns of fish species richness in Ontario lakes in relation to historical and environmental factors. Can. J. Fish. Aquat. Sci., **52**, 1462–74.

Mandrak, N. E. and E. J. Crossman. (1992) Postglacial dispersal of freshwater fishes into Ontario. Can. J. Zool., **70**, 2247–59.

Maness, J. D. and V. H. Hutchison. (1980) Acute adjustment of thermal tolerance in vertebrate ectotherms following exposure to critical thermal maxima. J. Therm. Biol., **5**, 225–33.

Mann, R. H. K. and D. R. O. Orr. (1969) A preliminary study of the relationships of fish in a hard-water and a soft-water stream in England. J. Fish. Biol., **1**, 31–44.

Mann, R. H. K. and T. Penczak. (1986) Fish production in rivers: a review. Pol. Arch. Hydrobiol., **33**, 233–47.

Mantel, N. (1967) The detection of disease clustering and a generalized regression approach. Cancer Res., **27**, 209–20.

Marconato, A., A. Bisazza, and M. Fabris. (1993) The cost of parental care and egg cannibalism in the river bullhead, *Cottus gobio* L. (Pisces, Cottidae). Behav. Ecol. Sociobiol., **32**, 229–37.

Marsh, E. (1977) Structural modifications of the pectoral fin rays in the order Pleuronectiformes. Copeia, **1977**, 575–78.

Marsh, E. (1980) The effects of temperature and photoperiod on termination of spawning in the orangethroat darter (*Etheostoma spectabile*) in central Texas. TX J. Sci., **XXXII**, 129–42.

Marsh, E. (1984) Egg size variation in central Texas populations of *Etheostoma spectabile* (Pisces: Percidae). Copeia, **1984**, 291–301.

Marsh, E. (1986) Effects of egg size on offspring fitness and maternal fecundity in the orangethroat darter, *Etheostoma spectabile* (Pisces: Percidae). Copeia, **1986**, 18–30.

Marsh, P. C. and D. R. Langhorst. (1988) Feeding and fate of wild larval razorback sucker. Environ. Biol. Fish., **21**, 59–67.

Marshall, B. E. (1988) Seasonal and annual variations in the abundance of pelagic sardines in Lake Kariba, with special reference to the effects of drought. Arch. Hydrobiol., **112**, 399–409.

Marshall, T. R. and P. A. Ryan. (1987) Abundance patterns and community attributes of fishes relative to environmental gradients. Can. J. Fish. Aquat. Sci., **44**, 198–215.

Martin, F. D. (1984) Diets of four sympatric species of *Etheostoma* (Pisces: Percidae) from southern Indiana: interspecific and intraspecific multiple comparisons. Environ. Biol. Fish., **11**, 113–20.

Martin, T. E. (1981) Species–area slopes and coeffifients: a caution on their interpretation. Am. Nat., **118**, 823–37.

Marx, J. L. (1983) Surviving heat shock and other stresses. Science, **221**, 251–53.

Matheney, M. P. and C. F. Rabeni. (1995) Patterns of movement and habitat use by northern hog suckers in and Ozark stream. Trans. Am. Fish. Soc., **124**, 886–97.

Mather, M. E. and R. A. Stein. (1993) Direct and indirect effects of fish predation on the replacement of a native crayfish by an invading congener. Can. J. Fish. Aquat. Sci., **50**, 1279–88.

Matheson, R. E., Jr. and G. R. Brooks, Jr. (1983) Habitat segregation between *Cottus bairdi* and *Cottus girardi*: an example of complex inter- and intraspecific resource partitioning. Am. Midl. Nat., **110**, 165–76.

Mathur, D. (1973) Some aspects of life history of the blackbanded darter, *Percina nigrofasciata* (Agassiz) in Halawakee Creek, Alabama. Am. Midl. Nat., **89**, 381–93.

Matthews, K. R. and N. H. Berg. (1997) Rainbow trout responses to water temperature and dissolved oxygen stress in two southern California stream pools. J. Fish. Biol., **50**, 50–67.

Matthews, W. J. (1973) The fishes of Piney Creek–an Ozark mountain stream in north central Arkansas. Master's thesis, Arkansas State University, Jonesboro.

Matthews, W. J. (1977) Influence of physico-chemical factors on habitat selection by red shiners, *Notropis lutrensis* (Pisces: Cyprinidae). Ph.D. dissertation, University of Oklahoma, Norman.

Matthews, W. J. (1982) Small fish community structure in Ozark streams: structured assembly patterns or random abundance of species? Am. Midl. Nat., **107**, 42–54.

Matthews, W. J. (1984) Influence of turbid inflows on vertical distribution of larval shad and freshwater drum. Trans. Am. Fish. Soc., **113**, 192–98.

Matthews, W. J. (1985a) Critical current speeds and microhabitats of the benthic fishes *Percina roanoka* and *Etheostoma flabellare*. Environ. Biol. Fish., **12**, 303–08.

Matthews, W. J. (1985b) Summer mortality of striped bass in reservoirs of the United States. Trans. Am. Fish. Soc., **114**, 62–6.

Matthews, W. J. (1985c) Distribution of midwestern fishes on multivariate environmental gradients, with emphasis on *Notropis lutrensis*. Amer. Midl. Nat. 113:225–237.

Matthews, W. J. (1986a) Diel differences in gill net and seine catches of fish in winter in a cove of Lake Texoma, Oklahoma–Texas. TX J. Sci., **38**, 153–58.

Matthews, W. J. (1986b) Fish faunal 'breaks' and stream order in the eastern and central United States. Environ. Biol. Fish., **17**, 81–92.

Matthews, W. J. (1986c) Fish faunal structure in an Ozark stream: stability, persistence and a catastrophic flood. Copeia, **1986**, 388–97.

Matthews, W. J. (1986d) Geographic variation in thermal tolerance of a widespread minnow *Notropis lutrensis* of the North American mid-west. J. Fish. Biol., **28**, 404–17.

Matthews, W. J. (1987a) Physicochemical tolerance and selectivity of stream fishes as related to their geographic ranges and local distributions, in *Community and Evolutionary Ecology of North American Stream Fishes* (eds. W. J. Matthews and D. C. Heins). University of Oklahoma Press, Norman, pp. 111–20.

Matthews, W. J. (1987b) Geographic variation in *Cyprinella lutrensis* (Pisces: Cyprinidae) in the United States, with notes on *Cyprinella lepida*. Copeia, **1987**, 616–37.

Matthews, W. J. (1988) North American prairie streams as systems for ecological study. J. North Am. Benthol. Soc., **7**, 387–409.

Matthews, W. J. (1990) Fish community structure and stability in warmwater midwestern streams, in *Ecology and Assessment of Warmwater Streams: Workshop Synopsis* (ed. M. Bain). U.S. Fish Wildl. Serv. Biol. Rep., **90**, 16–7.

Matthews, W. J. and F. P. Gelwick. (1990) Fishes of Crutcho Creek and the North Canadian River near Oklahoma City: urbanization and temporal variability. Southwest. Nat., **35**, 403–10.

Matthews, W. J. and G. L. Harp. (1974) Preimpoundment ichthyofaunal survey of the Piney Creek watershed. Izard County, Arkansas. Proc. AK Acad. Sci., **28**, 39–43.

Matthews, W. J. and L. G. Hill. (1979a) Influence of physico-chemical factors on habitat selection by red shiners, *Notropis lutrensis* (Pisces: Cyprinidae). Copeia, **1979**, 70–81.

Matthews, W. J. and L. G. Hill. (1979b) Age-specific differences in the distribution of red shiners, *Notropis lutrensis*, over physico-chemical ranges. Am. Midl. Nat., **101**, 366–72.

Matthews, W. J. and L. G. Hill. (1980) Habitat partitioning in the fish community of a southwestern river. Southwest. Nat., **25**, 51–66.

Matthews, W. J. and L. G. Hill. (1988) Physical and chemical profiles in Lake Texoma (Oklahoma–Texas) in summer 1982 and 1983. Proc. OK Acad. Sci., **68**, 33–8.

Matthews, W. J., L. G. Hill, D. R. Edds, J. J. Hoover, and T. G. Heger. 1988b. Trophic ecology of striped bass, *Morone saxatilis*, in a freshwater reservoir (Lake Texoma, U.S.A.). J. Fish. Biol. 33:273–88.

Matthews, W. J. and D. C. Heins. (1987) *Community and Evolutionary Ecology of North American Stream Fishes*. University of Oklahoma Press, Norman.

Matthews, W. J. and J. Maness. (1979) Critical thermal maxima, oxygen tolerances and success of cyprinid fishes in a southwestern river. Southwest. Nat., **24**. 374–77.

Matthews, W. J. and H. W. Robison. (In Press) Influence of drainage connectivity, drainage area, and regional species richness on fishes of the Interior Highlands in Arkansas. Amer. Midl. Nat.

Matthews, W. J. and H. W. Robison. (1988) The distribution of the fishes of Arkansas: a multivariate analysis. Copeia, **1988**, 358–74.

Matthews, W. J. and J. T. Styron, Jr. (1981) Tolerance of headwater versus mainstream fishes for abrupt physicochemical change. Am. Midl. Nat., **105**, 149–58.

Matthews, W. J. and E. G. Zimmerman. (1990) Potential effects of global warming on native fishes of the southern Great Plains and the Southwest. Fisheries, **15**, 26–32.

Matthews, W. J., J. R. Bek, and E. Surat. (1982b) Comparative ecology of the darters, *Etheostoma podostemone, E. flabellare,* and *Percina roanoka* in the Upper Roanoke River Drainage Virginia. Copeia, **1982**, 805–14.

Matthews, W. J., R. C. Cashner, and F. P. Gelwick. (1988a) Stability and persistence of fish faunas and assemblages in three midwestern streams. Copeia, **1988**, 945–55.

Matthews, W. J., F. P. Gelwick, and J. J. Hoover. (1992a) Food of and habitat use by juveniles of species of *Micropterus* and *Morone* in a southwestern reservoir. Trans. Am. Fish. Soc., **121**, 54–66.

Matthews, W. J., B. C. Harvey, and M. E. Power. (1994) Spatial and temporal patterns in the fish assemblages of individual pools in a midwestern stream (USA). Environ. Biol. Fish., **39**, 381–97.

Matthews, W. J., L. G. Hill, and S. M. Schellhaass. (1985) Depth distribution of striped bass and other fish in Lake Texoma (Oklahoma–Texas) during summer stratification. Trans. Am. Fish. Soc., **114**, 84–91.

Matthews, W. J., D. J. Hough, and H. W. Robison. (1992b) Similarities in fish distribution and water quality patterns in streams of Arkansas: congruence of multivariate analyses. Copeia, **1992**, 296–305.

Matthews, W. J., R. E. Jenkins, and J. T. Styron, Jr. (1982a) Systematics of two forms of blacknose dace, *Rhynichthys atratulus* (Pisces: Cyprinidae) in a zone of syntopy, with a review of the species group. Copeia, **1982**, 902–20.

Matthews, W. J., M. E. Power, and A. J. Stewart. (1986) Depth distribution of *Campostoma* grazing scars in an Ozark stream. Environ. Biol. Fish., **17**, 291–97.

Matthews, W. J., W. D. Shepard, and L. G. Hill. (1978) Aspects of the ecology of the duskystripe shiner, *Notropis pilsbyri* (Cypriniformes, Cyprinidae), in an Ozark stream. Am. Midl. Nat., **100**, 247–52.

Matthews, W. J., A. J. Stewart, and M. E. Power. (1987) Grazing fishes as components of North American stream ecosystems: effects of *Campostoma anomalum,* in *Community and Evolutionary Ecology of North American Stream Fishes* (eds. W. J. Matthews and D. C. Heins). University of Oklahoma Press, Norman, pp. 128–35.

Matthews, W. J., E. Surat, and L. G. Hill. (1982c) Heat death of the orangethroat darter *Etheostoma spectabile* (Percidae) in a natural environment. Southwest. Nat., **27**, 216–17.

Matthews, W. J., L. G. Hill, D. R. Edds, and F. P. Gelwick. (1989) Influence of water quality and season on habitat use by striped bass in a large southwestern reservoir. Trans. Am. Fish. Soc., **118**, 243–50.

Mattingly, H. T. and M. J. Butler IV. (1994) Laboratory predation on the Trinidadian guppy: implications for the size-selective predation hypothesis and guppy life history evolution. Oikos, **69**, 54–64.

Matuszek, J. E. and G. L. Beggs. (1988) Fish species richness in relation to lake area, pH, and other abiotic factors in Ontario lakes. Can. J. Fish. Aquat. Sci., **45**, 1931–41.

Maturakis, E. G., W. S. Woolcott, and R. E. Jenkins. (1987) Physiographic analyses of the longitudinal distribution of fishes in the Rappahannock River, Virginia. Assoc. Southeast. Biol. Bull., **34**, 1–14.

May, R. M. (1975) Patterns of species abundance and diversity, in *Ecology and Evolution of Communities* (eds. R. L. Cody and J. M. Diamond). Harvard University Press, (Belknap Press), Cambridge, MA, pp. 81–120.

May, R. M. (1976) Patterns in multi-species communities, in *Theoretical Ecology Principles and Applications* (ed. R. M. May). W. B. Saunders Company, Philadelphia, pp. 142–62.

May, R. M. (1988) How many species are there on Earth? Science, **241**, 1441–49.

Mayden, R. L. (1985) Biogeography of Ouachita Highland fishes. Southwest. Nat., **30**, 195–211.

Mayden, R. L. (1987) Historical ecology and North American highland fishes: a research program in community ecology, in *Community and Evolutionary Ecology of North American Stream Fishes* (eds. W. J. Matthews and D. C. Heins). University of Oklahoma Press, Norman, pp. 210–22.

Mayden, R. L. (1988a) Systematics of the *Notropis zonatus* species group, with description of a new species from the Interior Highlands of North America. Copeia, **1988**, 153–83.

Mayden, R. L. (1988b) Vicariance biogeography, parsimony, and evolution in North American freshwater fishes. System. Zool., **37**, 329–55.

Mayden, R. L. (1989) Phylogenetic studies of North American minnows, with emphasis on the genus *Cyprinella* (Teleostei: Cypriniformes). University of Kansas Museum Nat. Hist., Misc. Publ. **80**, 1–189.

Mayden, R. L. (1992a) *Systematics, Historical Ecology and North American Freshwater Fishes*. Stanford University Press, Stanford, CA.

Mayden, R. L. (1992b) An emerging revolution in comparative biology and the evolution of North American freshwater fishes, in *Systematics, Historical Ecology and North American Freshwater Fishes* (ed. R. L. Mayden). Stanford University Press, Stanford, CA., pp. 864–90.

Mayden, R. L. and S. J. Walsh. (1984) Life history of the least madtom *Noturus hildebrandi* (Siluriformes: Ictaluridae) with comparisons to related species. Am. Midl. Nat., **112**, 349–68.

Mazumder, A., W. D. Taylor, D. J. McQueen, D. R. S. Lean, and N. R. Lafontaine. (1990) A comparison of lakes and lake enclosures with contrasting abundances of planktivorous fish. J. Plank. Res., **12**, 109–24.

McAllister, D. E., S. P. Platania, F. W. Schueler, M. E. Baldwin, and D. S. Lee. (1986) Ichthyofaunal patterns on a geographic grid, in *The Zoogeography of North American Freshwater Fishes* (eds. C. H. Hocutt and E. O, Wiley). John Wiley and Sons, New York, pp. 17–51.

McCabe, G. J., Jr. (1995) Relations between winter atmospheric circulation and annual streamflow in the western United States. Clim. Res., **5**, 139–48.

McCammon, G. W. and C. von Geldern, Jr. (1979) Predator–prey systems in large reservoirs, in *Predator–Prey Systems in Large Reservoirs* (ed. H. Clepper). Sport Fishing Institute, Washington, DC, pp. 431–42.

McCarraher, D. B., M. L. Madsen, and R. E. Thomas. (1971) Ecology and fishery management of McConaughy Reservoir, Nebraska, in *Reservoir Fisheries and Limnology* (ed. G. E. Hall). Special Publication No. 8, American Fisheries Society, Washington, DC, pp. 299–311.

McCauley, R. W. (1958) Thermal relations of geographic races of *Salvelinus*. Can. J. Zool., 36, 655–62.

McCauley, R. W. and J. M. Casselman. (1981) The final preferendum as an index of the temperature for optimum growth in fish. Proc. World Symp. Aquacult., Heated Effluents and Recirc. Systs. II, 82–93.

McCauley, R. W. and N. W. Huggins. (1979) Ontogenetic and non-thermal seasonal effects on thermal preferenda of fish. Am. Zool., 19, 267–72.

McComas, S. R. and R. W. Drenner. (1982) Species replacement in a reservoir fish community: silverside feeding mechanics and competition. Can. J. Fish. Aquat. Sci., 39, 815–21.

McCormick, F. H. and N. Aspinwall. (1983) Habitat selection in three species of darters. Environ. Biol. Fish., 8, 279–82.

McCormick, P. V. (1990) Direct and indirect effects of consumers on benthic algae in isolated pools of an ephemeral stream. Can. J. Fish. Aquat. Sci., 47, 2057–65.

McCune, A. R. (1981) Quantitative description of body form in fishes: implications for species level taxonomy and ecological inference. Copeia, 1981, 897–901.

McCune, A. R., K. S. Thomson, and P. E. Olsen. (1984) Semionotid fishes from the Mesozoic great lakes of North America, in *Evolution of Fish Species Flocks* (eds. A. A. Echelle and I. Kornfeld). University of Maine Press, Orono, pp. 27–44.

McDonald, D. G. (1983) The effects of H^+ upon the gills of freshwater fish. Can. J. Zool., 61, 691–703.

McDonald, M. E. and A. E. Hershey. (1992) Shifts in abundance and growth of slimy sculpin in response to changes in the predator population in an arctic Alaskan lake. Hydrobiology, 240, 219–23.

McDonald, M. E., A. E. Hershey, and W. J. O'Brien. (1992) Cost of predation avoidance in young-of-year lake trout, *Salvelinus namaycush*: growth differential in sub-optimal habitats. Hydrobiology, 240, 213–18.

McDowall, R. M. (1968) Interactions of the native and alien faunas of New Zealand and the problem of fish introductions. Trans. Am. Fish. Soc., 97, 1–11.

McDowall, R. M. (1978) *New Zealand Freshwater Fishes—A Guide and Natural History*. Heinemann Educational Books, Auckland.

McDowall, R. M. (1984) Exotic fishes: the New Zealand experience, in *Distribution, Biology and Management of Exotic Fishes* (eds. W. R. Courtney, Jr. and J. R. Stauffer, Jr.). John Hopkins University Press, Baltimore, MD, pp. 200–14.

McDowall, R. M. (1987) The occurence and distribution of diadromy among fishes. Am. Fish. Soc. Symp., 1, 1–13.

McDowall, R. M. (1988) *Diadromy in Fishes: Migrations Between Freshwater and Marine Environments*. Timber Press, Portland, OR.

McDowall, R. M. (1990) Freshwater fishes and fisheries of New Zealand—the angler's Eldorado. Aquat. Sci., **2**, 281–34.

McDowall, R. M. and A. H. Whitaker. (1975) The freshwater fishes, in *Biogeography and Ecology in New Zealand* (ed. G. Kuschel). Dr. W. Junk, The Hague.

McFarland, W. N. and E. R. Loew. (1983) Wave produced changes in underwater light and their relations to vision. Environ. Biol. Fish., **8**, 173–84.

McFarland, W. N. and N. M. Kotchian. (1982) Interaction between schools of fish and mysids. Behav. Sociobiol., **11**, 71–6.

McFarland, W. N., F. H. Pough, T. J. Cade, and J. B. Heiser. (1979) *Vertebrate Life*. MacMillan Publishing, New York.

McFarlane, R. W., B. C. Moore, and S. E. Williams. (1976) Thermal tolerance of stream cyprinid minnows, in *Thermal Ecology II* (eds. G. W. Esch and R. W. McFarlane). Energy Research and Development Administration (ERDA) Symposium Series No. 40, pp. 141–44.

McIntosh, A. R., T. A. Crowl, and C. R. Townsend. (1994) Size-related impacts of introduced brown trout on the distribution of native common river galaxias. NZ J. Mar. Freshwater Res., **28**, 135–44.

McIntosh, A. R. and C. R. Townsend. (1996) Interactions between fish, grazing invertebrates and algae in a New Zealand stream: a trophic cascade mediated by fish-induced changes to grazer behaviour? Oecologia 108:174–81.

McIntosh, R. P. (1967) The continuum concept of vegetation. Botan. Rev., **33**, 130–87.

McIntosh, R. P. (1995) H. A. Gleason's "individualistic concept" and theory of animal communities: a continuing controversy. Biol. Rev., **70**, 317–57.

McKaye, K. R. and M. K. Oliver. (1980) Geometry of a selfish school: defence of cichlid young by a bagrid catfish in Lake Malawi, Africa. Animal Behav., **28**, 1278–90.

McLennan, D. A. (1993) Temporal changes in the structure of the male nuptial signal in the brook stickleback, *Culaea inconstans* (Kirtland). Can. J. Zool., **71**, 1111–19.

McLennan, D. A. (1995) Male mate choice based upon female nuptial coloration in the brook stickleback, *Culaea inconstans* (Kirtland). Animal Behav., **50**, 213–21.

McNeely, D. L. (1986) Longitudinal patterns in the fish assemblages of an Ozark stream. Southwest. Nat., **31**, 375–80.

McNeely, D. L. (1987) Niche relations within an Ozark stream cyprinid assemblage. Environ. Biol. Fish., **18**, 195–208.

McPeek, M. A. (1990) Determination of species composition in the *Enallagma* damselfly assemblages of permanent lakes. Ecology, **71**, 83–98.

McPhail, J. D. and C. C. Lindsey. (1986) Zoogeography of the freshwater fishes of Cascadia (the Columbia system and rivers north to the Stikine), in *The Zoogeography of North American Freshwater Fishes* (eds. C. H. Hocutt and E. O. Wiley). John Wiley and Sons, New York, pp. 615–37.

McPhee, J. (1980) *Basin and Range*, Farrar Straus Giroux, New York.

McPhee, J. (1987) The control of nature—Atchafalaya. The New Yorker, Feb. 23, 1987, 39–100.

McQueen, D. J. (1990) Manipulating lake community structure: where do we go from here? Freshwater Biol., **23**, 613–20.

McQueen, D. J. and J. R. Post. (1988) Cascading trophic interactions: uncoupling at the zooplankton–phytoplankton link. Hydrobiologia, **159**, 277–96.

McQueen, D. J., J. R. Post, and E. L. Mills. (1986) Trophic relationships in freshwater pelagic ecosystems. Can. J. Fish. Aquat. Sci., **43**, 1571–81.

Meador, M. R. and W. E. Kelso. (1990) Growth of largemouth bass in low-salinity environments. Trans. Am. Fish. Soc., **119**, 545–52.

Meek, A. (1930) *The Progress of Life: A Study in Phycho-genetic Evolution*, Longmans, New York. [Original not seen, cited in Allee et al., (1949)].

Meek, S. E. (1891) Report of explorations made in Missouri and Arkansas during 1889, with an account of the fishes observed in each of the river basins examined. Bull. U.S. Fish Comm., **9**, 113–41.

Meek, S. E. (1894) Report of investigations respecting the fishes of Arkansas, conducted during 1891, 1892, and 1893, with a synopsis of previous explorations in the same state. Bull. U.S. Fish Comm., **14**, 67–94.

Meffe, G. K. (1984) Effects of abiotic disturbance on coexistence of predator–prey fish species. Ecology, **65**, 1525–34.

Meffe, G. K. (1985) Predation and species replacement in American southwestern fishes: a case study. Southwest. Nat., **30**, 173–87.

Meffe, G. K. (1987) Embryo size variation in mosquitofish: optimality vs. plasticity in propagule size. Copeia 1987:762–68.

Meffe, G. K. (1990) Offspring size variation in eastern mosquitofish (*Gambusia holbrooki*: Poeciliidae) from contrasting thermal environments. Copeia, **1990**, 10–8.

Meffe, G. K. (1991) Life history changes in eastern mosquitofish (*Gambusia holbrooki*) induced by thermal elevation. Can. J. Fish. Aquat. Sci., **48**, 60–6.

Meffe, G. K. and T. M. Berra. (1988) Temporal characteristics of fish assemblage structure in an Ohio stream. Copeia, **1988**, 684–90.

Meffe, G. K. and W. L. Minckley (1987) Persistence and stability of fish and invertebrate assemblages in a repeatedly disturbed Sonoran Desert stream. Am. Midl. Nat., **117**, 177–91.

Meffe, G. K. and A. L. Sheldon. (1990) Post-defaunation recovery of fish assemblages in southeastern blackwater streams. Ecology, **71**, 657–67.

Meffe, G. K., D. A. Hendrickson, and W. L. Minckley. (1983) Factors resulting in decline of the endangered sonoran topminnow *Poeciliopsis occidentalis* (Atheriniformes: Poeciliidae) in the United States. Biol. Conserv., **25**, 135–59.

Mendelson, J. (1975) Feeding relationships among species of *Notropis* (Pisces: Cyprinidae) in a Wisconsin stream. Ecol. Monogr., **45**, 199–230.

Menendez, R. (1976) Chronic effects of reduced pH on brook trout (*Salvelimus fontinalis*). J. Fish. Res. Bd. Can., **33**, 118–23.

Menge, B. A. (1995) Indirect interactions in marine rocky intertidal interaction webs: patterns and importance. Ecol. Monogr. **65**, 21–74.

Menon, A. G. K. (1954) Fish geography of the Himalayas. Proc. Nat. Inst. Sci. India, **20**, 467–93.

Merrick, J. R. and G. E. Schmida (eds.). (1984) *Australian Freshwater Fishes—Biology and Management*. Griffin Press, Ltd., Netley, South Australia.

Metcalf, A. L. (1959) Fishes of Chautauqua, Cowley and Elk Counties, Kansas. Univ. Kansas Publ., Mus. Nat. Hist., **11**, 345–400.

Metcalf, A. L. (1966) Fishes of the Kansas River system in relation to zoogeography of the Great Plains. Univ. Kansas Publ., Mus. Nat. Hist., **17**, 23–189.

Metcalfe, I. (1994) Late Paleozoic and Mesozoic Paleogeography of eastern Pangea and Tethys, in *Pangea: Global Environments and Resources* (eds. A. F. Embry, B. Beauchamp, and D. J. Glass). Memoir 17, Canadian Society Petroleum Geology, Calgary, Alberta, pp. 97–111.

Meyer, A. (1993) Phylogenetic relationships and evolutionary processes in East African cichlid fishes. Trends Ecol. Evol., **8**, 279–84.

Meyer, J. L., E. T. Schultz, and G. S. Helfman. (1983) Fish schools: an asset to corals. Science, **220**, 1047–48.

Meyer, J. L. and E. T. Schultz. (1985) Migrating haemulid fishes as a source of nutrients and organic matter on coral reefs. Limnol. Oceanogr., **30**, 146–56.

Meyerhoff, R. D. and O. T. Lind. (1987) Factors affecting the benthic community structure of a discontinuous stream in Guadalupe Mountains National Park, Texas. Int. Rev. Hydrobiol., **72**, 283–96.

Milinski, M. (1986) Constraints placed by predators on feeding behavior, in *The Behavior of Teleost Fishes* (ed. T. J. Pitcher). The Johns Hopkins University Press, Baltimore, MD., pp. 236–52.

Milinski, M. and T. C. M. Bakker. (1990) Female sticklebacks use male coloration in mate choice and hence avoid parasitized males. Nature, **344**, 330–33.

Miller, A. M. and S. W. Golladay. (1996) Effects of spates and drying on macroinvertebrate assemblages of an intermittent and a perennial prairie stream. J. North Amer. Benthol. Soc. 15:670–89.

Miller, D. J. (1989) Introductions and extinctions of fish in the African Great Lakes. Trends Ecol. Evol., **4**, 56–9.

Miller, G. L. (1983) Trophic resource allocation between *Percina sciera* and *P. ouachitae* in the Tombigbee River, Mississippi. Am. Midl. Nat., **110**, 299–313.

Miller, P. J. (1996) The functional ecology of small fish: Some opportunities and consequences. Symposium Zool. Soc. London 69:175–99.

Miller, R. J. (1979) Relationships between habitat and feeding mechanisms in fishes, in *Predator–Prey Systems in Fisheries Management* (ed. H. Clepper). Sport Fishing Institute, Washington, DC.

Miller, R. R. (1966) Geographical distribution of Central American freshwater fishes. Copeia, **1966**, 773–802.

Miller, R. R. (1982a) First fossil record (Plio-Pleistocene) of threadfin shad, *Dorosoma petenense*, from the Gatuna Formation of southeastern New Mexico. J. Paleontol., **56**, 423–25.

Miller, R. R. (1982b) Pisces, in *Aquatic Biota of Mexico, Central America and the West Indies* (eds. S. H. Hurlbert and A. Villalobos-Figueroa). San Diego State University, San Diego, CA, pp. 486–501.

Millinski, M. and R. Heller. (1978) Influence of a predator on the optimal foraging behaviour of sticklebacks (*Gasterosteus aculeatus* L.). Nature, **275**, 642–44.

Mills, C. A. (1988) The effect of extreme northerly climatic conditions on the life history of the minnow, *Phoxinus phoxinus* (L.). J. Fish. Biol., **33**, 545–61.

Mills, H. B., W. C. Starrett, and F. C. Bellrose. (1966) Man's effect on the fish and wildlife of the Illinois River. Illinois Natural History Survey Biological Notes 57:1–24.

Minckley, W. L. (1963) The ecology of a spring stream, Doe Run, Meade County, Kentucky. Wildlife Monographs, No. 11.

Minckley, W. L. (1984) Cuatro Cienegas fishes: Research review and a local test of diversity versus habitat size. J. Arizona-Nevada Acad. Sci. 19:13–21.

Minckley, W. L. (1991) Native fishes of the Grand Canyon region: an obituary, in *Colorado River Ecology and Dam Management*. National Academy Press, Washington, DC, pp. 124–77.

Minckley, W. L. and W. E. Barber. (1971) Some aspects of biology of the longfin dace, a cyprinid fish characteristic of streams in the Sonoran Desert. Southwest. Nat., **15**, 459–64.

Minckley, W. L. and J. E. Deacon (eds). (1991) *Battle Against Extinction: Native Fish Management in the American West*. University of Arizona Press, Tucson, AZ.

Minckley, W. L. and G. K. Meffe. (1987) Differential selection by flooding in stream-fish communities of the arid American southwest, in *Community Evolutionary and Ecology of North American Stream Fishes* (eds. W. J. Matthews and D. C. Heins). University of Oklahoma Press, Norman, pp. 93–104.

Minckley, W. L., D. A. Hendrickson, and C. E. Bond. (1986) Geography of western North American freshwater fishes: description and relationships to intra-continental tectonism, in *The Zoogeography of North American Freshwater Fishes* (eds. C. H. Hocutt and E. O. Wiley). John Wiley and Sons, New York, pp. 519–613.

Minns, C. K. (1989) Factors affecting fish species richness in Ontario lakes. Trans. Am. Fish. Soc., **118**, 533–45.

Minns, C. K. (1990) Patterns of distribution and association of freshwater fish in New Zealand. NZ J. Mar. Freshwater Res., **24**, 31–44.

Minns, C. K. (1995) Allometry of home range size in lake and river fishes. Can. J. Fish. Aquat. Sci., **52**, 1499–1508.

Minshall, G. W. (1988) Stream ecosystem theory: a global perspective. J. North Am. Benthol. Soc., **7**, 263–88.

Mire, J. B. and L. Millett. (1994) Size of mother does not determine size of eggs or fry in the owens pupfish. *Cyprinodon radiosus*. Copeia, **1994**, 100–07.

Mittlebach, G. G. (1981a) Patterns in invertebrate size and abundance in aquatic habitats. Can. J. Fish. Aquat. Sci., **38**, 896–904.

Mittlebach, G. G. (1981b) Foraging efficiency and body size: a study of optimal diet and habitat use by bluegills. Ecology, **62**, 1370–86.

Mittlebach, G. G. (1983) Optimal foraging and growth in bluegills. Oecologia, **59**, 157–62.

Mittlebach, G. G. (1984) Predation and resource partitioning in two sunfishes Centrarchidae. Ecology, **65**, 499–513.

Mittlebach, G. G. (1986) Predator-mediated habitat use: some consequences for species interaction. Environ. Biol. Fish., **16**, 159–69.

Mittlebach, G. G. (1988) Competition among refuging sunfishes and effects of fish density on littoral zone invertebrates. Ecology, **69**, 614–23.

Mittlebach, G. G., A. M. Turner, D. J. Hall, and J. E. Rettig. (1995) Perturbation and resilience: a long-term, whole-lake study of predator extinction and reintroduction. Ecology, **76**, 2347–60.

Montgomery, D. R., J. M. Buffington, R. D. Smith, K. M. Schmidt, and G. Pess. (1995) Pool spacing in forest channels. Water Res. Resr., **31**, 1097–105.

Montgomery, D. R., T. B. Abbe, J. M. Buffington, N. P. Peterson, K. M. Schmidt, and J. D. Stock. (1996) Distribution of bedrock and alluvial channels in forested mountain drainage basins. Nature, **381**, 587–88.

Montgomery, W. L. (1975) Interspecific associations of sea-basses (Serranidae) in the Gulf of California. Copeia, **1975**, 785–b97.

Montgomery, W. L. (1981) Mixed-species schools and the significance of vertical territories of damselfishes. Copeia, **1981**, 477–81.

Montgomery, W. L., G. E. Glasgow, Jr., K. B. Staley, and J. R. Mills. (1987) Alternative mating behaviors of male Atlantic salmon (*Salmo salar*), with special reference to mature male parr, in *Community and Evolutionary Ecology of North American Stream Fishes*, (eds. W. J. Matthews and D. C. Heins). University of Oklahoma Press, Norman, pp. 232–38.

Moodie, G. E. E. and C. C. Lindsey. (1972) Life history of a unique cyprinid fish, the chiselmouth (*Acrocheilus alutaceus*), in British Columbia. Syesis 5:55–61.

Moore, G. A. (1950) The cutaneous sense organs of barbeled minnows adapted to life in the muddy waters of the Great Plains region. Trans. Am. Microsc. Soc., **LXIX**, 69–95.

Moore, G. A. (1956) The cephalic lateral line system in some sunfishes (*Lepomis*). J. Comp. Neurol., **104**, 49–55.

Moore, J. A. (1949) Geographic variation of adaptive characters in *Rana pipiens* Schreber. Evolution, **3**, 1–24.

Moore, W. G. (1942) Field studies on the oxygen requirements of certain freshwater fishes. Ecology, **23**, 319–29.

Morell, V. (1994) New African dinosaurs give an Old World a novel look. Science, **266**, 219–20.

Morin, P. J. (1984a) The impact of fish exclusion on the abundance and species composition of larval odonates: results of short-term experiments in a North Carolina farm pond. Ecology, **65**, 53–60.

Morin, P. J. (1984b) Odonate guild composition: experiments with colonization history and fish predation. Ecology, **65**, 1866–73.

Morin, P. J. (1988) Effects of vertebrate predation on zooplankton community composition. ISI Atlas Sci., Animal Plant Sci., **1988**, 5–8.

Morin, R. and R. J. Naiman. (1990) The relation of stream order to fish community dynamics in boreal forest watersheds. Pol. Arch. Hydrobiol., **37**, 135–50.

Morisita, M. (1959) Measuring of interspecific association and similarity between communities. Mem. Fac. Sci. Kyushu University, Ser. E. Biology, **3**, 65–80. (Original not seen.)

Morse, D. H. (1970) Ecological aspects of some mixed-species foraging flocks of birds. Ecol. Monogr., **40**, 119–68.

Mortensen, E. (1977) Fish production in small Danish streams. Folia Limnol. Scand., **17**, 21–6.

Moses, B. S. (1987) The influence of flood regime on fish catch and fish communities of the Cross River floodplain ecosystem, Nigeria. Environ. Biol. Fish. **18**, 51–65.

Moshenko, R. W. and J. H. Gee. (1973) Diet, time and place of spawning, and environments occupied by creek chub *Semotilus atromaculatus* in the Mink River, Manitoba. J. Fish. Res. Bd. Can., **30**, 357–62.

Motta, P. J., S. F. Norton, and J. J. Luczkovich. (1995) Perspectives on the ecomorphology of bony fishes. Environ. Biol. Fish., **44**, 11–20.

Moyle, J. B. (1946) Some indices of lake productivity. Trans. Am. Fish. Soc., **76**, 322–33.

Moyle, J. B. (1949) Fish-population concepts and management of Minnesota lakes for sport fishing. Trans. North Am. Wildl. Conf., **14**, 283–93.

Moyle, P. B. (1973) Ecological segregation among three species of minnows (Cyprinidae) in a Minnesota lake. Trans. Am. Fish. Soc., **4**, 795–805.

Moyle, P. B. (1976) *Inland Fishes of California*. University of California Press, Berkeley.

Moyle, P. B. (1977) In defense of sculpins. Fisheries, **2**, 20–3.

Moyle, P. B. (1994) Biodiversity, biomonitoring, and the structure of stream fish communities, in *Biological Monitoring of Aquatic Systems* (eds. S. L. Loeb and A. Spacie). Lewis Publ., Boca Raton, FL, pp. 171–86.

Moyle, P. B. and D. M. Baltz. (1985) Microhabitat use by an assemblage of California stream fishes: developing criteria for instream flow determinations. Trans. Am. Fish. Soc., **114**, 695–704.

Moyle, P. B. and J. J. Cech, Jr. (1988) *Fishes—An Introduction to Ichthyology*, 2nd ed. Prentice-Hall, Englewood Cliffs, NJ.

Moyle, P. B. and B. Herbold. (1987) Life-history patterns and community structure in stream fishes of Western North America. Comparisons with Eastern North America and Europe, in *Community and Evolutionary Ecology of North American Stream Fishes*, (eds. W. J. Matthews and D. C. Heins). University of Oklahoma Press, Norman, pp. 25–32.

Moyle, P. B. and H. W. Li. (1979) Community ecology and predator–prey relations in warmwater streams, in *Predator–Prey Systems in Fisheries Management* (ed. H. Clepper). Sport Fishing Institute, Washington, DC, pp. 171–80.

Moyle, P. B. and R. Nicols. (1974) Decline of the native fish fauna of the Sierra-Nevada foothills in central California. Am. Midl. Nat., **92**, 72–83.

Moyle, P. B. and F. R. Senanayake. (1984) Resource partitioning among the fishes of rainforest streams in Sri Lanka. J. Zool. Lond., **202**, 195–223.

Moyle, P. B. and B. Vondracek. (1985) Persistence and structure of the fish assemblage in a small California stream. Ecology, **66**, 1–13.

Moyle, P. B., R. A. Daniels, B. Herbold, and D. M. Baltz. (1985) Patterns in distribution and abundance of a noncoevolved assemblage of estuarine fishes in California. Fish. Bull. **84**, 105–17.

Moyle, P. B., J. J. Smith, R. A. Daniels, T. L. Taylor, D. G. Price, and D. M. Baltz. (1982) *Distribution and Ecology of Stream Fishes of the Sacramento–San Joaquin Drainage System*, California. University of California Press, Berkeley, pp. 1–256.

Mummert, J. R. and R. W. Drenner. (1986) Effect of fish size on the filtering efficiency and selective particle ingestion of a filter-feeding clupeid. Trans. Am. Fish. Soc., **115**, 522–28.

Mundahl, N. D. and C. G. Ingersoll. (1983) Early autumn movements and densities of johnny (*Etheostoma nigrum*) and fantail (*E. flabellare*) darters in a southwestern Ohio stream. Ohio Acad. Sci., **83**, 103–08.

Mundy, P. R. and H. T. Boschung. (1981) An analysis of the distribution of lotic fishes with application to fisheries management. Am. Fish. Soc. Warmwater Streams Symp., pp. 266–75.

Muniz, I. P., H. M. Seip, and I. H. Sevaldrud. (1984) Relationship between fish populations and pH for lakes in southernmost Norway. Water Air Soil Pollut., **23**, 97–113.

Murdoch, W. W., S. Avery, and M. E. Smythe. (1975) Switching in a predatory fish. Ecology, **56**, 1094–105.

Murphy, G. I. (1968) Pattern in life history and the environment. Am. Nat., **102**, 391–403.

Murphy, M. L. and J. D. Hall. (1981) Varied effects of clear-cut logging on predators and their habitats in small streams of the Cascade Mountains, Oregon. Canadian J. Fish. Aquat. Sci. **38**, 137–145.

Murphy, M. L., J. Heifetz, S. W. Johnson, K. V. Koski, and J. F. Thedinga. (1986) Effects of clear-cut logging with and without buffer strips on juvenile salmonids in Alaska streams. Can. J. Fish. Aquat. Sci., **43**, 1521–33.

Murray, N. L. (1996) *Oklahoma's Biodiversity Plan: A Shared Vision for Conserving Our Natural Heritage*. OK Department of Wildlife Conservation, Oklahoma City.

Myers, G. S. (1938) Fresh-water fishes and West Indian zoogeography. Annual Report Smithsonian Institution for 1937, pp. 339–64.

Myers, G. S. (1949) Salt-tolerance of fresh-water fish groups in relation to zoogeographical problems. Bijdr. Dierk. 28:315–22 [original not seen, cited from Berra, (1981)].

Myers, G. S. (1951) Freshwater fishes and East Indian zoogeography. Stanford Icthyological Bulletin 4:11–21.

Myers, G. S. (1966) Derivation of the freshwater fish fauna of Central America. Copeia, **1966**, 766–73.

Naesje, T. F., B. Jonsson, O. T. Sandlund, and G. Kjellberg. (1991) Habitat switch and niche overlap in coregonid fishes: effects of zooplankton abundance. Can. J. Fish. Aquat. Sci., **48**, 2307–315.

Naiman, R. J., H. DéCamps, J. Pastor, and C. A. Johnston. (1988) The potential importance of boundaries to fluvial ecosystems. J. North Am. Benthol. Soc., **7**, 289–306.

Naiman, R. J., J. M. Melillo, M. A. Lock, T. E. Ford, and S. E. Reice. (1987) Longitudinal patterns of ecosystem processes and community structure in a subarctic river continuum. Ecology, **68**, 1139–56.

Nakatsuru, K. and D. L. Kramer. (1982) Is sperm cheap? limited male fertility and female choice in the lemon tetra (Pisces, Characidae). Science, **216**, 753–55.

Nance, R. D. and J. B. Murphy. (1994) Orogenic style and the configuration of supercontinents, in *Pangea: Global Environments and Resources* (eds. A. F. Embry, B. Beauchamp, and D. J. Glass). Memoir 17, Canadian Society of Petroleum Geology, Calgary, Alberta, pp. 49–66.

Nash, R. D. M. (1988) The effects of disturbance and severe seasonal fluctuations in environmental conditions on north temperate shallow-water fish assemblages. Estuar. Coastal Shelf Sci., **26**, 123–35.

Neel, J. K. (1951) Interrelations of certain physical and chemical features in a headwater limestone stream. Ecology, **32**, 368–91.

Neill, W. H. (1979) Mechanisms of fish distribution in heterothermal environments. Am. Zool., **19**, 305–18.

Neill, W. H. and J. J. Magnuson. (1974) Distributional ecology and behavioral thermoregulation of fishes in relation to heated effluent from a power plant at Lake Monona, Wisconsin. Trans. Am. Fish. Soc., **103**, 663–710.

Neill, W. H., J. J. Magnuson, and G. D. Chipman. (1972) Behavioral thermoregulation by fishes: a new experimental approach. Science, **176**, 1443–45.

Nelson, D. O. and C. L. Prosser. (1981) Intracellular recordings from thermosensitive preoptic neurons. Science, **213**, 787–89.

Nelson, G. and N. I. Platnick. (1980) A vicariance approach to historical biogeography. BioScience, **30**, 339–43.

Nelson, J. S. (1976) *Fishes of the World*. John Wiley and Sons, New York.

Nelson, J. S. (1984) *Fishes of the World*, 2nd ed. John Wiley and Sons, New York.

Nelson, J. S. (1994) *Fishes of the World*, 3rd ed. John Wiley and Sons, New York.

Nelson, R. L., W. S. Platts, D. P. Larsen, and S. E. Jensen. (1992) Trout distribution and habitat in relation to geology and geomorphology in the North Fork Humboldt River drainage, northeastern Nevada. Trans. Am. Fish. Soc., **121**, 405–26.

Neverman, D. and W. A. Wurtsbaugh. (1994) The thermoregulatory function of diel vertical migration for a juvenile fish, *Cottus extensus*. Oecologia 98:247–256.

Nicholson, A. J. (1933) The balance of animal populations. J. Animal Ecol., **2**, 132–78.

Nicholson, A. J. (1954) An outline of the dynamics of animal populations. Austral. J. Zool., **2**, 9–65.

Nigro, A. A. and J. J. Ney. (1982) Reproduction and early-life accommodations of land-locked alewives to a southern range extension. Trans. Am. Fish. Soc., **111**, 559–69.

Nikolsky, G. V. (1963) *The Ecology of Fishes*. Academic Press, London.

Nilsson, N. A. (1960) Seasonal fluctuations in the food segregation of trout, char and whitefish in 14 north-Swedish lakes. Rep. Inst. Freshwater Res., Drottningham, **41**, 185–205.

Nilsson, N. A. (1967) Interactive segregation between fish species, in *The Biological Basis for Freshwater Fish Production* (ed. S. D. Gerking). Blackwell Scientific, Oxford, pp. 295–313.

Nilsson, N. A. and T. G. Northcote. (1981) Rainbow trout *Salmo gairdneri* and cutthroat trout *S. clarki* interactions in coastal British Columbia lakes. Can. J. Fish. Aquat. Sci., **38**, 1228–46.

Northcote, T. G. (1954) Observations on the comparative ecology of two species of fish, *Cottus asper* and *Cottus rhotheus*, in British Colombia. Copeia, **1954**, 25–8.

Northcote, T. G. (1988) Fish in the structure and function of freshwater ecosystems: a top-down view. Can. J. Fish. Aquat. Sci., **45**, 361–79.

Northcote, T. G., M. S. Arcifa, and K. A. Munro. (1990) An experimental study of the effects of fish zooplanktivory on the phytoplankton of a Brazilian reservoir. Hydrobiologia, **194**, 31–45.

Norton, S. F. (1995) A functional approach to ecomorphological patterns of feeding in cottid fishes. Environ. Biol. Fish., **44**, 61–78.

Norton, S. F., J. J. Luczkovich, and P. J. Motta. (1995) The role of ecomorphological studies in the comparative biology of fishes. Environ. Biol. Fish., **44**, 287–304.

Novacek, M. J. and L. G. Marshall. (1976) Early biogeographic history of Ostariophysian fishes. Copeia, **1976**, 1–12.

Nriagu, J. O. (1983) Rapid decomposition of fish bones in Lake Erie sediments. Hydrobiologia, **106**, 217–22.

Nurnberg, G. K. (1995a) Quantifying anoxia in lakes. Limnol. Oceanogr., **40**, 1100–11.

Nurnberg, G. K. (1995b) The anoxic factor, a quantitative measure of anoxia and fish species richness in central Ontario lakes. Trans. Am. Fish. Soc., **124**, 677–86.

Nyberg, D. W. (1971) Prey capture in the largemouth bass. Amer. Midland Nat. 86:128–44.

Oberdorff, T., E. Guilbert, and J. C. Lucchetta. (1993) Patterns of fish species richness in the Seine River basin, France. Hydrobiologia, **259**, 157–67.

Oberdorff, T., J. F. Guegan, and B. Hugueny. 1995. Global scale patterns of fish species richness in rivers. Ecography 18:345–52.

Oberdorff, T., B. Hugueny, and J. F. Guegan. (in press). Is there an influence of historical events on contemporary fish species richness in rivers? Comparisons between Western Europe and North America. Journal of Biogeography.

O'Brien, W. J., B. I. Evans, and G. L. Howick. (1986) A new view of the predation cycle of a planktivorous fish, white crappie (*Pomoxis annularis*). Can. J. Fish. Aquat. Sci., **43**, 1894–99.

O'Brien, W. J., B. Evans, and C. Luecke. (1985) Apparent size choice of zooplankton by planktivorous sunfish: exceptions to the rule. Environ. Biol. Fish., **13**, 225–33.

O'Brien, W. J., A. E. Hershey, J. E. Hobbie, M. A. Hullar, G. W. Kipphut, M. C. Miller, B. Moller, and J. R. Vestal. (1992) Control mechanisms of arctic lake ecosystems: a limnocorral experiment. Hydrobiologia, **240**, 143–88.

Ogden, J. C. and J. P. Ebersole. (1981) Scale and community structure of coral reef fishes: a long-term study of a large artificial reef. Mar. Ecol. Prog. Ser., **4**, 97–103.

Ogutu-Ohwayo, R. (1990) The decline of the native fishes of Lake Victoria and Kyoga (East Africa) and the impact of introduced species, especially the Nile perch, *Lates niloticus*, and the Nile tilapia, *Oreochromis niloticus*. Environ. Biol. Fish., **27**, 81–96.

Oliver, M. K. and K. R. McKaye. (1982) Floating islands: a means of fish dispersal in Lake Malawi, Africa. Copeia, **1982**, 748–54.

Olmsted, L. L. and D. G. Cloutman. (1974) Repopulation after a fish kill in Mud Creek, Washington County, Arkansas following pesticide pollution. Trans. Am. Fish. Soc., **103**, 79–87.

Omernik, J. M. (1986) *Ecoregions of the Conterminous United States (Map)*. U.S. Environ. Protect. Agency, Corvallis, OR.

Orr, R. T. (1971) *Vertebrate Biology*, 3rd ed. W. B. Saunders Co., Philadelphia.

Orth, D. J. and O. E. Maughan. (1984) Community structure and seasonal changes in standing stocks of fish in a warm-water stream. Amer. Midl. Nat. 112:369–78.

Osborne, L. L. and M. J. Wiley. (1992) Influence of tributary spatial position on the structure of warmwater fish communities. Can. J. Fish Aquat. Sci., **49**, 671–81.

Osenberg, C. W. and G. G. Mittlebach. (1989) Effects of body size on the predator–prey interaction between pumpkinseed sunfish and gastropods. Ecol. Monogr., **59**, 405–32.

Osenberg, C. W., E. E. Werner, G. G. Mittlebach, and D. J. Hall. (1988). Growth patterns in bluegill (*Lepomis macrochirus*) and pumpkinseed (*L. gibbosus*) sunfish: environmental variation and the importance of ontogenetic niche shifts. Can. J. Fish. Aquat. Sci., **45**, 17–26.

Ottaway, E. M. and A. Clarke. (1981) A preliminary investigation into the vulnerability of young trout (*Salmo trutta* L.) and Atlantic salmon (*S. salar* L.) to downstream displacement by high water velocities. J. Fish. Biol., **19**, 135–45.

Otto, R. G. (1973) Temperature tolerance of the mosquitofish, *Gambusia affinis* (Baird and Girard). J. Fish. Biol., **5**, 575–85.

Otto, R. G. and J. O'H. Rice. (1977) Response of a freshwater sculpin (*Cottus cognatus gracilis*) to temperature. Trans. Am. Fish. Soc., **106**, 89–94.

Packer, R. K. and W. A. Dunson. (1970) Effects of low environmental pH on blood pH and sodium balance of brook trout. J. Expt. Zool., **174**, 65–72.

Page, L. M. (1977) The lateralis system of darters (Etheostomatini). Copeia, **1977**, 472–75.

Page, L. M. (1983) *Handbook of Darters*. THF Publ., Neptune City, NJ.

Page, L. M. (1993) Are there as many species of darters as minnows in North America? Abstracts Annual Meet. ASIH, Austin, TX.

Page, L. M. and R. L. Mayden. (1979) Nesting site of the lollypop darter *Etheostoma neopterum*. Trans. KY Acad. Sci., **40**, 56–7.

Page, L. M. and D. W. Schemske. (1978) The effect of interspecific competition of the distribution and size of darters of the subgenus *Catonotus* (Percidae: *Etheostoma*). Copeia, **1978**, 406–12.

Page, L. M. and P. W. Smith. (1971) The life history of the slenderhead darter, *Percina phoxocephala*, in the Embarras River, Illinois. IL Nat. Hist. Notes, **74**, 1–14.

Page, L. M. and D. L. Swofford. (1984) Morphological correlates of ecological specialization in darters. Environ. Biol. Fish., **11**, 139–59.

Pagel, M. D., P. H. Harvey, and H. C. J. Godfray. (1991) Species-abundance, biomass, and resource-use distributions. Amer. Nat. 138:836–50.

Paine, M. D. (1984) Ecological and evolutionary consequences of early ontogenies of darter (*Etheostomatini*). Environ. Biol. Fish., **11**, 97–106.

Paine, M. D. (1986) Developmental allometry in three species of darters (Percidae: *Etheostoma*). Can. J. Zool., **64**, 347–52.

Paine, M. D. (1990) Life history tactics of darters (Percidae: Etheostomatiini) and their relationship with body size, reproductive behaviour, latitude and rarity. J. Fish. Biol., **37**, 473–88.

Paine, M. D. and E. K. Balon. (1986) Early development of johnny darter, *Etheostoma nigrum*, and fantail darter, *E. flabellare*, with a discussion of its ecological and evolutionary aspects. Environ. Biol. Fish., **15**, 191–220.

Paine, M. D., J. J. Dodson, and G. Power. (1982) Habitat and food resource partitioning among four species of darters Percidae: *Etheostoma* in a southern Ontario stream. Can. J. Zool., **60**, 1635–41.

Paine, R. T. (1969) A note on trophic complexity and community stability. Am. Nat., **103**, 91–93.

Paine, R. T. (1974) Intertidal community structure: experimental studies on the relationship between a dominant competitor and its principal predator. Oecologia, **15**, 93–120.

Paine, R. T. (1980) Food webs: linkage, interaction strength and community infrastructure. J. Animal Ecol., **49**, 667–85.

Paine, R. T. (1995) A conversation on refining the concept of a keystone species. Conservation Biology 9:962–64.

Paller, M. C. (1994) Relationships between fish assemblage structure and stream order in South Carolina coastal plain streams. Trans. Am. Fish. Soc., **123**, 150–61.

Palmer, M. W. and P. S. White. (1994) Scale dependence and the species area relationship. Am. Nat., **144**, 717–40.

Paloumpis, A. A. (1957) The effects of drought conditions on the fish and bottom organisms of two small oxbow ponds. Trans. IL State Acad. Sci., **50**, 60–4.

Paloumpis, A. A. (1958) Response of some minnows to flood and drought conditions in an intermittent stream. Iowa State College J. Sci., **32**, 547–61.

Paradis, A. R., P. Pepin, and J. A. Brown. (1996) Vulnerability of fish eggs and larvae to predation: review of the influence of the relative size of prey and predator. Can. J. Fish. Aquat. Sci., **53**, 1226–35.

Parker, G. A. (1992) The evolution of sexual size dimorphism in fish. J. Fish. Biol., **41** (Suppl. B), 1–20.

Parenti, L. (1981) A phylogenetic and biogeographic analysis of cyprinodontiform fishes (Teleostei, Atherinomorpha). Bull. Am. Museum Nat. Hist., **168**, 341–557.

Parrish, D. L. and F. J. Margraf. (1990) Interactions between white perch *Merone americana* and yellow perch *Perca flavescens* in Lake Erie as determined from feeding and growth. Can. J. Fish. Aquat. Sci., **47**, 1779–87.

Partridge, L. and P. H. Harvey. (1988) The ecological context of life history evolution. Science, **241**, 1449–55.

Partridge, W. D. (1991) Effects of *Campostoma anomalum* on the export of various size fractions of particulate organic matter. Master's thesis, University of Oklahoma, Norman.

Paszkowski, C. A. (1985) The foraging behavior of the central mudminnow and yellow perch; the influence of foraging site intraspecific and interspecific competition. Oecologia, **66**, 271–79.

Paszkowski, C. A. (1986) Foraging site use and interspecific competition between bluegills and golden shiners. Environ. Biol. Fish., **17**, 227–33.

Patrick, D. M., L. Mao, and S. T. Ross. (1991) The impact of geomorphic change on the distribution of bayou darters in the Bayou Pierre System. Mississippi Dept. Wildlife, Fisheries and Parks; Museum of Natural Science (Jackson, MS), Museum Technical Report No. 18.

Patten, B. C. (1975) A reservoir cove ecosystem model. Trans. Am. Fish. Soc., **104**, 596–619.

Patten, R. S. and J. E. Ellis. (1995) Patterns of species and community distributions related to environmental gradients in an arid tropical ecosystem. Vegetatio, **117**, 69–79.

Patterson, C. and A. E. Longbottom. (1989) An Eocene amiid fish from Mali, West Africa. Copeia 1989:827–36.

Pauly, D. (1989) Food consumption by tropical and temperate fish populations: some generalizations. J. Fish. Biol., **35** (Suppl. A), 11–20.

Pearl, R. (1928) *The Rate of Living*. Knopf, New York.

Pearse, A. S. (1916) The food of the shore fishes of certain Wisconsin lakes. Bull. U.S. Fish. Comm., **35**, 281–82.

Pearse, A. S. (1920) Distribution and food of the fishes of Green Lake, Wisc., in summer. Bull. US Bureau of Fish., **37**, 253–72.

Pearsons, T. N., H. W. Li, and G. A. Lamberti. (1992) Influence of habitat complexity on resistance to flooding and resilience of stream fish assemblages. Trans. Am. Fish. Soc., **121**, 427–36.

Peckarsky, B. L. (1983) Biotic interactions or abiotic limitations? A model of lotic community structure, in *Dynamics of Lotic Ecosystems* (eds. T. D. Fontaine III and S. M. Bartell). Ann Arbor Science, Ann Arbor, MI, pp. 303–23.

Peckarsky, B. L. (1984) Predator-prey interactions among aquatic insects, in (eds. V. H. Resh and D. M. Rosenberg) *The Ecology of Aquatic Insects*, Praeger, New York, pp. 196–254.

Penczak, T. (1985) Phosphorus, nitrogen, and carbon cycling by fish populations in two small lowland rivers in Poland. Hydrobiologia, **120**, 159–65.

Penczak, T. and R. H. K. Mann. (1990) The impact of stream order on fish populations in the Pilica drainage basin, Poland. Pol. Arch. Hydrobiol., **37**, 243–61.

Penczak, T. and I. Tátrai. (1985) Contribution of bream, *Abramis brama* (L.), to the nutrient dynamics of Lake Balaton. Hydrobiologia, **126**, 59–64.

Penczak, T., I. Forbes, T. Atkin, and T. Hill. (1991) Fish community structure in the rivers of Lincolnshire and South Humberside, England. Hydrobiologia, **211**, 1–9.

Penczak, T., J. Lobón-Cerviá, K. O'Hara, and H. Jakubowski. (1986) Production and food consumption by fish populations in the Pilawa and Dobrzyca Rivers, North Poland. Pol. Arch. Hydrobiol., **33**, 345–72.

Persson, L. (1983a) Effects of intra-and interspecific competition on dynamics and size structure of a perch *Perca fluviatilis* and a roach *Rutilus rutilus* population. Oikos, **41**, 126–32.

Persson, L. (1983b) Food consumption and competition between age classes in a perch *Perca fluviatilis* population in a shallow eutrophic lake. Oikos, **40**, 197–202.

Persson, L. (1983c) Food consumption and the significance of detritus and algae to intraspecific competition in roach *Rutilus rutilus* in a shallow eutrophic lake. Oikos, **41**, 118–25.

Persson, L. (1986) Temperature-induced shift in foraging ability in two fish species, roach (*Rutilus rutilus*) and perch (*Perca fluviatilis*): implications for coexistence between poikilotherms. J. Animal Ecol., **55**, 829–39.

Persson, L. (1987) Competition-induced switch in young of the year perch, *Perca fluviatilis*, an experimental text of resource limitation. Environ. Biol. Fish., **19**, 235–39.

Persson, L. (1993) Predator-mediated competition in prey refuges: the importance of habitat dependent prey resources. Oikos, **68**, 12–22.

Persson, L. and P. Eklöv. (1995) Prey refuges affecting interactions between piscivorous perch and juvenile perch and roach. Ecology, **76**, 70–81.

Persson, L., J. Bengtsson, B. A. Menge, and M. E. Power. (1995) Productivity and consumer regulation—concepts, patterns, and mechanisms, in *Food Webs* (eds. G. A. Polis and K. O. Winemiller). Chapman & Hall, New York, pp. 396–434.

Persson, L., S. Diehl, L. Johansson, G. Andersson, and S. F. Hamrin. (1992) Trophic interactions in temperate lake ecosystems: a test of food chain theory. Am. Nat., **140**, 59–84.

Persson, L., L. Johansson, G. Anderson, S. Diehl, and S. F. Hamrin. (1993) Density dependent interactions in lake ecosystems: whole lake perturbation experiments. Oikos, **66**, 193–208.

Peters, R. H. (1991) *A Critique for Ecology* (ed. R. H. Peters). Cambridge University Press, Cambridge.

Peters, R. H. and J. V. Raelson. (1984) Relations between individual size and mammalian population density. Am. Nat., **124**, 498–517.

Peterson, C. H. and R. Black. (1988) Density-dependent mortality caused by physical stress interacting with biotic history. Am. Nat., **131**, 257–70.

Peterson, J. T. and C. F. Rabeni. (1996) Natural thermal refugia for temperate warmwater stream fishes. North Am. J. Fish. Managmt., **16**, 738–46.

Petranka, J. W. (1983) Fish predation: a factor affecting the spatial distribution of a stream-breeding salamander. Copeia, **1983**, 624–28.

Pflieger, W. L. (1966) Young of the orangethroat darter (*Etheostoma spectabile*) in nests of the smallmouth bass (*Micropterus dolomieui*). Copeia, **1966**, 139–40.

Pflieger, W. L. (1971) A distributional study of Missouri fishes. Univ. Kansas Museum Nat. Hist., **20**, 225–570.

Pflieger, W. L. (1975) *The Fishes of Missouri*. Missouri Department of Conservation, Jefferson City.

Pflieger, W. L. and T. B. Grace. (1987) Changes in the fish fauna of the lower Missouri River, 1940–1983, in *Community and Evolutionary Ecology of North American Stream Fishes* (eds. W. J. Matthews and H. W. Robison). University of Oklahoma Press, Norman, pp. 166–77.

Pflug, D. E. and G. B. Pauley. (1983) The movement and homing of smallmouth bass, *Micropterus dolomieui*, in Lake Sammamish, Washington. Calif. Fish Game, **69**, 207–16.

Phillips, E. C. and R. V. Kilambi. (1996) Food habits of four benthic fish species (*Etheostoma spectibile, Percina caprodes, Noturus exilis, Cottus carolinae*) from northwest Arkansas streams. Southwest. Nat., **41**, 69–73.

Pianka, E. R. (1970) On r and K selection. Am. Nat., **104**, 592–97.

Pianka, E. R. (1974) Niche overlap and diffuse competition. Proc. Nat. Acad. Sci. USA, **71**, 2141–45.

Pianka, E. R. (1978) *Evolutionary Ecology*, 2nd ed. Harper and Row, New York.

Pianka, E. R. and W. S. Parker. (1975) Age-specific reproductive tactics. Am. Nat., **109**, 453–64.

Picard, C. R., R. Freitag, and E. P. Iwachewski. (1993) Aspects of smallmouth bass, *Micropterus dolomieui*, life history in northwestern Ontario, Canada. J. Freshwater Ecol., **8**, 355–61.

Pielou, E. C. (1975) *Ecological Diversity*. John Wiley and Sons, New York.

Pierce, C. L., J. B. Rasmussen, and W. C. Leggett. (1994) Littoral fish communities in southern Quebec lakes: relationships with limnological and prey resource variables. Can. J. Fish. Aquat. Sci., **51**, 1128–38.

Pigg, J. and L. G. Hill. (1974) Fishes of the Kiamichi River. Proc. OK Acad. Sci., **54**, 124–30.

Pilette, R. (1989) Evaluating direct and indirect effects in ecosystems. Am. Nat., **133**, 303–07.

Pitcher, T. J. (1986) Functions of shoaling in teleosts, in *The Behavior of Teleost Fishes* (ed. J. T. Pitcher). John Hopkins University Press, Baltimore, MD, pp. 294–337.

Pitcher, T. J. and P. J. B. Hart. (1982) *Fisheries Ecology*. AVI Publishing, Westport, CT, pp. 77–108.

Pitcher, T. J., A. E. Magurran, and J. R. Allan. (1985) Size-segregative behavior in minnow shoals. J. Fish. Biol. **29**, 83–95.

Pitcher, T. J., A. E. Magurran, and I. J. Winfield. (1982) Fish in larger shoals find food faster. Behav. Ecol. Sociobiol., **10**, 149–51.

Plante, C. and J. A. Downing. (1993) Relationship of salmonine production to lake trophic status and temperature. Can. J. Fish. Aquat. Sci., **50**, 1324–28.

Platania, S. (1991) Fishes of the Rio Chama and upper Rio Grande, New Mexico, with preliminary comments on their longitudinal distribution. Southwest. Nat., **36**, 186–93.

Poff, N. L. (1992) Why disturbances can be predictable: A perspective on the definition of disturbance in streams. J. North Amer. Benthol. Soc. 11:86–92.

Poff, N. L. and J. D. Allan. (1995) Functional organization of stream fish assemblages in relation to hydrological variability. Ecology, **76**, 606–27.

Poff, N. L. and J. V. Ward. (1990) Physical habitat template of lotic systems: recovery in the context of historical pattern of spatiotemporal heterogeneity. Environ. Mangmt., **14**, 629–45.

Poff, N. L., M. A. Palmer, P. L. Angermeier, R. L. Vadas, Jr., C. C. Hakenkamp, A. Bely, P. Aresnburger, and A. P. Martin. (1993) Size structure of the metazoan community in a Piedmont stream. Oecologia, **95**, 202–09.

Polis, G. A. (1984) Age structure component of niche width and intraspecific resource partitioning: can age groups function as ecological species? Am. Nat., **123**, 541–64.

Polis, G. A. and K. O. Winemiller. (1995). *Food Webs: Integration of Patterns and Dynamics*, Chapman & Hall, New York.

Polivka, K. M. (1997) The riverine landscape and microhabitat use by the Arkansas River shiner, *Notropis girardi*. Unpublished M.S. thesis, University of Oklahoma, Norman, OK.

Popiel, S. A., A. Perez-Fuentetaja, D. J. McQueen, and N. C. Collins. (1996) Determinants of nesting success in the pumpkinseed (*Lepomis gibbosus*): a comparison of two populations under different risks from predation. Copeia, **1996**, 649–56.

Post, J. R. and D. Cucin. (1984) Changes in the benthic community of a small Precambrian lake following the introduction of yellow perch, *Perca flavescens*. Can. J. Fish. Aquat. Sci., **41**, 1496–1501.

Poulin, R., N. G. Wolf, and D. L. Kramer. (1987) The effects of hypoxia on the vulnerability of guppies *Poecilia reticulata* (Poeciliidae) to an aquatic predator, *Astronotus acellatus* (Cichlidae). Environ. Biol. Fish., **20**, 285–92.

Power, G. (1990) Salmonid communities in Quebec and Labrador; temperature relations and climate change. Pol. Arch. Hydrobiol., **37**, 13–28.

Power, M. E. (1983) Grazing responses of tropical freshwater fishes to different scales of variation in their food. Environ. Biol. Fish., **9**, 103–15.

Power, M. E. (1984a) Depth distributions of armored catfish: predator-induced resource avoidance? Ecology, **65**, 523–28.

Power, M. E. (1984b) Habitat quality and the distribution of algae-grazing catfish in a Panamanian stream. J. Animal Ecol., **53**, 357–74.

Power, M. E. (1984c) The importance of sediment in the grazing ecology and size class interactions of an armored catfish, *Ancistrus spinosus*. Environ. Biol. Fish., **10**, 173–81.

Power, M. E. (1987) Predator avoidance by grazing fishes in temperate and tropical streams: importance of stream depth and prey size, in *Predation*, (eds. W. C. Kerfoot and A. Sih). University Press of New England, Hanover, NH, pp. 333–51.

Power, M. E. (1990a) Resource enhancement by indirect effects of grazers: armored catfish, algae, and sediment. Ecology, **71**, 897–904.

Power, M. E. (1990b) Effects of fish in river food webs. Science, **250**, 811–14.

Power, M. E. (1992a) Habitat heterogeneity and the functional significance of fish in river food webs. Ecology, **73**, 1675–88.

Power, M. E. (1992b) Top-down and bottom-up forces in food webs: do plants have primacy? Ecology, **73**, 733–46.

Power, M. E. (1995) Floods, food chains, and ecosystem processes in rivers, in *Linking Species and Ecosystems* (eds. C. G. Jones and J. H. Lawton). Chapman & Hall, New York, pp. 52–60.

Power, M. E. and W. J. Matthews. (1983) Algae-grazing minnows (*Campostoma anomalum*), piscivorous bass (*Micropterus* spp.), and the distribution of attached algae in a small prairie-margin stream. Oecologia, **60**, 328–32.

Power, M. E. and A. J. Stewart. (1987) Disturbance and recovery of an algal assemblage following flooding in an Oklahoma stream. Am. Midl. Nat., **117**, 333–45.

Power, M. E., T. L. Dudley, and S. D. Cooper. (1989) Grazing catfish, fishing birds, and attached algae in a Panamanian stream. Environ. Biol. Fish., **26**, 285–94.

Power, M. E., J. C. Marks, and M. S. Parker. (1992) Variation in the vulnerability of prey to different predators: community level consequences. Ecology, **73**, 2218–23.

Power, M. E., W. J. Matthews, and A. J. Stewart. (1985) Grazing minnows, piscivorous bass and stream algae: dynamics of a strong interaction. Ecology, **66**, 1448–56.

Power, M. E., M. S. Parker, and J. T. Wootton. (1995b) Disturbance and food chain lengths in rivers, in *Food Webs* (eds. G. A. Polis and K. O. Winemiller). Chapman & Hall, New York, pp. 286–97.

Power, M. E., A. J. Stewart, and W. J. Matthews. (1988) Grazer control of algae in an Ozark mountain stream: effects of short-term exclusion. Ecology, **69**, 1894–98.

Power, M. E., A. Sun, G. Parker, W. E. Dietrich, and J. T. Wootton. (1995a) Hydraulic food-chain models. An approach to the study of food-web dynamics in large rivers. BioScience, **45**, 159–67.

Power, M. E., A. Sun, G. Parker, W. E. Dietrich, and J. T. Wootton. (1995c) How does floodplain width affect floodplain river ecology? A preliminary exploration using simulations. Geomorphology, **13**, 301–17.

Power, M. E., D. Tilman, J. A. Estes, A. Menge, W. J. Bond, L. S. Mills, G. Daily, J. C. Castillo, J. Lunchenco, and R. T. Paine. (1996) Challenges in the quest for keystones. BioScience, **46**, 609–20.

Prejs, A. (1978) Lake macrophytes as the food of roach (*Rutilus rutilus* L.) and rudd (*Scardinius erythrophthals* L.). II. Daily intake of macrophyte food in relation to body size of fish. Ekol. Pol., **26**, 537–53.

Prejs, A. (1984) Herbivory by temperate freshwater fishes and its consequences. Environ. Biol. Fish., **10**, 281–96.

Prejs, A. and K. Prejs. (1987) Feeding of tropical freshwater fishes: seasonality in resource availability and resource use. Oecologia, **71**, 397–404.

Preston, F. W. (1948) The commonness and rarity of species. Ecology, **29**, 254–83.

Preston, F. W. (1962) The canonical distribution of commonness and rarity, Parts I and II. Ecology, **43**, 185–215, 410–32.

Pringle, C. M., R. J. Naiman, G. Bretschko, J. R. Karr, M. W. Oswood, J. R. Webster, R. L. Welcomme, and M. J. Winterbourn. (1988) Patch dynamics in lotic systems: the stream as a mosaic. J. North Am. Benthol. Soc., **7**, 503–24.

Propst, D. L. and K. R. Bestgen. (1991) Habitat and biology of the loach minnow, *Tiaroga cobitis*, in New Mexico. Copeia, **1991**, 29–38.

Propst, D. L. and J. A. Stefferud. (1994) Distribution and status of the Chihuahua chub (Teleostei: Cyprinidae: *Gila nigrescens*), with notes on its ecology and associated species. Southwest. Nat., **39**, 224–34.

Pusey, B. J., A. H. Arthington, and M. G. Read. (1993) Spatial and temporal variation in fish assemblage structure in the Mary River, south-eastern Queensland: the influence of habitat structure. Environ. Biol. Fish., **37**, 355–80.

Pusey, B. J., A. H. Arthington, and M. G. Read. (1995) Species richness and spatial variation in fish assemblage structure in two rivers of the wet tropics of northern Queensland, Australia. Environ. Biol. Fish., **42**, 181–99.

Pyke, G. H. (1978) Optimal foraging in hummingbirds: testing the marginal value theorem. Am. Zool., **18**, 739–52.

Pyron, M. (1995) Mating patterns and a test for female mate choice in *Etheostoma spectabile* (Pisces, Percidae). Behav. Ecol. Sociobiol., **36**, 407–12.

Pyron, M. (1996a) Sexual size dimorphism and phylogeny in North American minnows. Biol. J. Linn. Soc., **57**, 327–42.

Pyron, M. (1996b) Male orangethroat darters, *Etheostoma spectabile*, do not prefer larger females. Environ. Biol. Fish., **47**, 407–10.

Pyron, M. and C. M. Taylor. (1993) Fish community structure of Oklahoma Gulf Coastal Plains. Hydrobiology, **257**, 29–35.

Quinn, T. P. and N. P. Peterson. (1996) The influence of habitat complexity and fish size on over-winter survival and growth of individually marked juvenile coho salmon

(*Oncorhynchus kisutch*) in Big Beef Creek, Washington. Can. J. Fish. Aquat. Sci., **53**, 1555–64.

Rader, R. B. and C. L. Richardson. (1994) Response of microinvertebrates and small fish to nutrient enrichment in the northern Everglades. Wetlands, **14**, 134–46.

Raffetto, N. S., J. R. Baylis, and S. L. Serns. (1990) Complete estimates of reproductive success in a closed population of smallmouth bass (*Micropterus dolomieui*). Ecology, **71**, 1523–35.

Rahel, F. J. (1984) Factors structuring fish assemblages along a bog lake successional gradient. Ecology, **65**, 1276–89.

Rahel, F. J. (1986) Biogeographic influences on fish species composition of northern Wisconsin lakes with applications for lake acidification studies. Can. J. Aquat. Sci., **43**, 124–34.

Rahel, F. J. (1989) Nest defense and aggressive interactions between a small benthic fish (the johnny darter *Etheostoma nigrum*) and crayfish. Environ. Biol. Fish., **24**, 301–06.

Rahel, F. J. and W. A. Hubert. (1991) Fish assemblages and habitat gradients in a Rocky Mountain-Great Plains stream: biotic zonation and additive patterns of community change. Trans. Am. Fish. Soc., **120**, 319–32.

Rahel, F. J. and J. J. Magnuson. (1983) Low pH and the absence of fish species in naturally acidic Wisconsin lakes: inferences for cultural acidification. Can. J. Fish. Aquat. Sci., **40**, 3–9.

Rahel, F. J. and J. W. Nutzman. (1994) Foraging in a lethal environment: fish predation in hypoxic waters of a straitified lake. Ecology, **75**, 1246–53.

Rahel, F. J. and R. A. Stein. (1988) Complex predator–prey interactions and predator intimidation among crayfish, piscivorous fish, and small benthic fish. Oecologia, **75**, 94–8.

Rahel, F., J. D. Lyons, and P. A. Cochran. (1984) Stochastic or deterministic regulation of fish assemblage structure? It may depend on how the assemblage is defined. Am. Nat., **124**, 583–89.

Rainbow, W. J. (1991) Cyprinids of South East Asia, in *Systematics, Biology and Exploitation* (eds. I. J. Whitfield and J. S. Nelson). Chapman & Hall, New York, pp. 156–210.

Ramcharan, C. W., D. J. McQueen, E. Demers, S. A. Popiel, A. M. Rocchi, N. D. Yan, A. H. Wong, and K. D. Hughes. (1995) A comparative approach to determining the role of fish predation in structuring limnetic ecosystems. Arch. Hydrobiol., **133**, 389–416.

Rand, P. S., D. J. Stewart, B. F. Lantry, L. G. Rudstam, O. E. Johannsson, A. P. Goyke, S. B. Brandt, R. O'Gorman, and G. W. Eck. (1995) Effect of lake-wide planktivory by the pelagic prey fish community in Lakes Michigan and Ontario. Can. J. Fish. Aquat. Sci., **52**, 1546–63.

Randall, R. G., J. R. M. Kelson, and C. K. Minns. (1995) Fish production in freshwaters: are rivers more productive than lakes? Can. J. Fish. Aquat. Sci., **52**, 631–43.

Rawson, D. S. (1952) Mean depth and the fish production of large lakes. Ecology, **33**, 513–21.

Reash, R. J. and J. Pigg. (1990) Physicochemical factors affecting the abundance and species richness of fishes in the Cimmaron River. Proc. OK Acad. Sci., **70**, 23–8.

Reeves, G. H., F. H. Everest, and J. D. Hall. (1987) Interactions between the redside shiner (*Richardsonius balteatus*) and the steelhead trout (*Salmo gairdneri*) in western Oregon: the influence of water temperature. Can. J. Fish. Aquat. Sci., **44**, 1603–13.

Reeves, G. H., L. E. Benda, K. M. Burnett, P. A. Bisson, and J. R. Sedell. (1995) A disturbance-based ecosystem approach to maintaining and restoring freshwater habitats of evolutionarily significant units of anadromous salmonids in the Pacific northwest, in *Evolution and the Aquatic Ecosystem: Defining Unique Units in Population Conservation* (eds. J. L. Nielson and D. A. Power). Am. Fish. Soc. Symp., **17**, 334–49.

Regan, C. T. (1906–1908) Pisces. Biologia Centrali-Americana, **8**, 1–203. [Original not seen, cited by Echelle and Echelle (1984).]

Regan, C. T. (1909) The classification of teleost fishes. Annual Mag. Nat. Hist., **8**, 75–86.

Reice, S. R. (1983) Predation and substratum: factors in lotic community structure, in *Dynamics of Lotic Ecosystems* (eds. T. D. Fontaine III and S. M. Bartell). Ann Arbor Science, Ann Arbor, MI.

Reice, S. R. and R. L. Edwards. (1986) The effect of vertebrate predation on lotic macroinvertebrate communities in Québec, Canada. Can. J. Zool., **64**, 1930–36.

Reighard, J. (1908) Two-foot water glass supported on four legs and provided with screen, as used for studying and photographing lampreys (*Lampetra wilderi*). Bull. U.S. Fish. Comm., **28**, 1137.

Reighhard, J. (1920) The breeding behavior of the suckers and minnows. Biol. Bull., **XXXVIII**, 1–32.

Reinertsen, H., A. Jensen, J. I. Koksvik, A. Langeland, and Y. Olsen. (1990) Effects of fish removal on the limnetic ecosystem of a eutrophic lake. Can. J. Fish. Aquat. Sci., **47**, 166–73.

Reinthal, P. (1993) Evaluating biodiversity and conserving Lake Malawi's chiclid fauna. Conserv. Biol., **7**, 712–18.

Reinthal, P. N. (1990a) Morphological analyses of the neurocranium of a group of rock-dwelling cichlid fishes (Cichlidae: Perciformes) from Lake Malawi, Africa. Zool. J. Linn. Soc., **98**, 123–39.

Reinthal, P. N. (1990b) The feeding habits of a group of herbivorous rock-dwelling cichlid fishes (Cichlid: Perciformes) from Lake Malawi, Africa. Environ. Biol. Fish., **27**, 215–33.

Reinthal, P. N. and G. W. Kling. (1994) Exotic species, tropic interactions, and ecosystem dynamics: a case study of Lake Victoria, in *Theory and Application in Fish Feeding Ecology* (eds. D. J. Strouder, K. L. Fresh, R. J. Feller, and M. Duke). University of South Carolina Press, Columbia, SC, pp. 298–313.

Reisen, W. K. (1972) The influence of organic drift on the food habits and life history of the yellowfin shiner, *Notropis lutipinnis* (Jordan and Brayton). Am. Midl. Nat., **88**, 376–83.

Reist, J. D. (1980) Selective predation upon pelvic phenotypes of brook stickleback, *Culàea inconstans*, by northern pike, *Esox lucius*. Can. J. Zool., **58**, 1245–52.

Reite, O. B., G. M. O. Malory, and B. Aasehaug. (1974) pH, salinity and temperature tolerance of Lake Magadi *Tilapia*. Naturalist, **247**, 315.

Renfro, J. L. and L. G. Hill. (1971) Osmotic acclimation in the Red River pupfish, *Cyprinodon rubrofluviatilis*. Compar. Biochem. Physiol., **49A**, 711–14.

Resetarits, W. J., Jr. (1991) Ecological interactions among predators in experimental stream communities. Ecology, **72**, 1782–93.

Resetarits, W. J., Jr. (1995) Competitive asymmetry and coexistence in size-structured populations of brook trout and spring salamanders. Oikos, **73**, 188–98.

Resh, V. H., A. V. Brown, A. P. Covich, M. E. Gurtz, H. W. Li, G. W. Minshall, S. R. Reice, A. A. Sheldon, J. B. Wallace, and R. C. Wissmar. (1988) The role of disturbance in stream ecology. J. North Am. Benthol. Soc., **7**, 433–55.

Reshetnikov, Y. S. and F. M. Shakirova. (1993) A zoogeographical analysis of the ichthyofauna of central Asia including a list of freshwater fishes. J. Ichthyol., **33**, 99–110.

Reynolds, L. F. (1883) Migration patterns of five fish species in the Murray–Darling River System. Aust. J. Mar. Freshwater Res., **34**, 857–71.

Reynolds, W. W. (1977) Thermal equilibration rates in relation to heartbeat and ventilatory frequencies in largemouth blackbass, *Micropterus salmoides*. Compar. Biochem. Physiol., **56A**, 195–201.

Reynolds, W. W. and M. E. Casterlin. (1979a) Behavioral thermoregulation and the "final preferendum" paradigm. Am. Zool., **19**, 211–24.

Reynolds, W. W. and M. E. Casterlin. (1979b) Effect of temperature on locomotor activity in the goldfish (*Carassius auratus*) and the bluegill (*Lepomis macrochirus*): presence of an "activity well" in the region of the final referendum. Hydrobiologia, **65**, 3–5.

Reznick, D. (1981) "Grandfather effects": the genetics of interpopulation differences in offspring size in the mosquitofish. Evolution, **35**, 941–53.

Reznick, D. (1982) The impact of predation on life history evolution in Trinidadian guppies: genetic basis of observed life history patterns. Evolution, **36**, 1236–50.

Reznick, D. (1983) The structure of guppy life histories: the tradeoff between growth and reproduction. Ecology, **64**, 862–73.

Reznick, D. (1989) Life-history evolution in guppies: 2. Repeatability of field observations and the effects of season on life histories. Evolution, **43**, 1285–97.

Reznick, D. and B. Braun. (1987) Fat cycling in the mosquitofish (*Gambusia affinis*): fat storage as a reproductive adaptation. Oecologia, **73**, 401–13.

Reznick, D. and H. Bryga. (1987) Life-history evolution in guppies (*Poecilia reticulata*): 1. Phenotypic and genetic changes in an introduction experiment. Evolution, **41**, 1370–85.

Reznick, D. and J. A. Endler. (1982) The impact of predation on life history evolution in Trinidadian guppies (*Poecilia reticulata*). Evolution, **36**, 160–77.

Reznick, D. and D. B. Miles. (1989) A review of life history patterns in poeciliid fishes, in *Ecology and Evolution of Livebearing Fishes (Poeciliidae)* (eds. G. K. Meffe and F. F. Snelson, Jr.). Prentice-Hall, Englewood Cliffs, NJ, pp. 125–48.

Reznick, D., H. Bryga, and J. A. Endler. (1990) Experimentally induced life-history evolution in a natural population. Nature, **346**, 357–59.

Richards, C., L. B. Johnson, and G. E. Host. (1996) Landscape-scale influences on stream habitats and biota. Can. J. Fish. Aquat. Sci., **53**, 295–311.

Richards, J. S. (1976) Changes in fish species composition in the Au Sable River, Michigan from the 1920's to 1972. Trans. Am. Fish. Soc., **1**, 32–40.

Richardson, W. B. and S. T. Threlkeld. (1993) Complex interactions of multiple aquatic consumers: an experimental mesocosm manipulation. Can. J. Fish. Aquat. Sci., **50**, 29–42.

Richey, J. E., M. A. Perkins, and C. R. Goldman. (1975) Effects of the kokanee salmon (*Oncorhynchus nerka*) decomposition on the ecology of a subalpine stream. J. Fish. Res. Bd. Can., **32**, 817–20.

Ricklefs, R. E. (1987) Community diversity: relative roles of local and regional processes. Science, **235**, 167–71.

Ricklefs, R. E. and G. W. Cox. (1977) Morphological similarity and ecological overlap among passerine birds on St. Kitts, British West Indies. Oikos, **29**, 60–6.

Ricklefs, R. E., D. Cochran, and E. R. Pianka. (1981) A morphological analysis of the structure of communities of lizards in desert habitats. Ecology, **62**, 1474–83.

Ridgway, M. S. and T. G. Friesen. (1992) Annual variation in parental care in smallmouth bass, *Micropterus dolomieu*. Environ. Biol. Fish., **35**, 243–55.

Ridgeway, M. S. and J. D. McPhail. (1984) Ecology and evolution of sympatric sticklebacks (*Gasterosteus*): mate choice and reproductive isolation in the Enos Lake species pair. Can. J. Zool., **62**, 1813–18.

Rigler, F. H. (1982) The relation between fisheries management and limnology. Trans. Am. Fish. Soc., **111**, 121–32.

Riley, S. C., K. D. Fausch, and C. Gowan. (1992) Movement of brook trout (*Salvelinus fontinalis*) in four small subalpine streams in northern Colorado. Ecol. Freshwater Fish, **1**, 112–22.

Rincón, P. A. and J. Lobón-Cerviá. (1995) Use of an encounter model to predict size-selective predation by a stream-dwelling cyprinid. Freshwater Biol., **33**, 181–91.

Rincón, P. A., J. C. Velasco, N. Gonzales-Sanchez, and C. Pollo. (1990) Fish assemblage in small streams in western Spain: the influence of an introduced predator. Arch. Hydrobiol., **118**, 81–91.

Rinne, J. N. (1975) Changes in minnow populations in a small desert stream resulting from naturally and artificaly induced factors. Southwest. Nat., **20**, 185–95.

Rinne, J. N. (1992) Physical habitat utilization of fish in a Sonoran Desert stream, Arizona, Southwestern United States. Ecology of Freshwater Fish 1:35–41.

Rister, P. W. (1994) Inventory and classification of streams in the lower Cumberland River and Tennessee River drainages. Bull. KY Dept. Fish. Wildl. Res. Fish., **92**:1–211.

Ritossa, F. M. (1962) A new puffing pattern induced by heat shock land DNP in Drosophila. Experientia, **18**, 571–73.

Ritossa, F. M. (1964) Experimental activation of specific loci in polytene chromosomes of Drosophilia. Expt. Cell Res., **36**, 515–23.

Ritter, D. F., R. C. Kochel, and J. R. Miller. (1995) *Process Morphology*, 3rd ed., W. C. Brown Publ., Boston.

Roberts, T. R. (1972) Ecology of fishes in the Amazon and Congo basins. Bull. Museum Compar. Zool., **143**, 117–47.

Roberts, T. R. (1975) Geographical distribution of African freshwater fishes. Zool. J. Linn. Soc., **57**, 249–319.

Robinson, B. W. and D. S. Wilson. (1994) Character release and displacement in fishes: a neglected literature. Am. Nat., **144**, 596–627.

Robinson, B. W., D. S. Wilson, A. S. Margosian, and P. T. Lotito. (1993) Ecological and morphological differentiation of pumpkinseed sunfish in lakes without bluegill sunfish. Evol. Ecol., **7**, 451–64.

Robinson, C. L. K. and W. M. Tonn. (1989) Influence of environmental factors and piscivory in structuring fish assemblages of small Alberta lakes. Can. J. Fish. Aquat. Sci., **46**, 81–9.

Robinson, G. D., W. A. Dunson, J. E. Wright, and G. E. Mamolito. (1976) Differences in low pH tolerance among strains of brook trout (*Salvelinus fontinalis*). J. Fish. Biol., **8**, 5–17.

Robinson, J. V. and G. A. Wellborn. (1988) Ecological resistance to the invasion of a freshwater clam, *Corbicula fluminea*: fish predation effects. Oecologia, **77**, 445–52.

Robison, H. W. (1979) Additions to the Strawberry River ichthyofauna. Proc. AK Acad. Sci., **33**, 89–90.

Robison, H. W. (1986) Zoogeographic implications of the Mississippi River basin, in *The Zoogeography of North American Freshwater Fishes* (eds. C. H. Hocutt and E. O. Wiley). John Wiley and Sons, New York, pp. 267–85.

Robison, H. W. and J. K. Beadles. (1974) Fishes of the Strawberry River system in northcentral Arkansas. Proc. AR Acad. Sci., **28**, 65–70.

Robison, H. W. and T. M. Buchanan. (1988) *Fishes of Arkansas*. University of Arkansas Press, Fayetteville.

Robison, H. W. and G. L. Harp. (1971) A pre-impoundment limnological study of the Strawberry River in northeastern Arkansas. Proc. AR Acad. Sci., **25**, 70–9.

Rodd, F. H. and D. N. Reznick. (1991) Life history evolution in guppies: III. the impact of prawn predation on guppy life histories. Oikos, **62**, 13–9.

Rodrighez, M. A. and W. M. Lewis, Jr. (1997) Structure of fish assemblages along environmental gradients in floodplain lakes of the Orinoco River. Ecological Monographs 67:109–28.

Roell, M. J. and D. J. Orth. (1994) The roles of predation, competition, and exploitation in the trophic dynamics of a warmwater stream: a model synthesis, analysis, and application. Hydrobiologia, **291**, 157–78.

Rogers, C. S. (1993) Hurricanes and coral reefs: the intermediate disturbance hypotheses revisited. Coral Reefs, **12**, 127–37.

Rohlf, F. J. (1990) *NTSYS-pc: Numerical Taxonomy and Multivariate Analysis System*, Exeter Software, Setauket, NY.

Rohm, C. M., J. W. Giese, and C. C. Bennett. (1987) Evaluation of an aquatic ecoregion classification of streams in Arkansas. J. Freshwater Ecol., **4**, 127–40.

Romer, A. S. 1962. *The Vertebrate Body* (3rd Edition). W. B. Saunders, Philadelphia, PA.

Romer, A. S. and T. S. Parsons. (1977) *The Vertebrate Body*, 5th ed., W. B. Saunders Philadelphia, pp. 624.

Romo, S., E. Van Donk, R. Gylstra, and R. Gulati. (1996) A multivariate analysis of phytoplankton and food web changes in a shallow biomanipulated lake. Freshwater Biol., **36**, 683–96.

Root, R. W. (1931) The respiratory function of the blood of marine fishes. Biol. Bull. Mar. Lab., Woods Hole, **61**, 427–56.

Root, T. L. (1993) Effects of global climate change on North American birds and their communities, in *Biotic Interactions and Global Change* (eds. P. M. Kareva, J. G. Kingsolver, and R. B. Huey). Sinauer Associates, Sunderland, MA, pp. 280–92.

Rosen, D. E. (1978) Vicariant patterns and historical explanation in biogeography. Syst. Zool., **27**, 159–88.

Rosenzweig, M. L. (1995) *Species Diversity in Space and Time*. Cambridge University Press, Cambridge.

Ross, D. H. and J. B. Wallace. (1983) Longitudinal patterns of production, food consumption, and seston utilization by net-pinning caddisflies (Trichoptera) in a southern Appalachian stream (USA). Holar. Ecol., **6**, 270–84.

Ross, R. D. (1969) Drainage evolution and fish distribution problems in the southern Appalachians of Virginia, in *The Distributional History of the Biota of the Southern Appalachians. Part I: Invertebrates* (ed. P. C. Holt). Research Division Monograph No. 1, Virginia Polytechnic Institute and State University, Blacksburg. pp. 277–92.

Ross, S. T. (1986) Resource partitioning in fish assemblages: a review of field studies. Copeia, **1986**, 352–88.

Ross, S. T. and J. A. Baker. (1983) The response of fishes to periodic spring floods in a southeastern stream. Am. Midl. Nat., **109**, 1–14.

Ross, S. T., J. A. Baker, and K. E. Clark. (1987) Microhabitat partitioning of southeastern stream fishes: temporal and spatial predictability, in *Community and Evolutionary Ecology of North American Stream Fishes* (eds. W. J. Matthews and D. C. Heins). University of Oklahoma Press, Norman, pp. 43–51.

Ross, S. T., J. G. Knight, and S. D. Wilkins. (1990) Longitudinal occurrence of the bayou darter (Percidae: *Etheostoma rubrum*) in Bayou Pierre—a response to stream order or habitat availability? Pol. Arch. Hydrobiol., **37**, 221–33.

Ross, S. T., W. J. Matthews, and A. A. Echelle. (1985) Persistence of stream fish assemblages: effects of environmental change. Am. Nat., **126**, 24–40.

Rosser, Z. C. and R. G. Pearson. (1995) Responses of rock fauna to physical disturbance in two Australian tropical rainforest streams. J. North Am. Benthol. Soc., **14**, 183–96.

Rossi, R. E., D. J. Mulla, A. G. Journel, and E. H. Franz. (1992) Geostatistical tools for modeling and interpreting ecological spatial dependence. Ecol. Monogr., **62**, 277–314.

Roubach, R. and U. Saint-Paul. (1994) Use of fruits and seeds from Amazonian inundated forests in feeding trials with *Colossoma macropomum* (Cuvier, 1818) (Pisces, Characidae). J. Appl. Ichthyol., **10**, 134–40.

Roughgarden, J. (1972) Evolution of niche width. Nat., **106**, 683–718.

Roughgarden, J. (1983) Competition and theory in community ecology. Am. Nat., **122**, 583–601.

Rowe, D. K. (1994) Vertical segregation and seasonal changes in fish depth distributions between lakes of contrasting trophic status. J. Fish. Biol., **45**, 787–800.

Rozas, L. P. and W. E. Odum. (1987) Use of tidal freshwater marshes by fishes and macrofaunal crustaceans along a marsh stream-order gradient. Estuaries, **10**, 36–43.

Rubenstein, D. I. (1981) Individual variation and competition in the Everglades pygmy sunfish. J. Animal Ecol., **50**, 337–50.

Ruby, S. M., J. Aczel, and G. R. Craig. (1977) The effects of depressed pH on oogenesis in flagfish *Jordanella floridae*. Water Res., **11**, 757–62.

Ruppert, J. B., R. T. Muth, and T. P. Nesler. (1993) Predation on fish larvae by adult red shiner, Yampa and Green Rivers, Colorado. Southwest. Nat., **38**, 397–99.

Rutherford, D. A., A. A. Echelle, and O. E. Maughn. (1987) Changes in the fauna of the Litte River drainage, Southeastern Oklahoma, 1948–1955 and 1981–1982: A test of the hypothesis of environmental degradation, in *Community and Evolutionary Ecology of North American Stream Fishes* (eds. W. J. Matthews and D. C. Heins). University of Oklahoma Press, Norman, pp. 178–83.

Rutherford, D. A., W. E. Kelso, C. F. Bryan, and G. C. Constant. (1995) Influence of physicochemical characteristics on annual growth increments of four fishes from the lower Mississippi River. Trans. Am. Fish. Soc., **124**, 687–97.

Rutledge, C. J., E. G. Zimmerman, and T. L. Beitinger. (1990) Population genetic responses of two minnow species (Cyprinidae) to seasonal stream intermittency. Genetics, **80**, 209–19.

Ryder, R. A. (1965) A method of estimating the potential fish production of north-temperate lakes. Trans. Am. Fish. Soc., **94**, 214–18.

Ryder, R. A. (1982) The morphoedaphic index—use, abuse, and fundamental concepts. Trans. Am. Fish. Soc., **111**, 154–64

Ryder, R. A. and S. R. Kerr. (1978) The adult walleye in the percid community—a niche definition on feeding behavior and food specificity, in *Selected Coolwater Fishes of North America* (ed. R. Kendall), Special Publication No. 11, American Fisheries Society, Washington, DC, pp. 39–51.

Sabat, A. M. (1994a) Mating success in brood-guarding male rock bass, *Amblopites rupestris*: the effect of body size. Environ. Biol. Fish., **39**, 411–15.

Sabat, A. M. (1994b) Costs and benefits of parental effort in a brood-guarding fish (*Ambloplites rupestris*, Centrarchidae). Behav. Ecol., **5**, 195–201.

Saiki, M. K. (1984) Environmental conditions and fish faunas in low elevation rivers on the irrigated San Joaquin Valley floor, California. Calif. Fish Game, **70**, 145–57.

Saint-Paul, U. and G. M. Soares. (1987) Diurnal distribution and behavioral responses of fishes to extreme hypoxia in an Amazon floodplain lake. Environ. Biol. Fish., **20**, 91–104.

Sale, P. F. (1979) Habitat partitioning and competition in fish communities, in *Predator–*

Prey Systems in Fishery Management (ed. H. Clepper). Sport Fishing Institute, Washington, DC, pp. 323–31.

Sal'nikov, V. B. and Y. S. Reshetnikov. (1992) Formation of fish populations in artificial waters in Turkmenistan. J. Ichthyol., **31**, 82–95.

Sato, M., H. Mitani, and A. Shima. (1990) Eurythermic growth and synthesis of heat shock proteins of primary cultured goldfish cells. Zool. Sci., **7**, 395–99.

Savino, J. F. and R. A. Stein. (1982) Predator–prey interaction between largemouth bass and bluegills as influenced by simulated, submersed vegetation. Trans. Am. Fish. Soc., **111**, 255–66.

Savino, J. F. and R. A. Stein. (1989) Behavioral interactions between fish predators and their prey: effects of plant density. Animal Behav., **37**, 311–21.

Sazima, I. and F. A. Machado. (1990) Underwater observations of piranhas in western Brazil. Environ. Biol. Fish., **28**, 17–31.

Scalet, C. G. (1973) Stream movements and population density of the orangebelly darter, *Etheostoma radiosum cyanorum* (Osteichthyes: Percidae). Southwest. Nat., **17**, 381–87.

Schaefer, J. (1995) Stream microhabitat use by two centrarchids: (*Lepomis megalotis*) and (*Lepomis macrochirus*). Unpublished MS thesis, University of Oklahoma, Norman, OK.

Schaefer, J. F., S. T. Heullett, and T. M. Farrell. (1994) Interactions between two poeciliid fishes *Gambusia holbrooki* and *Heterandria formosa* and their prey in a Florida marsh. Copeia, **1994**, 516–20.

Schaeffer, N. and D. E. Rosen. (1961) Major adaptive levels in the evolution of the actinopterygean feeding mechanism. Am. Zool., **1**, 187–204.

Schindler, D. E. (1992) Nutrient regeneration by sockeye salmon (*Oncorhynchus nerka*) fry and subsequent effects on zooplankton and phytoplankton. Can. J. Fish. Aquat. Sci., **49**, 2498–2506.

Schindler, D. E., et al. (1993) Food web structure and phosphorus cycling in lakes. Trans. Am. Fish. Soc., **122**, 756–72.

Schindler, D. E., S. R. Carpenter, K. L. Cottingham, X. He, J. R. Hodgson, J. F. Kitchell, and P. A. Soranno. (1995) Food web structure and littoral zone coupling to pelagic trophic cascades, in *Food Webs: Integration of Patterns and Dynamics*, (eds. G. A. Polis and K. O. Winemiller). Chapman & Hall, New York, pp. 96–105.

Schindler, D. E., J. R. Hodgson, and J. F. Kitchell. (1997) Density-dependent changes in individual foraging specialization of largemouth bass. Oecologia 110:592–600.

Schlesinger, D. A. and H. A. Regier. (1982) Climatic and morphoedaphic indices of fish yields from natural lakes. Trans. Am. Fish. Soc., **111**, 141–50.

Schloemer, C. L. (1947) Reproductive cycles of five species of Texas centrarchids. Science, **106**, 85.

Schlosser, I. and P. L. Angermeier. (1990) The influence of environmental variability, resource abundance, and predation on juvenile cyprinid and centrarchid fishes. Pol. Arch. Hydrobiol., **37**, 265–84.

Schlosser, I. and L. A. Toth. (1984) Niche relationships and population ecology of rainbow

Etheostoma caeruleum and fantail *E. flabellare* darters in a temporary variable environment. Oikos, **42**, 229–38.

Schlosser, I. J. (1982) Fish community structure and function along two habitat gradients in a headwater stream. Ecol. Monogr., **52**, 395–414.

Schlosser, I. J. (1985) Flow regime, juvenile abundance, and the assemblage structure of stream fishes. Ecology, **66**, 1484–90.

Schlosser, I. J. (1987a) The role of predation in age- and size-related habitat use by stream fishes. Ecology, **66**, 651–59.

Schlosser, I. J. (1987b) A conceptual framework for fish communities in small warmwater streams, in *Community and Evolutionary Ecology of North American Stream Fishes* (eds. W. J. Matthews and D. C. Heins). University of Oklahoma Press, Norman, pp. 17–24.

Schlosser, I. J. (1988a) Predation rates and the behavioral response of adult brassy minnows, *Hybognathus hankinsoni* to creek chub and smallmouth bass predators. Copeia, **1988**, 691–97.

Schlosser, I. J. (1988b) Predation risk and habitat selection by two size classes of a stream cyprinid: experimental test of a hypothesis. Oikos, **52**, 36–40.

Schlosser, I. J. (1990) Environmental variation, life history attributes, and community structure in stream fishes: implications for environmental management and assessment. Environ. Mangmt., **14**, 621–28.

Schlosser, I. J. (1991) Stream fish ecology: a landscape perspective. BioScience, **41**, 704–12.

Schlosser, I. J. (1995) Critical landscape attributes that influence fish population dynamics in headwater streams. Hydrobiologia, **303**, 71–81.

Schlosser, I. J. and K. K. Ebel. (1989) Effects of flow regime and cyprinid predation on a headwater stream. Ecol. Monogr., **59**, 41–57.

Schluter, D. (1993) Adaptive radiation in sticklebacks: size, shape, and habitat use efficiency. Ecology, **74**, 699–709.

Schluter, D. (1994) Experimental evidence that competition promotes divergence in adaptive radiation. Science, **266**, 798–801.

Schmidt, R. E. (1986) Zoogeography of the Northern Appalachians, in *The Zoogeography of North American Freshwater Fishes* (eds. C. H. Hocutt and E. O. Wiley). John Wiley and Sons, New York, pp. 137–59.

Schmidt, R. J. and S. J. Holbrook. (1984) Gape-limitation, foraging tactics and prey size selectivity of two microcarnivorous species of fish. Oecologia, **63**, 6–12.

Schmidt-Neilsen, K. (1975) *Animal Physiology—Adaption and Environment*, Cambridge University Press, London.

Schmitz, E. H. and C. D. Baker. (1969) Digestive anatomy of the gizzard shad, *Dorosoma cepedianum* and the threadfin shad, *D. petenense*. Trans. Am. Microsc. Soc., **88**, 525–46.

Schneider, D. C. and R. L. Haedrich. (1989) Prediction limits of allometric equations: a reanalysis of Ryder's morphoedaphic index. Can. J. Fish. Aquat. Sci., **46**, 503–08.

Schoener, T. W. (1970) Size patterns in West Indian *Anolis* lizards II: correlations with the sizes of particular sympatric species-displacement and convergence. Am. Nat., **104**, 155–74.

Schoener, T. W. (1971) Theory of feeding strategies. Annu. Rev. Ecol. Syst., **2**, 369–404.

Schoener, T. W. (1974) Resource partitioning in ecological communities. Science, **185**, 27–39.

Schoener, T. W. (1982) The controversy over interspecific competition. Am. Sci., **70**, 586–95.

Schoenherr, A. A. (1979) Niche separation within a population of freshwater fishes in an irrigation drain near the Salton Sea, California. Bull. S. CA Acad. Sci., **78**, 46–55.

Schofield, K., C. R. Townsend, and A. G. Hildrew. (1988) Predation and the prey community of a headwater stream. Freshwater Biol., **20**, 85–95.

Schramm, H. L., Jr. and A. V. Zale. (1985) Effects of cover and prey size on preferences of juvenile largemouth bass for blue tilapias and bluegills in tanks. Trans. Am. Fish. Soc., **114**, 725–31.

Schriver, P., J. Bøgestrand, E. Jeppesen, and M. Søndergaard. (1995) Impact of submerged macrophytes on fish–zooplankton–phytoplankton interactions: large-scale enclosure experiments in a shallow eutrophic lake. Freshwater Biol., **33**, 255–70.

Schut, J., S. S. DeSilva, and K. Kortmulder. (1984) Habitat, associations and competition of eight *Barbus (=Puntius)* species (Pisces, Cyrinidae) indigenous to Sri Lanka. Netherlands. J. Zool., **34**, 159–81.

Schwartz, F. J. (1988) Pre- and post-drought fish surveys of selected freshwater ponds located in Nags Head, North Carolina. ASB Bull., **35**, 189–98.

Scott, D. (1995) Vegetation: a mosaic of discrete communities, or a continnum? NZ J. Ecol., **19**, 47–52.

Scrimgeour, G. J. and J. M. Culp. (1994a) Foraging and evading predators: the effect of predator species on a behavioural trade-off by a lotic mayfly. Oikos, **69**, 71–9.

Scrimgeour, G. J. and J. M. Culp. (1994b) Feeding while evading predators by a lotic mayfly: linking short-term foraging behaviors to long-term fitness consequences. Oecologia, **100**, 128–34.

Scrimgeour, G. J. and M. J. Winterbourn. (1987) Diet, food resource partitioning and feeding periodicity of two riffle-dwelling fish species in a New Zealand river. J. Fish. Biol., **31**, 309–24.

Scrimgeour, G. J., J. M. Culp, and K. J. Cash. (1994a) Anti-predator responses of mayfly larvae to conspecific and predator stimuli. J. North Am. Benthol. Soc., **13**, 299–309.

Scrimgeour, G. J., J. M. Culp, and F. J. Wrona. (1994b) Feeding while avoiding predators: evidence for a size-specific trade-off by a lotic mayfly. J. North Am. Benthol. Soc., **12**, 368–78.

Sechnick, C. W., R. F. Carline, R. A. Stein, and E. T. Rankin. (1986) Habitat selection by smallmouth bass in response to physical characteristics of a simulated stream. Trans. Am. Fish. Soc., **115**, 314–21.

Seda, J. and A. Duncan. (1994) Low fish predation pressure in London reservoirs: II. consequences to zooplankton community structure. Hydrobiologia, **291**, 179–91.

Seegrist, D. W. and R. Gard. (1972) Effects of floods on trout in Sagehen Creek, California. Trans. Am. Fish. Soc., **101**, 478–82.

Seghers, B. (1973) An analysis of geographic variations in the antipredator adaptions of the guppy, *Poecilia reticulata*, Ph.D. dissertation, University of British Columbia.

Seifert, R. P. (1984) Does competition structure communities? Field studies on neotropical *Heliconia* insect communities, in *Ecological Communities: Conceptual Issues and the Evidence*, (eds. D. R. Strong, Jr., D. Simberloff, L. F. Abele, and A. B. Thistle). Princeton University Press, Princeton, NJ, pp. 54–63.

Semlitsch, R. D. (1987) Interactions between fish and salamander larvae-costs of predator avoidance or competition? Oecologia, **72**, 481–86.

Semlitsch, R. D. (1988) Allotopic distribution of two salamanders: effects of fish predation and competitive interactions. Copeia, **1988**, 290–98.

Semlitsch, R. D. and J. W. Gibbons. (1988) Fish predation in size-structured populations of treefrog tadpoles. Oecologia (Berlin), **75**, 321–26.

Serafy, J. E. and R. M. Harrell. (1993) Behavioural response of fishes to increasing pH and dissolved oxygen: field and laboratory observations. Freshwater Biol., **30**, 53–61.

Settles, W. H. and R. D. Hoyt. (1978) The reproductive biology of the southern redbelly dace, *Chrosomus erythrogaster* Rafinesque, in a spring-fed stream in Kentucky. Am. Midl. Nat., **99**, 290–98.

Sexton, O. J. and C. Phillips. (1986) A qualitative study of fish–amphibian interactions in 3 Missouri ponds. Trans. MO Acad. Sci., **20**, 25–35.

Sexton, O. J., C. Phillip, and E. Routman. (1994) The response of naive breeding adults of the spotted salamander to fish. Behaviour, **130**, 113–21.

Shafland, P. L. and J. M. Pestrak. (1982) Lower lethal temperatures for fourteen non-native fishes in Florida. Environ. Biol. Fish., **7**, 149–56.

Shannon, J. P., D. W. Blinn, P. L. Benenati, and K. P. Wilson. (1996) Organic drift in a regulated desert river. Can. J. Fish. Aquat. Sci., **53**, 1360–69.

Shapiro, J. and D. I. Wright. (1984) Lake restoration by biomanipulation: Round Lake, Minnesota, the first two years. Freshwater Biol., **14**, 371–83.

Shaw, E. (1962) The schooling of fishes. Sci. Am., **206**, 128–38.

Shaw, E. (1970) Schooling in fishes: critique and review, in *Development and Evolution of Behavior* (eds. L. R. Aronson, E. T. Toboch, D. S. Lehrman, and J. S. Rosenblatt). W. H. Freeman, San Francisco, CA, pp. 452–80.

Shaw, E. (1978) Schooling fishes. Am. Sci., **66**, 166–75.

Sheldon, A. L. (1980) Coexistance of perlid stoneflies (*Plecoptera*): predictions from multivariate morphometrics. Hydrobiologia, **71**, 99–105.

Sheldon, A. L. (1987) Rarity: patterns and consequences for stream fishes, in *Community and Evolutionary Ecology of North American Stream Fishes* (eds. W. J. Matthews and D. C. Heins). University of Oklahoma Press, Norman, pp. 203–09.

Sheldon, A. L. and G. K. Meffe. (1995) Short-term recolonization by fishes of experimentally defaunated pools of a coastal plain stream. Copeia, **1995**, 828–37.

Sheldon, A. S. (1968) Species diversity and longitudinal succession in stream fishes. Ecology, **49**, 193–98.

Shelford, V. E. (1911) Ecological succession. I. Stream fishes and the method of physiographic analysis. Biol. Bull., **21**, 9–35.

Shelford, V. E. and W. C. Allee. (1913) The reactions of fishes to gradients of dissolved atmospheric gases. J. Expt. Zool., **14**, 207–66.

Shelton, W. L., A. Soliman, and S. Rothbard. (1995) Experimental observations on feeding biology of black carp (*Mylopharyngodon piceus*). Israeli J. Aquacult.—Bamidgeh, **47**, 59–67.

Shetter, D. S., O. H. Clark, and A. S. Hazzard. (1946) The effects of deflectors in a section of a Michigan trout stream. Trans. Am. Fish. Soc., **76**, 248–78.

Shirley, K. E. and A. K. Andrews. (1977) Growth, production, and mortality of largemouth bass during the first year of life in Lake Carl Blackwell, Oklahoma. Trans. Am. Fish. Soc., **106**, 590–95.

Shirvell, C. S. (1990) Role of instream rootwads as juvenile coho salmon (*Oncorhynchus kisutch*) and steelhead trout (*O. mykiss*) cover habitat under varying streamflows. Can. J. Fish. Aquat. Sci., **47**, 852–61.

Shirvell, C. S. (1994) Effect of change in streamflow on the microhabitat use and movements of sympatric juvenile coho salmon (*Oncorhynchus kisutch*) and chinook salmon (*O. tshawytscha*) in a natural stream. Can. J. Fish. Aquat. Sci., **51**, 1644–52.

Shute, P. W., J. R. Shute, and D. G. Lindquist. (1982) Age, growth and early life history of the waccamaw darter, *Etheostoma perlongum*. Copeia, **1982**, 561–67.

Sigler, W. F., S. Vigg, and M. Bres. (1985) Life history of the cui-ui, *Chasmistes cujus* Cope, in Pyramid Lake, Nevada: a review. Great Basin Nat., **45**, 571–603.

Sih, A. (1980) Optimal behavior: Can foragers balance two conflicting demands? Science, **210**, 1041–43.

Sih, A. (1982) Foraging strategies and the avoidance of predation by an aquatic insect, *Notonecta hoffmani*. Ecology, **63**, 786–96.

Sih, A. (1987) Predators and prey lifestyles: an evolutionary and ecological overview, in *Predation* (eds. W. C. Kerfoot and A. Sih). University Press of New England, Hanover, NH, pp. 203–24.

Sih, A., L. B. Kats, and R. D. Moore. (1992) Effects of predatory sunfish on the density, drift, and refuge use of stream salamander larvae. Ecology, **73**, 1418–30.

Simberloff, D. (1974) Equilibrium theory of island biogeography and ecology. Annu. Rev. Ecol. Syst., **5**, 161–91.

Simberloff, D. (1976) Experimental zoogeography of islands: effects of island size. Ecology, **57**, 629–48.

Simonsen, J. F. and P. Harremoës. (1978) Oxygen and pH fluctuations in rivers. Water Res., **12**, 477–89.

Simonson, T. D. and J. Lyons. (1995) Comparison of catch per effort and removal procedures for sampling stream fish assemblages. North Am. J. Fish. Mangmt., **15**, 419–27.

Simonson, T. D. and W. A. Swenson. (1990) Critical stream velocities for young-of-year smallmouth bass in relation to habitat use. Trans. Am. Fish. Soc., **119**, 902–09.

Simpson, G. G. (1953) Evolution and geography—an essay on historical biogeography with special reference to mammals. Condon Lectures, Oregon State System of Higher Education, Eugene.

Sioli, H. (1975) Tropical river: the Amazon, in *River Ecology*, (ed. B. A. Whitton). University of California Press, Berkeley, pp. 461–88.

Skinner, W. D. (1985) Night–day drift patterns and the size of larvae of two aquatic insects. Hydrobiologia, **124**, 283–85.

Skud, B. E. (1982) Dominance in fishes: the relation between environment and abundance. Science, **216**, 144–49.

Skulason, S., D. L. G. Noakes, and S. S. Snorrason. (1989) Ontogeny of trophic morphology in four sympatric morphs of arctic charr *Salvelinus alpinus* in Thingvallavatn, Iceland. Biol. J. Linn. Soc., **38**, 281–301.

Skulason, S., A. S. Snorrason, D. Ota, and D. L. G. Noakes. (1993) Genetically based differences in foraging behavior among sympatric morphs of arctic charr (Pisces: Salmonidae). Animal Behav., **45**, 1179–92.

Smale, M. A. and C. F. Rabeni. (1995a) Influences of hypoxia and hyperthermia on fish species composition in headwater streams. Trans. Am. Fish. Soc., **124**, 711–25.

Smale, M. A. and C. F. Rabeni. (1995b) Hypoxia and hyperthermia tolerances of headwater stream fishes. Trans. Am. Fish. Soc., **124**, 698–710.

Smart, R. A. (1978) A comparison of morphological and ecological overlaps in a *Peromyscus* community. Ecology, **59**, 216–20.

Smith, C. L. (1995) *Fishwatching: An Outdoor Guide to Freshwater Fishes*. Cornell Univ. Press, Ithaca, NY.

Smith, C. L. and C. R. Powell. (1971) The summer fish communities of Brier Creek, Marshall County, Oklahoma. Am. Museum Novitates, **2458**, 1–30.

Smith, C. L. and J. C. Tyler. (1972) Space resource sharing in a coral reef fish community, in *Results of the Tektite Program: Ecology of Coral Reef Fishes* (eds. B. C. Collette and S. A. Earle). Nat. Hist. Museum Los Angeles County Sci. Bull., **14**, pp. 125–70.

Smith, G. R. (1981) Effects of habitat size on species richness and adult body sizes of desert fishes, in *Fishes in North American Deserts* (eds. R. J. Naiman and D. L. Soltz). John Wiley and Sons, New York, pp. 125–71.

Smith, G. R. (1992) Phylogeny and biogeography of the Catostomidae, freshwater fishes from North America and Asia, in (ed. R. L. Mayden) *Systematics, Historical Ecology, & North American Freshwater Fishes*, Stanford University Press, pp. 778–826.

Smith, G. R. and D. R. Fisher. (1970) Factor analysis of distribution patterns of Kansas fishes, in (ed. not given) *Pleistocene and recent environments of the central Great Plains*. Special Publication 3, University of Kansas Press, Lawrence, KS.

Smith, H. M. (1945) The freshwater fishes of Siam, or Thailand. Bull. Smithsonian Instit. U.S. Nat. Museum, **188**, 1–38.

Smith, M. L. and R. R. Miller. (1986) The evolution of the Rio Grande basin as inferred from its fish fauna, in *The Zoogeography of North American Freshwater Fishes* (eds. C. H. Hocutt and E. O. Wiley). John Wiley and Sons, New York, pp. 457–85.

Smock, L. A., L. C. Smith, J. B. Jones, Jr., and S. M. Hooper. (1994) Effects of drought and hurricane on a coastal headwater stream. Arch. Hydrobiol., **131**, 25–38.

Sneath, P. H. A. and R. R. Sokal (eds.). (1973) *Numerical Taxonomy—The Principles and Practice of Numerical Systematics*. W. H. Freeman, San Francisco.

Snelson, F. F., Jr. (1982) Indeterminate growth in males of the sailfin molly, *Poecilia latipinna*. Copeia, **1982**, 296–304.

Snelson, F. F., Jr. (1984) Seasonal maturation and growth of males in a natural population of *Poeculia latipinna*. Copeia, **1984**, 252–55.

Snyder, G. K. and W. W. Weathers. (1975) Temperature adaptations in amphibians. Am. Nat., **109**, 93–101.

Sokal, R. R. and F. J. Rohlf. (1969) *Biometry* (1st edition). W. H. Freeman and Co., San Francisco, CA.

Sokolov, L. I., E. L. Sokolov, V. A. Pegasov, I. M. Shatunovskiy, and A. N. Kistenev. (1994) The ichthyofauna of the Moscow River within the boundaries of the City of Moscow. J. Ichthyol., **34**, 141–51.

Soluk, D. A. (1993) Multiple predator effects: predicting combined functional response of stream fish and invertebrate predators. Ecology, **74**, 219–25.

Soluk, D. A. and N. C. Collins. (1988a) Synergistic interactions between fish and stoneflies: facilitation and interfacing among stream predators. Oikos, **52**, 94–100.

Soluk, D. A. and N. C. Collins. (1988b) Balancing risks? responses and nonresponses of mayfly larvae to fish and stonefly predators. Oecologia, **77**, 370–74.

Soule, P. T. (1993a) Hydrologic drought in the contiguous United States, 1900–1989: spatial patterns and multiple comparison of means. Geophys. Res. Lett., **20**, 2367–70.

Soule, P. T. (1993b) Spatial patterns and teleconnections of Utah-based droughts in the contiguous USA. J. Arid Environ., **24**, 217–29.

Spangler, G. R. and J. J. Collins. (1992) Lake Huron fish community structure based on gill-net catches corrected for selectivity. North Am. J. Fish. Mangmt., **12**, 585–97.

Spellerberg, I. F. (1973) Critical minimum temperature of reptiles, in *Effects of Temperature on Ectothermic Organisms* (ed. W. Wieser). Springer-Verlag, Berlin, pp. 239–47.

Spencer, C. N. and D. L. King. (1984) Role of fish in regulation of plant and animal communities in eutrophic ponds. Can. J. Fish. Aquat. Sci., **41**, 1851–55.

Spigarelli, S. A., R. M. Goldstein, W. Prepejchal, and M. M. Thommes. (1982) Fish abundance and distribution near three heated effluents to Lake Michigan. Can. J. Fish. Aquat. Sci., **39**, 305–15.

Spigarelli, S. A., M. M. Thommes, W. Prepejchal, and R. M. Goldstein. (1983) Selected temperatures and thermal experience of brown trout, *Salmo trutta*, in a steep thermal gradient in nature. Environ. Biol. Fish., **2**, 137–49.

Spoor, W. A. (1977) Oxygen requirements of embryos and larvae of the largemouth bass, *Micropterus salmonides* (Lacépède). J. Fish. Biol., **11**, 77–86.

Sprules, W. G., S. B. Brandt, D. J. Stewart, M. Munawar, E. H. Jin, and J. Love. (1991)

Biomass size spectrum of the Lake Michigan pelagic food web. Can. J. Fish. Aquat. Sci., **48**, 105–15.

Stager, J. C., P. N. Reinthal, and D. A. Livingstone. (1986) A 25,000-year history for Lake Victoria, East Africa, and some comments on its significance for the evolution of cichlid fishes. Freshwater Biology 16:15–19.

Stanford, J. A. and J. V. Ward. (1983) Insect species diversity as a function of environmental variability and disturbance in stream systems, in *Stream Ecology—Application and Testing of General Ecological Theory* (eds. J. R. Barnes and G. W. Minshall). Plenum Press, New York, pp. 265–78.

Stanford, J. A. and J. V. Ward. (1993) An ecosystem perspective of alluvial rivers: connectivity and the hyporheic corridor. J. North Am. Benthol. Soc., **12**, 48–60.

Stangel, P. W. and R. D. Semlitsch. (1987) Experimental analysis of predation on the diel vertical migrations of a larval salamander. Can. J. Zool., **65**, 1554–58.

Stanley, E. H. and A. J. Boulton. (1995) Hyporeic processes during flooding and drying in a Sonoran Desert stream. Arch. Hydrobiol., **134**, 1–26.

Starrett, W. C. (1950a) Distribution of the fishes of Boone County, Iowa, with special reference to the minnows and darters. Am. Midl. Nat., **43**, 112–27.

Starrett, W. C. (1950b) Food relationships of the minnows of the Des Moines River, Iowa. Ecology, **31**, 261–33.

Starrett, W. C. (1951) Some factors affecting the abundance of minnows in the Des Moines River, Iowa. Ecology, **32**, 13–27.

Statzner, B. and B. Higler. (1986) Stream hydraulics as a major determinant of benthic invertebrate zonation patterns. Freshwater Biol., **16**, 127–39.

Statzner, B. and T. F. Holm. (1989) Morphological adaption of shape to flow: microcurrents around lotic macroinvertebrates with Reynolds numbers at quasi-natural flow. Oecologia, **78**, 145–57.

Stauffer, J. R., Jr., E. L. Melisky, and C. Hocutt. (1984) Interrelationships among preferred, avoided, and lethal temperatures of three fish species. Arch. Hydrobiol., **100**, 159–69.

Stauffer, J. R., Jr., K. L. Dickson, J. Cairns, Jr., and D. S. Cherry. (1976) The potential and realized influences of temperature on the distribution of fishes in the New River, Glen Lyn, Virginia. Wildl. Monogr., **50**, 1–40.

Stauffer, J. R., Jr., C. H. Hocutt, M. T. Masnik, and J. E. Reed, Jr. (1975) The longitudinal distribution of the fishes of the East River, West Virginia–Virginia. VA J. Sci., **26**, 121–25.

Stearns, S. C. (1983) The evolution of life-history traits in mosquitofish since their introduction to Hawaii in 1905: rates of evolution, heritabilities, and developmental plasticity. Am. Zool., **23**, 65–75.

Stearns, S. C. and R. E. Crandall. (1984) Plasticity for age and size at sexual maturity: A life history response to unavoidable stress, in *Fish Reproduction: Strategies and Tactics*, (eds. G. Potts and R. J. Wootton). Academic Press, London, UK, pp. 13–33.

Stefan, H. G., M. Hondzo, J. G. Eaton, and J. H. McCormick. (1995) Validation of a fish habitat model for lakes. Ecol. Model., **82**, 211–24.

Stehr, W. C. and J. W. Branson. (1938) An ecological study of an intermittent stream. Ecology, **19**, 294–310.

Stein, R. A. (1979) Behavioral response of prey to fish predators, in *Predator–Prey Systems in Fisheries Management* (ed. H. Clepper). Sport Fishing Institute, Washington, DC, pp. 343–53.

Stein, R. A. and J. J. Magnuson. (1976) Behavioral response of crayfish to a fish predator. Ecology 58:571–81.

Stenson, J. A. E. (1980) Predation pressure from fish on two *Chaoborus* species as related to their visibility, in *Evolution and Ecology of Zooplankton Communities* (ed. C. W. Kerfoot). University Press of New England, Hanover, NH, pp. 618–22.

Stevens, E. D. and A. M. Sutterlin. (1976) Heat transfer between fish and ambient water. J. Expt. Biol., **65**, 131–45.

Stevenson, M. M. (1992) Food habits within the Laguna Chichancanab *Cyprinodon* (Pisces: Cyprinodontidae) species flock. Southwest. Nat., **37**, 337–43.

Stevenson, M. M., G. D. Schnell, and R. Black. (1974) Factor analysis of fish distribution patterns in Western and Central Oklahoma. Syst. Zool. 23:202–218.

Stewart, A. J. (1987) Responses of stream algae to grazing minnows and nutrients: a field test for interactions. Oecologia, **72**, 1–7.

Stewart, B. G., J. G. Knight, and R. C. Cashner. (1992) Longitudinal distribution and assemblages of fishes of Byrd's Mill Creek, A southern Oklahoma Arbuckle Mountain stream. Southwest. Nat., **37**, 138–47.

Stewart, D. J., J. F. Kitchell, and L. B. Crowder. (1981) Forage fishes and their salmonid predators in Lake Michigan. Trans. Am. Fish. Soc., **110**, 751–63.

Stiassny, M. L. J. (1996) An overview of freshwater biodiversity: with some lessons from African fishes. Fisheries, **21**, 7–13.

Stine, S. (1994) Extreme and persistent drought in California and Patagonia during medieval time. Nature 369:546–49.

Stock, J. D. and I. J. Schlosser. (1991) Short-term effects of a catastrophic beaver dam collapse on a stream fish community. Environ. Biol. Fish., **31**, 123–29.

Stoermer, E. F., G. Emmert, M. L. Julius, and C. L. Schelske. (1996) Paleolimnologic evidence of rapid recent change in Lake Erie's trophic status. Canadian J. Fish. Aquat. Sci. 53:1451–58.

Stone, L. and A. Roberts. (1991) Conditions for a species to gain advantage from the presence of competitors. Ecology, **72**, 1964–72.

Stoneman, C. L. and M. L. Jones. (1996) A simple method to classify stream thermal stability with single observations of daily maximum water and air teperatures. North Am. J. Fish. Mangmt., **16**, 728–37.

Stoner, G. and F. Breden. (1988) Phenotypic differentiation in female preference related to geographic variation in male predation risk in the Trinidad guppy (*Poecilia reticulata*). Behav. Ecol. Sociobiol., **22**, 285–91.

Strahler, A. N. (1957) Quantitative analysis of watershed geomorphology. Trans. Am. Geophysi. Union, **38**, 913–20.

Strange, E. M. (1995) Pattern and process in stream fish community organization: field study and simulation modeling. Ph.D. thesis, University of California, Davis.

Strange, E. M., P. B. Moyle, and T. C. Foin. (1992) Interactions between stochastic and deterministic processes in stream fish community assembly. Environ. Biol. Fish., **36**, 1–15.

Strauss, R. E. (1987) The importance of phylogenetic constraints in comparisons of morphological structure among fish assemblages, in *Community and Evolutionary Ecology of North American Stream Fishes* (eds. W. J. Matthews and D. C. Heins). University of Oklahoma Press, Norman, pp. 136–43.

Strauss, R. E. (1990) Predation and life-history variation in *Poecilia reticulata* (Cyprinodontiformes: Poeciliidae). Environ. Biol. Fish., **27**, 121–30.

Strauss, R. E. and F. L. Bookstein. (1982) The truss: body form reconstructions in morphometrics. Syst. Zool. 31:113–35.

Stromberg, J. C., B. D. Richter, D. T. Patten, and L. G. Wolden. (1993) Response of a Sonoran riparian forest to a 10-year return flood. Great Basin Nat., **53**, 118–30.

Sturmbauer, C., W. Mark, and R. Dallinger. (1992) Ecophysiology of Aufwuchs-eating cichlids in Lake Tanganyika: niche separation by trophic specialization. Environ. Biol. Fish., **35**, 283–90.

Sublette, J. E., M. D. Hatch, and M. Sublette. (1990) *The Fishes of New Mexico*. University of New Mexico Press, Albuquerque.

Sugihara, G. (1980) Minimal community structure: an explanation of species abundance patterns. Am. Nat., **116**, 770–87.

Surat, E. M., W. J. Matthews, and J. R. Bek. (1982) Comparative ecology of *Notropis albeolus, N. ardens,* and *N. cerasinus* (Cyprinidae) in the upper Roanoke River drainage, Virginia. Am. Midl. Nat., **107**, 13–24.

Suthers, I. M. and J. H. Gee. (1986) Role of hypoxia in limiting diel spring and summer distribution of juvenile yellow perch (*Perca flavescens*) in a prairie marsh. Can. J. Fish. Aquat. Sci., **43**, 1562–70.

Svardson, G. (1949) Competition between trout and char *Salmo trutta* and *S. alpinus*. Rep. Inst. Freshwater Res., Drottningham, **29**, 108–11.

Swaidner, J. E. and T. M. Berra. (1979) Ecological analysis of the fish distribution in Green Creek, a spring-fed stream in northern Ohio. Ohio J. Sci., **79**, 84–92.

Swarts, F. A., W. A. Dunson, and J. E. Wright. (1978) Genetic and environmental factors involved in increased resistance of brook trout to sulfuric acid solutions and mine acid-polluted waters. Trans. Am. Fish. Soc., **107**, 651–77.

Swift, C. R., C. R. Gilbert, S. A. Bortone, G. H. Burgess, and R. W. Yerger. (1986) Zoogeography of the freshwater fishes of the southeastern United States: Savannah River to Lake Pontchartrain, in *Zoogeography of North American Freshwater Fishes* (eds. C. Hocutt and E. O. Wiley). John Wiley and Sons, New York, pp. 213–65.

Swingle, H. S. (1950) Relationships and dynamics of balanced and unbalanced fish populations. AL Agri. Expt. Statn. Bull., **274**, 1–74.

Sylvester, J. R. and J. D. Broughton. (1983) Distribution and relative abundance of fish in Pool 7 of the upper Mississippi River. North Am. J. Fish. Mangmt., **3**, 67–71.

Symons, P. E. K. and M. Heland. (1978) Stream habitat and behavioral interactions of underyearling and yearling Atlantic salmon *Salmo salar*. J. Fish. Res. Bd. Can., **35**, 175–83.

Tallman, R. F. and J. H. Gee. (1982) Intraspecific resource partitioning in a headwaters stream fish, the pearl dace *Semotilus margarita* (Cyprinidae). Environ. Biol. Fish., **3**, 243–49.

Tarboton, David G. (1995) Hydrologic scenarios for severe sustained drought in the southwestern United States. Water Res. Bull., **31**, 803–13.

Tátrai, I. and V. Istvánovics. (1986) The role of fish in the regulation of nutrient cycling in Lake Balaton, Hungary. Freshwater Biol., **16**, 417–24.

Tátrai, I., L. G.-Tóth, and J. E. Ponyi. (1985) Effects of bream (*Abramis brama* L.) on the lower trophic level and on the water quality in Lake Balaton. Arch. Hydrobiol., **105**, 205–17.

Taylor, B. E. (1980) Size-selective predation on zoolankton, in *Evolution and Ecology of Zooplankton Communities* (ed. C. W. Kerfoot). University Press of New England, Hanover, NH, pp. 377–87.

Taylor, C. M. (1996) Abundance and distribution within a guild of benthic stream fishes: local processes and regional patterns. Freshwater Biol., **36**, 385–96.

Taylor, C. M. (1997) Fish species richness and incidence patterns in isolated and connected stream pools: effects of pool volume and spatial position. Oecologia **110**, 560–566.

Taylor, C. M. and N. J. Gotelli. (1994) The macroecology of *Cyprinella*: correlates of phylogeny, body size, and geographical range. Am. Nat., **144**, 549–69.

Taylor, C. M. and P. W. Lienesch. (1996) Regional parapatry of the congeneric cyprinids *Lythrurus snelsoni* and *L. umbratilis*: species replacement along a complex environmental gradient. Copeia, **1996**, 493–97.

Taylor, C. M., M. R. Winston, and W. J. Matthews. (1991) Distribution and abundance of sport and forage fishes of the upper Red River drainage (above Lake Texoma) in Oklahoma. Final Report, Federal Aid Project No. F-48-R. Oklahoma Department of Wildlife Conservation.

Taylor, C. M., M. R. Winston, and W. J. Matthews. (1993) Fish species-environment and abundance relationships in a Great Plains river system. Ecography, **16**, 16–23.

Taylor, C. M., M. R. Winston, and W. J. Matthews. (1996) Temporal variation in tributary and mainstream fish assemblages in a Great Plains stream system. Copeia, **1996**, 280–89.

Taylor, E. B. (1990) Environmental correlates of life-history variation in juvenile chinook salmon, *Oncorhynchus tshawytscha* (Walbaum). J. Fish. Biol., **37**, 1–17.

Taylor, F. and A. C. Hendricks. (1987) The influence of fish on leaf breakdown in a Virginia pond. Freshwater Biol., **18**, 45–51.

Terborgh, J. W. and J. Faaborgh. (1980) Saturation of bird communities in the West Indies. Am. Nat., **116**, 178–95.

ter Braak, C. J. F. (1986) Canonical correspondence analysis: A new eigenvector technique for multivariate direct gradient analysis. Ecology 67:1167–1179.

Theodorakis. C. W. (1989) Size segregation and the effects of oddity on predation risk in minnow schools. Animal Behav., **38**, 496–502.

Thiebaux, M. L. and L. M. Dickie. (1993) Structure of the body-size spectrum of the biomass in aquatic ecosystems: a consequence of allometry in predator–prey interactions. Can. J. Fish. Aquat. Sci., **50**, 1308–17.

Thomas, D. L. (1970) An ecological study of four darters of the genus *Percina* (Percidae) in the Kaskaskia River, Illinois. Ill. Nat. Hist. Surv. Biol. Notes, **70**, 2–18.

Thompson, D. B. A. and M. L. P. Thompson. (1985) Early warning and mixed species association: the "Plover's Page" revisited. IBIS. **127**, 559–62.

Thompson, D. H. (1925) Some observations on the oxygen requirements of fishes in the Illinois River. Ill. Nat. Hist. Sur., **15**, 423–37.

Thompson, D. H. (1933) *The Migration of Illinois Fishes*, Biol. Notes 1. Illinois Natural History Survey, Urbana, IL, p. 26.

Thompson, D. H. and F. D. Hunt. (1930) The fishes of Champaign County: a study of the distribution and abundance of fishes in small streams. Ill. Nat. Hist. Bull., **XIX**, 5–71.

Thonny, J. -P. and R. J. Gibson. (1989) Feeding strategies of brook trout, *Salvelinus fontinalis*, and juvenile Atlantic salmon, *Salmo salar*, in a Newfoundland river. Can. Field-Nat., **103**, 48–56.

Thornton, R. G., B. L. Kimmel, and F. E. Payne (eds.). (1990) *Reservoir Limnology: Ecological Perspectives*. John Wiley and Sons, New York.

Thorp, J. H. and E. A. Bergey. (1981) Field experiments on responses of a freshwater, benthic macroinvertebrate community to vertebrate predators. Ecology, **62**, 365–75.

Thorpe, J. E. (1994) Performance thresholds and life-history flexibility in salmonids. Conserv. Biol., **8**, 877–79.

Threlkeld, S. T. (1987) Experimental evaluation of trophic-cascade and nutrient-mediated effects of planktivorous fish on plankton community structure, in *Predation* (eds. W. C. Kerfoot and A. Sih). University Press of New England, Hanover, NH, pp. 161–73.

Threlkeld, S. T. (1988) Planktivory and planktivore biomass effects on zooplankton, phytoplankton, and the trophic cascade. Limnol. Oceanogr., **33**, 1362–75.

Threlkeld, S. T. and D. M. Søballe. (1988) Effects of mineral turbidity on freshwater plankton communities: three exploratory tank experiments of factorial design. Hydrobiologia, **159**, 223–36.

Threlkeld, S. T. and R. W. Drenner. (1987) An experimental mesocosm study of residual and contemporary effects of an omnivorous, filter-feeding, clupeid fish on plankton community structure. Limnol. Oceanogr., **32**, 1331–41.

Titus, R. G. and H. Mosegaard. (1992) Fluctuating recruitment and variable life history of migratory brown trout, *Salmo trutta* L., in a small, unstable stream. J. Fish. Biol., **41**, 239–55.

Todd, C. S. and K. W. Stewart. (1985) Food habits and dietary overlap of nongame insectivorous fishes in Flint Creek, Oklahoma, a western Ozark foothills stream. Great Basin Nat., **45**, 721–33.

Toline, C. A. and A. J. Baker. (1993) Foraging tactic as a potential selection pressure influencing geographic differences in body shape among populations of dace (*Phoxinus eos*). Can. J. Zool., **71**, 2178–84.

Tonn, W. M. (1985) Density compensation in *Umbra-Perca* fish assemblages of northern Wisconsin lakes. Ecology, **66**, 415–29.

Tonn, W. M. (1990) Climate change and fish communities: a conceptual framework. Trans. Am. Fish. Soc., **119**, 337–52.

Tonn, W. M. and J. J. Magnuson. (1982) Patterns in the species composition and richness of fish assemblages in northern Wisconsin lakes. Ecology, **63**, 1149–66.

Tonn, W. M., J. J. Magnuson, M. Rask, and J. Toivonen. (1990) Intercontinental comparison of small-lake fish assemblages: the balance between local and regional processes. Am. Nat., **136**, 345–75.

Townsend, C. R. (1989) The patch dynamics concept of stream community ecology. J. N. Amer. Benthol. Soc. 8:36–50.

Townsend, C. R. and T. A. Crowl. (1991) Fragmented population structure in a native New Zealand fish: an effect of introduced brown trout? Oikos, **61**, 347–54.

Townsend, C. R., A. G. Hildrew, and J. Francis. (1983) Community structure in some southern English streams: the influence of physicochemical factors. Freshwater Biol., **13**, 521–44.

Tramer, E. J. (1978) Catastrophic mortality of stream fishes trapped in shrinking pools. Am. Midl. Nat., **97**, 469–78.

Tramer, E. J. and P. M. Rogers. (1973) Diversity and longitudinal zonation in fish populations of two streams entering a metropolitan area. Am. Midl. Nat., **90**, 366–74.

Trautman, M. B. (1942) Fish distribution and abundance correlated with stream gradients as a consideration in stocking programs. Trans. North Am. Wildl. Conf., **7**, 211–33.

Trautman, M. B. and D. K. Gartman. (1974) Re-evaluation of the effects of man-made modifications on Gordon Creek between 1887 and 1973 and especially as regards its fish fauna. Ohio J. Sci., **74**, 162–73.

Trautman, M. B. (1957) *The Fishes of Ohio*, Ohio State University Press, Columbus, OH.

Travnichek, V. H., M. B. Bain, and M. J. Maceina. (1995) Recovery of a warm-water fish assemblage after the initiation of a minimum-flow release downstream from a hydroelectric dam. Trans. Am. Fish. Soc., **124**, 836–44.

Trenberth, K. E., G. W. Branstator, and P. A. Arkin. (1988) Origins of the 1988 North American drought. Science, **242**, 1640–45.

Trippel, E. A. and F. W. H. Beamish. (1993) Multiple trophic level structuring in *Salvelinus–Coregonus* assemblages in Boreal Forest lakes. Can. J. Fish Aquat. Sci., **50**, 1442–55.

Tuomi, J., T. Hakala, and E. Haukioja. (1983) Alternative concepts of reproductive effort, costs of reproduction, and selection in life-history evolution. Am. Zool., **23**, 25–34.

Turner, C. L. (1921) Food of the common Ohio darters. Ohio J. Sci., **22**, 41–62.

Turner, T. F., J. C. Trexler, G. L. Miller, and K. E. Toyer. (1994) Temporal and spatial dynamics of larval and juvenile fish abundance in a temperate floodplain river. Copeia, **1994**, 174–83.

Uehlinger, U., H. Bührer, and P. Reichert. (1996) Periphyton dynamics in a floodprone prealpine river: evaluation of significant processes by modelling. Freshwater Biol., **36**, 249–63.

Ultsch, G. R., H. Boschung, and M. J. Ross. (1978) Metabolism, critical oxygen tension, and habitat selection in darters (*Etheostoma*). Ecology, **59**, 99–107.

Underhill, J. C. (1986) The fish fauna of the Laurentian Great Lakes, the St. Lawrence lowlands, Newfoundland and Labrador, in (eds. C. H. Hocutt and E. O. Wiley) *The Zoogeography of North American Freshwater Fishes*, John Wiley and Sons, New York, pp. 105–36.

Unger, P. A. and W. M. Lewis, Jr. (1991) Population ecology of a pelagic fish, *Xenomelaniris venezuelae* (Atherinidae), in Lake Valencia, Venezuela. Ecology, **72**, 440–56.

Vadas, R. L., Jr. (1990) The importance of omnivory and predator regulation of prey in freshwater fish assemblages of North America. Environ. Biol. Fish., **27**, 285–302.

Valett, H. Maurice, C. C. Hakenkamp, and A. J. Bolton. (1993) Perspectives on the hyporheic zone: integrating hydrology and biology. Intro. J. North Am. Benthol. Soc., **12**, 40–3.

VanCouvering, J. A. H. (1977) Early records of freshwater fishes in Africa. Copeia, **1977**, 163–66.

Van der Meer, H. J., G. C. Anker, and C. D. N. Barel. (1995) Ecomorphology of retinal structures in zooplanktivorous haplochromine cichlids (Pisces) from Lake Victoria. Environ. Biol. Fish., **44**, 115–32.

Van Oijen, M. P. J. (1982) Ecological differentiation among the piscivorous haplochromine cichlids of Lake Victoria (East Africa). Netherlands J. Zool., **32**, 336–63.

Van Staaden, M. J., R. Huber, L. S. Kaufman, and K. F. Liem. (1994/95) Brain evolution in cichlids of the African Great Lakes: brain and body size, general patterns, and evolutionary trends. Zoology, **98**, 165–78.

Vanni, M. J. (1986) Fish predation and zooplankton demography: indirect effects. Ecology, **67**, 337–54.

Vanni, M. J. and D. L. Findlay. (1990) Trophic cascades and phytoplankton community structure. Ecology, **71**, 921–37.

Vanni, M. J., C. Leucke, J. F. Kitchell, Y. Allen, J. Temte, and J. J. Magnuson. (1990) Effects on lower trophic levels of massive fish mortality. Nature, **344**, 333–35.

Vannote, R. L., G. W. Minshall, K. W. Cummins, J. R. Sedell, and C. E. Cushing. (1980) The river continuum concept. Can. J. Fish. Aquat. Sci., **37**, 130–37.

Vaughn, C. C., F. P. Gelwick, and W. J. Matthews. (1993) Effects of algivorous minnows on production of grazing stream invertebrates. Oikos, **66**, 119–28.

Vaux, P., W. Wurtsbaugh, H. Trevino, L. Marino, E. Bustamante, J. Torres, P. Richerson, and R. Alfaro. (1988) Ecology of the pelagic fishes of Lake Titicaca, Peru-Bolivia. Biotropica, **20**, 220–29.

Venugopal, M. N. and I. J. Winfield. (1993) The distribution of juvenile fishes in a hypereutrophic pond: can macrophytes potentially offer a refuge for zooplankton? J. Freshwater Ecol., **8**, 389–96.

Victor, B. C. and E. B. Brothers. (1982) Age and growth of the fallfish, *Semotilus corparalis* with daily otolith increments as a method of annulus verification. Can. J. Zool., **60**, 2543–50.

Vigg, S. and T. Hassler. (1982) Distribution and relative abundance of fish in Ruth Reservoir, California, in relation to environmental variables. Great Basin Nat., **42**, 529–40.

Vinyard, G. L., R. W. Drenner, M. Gophen, U. Pollinger, D. L. Winkleman, and K. D. Hambright. (1988) an experimental study of the plankton community impacts of two omnivorous filter-feeding cichlids, *Tilapia galilaea* and *Tilapia auera*. Can. J. Fish. Aquat. Sci., **45**, 685–690.

Visman, V., D. J. McQueen, and E. Demers. (1994) Zooplankton spatial patterns in two lakes with contrasting fish community structure. Hydrobiologia, **284**, 177–91.

Vitt, L. J. (1996) Biodiversity of Amazonian lizards, in *Neotropical Biodiversity and Conservation* (ed. A. C. Gibson). University of California, Los Angeles, pp. 89–108.

Vitousek, P. M. (1977) The March 19, 1997 regulation of element concentrations in mountain streams in the northeastern United States. Ecol. Monogr., **47**, 65–87.

Vives, S. P. (1990) Nesting ecology and behaviour of hornyhead chub *Nocomis biguttatus*, a keystone species in Allequash Creek, Wisconsin. Am. Midl. Nat., **124**, 46–56.

Vogel, S. (1981) *Life in Moving Fluids: The Physical Biology of Flow*. Willard Grant Press, Boston, MA.

Vogt, G. F., Jr. and T. G. Coon. (1990) A comparison of the foraging behavior of two darter (*Etheostoma*) species. Copeia, **1990**, 41–9.

Vøllestad, L. A. (1985) Resource partitioning of roach *Rutilus rutilus* and bleak *Alburnus alburnus* in two eutrophic lakes in SE Norway. Holar. Ecol., **8**, 88–92.

Vøllestad, L. A. and R. Andersen. (1985) Resource partitioning of various age groups of brown trout *Salmo trutta* in the littoral zone of Lake Selura, Norway. Arch. Hydrobiol., **105**, 177–85.

Vrijenhoek, R. C., G. Marteinsdottir, and R. A. Schenck. (1987) Genotypic and phenotypic aspects of niche diversification in fishes, in *Community and Evolutionary Ecology of North American Stream Fishes* (eds. W. J. Matthews and D. C. Heins). University of Oklahoma Press, Norman, pp. 245–50.

Wahl, D. H. and R. A. Stein. (1989) Comparative vunerability of three esocids to largemouth bass *Micropterus salmoides* predation. Can. J. Fish. Aquat. Sci., **46**, 2095–103.

Wainwright, P. C. (1987) Biomechanical limits to ecological performance: mollusk crushing by the Caribbean hogfish, *Lachnolaimus maximus* (Labridae). J. Zool. Lond., **213**, 283–97.

Wainwright, P. C. (1989) Prey processing in haemulid fishes: patterns of variation in pharyngeal jaw muscle activity. J. Expt. Biol., **141**, 359–75.

Wainwright, P. C. and G. V. Lauder. (1992) The evolution of feeding biology in sunfishes (Centrarchidae), in *Systematics, Historical Ecology, and North American Freshwater Fishes* (ed. R. L. Mayden). Stanford University Press, Stanford, CA, pp. 472–91.

Wainwright, P. C. and B. A. Richard. (1995) Predicting patterns of prey use from morphology of fishes. Environ. Biol. Fish., **44**, 97–113.

Walch, L. A. and E. P. Bergersen. (1982) Home range and activity patterns of lake trout in central Colorado. Fish. Res., **1**, 311–18.

Wallace, J. B. and A. C. Benke. (1984) Quantification of wood habitat in subtropical coastal plain streams. Can. J. Fish. Aquat. Sci., **41**, 1643–52.

Wallace, R. K., Jr. and J. S. Ramsey. (1981) Reproductive behaviour and biology of the bluestripe shiner (*Notropis callitaenia*) in Uchee Creek, Alabama. Am. Midl. Nat., **106**, 197–200.

Wallin, J. E. (1989) Bluehead chub *Nocomis leptocephalus* nests used by yellow-fin shiners *Notropis lutipinnis*. Copeia, **1989**, 1077–80.

Wallin, J. E. (1992) The symbiotic nest association of yellowfin shiners, *Notropis lutipinnis*, and bluehead chubs, *Nocomis leptocephalus*. Environ. Biol. Fish., **33**, 287–92.

Wallus, R., B. L. Yeager, and T. P. Simon. (1990) *Reproductive Biology and Early History of Fishes in the Ohio River Drainage* (eds. R. Wallus, J. P. Buchanan, H. J. Cathey, G. E. Hall, R. J. Pryor, T. P. Simon, C. T. Swor, and C. W. Voigtlander). Tennessee Valley Authority, Chattanooga, TN, Vol. 1.

Walch, L. A. and E. P. Bergersen. (1982) Home range and activity patterns of lake trout in central Colorado. Fish. Res., **1**, 311–18.

Walsh, G. and G. J. FitzGerald. (1984) Resource utilization and coexistance of three species of sticklebacks Gasterosteidae in tidal salt-marsh pools. J. Fish Biol., **25**, 405–20.

Ward, J. V. (1983) The serial discontinuity concept of lotic ecosystems, in *Dynamics of Lotic Ecosystems* (eds. T. D. Fontaine and S. M. Bartell). Ann Arbor Science, Ann Arbor, MI, pp. 29–42.

Ward, J. V. (1989) The four-dimensional nature of lotic ecosystems. J. North Am. Benthol. Soc., **8**, 2–8.

Ward, J. V. and J. A. Stanford. (1983) The intermediate-disturbance hypothesis: an explanation for biotic diversity patterns in lotic ecosystems, in *Dynamics of Lotic Ecosystems* (eds. T. D. Fontaine III and S. M. Bartell). Ann Arbor Science, Ann Arbor, MI, pp. 347–56.

Ward, J. V. and J. A. Stanford. (1995a) The serial discontinuity concept; extending the model to floodplain rivers. Regul. Rivers: Res. Mangmt., **10**, 159–68.

Ward, J. V. and J. A. Stanford. (1995b) Ecological connectivity in alluvial river ecosystems and its disruption by flow regulation. Regul. Rivers: Res. Mangmt., **11**, 105–19.

Warren, C. E., J. H. Wales, G. E. Davis, and P. Doudoroff. (1964) Trout production in an experimental stream enriched with sucrose. J. Wildl. Mangmt., **28**, 617–60.

Warren, M. A., R. C. Cashner, and R. D. Suttkus. (1994) Fishes of the Buffalo River system, Wilkinson County, southwestern Mississippi. Tulane Studies Zool. Bot., **29**, 127–39.

Waters, T. F. (1988) Fish production-benthos production relationships in trout streams. Polish Archiv. Hydrobiol. 35:545–61.

Waters, T. F. (1995) Sediment in streams. Am. Fish. Soc. Monogr., **7**, 79–118.

Watson, D. J. and E. K. Balon. (1984) Ecomorphological analysis of fish taxocenes in rainforest streams of northern Borneo. J. Fish. Biol., **25**, 371–84.

Weatherley, A. H. (1972) *Growth and Ecology of Fish Populations*. Academic Press, London.

Webb, P. W. (1975) Hydrodynamics and energetics of fish propulsion. Bull. Fish. Res. Bd. Canada **190**, 1–159.

Webb, P. W., C. L. Gerstner, and S. T. Minton. (1996) Station-holding by the mottled sculpin, *Cottus bairdi* (Teleostei: Cottidae), and other fishes. Copeia, **1996**, 488–93.

Weber, J. -M. and D. L. Kramer. (1983) Effects of hypoxia and surface access on growth, mortality, and behavior of juvenile guppies, *Poecilia reticulata*, Can. J. Fish. Aquat. Sci., **40**, 1583–88.

Weddle, G. K. and B. M. Burr. (1991) Fecundity and the dynamics of multiple spawning in darters: an in-stream study of *Etheostoma rafinesquei*. Copeia, **1991**, 419–33.

Weeks, S. C. and O. E. Gaggiotti. (1993) Patterns of offspring size at birth in clonal and sexual strains of *Poeciliopsis (Poeciliidae)*. Copeia, **1993**, 1003–09.

Weihs, D. (1980) Hydrodynamics of suction feeding of fish in motion. J. Fish. Biol., **16**, 425–33.

Weihs, D. (1989) Design features and mechanics of axial locomotion in fish. Am. Zool., **29**, 151–60.

Weitzman, S. H. (1962) The osteology of *Brycon meeki*, a generalized characid fish, with an osteological definition of the family. Stanford Ichthy. Bull., **8**, 1–77.

Welcomme, R. L. (1979) *Fisheries Ecology of Floodplain Rivers*, Longman, London.

Welcomme, R. L. (1985) River fisheries. FAO Fish. Tech. Paper, **262**, 1–330.

Welcomme, R. L. (1988) Concluding remarks I: on the nature of large tropical rivers, floodplains, and future research directions. J. North Am. Benthol. Soc., **7**, 525–26.

Welcomme, R. L. (1995) Relationships between fisheries and the integrity of river systems. Regulated Rivers Research and Management 11:121–36.

Wellborn, G. A. (1994) Size-biased predation and prey life histories: a comparative study of freshwater amphipod populations. Ecology, **75**, 2104–17.

Wellborn, G. A. (1995a) Determinants of reproductive success in freshwater amphipod species that experience different mortality regimes. Animal Behav., **50**, 353–63.

Wellborn, G. A. (1995b) Predator community composition and patterns of variation in life history and morphology among *Hyalella* (Amphipoda) populations in southeast Michigan. Am. Midl. Nat., **133**, 322–32.

Wellborn, G. A. and J. V. Robinson. (1991) The influence of fish predation in an experienced prey community. Can. J. Zool., **69**, 2515–22.

Wellborn, G. A., D. K. Skelly, and E. E. Werner. (1996) Mechanisms creating community structure across a freshwater habitat gradient. Annu. Rev. Ecol. Syst., **27**, 337–63.

Welton, J. S., C. A. Mills, and E. L. Rendle. (1983) Food and habitat partitioning in two small benthic fishes, *Noemacheilus barbatulus* (L.) and *Cottus gobio* L. Arch. Hydrobiol., **97**, 434–57.

Werner, E. E. (1974) The size, prey size, handling time relation in several sunfishes and some implications. J. Fish. Res. Bd. Can., **31**, 1531–36.

Werner, E. E. (1977) Species packing and niche complementarity in three sunfishes. Am. Nat., **3**, 553–78.

Werner, E. E. (1984) The mechanisms of species interactions and community organization in fish, in *Ecological Communities: Conceptual Issues and the Evidence* (eds. D. S. Strong, D. Simborloff, L. G. Able, and A. B. Thistle). Princeton University Press, Princeton, NJ, pp. 360–82.

Werner, E. E. and J. F. Gilliam. (1984) The ontogenetic niche and species interractions in size-structured populations. Annu. Rev. Ecol. Syst., **15**, 393–425.

Werner, E. E. and D. J. Hall. (1974) Optimal foraging and the size selection of the bluegill sunfish *Lepomis macrochirus*. Ecology, **55**, 1042–52.

Werner, E. E. and D. J. Hall. (1976) Niche shifts in sunfishes: experimental evidence and significance. Science, **191**, 404–06.

Werner, E. E. and D. J. Hall. (1977) Competition and habitat shift in two sunfishes (Centrarchidae). Ecology, **58**, 869–76.

Werner, E. E. and D. J. Hall. (1979) Foraging efficiency and habitat switching in competing sunfish. Ecology, **60**, 256–64.

Werner, E. E. and D. J. Hall. (1988) Ontogenetic habitat shifts in bluegill: the foraging rate-predation risk trade-off. Ecology, **69**, 1352–66.

Werner, E. E., D. J. Hall, and M. D. Werner. (1978) Littoral zone fish communities of two Florida lakes and a comparison with Michigan lakes. Environ. Biol. Fish., **3**, 163–72.

Werner, E. E., G. G. Mittlebach, and D. J. Hall. (1981) The role of foraging profitability and experience in habitat use by the bluegill sunfish. Ecology, **62**, 116–25.

Werner, E. E., G. G. Mittlebach, D. J. Hall, and F. J. Gilliam. (1983a) Experimental tests of optimal habitat use in fish: the role of relative habitat profitability. Ecology, **64**, 1525–39.

Werner, E. E., J. F. Gilliam, D. J. Hall, and G. G. Mittlebach. (1983b) An experimental test of the effects of predation risk on habitat use in fish. Ecology, **64**, 1540–48.

Werner, E. E., D. J. Hall, D. R. Laughlin, D. J. Wagner, L. A. Wilsmann, and F. C. Funk. (1977) Habitat partitioning in a freshwater fish community. J. Fish. Res. Bd. Can., **34**, 360–70.

Westneat, M. W. (1994) Transmission of force and velocity in the feeding mechanisms of labrid fishes (Teleostei, Perciformes). Zoomorphology, **114**, 103–18.

Westneat, M. W. (1995) Phylogenetic systematics and biomechanics in ecomorphology. Environ. Biol. Fish., **44**, 263–83.

Wetzel, R. G. (1975) *Limnology* (1st edition). W. B. Saunders, Philadelphia, PA.

Wetzel, R. G. (1983) *Limnology*, 2nd ed. Saunders College Publ., Philadelphia.

Wetzel, R. G. (1995) Death, detritus, and energy flow in aquatic ecosystems. Freshwater Biol., **33**, 83–89.

Wetzel, R. G. and G. Likens. (1991) *Limnological Analyses*, 2nd ed. Springer-Verlag, New York.

Whitaker, J. O., Jr. (1976) Fish community changes at one Vigo County locality over a twelve-year period. Proc. Indiana Acad. Sci. 85:191–207.

White, C. N., L. E. Hightower, and J. R. Schultz. (1994) Variation in heat-shock proteins among species of desert fishes (Poeciliidae, Poeciliopsis). Molec. Biol. Evol., **11**, 106–19.

White, D. S. (1993) Perspectives on defining and delineating hyporheic zones. J. Am. Benthol. Soc., **12**, 61–9.

Whiteside, B. G. and R. M. McNatt. (1972) Fish species diversity in relation to stream order and physicochemical conditions in the Plum Creek Drainage Basin. Am. Midl. Nat., **88**, 90–101.

Whitmore, C. M., C. E. Warren, and P. Doudoroff. (1960) Avoidance reactions of salmonid and centrarchid fishes to low oxygen concentrations. Trans. Am. Fish. Soc., **89**, 17–26.

Whitney, R. J. (1942) Diurnal fluctuations of oxygen and pH in two small ponds and a stream. J. Expt. Biol., **19**, 92–9.

Whittaker, R. H. (1956) Vegetation of the Great Smokey Mountains. Ecol. Monogr., **22**, 1–44.

Whittaker, R. H. (1967) Gradient analysis of vegetation. Biol. Rev., **42**, 207–64.

Whittaker, R. H. (1970) The population structure of vegetation. Reprinted in, *Phytosociology* (ed. R. P. McIntosh), Benchmark Papers in Ecology, Volume 6. Dowden, Hutchinson and Ross, Stroudsburg, PA, pp. 360–80.

Whittaker, R. H. (1972) Evolution and measurement of species diversity. Taxonomy, **21**, 213–51.

Whittier, T. R., R. M. Hughes, and D. P. Larsen. (1988) Correspondence between ecoregions and spatial patterns in stream ecosystems in Oregon. Can. J. fish. Aquat. Sci. 45:1264–78.

Whittier, T. R. and D. L. Miller. (1986) Stream fish communities revisited: a case of mistaken identity. Am. Nat., **128**, 433–37.

Whitton, B. A. (1975) *River Ecology.* University of California Press, Berkeley.

Whoriskey, F. G. and G. J. FitzGerald. (1987) Intraspecific competition in sticklebacks (Gasterosteidae; Pisces): does Mother Nature concur? J. Animal Ecol., **56**, 939–47.

Wickliff, E. L. (1945) Some effects of droughts and floods on stream fish. OH Div. Conser. Nat. Res., **17**, 23–31.

Wiebe, A. H. (1931) Diurnal variations in the amount of dissolved oxygen, alkalinity, and free ammonia in certain fish ponds at Fairport (Iowa). Ohio J. Sci., **31**, 120–26.

Wiener, J. G., P. J. Rago, and J. M. Eilers (eds.). (1984) *Early Biotic Responses to Advancing Lake Acidification.* Butterworths, Boston.

Wiens, J. A. (1977) On competition and variable environments. Am. Sci., **65**, 590–97.

Wiens, J. A. (1984) Conclusions: non-equilibrium, reality, and myths, in *Ecological Communities—Conceptual Issues and the Evidence*, (eds. D. R. Strong, Jr., D. Simberloff, L. G. Abele, and A. B. Thistle). Princeton University Press, Princeton, NJ, pp. 451–57.

Wiens, J. A. and J. T. Rotenberry. (1981) Morphological size ratios and competition in ecological communities. Am. Nat., **117**, 592–99.

Wikramanayake, E. D. (1990) Ecomorphology and biogeography of a tropical stream fish assemblage: evolution of assemblage structure. Ecology, **71**, 1756–64.

Wikramanayake, E. D. and P. B. Moyle. (1989) Ecological structure of tropical fish assemblages in wet-zone streams of Sri Lanka. J. Zool. Lond., **218**, 503–26.

Wiley, E. O. (1976) The phylogeny and biogeography of fossil and recent gars (Actinopterygii: Lepisosteidae). Univ. Kansas Museum Nat. Hist. Misc. Publ., **64**, 1–111.

Wiley, E. O. (1983) *Phylogenetics: The Theory and Practice of Phylogenetic Systematics*. John Wiley and Sons, New York.

Wiley, E. O. (1988a) Vicariance biogeography. Annu. Rev. Ecol. Syst., **19**, 513–42.

Wiley, E. O. (1988b) Parsimony analysis and vicariance biogeography. Syst. Zool., **37**, 271–90.

Wilkie, M. P., P. A. Wright, G. K. Iwama, and C. M. Wood. (1993) The physiological responses of the Lahontan cutthroat trout (*Oncorhynchus clarki henshawi*), a resident of highly alkaline Pyramid Lake (pH 9.4), to challenge at pH 10. J. Expt. Biol., **175**, 173–94.

Williams, C. B. (1964) *Patterns in the Balance of Nature*. Academic Press, New York.

Williams, G. C. (1959) Ovary weights of darters; a test of the alleged association of parental care with reduced fecundity in fishes. Copeia, **1959**, 18–24.

Williams, G. C. (1966) *Adaptation and Natural Selection—A Critique of Some Current Evolutionary Thought*. Princeton University Press, Princeton, NJ.

Williams, G. C. (1967) Identification and seasonal size changes of eggs of the labrid fishes, *Tautogolabrus adspersus* and *Tautoga onitis*, of Long Island Sound. Copeia, **1967**, 452–53.

Williams, J. E. and C. E. Bond. (1983) *Status and Life History Notes on the Native Fishes of the Alvord Basin, Oregon and Nevada*. Department of Fisheries and Wildlife, Oregon State University, Corvalis, OR, pp. 409–20.

Wilson, D. S. (1975) The adequacy of body size as a niche difference. Am. Nat., **109**, 769–84.

Wilson, M. V. H. (1992) Importance for phylogeny of single and multiple stem-group fossil species with examples from freshwater fishes. Syst. Biol., **41**, 462–70.

Wilson, M. V. H. and R. R. G. Williams. (1992) Phylogenetic biogeographic, and ecological significance of early fossil records of North American teleostean fishes in *Systematics, Historical Ecology and North American Freshwater Fishes* (ed. R. L. Mayden). Stanford University Press, Stanford, CA, pp. 224–44.

Wilson, S. and C. Hubbs. (1972) Developmental rates and tolerances of the plains killifish, *Fundulus kansae*, and comparison with related fishes. TX J. Sci., **XXIII**, 371–379.

Wilzbach, M. A. (1985) Relative roles of food abundance and cover in determining habitat distribution of stream-dwelling cutthroat trout (*Salmo clarki*). Can. J. Fish. Aquat. Sci., **42**, 1668–72.

Winemiller, K. O. (1989) Ontogenetic diet shifts and resource partitioning among piscivorous fishes in the Venezuelan llanos. Environ. Biol. Fish., **26**, 177–99.

Winemiller, K. O. (1990) Spatial and temporal variation in tropical fish trophic networks. Ecol. Monogr., **60**, 331–67.

Winemiller, K. O. (1991a) Comparative ecology of Serranochromis species (Teleostei: Cichlidae) in the upper Zambezi River floodplain. J. Fish. Biol., **39**, 617–39.

Winemiller, K. O. (1991b) Ecomorphological diversification in lowland freshwater fish assemblages from five biotic regions. Ecol. Monogr., **61**, 343–65.

Winemiller, K. O. (1992a) Ecomorphology of freshwater fishes. Nat. Geogr. Res. Explor., **8**, 308–27.

Winemiller, K. O. (1992b) Life-history strategies and the effectiveness of sexual selection. Oikos, **63**, 318–27.

Winemiller, K. O. (1993) Seasonality of reproduction by livebearing fishes in tropical rainforest streams. Oecologia, **95**, 266–76.

Winemiller, K. O. (1995) Factors driving temporal and spatial variation in aquatic floodplain food webs, in *Food Webs* (eds. G. A. Polis and K. O. Winemiller). Chapman & Hall, New York, pp. 298–312.

Winemiller, K. O. and M. A. Leslie. (1992) Fish assemblages across a complex tropical freshwater/marine ecotone. Environ. Biol. Fish., **34**, 29–50.

Winemiller, K. O. and G. A. Polis. (1995) Food webs: what can they tell us about the world? in *Food Webs* (eds. G. A. Polis and K. O. Winemiller). Chapman & Hall, New York, pp. 1–22.

Winemiller, K. O. and K. A. Rose. (1992) Patterns of life-history diversification in North American fishes: implications for population regulation. Can. J. Fish. Aquat. Sci., **49**, 2196–218.

Winemiller, K. O. and R. A. Rose. (1993) Why do most fish produce so many tiny offspring? Am. Nat., **142**, 585–603.

Winemiller, K. O. and D. H. Taylor. (1982) Smallmouth bass nesting behaviour and nest site selection in a small Ohio stream. Ohio J. Sci., **82**, 266–73.

Winemiller, K. O., L. C. Kelso-Winemiller, and A. L. Brenkert. (1995) Ecomorphological diversification and convergence in fluvial cichlid fishes. Environ. Biol., **44**, 235–61.

Winfield, I. J. (1991) Fishes, waterfowl and eutrophied ecosystems: a perspective from a European vertebrate ecologist. Mem. Ist. Ital. Idrobiol., **48**, 113–26.

Winfield, I. J. and C. R. Townsend. (1991) The role of cyprinids in ecosystems, in *Cyprinid Fishes—Systematics, Biology and Exploitation* (eds. I. J. Winfield and J. S. Nelson). Chapman & Hall, London, pp. 552–71.

Winkler, P. (1979) Thermal preference of *Gambusia affinis affinis* as determined under field and laboratory conditions. Copeia, **1979**, 60–4.

Winklemann, D. L. (1996) Reproduction under predatory threat: trade-offs between nest guarding and predator avoidance in male dollar sunfish. Copeia, **1996**, 845–51.

Winkleman, D. L. and J. M. Aho. (1993) Direct and indirect effects of predation on mosquitofish behavior and survival. Oecologia, **96**, 300–03.

Winn, H. E. (1958) Comparative reproductive behavior and ecology of fourteen species of darters (Pisces-Percidae). Ecol. Monogr., **28**, 155–91.

Winn, H. E. (1958) Observations on the reproductive habits of darters (*Pisces–Percidae*). Am. Midl. Nat., **59**, 190–211.

Winston, M. R. (1995) Co-occurrence of morphologically similar species of stream fishes. Am. Nat., **145**, 527–45.

Winston, M. R., C. M. Taylor, and J. Pigg. (1991) Upstream extirpation of four minnow species due to damming of a prairie stream. Trans. Am. Fish. Soc., **120**, 98–105.

Wiseman, S. W., S. D. Cooper, and T. L. Dudley. (1993) The effects of trout on epibenthic odonate naiads in stream pools. Freshwater Biol., **30**, 133–45.

Wismer, D. A., D. L. DeAngelis, and B. J. Shuter. (1985) An empirical model of size distributions of smallmouth bass. Trans. Am. Fish, Soc., **114**, 737–42.

Witte, F. (1984) Ecological differentiation in Lake Victoria haplochromines: comparison of cichlid species flocks in African lakes, in *Evolution of Fish Species Flocks* (eds. A. A. Echelle and I. Kornfeld). University of Maine Press, Orono, pp. 155–67.

Witte, F., T. Goldschmidt, J. H. Wanink, M. J. P. Van Oiveh, P. C, Goudswaard, E. L. M. Whitte-Mass, and N. Bouton. (1992) The destruction on an endemic species flock: quantitative data on the decline of the haplochromine species of the Mwanza Gulf of Lake Victoria. Environ. Biol. Fish., **34**, 1–8.

Wolda, H. (1981) Similarity indices, sample size, and diversity. Oecologia, **50**, 296–302.

Wood, B. M. and M. B. Bain. (1995) Morphology and microhabitat use in stream fish. Can. J. Fish. Aquat. Sci., **52**, 1487–98.

Woodward, B. D. (1983) Predator-prey interactions and breeding-pond use of temporary-pond species in a desert anuran community. Ecology, **64**, 1549–55.

Wootton, J. T. and M. P. Oemke. (1992) Latitudinal differences in fish community trophic structure, and the role of fish herbivory in a Costa Rican stream. Environ. Biol. Fish., **35**, 311–19.

Wootton, J. T. and M. E. Power. (1993) Productivity, consumers, and the structure of a river food chain. Proc. Nat. Acad. Sci. USA, **90**, 1384–87.

Wootton, J. T., M. S. Parker, and M. E. Power. (1996) Effects of disturbance on river food webs. Science, **273**, 1558–61.

Wootton, R. J. (1984) Introduction: tactics and strategies in fish reproduction, in *Fish Reproduction: Strategies and Tactics*, (eds. G. W. Potts and R. J. Wootton). Academic Press, London, pp. 1–12.

Wootton, R. J. (1990) *Ecology of Teleost Fishes*. Chapman & Hall, London.

Wootton, R. J. (1992) Constraints in the evolution of fish life histories. Netherlands J. Zool., **42**, 291–303.

Work, K. A. (1997) The ecology of the exotic cladoceran, *Daphnia lumholtzi* (Sars), in Lake Texoma, Oklahoma–Texas. Ph.D. dissertation, University of Oklahoma, Norman.

Work, K. A. and M. Gophen. (1995) The invasion of *Daphnia lumholtzi* (Sars) into Lake Texoma (USA). Arch. Hydrobiol., **133**, 287–303.

Wright, D., W. J. O'Brien, and G. L. Vinyard. (1980) Adaptive value of vertical migration: a simulation model argument for the predation hypothesis, in *Evolution and Ecology of Zooplankton Communities* (ed. W. C. Kerfoot). University Press of New England, Hanover, NH.

Wright, D. I. and W. J. O'Brien. (1982) The development and field test of a tactical model of the planktivorous feeding of white crappie (*Pomoxis annularis*). Ecol. Monogr., **54**, 65–98.

Wright, R. F. (1976) The impact of forest fire on the nutrient influxes to small lakes in northeastern Minnesota. Ecology, **57**, 649–63.

Wuerthrich, B. (1996) Deliberate flood renews habitats. Science, **272**, 344–45.

Wurtsbaugh, W. and H. Li. (1985) Diel migration of a zooplanktivorous fish (*Menidia beryllina*) in relation to the distribution of its prey in a large eutrophic lake. Limnol. Oceanogr., **30**, 565–75.

Wynes, D. L. and T. E. Wissing. (1982) Resource sharing among darters in an Ohio stream. Am. Midl. Nat., **107**, 294–304.

Yanagisawa, Y. (1986) Parental care in a monogamous mouthbrooding cichlid *Xenotilapia flavipinnis* in Lake Tanganyika. Jpn. J. Ichthyol., **33**, 249–61.

Yanagisawa, Y. (1987) social organization of a polygynous cichlid *Lamprologus furcifer* in Lake Tanganyika. Jpn. J. Ichthyol., **34**, 82–90.

Yant, P. R., J. A. Karr, and P. L. Angermeier. (1984) Stochasticity in stream fish communities: an alternative interpretation. Am. Nat., **124**, 573–82.

Yen, M. D. (1985) Species composition and distribution of the freshwater fish fauna of the north of Vietnam. Hydrobiologia, **121**, 281–86.

Young, M. K. (1996) Summer movements and habitat use by Colorado River cutthroat trout (*Oncorhynchus clarki pleuricus*) in small, montane streams. Can. J. Fish. Aquat. Sci., **53**, 1403–08.

Yule, D. L. and C. Luecke. (1993) Lake trout consumption and recent changes in the fish assemblage of Flaming Gorge Reservoir. Trans. Am. Fish. Soc., **122**, 1058–69.

Zakaria-Ismail, M. (1994) Zoogeography and biodiversity of the freshwater fishes of southeast Asia. Hydrobiologia, **285**, 41–8.

Zalewski, M. and R. J. Naiman. (1985) The regulation of riverine fish communities by a continuum of abiotic-biotic factors, in *Habitat Modification and Freshwater Fisheries* (ed. J. S. Alabaster). Butterworths, London, pp. 3–9.

Zalewski, M., P. Frankiewicz, M. Przybylski, J. Banbura, and M. Nowak. (1990) Structure and dynamics of fish communities in temperate rivers in relation to the abiotic-biotic regulatory continuum concept. Pol. Arch. Hydrobiol., **37**, 151–76.

Zar, J. H. (1984) *Biostatistical Analysis* (2nd Edition). Prentice-Hall, Englewood Cliffs, NJ.

Zaret, T. M. (1979) Predation in freshwater fish communities, in *Predator-Prey Systems in Fisheries Management* (ed. H. Klepper). Sport Fishing Institute, Washington, DC, pp. 135–43.

Zaret, T. M. and A. S. Rand. (1971) Competition in tropical stream fishes: support for the competitive exclusion principle. Ecology, **52**, 336–42.

Zimmerman, E. G. (1987) Relationships between genetic parameters and life-history characteristics of stream fish, in *Community and Evolutionary Ecology of North American Stream Fishes* (eds. W. J. Matthews and D. C. Heins). University of Oklahoma Press, Norman, OK, pp. 239–44.

Zimmerman, E. G. and M. C. Richmond. (1981) Increased heterozygosity at the MDH-B locus in fish inhabitating a rapidly fluctuating thermal environment. Trans. Am. Fish. Soc., **110**, 410–16.

Zimmerman, L. C., E. A. Standora, and J. R. Spotila. (1989) Behavioural thermoregulation of largemouth bass (*Micropterus salmoides*): response of naive fish to the thermal gradient in a nuclear reactor cooling reservoir. J. Therm. Biol., **14**, 123–32.

Subject Index

Abiotic factors, 15, 264, 318, 604
Abundance, 32, 73–77, 104
  hollow-curve, 73
Accessory organs, 359
Adaptation, 373, 385, 395, 412, 420
Age, first reproduction, 425
Age-classes, 502
Air breathing, 359
Algae, 14, 22, 23, 68, 145, 349, 504, 551, 587, 588, 596, 605, 611–616
  as shelter, 348, 351, 416, 551, 552
  eating, 19, 144, 415, 499, 551, 583–589, 611–616
Algivorous fishes, 19, 23, 281, 490, 499, 583–589, 611–616
Algivory, 144, 415, 499, 551, 583–589, 611–616
Alkalinity, 140
Allen's paradox, 567, 576
Amphibians, 575
Amphidromy, 453
Amphipods, 573
Anadromy, 453
Ancestral Plains Stream, North America, 172
Anoxia, 136, 138, 155, 161, 273, 292, 345, 349, 351, 540
Anoxic factor, 358
Anthropogenic effects, 15, 19, 37, 157, 162, 188–189, 370, 417
Aquatic surface respiration (ASR), 140, 359, 395
Area, 264–268
Artificial streams, 23, 45, 504, 518, 544, 547–548, 575, 587, 588, 600
Aspect ratio, 388, 399

Assemblage, 2, 93
  composition, 15, 18, 109
  local, 191, 219, 255, 377, 534, 541, 564
  of fishes, 103, 131
  structure, 12–15, 30–85, 105, 264, 319
Assemblages
  discrete, 86–104, 366
  overlapping, 86–104
  plant, 87
Assembly rules, 262
Aufwuchs, 330
Autecology, 15, 18, 381

Bacteria, 22
Barriers, 45
  to movement, 191, 192, 304, 545, 560
Base flow, 185
Bayou, 200
Behavior, 19
  foraging, 546–549, 552–557
  reproductive, 80
Benthic fishes, 105, 333, 335, 393–395, 405
Bernoulli effect, 382
Bicarbonate buffer system, 140
Big river fauna, 31, 269
Big rivers, 151
Bigger-deeper hypothesis, 542
Biodiversity, 1, 33, 315
Bioenergetics, 19, 414, 415, 468, 504, 602
Bioengineers, 566
Biofilm, 68, 592
Biogeography
  descriptive, 245
  ecological, 245
  global, 192

733

historical, 18, 245–249
interpretive, 245
regional, 192
subcontinental, 192
Biomanipulation, 602, 602–610
Biomechanics, 382, 385
Biotic interactions, 15, 604
Bioturbation, 593
Body, 384
caudal peduncle, 391
depth, 384, 391
form, 7
size, 10, 15, 32, 77–81, 260, 296, 314, 389, 441
Body shape, 10, 387, 390, 391
Bogs, 149
Bohr shift, 142, 360
Bottom-up control, 19, 60, 133, 496, 578, 600, 602
Boundary-layer, 382
Brain, 409
Breaks, in fish faunas, 309
Broken-stick model, 74
Burst speed, 390

Calcium carbonate, 140
Cannibalism, 411, 441, 569, 603
Canonical correspondence analysis (CCA), 102
Carbon dioxide, 135
Carcasses, of fish, 135
Carolina Bays, 155
Carrying capacity, 461
Cascading effects, 19, 22, 551, 566, 577, 594, 595, 597–607, 607–610, 612
Catadromy, 453
Cenozoic, the, 220
Central Highlands biogeographic track, 251
Change, long-term, 106, 107
Channel
formation, 326, 331
morphometry, 182–185, 188
Chemoreception, 580
Clades, 215
monophyletic, 27
paraphyletic, 27
polyphyletic, 27
Classification, 26
Climate, 133
Climate change, 14, 19
Cluster analysis, 91, 195, 315, 339

Clutch
parameters, 432–433
volume, 433
Coefficient of variation (CV), 113
Coevolution, 24, 225, 235, 248, 457, 487, 525–531, 580, 589
coupled, 526, 535
diffuse, 25, 526
stepwise, 25
Cold, 357, 415
Colonization, 11, 155
postglacial, 157
Come-and-go species, 104, 110
Community, 2
abiotic vs. biotic control, 319–320
Clementsian, 87, 93, 103, 457, 497
deterministic, 108
equilibrium, 108, 455
Gleasonian, 87, 93, 103, 457, 497
stability, 104
stochastic, 108
structure, 30–85
Community importance (CI), 610
Community unit, 86, 87, 93
Competition, 19, 87, 88, 91, 153, 192, 278, 455–497, 530, 534, 602, 615
among age-classes, 507–510
asymmetrical, 505, 511
interspecific, 320, 455–497, 558
intraspecific, 464, 497–507, 558
Competitive exclusion, 89, 456
Continental movements, 12, 14, 24, 194, 202
Continuum, of stream fishes, 305–307
Cordilleran Ice Sheet, 232
Core-satellite hypothesis, 90
Correspondence Analysis (CA), 103
Cost curve, 472
Cover, 279, 283, 504
Coves, 292–296
Crayfish, 23, 550, 551, 615
Cretaceous, the, 205, 206, 208, 212, 215–217, 221, 222
Critical Thermal Maximum (CTM), 355, 371
Crunch, ecological, 456–457, 463
Current speed, 182,
critical, 394
Currents, water, 314
Cyprinid origins, 213–219

D-link, 290, 311, 312
Dams, 188, 417

Defaunation, 108
Delta formation, 295
Density, 133
  of fishes, 297, 495
Density-dependent effects, 104, 499–502
Depositional zone, 149, 182
Depth, water, 91, 271–274
Deserts, 38, 283, 343, 346, 538, 585
Detrended Correspondence Analysis (DCA), 95, 99, 103, 105, 116, 121, 316–317
Detritivorous fishes, 566, 584, 585, 593, 616
Detritivory, 22, 584, 602
Detritus, 56, 68, 504, 585, 615
Development, 437
Diadromy, 453–454
Diel patterns, 31
Diet overlap, intraspecific, 502–503, 506
Direct effects of fish in ecosystems, 567–575
Directional affinities, 249–254
Discharge, 276–278, 435
Discriminant Functions Analysis (DFA), 91, 316
Dispersal, 24
Displacement, 467, 469, 473, 512
  habitat, 466–469
Dissolved organic carbon, 22
Distribution
  African-Oriental, 210
  ancient families, 212
  breaks in, 88, 91
  broad tropical, 211
  endemics, 100
  Eurasian-Oriental, 209
  Holarctic, 208
  Nearctic-Neotropical, 211
  Neotropical-African, 210
  South Circumpolar, 212
  Transatlantic, 209
  Transpacific, 209
Distributional limits (DL), 306
Distributions
  of families, worldwide, 207
  multivariate analyses of, 315–317
Disturbance, 19, 313, 318, 320–326, 377–379, 456, 593, 603
  definitions of, 320–321
  mechanical, 68
  press, 322
  pulse, 322
  time scales of, 323–326
Diversity, 2, 32, 35, 36
  continental, 35
  morphological, 7, 46
  of fishes, 2–7
  regional, 36
Dominance, 503–505
Downstream link (d-link), 290, 311, 312
Drag, 381–384
  friction, 382, 388
  pressure, 382, 383, 389
Drainage, 173, 191, 196
  area, 264–268
  classification, Jenkins, 175–180
  density, 174, 176
  formation, 24, 172–174
  networks, 14
Drainage patterns, 175–180
  dendritic, 175
  trellised, 175
Drift, 45
  invertebrate, 45, 413, 494, 567, 569, 612
  larval fish, 559
Drift feeding, 58, 314, 468
Drought, 14, 19, 301, 320, 321, 341–353, 377–379, 603
  effects on fish, 105, 108, 111
  historic record, 341–342
  phases of, 349–353
  severity, 109, 130
Dystrophy, 159

Early life history, 437–439
Ecological traits, 25
Ecomorphology, 385, 386, 403–410
Ecoregions, 192, 254–255
Ecospecies, 55, 80, 500, 513
Ecosystem engineering, 593–594
Ecosystems, 12, 19, 60, 343
  fish effects in, 12, 70, 565–616
  stream, 171–172
Egg, 422, 522, 557
  number of, 433
  predation, 538
  size, 360, 433–435
Electromyography, 401
Energetics, 15
Environment, effects on distribution, 315, 317
Eocene, the, 206, 209, 212, 217, 223
Epilimnion, 292
Epilimnion/hypolimnion ratio, 158, 273
Erosional zone, 149, 182
Eutrophication, 157–159, 600, 608

Evenness, 48, 49
Evolution, 10, 18, 155, 412
Evolutionary past
  deep, 12, 18, 24, 45, 46, 192, 525
  recent, 24, 46
Evolutionary stable stragies (ESS), 424, 432
Exotic species, 531, 541–542, 560, 605
Experiments, 23, 444, 464
  competition, 464–465, 467–468, 470, 476–477, 490, 496, 506
  manipulative, 465, 485, 490
  natural, 480, 485, 490
  predator-prey, 542–543, 547–553, 581
Explanations
  ecological, 24
  evolutionary, 24
  proximate, 24
  ultimate, 24
Extinction, 265
Eyes, 395

Factor analysis, 315
Families, 32, 45, 48, 406
  distributions worldwide, 207
  of fishes, 3, 6, 10, 24, 194, 196–197, 225
Fauna, regional, 379
Faunal provinces, fish, 89
Faunal screens, 192
Feces, 72, 591, 597, 615
Fecundity, 433
Feeding, 9, 19, 54, 325, 530
  benthic, 8
  particulate, 68, 576
  pump filter, 576
  specialists, 54
Female mate choice, 430
Fetch, 152, 164, 166
Filter, faunal, 15, 192, 347, 365, 379
Filter feeder, 576, 580, 582
Fineness ratio, 369
Fins, 7, 10, 384, 391–394
  anal, 392
  as keels, 392
  as rudders, 392
  as stabilizers, 392
  caudal, 388, 390, 393, 395
  dorsal, 392–393
  pectoral, 387, 393, 395, 593
  pelvic, 393, 395, 593
Fish effects
  in ecosystems, 19, 66, 68, 565–616

  in food webs, 22
  on algae, 22, 23, 577, 584, 587–588, 611–616
  on bacteria, 588
  on crayfish, 574, 588
  on life history traits, 573–574
  on macroinvertebrates, 567–574, 584, 588, 612
  on nutrients, 22, 566, 576, 581–582
  on other vertebrates, 587–588, 590–593, 601
  on phytoplankton, 577, 581–583
  on POM, 22, 587–588, 596, 615
  on primary production, 582, 584, 587
  on production, 576, 582
  on vascular plants, 589–590
  on water quality, 582
  on zooplankton, 19
  second-order, 594–597
Fish faunal provinces
  of Africa, 240
  Arctic Archipelago, 236, 244
  Cascadia, 231, 234, 236
  Central Appalachian, 236
  Central Asia-Afganistan, 238
  Central Mexican, 237, 244
  Colorado, 237, 244
  of Eurasia, 238–239
  Great Basin-West Coast, 237
  Great Lakes/St. Lawrence, 236
  Hudson Bay, 236, 244
  of India, 238
  Mississippi, 236
  Northern Appalachian, 236
  Rio Grande, 237, 244
  of Russia, 238
  Sonoran-Sinaloan, 237
  of South America, 240
  of Southeast Asia, 239
  Southeastern/Gulf Slope, 237
  Western Gulf Slope, 237
  Yukon-Mackenzie, 234, 236, 244
Fish faunal regions, 89, 195
Fish zones, 289–300, 303–305
  barbel, 303
  bream, 303
  grayling, 303
  trout, 303
Flood pulse, 328
Flood-exploitative species, 378, 450
Flooded forest, 589
Floodplain, 42, 278, 303, 326, 327, 378, 589

Floods, 14, 19, 42, 185–188, 320–321, 325–326, 377–379, 481, 495, 603
  effects on fish, 42, 105, 108, 110–112, 277, 301, 326–340, 416, 437
  erosive, 126, 130, 131, 326, 327, 334, 378
  long-term effects of, 331–333, 339
  non-erosive, 327
Flow, 18, 91, 311
Fluid dynamics, 382
Food availability, 14
Food webs, 22, 63, 278, 328, 343, 564–566, 593–595, 602, 604–610
  kinds of, 602
Foraging, 476, 554
  optimal, 417
Foreshortening (of fin rays), 387, 394
Fossils, 24, 214, 215, 225, 246, 265
  minimum age, 214, 215
Freshwater bridges, postglacial, 202
Freshwater dispersant, 201
Freshwater fishes
  primary, 194, 195, 197–202
  secondary, 194, 195, 197–202
Fretwell-Lucas model, 500
Functional groups, 55, 66–73
Fuselage, 7, 384, 388

Gape width, 81
Generalists, 278
Generalized tracks, 245
Geology, 14, 18, 133, 140, 191, 269, 312
Ghost of competition past, 457
Gill-raker, 510, 582
  retention of particles, 582
Glacial lakes, 11, 155, 228
Glacial periods, 227–233
  Kansan, 229
  Nebraskan, 229
  Wisconsinian, 227, 231
Glaciation, 18
  Pleistocene, 24, 227, 228, 232
Global warming, 469
Glochidia, 572
Grazers, 577
Grazing fishes, 67, 584
Grazing scars, 586
Groundwater, 181
Growth, 437–440, 441, 502, 545, 557, 563
  indeterminate, 497
Guarding, 436, 522

Guilds, 54, 55, 519, 520
Gut, 10, 408

Habitat, 10, 11, 88, 89, 255–260, 456, 593, 595
  complementation, 314
  complexity, 18, 265, 279
  connectivity of units, 287–289
  diversity, 37, 283
  heterogeneity, 272, 279, 283, 535
  neighborhoods, 314
  patches, 32, 166, 314, 418
  preference, 18
  selection, 91, 137, 146–149, 162, 487
  size, 36, 38, 268–278
  structure, 18, 131
  supplementation, 314
  volume, 274–276
  zonation, 18
Habitat use, 417, 502, 527
  predator effect on, 542–544
Habitats
  backwater, 145, 293, 416–417
  benthic, 7
  canos, 298
  cove, 293–296
  lagoons, 298
  lentic, 42, 149–151
  lotic, 42, 149–151
  sand ripples, 413
  sand-ridge, 413
  water column, 7, 137
Harsh environment, 272, 341, 353–354, 365, 373
Harsh-to-benign hypothesis, 115, 368
Head, 384, 396
Headcutting; of stream channels, 180–181
Headwater fauna, 301
Headwaters (of streams), 11, 35, 39, 259, 278, 296, 306, 348, 359, 369
Heat, 136–137
Heat budget, 136
Heat hardening, 356
Heat-shock proteins (HSP), 143, 355–357
Herbivory, 19, 22
Herbivorous fishes, 566, 589
Hierarchies, of individuals, 503–504
Historical biogeography, 245–249
Historical ecology, 191, 192, 245, 247
Holocene, the, 220
Home range, 314, 448–450

Horton-Strahler system for stream order, 175, 307, 310–312
Hotspots, 279
HSS hypothesis, 597
Hydraulic heterogeneity, 284
Hydraulic scour, 437
Hydrodynamics, 7, 9, 18, 384, 390, 394, 412, 435, 437, 507
Hydrofoils, 395
Hydrograph, 185
Hypolimnion, 138, 292
Hyporheic zone, 172, 343, 569
Hypoxia, 139–140, 345–346
Hypsographic curves, 170

Ice, 133–134, 160, 345–346
Ideal Free Distribution (IFD), 375, 417, 418, 497, 500, 501, 547
Indirect effects, of fish, 566, 594–610
Individualism, 86, 88, 89, 90, 93, 149, 380, 526
Infiltration rates, 182
Information transfer, 514
Insectivory, 22, 144, 584, 599
Interactive segregation, 457, 458, 472–474, 485, 491
Interbrood interval, 443
Interglacial periods, 230
Intermediate disturbance hypothesis, 322, 328, 457, 534
Invasion, 15
Invertebrates, 14, 22, 63, 67, 83, 142, 281, 323, 330, 343, 346, 414, 496, 567, 569, 570, 593, 596, 599, 612, 615
Island
  floating, 448
  size, 265
Iteroparity, 421, 427–429

Jaccard's index, 116, 119, 121, 125, 195
Jaws, 7, 8, 396–399
  crushing, 8, 398
  evolution of, 397–399
  maxilla, 397
  premaxilla, 397
  protrusable, 397, 400
Jurassic, the, 221
Juveniles, in assemblages, 495

Karst regions, 181
Kendall's W, 111

Kettle lakes, 32, 156, 228
Keystone species, 19, 278, 457, 534, 564, 565, 609–611

Lagoons, 43, 298
Lake formation, 131, 152–157
Lakes
  dimictic, 160
  drainage, 266
  eutrophic, 92, 157
  fault block, 273
  fluviatile, 155
  glacial, 155–157
  kettle, 156, 228
  landslide, 155
  meromictic, 160
  moraine, 228
  oligotrophic, 92, 157
  oxbow, 131, 155
  seepage, 266
  tropical, 35
  warm monomictic, 134, 160, 604
Land bridges, 15, 192, 194, 206, 212, 228
Landscape, 328
Landscape ecology, 287, 312–315
Langmuir circulation, 163
Larval fishes, 151, 166, 277, 295, 324, 330, 416–417, 433–434, 580–581
Larval success, 19, 433–434
Life history, 10, 80, 420–448, 603
  alternative, 431–432
  bet-hedging, 427–428
  equilibrium, 423
  heritability, 444–445
  opportunistic, 423
  optimality, 424
  periodic, 423
  trade-offs, 423–424, 439, 443–444
Life history strategies, 423, 442, 445
Life history traits, 10, 296, 419–424, 443
  intraspecific variation, 434–435, 439–442
Life table, 420–421
Light, 135–136, 323, 553, 600
Limnetic zone, 31, 290
Link number, 290
Little Ice Age, 231
Littoral zone, 31, 274, 290, 477, 554
Locality, 30
Longitudinal zonation, 296–307, 369
Lotka-Volterra, 278, 456

Macroecology, 15, 22, 255, 260
Macrophytes, 138, 169, 279, 282, 292, 329, 330, 585, 589, 592
Male mate choice, 430
Male-male competition, 430
Mantel test, 243, 339, 408
Marginal Value Theorem, 418
Mate choice, 430–431
Materials
　allochthonous, 135
　autochthonous, 135
Mating systems, 429–432
Medieval Warm Period, 231
Mesozoic, the, 220
Metalimnion, 292
Metapopulations, 328
Microhabitat, 7, 9, 18, 32, 45, 91, 149, 174, 299, 310, 316, 331–332, 367, 375, 413, 415, 465, 475, 490–491, 493, 507, 513, 517, 519–520, 530, 554
Microhydraulics, 14, 185, 412, 486
Microlayer, of flowing water, 413
Micropiscivory, 538
Migration, 11, 105, 274, 314, 342, 346, 347, 448–454, 590, 592
　vertical, 511
Miocene, the, 207, 212, 215, 218, 219, 221, 223, 224
Mixed-species effects, 512–525
Mixed-species, number of species, 524–525
Mollusk-crushing, 402, 570
Monophyly, 26
Morphoedaphic index (MEI), 22, 170, 273–274
Morphology, 7, 24, 25, 45, 380, 384–412
　descriptive, 385, 386–400
　functional, 10, 385, 400–403
　intraspecific variation in, 410–411
　trophic, 396–403
Morphometry, of lakes, 157, 166–171
Mortality, 446–447, 495, 554
　curves, 19, 446–447
Mouth, 7, 8, 81, 395, 437
　position, 7, 395, 408
Movement, 30, 31, 314, 325, 375, 448–454, 560
　among lakes, 11
Movements
　daily, 31, 450–451, 591
　local, 375, 450–451, 560
　seasonal, 451–453

Mud-eaters, 565
Multivariate
　analysis, 90, 91
　space, 93–104, 116, 122–126
Mutualism, 19, 90, 192, 263

Nest-building, 69, 324, 436, 594
Nestedness, 88
Nests, 90, 263, 323, 330, 437, 438, 445, 522
Newark Supergroup of lakes, 153
Niche, 23, 89, 455–459
　food, 456, 490–491
　fundamental, 456
　habitat, 456
　overlap, 459, 464, 482, 487, 493
　realized, 456
　segregation, 458–460, 478–479, 491
　thermal, 137, 479
　time, 480
Niche preemption model, 52, 53
Nickpoint migration, 105, 302
Nitrogen budget, 585
Null models, 25, 93,
Nutrient, 14, 68, 72, 169, 327, 433, 566, 578, 600–601, 604–605, 607
　availability, 14, 19, 157, 159, 585, 605
　cycling, 22, 585, 590
　transport of, 20, 22, 566–567, 590–591

Oligocene, the, 206
Omnivores, 581
Omnivory, 144, 603
Ontogeny, 54, 55, 69, 111, 437, 498, 502, 507, 509, 576, 597, 602
Optic nerve, 403
Optimal foraging theory, 419
Organs of attachment, 387, 395
Orogeny, 14, 133
Ostariophysan origins, 213–219
Oxygen, 15, 92, 138–140, 158–159
　depletion, 161, 162
　dissolved, 144, 145, 345, 349, 351, 357–360, 540
　effects of low, 140
　tolerance, 371–372
Ozone, 135

Paleocene, the, 215, 216, 222
Parental care, 436–437
Parental investment, 435
Particulate organic matter (POM), 22, 68, 615

Patch, 313
Patch choice, 417–419
Patchiness, 584, 588
Percent organic matter, 159
Percent Similarity Index (PSI), 113, 116, 117, 119, 120, 125
Periphyton, 19, 58, 329, 343, 344, 585, 586, 589, 611–616
Persistence, 104
pH, 11, 92, 140–142, 145, 159, 360–361, 540, 601
Pharyngeal jaw apparatus (PJA), 384, 396, 399, 402
Pharyngeal teeth, 55, 399, 570, 582
Phosphorus, 590–592
Photosynthesis, 138, 141, 144, 145, 147, 162, 359, 361
Photosynthetically active radiation (PAR), 136
Phylogenetic baggage, 9, 483
Phylogeny, 25, 27, 386, 402, 412, 420, 441, 483–484
  effect on tolerance, 367, 370–374
Physicochemical selectivity, 361
Physicochemical stress, 60, 353–354, 377–379
Physicochemical tolerance, 14, 60, 530
Physiographic succession, 302
Physiological tolerance, 19, 353–357, 359
Phytoplankton, 138, 292, 566, 600–601, 604–610
Piscivore-prey ratios, 60–66
Piscivores, 60, 62, 92, 273, 342, 358, 419, 495, 534, 589, 604–610
  avian, 536, 542, 548, 563, 584
  effects in lakes, 92, 539–542, 604, 609
  effects in streams, 495, 536–539, 542–549, 559–563
Piscivory, 23, 56, 349, 532–547
Planktivores, 604–610
Planktivory, 68, 144, 169, 575–581
Playas, 11, 149
Pleistocene, the, 15, 25, 152, 157, 172, 173, 218, 227, 228, 230, 235
Pliocene, the, 172, 173, 209, 210
Plunge point, 160
Poisson distribution, 95–100
Polar coordinates analysis, 316
Pollution, 359
Ponds, 11, 43
Pool development, 18, 60, 91, 272–273
Pools, 31, 44, 57, 288, 309, 314, 344, 349, 418, 543–544

Population
  fragmentation, 559
  size-structured, 78, 497, 554
Position-holding, 387, 394, 414, 468
Postglacial colonization, 24, 235
Predation, 19, 462, 527, 573, 597
Predation-risk, 430, 440, 514, 547–549, 552–557
Predator, 456
  gape-limited, 499, 563, 576
  model of effects, 533–536
  size-selective, 569, 573
Predator avoidance, 513–514, 545, 558, 579
Predator threat, 23, 349, 554–558, 597, 599
Predator-mediated effects, on life history, 443–444
Predator-prey
  interactions, 535, 542–544, 549–550
  relationships, 32
Predator-swamping, 556
Prey detection, 576
Prey size
  gape-limitation, 78, 80, 576
  optimal, 78
Prey-piscivore ratios, 32
Primary productivity, 19, 135, 136, 157, 159, 161, 289–290, 293, 496, 576–577, 584, 607, 612
Principal Components Analysis (PCA), 99, 112, 116, 122, 404, 411, 423, 609
Profundal zone, 290
Proportional sampling, 23, 256

Q10, 143, 375

Range size, 245, 260
Recolonization, 92, 108, 322, 342, 346
Redundancy of species, 602
Refugia, 23, 24, 92, 131, 322, 331, 342, 352, 354
  from predators, 549, 557, 559
  from stress, 92, 162, 354, 374–376, 495
  glacial, 24, 157, 230, 232
Regional patterns, 15, 192, 194, 255
Regional-local models, 255–260
Relative depth, of lakes, 170
Reproduction, 19, 360, 416, 420, 425–431
  frequency of, 427–429
Reproductive
  modes, 422
  season, 426–427

Reservoirs, 43, 111, 295, 296, 342, 606–607
Resource
  availability, 456–457, 459, 474, 493
  axes, 458
Resource partitioning, 19, 45, 89, 137, 192, 458–460, 463, 470–476, 478–497, 526
  intraspecific, 507–512
Respiration, 138, 144
Reynolds number, 384
Rheotaxis, 451
Ridge-and-valley, 175, 311
Riffle assemblages, 31, 309, 349
Riffle-pool, 14, 30, 31, 112, 124, 149, 174, 328
Riffles, 57, 288, 302, 493
Riparian zone, 126, 332, 334, 337, 378
Risk-balancing, 552–554
River basins, 14, 18, 89, 126, 173, 241–243
River Continuum Concept (RCC), 20, 67, 172, 297, 299, 307, 566
Rock, 282
Root effect, 142, 360
Runoff, 181

Salinity, 11, 90, 103, 365–366
Saltwater dispersants, 201
Sampling, 118–119
Scattering, of light, 135
Schooling
  mixed-species, 513
  monospecific, 498, 513
Scour, 130, 325, 327, 329, 331, 337
Screens, faunal, 15
Sea level lowering, 228
Seasonal effects, 274
Seasonality, 32
Secondary production, 576
Seiches, 136, 152, 164–166
Selectivity, 366–370
Semelparity, 427–429
Sexual dimorphism, 437
Sexual selection, 429–431
Shade, 279–348
Shelter, 279
Shoals, mixed-species, 513
Shoreline development, of lakes, 169
Silt, 295, 332, 587, 593, 607, 615
Size efficiency hypothesis, 577
Size-selective hypothesis, 80
Snags, 281
Sneaker, 432

Spawning, 90, 263, 279, 326, 330, 417, 422, 426, 427, 432–433, 448, 451–452
Specialists, 278, 462
Speciation, 304
  allopatric, 153
  microallopatric, 153, 248
  sympatric, 153
Species
  additions, 298
  associations, 89–91, 486, 517, 520
  density, 272
  endemic, 5, 197, 207, 236, 243–245
  flocks, 11, 33, 35, 43, 153, 237, 528
  number of, 1
  packing, 194
  per family, 32, 45–54
  rare, 73, 75 77, 110, 117
  regional pool, 225
  replacements, 298
Species richness, 32, 33, 272, 288, 323, 338
  continental, 255,
  local, 3, 256, 259, 260
  regional, 256, 259, 260
  saturation, 255, 256
Species-area curve, 22, 23, 265–267
Species-area effects, 264–268
Species-stacking, 83
Specific gravity, 134
Springs, 38, 149, 152, 181, 277, 300, 301, 310, 354
Stability of assemblages, 14, 38, 44, 92, 104–127
Starvation, 357, 433, 438–439, 503, 603
Stratification
  direct, 134
  of lakes, 134–136, 139, 143, 159–163
  of stream pools, 149, 150
  reverse, 134
Stream
  gradient, 14
  width, 18, 268–270
Stream capture, 173, 177–180
Stream ecology, 131
Stream order, 175, 290, 307–312
Streamlining, 7, 14, 384, 414
Streams
  intermittent, 310, 343
  tropical, 33, 35, 39, 51, 75, 278, 396, 405, 474, 493, 566, 584–585
Stress, 318, 347
  abiotic, 353–354, 456

Strong interactors, 19, 497, 564, 565, 611–616
Structure, 279, 292, 328, 413, 549–550
Studies, long-term, 107–121
Sublethal stress, 363, 376
Substrate
  bedrock, 131
  cobble, 413
  gravel, 413
  sand, 413
Supercontinent, 202, 203
Survivorship, 447
Swim bladder, 395
Swimming
  tail, 389–390
  undulation, 389
Switching, 54, 55, 388–390, 564–603
Systematics, 26, 27

Taxon-cycle, 105
Tectonic lakes, 155,
Teeth, oral, 217, 396
Temperature, 18, 121, 137, 143, 148, 341, 343–344, 349, 351, 354–357
  final preferendum, 147
  fluctuation, 368
  selection, 147, 148
Temperature-oxygen squeeze, 363
Terminology, of Jenkins, 174
Terraces, riparian, 332
Tertiary, the, 222, 224
Tetrapods, 3–5, 25
Thalweg, 288
Thermal
  refugia, 354
  tolerance, 162, 253, 351, 354–357
Thermoregulation, behavioral, 137, 147, 558
Tolerance, 345, 347, 353–354, 371, 530
  intraspecific variation in, 361–363
  oxygen, 139, 253
  pH, 142
  salinity, 197, 200, 219
Tolerance versus selectivity, 362, 366–370
Top-down control, 19, 60, 133, 496, 565, 578, 600–603
Top-down versus bottom-up regulation, 600–603
Total dissolved solids (TDS), 171, 273–274
Triassic, the, 203
Trophic cascades (*see* "cascading effects")
Trophic categories

herbivore, 54, 56, 60
invertivore, 56
omnivore, 54, 56, 581
planktivore, 56
Trophic groups, 32, 54–60, 67
Trophic potential, 32
Trophic species, 602
Tropics, 35, 43, 327
Truss system, 404, 407, 411
Turbidity, 131, 142, 269, 295, 325, 330
TWINSPAN, 91

Upper extreme habitat temperature (HEHT), 354
Urbanization, 189

Variation
  interspecific, 14
  intraspecific, 14
Vertebrates, 2–5, 202, 401
Vicariance, 24, 153, 221, 304
Vicariance biogeography, 245
Vigilance, 518
Volcanic lakes, 155, 156

Wallace's Line, 194, 211
Water
  chemistry, 140–146, 317
  clarity, 607–609
  density, 133–135
  depth, 18
  molecule, 133–134
  movement, 163–166
  quality, 317, 566, 583, 604–610
  sources, 181–182
Water-column fishes, 58, 396, 418
Waterfalls, 11, 304
Watershed, 171, 174
Waterspout, 11
Waves, 163, 164, 292, 438
Weberian apparatus, 213, 222
Wind, 134, 136, 152, 160, 163, 292, 323–324
Winter, 162, 272, 311, 333, 334, 345,–347, 357, 415, 495
  aggregations, 415
Winterkill, 92, 139, 155, 162, 357–358, 608
Woody debris, 14, 187–188, 279–283, 314, 331, 550

z, 265–268
Zonation

in lakes, 143, 290–293
  in streams, 296–307, 365–366, 369
  vertical, 290–292
Zoogeographic realms, 5, 192–196
  African, 6, 196
  Australian, 6, 194, 196, 197
  Holarctic, 6, 203, 208
  Nearctic, 6, 196, 197
  Neotropical, 6, 194, 196, 197
  Oriental, 6, 196, 197, 208, 211, 239
  Palearctic, 6, 196, 197
Zoogeography, 12, 15, 18, 191
Zooplankton, 166, 292, 566, 575–581
  size-structure, 566, 577

Taxonomic Index

Abramis brama, 592
Acanthopterygii, 26
Acipenser fulvescens, 220
Acipenseridae, 220
Acrocheilus alutaceus, 586
African reedfish, 147
Agosia chrysogaster, 343, 348, 585, 611
Alewife, 425, 539, 576, 580, 590
Alligator gar, 546
Alosa pseudoharengus, 425, 539, 576, 580, 590
   A. sapidisima, 428
Ambloplites, 281, 373, 472
   A. rupestris, 356, 405, 436, 479, 507
Amblyopsidae, 224
American shad, 428
Amia, 397
   A. calva, 212
Amieurus, 92, 207, 223, 386
   A. melas, 71
   A. natalis, 356
   A. nebulosus, 405, 426
Amiidae, 212, 221
Ammocrypta, 403
Anabantid, 401
Anabantidae, 210
Ancistrus, 584
Anguilla rostrata, 454
Anguillids, 26
Anguilliformes, 6
Anotopysi, 213, 218
Aphredoderidae, 224
Aphredoderus sayanus, 281, 283
Aplocheilidae, 196, 211

Aplodinotus grunniens, 452, 570
Arctic char, 411, 439, 444
Arctic grayling, 504
Ariid catfishes, 217
Astronotus ocellatus, 550
Astyanax, 237
   A. mexicanus, 544
Atherinidae, 207, 224
Atherinids, 62, 169, 331
Atheriniformes, 7
Atlantic salmon, 432, 503, 505
Atractosteus, 221, 269
   A. spathula, 546
Aulopiformes, 6
Australian grayling, 424

Bagridae, 210, 218, 223
Banded darter, 435
Banded sculpin, 469
Barb, 419
Barbus, 486
   B. anoplus, 419
Bass, 130, 224
Bayou darter, 426
Bigeye shiner, 427, 551
Black carp, 570
Blacknose dace, 179, 500, 547, 548
Blacktail shiner, 281, 414, 435
Bleak (or sardine), 578
Blennies, 468
Blue catfish, 10, 269, 325, 465
Blue sucker, 269, 271
Blue tilapia, 550
Bluegill, 55, 281, 391, 392, 411, 414, 417,

432, 439, 460, 476, 477, 495, 507, 549, 550, 552, 553, 558, 575, 578
Bluehead shiner, 432
Bluntnose minnow, 515, 513
Bonytongues, 7
Bowfins, 220, 221, 236
Brassy minnow, 543, 547, 548
Bream, 592
Brook silverside, 466
Brook stickleback, 540
Brook trout, 60, 147, 281, 449, 470, 504, 509, 575
Brown bullhead, 426
Brown trout, 315, 347, 449, 475, 504, 505, 536, 538, 560
Buffalo, 269
Bullhead, 362
Bullhead catfish, 159
Burbot, 274, 466

California roach, 589
Campostoma, 19, 23, 24, 67, 263, 399, 499, 545, 546, 557, 561, 563, 599, 611–616
  C. anomalum, 10, 23, 45, 58, 60, 108, 109, 281, 287, 289, 335, 336, 415, 419, 450, 500, 514, 543, 551–552, 585–589, 591–593, 596–599
Caribbean hogfish, 402
Carp, 62, 213, 565, 589, 593, 596
Carpiodes, 37, 269
Carpsuckers, 269
Catfishes, 7, 10, 26, 37, 45, 197, 208, 210, 213, 215, 216, 218, 222, 223, 281, 283, 386, 522, 584
Catostomidae, 67, 111, 209, 223, 277, 333, 387, 399, 413, 452
Catostomids, 62, 201, 236, 237, 267, 352, 394
Catostomus catostomus, 209
  C. commersoni, 92, 515, 544
  C. occidentalis, 470, 511
Cavefishes, 224
Central mudminnow, 139, 358, 505
Centrarchidae, 44, 46, 92, 207, 224, 277, 349, 371, 373, 409, 477
Centrarchids, 92, 201, 263, 272, 436, 449, 481, 508, 539, 573, 576, 589
Chaetostoma, 584
Chain pickerel, 542
Channel catfish, 325, 347, 465
Channidae, 210, 213, 218

Channiformes, 210
Chanoides, 213
Characidae, 27, 237
Characiformes, 67, 197, 213, 215
Characiforms, 216, 217
Characins, 7, 26, 27, 45, 197, 213, 222, 453
Charcoids, 213, 216
Chasmistes, 223
Chihuahua chub, 280
Chinook salmon, 425, 435
Chirostoma, 237
Chiselmouth, 586
Chologaster agassizi, 147
Chum salmon, 553
Cichla ocellaris, 541
Cichlidae, 5, 196, 211, 237
Cichlids, 54, 153, 211, 248, 399, 401, 403, 429, 482, 490, 491, 522, 541, 550, 583, 596, 600
Cisco, 460, 576, 580
Ciscoes, 291
Clariidae, 210
Climbing perches, 210
Clinostomus, 499
  C. funduloides, 314, 336, 506, 519
Clupeidae, 206, 222
Clupeids, 26, 31, 62
Clupeiformes, 7, 26
Cobitidae, 197, 209
Cod, 468
Coho salmon, 281, 432, 468, 487
Collosoma, 8, 398
Colorado squawfish, 188, 237
Common shiner, 515–516
Copper redhorse, 570
Coregonus, 291, 460
  C. albula, 505, 511
  C. artedii, 576, 580
Cottidae, 225, 349, 490
Cottids, 236
Cottocomephoridae, 208
Cottus, 62, 268
  C. asper, 460
  C. bairdi, 405, 426, 485, 506, 569
  C. carolinae, 269, 301, 469, 499
  C. cognatus, 557
  C. girardi, 485, 569
  C. gobio, 362, 436
  C. hypselurus, 269, 301
  C. rhotheus, 460

Crappie, 296
Creek chub, 62, 500, 509, 518, 543, 547, 548
Creek chubsucker, 69
Crenicichla alta, 443
Ctenopharyngodon idella, 589
Culaea inconstans, 540
Cutthroat trout, 283, 449
Cycleptus elongatus, 269, 271
Cyprinella, 78, 260, 371, 475, 499, 516
   C. galactura, 82, 251, 301, 529
   C. lutrensis, 90, 146, 169, 229, 345, 352, 356, 362, 366, 426, 427, 451, 469, 497, 500, 528, 538
   C. spiloptera, 109
   C. venusta, 45, 108, 115, 169, 281, 345, 414, 435
   C. whipplei, 281
Cyprinidae, 3, 10, 44, 46, 50, 58, 74, 197, 211, 213, 219, 223, 239, 256, 277, 333, 349, 371, 407, 409, 490, 499
Cyprinids, 10, 92, 201, 202, 211, 217, 236, 267, 404, 519, 539, 578
Cypriniformes, 6, 7, 26, 197, 210, 213, 215, 216
Cypriniforms, 7
Cyprinodon, 147, 153, 393, 400, 451
   C. macularius, 431
   C. pecosensis, 431
   C. rubrofluviatilis, 143, 365
   C. variegatus, 355
Cyprinodontidae, 195, 196, 207, 208, 211, 223, 349, 371
Cyprinodontiformes, 7
Cyprinodonts, 7
Cyprinoids, 216
Cyprinus carpio, 62, 71, 565, 589, 593

Dace, 38, 439, 440, 518
Darters, 3, 7, 10, 60, 62, 84, 89, 92, 139, 147, 209, 224, 256, 277, 282, 300, 302, 333, 387, 389, 391, 393, 394, 399, 403, 405, 409, 413, 414, 422, 427, 470, 481, 490, 507, 594
Desert pupfish, 147, 451
Diplomystidae, 217
Dollar sunfish, 436
Dolly Varden, 487
Dorosoma, 159, 149, 225, 296, 399, 439, 607
   D. cepedianum, 71, 357, 575–576, 581–583
   D. petenense, 467

Eastern mosquitofish, 467
Eels, 389, 454
Elassoma evergladei, 433
   E. zonatum, 115
Electric eels, 207
Electrophoridae, 207
Elephantfishes, 7
Elopomorphs, 26
Embiotocidae, 401
Eohiodon, 222
Erimystax dissimilis, 337, 338
Erimyzon oblongus, 69
Erpetoichthys clalbaricus, 147
Esocidae, 26, 206, 208, 222
Esocids, 26, 27, 62, 550
Esociformes, 27
Esox, 7, 24, 25, 92, 538, 539
   E. americanus, 389
   E. lucius, 291, 540, 597, 608
   E. niger, 542
Etheostoma, 58, 82, 139, 149, 225, 389, 393, 399, 403, 441
   E. blennioides, 420, 485
   E. caeruleum, 268, 301, 465, 478, 485, 486
   E. flabellare, 109, 334–336, 347, 369, 405, 448, 466, 478, 482, 485, 486, 507, 508
   E. lepidum, 108, 438
   E. nigrum, 448, 550
   E. olmstedi, 436
   E. perlongum, 438
   E. podostemone, 335–336
   E. punctulatum, 283
   E. radiosum, 448
   E. rubrum, 310, 426
   E. spectabile, 108, 263, 268, 301, 347, 355, 362, 430, 431, 434, 435, 452, 465, 469, 495, 499, 508, 523, 528
   E. whipplei, 426, 434
   E. zonale, 435, 478
European minnow, 518
Euteleostei, 26
Everglades pigmy sunfish, 433

Fallfish, 440
Fantail darter, 369, 448, 486
Fathead minnow, 540, 547, 548
Featherbacks, 210
Flathead catfish, 223
Freshwater drum, 570
Fundulidae, 58, 184, 195, 224

Fundulus, 207, 223, 351, 393, 395, 478, 508
  F. catenatus, 82, 252, 529
  F. notatus, 45, 169, 500, 562
  F. olivaceus, 82
  F. waccamensis, 516
  F. zebrinus, 365

Gadiformes, 7
Galaxiidae, 212
Galaxiids, 346, 560
Galaxis vulgaris, 475
Gambusia, 351, 425, 438, 444, 538, 576
  G. affinis, 90, 169, 282, 301, 365, 426, 435, 441
  G. geiseri, 108, 282, 301, 345, 441, 544
  G. holbrooki, 542
  G. luna, 484
  G. puncticulata, 484
  G. sexradiata, 484
Gambusias, 224, 300
Gar, 37, 62, 212, 220, 221, 236, 269, 352
Gasterosteiformes, 7
Gasterosteus, 505
  G. aculeatus, 147, 411, 469
Giant rivulus, 557
Gila atraria, 546
  G. bicolor, 510
  G. bicolor mohavensis, 353
  G. cypha, 188
  G. nigrescens, 280
  G. orcutta, 353
Gizzard shad, 576–577, 581–583
Gobio gobio, 518
Gobiomorphus breviceps, 475
Golden perch, 452
Golden shiner, 450, 558
Goldcyc, 221
Gonorhynchidae, 213
Gonorhynchiformes, 213, 218
Goodeidae, 237
Grass carp, 589
Green sunfish, 60, 148, 159, 282, 403, 445, 449, 476, 477, 494, 544, 551–553, 574
Greenthroat darter, 438
Gudgeon, 518, 578
Guppy, 359, 430, 443, 543, 550, 555–557
Gymnarchidae, 207
Gymnarchids, 207
Gymnocephalus cernua, 469
Gymnotids, 216

Gymnotiformes, 7, 213
Gymnotiforms, 144, 215, 218

Haplochromines, 399, 403, 482, 541, 600
Hesperoleucas symmetricus, 589
Heterandria, 538
Hillstream loaches, 207, 394
Hiodon, 26, 221
Hiodontidae, 206, 212, 221, 222
Hiodontids, 7
Homalopteridae, 207
Hoplias, 542–543, 560
  H. malabaricus, 557
Hornyhead chub, 548
Humpback chub, 188, 237, 452
Hybognathus, 10
  H. hankinsoni, 543, 547, 548
  H. placitus, 269
Hydrocynus vittatus, 537
Hypentelium nigricans, 82, 334–336, 394, 405, 413, 449, 450, 514, 596
  H. roanokense, 335
Hysidoridae, 217
Hysterocarpus traski, 425

Ictaluridae, 10, 206, 223, 281, 349, 352, 371
Ictalurus, 37, 62, 207, 223, 269
  I. furcatus, 10, 465
  I. punctatus, 465
Ictiobus, 37, 207, 223, 269
Ide, 589
Inland silversides, 324, 466

Johnny darter, 448, 550

Killifishes, 7, 224, 359, 478, 542
Kneria paucusquamata, 303
Kneriidae, 213
Knifefishes, 7, 26
Knightia eocaena, 222
  K. vetusta, 222
Kokanee, 546

Labeo, 586
Labidesthes, 7
  L. sicculus, 331, 399, 466
Lachnolaimus maximum, 402
Lahontan redside, 124
Lahonton cutthroat trout, 11
Lake trout, 274, 449, 460, 557
Lake whitefish, 274

Lampreys, 106, 389, 413, 429
Largemouth bass, 23, 159, 169, 201, 390, 449, 450, 476, 477, 502, 503, 543, 545–547, 549–553, 558, 607, 609, 611
Largespring gambusia, 544
Lates niloticus, 153, 541, 600
Leaffishes, 221
Leopard dace, 471
Lepidomeda vittata, 538
Lepisosteidae, 37, 62, 208, 212, 221
Lepisosteus, 221, 269
Lepomis, 62, 81, 82, 92, 169, 281, 282, 295, 323, 335, 351, 373, 386, 394, 406, 424, 440, 472, 473, 494, 543, 562, 580, 594, 599
   L. auritus, 263
   L. cyanellus, 60, 90, 148, 275, 282, 287, 403, 445, 449, 476, 550, 544, 551–553, 558, 574–575
   L. gibbosus, 78, 405, 411, 477, 511, 570
   L. humilus, 62, 403
   L. macrochirus, 55, 281, 392, 411, 414, 432, 463, 476, 507, 552, 554, 575
   L. marginatus, 436
   L. megalotis, 62, 90, 275, 311, 414, 429, 432, 440, 445, 449, 500
   L. microlophus, 401, 570
Leuciscus idus, 589
   L. leuciscus, 439, 518
Limnothrissa miodon, 348
Ling, 460
Little Colorado spinedace, 538
Live-bearers, 7, 211
Loaches, 7, 197, 209, 210
Logperch, 147, 514, 515
Longear sunfish, 311, 414, 416, 429, 432, 440, 445, 449, 494
Longnose dace, 471, 518
Lophiiformes, 7
Loricariid catfishes, 281, 584, 593, 596
Lota lota, 460
Lungfishes, 8, 401
Luxilus, 373
   L. albeolus, 178, 334–336, 483
   L. cardinalis, 27, 561, 587
   L. cerasinus, 334–336, 483
   L. chrysocephalus, 28, 178, 413
   L. cornutus, 178, 413, 452, 474, 515
   L. pilsbryi, 27, 54, 82, 83, 183, 184, 301, 450, 452, 460, 506, 508, 570, 596
Lythrurus, 269, 449

L. ardens, 334–336, 483
L. umbratilis, 301

Macquaria ambigua, 452
Macrhybopsis meeki, 269
   M. storeriana, 269
Madtom catfish, 10, 452, 481
Masticembelidae, 210
Meda fulgida, 469
Menidia, 224, 296, 426
   M. beryllina, 163, 166, 169, 295, 324, 399, 427, 438, 466, 580
   M. extensa, 516
   M. menidia, 576
Mexican tetra, 544
Micropterus, 62, 81, 82, 169, 276, 282, 373, 388, 439, 500, 537, 586, 599
   M. dolomieu, 174, 281, 283, 331, 334, 355, 357, 415, 428, 429, 436, 479, 515, 523, 543, 544, 547, 548, 557, 559, 596, 611
   M. punctulatus, 60, 347, 563, 611
   M. salmoides, 23, 60, 92, 169, 287, 351, 390, 398, 439, 449, 450, 476, 502, 543, 546, 547, 609, 611
Mimic shiner, 515, 516
Minnow, 23, 89, 111, 211
Minnows, 3, 7, 10, 26, 44, 45, 54, 58, 59, 60, 62, 90, 92, 112, 130, 174, 213, 222, 223, 237, 259, 263, 269, 277, 280–282, 300, 333, 359, 368, 404, 406, 407, 409, 413, 429, 448, 481, 490, 520, 523, 527, 543, 594, 597, 599
Mirogrex terraesanctae, 578
Molly, 542
Mormyriformes, 26
Morone, 62, 207, 393
   M. chrysops, 31, 169, 467
   M. mississippiensis, 62
   M. saxatilis, 54, 169, 291, 363, 453, 467, 502, 542, 603, 606
Moronidae, 224
Mosquitofish, 282, 538
Mottled sculpin, 506
Moxostoma, 82, 184, 452, 587
   M. carinatum, 570
   M. cervinum, 335
   M. erythrurum, 336
   M. hubbsi, 570
   M. rhothoecum, 335, 336, 538
Mudminnow, 92, 208

Taxonomic Index / 749

Mylopharyngodon piceus, 570
Myxocyprinus asiaticus, 209

Nanidae, 211
Neosho madtom, 347, 541
Nile perch, 60, 153, 600
Nocomis, 332, 522, 538, 594
   N. biguttatus, 82, 514, 548
   N. leptocephalus, 334–336
Noemacheilus barbatulus, 362
Northern hogsucker, 334, 394, 413, 449, 450, 515, 596
Northern pike, 274, 540, 597, 608
Northern redbelly dace, 547, 548
Notemigonus, 223
   N. chrysoleucas, 92, 371, 405, 558
Notopteridae, 210
Notropis, 10, 263, 373, 406, 472, 475, 508, 515–517
   N. amabilis, 108, 345
   N. atherinoides, 345
   N. bairdi, 269
   N. boops, 45, 287
   N. buccatus, 481
   N. girardi, 184, 269, 356
   N. heterolepis, 405
   N. longirostris, 435, 481
   N. lutipinnis, 429
   N. nubilus, 58, 82, 230, 301, 514, 561, 593
   N. petersoni, 516
   N. potteri, 269
   N. rubellus, 27, 73, 82, 405, 414, 494, 570
   N. snelsoni, 301
   N. stramineus, 109
   N. suttkusi, 27
   N. telescopus, 82, 251, 301, 529
   N. volucellus, 474, 515
Noturus, 10, 481
   N. flavus, 405
   N. gyrinus, 283
   N. placidus, 347

Oncorhynchus, 432
   O. clarki, 354, 449
   O. gorbuscha, 553, 590
   O. keta, 553
   O. kisutch, 281, 432, 468, 487
   O. mykiss, 147, 452, 470, 505, 506, 511, 519, 538
   O. nerka, 546, 590
   O. tshawytscha, 425, 435

Orangebelly darter, 315, 448
Orangespotted sunfish, 495
Orangethroat darter, 431, 434, 435, 469, 523
Orthodon microlepidotus, 139, 359
Osmeridae, 26
Osmeriformes, 6, 7, 27
Osmerus mordax, 539
Ostariophysi, 26, 206, 213–219
Osteoglossidae, 210, 211
Osteoglossids, 7, 26, 210
Osteoglossiformes, 210
Osteoglossum, 8
Otophysi, 213, 215, 218, 219, 222
Ozark minnow, 230

Pacific salmon, 432
Paddlefishes, 209, 220, 236, 269
Paleopsephurus, 221
Pantadon, 8
Peacock bass, 60, 541
Pearl dace, 508
Perca, 225, 472, 479, 559
   P. flavescens, 92, 139, 291, 355, 505, 540, 576, 580
   P. fluviatilus, 469, 550, 559, 600
Perch, 7, 209, 224, 355, 469, 504, 550
Percichthyidae, 224
Percidae, 3, 48, 60, 74, 84, 89, 209, 224, 259, 277, 349, 371, 389, 403, 405, 409, 413, 422, 427, 437, 490
Percids, 62, 267
Perciformes, 7, 499
Perciforms, 7, 26, 210
Percina, 225, 333, 399, 413, 485
   P. burtoni, 514
   P. caprodes, 147, 514
   P. evides, 514
   P. rex, 334
   P. roanoka, 335–336
   P. sciera, 108, 345
Percopsidae, 224
Petromyzontidae, 220
Phenacobius mirabilis, 73, 108
Phoxinus, 159, 263
   P. eos, 411, 547, 548
   P. erythrogaster, 58, 300, 338
   P. oreas, 336
   P. phoxinus, 426, 518
Phractolaimidae, 213
Pickerels, 24, 26, 208, 389
Pikes, 24, 26, 89, 208, 222, 291, 538

Pimephales, 7, 371, 472
  P. notatus, 82, 109, 334–336, 474, 515
  P. promelas, 90, 356, 540, 547, 548
  P. vigilax, 90, 169
Pink salmon, 553, 590
Piranha, 547
Pirate perch, 224, 281, 283
Plaice, 468
Plains killifish, 365, 366
Pleuronectiformes, 7
Poecilia, 543
  P. latipinna, 426
  P. reticulata, 430, 443, 550, 555–557
Poeciliidae, 207, 211, 212, 224, 237
Poeciliopsis, 356, 441, 444, 538
Polyodon, 62, 221
  P. spathula, 209, 220, 269
Polyodontidae, 209, 220
Pomoxis, 296, 373, 439, 472
Prochilodus mariae, 584, 585, 593, 596, 611
Protroctes maraena, 424
Psephurus gladius, 209, 220
Pternotropis hubbsi, 432
Ptychocheilus lucius, 188
Pumpkinseed sunfish, 78, 411, 477, 512, 570
Pupfishes, 7, 153, 223, 237, 367, 425, 430, 431
Pydocentrus notatus, 547
Pylodictis, 62, 269

Rainbow darter, 486
Rainbow smelt, 539
Rainbow trout, 103, 147, 452, 470, 505, 506, 511, 519, 538
Razorback sucker, 148, 188, 237, 452
Red drum, 546
Red River pupfish, 143, 365, 366
Red shiner, 229, 404, 426, 427, 451, 469, 497, 538
Redear sunfish, 401, 495, 570
Redside shiner, 470
Rhinichthys atratulus, 179, 335, 500, 547, 548
  R. cataractae, 426, 471, 518
  R. falcatus, 471
Richardsonius balteatus, 470
  R. egregius, 124
River bullhead, 436
River redhorse, 570
Rivulines, 211
Rivulus, 543, 555–556, 560
  R. harti, 443, 555–557, 560

Roach, 355, 439, 440, 469, 504, 550, 559, 570, 589, 600, 601, 608
Rock bass, 281, 436, 479, 507
Rock-dwelling cichlids, 482
Rosyside dace, 314, 506, 519
Rudd, 589, 592
Ruffe, 469
Rutilus rutilus, 355, 439, 469, 550, 559, 570, 589, 608

Sacramento blackfish, 139, 359
Sacramento sucker, 470, 511
Sailfin molly, 426
Salmo clarki, 283
  S. salar, 432, 503, 505
  S. trutta, 147, 315, 347, 356, 449, 468, 470, 475, 505, 509, 538
Salmonidae, 26, 92, 206, 222
Salmonids, 5, 26, 37, 92, 158, 236, 328, 333, 354, 413, 419, 422, 429, 449, 453, 469, 499, 502, 508, 566, 594
Salmoniformes, 6, 7, 26, 27
Salmoniforms, 7
Salmons, 26, 222
Salvelinus alpinus, 411, 439, 468
  S. fontinalis, 282, 354, 575
  S. malma, 487
  S. namaycush, 449, 460, 557
Scardinus eryopthalmus, 589, 592, 601
Schilbeidae, 210
Sciaenops ocellatus, 546
Scorpaeniformes, 7, 499
Sculpins, 38, 92, 158, 225, 268, 333, 393, 394, 448, 470, 490, 569, 574
Semionotids, 153
Semotilus atromaculatus, 62, 275, 500, 509, 518, 543, 547, 548, 553
  S. corporalis, 440
  S. margarita, 508
Shads, 31, 159, 166, 169, 222, 295, 296, 399, 439, 607
Sheepshead minnow, 355
Siluridae, 210
Siluriformes, 6, 7, 197, 208, 210, 213
Siluriforms, 7, 215, 216
Siluriphysi, 213, 215, 218
Silversides, 224, 295
Slimy sculpin, 557
Smallmouth bass, 184, 263, 277, 280, 281, 283, 331, 355, 357, 415, 425, 428, 429,

436, 479, 495, 559, 560–561, 563, 596, 611
Smelts, 26
Snakeheads, 210
Sockeye salmon, 137, 536, 590
Southern duskystripe shiner, 450, 506
Spikedace, 469
Spiny eels, 210
Spotted bass, 347, 563, 611
Spring cavefish, 147
Squawfish, 452
Steelcolor shiner, 91
Steelhead trout, 470, 600
Stickleback, 431, 433, 436, 460, 505, 553, 573
Stippled darter, 283
Stizostedion, 62, 479, 608
  S. vitreum, 460
Stone loach, 362
Stonerollers, (*see Campostoma anomalum*)
Striped bass, 80, 169, 291, 325, 354, 363, 453, 467, 502, 542, 603, 606
Sturgeons, 62, 220, 454
Suckers, 7, 26, 37, 38, 67, 111, 184, 209, 222, 223, 263, 277, 333, 352, 387, 399, 413
Sunfishes, 7, 44, 59, 60, 92, 111, 130, 169, 207, 224, 263, 277, 280–282, 323, 386, 394, 406, 409, 424, 428, 430, 431, 494, 498, 500, 543, 562, 580, 599
Surfperches, 401
Swordtails, 10
Synbranchiformes, 7

Tadpole madtom, 283
Tarpons, 26

Tessellated darter, 436
Threadfin shad, 357, 467
Threespine stickleback, 411, 469
Thymallus arcticus, 504
Tiaroga cobitis, 283
Tigerfish, 537
Tilapia aurea, 582
  T. esculenta, 582
  T. galilaea, 583
  T. grahami, 11
Topminnows, 7, 58, 59, 62, 184, 351
Torrent sucker, 538
Trout, 26, 38, 124, 222, 334, 417, 536, 567
Trout perch, 224
Tui chub, 510
Tule perch, 425

Umbra lima, 139, 358, 505, 539
Umbridae, 208
Utah chub, 546

Vendace, 505, 511
Vertebrates, 401

Walleye, 274, 460
Weakly electric fishes, 213
White bass, 31, 169, 467
White sucker, 274, 515, 516, 544

Xeomelanaris venezuelae, 438
Xyrauchen texanus, 148, 188

Yellow perch, 92, 139, 274, 291, 505, 540, 576, 580, 600
Yellowfin shiner, 429

Locality Index

Afganistan, 238
Africa, 7, 35, 39, 48, 49, 152, 153, 194, 196, 203, 205–208, 210–212, 215, 216, 218, 219, 221, 243, 259, 266, 303, 387, 399, 407, 480, 537, 541, 603
Alabama, 308, 321, 409
Alabama River, 242
Alaska, 206, 232, 234, 289, 407, 590
Alberta, 221–223, 540
Alps, the, 227
Amazon River, 5, 35, 42, 140, 144, 172, 228, 240, 348, 448, 453, 589
Amur River basin, 570
Anadyr River basin, Russia, 238
Andes Mountains, 14, 42, 172, 240
Antarctica, 205
Appalachian Mountains, 5, 14, 112, 234, 243, 247, 251, 281, 506, 519, 529
Appalachicola basin, 242
Apure River, Venezuela, 42
Arabia, 207
Aravaipa Creek, Arizona, 46, 112, 120, 121, 124
Arbuckle Mountains, Oklahoma, 254, 309
Arctic Archipelago, 234
Arctic Ocean, 232
Arizona, 112, 120, 121, 124, 224, 230, 348, 538
Arkansas, 27, 33, 107, 110, 111, 121, 145, 169, 186, 187, 223, 243, 244, 249–252, 259, 267, 269, 301, 305, 344, 407, 450
Arkansas River, 172, 242, 247, 252, 262, 305, 308
Asia, 205, 207, 215, 227, 238, 239, 299, 301, 387, 394, 484, 570, 589

Asia northern, 39, 297, 203, 224, 454
Asia southeast, 206, 209, 217–219, 227, 238
Atchafalaya River Basin, Louisiana, 278
Atlantic Divide, North America, 178
Australia, 6, 39, 49, 194, 196, 197, 205, 208, 211, 240, 299, 346, 347, 452, 454

Baja California, 237
Balkans, 227
Baltic, the, 347
Baron Fork (of Illinois River), Oklahoma, 330, 415, 545, 557, 559, 560, 563, 586–589, 612
Battle Branch, Oklahoma, 316
Bayou Sara, Louisiana, 91
Belize, 484
Bering Sea, 232
Bering Strait, 206, 232
Beringia, 194, 205, 206, 209, 210, 222, 225, 228, 232, 234
Big Bend National Park, Texas, 189
Big Sandy Creek, Texas, 309
Big Vermillion River, Indiana, 329
Black Creek, Mississippi, 112, 120, 122
Black River, Arkansas-Missouri, 252, 305, 306
Black Sea, 5, 227
Blue Ridge divide, Virginia, 178
Blue River, Oklahoma, 521, 551
Boone County, Iowa, 109
Borneo, 35, 405
Brazil, 211, 218, 221, 240, 547
Brier Creek, Oklahoma, 11, 23, 44, 48, 57, 59, 60, 70, 81, 90, 104, 110, 111, 120, 121, 124, 130, 139, 143, 185, 186, 188, 275, 282, 285, 288, 301, 315, 324, 326, 329,

Locality Index / 753

348, 355, 368, 370, 416, 424, 428, 438, 446–447, 450, 500, 501, 508, 538, 543, 545, 561–562, 586–587, 599
British Columbia, 222, 449
British Isles, 39, 56, 361, 362, 439
Broken Bow Lake, Oklahoma, 269
Bruce Lakes District, Ontario, 99
Buffalo River, Mississippi, 75, 77
Bull Shoals Lake, Arkansas, 169
Byrds Mill Creek, Oklahoma, 46

California, 224, 342, 354, 511, 600
Canada, 7, 24, 80, 92, 157, 206, 220, 221, 234, 266, 290, 422, 453, 508, 604, 608
Canadian River, Oklahoma, 279, 479
Caribbean, 212, 246
Caspian Sea, 5, 227
Cedar Fork River, Ohio, 112
Central America, 4, 6, 48, 192, 196, 200, 207, 210–212, 220, 224, 237
Chagres River, Panama, 541
China, 209, 218, 220, 221, 223
Chukostsk Sea, Russia, 238
Clear Boggy River, Oklahoma, 521
Coastal Plain, eastern North America, 242
Colorado, 109, 308, 347, 449, 569
Colorado River basin, 5, 126, 148, 267, 367, 452
Colorado River, Texas, 344
Colorado River, western United States, 5, 188, 341, 417, 452, 469
Columbia, 297
Congo River, 42, 228, 348
Conowingo Creek, Maryland-Pennsylvania, 270
Continental Divide, North America, 208, 223, 224, 227
Costa Rica, 35, 49, 298, 407, 585, 603
Crater Lake, Oregon, 170
Cross River, Nigeria, 327
Cuatro Cienegas, Mexico, 268
Cuba, 212
Cucumber Creek, Oklahoma, 288

Death Valley, California, 223
Denmark, 144
Des Moines River, Iowa, 91, 104, 345, 460, 463
Devil's Lake, North Dakota, 231
Devils River, Texas, 333, 479

Doe Run, Kentucky, 142
Driftless Area, Wisconsin, 230

Eastern Chukotka, Russia, 238
Ecuador, 35, 297
Edisto River, 242
Edwards Aquifer, Texas, 237
Edwards Plateau, Texas, 244
Embarras River, Illinois, 267, 308
Esla River basin, Spain, 538
Ethiopia, 196, 197
Euramerica, 205
Eurasia, 6, 202, 208, 216
Europe, 6, 7, 37, 38, 39, 48, 149, 203, 205, 216, 227, 298, 303, 454, 510, 589
Europe, streams of, 5, 267
Everglades, Florida, 289

Fall Line, the (North America), 237, 242
Finland, 258, 581
Flaming Gorge Reservoir, Utah-Wyoming, 545
Fly River, New Guinea, 278
France, 267, 297

Ganges River, 227
Gangetic Plain, 36, 308
Gatun Lake, Panama, 541
Ghana, 586
Glen Canyon Dam, 188, 417
Glenfinish River, Ireland, 509
Gondwanaland, 203, 205
Great Basin, North America, 152, 155, 228, 243
Great Britain, 471
Great Lakes, of Africa, 2, 5, 11–12, 33, 54, 106, 248, 399, 409, 482, 583
Great Lakes, of North America, 5, 31, 106, 259
Great Plains (North America), 11, 109, 172, 189, 220, 224, 228, 229, 231, 250, 251, 283, 299, 342, 358, 359, 368, 372, 373, 451, 480
Green River formation, Wyoming, 56, 209, 222, 265
Greenland, 221
Guadalupe River, Texas, 345
Gulf Coast (North America), 244
Gulf of St. Lawrence, 234

Hainan Island, 239
Hawaii, 330, 425, 454

Hickory Creek, Oklahoma, 295
Himalaya Mountains, 6, 36, 206, 227, 308, 387
Hudson Bay, 229, 234
Hudson-James Bays, 259

Iberian Peninsula, 227
Iceland, 510
Illinois, 55, 89, 90, 111, 232, 333, 346, 347, 450, 452, 470
Illinois River, Illinois, 144
India, 205, 206, 210, 212, 238
Indian subcontinent, 210
Indiana, 111, 120, 121, 122, 232, 344
Indo-Chinese Peninsula, 239
Indo-Malaysian archipelago, 211
Indonesia, 239
Interior Highlands, North America, 33, 172, 247, 359, 372
Iowa, 450
Iran, 207
Ireland, 509
Israel, 480

James River, Virginia, 178, 179, 243
Janes Creek, Arkansas, 37, 48, 51
Japan, 238
Java, 210
Jordan Creek, Illinois, 277, 438, 485, 544

Kali Gandaki River, Nepal, 36, 99, 102, 308
Kamchatka, 590
Kankakee River, Illinois, 327
Kansas, 103, 110, 251, 315, 347, 362, 450
Kenya, 218
Keystone Lake, Oklahoma, 453
Kiamichi River, Oklahoma, 90, 110, 112, 281, 315
Kretam Kechil River, Borneo, 35

Lake Agassiz, glacial, 233
Lake Baikal, 5, 208
Lake Chad, 5
Lake Chapala, Mexico, 237
Lake Chichancanab, Mexico, 152, 153, 400
Lake Erie, 5
Lake Gatun, Panama, 60
Lake Kariba, Africa, 348
Lake Kenneret, Israel, 171, 358, 576, 578, 582
Lake Magidi, Kenya, 11
Lake Malawi, 5, 409, 448, 482, 522

Lake Mendota, Wisconsin, 171, 601
Lake Michigan, 5, 80, 137, 539, 576
Lake Myrkdalsvatnet, Norway, 509
Lake Nipigon, Canada, 460
Lake Norfork, Arkansas, 169
Lake Okeechobee, Florida, 155
Lake Ontario, 5, 137, 576
Lake Opinicon, Ontario, 463, 471, 472, 507, 512
Lake Pontchartrain, Louisiana, 200, 478
Lake Selura, Norway, 509
Lake Sobygard, Denmark, 601
Lake Tahoe, Nevada, 590
Lake Tanganyika, 5, 43
Lake Texoma, Oklahoma-Texas, 80, 139, 141, 160, 163–166, 169, 170, 292–294, 296, 324–326, 354, 363, 365, 439, 450–453, 465–467, 502, 521, 542, 581, 606, 607
Lake Valencia, Venezuela, 438
Lake Victoria, 5, 60, 152–155, 248, 403, 482, 491, 541, 600
Laurasia, 203, 216
Little Auglaize River, Ohio, 346
Little Colorado River, 452
Little Miami River, Ohio, 481
Little Missouri River, Arkansas, 553
Little River, Oklahoma, 451
Little Tennessee River, 491
Louisiana, 110, 115, 120, 122, 124, 200, 251, 316, 404
Luongo River, Africa, 35, 49, 303, 304

Madagascar, 196, 211
Magdalena River, Columbia, 240, 297, 322
Malay Peninsula, 238–239
Marais des Cygnes River, Kansas, 347
Martis Creek, California, 111, 120, 121, 124
Mary River, Australia, 49
McKittrick Creek, Texas, 347
Meadow Creek, Virginia, 179
Mediterranean, 208, 211
Mesa Central, Mexico, 192, 237
Mexico, 224, 237, 268, 277, 280, 353
Michigan, 282, 536
Midcontinental sea, North America, 205
Midwest (of the United States), 100, 215, 286, 324, 354, 516, 521, 544, 551, 558, 593
Mill Creek, Oklahoma, 37, 62
Mindanao, Philippines, 239
Minnesota, 358, 515
Mississippi, 308, 404

Locality Index / 755

Mississippi River, 174, 252, 305, 327, 529
Mississippi River Basin, 5, 39, 242, 243, 251, 326, 452
Missouri, 251, 269, 289, 334, 367, 371, 495
Missouri River, 39, 174, 229, 232
Mobile Bay drainage, 237
Mojave River, California, 353
Mongolia, 238
Mono Lake, California, 11
Montana, 220, 224
Morocco, 210
Mulgrave River, Australia, 299

Napo River, Ecuador, 35, 73, 297
Nelson River, 259
Neosho River, Kansas, 109, 281, 347
Neotropics, 407
Nepal, 36, 39, 99, 102
Neuse River, 242
Nevada, 224
Newark Supergroup of lakes, 153
New England, 232
New Guinea, 240, 278
New Mexico, 280
New River, Virginia, 178, 179
New York, 232, 266, 271, 500, 547
New Zealand, 6, 39, 60, 196, 212, 240, 453, 454, 475, 491, 560, 567, 569, 594
Nida River, Poland, 5
Nigara River, New York, 479
Nile River, 240
North America, 7, 27, 39, 50, 205, 208
   eastern, 5, 35, 36, 39, 46, 62, 203, 209, 211, 230, 247, 612, 615
   western, 5, 11, 37, 39, 206, 209, 222, 299, 354, 469
North Carolina, 348, 448, 591
Norway, 361, 449, 563, 509, 510
Nysa Klodzka River, 300

Ohio, 109, 112, 232, 250, 344, 470, 478, 485
Ohio River, 39, 178, 232, 242, 251
Oklahoma, 27, 33, 90, 99, 102, 104, 110–112, 130, 139, 142, 145, 160, 169, 186, 188, 189, 224, 225, 229, 251, 269, 281, 301, 315, 375, 404, 407, 448, 495, 498, 520, 521, 551, 585, 611
Ontario, 92, 99, 103, 233, 235, 256, 266, 273, 358, 405, 604, 608
Oregon, 99, 103, 225, 307, 316
Orinoco River, 144, 240, 298, 317

Oswego Creek, New York, 73
Otter Creek, Indiana, 111, 120–122
Ouachita Mountains, Arkansas-Oklahoma, 31, 33, 145, 172, 175, 243, 247, 251, 259, 267, 269, 275, 288, 299, 553
Ozark Mountains, 5, 33, 145, 172, 175, 188, 230, 235, 242, 243, 247, 251, 252, 259, 267, 268, 281, 305, 306, 336, 516, 519, 527, 528, 596

Panama, 35, 48, 275, 281, 283, 286, 474, 584
Pangaea, 24, 202, 203, 205, 221
Paraguay, 406
Parana River, 240, 327
Patagonia, 240
Pearl River, Louisiana-Mississippi, 110, 115, 120–122, 124
Pennington Creek, Oklahoma, 281, 414
Pennsylvania, 220, 230
Penzhina River, Russia, 238
Penzhina Bay, Russia, 238
Phillipines, The, 11, 239
Piedmont, eastern North America, 242, 243
Pilica River, Poland, 297
Piney Creek, Arkansas, 48, 57, 58, 70, 75, 82, 107, 110–112, 120, 121, 124, 125, 183, 186, 187, 252, 268, 297, 301, 326, 327, 336, 450, 452, 506, 508, 596
Plum Creek, Colorado, 332
Poland, 405, 592
Potomac River, 242
Pyramid Lake, Nevada, 11, 510

Quebec, 309

Rappahannock River, 242
Red River, Minnesota, 308
Red River, Oklahoma-Texas, 90, 99, 102, 229, 269, 317, 325, 331, 365, 366, 407
Rhone River, 329
Rio Frijolito, Panama, 49, 584
Rio Grande basin, 269, 277
Rio Juan Grande, Panama, 49
Rio Tortuguero, Costa Rica, 49, 75, 77
River Cam, 471
River Havelse, Denmark, 144
Roanoke River, Virginia, 75, 84, 109, 178, 272, 334, 365, 369, 391, 399, 408, 409, 450, 458, 466, 483, 484, 495, 521, 538
Rocky Mountains, 224, 342
Russia, 208, 605

Sagehen Creek, California, 110, 188
Saltilla River, Georgia, 281
Salween Basin, 239
San Juan River, New Mexico, 105
Saskatchewan, 222
Sea of Okhotsk, Russia, 238
Siberia, 209, 221, 232
Sinking Creek, Virginia, 179
Sister Grove Creek, Texas, 188
Smokey Hill River, Kansas, 110
Solomon River, Kansas, 110
Sonoran Desert, 348
South America, 4, 6, 7, 8, 27, 35, 39, 42, 194, 196, 197, 203, 205–207, 211, 215, 267, 342, 387, 406, 589, 596, 603, 611
South Canadian River, Oklahoma, 91, 145, 316, 317, 375, 379, 451, 521
South Carolina, 99, 102, 270
South Concho River, Texas, 282, 301, 345, 544
South Dakota, 224, 225
South Johnstone River, Australia, 299
Soviet Union, 238
Spain, 205, 206, 328, 538
Spring Creek, Oklahoma, 286
Squaw Creek, Iowa, 345
Sri Lanka, 35, 406, 408, 486, 527
Strawberry River, Arkansas, 305–307
Sulphur River, Texas, 91
Sunda Shelf, 239
Sweden, 575
Sycamore Creek, Arizona, 343, 348, 611

Taiwan, 238
Tallapoosa River, Alabama, 321
Tar River, 242
Teays River, Virginia-Ohio, 178
Tennessee River, 242
Terlingua Creek, Texas, 189
Texas, 91, 100, 104, 107, 108, 139, 160, 188, 189, 225, 228, 229, 244, 251, 271, 317, 345, 347, 363, 365, 434, 452, 544, 551, 603
Thailand, 239
Tibet, 239

Toledo, Ohio, 272
Tombigbee River, Alabama-Mississippi, 308, 485
Tonkin Gulf, 239
Toolik Lake, Alaska, 557
Tornillo Creek, Texas, 289
Trans-Pecos region, Texas, 244
Trinidad, 444, 542–544, 555–557, 559
Turgai Sea, 205
Tyner Creek, Oklahoma, 289, 415

United States
  eastern, 4, 34, 38, 48, 149
  midwest, 100, 215, 286, 324, 354, 516, 521, 544, 551, 558, 593
  western, 31, 38, 48, 152, 353
Upper Klamath Lake, California, 358
Utah, 342

Vangsvatnet Lake, Norway, 510
Venezuela, 42, 43, 297, 348, 407, 584, 596, 603
Vietnam, 238, 239
Virginia, 175, 178, 179, 485, 538

Wabash River, Indiana, 329
Wales, 346, 475
Washington, 222
West Indies, 196
White River, Arkansas-Missouri, 27, 51, 242, 244, 262, 305, 306, 452
Willamette River, Oregon, 307
Wintergreen Lake, Michigan, 609
Wisconsin, 75, 92, 171, 255, 258, 291, 317, 357, 361, 475, 516, 518, 539
Wisconsin River, 358
Wrangell-St. Elias Icefield, Alaska, 234
Wyoming, 217, 220, 518

Yangtze River, China, 209, 220
York River, 242, 243

Zaire, 218, 303, 304
Zambesi River, 240
Zambia, 42, 407

DATE DUE			
NOV 2 0 1999			
APR 0 1 2000			
AUG 0 4 2000			
NOV 2 6 2000			
DEC 1 4 2000			
JAN 1 5 2001			
FEB 1 5 2001			
JUL 1 0 2001			
NOV 0 1 2003			
NOV 0 - 2003			
NOV 0 5 2003			
GAYLORD			PRINTED IN U.S.A.